Analog MOS
Integrated Circuits, II

OTHER IEEE PRESS BOOKS

Instrumentation and Techniques for Radio Astronomy, *Edited by P. F. Goldsmith*
Network Interconnection and Protocol Conversion, *Edited by P. E. Green, Jr.*
VLSI Signal Processing, III, *Edited by R. W. Brodersen and H. S. Moscovitz*
Microcomputer-Based Expert Systems, *Edited by A. Gupta and B. E. Prasad*
Principles of Expert Systems, *Edited by A. Gupta and B. E. Prasad*
High Voltage Integrated Circuits, *Edited by B. J. Baliga*
Microwave Digital Radio, *Edited by L. J. Greenstein and M. Shafi*
Oliver Heaviside: Sage in Solitude, *By P. J. Nahin*
Radar Applications, *Edited by M. I. Skolnik*
Principles of Computerized Tomographic Imaging, *By A. C. Kak and M. Slaney*
Selected Papers on Noise in Circuits and Systems, *Edited by M. S. Gupta*
Spaceborne Radar Remote Sensing: Applications and Techniques, *By C. Elachi*
Engineering Excellence, *Edited by D. Christiansen*
Selected Papers on Logic Synthesis for Integrated Circuit Design, *Edited by A. R. Newton*
Planar Transmission Line Structures, *Edited by T. Itoh*
Introduction to the Theory of Random Signals and Noise, *By W. B. Davenport, Jr. and W. L. Root*
Teaching Engineering, *Edited by M. S. Gupta*
Selected Papers on Computer-Aided Design of Very Large Scale Integrated Circuits, *Edited by A. L. Sangiovanni-Vincentelli*
Robust Control, *Edited by P. Dorato*
Writing Reports to Get Results: Guidelines for the Computer Age, *By R. S. Blicq*
Multi-Microprocessors, *Edited by A. Gupta*
Advanced Microprocessors, II, *Edited by A. Gupta*
Adaptive Signal Processing, *Edited by L. H. Sibul*
Selected Papers on Statistical Design of Integrated Circuits, *Edited by A. J. Strojwas*
System Design for Human Interaction, *Edited by A. P. Sage*
Microcomputer Control of Power Electronics and Drives, *Edited by B. K. Bose*
Selected Papers on Analog Fault Diagnosis, *Edited by R. Liu*
Advances in Local Area Networks, *Edited by K. Kümmerle, J. O. Limb, and F. A. Tobagi*
Load Management, *Edited by S. Talukdar and C. W. Gellings*
Computers and Manufacturing Productivity, *Edited by R. K. Jurgen*
Selected Papers on Computer-Aided Design of Analog Networks, *Edited by J. Vlach and K. Singhal*
Being the Boss, *By L. K. Lineback*
Effective Meetings for Busy People, *By W. T. Carnes*
Selected Papers on Integrated Analog Filters, *Edited by G. C. Temes*
Electrical Engineering: The Second Century Begins, *Edited by H. Freitag*
VLSI Signal Processing, II, *Edited by S. Y. Kung, R. E. Owen, and J. G. Nash*
Modern Acoustical Imaging, *Edited by H. Lee and G. Wade*
Low-Temperature Electronics, *Edited by R. K. Kirschman*
Undersea Lightwave Communications, *Edited by P. K. Runge and P. R. Trischitta*
Multidimensional Digital Signal Processing, *Edited by the IEEE Multidimensional Signal Processing Committee*
Adaptive Methods for Control System Design, *Edited by M. M. Gupta*
Residue Number System Arithmetic, *Edited by M. A. Soderstrand, W. K. Jenkins, G. A. Jullien, and F. J. Taylor*
Singular Perturbations in Systems and Control, *Edited by P. V. Kokotovic and H. K. Khalil*
Getting the Picture, *By S. B. Weinstein*
Space Science and Applications, *Edited by J. H. McElroy*
Medical Applications of Microwave Imaging, *Edited by L. Larsen and J. H. Jacobi*
Modern Spectrum Analysis, II, *Edited by S. B. Kesler*
The Calculus Tutoring Book, *By C. Ash and R. Ash*
Imaging Technology, *Edited by H. Lee and G. Wade*
Phase-Locked Loops, *Edited by W. C. Lindsey and C. M. Chie*
VLSI Circuit Layout: Theory and Design, *Edited by T. C. Hu and E. S. Kuh*
Monolithic Microwave Integrated Circuits, *Edited by R. A. Pucel*
Next-Generation Computers, *Edited by E. A. Torrero*
Analog MOS Integrated Circuits, *Edited by P. R. Gray, D. A. Hodges, and R. W. Brodersen*

A complete listing of IEEE PRESS books is available upon request.

Analog MOS Integrated Circuits, II

Edited by
Paul R. Gray
Professor of Electrical Engineering and Computer Sciences
University of California, Berkeley

Bruce A. Wooley
Professor of Electrical Engineering
Stanford University

Robert W. Brodersen
Professor of Electrical Engineering and Computer Sciences
University of California, Berkeley

A volume in the IEEE PRESS Selected Reprint Series,
prepared under the sponsorship of the IEEE Solid-State Circuits Council.

IEEE
PRESS

The Institute of Electrical and Electronics Engineers, Inc., New York

Library of Congress Cataloging-in-Publication Data

Analog MOS integrated circuits, II.

(IEEE Press selected reprint series)
''Prepared under the sponsorship of the IEEE Solid-State
Circuits Council.''
Includes indexes.
1. Linear integrated circuits. 2. Metal oxide semiconduc-
tors. 3. Integrated circuits—Large scale integration. I. Gray,
Paul R., 1942–   . II. Wooley, Bruce A., 1943–   .
III. Brodersen, Robert W., 1945–   . IV. IEEE Solid-State
Circuits Council. V. Title: Analog MOS integrated circuits 2.
VI. Title: Analog MOS integrated circuits two.
TK7874.A575 1988   621.381'73   88-32043
ISBN 0-87942-246-7

Contents

Preface

SINCE its inception in the mid-1970s, the field of analog metal–oxide–semiconductor (MOS) integrated circuit design has continued to become of increasing importance in a steadily expanding array of applications. This growth has been fueled by the continued scaling of MOS very large scale integration (VLSI) technology, which has made the digital, rather than analog, processing of signals increasingly attractive. As digital VLSI circuits become pervasive in systems that were traditionally analog in nature, the primary role of analog circuits is becoming that of providing the interfaces between the analog environment of the physical world and a digital environment wherein information is represented in a form that is quantized into discrete intervals of both time and amplitude. Specific functions required in such interfaces include amplification, filtering, and both analog-to-digital and digital-to-analog conversion. Realization of these functions in turn requires building block circuits such as amplifiers, integrators, comparators, and reference networks.

The benefits of integrating the analog interface electronics in the same technology used to implement the digital signal and data processing functions stem from both economic and performance considerations. Traditional analog circuit design using bipolar technology requires that interface circuits be realized in a technology that differs greatly from that in which most of the system itself is implemented, i.e., complementary MOS (CMOS). As a consequence, although the interface circuits may represent a steadily decreasing proportion of an entire system, they have often come to represent an increasingly burdensome share of the system cost. It is also true that the performance of digital systems is typically limited by the capabilities of the interface circuits. Realization of the interface and data processing circuits in separate technologies implies the existence of an additional electronic and packaging interface that may significantly degrade the overall interface performance.

The above considerations have provided a strong motivation to implement analog interface functions in MOS technology, and the contributions represented in this reprint volume are a result of that motivation. For high-performance analog circuit design, the inherent limitations of MOS devices present a difficult challenge. Compared with bipolar devices, the disadvantages of MOS transistors include low transconductance, large device mismatch, and high noise levels. However, MOS technology also offers some considerable advantages for analog circuit design, namely, the availability of capacitive charge storage and fast, zero-offset analog switches. These advantages allow for the use of techniques such as offset storage and cancellation, dynamic biasing, analog pipelining, chopper stabilization, and double-correlated sampling to overcome the device limitations. Although specifically developed analog technologies have traditionally been regarded as essential to achieving the highest possible levels of performance, accomplishments of the past decade suggest that in on-chip interface applications the performance achievable with MOS circuits can often rival, and may even surpass, that obtainable with separate chips integrated in a dedicated analog bipolar technology.

Analog MOS Integrated Circuits edited by P. R. Gray, D. A. Hodges, and R. W. Brodersen, a volume in the IEEE PRESS Selected Reprint Series, appeared in 1980. Since that time, dramatic progress has been made both in the development of new techniques to allow MOS technology to realize high-performance analog interface functions, and in the use of the technology to address applications of economic importance. The present volume is intended to supplement the 1980 volume, and compiles work that has, for the most part, been published since that time.

This volume is divided into two major parts. The first is devoted to advances in circuit techniques for filtering, amplification, A/D conversion, and so forth. The second is devoted to representative applications in the areas of data acquisition, telecommunications, and data communications. Due to limited space, many important and innovative contributions in these areas could not be included, but we believe that those included here are representative of progress in the area since 1980.

PAUL R. GRAY
BRUCE A. WOOLEY
ROBERT W. BRODERSEN

Part I
Basic Concepts

The Design of High-Performance Analog Circuits on Digital CMOS Chips

ERIC A. VITTOZ, MEMBER, IEEE

Abstract —Devices available in digital oriented CMOS processes are reviewed, with emphasis on the various modes of operation of a standard transistor and their respective merits, and on additional specifications required to apply devices in analog circuits. Some basic compatible analog circuit techniques and their related tradeoffs are then surveyed by means of typical examples. The noisy environment due to cohabitation on the chip with digital circuits is briefly evoked.

I. INTRODUCTION

THE EVOLUTION of scaled-down digital processes will shift the boundary between digital and analog parts of systems [1]. However, analog circuits will remain irreplaceable components of systems-on-a-chip. Besides A/D conversion, they will always be needed to perform a variety of critical tasks required to interface digital with the external world, such as amplification, prefiltering, demodulation, signal conditioning for line transmission, for storage, and for display, generation of absolute values (voltages, currents, frequencies), and to implement compatible sensors on chip. In addition, analog will retain for a long while its advantage over digital when very high frequency or very low power is required.

Most of the limitations of analog circuits are due to the fact that they operate with electrical variables and not simply with numbers. Therefore, their accuracy is fundamentally limited by unavoidable mismatches between components, and their dynamic range is limited by noise, offset, and distortions.

For economical reasons, the analog part of a system-on-a-chip must be fully compatible with a process basically tailored for digital requirements, and this with a minimum number of additional specifications. Section II will review all active and passive devices available in digital CMOS processes, together with the additional specifications needed to use them for implementing analog circuits. Some basic analog circuit techniques will then be described in Section III by means of typical examples. Finally, the problems related to the noisy environment due to cohabitation on the chip with large digital circuits will be briefly evoked in Section IV.

Manuscript received November 2, 1984; revised December 24, 1984.
The author is with Centre Suisse d'Electronique et de Microtechnique. (CSEM, formerly CEH) Maladiere 71, 2000 Neuchâtel 7, Switzerland.

II. DEVICES AVAILABLE FOR ANALOG CIRCUITS

A. Transistors

A clear understanding of the various ways of biasing a normal MOS transistor, and of their respective merits, is a key factor in the design of optimum analog subcircuits. Fig. 1 illustrates the complete transfer characteristics $I_D(V_G)$ of a n-channel MOS transistor in saturation for various possible modes of operation. For the sake of symmetry, all potentials are defined with respect to that of the local substrate, in this case the p-well.

The general behavior of drain current in saturation I_D in the two basic modes of field effect operation can be described by two separate approximative models [2], [3] which sacrifice accuracy to clarity and simplicity:

Strong inversion ($I_D \gg \beta U_T^2$)

$$I_D = \frac{\beta}{2n}(V_G - V_{T0} - nV_S)^2,$$

$$\text{for } V_D > V_{D\,\text{sat}} = (V_G - V_{T0})/n. \quad (1)$$

Weak inversion ($I_D \ll \beta U_T^2$)

$$I_D = K\beta U_T^2 \exp\left((V_G - V_{T0} - nV_S)/nU_T\right),$$

$$\text{for } V_D > V_{D\,\text{sat}} = 3 \text{ to } 4U_T. \quad (2)$$

These models only include the three most important device parameters required for circuit design

$\beta = \mu C_{ox} W/L$ transfer parameter for strong inversion

V_{T0} gate threshold voltage for $V_S = 0$

n slope factor in weak inversion, which also describes approximately the effect of fixed charges in the channel in strong inversion. Its value depends slightly on V_S [3] and ranges usually from 1.3 to 2.

K is a factor somewhat larger than 1, which connects weak and strong inversion. Its exact value has no importance in circuit design, since transistors in weak inversion must be biased at fixed drain current I_D to avoid the very high sensitivity to $U_T = kT/q$ and V_{T0} for fixed gate voltage V_G.

In CMOS logic circuits, transistors usually operate with $V_S = 0$ as shown in heavy lines in Fig. 1. Their role is to

Reprinted from *IEEE J. Solid-State Circuits*, vol. SC-20, no. 3, pp. 657–665, June 1985.

3

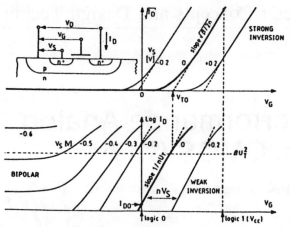

Fig. 1. General transfer characteristics $I_D(V_G)$ of a MOS transistor in saturation. Different modes of operation can be identified, namely strong inversion, weak inversion, and bipolar.

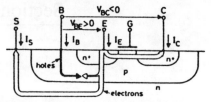

Fig. 2. Flows of carriers in bipolar operation.

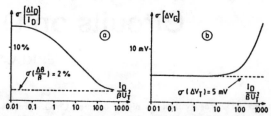

Fig. 3. Matching of a pair of MOS transistors as a function of drain current, for (a) same gate voltage and (b) same drain current. Uncorrelated components with mean standard deviations of 2 percent for $\Delta\beta/\beta$ and 5 mV for ΔV_T are assumed in this example.

provide maximum drain current in the "on" state for $V_G = V_{cc}$, and minimum residual current I_{D0} in the "off" state for $V_G = 0$. The only requirements for digital circuits are thus a maximum possible value of transfer parameter β, and a value of threshold voltage V_{T0} as low as possible while ensuring acceptable value of residual current for $V_G = 0$

$$I_{D0} = K\beta U_T^2 \exp(-V_{T0}/nU_T). \qquad (3)$$

This is only possible if slope factor n of weak inversion is not too large. These requirements are also favorable to analog circuits, since they allow a maximum value of transconductance g_m which can be easily derived from (1) and (2) as

$$g_m = ((2\beta I_D)/n)^{1/2} = 2I_D/(V_G - V_{T0} - nV_S)$$
$$\text{(strong inversion)} \qquad (4)$$

$$g_m = I_D/nU_T \qquad \text{(weak inversion)}. \qquad (5)$$

However, specifications on the maximum range of variation of β, V_{T0}, and n are usually necessary.

Transconductance g_m is proportional to drain current I_D in weak inversion, but only to the square root of I_D in strong inversion. If source voltage V_S is not zero, the gate voltage for constant drain current is shifted by nV_S for both modes of operation. Thus transconductance g_{ms} from source to drain is given by

$$g_{ms} = ng_m. \qquad (6)$$

Fig. 1 also shows that when gate voltage V_G is sufficiently negative, it has not more effect on drain current, which means that gate transconductance g_m decreases to zero. However, I_D can still be controlled by negative values of source voltage V_S which corresponds to a forward-biased source junction. The device then operates as a lateral bipolar, with the flow of carriers pushed away from the surface by the negative gate potential [4]. The various flows of carriers in this mode of operation are shown in Fig. 2. Source, drain, and p-well have been renamed emitter E,

collector C, and base B. Since a large fraction of emitter current I_E flows to substrate, the maximum alpha-gain of this lateral bipolar is only 0.2 to 0.6, depending on the process. However, owing to the low rate of recombination in the well, the β-gain can reach quite acceptable values ranging from 20 to 500. This high value of current gain is only obtained for the transistor implemented in the well, in this case a n-p-n. The n-p-n to substrate can be used without lateral collector, but only in common collector configurations.

For implementing analog circuits, it is necessary to specify the matching properties of similar adjacent transistors. Matching must be characterized by two independent statistical values: threshold mismatch ΔV_T, which may have in practice a mean standard deviation ranging from 1 to 20 mV, and $\Delta\beta/\beta$ mismatch which is usually in the range of 0.5–5 percent. Fig. 3 shows that when two transistors have the same gate voltage, as in a current mirror, the mismatch of their drain currents

$$\Delta I_D/I_D = \Delta\beta/\beta - (g_m/I_D)\Delta V_T \qquad (7)$$

is maximum in weak inversion, for which g_m/I_D is maximum, and only comes down to $\Delta\beta/\beta$ when the transistors operate deeply in strong inversion. On the contrary, when they have the same drain current, as in a differential pair, the mismatch of their gate voltages

$$\Delta V_G = \Delta V_T - (I_D/g_m)\cdot\Delta\beta/\beta \qquad (8)$$

is just ΔV_T in weak inversion, and increases in strong inversion where g_m/I_D is reduced.

Noise is a very important limitation of most analog circuits. As shown in Fig. 4, the noise of a transistor must also be characterized by at least two independent sources: White channel noise is independent of the process and corresponds to an equivalent input noise resistance R_N approximately equal to the inverse of transconductance g_m [5]. Gate interface $1/f$ noise dominates at low frequencies

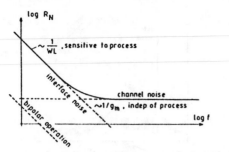

Fig. 4. Contributions to equivalent input noise resistance R_N of a MOS transistor.

TABLE I
SPECIFICATIONS ON TRANSISTORS FOR ANALOG AND DIGITAL APPLICATIONS.

(Analog Requires Specifications on Additional Parameters, and on Maximum and Minimum Values of Some Parameters.)

Parameter	Digital	Analog
β	min.	min.–max.
v_T	max.	min.–max.
n	max.	min.–max.
I_{DO}	max.	max.
β –bipolar	——	min.
mismatch	——	max.
1/f noise	——	(max.)
output resist.	——	min.

and is approximately independent of drain current. It is inversely proportional to gate area, and very sensitive to process quality. It should therefore be eliminated by circuit techniques such as chopping or autozeroing [6]–[8].

Both flicker noise and threshold mismatch are drastically reduced when the transistor operates in the lateral bipolar mode [4]. This is because the device is then shielded from all surface effects.

The respective qualitative specifications on transistors for digital and analog applications are summarized in Table I. An additional requirement for analog is a high value of output resistance which is approximately proportional to channel length. Designs should be made independent of the exact value of this parameter.

B. Passive Components

In digital CMOS circuits, passive components, namely capacitors and resistors, are only present as parasitics and should therefore by minimized. On the contrary, functional passive components of reasonable values and acceptable quality are required in most analog subcircuits.

Excellent precision capacitors can be implemented in a compatible way by using the silicon dioxide dielectric, provided both electrodes have a sufficiently low resistivity. Thin oxide gate capacitors are available in metal gate technologies, but they cannot be implemented in Si-gate processes without additional steps. For processes with a single polysilicon layer, the only reasonable choice is the capacitor between aluminum and polysilicon layers [9], which usually achieves rather low specific values. Many modern technologies provide two layers of polysilicon that can be used as electrodes for the capacitors [10].

Good resistors of less than 100 Ω/sq. can be obtained in the polysilicon layer. Higher values of few kiloohms per square are possible by using the well diffusion, but these resistors are slightly voltage dependent, and they are always associated with a large parasitic capacitance. Lightly doped polysilicon resistors such as those used to implement quasi-static RAM's [11] achieve very high values but they have a very poor accuracy.

Most of the modern design techniques for analog circuits are based on ratios of capacitances or resistances, and therefore only require specifications on matching and linearity of passive devices. If absolute values are needed as well, data on spread, temperature behavior, and aging must be available, and must be ensured by periodic statistical measurements.

No floating diode is usually available, except the base–emitter junction of the bipolar transistor to substrate. Some special micropower processes offer a lateral diode in the polysilicon layer [12].

III. BASIC ANALOG CIRCUIT TECHNIQUES

A. Optimum Matching

Most analog circuit techniques are based on the matching properties of similar components. For a given process, matching of critical devices may be improved by enforcing the set of rules that are summarized in Table II. These rules are not specific to CMOS and are applicable to all kinds of IC technologies. The relevancy and the quantitative importance of each of these rules depend on the particular process and on the particular device under consideration.

1) Devices to be matched should have the same structure. For instance, a junction capacitor cannot be matched with an oxide capacitor. This also means that the error due to parasitic junction capacitors cannot be compensated by adjusting the value of functional oxide capacitors.

2) They should have same temperature, which is no problem if power dissipated on chip is very low. Otherwise, devices to be matched should be located on the same isotherm, which can be obtained by a symmetrical implementation with respect to the dissipative devices.

3) They should have same shape and same size. For example, matched capacitors should have same aspect ratios, and matched transistors or resistors should have same width and same length, and not simply same aspect ratios.

4) Minimum distance between matched devices is necessary to take advantage of spatial correlation of fluctuating physical parameters.

5) Common-centroid geometries should be used to cancel constant gradients of parameters. Good practical examples

TABLE II
RULES FOR OPTIMUM MATCHING

1. Same structure
2. Same temperature
3. Same shape, same size
4. Minimum distance
5. Common—centroid geometries
6. Same orientation
7. Same surroundings
8. Non minimum size

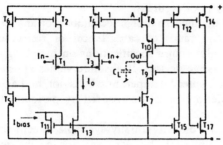

Fig. 5. Single-stage cascode operational transconductance amplifier (OTA) [13].

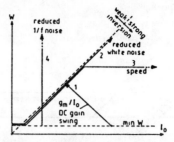

Fig. 6. Optimization of width W and tail current I_0 of input pair $T_1 - T_3$. Arrowed paths 1 to 4 indicate displacements in plane (W, I_0) for maximum improvement of various features of the amplifier.

are the quad configuration used to implement a pair of transistors, and common-centroid sets of capacitors.

6) The same orientation on chip is necessary to eliminate dissymmetries due to unisotropic steps in the process, or to the unisotropy of the silicon substrate itself. In particular, the source to drain flows of current in matched transistors should be strictly parallel.

7) Devices to be matched should have the same surroundings in the layout. This to avoid for instance the end effect in a series of current sources implemented as a line of transistors, or the street effect in a matrix of capacitors.

8) Using nonminimum size is an obvious way of reducing the effect of edge fluctuations, and to improve spatial averaging of fluctuating parameters.

Matching can be extended to the realization of non-unity n/m ratios by separately grouping m and n matched devices. A slight alteration of one or many devices is necessary when intermediate ratios are required.

B. Amplifiers

The basic configurations and tradeoffs related to the realization of amplifiers can be discussed with the example of a single-stage cascode OTA represented in Fig. 5 [13].

Differential pair $T_1 - T_3$ converts the differential input voltage into a difference of currents which is integrated in load capacitance C_L. These transistors can have minimum channel length since they are loaded by the high input conductance of current mirrors. Remaining design parameters are then channel width W and value of tail current I_0. Optimization of this input pair therefore amounts to choosing the best possible point in the (W, I_0) plane, with respect to conflicting requirements. This plane is represented in Fig. 6, with the limit between weak and strong inversion which corresponds to a given value of W/I_0.

Displacements in the plane for maximum improvement of various important features of the whole amplifier are represented in a qualitative way.

Transconductance for a given current, dc gain, and maximum possible swing are all improved by increasing W/I_0 to approach weak inversion, where they reach their maximum values (path 1). This also reduces input offset voltage.

The white noise spectral density is inversely proportional to transconductance g_m, which increases linearly with current I_0 up to the upper limit of weak inversion. To keep the advantages of weak inversion, a further increase of g_m requires a parallel increase of width W and current I_0 (path 2), which is only limited by size.

Speed (path 3) is proportional to g_m as long as parasitic capacitances of the transistors (proportional to W) are constant or negligible. A further increase in speed requires a progressive incursion into strong inversion, which results in progressive degradations of dc gain and of maximum possible swing. Speed in strong inversion only increases with the square root of current.

Low frequency $1/f$ noise is reduced by increasing channel width (path 4) and by choosing the better type of transistor, which is usually a p-channel.

If input current can be tolerated, very low $1/f$ noise and very high speed can be achieved by using transistors operated as lateral bipolars [4], [14].

The maximum differential current available from the pair is equal to tail current I_0, which puts a fundamental limit to slew rate. This problem can be circumvented by momentarily increasing current I_0 by a fixed amount each time an input step is anticipated, which yields a dynamic amplifier [15]. It can also be increased by an amount proportional to the difference of drain currents to realize an adaptive bias [16], [17], which provides operation in class AB.

Complementary pairs $T_1 - T_2$ and $T_3 - T_4$ can be viewed as common emitter amplifiers, each of which amplifies half of the total differential input voltage. Gain $g_{m1,3}/g_{m2,4}$ must be high enough, of the order of 3 to 10, to have noise and offset voltage limited to the only contribution of the differential input pair. This requires operation of transistors T_2 and T_4 deep enough into strong inversion. Too much gain reduces the stability phase margin.

As was illustrated by Fig. 3, the mismatch of current mirrors is maximum in weak inversion. It can be shown

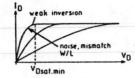

Fig. 7. Optimization of W/L of current mirrors.

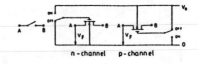

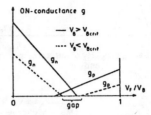

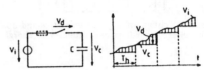

Fig. 8. Realization of an analog switch. On-conductances g_n and g_p of n-channel and p-channel transistors depend on floatation voltage V_F.

Fig. 9. Elementary sample-and-hold.

that this is true as well for both white and $1/f$ noise [18]. However, according to relations (1) and (2), weak inversion provides the minimum possible value of drain saturation voltage of the order of 100 mV. Therefore, the optimization of W/L of current mirrors $T_{11} - T_{13} - T_{15}$, $T_2 - T_6$, $T_4 - T_8$, $T_5 - T_7$ amounts to an acceptable compromise between small mismatch and low noise ($V_G - V_{T0}$ large), and small saturation voltage ($V_G - V_{T0}$ small) to permit large-signal swing (Fig. 7).

The overall transconductance of the amplifier can be multiplied by ratio A of mirror $T_4 - T_8$, at the expense of a reduction in phase margin. Some of the mirrors can be avoided by using the folded cascode scheme [20].

Cascode transistors T_9 and T_{10} decrease the output conductance by a factor equal to g_{ms}/g_o. This is obtained without any noise penalty, and with only a very small reduction of phase margin. The resulting dc gain is thus higher than that of a two-stage noncascode amplifier which requires internal compensation. Gain may be further boosted by using double or triple cascode [19], until it becomes limited by the direct conductance to ground due to impact ionization in the drain depletion layers.

The reduction of maximum output swing due to the cascode transistors can be minimized by careful design of the bias circuitry $T_{12} - T_{14} - T_{15} - T_{17}$ [13], [20]. Drain voltages of transistors T_7 and T_8 can be made equal to their limit value V_{Dsat} for saturation, independently of bias current. Maximum output swing is then only reduced by $4V_{Dsat}$ with respect to total supply voltage, which only amounts to about 400 mV in weak inversion.

The circuit can be modified to provide differential output [20]. This doubles the maximum output swing, but requires a common mode feedback scheme.

All amplifiers based on a differential input pair suffer noise and speed penalties with respect to a simple CMOS inverter used as an amplifier with an adequate biasing scheme [21]. This kind of amplifier is furthermore free from any slew rate limitation. It represents a very attractive solution for very low power [7] or very high speed [22] applications, in spite of its poor intrinsic PSRR.

C. Switch and Sample-and-Hold

The realization of the analog switch, which is a very important component of CMOS analog circuits, is illustrated in Fig. 8. A n-channel transistor is switched on by connecting its gate to the positive power line V_B. However, its on-conductance g_n comes to zero if potential V_F at which the device floats is too high. The same is true if V_F is too low for on-conductance g_p of the p-channel transistor

with gate connected to zero. If total supply voltage V_B is larger than a critical value V_{Bcrit}, the conductance of the switch can be ensured independently of V_F by connecting both types of transistors in parallel. Below this critical value, a gap of conduction appears at intermediate levels of V_F. Because of substrate effects, this critical supply voltage may widely exceed the sum $V_{T0p} + V_{T0n}$ of p and n thresholds for zero source voltages. For example, $V_{T0p} = 0.6$ V and $V_{T0n} = 0.7$ V may correspond to $V_{Bcrit} = 2.3$ V [18], [23]. This very severe limitation to low-voltage operation of analog circuits may be circumvented by on-chip clock voltage multiplication [8], [24].

Leakage in the off state is due to residual channel current and to reverse currents of junctions. Care must be taken not to bootstrap the switch potential beyond that of the power supply lines, which would forward bias these junctions [25].

The combination of a switch and a capacitor provides a basic sample-and-hold shown in Fig. 9. Voltage V_C across capacitor C keeps a constant value equal to that of input voltage V_i at the last sampling instant. The value the of noise voltage at the sampling instant is also frozen in capacitor C; therefore, the total noise power is concentrated below the clock frequency. Voltage V_d across the switch is readjusted to zero each time the switch is closed, which corresponds to the transfer function for the fundamental signal (component of output signal V_d at the frequency of input signal V_i) shown in Fig. 10 [26]. At low frequency, this autozeroing by means of a sample-and-hold amounts to a differentiation, with a time constant equal to half the value of hold duration T_h. It may be used to cancel offset and to reduce low-frequency noise components generated in a circuit [6], [27].

Another source of sampling error is caused by the charge which is released from the channel into holding capacitor C when the transistor of the switch is blocked [28]. This problem has been analyzed in the general case shown in

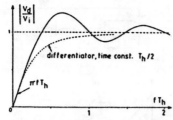

Fig. 10. Differentiating property of autozeroing obtained by sample-and-hold.

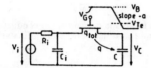

Fig. 11. Equivalent circuit of a practical sample-and-hold. Finite fall time of gate voltage V_G allows redistribution of charge through the transistor.

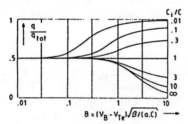

Fig. 12. Calculated fraction q of total charge q_{tot} left in holding capacitor C after switch-off [23].

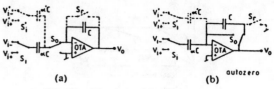

(a)   (b)

Fig. 13. Stray insensitive SC integrators.

Fig. 11 where the source of the signal is assumed to have an internal capacitance C_i [23], [26]. Finite fall time of V_G (slope $-a$) from initial value V_B to effective threshold voltage V_{Te} allows a redistribution through the transistor of total charge $q_{tot} = C_{gate}(V_B - V_{Te})$ between C and C_i. The result obtained by numerical integration of a normalized nonlinear equation describing this process, for time constant $R_i C_i$ much larger than the switching time, is represented in Fig. 12. This figure shows the fraction q of total channel charge q_{tot} which goes into holding capacitor C for various values of ratio C_i/C. This fraction is a function of an intermediate parameter B which combines clock amplitude, clock slope, β of transistor, and value of holding capacitor C. These curves suggest various strategies for minimizing parasitic charge q.

A first possibility is to choose C_i very large and B much larger than 1. All charges released into C flow back into C_i during the decay of gate voltage, and q tends to zero (some easily calculable additional charge is due to the coupling through overlap capacitors after switching off). The drawback is the long period of time needed for switching off.

A second solution is to equilibrate the values of both capacitors [29]. By symmetry, half of the channel charge flows in each capacitor, and can be compensated by half-sized dummy switches that are switched on when the main switch is blocked [28].

The need for equal values of capacitors may be eliminated by choosing a value of B much smaller than 1 which also

ensures equipartition of the total charge. Charge q is then compensated by a single dummy switch.

When complementary transistors are used to implement the switch, they partially compensate each other, although matching is very poor. The effect of charge injection can be drastically reduced by appropriate circuit techniques such as differential implementation [25], and active compensation by a low sensitivity auxiliary input [26], [30].

D. Switched Capacitor Integrators

Switched capacitor integrators are the building blocks of all kinds of circuits, in particular SC filters. Two different implementations that are insensitive to parasitic capacitances to ground are shown in Fig. 13. Both provide a differential input and a time constant $1/\alpha f_c$, which only depends on clock frequency f_c and ratio of capacitors α. Version b includes autozeroing, which reduces low-frequency noise and compensates offset. It can therefore be realized with a nondifferential amplifier such as a CMOS inverter [7].

Output resetting can be obtained by additional switch S_r, and many differential signals can be separately weighted and summed by repeating the input circuitry, as shown in dotted lines. These integrators may be damped at will by connecting one of these additional inputs to output.

They can be transformed into amplifiers with controlled gain, either by resetting the output at every clock cycle, or by deleting integrating capacitor C in a damped configuration.

E. Comparators

Comparators must usually achieve a very low value of input offset voltage. An excellent solution is obtained by removing integrating capacitor C in the basic integrator of Fig. 13b [31]. Any difference $V_{i+} - V_{i-}$ will cause an output current to charge (or discharge) the parasitic output capacitance. The comparator thus behaves as an integrator of input error voltage, and sensitivity is proportional to the time allotted for comparison. It is ultimately limited by the finite dc gain of the amplifier. The speed-sensitivity ratio may be increased by achieving nth order integration along a cascade of n stages [26], as shown in Fig. 14.

The effects of charge injection and switching noise may be virtually cancelled by sequentially opening switches S_1 to S_n before toggling switch S_0: When S_1 is opened first, charge injection and sampled noise cause an error voltage across C_1. Since switch S_2 is still closed, a compensation voltage appears across C_2 after equilibration. The same is true when S_2 to S_{n-1} are then opened sequentially. The

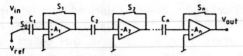

Fig. 14. Multistage comparator.

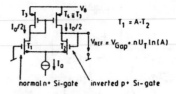

Fig. 17. Extraction of bandgap voltage V_{gap} by MOS transistors [36].

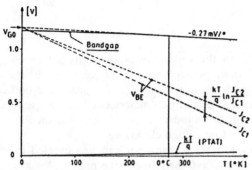

Fig. 15. "Built-in" voltages available in silicon.

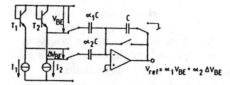

Fig. 18. Principle of SC-weighted bandgap reference.

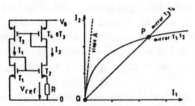

Fig. 16. Extraction of $U_T = kT/q$ with MOS transistors $T_1 - T_2$ operated in weak inversion.

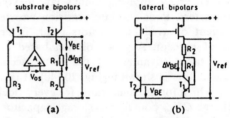

Fig. 19. Principle of R-weighted bandgap reference.

only residual error at the input is thus that due to switch S_n divided by the gain of the $n-1$ first stages [32], [33]. Accuracy is further improved by using a fully differential implementation for which values of offset as low as 5 µV have been reported [34].

F. Voltage and Current References

The realization of absolute references must be based on intrinsic physical values, in order to reduce their sensitivity to process variations.

The various "built-in" voltages provided by silicon are represented in Fig. 15. They can be extracted by adequate circuits to implement voltage references.

Thermal voltage $U_T = kT/q$, proportional to absolute temperature (PTAT), can be extracted by two MOS transistors operated in weak inversion with different current densities, as shown in Fig. 16 [3], [35]. If $T_3 = T_4$, application of weak inversion model (2) to transistors T_1 and $T_2 = AT_1$ yields

$$V_{ref} = U_T \ln(A) \qquad (9)$$

which in turn imposes current I_2 in the circuit.

Bandgap voltage V_{gap} decreases approximately linearly with temperature from extrapolated value V_{G0}, with a slight curvature. A possible technique for direct extraction of V_{gap} is shown in Fig. 17 [36]. Transistor T_1 is n-channel with a normal n$^+$-doped silicon gate. Transistor T_2 is also

n-channel but with a p$^+$-doped gate, as is possible in some technologies. Their threshold voltages therefore differ by approximately V_{gap} which appears at the output of this simple amplifier connected in unity gain configuration. After compensation by a small PTAT voltage obtained by operating T_1 and T_2 in weak inversion at different current densities, a temperature coefficient lower than 30 ppm °C can be obtained. All references based on MOS operation have their accuracy degraded by large uncontrolled offset components due to surface effects.

Base–emitter voltage V_{BE} of bipolar transistors is not really a physical parameter, but it only depends very slightly on the process. The difference ΔV_{BE} of two bipolars operated at different current densities is strictly proportional to kT/q. After multiplication by an adequate factor, it can be added to V_{BE} to obtain a voltage of value V_{G0} independent of temperature. This principle of bandgap reference is well known in bipolar technology, and can be applied to the bipolars available in CMOS technology.

Weighting and summing V_{BE} and ΔV_{BE} can be achieved by the SC circuit shown in Fig. 18, which is derived from the integrator of Fig. 13(b) [23], [37], [38]. Transistors T_1 and T_2 are bipolars to substrate. Accuracy is mainly limited by charge injection from the feedback switch.

Another solution consists in using a resistive divider, as depicted in Fig. 19. Version a uses substrate bipolars and a CMOS amplifier [39]. The offset of this amplifier, which is multiplied by R_2/R_1 of the order of 10, causes independent errors of the value of V_{ref} and of its temperature coefficient. Adjustment at two different temperatures would thus be required to achieve an accuracy of a few millivolts. Version b uses compatible lateral bipolars and avoids any MOS amplifier [4]. The error due to the p-channel mirror is

9

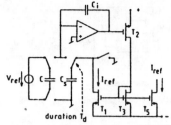

Fig. 20. SC voltage-to-current converter [40].

lowered by operating deeply in strong inversion. The offset of bipolars is small and only causes an error PTAT, which can be corrected at a single temperature.

Semiconductor physics do not provide any "built-in" current. Current references must thus be derived from voltage references by applying Ohm's law in voltage to current converters. Poor absolute precision and temperature coefficient of available resistors result in errors of many tens of percent. A good solution based on a SC scheme that takes advantage of the better accuracy of available capacitors is shown in Fig. 20 [40]. It is a closed loop system which forces equilibrium, for every clock cycle, between charge CV_{ref} poured into storage capacitor C_s and charge $I_{ref}T_d$ withdrawn from the same capacitor. Thus

$$I_{ref} = CV_{ref}/T_d. \qquad (10)$$

Generation of an independent frequency on chip is not possible with a precision better than a few tens of percent. It is normally not required since an accurate clock frequency is usually provided by the system, from which synchronous signals of any frequency can be derived in a digital way. Totally asynchronous signals of accurate frequency can be produced by quasi-sinusoidal SC oscillators [41]–[43].

IV. ANALOG CIRCUITS IN A DIGITAL ENVIRONMENT

If no precaution is taken, the dynamic range of analog circuits will be limited by the noise generated by digital circuits operating on the same chip. This problem is specially accute for sampled circuits which fold down high-frequency noise components by undersampling.

Coupling may occur through power lines, current in the common substrate, capacitive links, and possibly by minority carriers that are released when digital transistors are being blocked.

Various provisions can be suggested to improve the situation: Utilization of separate power lines, including pads, bondings, and possibly pins. Implementation of power supply filters or regulators. High value of PSRR, also at high frequencies for sampled circuits; care must be taken not to destroy the PSRR of amplifiers by the additional circuitry. Systematic implementation of fully differential structures. Avoidance of large current spikes in digital circuits, and of any digital transition during critical analog tasks. Provision of maximum distance on chip to digital lines, and of separate clean clocks for analog cir-

cuits. Critical nodes should be shielded from substrate and from digital lines by adequate layers, and analog circuits can be separated by special wells that collect parasitic minority carriers. Processes that provide an epitaxial layer on a highly doped substrate, to improve immunity to latch-up, allow to drain all parasitic currents to substrate.

V. CONCLUSION

Thanks to the versatility of the CMOS technology, all kinds of analog circuits can be combined on the same chip with digital circuits, without any process modification. However, additional parameters need to be specified and guaranteed, the number of which depends on the type of function and on design cleverness.

A standard MOS transistor can be operated in various modes which have their respective merits. Weak inversion provides maximum values of gain and signal swing, and minimum offset and noise voltages. Strong inversion is required to achieve high speed, and provides minimum relative values of offset and noise currents. The lateral bipolar mode exhibits excellent matching properties and very low $1/f$ noise, and can be used to implement a variety of schemes previously developed for normal bipolars transistors.

Mismatch of active and passive devices represents a major limitation to the accuracy of analog circuits. It can be minimized by respecting a set of basic rules. Designs must be based on sound concepts that take maximum advantage of all available components.

Single-stage cascoded OTA's should be preferred to multistage amplifiers. Their optimization amounts to choosing the best possible compromise with respect to various conflicting requirements. The excellent performance of the MOS transistor as a switch and the availability of high-quality capacitors are key elements in the implementation of a variety of analog functions. The elementary sample-and-hold that is made possible by the absence of any dc gate control current can be used to reduce low-frequency noise and to compensate offset. Its major limitation is due to charge injection from the switch, which can be evaluated and partially compensated by an adequate strategy.

Accurate absolute references are very critical circuits which must be based on intrinsic physical values to achieve low sensitivity to process. This is possible for voltage references, from which current references can be derived by applying Ohm's law.

Precautions must be taken to avoid degradation of analog performances by the noise generated by the digital part of the chip.

REFERENCES

[1] R. W. Brodersen and P. R. Gray, "The role of analog circuits in future VLSI technologies," in *ESSCIRC '83 Dig. Tech. Papers,* Sept. 1983, pp. 105–110.
[2] J. D. Chatelain, "Dispositifs à semiconducteurs," *Traité d'Electricité EPFL,* vol. VII (Georgi, St-Saphorin, Switzerland), 1979.

[3] E. Vittoz and J. Fellrath, "CMOS analog integrated circuits based on weak inversion operation," *IEEE J. Solid-State Circuits*, vol. SC-12, pp. 224–231, June 1977.

[4] E. Vittoz, "MOS transistors operated in the lateral bipolar mode and their applications in CMOS technology," *IEEE J. Solid-State Circuits*, vol. SC-18, pp. 273–279, June 83.

[5] J. Fellrath and E. Vittoz, "Small signal model of MOS transistors in weak inversion," in *Proc. Journées d'Electronique 1977* (EPF-Lausanne), pp. 315–324.

[6] H. W. Klein and W. Engl, "Design techniques for low noise CMOS operational amplifiers," in *ESSCIRC '84 Dig. Tech. Papers* (Edinburgh), Sept. 1984, pp. 27–30.

[7] F. Krummenacher, "Micropower switched capacitor biquadratic cell," *IEEE J. Solid-State Circuits*, vol. SC-17, pp. 507–512, June 1982.

[8] M. Degrauwe and F. Salchli, "A multipurpose micropower SC-filter," *IEEE J. Solid-State Circuits*, vol. SC-19, pp. 343–348, June 1984.

[9] E. Vittoz and F. Krummenacher, "Micropower SC filters in Si-gate technology," in *Proc. ECCTD '80* (Warshaw), Sept. 1980, pp. 61–72.

[10] J. L. McCreary, "Matching properties, and voltage and temperature dependence of MOS capacitors," *IEEE J. Solid-State Circuits*, vol. SC-16, pp. 608–616, Dec. 1981.

[11] Y. Uchida *et al.*, "A low power resistive load 64 kbit CMOS RAM," *IEEE J. Solid-State Circuits*, vol. SC-17, pp. 804–909, Oct. 1982.

[12] M. Dutoit and F. Sollberger, "Lateral polysilicon p-n diodes," *J. Electrochem. Soc.*, vol. 125, pp. 1648–1651, Oct. 1978.

[13] F. Krummenacher, "High voltage gain CMOS OTA for micropower SC filters," *Electron. Lett.*, vol. 17, pp. 160–162, 1981.

[14] S. Gustafsson *et al.*, "Low-noise operational amplifiers using bi-polar input transistors in a standard metal gate CMOS process," *Electron. Lett.*, vol. 20, pp. 563–564, 1984.

[15] B. J. Hosticka, "Dynamic CMOS amplifiers," *IEEE J. Solid-State Circuits*, vol. SC-15, pp. 887–894, Oct. 1980.

[16] M. G. Degrauwe *et al.*, "Adaptive biasing CMOS amplifiers," *IEEE J. Solid-State Circuits*, vol. SC-17, pp. 522–528, June 1982.

[17] E. Seevinck and R. F. Wassenaar, "Universal adaptive biasing principle for micropower amplifiers," in *ESSCIRC '84 Dig. Tech. Papers* (Edinburgh), Sept. 1984, pp. 59–62.

[18] E. Vittoz, "Micropower techniques," *Advanced Summer Course on Design of MOS-VLSI Circuits for Telecommunications*, L'Aquila, Italy, June 1984, to be published by Prentice-Hall, 1985.

[19] P. W. Li *et al.*, "A ratio-independent algorithmic ADC technique," in *ISSCC Dig. Tech. Papers*, Feb. 1984, pp. 62–63.

[20] T. C. Choi *et al.*, "High-frequency CMOS switched-capacitor filters for communications application," *IEEE J. Solid-State Circuits*, vol. SC-18, pp. 652–664, Dec. 1983.

[21] F. Krummenacher, E. Vittoz, and M. Degrauwe, "Class *AB* CMOS amplifiers for micropower SC filters," *Electron. Lett.*, vol. 17, pp. 433–435, June 1981.

[22] S. Masuda *et al.*, "CMOS sampled differential push–pull cascode operational amplifier," in *Proc. ISCAS '84* (Montreal, Ont., Canada), May 1984, p. 1211.

[23] E. Vittoz, "Microwatt SC circuit design," "Summer course on SC circuits," KU-Leuven, Belgium, June 1981, republished in *Electro-component Sci. Technol.*, vol. 9, pp. 263–273, 1982.

[24] F. Krummenacher, H. Pinier, and A. Guillaume, "Higher sampling rates in SC circuits by on-chip clock-voltage multiplication," in *ESSCIRC '83 Dig. Tech. Papers* (Lausanne), pp. 123–126, Sept. 1983.

[25] D. J. Alstott, "A precision variable-supply CMOS comparator," *IEEE J. Solid-State Circuits*, vol. SC-17, pp. 1080–1087, Dec. 1982.

[26] E. Vittoz, "Dynamic analog techniques," *Advanced Summer Course on Design of MOS-VLSI Circuits for Telecommunications*, L'Aquila, Italy, June 1984, to be published by Prentice-Hall, 1985.

[27] C. Enz, "Analysis of the low-frequency noise reduction by autozero technique," *Electron. Lett.*, vol. 20, pp. 959–960, Nov. 1984.

[28] E. Suarez *et al.*, "All-MOS charge redistribution analog-to digital conversion techniques," *IEEE J. Solid-State Circuits*, vol. SC-10, pp. 379–385, Dec. 1975.

[29] L. Bienstman and H. J. De Man, "An eight-channel 8-bit micro-processor compatible NMOS converter with programmable scaling," *IEEE J. Solid-State Circuits*, vol. SC-15, pp. 1051–1059, Dec. 1980.

[30] M. G. Degrauwe, E. Vittoz, and I. Verbauwhede, "A micropower CMOS instrumentation amplifier," in *ESSCIRC '84 Dig. Tech. Papers* (Edinburgh), Sept. 1984, pp. 31–34.

[31] Y. S. Lee *et al.*, "A 1 mV MOS comparator," *IEEE J. Solid-State Circuits*, vol. SC-13, pp. 294–297, June 1978.

[32] R. Poujois *et al.*, "Low-level MOS transistor amplifier using storage techniques," in *ISSCC Dig. Tech. Papers*, 1973, pp. 152–153.

[33] A. R. Hamade, "A single chip all-MOS 8-bit ADC," *IEEE J. Solid-State Circuits*, vol. SC-14, pp. 785–791, Dec. 1978.

[34] R. Poujois and J. Borel, "A low drift fully integrated MOSFET operational amplifier," *IEEE J. Solid-State Circuits*, vol. SC-13, pp. 499–503, Aug. 1978.

[35] E. Vittoz and O. Neyroud, "A low-voltage CMOS bandgap reference," *IEEE J. Solid-State Circuits*, vol. SC-14, pp. 573–577, June 1979.

[36] H. Oguey and B. Gerber, "MOS voltage reference based on poly-silicon gate work function difference," *IEEE J. Solid-State Circuits*, vol. SC-15, pp. 264–269, June 1980.

[37] O. Leuthold, "Integrierte Spannungsüberwachungsschaltung," presented at Meet. Swiss Chapter Solid-State Devices Circuits (Bern, Switzerland), Oct. 1981.

[38] B. S. Song and P. R. Gray, "A precision curvature-compensated CMOS bandgap reference," *IEEE J. Solid-State Circuits*, vol. SC-18, pp. 634–643, Dec. 1983.

[39] R. Ye and Y. Tsividis, "Bandgap voltage reference sources in CMOS technology," *Electron. Lett.*, vol. 18, pp. 24–25, 1982.

[40] H. W. Klein and W. L. Engl, "A voltage–current-converter based on a SC-controller," in *ESSCIRC '83 Dig. Tech. Papers*, (Lausanne, Switzerland), Sept. 1983, pp. 119–122.

[41] E. Vittoz, "Micropower SC oscillator," *IEEE J. Solid-State Circuits*, vol. SC-14, pp. 622–624, June 1979.

[42] B. J. Hosticka *et al.*, "Switched-capacitor FSK modulator and demodulator in CMOS technology," in *ESSCIRC '83 Dig. Tech. Papers*, pp. 231–216 (Lausanne, Switzerland), Sept. 1983.

[43] F. Krummenacher, "A high resolution capacitance-to-frequency converter," in *ESSCIRC '84 Dig. Tech. Papers* (Edinburgh), Sept. 1984, pp. 95–98.

MOS Operational Amplifier Design—
A Tutorial Overview

PAUL R. GRAY, FELLOW, IEEE, AND ROBERT G. MEYER, FELLOW, IEEE

(Invited Paper)

Abstract—This paper presents an overview of current design techniques for operational amplifiers implemented in CMOS and NMOS technology at a tutorial level. Primary emphasis is placed on CMOS amplifiers because of their more widespread use. Factors affecting voltage gain, input noise, offsets, common mode and power supply rejection, power dissipation, and transient response are considered for the traditional bipolar-derived two-stage architecture. Alternative circuit approaches for optimization of particular performance aspects are summarized, and examples are given.

I. INTRODUCTION

THE rapid increase in chip complexity which has occurred over the past few years has created the need to implement complete analog–digital subsystems on the same integrated circuit using the same technology. For this reason, implementation of analog functions in MOS technology has become increasingly important, and great strides have been made in recent years in implementing functions such as high-speed DAC's, sampled data analog filters, voltage references, instrumentation amplifiers, and so forth in CMOS and NMOS technology [1]. These developments have been well documented in the literature. Another key technical development has been a maturing of the state of the art in the implementation of operational amplifiers (op amps) in MOS technology. These amplifiers are key elements of most analog subsystems, particularly in switched capacitor filters, and the performance of many systems is strongly influenced by op amp performance. Many of the developments in MOS operational amplifier design have not been as well documented in the literature, and the intent of this paper is to review the state of the art in this field. This paper is focused on the design of op amps for use within single-chip analog–digital LSI systems, and the particular problems of the design of stand-alone CMOS amplifiers are not addressed.

Manuscript received August 24, 1982; revised September 27, 1982. This work was supported by the Joint Services Electronics Program under Contract F49620-79-c-0178 and the National Science Foundation under Grant ENG79-07055.

The authors are with the Department of Electrical Engineering and Computer Sciences and the Electronics Research Laboratory, University of California, Berkeley, CA 94720.

In Section II, the important performance requirements and objectives for operational amplifiers within a monolithic analog subsystem are summarized. In Section III, the performance of the basic two-stage CMOS operational amplifier architecture is summarized. In Section IV, alternative circuit approaches for the improvement of particular performance aspects are considered. In Section V, the particular problems associated with NMOS depletion load amplifier design are considered, and in Section VI, the design of output stages is considered. Finally, a summary and discussion of the design of amplifiers in scaled technologies are presented in Section VII.

II. PERFORMANCE OBJECTIVES FOR MOS OPERATIONAL AMPLIFIERS

The performance objectives for operational amplifiers to be used within a monolithic analog subsystem are often quite different from those of traditional stand-alone bipolar amplifiers. Perhaps the most important difference is the fact that for many of the amplifiers in the system, the load which the output of the amplifier has to drive is well defined, and is often purely capacitive with values of a few picofarads. In contrast, stand-alone general-purpose amplifiers usually must be designed to achieve a certain level of performance independent of loading over capacitive loads up to several hundred picofarads and resistive loads down to 2 kΩ or less. Within a monolithic analog subsystem, only a few of the amplifiers must drive a signal off chip where the capacitive and resistive loads are significant and variable. In this paper, these amplifiers will be termed output buffers, and the amplifiers whose outputs do not go off chip will be termed internal amplifiers. The particular problems of the design of these output buffers are considered in Section VII.

A typical application of an internal operational amplifier, a switched capacitor integrator, is illustrated in Fig. 1. The basic function of the op amp is to produce an updated value of the output in response to a switching event at the input in which the sampling capacitor is charged from the source and discharged into the summing node. The output must assume the new updated value within the required accuracy, typically on the order of 0.1 percent, within one clock period, typically on the order of 1 μs for voiceband filters. Important performance

Reprinted from *IEEE J. Solid-State Circuits*, vol. SC-17, no. 6, pp. 969–982, Dec. 1982.

12

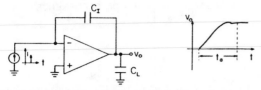

Fig. 1. Typical application of an internal MOS operational amplifier, a switched capacitor integrator.

TABLE I
TYPICAL PERFORMANCE, CONVENTIONAL TWO-STAGE CMOS
INTERNAL OPERATIONAL AMPLIFIER
(+/−5 V SUPPLY, 4 μm SI GATE CMOS)

dc gain (capacitive load only)	5000
Setting time, 1 V step, C_l = 5 pF	500 ns
Equiv. input noise, 1 kHz	100 nV/$\sqrt{\text{Hz}}$
PSRR, dc	90 dB
PSRR, 1 kHz	60 dB
PSRR, 50 kHz	40 dB
Supply capacitance	1 fF
Power dissipation	0.5 mW
Unity-gain frequency	4 MHz
Die area	75 mils^2
Systematic offset	0.1 mV
Random offset std. deviation	2 mV
CMRR	80 dB
CM range	within 1 V of supply

parameters are the power dissipation, maximum allowable capacitive load, open-loop voltage gain, output voltage swing, equivalent input flicker noise, equivalent input thermal noise, power supply rejection ratio, supply capacitance (to be defined later), and die area. In this particular application the input offset voltage, common-mode rejection ratio, and common-mode range are less important, but these parameters can be important in other applications. Because of the inherent capacitive sample/hold capability in MOS technology, dc offsets can often be eliminated at the subsystem level, making operational amplifier offsets less important. A typical set of values for the parameters given above for a conventional amplifier design in 4 μm CMOS technology are given in Table I. In the following section, the factors affecting the various performance parameters are evaluated for the most widely used amplifier architecture.

III. BASIC TWO-STAGE CMOS OPERATIONAL AMPLIFIER

Currently, the most widely used circuit approach for the implementation of MOS operational amplifiers is the two-stage configuration shown in Fig. 2(b). This configuration is also widely used in bipolar technology, and the bipolar counterpart is also illustrated in Fig. 2(a). The behavior of this circuit when implemented in bipolar technology has been reviewed in an overview article published earlier [2]. This circuit configuration provides good common mode range, output swing, voltage gain, and CMRR in a simple circuit that can be compensated with a single pole-splitting capacitor. While the implementation of this architecture in NMOS technology requires additional circuit elements because of the lack of a complementary device, many NMOS amplifiers commercially manufactured at the present time use a conceptually similar

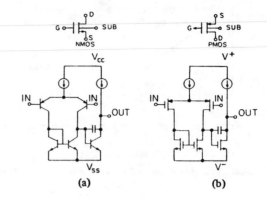

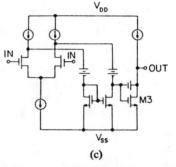

Fig. 2. Two-stage operational amplifier architecture. (a) Bipolar implementation. (b) CMOS implementation. (c) An example of an NMOS implementation with interstage coupling network.

configuration, as illustrated in Fig. 2(c) where a differential interstage level-shifting network composed of voltage and current sources has been inserted between the first and second stages so that both stages can utilize n-channel active devices and depletion mode devices as loads. The implementation of this circuit is discussed further in Section V.

In this section, we will analyze the various performance parameters of the CMOS implementation of this circuit, focusing particularly on the aspects which are different from the bipolar case.

Open Circuit Voltage Gain

An important difference between MOS and bipolar technology is the fact that the maximum transistor open circuit voltage gain g_m/g_o is much lower for MOS transistors than for bipolar transistors, typically by a factor on the order of 10–40 for typically used geometries and bias currents [3]. Under certain simplifying assumptions, voltage gain can be shown to be

$$g_m/g_o = \frac{2L}{V_{gs} - V_T} \left(\frac{dx_d}{dV_{ds}}\right)^{-1} \tag{1}$$

where x_d is the width of the depletion region between the end of the channel and the drain and L is the effective channel length. The expression illustrates several key aspects of MOS devices used as analog amplifiers. First, for constant drain current decreasing either the channel length or width results in a decrease in the gain, the latter because of the fact that V_{gs} increases. This fact, along with noise considerations, usually dictates the minimum size of the transistors that must be used in a given high-gain amplifier application. Usually, this is

13

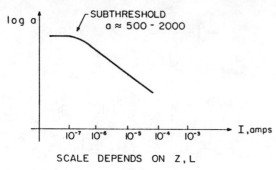

Fig. 3. Typical open circuit gain of an MOS transistor as a function of bias current.

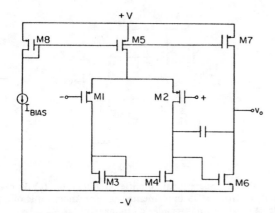

Fig. 4. Schematic of basic two-stage CMOS operational amplifier.

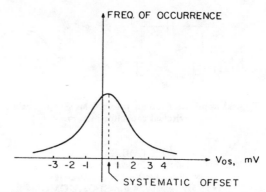

Fig. 5. Typical input offset distribution, MOS operational amplifier.

larger than the length and width used for digital circuits in the same technology.

Second, if the device geometry is kept constant, the voltage gain is inversely proportional to the square root of the drain current since $(V_{gs} - V_T)$ is proportional to the square root of the drain current. A typical variation of open circuit voltage gain as a function of drain current is shown in Fig. 3 [4]. The gain becomes constant at a value comparable to bipolar devices in the subthreshold range of current. This fact makes use of low current levels desirable, and at the same time complicates the design of high-speed amplifiers which must operate at high current.

Third, if device size and bias current are kept constant, the gain is an increasing function of substrate doping since dx_d/dV_{ds} decreases with increasing doping. Thus, devices which have received a channel implant to increase threshold voltage would display a higher open circuit gain than an unimplanted device whose channel doping was lower. Finally, the expression demonstrates that open circuit gain is not degraded by technology scaling in the constant field sense since all terms in the expression decrease in proportion. However, scaling in the quasi-constant field or constant voltage sense would result in a decrease in gain.

Turning to the operational amplifier, the voltage gain of the first stage of the circuit shown in Fig. 4 can be shown to be simply

$$A_{v1} = \frac{g_{m1}}{g_{o2} + g_{o4}} \qquad (2)$$

where g_m is the device transconductance and g_o is the small signal output conductance, and assuming that $M1$ and $M2$ are

identical and that $M3$ and $M4$ are identical. Similarly, the second stage voltage gain is

$$A_{v2} = \frac{g_{m6}}{g_{o6} + g_{o7}}. \qquad (3)$$

For switched capacitor filter applications, the overall voltage gain required is on the order of several thousand [5], implying a gain in each stage on the order of 50. In order to achieve this level of gain per stage, transistor bias currents and channel lengths and widths are usually chosen such that the transistor $(V_{gs} - V_T)$ is several hundred millivolts, and the drain depletion region is on the order of one fifth or less of the effective channel length at the typical drain bias of several volts. Circuit approaches to achieving more voltage gain or, alternatively, achieving the same voltage gain with smaller devices, are discussed in Section IV.

DC Offsets, DC Biasing, and DC Power Supply Rejection

The input offset voltage of an operational amplifier is composed of two components, the systematic offset and the random offset. The former results from the design of the circuit and is present even when all of the matched devices in the circuit are indeed identical. The latter results from mismatches in supposedly identical pairs of devices. A typical observed distribution of input offset voltages is shown in Fig. 5.

Systematic Offset Voltage

In bipolar technology, the comparatively high voltage gain per stage (on the order of 500) tends to result in a situation in which the input-referred dc offset voltage of an operational amplifier is primarily dependent on the design of the first stage. In MOS op amps, because of the relatively low gain per stage, the offset voltage of the differential to single-ended converter and second stage can play an important role. In Fig. 6, the operational amplifier of Fig. 4 has been split into two separate stages. Assuming perfectly matched devices, if the inputs of the first stage are grounded, then the quiescent output voltage at the drain of $M4$ is equal to the voltage at the drain of $M3$ ($M3$ and $M4$ have the same drain current and gate–source voltage, and hence must have the same drain–source voltage). However, the value of the gate voltage of $M6$ which is required to force the amplifier output voltage to zero may be different from the quiescent output voltage of the first stage. For a first stage gain of 50, for example, each 50 mV difference in these

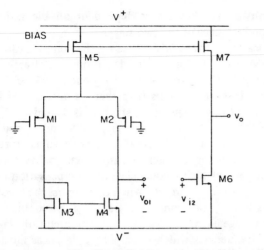

Fig. 6. Two-stage amplifier illustrating interstage coupling constraints.

voltages results in 1 mV of input-referred systematic offset. Thus, the W/L ratios of $M3$, $M4$, and $M6$ must be chosen so that the current density in these three devices are equal. For the simple circuit of Fig. 6, this constraint would take the form

$$\frac{(W/L)_3}{(W/L)_6} = \frac{(W/L)_4}{(W/L)_6} = \left(\frac{1}{2}\right)\frac{(W/L)_5}{(W/L)_7}. \tag{4}$$

In order that this ratio be maintained over process-induced variations in channel length, the channel lengths of $M3$, $M4$, and $M6$ usually must be chosen to be the same, and the ratios provided by properly choosing the channel widths. The use of identical channel lengths for the devices is at odds with the requirements (discussed later) that for low noise, $M3$ and $M4$ have low transconductance, and that for best frequency response under capacitive loading, $M6$ has high transconductance.

Systematic offset voltage is closely correlated with dc power supply rejection ratio. If a systematic offset exists, it is likely to display a dependence on power supply voltage, particularly if the bias reference source is such that the bias currents in the amplifier are not supply independent.

Random Input Offset Voltage

Source-coupled pairs of MOS transistors inherently display somewhat higher input offset voltage than bipolar pairs for the same level of geometric mismatch or process gradient. The reason for this is perhaps best understood intuitively by means of the conceptual circuit shown in Fig. 7. Here, a differential amplifier is made up of an identical pair of unilateral active devices biased at a current I and displaying a transconductance g_m. If the load elements, in this case assumed to be resistors, are assumed to mismatch by a percentage Δ, then in order for the output voltage of the differential amplifier to be zero, the absolute difference in the currents in the two devices must be equal to ΔI. This in turn requires that the dc input difference voltage applied to bring about this difference be

$$V_{gs} = \frac{I}{g_m}\Delta. \tag{5}$$

Thus, the input offset in this case depends on the I/g_m ratio of the active devices and the fractional mismatch in the

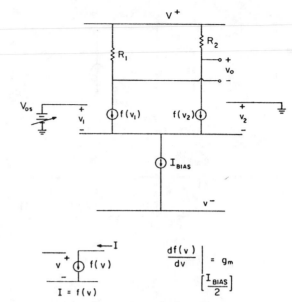

Fig. 7. Conceptual circuit for calculation of random offset voltage.

matched elements. A similar dependence is found for mismatches in many of the parameters of the active devices themselves, such as area mismatches in bipolar transistors and channel length and width mismatches in MOS transistors.

For bipolar devices, the I/g_m ratio is equal to kT/q or 0.026 V at room temperature. For MOS transistors, the ratio is $(V_{gs} - V_T)/2$, a bias-dependent quantity which is normally in the 100–500 mV range. While the offset voltage can be substantially improved by operating at low values of V_{gs}, the result is typically a somewhat larger offset voltage than in the bipolar case [2]. As discussed in a later section, the I/g_m ratio also directly effects the slew rate for class A input stages, so that often transient performance requirements place a lower limit on the allowable value of this parameter.

One mismatch component present in MOS devices which is not present in bipolar transistors is the mismatch in the threshold voltage itself. This component does not obey the above relationship, and results in a constant offset component which is bias current independent. Threshold mismatch is a strong function of process cleanliness and uniformity, and can be substantially improved by the use of common centroid geometries. Published data indicate that large-geometry common-centroid structures are capable of achieving threshold match distributions with standard deviations on the order of 2 mV in a silicon gate MOS process of current vintage [6].

Frequency Response, Compensation, Slew Rate, and Power Dissipation

The compensation of the two-stage CMOS amplifier can be carried out much as in the case of its bipolar equivalent using a pole-splitting capacitor C_c as shown in Fig. 2. However, important differences arise because of the much lower transconductance of the MOS transistor relative to bipolar devices [7]. The circuit can be approximately represented by the small-signal equivalent circuit of Fig. 8(a) if the nondominant poles due to the capacitances at the source of $M1$-2, the capacitance at the gate of $M3$, and any other nondominant poles which may exist on the circuit are neglected. This circuit has been analyzed by many authors [2], [8] because it occurs so fre-

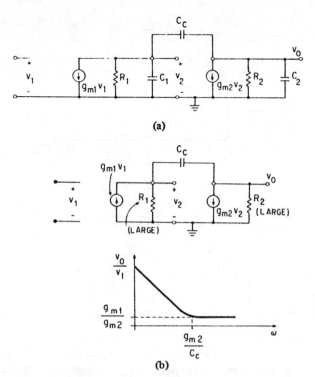

Fig. 8. (a) Small-signal equivalent circuit for two-stage amplifier. (b) Small-signal equivalent circuit with C_1 and C_2 set to zero, and gain of the circuit versus frequency.

quently in bipolar amplifiers. The circuit displays two poles and a right half-plane zero, which under the assumption that the poles are widely spaced, can be shown to be approximately located at

$$p_1 = \frac{-1}{(1 + g_{m2}R_2)C_cR_1} \qquad (6)$$

$$p_2 = \frac{-g_{m2}C_c}{C_2C_1 + C_2C_c + C_cC_1} \qquad (7)$$

$$z = +\frac{g_{m2}}{C_c}. \qquad (8)$$

Note that the pole due to the capacitive loading of the first stage by the second, p_1, has been pushed down to a very low frequency by the Miller effect in the second stage, while the pole due to the capacitance at the output node of the second stage, p_2, has been pushed to a very high frequency due to the shunt feedback. For this reason, the compensation technique is called pole splitting.

A unique problem arises when attempting to use pole splitting in MOS amplifiers. Analytically, the problem is illustrated by considering the location of the second pole p_2 and the right half-plane zero z relative to the unity-gain frequency g_{m1}/C_c. Here we make the simplifying assumption that the internal parasitic C_1 is much smaller than either the compensation capacitor C_c or the load capacitance C_2. This gives

$$\left|\frac{p_2}{\omega_1}\right| = \frac{g_{m2}C_c}{g_{m1}C_2} \qquad (9)$$

$$\left|\frac{z}{\omega_1}\right| = \frac{g_{m2}}{g_{m1}}. \qquad (10)$$

Note that the location of the right half-plane zero relative to

the unity-gain frequency is dependent on the ratio of the transconductances of the two stages.

Physically, the zero arises because the compensation capacitor provides a path for the signal to propagate directly through the circuit to the output at high frequencies. Since there is no inversion in that signal path as there is in the inverting path dominant at low frequencies, stability is degraded. The location of the zero can best be conceptually understood by considering a case in which C_1 and C_2 are zero as illustrated in Fig. 8(b). For low frequencies, this circuit behaves like an integrator, but at high frequencies, the compensation capacitor behaves like a short circuit. When this occurs, the second stage behaves like a diode-connected transistor, presenting a load to the first stage equal to $1/g_{m2}$. Thus, the circuit displays a gain at high frequencies which is simply g_{m1}/g_{m2}, as illustrated in Fig. 8(b). The polarity of this gain is opposite to that at low frequencies, turning any negative feedback that might be present around the amplifier into positive feedback.

In bipolar technology, the transconductance of the second stage is normally much higher than the first because it is operated at relatively high current and the transconductance of the bipolar device is proportional to current level. In MOS amplifiers, the transconductances of the two stages tend to be similar, in part because the transconductance varies only as the square root of the drain current. Also, the transconductance of the first stage must be kept reasonably high for thermal noise reasons.

Fortunately, two effective means have evolved for eliminating the effect of the right half-plane zero. One approach has been to insert a source follower in the path from the output back through the compensation capacitor to prevent the propagation of signals forward through the capacitor [7]. This works well, although it requires more devices and dc bias current. An even simpler approach is to insert a nulling resistor in series with the compensation capacitor as shown in Fig. 9 [9]. In this circuit, note that at high frequencies, the output current from the first stage must flow principally as drain current in the second stage transistor. This, in turn, gives rise to voltage variation at the gate of the second stage which is proportional to the small-signal current from the first stage and inversely proportional to the transconductance of the second stage. In the circuit of Fig. 8, this voltage appears directly at the output. However, if a resistor of value equal to $1/g_{m2}$ is inserted in series with the compensation capacitor, the voltage across this resistor will cancel the small-signal voltage appearing on the left side of the compensation capacitor, resulting in the cancellation of the feedthrough effect.

Using an analysis similar to that performed for the circuit of Fig. 8, one obtains pole locations which are close to those for the original circuit, and a zero location of

$$z = \frac{1}{C_c\left(\dfrac{1}{g_{m2}} - R_z\right)}. \qquad (11)$$

As expected, the zero vanishes when R_z is made equal to $1/g_{m2}$. In fact, the resistor can be further increased to move the zero into the left half-plane to improve the amplifier phase margin [10]. The movement of the zero for increasing values of R_z is illustrated in Fig. 10.

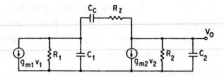

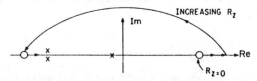

Fig. 9. Small-signal equivalent circuit of the basic amplifier with nulling resistor added in series with the compensation capacitor.

Fig. 10. Pole-zero diagram showing movement of the transmission zero for various values of R_z.

A second problem in compensation involves the effects of capacitive loading. From (9), the location of the nondominant pole due to capacitive loading on the output node relative to the unity-gain frequency is determined by the ratio of the second-stage transconductance to that of the first and the ratio of the load capacitance to the compensation capacitance. Since the stage transconductances tend to be similar, this implies that the use of load capacitances of the same order as the compensation capacitance will tend to degrade the unity-gain phase margin because of the encroachment of this nondominant pole. This is of considerable practical significance in switched capacitor filters where large capacitive loads must be driven, and the use of an output stage is undesirable for power dissipation and noise reasons.

Slew Rate

As in its bipolar counterpart, the CMOS op amp of Fig. 4 displays a relationship among slew rate, bandwidth, input stage bias current, and input device transconductance of

$$SR = \frac{I_{D1}}{g_{m1}} \omega_1 \qquad (12)$$

where g_{m1} is the input transistor transconductance, I_{D1} is the bias current of the input devices, and ω_1 is the unity-gain frequency of the amplifier. For the MOS case, this gives

$$SR = \frac{(V_{gs} - V_T)_1}{2} \omega_1. \qquad (13)$$

In effect, the $(V_{gs} - V_T)$ of the input stage is the range of differential input voltage for which the input stage stays in the active region. If the bandwidth is kept constant and this range is increased, the slew rate improves. Because this range is usually substantially higher in MOS amplifiers than in bipolar amplifiers, MOS amplifiers usually display relatively good slew rate. In micropower or precision applications where the input transistors are operated at very low $(V_{gs} - V_T)$, this may not be the case, however.

Power Dissipation

Even for the simple circuit of Fig. 4, the minimum achievable power dissipation is a complex function of the technology used and the particular requirements of the application. In

sampled data systems such as switched capacitor filters, the requirement is that the amplifier be able to settle in a certain time to a certain accuracy with a capacitive load of several picofarads. In this application, the factors determining the minimum power dissipation tend to be the fact that there must be enough standing current in the amplifier class A second stage such that the capacitance can be charged in the allowed time, and the fact that the amplifier must have sufficient phase margin to avoid degradation of the settling time due to ringing and overshoot. The latter requirement dictates a certain minimum g_m in transistor $M6$ for a given bandwidth and load capacitor. This, in turn, usually dictates a certain minimum bias current in $M6$ for a reasonable device size. If a class A source follower output stage is added, then the same comment would apply to its bias current since its g_m, together with the load capacitance, contribute a nondominant pole.

The preceding discussion is predicted on the use of class A circuitry (i.e., circuits whose available output current is not greater than the quiescent bias current). Quiescent power dissipation can be greatly reduced through the use of dynamic circuits and class B circuits, as discussed later.

Noise Performance

Because of the fact that MOS devices display relatively high $1/f$ noise, the noise performance is an important design consideration in MOS amplifiers. All four transistors in the input stage contribute to the equivalent input noise, as illustrated in Fig. 11. By simply calculating the output noise for each circuit and equating them (11),

$$V_{eq\,TOT}^2 = V_{eq1}^2 + V_{eq2}^2 + \left(\frac{g_{m3}}{g_{m1}}\right)(V_{eq3}^2 + V_{eq4}^2) \qquad (14)$$

where it has been assumed that $g_{m1} = g_{m2}$ and that $g_{m3} = g_{m4}$. Thus, the input transistors contribute to the input noise directly, while the contribution of the loads is reduced by the square of the ratio of their transconductance to that of the input transistors. The significance of this in the design can be further appreciated by considering the input-referred $1/f$ noise and the input-referred thermal noise separately.

Input-Referred $1/f$ Noise

The equivalent input noise spectrum of a typical MOS transistor is shown in Fig. 12. The dependence of the $1/f$ portion of the spectrum on device geometry and bias conditions has been studied by many authors [12]–[14]. Considerable discrepancy exists in the published data on $1/f$ noise, indicating that it arises from a mechanism that is strongly affected by details of device fabrication. Perhaps the most widely accepted model for $1/f$ noise is that for a given device, the gate-referred equivalent mean-squared voltage noise is approximately independent of bias conditions in saturation, and is inversely proportional to the gate capacitance of the device. The following analytical results are based on this model, but it should be emphasized that the actual dependence must be verified for each process technology and device type [12], [15]. Thus,

$$V_{1/f}^2 = \frac{K}{C_{ox}WL}\left(\frac{\delta f}{f}\right). \qquad (15)$$

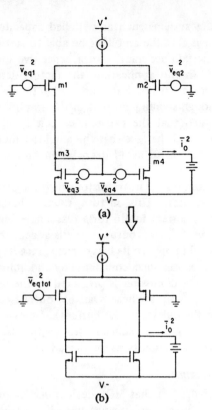

(a)

(b)

Fig. 11. CMOS input stage. (a) Device noise contributions. (b) Equivalent input noise.

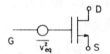

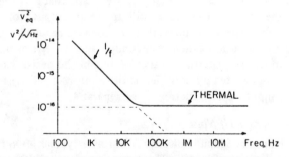

Fig. 12. Typical equivalent input noise, MOS transistor.

Utilizing this assumption, we obtain for the equivalent input noise

$$V_{1/f}^2 = \frac{2K_p}{W_1 L_1 C_{ox}} \left(1 + \frac{K_n \mu_n L_1^2}{K_p \mu_p L_3^2}\right)\left(\frac{\delta f}{f}\right) \qquad (16)$$

where K_n and K_p are the flicker noise coefficients for the n-channel and p-channel devices, respectively. Depending on processing details, these may be comparable or different by a factor of two or more. Note that the multiplying term in front is the input noise of the input transistors, and the second term is the increase in noise due to the loads. It is clear from this second term that the load contribution can be made small by simply making the channel lengths of the loads longer than

that of the input transistors by a factor of on the order of two or more. The input transistors can then be made wide enough to achieve the desired performance. It is interesting to note that increasing the width of the channel in the loads does not improve the $1/f$ noise performance.

Thermal Noise Performance

The input-referred thermal noise of an MOS transistor is given by [8]

$$V_{eq}^2 = 4kT\left(\frac{2}{3g_m}\right)\delta f. \qquad (17a)$$

Utilizing the same approach as for the flicker noise, this gives

$$V_{eq}^2 = 4kT \frac{4}{3\sqrt{2\mu_p C_{ox}(W/L)_1 I_D}} \left(1 + \sqrt{\frac{\mu_n(W/L)_3}{\mu_p(W/L)_1}}\right). \qquad (17b)$$

Again, the first term represents the thermal noise from the input transistors, and the term in parentheses represents the fractional increase in noise due to the loads. The term in parentheses will be small if the W/L's are chosen so that the transconductance of the input devices is much larger than that of the loads. If this condition is satisfied, then the input noise is simply determined by the transconductance of the input transistors.

Power Supply Rejection and Supply Capacitance

Power supply rejection ratio (PSRR) is a parameter of considerable importance in MOS amplifier design. One reason for this is that in complex analog–digital systems, the analog circuitry must coexist on the same chip with large amounts of digital circuitry. Even though separate analog and digital supply buses are often run on chip, it is hard to avoid some coupling of digital noise into the analog supplies. A second reason is that in many systems, switching regulators are used which introduce power supply noise into the supply voltage lines. If these high-frequency signals couple into the signal path in a sampled data system such as a switched capacitor filter or high-speed A/D converter, they can be aliased down into the frequency band where the signal resides and degrade the overall system signal-to-noise ratio. The parameters reflecting susceptibility to this phenomenon in the operational amplifier are the high-frequency PSRR and the supply capacitance.

The PSRR of an operational amplifier is simply the ratio of the voltage gain from the input to the output (open loop) to that from the supply to the output. It can be easily demonstrated that for frequencies less than the unity-gain frequency, if the operational amplifier is connected in a follower configuration and an ac signal is superimposed on one of the power supplies, the signal appearing at the output is equal to the applied signal divided by the PSRR for that supply. The basic circuit of Fig. 4 is particularly poor in terms of its high-frequency rejection from the negative power supply, as illustrated in Fig. 13. The primary reason is that as the applied frequency increases, the impedance of the compensation capacitor decreases, effectively shorting the drain of $M6$ to its gate for ac signals. Thus, the gain from the negative supply to the output

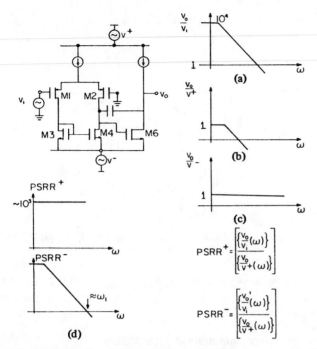

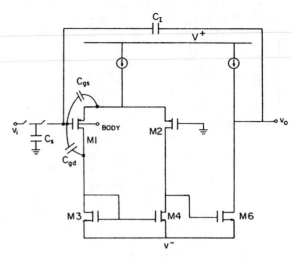

Fig. 14. Supply capacitance in a CMOS amplifier.

Fig. 13. High-frequency PSRR of bipolar-derived op amp. (a) Gain from input to output. (b) Gain from positive supply to output. (c) Gain from negative supply to output. (d) Positive and negative PSRR.

approaches unity and stays there out to very high frequencies. The same phenomenon causes the gain from the positive supply to fall with frequency as the open-loop gain does, so that the positive PSRR remains relatively flat with increasing frequency. The negative supply PSRR falls to approximately unity at the unity-gain frequency of the operational amplifier. Several alternative amplifier architectures have evolved which alleviate this problem, and they are discussed in a later section.

A second important contribution to coupling between the power supply and the signal path at high frequency is termed supply capacitance [10], [17]. This phenomenon manifests itself as a capacitive coupling between one or both of the power supplies and the operational amplifier input leads. The effect of this capacitance is illustrated in Fig. 14 for a switched capacitor integrator. Since the op amp input is connected to the summing node, then the power supply variations will appear at the integrator output attenuated by the ratio of the supply capacitance to the integrator capacitance. The result can be quite poor power supply rejection in switched capacitor filters and other sampled data analog circuits.

Supply capacitance effects can occur in several ways, but four important ones are given below.

1) Variation in drain voltage on $M1$, $M2$ with negative supply voltage. If the op amp inputs are grounded and the negative supply voltage changes, then a displacement current flows into the summing node because of the resulting change in voltage across the drain–gate capacitance of the input transistors. This is usually eliminated by the use of cascode transistors in series with the drains of the input transistors.

2) Variation of drain current in $M1$, $M2$ with supply voltage. Use of a bias reference which results in bias current variations with supply voltage will, in turn, cause the $V_{gs} - V_T$ of the input devices to change with supply voltage. This will cause a

displacement current to flow through the gate–source capacitance of $M1$, $M2$ onto the summing node. The usual solution to this problem is the use of a supply-independent bias reference.

3) Variation of body bias on $M1$, $M2$ with supply voltage variations. If the substrate terminal of the input transistors is tied to a supply or supply-related voltage, then as the supply voltage changes, the substrate bias changes. This, in turn, changes the threshold, which changes V_{gs}. The resulting displacement current in C_{gs} flows into the summing node. In CMOS operational amplifiers, the usual solution to this problem is to put the input transistors in a well and tie the well to the sources of the input transistors. This dictates, for example, that in a p-well process, the input devices be n-channel devices, and vice versa. In NMOS, the use of fully differential circuitry is probably the only way to fully alleviate the problem since the substrate must be tied to a supply. A second alternative is capacitive decoupling of the substrate so that it does not follow high-frequency supply variations [17].

4) Interconnect crossovers in the amplifier layout and in the system layout can produce undesired supply capacitance. This can usually be overcome with careful layout.

IV. ALTERNATIVE ARCHITECTURES FOR IMPROVED PERFORMANCE

The bipolar-derived amplifier discussed above is widely used at the present time, although with many variations, in a variety of applications. However, many alternative circuit approaches have been investigated and, in many cases, utilized in commercial products in order to achieve performance which is superior to that available from the basic circuit in some respect. In this section, we first consider variations on the basic circuit, and then alternative architectures.

Variations on the Basic Two-Stage Amplifier

Use of Cascodes for Improved Voltage Gain: In precision applications involving large values of closed-loop gain, the voltage gain available from the basic circuit shown in Fig. 4 may be inadequate. One approach to improving the voltage gain without adding an additional common-source stage with its

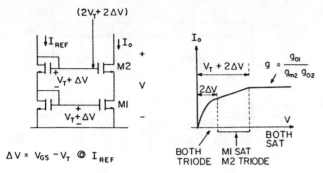

$$\Delta V = V_{GS} - V_T @ I_{REF}$$

Fig. 15.   Cascode current source.

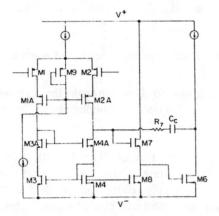

Fig. 16.   Two-stage amplifier with cascoded first stage.

associated high impedence node and pole is to add a common-gate, or cascode, transistor to increase the output resistance of the common-source transistors in the basic amplifier. The basic cascode circuit is shown in Fig. 15. It is easily demonstrated that the incremental output resistance of this current source is equal to

$$r_o = r_{o2}[1 + g_{m2}r_{o1}] + r_{o1}. \tag{18}$$

The output resistance is increased by an amount equal to the open circuit gain of the cascode transistor. This circuit may be directly applied to the basic two-stage amplifier in either the first stage, second stage, or in both stages. The circuit of Fig. 16 illustrates the use of cascodes in the first stage. One disadvantage of this circuit is a substantial reduction in input stage common mode range, but this can be alleviated by optimizing the biasing of the cascodes, to be discussed in Section V.

Improved PSRR Grounded-Gate Cascode Compensation: Read and Wieser [18] have recently described a technique for improving the negative supply PSRR of the circuit of Fig. 4. Conceptually, if the left end of the capacitor could be connected to a virtual ground, then the capacitor voltage would not have to change whenever the negative supply voltage changed in order to have the output remain constant. This was accomplished by inserting a cascode device in this loop with the gate connected to ground, as shown in Fig. 17. The displacement current from the capacitor flows into the source of this transistor and out the drain into the compensation point. An additional current source and current sink of equal values must be added to bias the common-gate device in the active region and so as not to contribute any systematic offset.

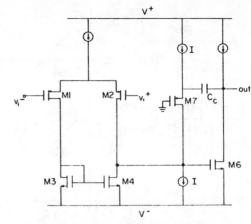

Fig. 17. Schematic of basic amplifier with cascode feedback compensation.

The resulting negative PSRR at high frequencies is greatly improved at the cost of a slight increase in complexity, random offset, and noise [16].

Common-Source Common-Gate Amplifiers

The basic amplifier considered thus far is actually a cascade of two common-source stages. An alternative approach is to use a cascade of a common-source stage and a common-gate stage, often called a cascode amplifier. An example of an amplifier utilizing this architecture is shown in Fig. 18. The voltage gain of this circuit at dc is approximately the same as that of the basic two-stage circuit. The small-signal impedance at the output node is increased by $g_m r_o$ relative to the output node of the two-stage circuit, and the voltage gain is simply the product of the transconductance of the input transistors and the impedance at the output mode:

$$A_v = \frac{g_{m1}}{\dfrac{g_{o2} + g_{o9}}{g_{m4}r_{o4}} + \dfrac{g_{o7}}{g_{m5}r_{o5}}}. \tag{19a}$$

The principal reasons for considering this architecture are twofold. First, the compensation capacitor and load capacitor are the same element in this circuit. The first nondominant pole comes from the g_m/C_{gs} time constant of the n-channel cascode devices, and gives a pole frequency approximately at the f_t of these devices. A second nondominant pole results from the differential to single-ended converter. However, the nondominant pole due to the load capacitance present in the two-stage circuit, is not present in this circuit. Thus, this circuit is capable of achieving higher stable closed-loop bandwidth with large capacitive load. The principal application of this architecture to date has been in high-frequency switched capacitor filters [20], [21].

An important advantage of this circuit is that it does not suffer from the degradation of the high-frequency power supply rejection problem inherent in the pole-split compensated two-stage architecture, assuming that the load capacitance or part of it is not tied to a power supply.

Because of the fact that cascode transistors are used at the output, the output swing of this circuit is lower than the common-source common-source amplifier. This problem can

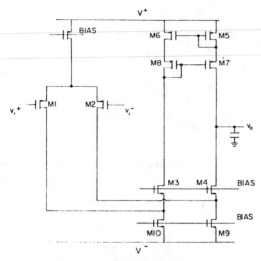

Fig. 18. One-stage amplifier schematic.

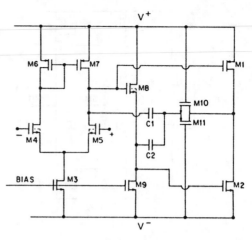

Fig. 19. Example circuit illustrating class AB second stage.

be minimized by modifying the bias generator such that the lower transistors in the cascode current source are biased on the edge of saturation (i.e., $V_{gd} = V_T$). This results in an available output voltage swing within $2(V_{gs} - V_T)$ of each supply, or perhaps 0.4 to 0.8 V in voiceband filters. MOS transistors actually display a rather indistinct transition from triode to saturation as the drain depletion region forms, and as a result, the bias point must actually be chosen so as to bias the lower transistor a few hundred millivolts into saturation if the predicted value of incremental output resistance is to be obtained.

A second disadvantage of this circuit is that more devices contribute to the input-referred voltage and input offset voltage. Assuming that transistors $M5$–$M8$ are biased at the same current as the input devices, the input-referred flicker noise can be shown to be

$$v_{eq}^2 = \frac{2K_p}{W_1 L_1} \left[1 + \frac{2K_n \mu_n}{K_p \mu_p} \left(\frac{L_1}{L_9}\right)^2 + \left(\frac{L_1}{L_s}\right)^2 \right] \frac{\delta f}{f}. \quad (19b)$$

In this case, the current sources $M9$ and $M10$ contribute an additional term not present in (16). However, as in the case of the common-source common-source amplifier, the equivalent input noise can be made almost equivalent to the noise of the input transistors alone by choosing the channel lengths of the input transistors to be short compared to those of $M5$, $M6$, $M9$, and $M10$. The same considerations apply for the thermal noise.

Class AB Amplifiers, Dynamic Amplifiers, and Dynamic Biasing

Many, if not all, MOS analog circuits commercially produced utilize class AB circuitry in some form. Here the term class AB is taken to mean a circuit which can source and sink current from a load which is larger than the dc quiescent current flowing in the circuit. The most widespread application is in output buffers, but if an important objective is the minimization of chip power, then the philosophy of using class AB operation can be extended to the internal amplifiers. The motivation for doing so is that one of the factors that dictates the value of the quiescent current with an MOS amplifier is the

value of current required to charge the load and/or the compensation capacitance in the required time. However, it is relatively rare that the operational amplifier outputs actually have to change the maximum amount in one clock cycle. Large power savings can be effected if only that current is drawn which is required to charge the capacitance on that particular cycle. An example of an amplifier utilizing a class A first stage and a class AB second stage is shown in Fig. 19 [22]. In a conventional circuit, the gate of $M2$ would be connected to a level-shifted version of the stage input voltage. Thus, when the first stage output swings positive, reducing the current in $M1$, the current in $M2$ increases above its quiescent dc value. An example of a single-stage amplifier that operates on this principle is shown in Fig. 20 [23]. This particular circuit can be used in the inverting mode only. With the input grounded, the quiescent current in the input transistors is determined by the bias voltages shown. Upon the application of a voltage to the input, the current in one side of the input stage increases monotonically with the applied voltage until the power supply is reached, while the other side of the input stage turns off. The amount of current available at the output is much larger than the quiescent current, and the circuit, as a result, does not follow the relationship of (12). In fact, the circuit does not display slew rate limiting in the usual sense. Another aspect of this circuit is the fact that the small-signal voltage gain in the quiescent mode can be quite high because of the low current level, and the fact that the voltage gain falls off during transients because of the high current levels is of little consequence. Similar circuits have been used extensively in bipolar technology [24].

Degrauwe et al. [25] have recently described a novel approach to the same objective. A conventional class A amplifier configuration is used, but an auxiliary circuit is used to detect the presence of large differential signals at the input. The bias current in the class A circuitry is then increased when such signals are present. Experimental versions of such amplifiers have yielded quiescent power dissipation of less than 10 μW.

A second class of amplifiers has been explored by several authors, beginning with Copeland [26], in which the quiescent current in the absence of signals is allowed to decay to zero. Such amplifiers are fully dynamic in the sense that no dc paths

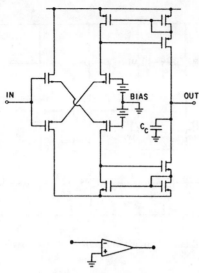

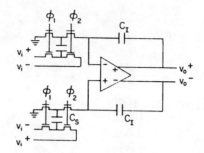

Fig. 21. Differential switched capacitor integrator.

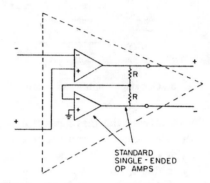

Fig. 22. Equivalent circuit for a differential output operational amplifier.

Fig. 20. Examples of a single-stage class *AB* op amp.

exist for current to flow from the supply. While very low values of power dissipation can be obtained, difficult problems of settling time and power supply rejection remain to be solved with these amplifiers.

Hostica [27] has described a third approach to micropower amplifier design for switched capacitor applications which utilizes a time-varying periodic bias current which is synchronous with the master clock in the filter. In contrast to the approaches described above, the power supply current is independent of signal amplitude, and is made large during the early part of the clock period for fast slewing and small during the later portion for high gain and power savings. The bias current waveform is generated by discharging a capacitor into the input of a current mirror. While this technique, in principle, dissipates more power than the other approaches under low signal conditions, it can be implemented with relatively simple circuitry, and it has demonstrated good experimental results for both one-stage and two-stage amplifiers [28].

V. DIFFERENTIAL OUTPUT AMPLIFIERS

As has been mentioned, power supply rejection is an important performance parameter for amplifiers to be used in complex analog/digital systems. In addition, one inevitable result of technology scaling is a reduction in power supply voltage with an accompanying reduction in internal signal swings and dynamic range. These two considerations make use of fully differential signal paths throughout the analog portions of the system attractive for some systems [29], [30]. The inherently differential nature of the circuit tends to give very high PSRR since the supply variations appear as a common mode signal. Also, the effective output swing is doubled, while the magnitude of the input-referred operational amplifier noise remains the same, giving a 6 dB improvement in operational amplifier noise-limited dynamic range.

A typical implementation of a differential switched capacitor integrator is shown in Fig. 21. The operational amplifier is required to produce two analog outputs which are symmetric about ground, in contrast to the single-ended case where only one is produced. An equivalent circuit for a differential op

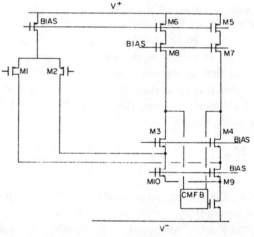

Fig. 23. Example of a differential output amplifier. The block labeled *CMFB* serves to keep the common-mode output voltage near ground.

amp is shown in Fig. 22. An example of a CMOS differential output operational amplifier is shown in Fig. 23.

An important problem in such amplifiers is the design of a feedback loop to force the common mode output voltage to be ground or some other internal reference potential. This feedback path can be implemented with transistors in a continuous-time circuit or can be implemented with switched capacitor circuitry. The continuous approach is potentially simpler, but presents a difficult design problem in making the common-mode output voltage independent of the differential mode signal voltage [21], [29]. Switched capacitor circuitry can make use of the linearity of MOS capacitors to achieve this goal [30]. The choice between the two techniques depends on the sensitivity of the particular application to variations in common-mode voltage.

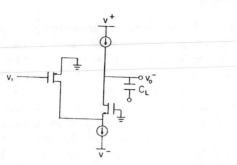

Fig. 24. Small-signal differential half circuit for the amplifier in Fig. 23.

Another important advantage of differential output amplifiers is that the differential single-ended converter with its associated nondominant poles is eliminated. The small-signal equivalent circuit for the circuit in Fig. 22, for example, is a simple common-source common-gate cascade, as shown in Fig. 24. This circuit has only one nondominant pole, at the f_t of the common-gate device. Thus the configuration is particularly well suited to the implementation of high-frequency switched capacitor filters. A configuration of this type has been used in recently reported work yielding high-Q switched capacitor filters clocked at 4 MHz with center frequencies of 250 kHz in a 4 μm silicon gate CMOS technology [21].

VI. NMOS OPERATIONAL AMPLIFIERS

The design of an operational amplifier of a given performance level in NMOS depletion load technology is a much more difficult task than in CMOS technology. The absence of a complementary device makes the implementation of level shifters which track supply voltage variations much more complex. The level of body effect found in most depletion load devices makes the realization of large gains per stage difficult. Assuming that the basic architecture is similar to that illustrated in Fig. 2(c), the small-signal properties, voltage gain, transient response and slew rate, input noise, and power supply rejection considerations are basically similar to the two-stage CMOS amplifier. The key additional considerations are the shunting effects of the incremental output conductance of the depletion load current sources and the impedance and power supply variation of the floating level-shifting voltage sources and the resulting degradation of power supply rejection ratio. Nonetheless, creative circuit design has resulted in NMOS amplifiers which nearly match CMOS amplifiers in most performance aspects, albeit at the cost of somewhat more complexity, die area, and power dissipation. Circuit techniques used to achieve this include replica biasing for tracking level shifters [9], [17], positive feedback for high voltage gain [31], [30], differential configurations for power supply rejection [29], [30], and others.

While there will no doubt always be a need for NMOS amplifiers for some applications, the emergence of CMOS as a key VLSI digital technology has resulted in the widespread adoption of CMOS for new mixed analog–digital designs.

VII. OUTPUT BUFFERS

In amplifier applications involving either a large capacitive or resistive load, an output stage must be added to the basic am-plifier to prevent the load from degrading the voltage gain or closed-loop stability. This situation most often arises when signals must be supplied off the chip to an external environment. The key requirements on such stages is that they be sufficiently broad band with heavy capacitive loading such that they do not degrade the loop stability of the operational amplifier, and such that the output is able to supply a large enough voltage swing to the load with the maximum load conductance. While class A source follower or emitter follower circuits can be used in some applications, quiescent power dissipation considerations usually dictate a class AB implementation of the circuit. This discussion is limited to class AB output buffers.

In bipolar operational amplifier design, the complementary emitter follower class AB configuration is used in the vast majority of cases. In contrast, class AB CMOS output stage implementations tend to vary widely, depending on the specific devices available in the particular technology used. The CMOS complementary source follower class AB output buffer stage shown in Fig. 25 is a direct analog of its bipolar counterpart. The primary drawback of this circuit is that the output voltage swing is limited by the gate–source voltage of the output transistors. This occurs because the transistors used for logic functions on the chip have thresholds in the 0.5–1 V range, so that the amount of swing lost due to threshold voltage plus the ($V_{gs} - V_T$) drop is too large for many applications. However, many technologies have an extra device type with very low threshold voltage, and in this case, this low threshold device can be used for one of the two output transistors. It is rare that both p-channel and n-channel low threshold devices are available in the same technology.

In many CMOS technologies, a bipolar transistor follower is available and can be used in place of one of the output followers. This provides very low output resistance and good output swing. In processes with light substrate doping, potential latchup problems can make the use of such devices in off-chip driver stages impractical because of the fact that the collector current of the transistor flows in the substrate and can cause voltage drops which cause a junction to be forward biased. An example of the use of a bipolar device in an MOS output stage together with a low threshold device is illustrated in Fig. 26.

A third alternative is the use of quasi-complementary configurations in which a common-source transistor together with an error amplifier is used in place of one or both of the follower devices. This circuit is shown conceptually in Fig. 27. The combination of the error amplifier and the common-source device mimics the behavior of a follower with high dc transconductance. Such quasi-complementary circuits provide excellent dc performance with voltage swings approaching the supply rails, but since the amplifier must be broad band to prevent crossover distortion problems, they present difficult problems in compensation of the local feedback loop in the presence of large capacitive loads. Proper control of the quiescent current is also a key design constraint.

Low threshold devices, bipolar devices, and quasi-complementary devices can be used in any combination, depending on what devices are available in the particular technology

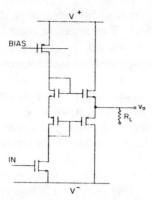

Fig. 25. Complementary source follower CMOS output stage based on the traditional bipolar implementation.

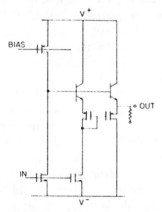

Fig. 26. Example of a CMOS output stage using a bipolar emitter follower and a low-threshold p-channel source follower.

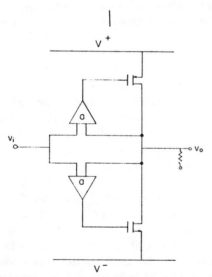

Fig. 27. Example of a complementary class B output stage using compound devices with imbedded common-source output transistors.

being used. Whereas in the bipolar case the vast majority of output stage applications can be satisfied using the traditional complementary class B emitter follower stage, no single circuit approach has yet emerged as the standard for CMOS output stages.

VIII. SUMMARY AND CONCLUSIONS

In this paper, we have attempted to summarize the various techniques and architectures which have been applied in the design of MOS operational amplifiers in the past several years. An important question is the extent to which these amplifier designs can be scaled as minimum feature sizes continue to decrease. As pointed out in a recent study [32], dc parameters such as voltage gain are generally unaffected for constant-field scaling, although they are degraded for quasi-constant voltage or constant voltage scaling. Perhaps the most difficult problem results from the fact that the effective dynamic range of the amplifier falls in scaled technologies. This occurs fundamentally because of the fact that analog signal swings fall with reductions in power supply voltage. Input-referred thermal noise remains constant because of the fact that the device transconductance remains constant under constant-field scaling. The input-referred $1/f$ noise increases, but this does not appear to be a fundamental limitation on system dynamic range because the signal can always be translated to a higher portion of the spectrum using techniques like chopper stabilization [29]. Also, newer technologies have demonstrated continuing reductions in $1/f$ noise as a result of better process control.

In sampled data analog amplifiers, filters and data converters, the primary limitation on dynamic range, assuming that $1/f$ noise has been removed, is the kT/C noise contributed by the analog switches making up the filter. The kT/C limited dynamic range also falls as the technology is scaled, and since for practical clock rates and capacitor sizes this noise source is dominant over op amp thermal noise, there appears to be no barrier to constant-field scaling of operational amplifiers for this application, assuming that $1/f$ noise is removed by circuit or technological means. Thus, the adaptation of the circuit approaches described in this paper to lower supply voltages and scaled devices, and the removal of $1/f$ noise from the signal path in such circuits, are important objectives in future work.

REFERENCES

[1] D. A. Hodges, P. R. Gray, and R. W. Broderson, "Potential of MOS technologies for analog integrated circuits," *IEEE J. Solid-State Circuits*, pp. 285–293, June 1978.

[2] J. E. Solomon, "The monolithic op amp, A tutorial study," *IEEE J. Solid-State Circuits*, vol. SC-9, pp. 314–332, Dec. 1974.

[3] Y. P. Tsividis, "Design considerations in single-channel MOS analog circuits—A tutorial," *IEEE J. Solid-State Circuits*, pp. 383–391, June 1978.

[4] P. R. Gray, "Basic MOS operational amplifier design—An overview," in *Analog MOS Integrated Circuits*. New York: IEEE Press, 1980, pp. 28–49.

[5] R. W. Broderson, P. R. Gray, and D. A. Hodges, "MOS switched capacitor filters," *Proc. IEEE*, pp. 61–75, Jan. 1979.

[6] O. H. Shade, Jr., "BiMOS micropower integrated circuits, *IEEE J. Solid-State Circuits*, vol. SC-13, pp. 791–798, Dec. 1978. See also O. H. Schade, Jr. and E. J. Kramer, "A low-voltage BiMOS op amp," *IEEE J. Solid-State Circuits*, vol. SC-16, pp. 661–668, Dec. 1981.

[7] Y. P. Tsividis and P. R. Gray, "An integrated NMOS operational amplifier with internal compensation," *IEEE J. Solid-State Circuits*, vol. SC-11, pp. 748–753, Dec. 1976.

[8] P. R. Gray and R. G. Meyer, *Analysis and Design of Analog Integrated Circuits*. New York: Wiley, 1977.

[9] D. Senderowicz, D. A. Hodges, and P. R. Gray, "A high-performance NMOS operational amplifier," *IEEE J. Solid-State Circuits*, vol. SC-13, pp. 760–768, Dec. 1978.

[10] W. C. Black and D. J. Allstott, "Low power CMOS channel filter," *IEEE J. Solid-State Circuits*, vol. SC-15, pp. 929–938, Dec. 1980.

[11] J. C. Bertails, "Low frequency noise considerations for MOS am-

plifier design," *IEEE J. Solid-State Circuits*, vol. SC-14, pp. 773–776, Aug. 1979.

[12] M. B. Das and J. M. Moore, "Measurements and interpretation of low-frequency noise in FET's," *IEEE Trans. Electron Devices*, vol. ED-21, Apr. 1974.

[13] N. R. Mantena and R. C. Lucas, "Experimental study of flicker noise in MIS transistors," *Electron. Lett.*, vol. 5, pp. 697–603, 1969.

[14] H. Mikoshiba, "1/f noise in n-channel silicon gate MOS transistors," *IEEE Trans. Electron Devices*, vol. ED-29, June 1962.

[15] E. Vittoz and J. Fellrath, "CMOS integrated circuits based on weak inversion operation," *IEEE J. Solid-State Circuits*, vol. SC-12, pp. 214–231, June 1977.

[16] B. Ahuja, Intel Corporation, private communication.

[17] H. Ohara, W. M. Baxter, C. F. Rahim, and J. L. McCreary, "A precision low power PCM channel filter with on-chip power supply regulation," *IEEE J. Solid-State Circuits*, vol. SC-15, pp. 1005–1013, Dec. 1980.

[18] R. Read, private communication.

[19] P. R. Gray and R. G. Meyer, "Recent advances in monolithic operational amplifier design," *IEEE Trans. Circuits Syst.*, pp. 317–327, May 1974.

[20] P. R. Gray, R. W. Broderson, D. A. Hodges, T. C. Choi, R. Kaneshiro, and K. C. Hsieh, "Some practical aspects of switched capacitor filter design," in *Dig. Tech. Papers, 1981 Int. Symp. Circuits Syst.*

[21] T. Choi, R. Kaneshiro, R. W. Broderson, and P. R. Gray, "High frequency CMOS switched capacitor filters for communications applications," in *Dig. Tech. Papers, 1983 Int. Solid-State Circuits Conf.*

[22] Y. A. Haque, R. Gregortan, D. Blasco, R. Mao, and W. Nicholson, "A two-chip PCM codec with filters," *IEEE J. Solid-State Circuits*, vol. SC-24, pp. 961–969, Dec. 1979.

[23] W. Black, personal communication.

[24] P. C. Davis and V. Saari, "A high slew rate monolithic op amp using compatible complementary PNPs," in *Dig. Tech. Papers, IEEE Int. Solid-State Circuits Conf.*, Philadelphia, PA, Feb. 1974.

[25] M. G. Degrauwe, J. Rijmenants, E. A. Vittoz, and H. J. De Man, "Adaptive biasing CMOS amplifiers," *IEEE J. Solid-State Circuits*, vol. SC-17, pp. 522–528, June 1982.

[26] M. A. Copeland and J. M. Rabaey, "Dynamic u amplifiers for MOS technology," *Electron. Lett.*, vol. 15, pp. 301–302, May 1979.

[27] B. J. Hosticka, "Dynamic CMOS amplifiers," *IEEE J. Solid-State Circuits*, vol. SC-15, pp. 887–894, Oct. 1980.

[28] B. J. Hosticka, D. Herbst, B. Hoefflinger, U. Kleine, J. Pandel, and R. Schweer, "Real-time programmable low-power SC bandpass filter," *IEEE J. Solid-State Circuits*, vol. SC-17, pp. 499–506, June 1982.

[29] K. C. Hsieh, P. R. Gray, D. Senderowicz, and D. Messerschmitt, "A low-noise differential chopper stabilized switched capacitor filtering technique," *IEEE J. Solid-State Circuits*, vol. SC-16, pp. 708–715, Dec. 1981.

[30] D. Senderowicz, S. F. Dreyer, J. M. Huggins, C. F. Rahim, and C. A. Laber, "Differential NMOS analog building blocks for PCM telephony," in *Dig. Tech. Papers, 1982 Int. Solid-State Circuits Conf.*, San Francisco, CA, Feb. 1982. Also appears in full length form in this issue, pp. 1014–1023.

[31] J. Guinea and D. Senderowicz, "High frequency NMOS switched capacitor filters using positive feedback techniques," this issue, pp. 1029–1038.

[32] S. Wong and C. A. T. Salama, "Scaling of MOS analog circuits for VLSI applications," in *Dig. Tech. Papers, 1982 Symp. VLSI Technology*, Tokyo, Japan, Sept. 1982.

Additional References on MOS Operational Amplifiers

[33] F. H. Musa and R. C. Huntington, "A CMOS monolithic 3½ digit A/D converter," in *Dig. Tech. Papers, 1976 Int. Solid-State Circuits Conf.*, Philadelphia, PA, Feb. 1976, pp. 144–145.

[34] A. G. F. Dingwall and B. D. Rosenthall, "Low-power monolithic COS/MOS dual-slope 11-bit A/D converters," in *Dig. Tech. Papers, 1976 Int. Solid-State Circuits Conf.*, Philadelphia, PA, Feb. 1976, pp. 146–147.

[35] Y. P. Tsividis and D. Fraser, "A process insensitive NMOS operational amplifier," in *Dig. Tech. Papers, 1979 Int. Solid-State Circuits Conf.*, Philadelphia, PA, Feb. 1979, pp. 188–189.

[36] S. Kelley and D. Ulmer, "A single-chip PCM codec," *IEEE J. Solid-State Circuits*, vol. SC-14, pp. 54–58, Feb. 1979.

[37] B. J. White, G. M. Jacobs, and G. Landsburg, "Monolithic dual tone multifrequency receiver," *IEEE J. Solid-State Circuits*, vol. SC-14, pp. 991–997, Dec. 1979.

[38] I. A. Young, "A high performance all-enhancement NMOS operational amplifier," *IEEE J. Solid-State Circuits*, vol. SC-14, pp. 1070–1076, Dec. 1979.

A Programmable CMOS Dual Channel Interface Processor for Telecommunications Applications

BHUPENDRA K. AHUJA, MEMBER, IEEE, PAUL R. GRAY, FELLOW, IEEE, WAYNE M. BAXTER, AND
GREGORY T. UEHARA, MEMBER, IEEE

Abstract —A CMOS analog VLSI chip for telecommunication applications has been designed with many desirable line card features, which are programmable through a unique digital interface from the central switching office. The paper emphasizes the circuit innovations of some key analog functions realized on the chip, specifically, the operational amplifier family, the precision bandgap reference circuit, and the line balancing function. The die size of the analog VLSI is approximately 50 000 mils^2, and the active power dissipation is 80 mW with a 1 mW standby mode.

Manuscript received June 15, 1984; revised August 3, 1984.
B. K. Ahuja, W. M. Baxter, and G. T. Uehara are with the Intel Corporation, Chandler, AZ 85224.
P. R. Gray is with the Department of Electrical Engineering and Computer Sciences, University of California, Berkeley, CA 94720.

I. INTRODUCTION

THE advent of single-chip codec/filter integrated circuits [1], [2] has greatly reduced the cost of per-line electronics in digital switching and transmission systems. However, one drawback of such chips has been the necessity of adding external components to the line card to perform functions such as line balancing. A second drawback has been the difficulty of adapting the line card to the specific requirements of different system applications. This paper describes a CMOS VLSI interface circuit incorporating all of the low-voltage functions for a subscriber line

Reprinted from *IEEE J. Solid-State Circuits*, vol. SC-19, no. 6, pp. 892–899, Dec. 1984.

card on the same monolithic chip. The circuit provides programming of all the features through a unique digital interface from the central switching office.

This paper does not attempt to describe each functional block on the chip, but gives an overview of the system architecture followed by details on the design of three analog blocks which incorporate some novel ideas, namely, the operational amplifier family, the precision bandgap reference circuit, and the programmable balance networks. In conclusion, some system attributes and device performance are also provided.

II. SYSTEM ARCHITECTURE

The most important objective in defining the architecture of this analog VLSI was to make all its features programmable through software control from the central switching office. Fig. 1 shows the internal organization of the chip. Along with the basic function of the codec/filter for the voice channel, the chip also provides gain control for transmit and receive directions, balance networks and 2- to 4-wire conversion, three-party conferencing, and a secondary analog channel which may be used for performing loop tests, loop monitor and control, or any other low-bandwidth application. For the transmit voice path, the input signal (VFX) first goes through an antialiasing filter (AAF), followed by a low-noise amplifier (LNA), which can be used to set the transmit gain with external resistors at TG_1 and TG_2. Another programmable gain stage (XPG) is provided in the voice signal path to allow gain control of $+6--6$ dB in 0.5 dB steps under software control. The signal is then band limited to 3.2 kHz in accordance with CCITT requirements by an eighth-order switched-capacitor transmit filter and an auto zero network (XF/AZ) prior to conversion into A or u law PCM words by an all capacitive charge redistribution A/D converter. This encoder is also used to perform another 8 bit PCM conversion for the secondary analog inputs SAI_1 and SAI_2, which can be encoded at an 8 kHz rate in single-ended or differential modes.

On the receive side, the D/A converter performs decoding of three A or u law PCM words every 125 μs. The primary and third-party voice PCM bytes are summed and filtered by a fifth-order switched-capacitor filter (RLPF). A receive programmable gain (RPG) circuit provides gain control of 0 to -12 dB in 0.5 dB steps under software control. Two on-chip power amplifiers (PA's) can drive up to 300 Ω transformer loads. Information decoded from the receive data byte appears at the secondary analog output (SAO), which is capable of driving up to a 10 kΩ load. No on-chip filtering is included for the secondary channels, and thus either of these should be externally filtered, or their applications should be limited to low bandwidths only.

The 2- to 4-wire conversion necessary for the subscriber interface is implemented on-chip. The option of using either internal or external balance networks (BNW's) provides flexibility for any application.

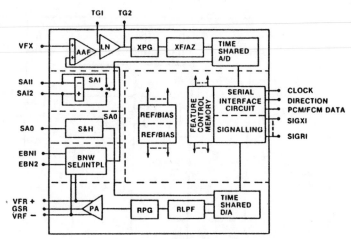

Fig. 1. Block diagram of the chip.

TABLE I
DESIGN OBJECTIVES OF THE OPERATIONAL AMPLIFIER FAMILY

Type	AVMIN	RLMIN	CLMAX	SWING	PDMAX	PSRR @ 100KHZ
Internal	5000	100K	10PF	± 4V	1mW	60dB
10k Buffer	5000	10K	10PF	± 4V	2mW	50dB
1k Buffer	5000	1K	100PF	± 4V	3mW	50dB
300u Buffer	5000	300	100PF	± 4V	6mW	50dB

Self and system test capabilities for diagnostic purposes have been integrated. A unique bidirectional serial interface to a line card controller chip provides further capabilities to the system designer by allowing features like time slot assignment through a microprocessor or an HDLC port and software control of the analog VLSI processor. This paper will describe the analog VLSI processor only.

III. OPERATIONAL AMPLIFIER FAMILY

The family of operational amplifiers used in the circuit is summarized in Table I. For internal applications involving purely capacitive loads, a conventional single-ended output two-stage amplifier is used, incorporating the cascode compensation scheme for improved high-frequency supply rejection from the negative supply. This amplifier has been discussed elsewhere [3] and will not be discussed further here.

For applications on the chip requiring the amplifier to drive finite values of load resistance such as resistive programmed internal gain stages and off-chip loads, a composite operational amplifier is used which consists of a common core preamplifier together with one of a family of three class AB output stages of different drive capability. The design of the output buffers was particularly challenging because of the requirements for low quiescent power dissipation together with an output voltage swing to within 1 V of each power supply while driving resistive loads.

Some key requirements of the core preamplifier are that it realize good power supply rejection at high frequencies, that its nondominant poles be at sufficiently high frequencies so that the excess phase budget can be used entirely by the output stage with its capacitive loading, and that its

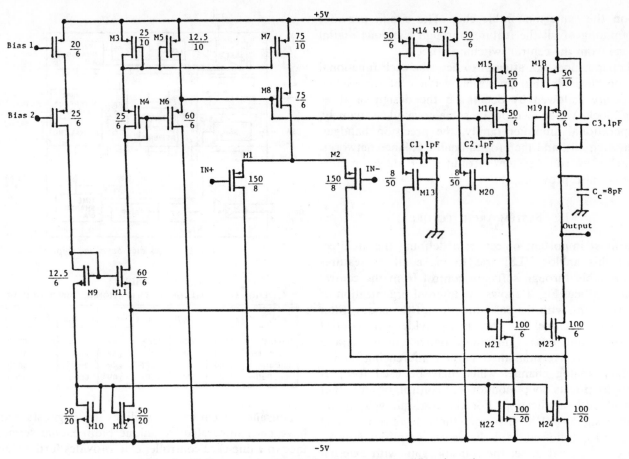

Fig. 2. Single-stage common-source–common-gate amplifier schematic.

power dissipation be kept to a minimum. In order to achieve this, the common-source–common-gate configuration, as shown in Fig. 2, was chosen. In this circuit, transistors M_9, M_{10}, M_{11}, and M_{12} are used to develop bias voltages for M_{21}, M_{23}, M_{22}, and M_{24} such that the latter two are biased on the edge of the triode region. The same function is provided by M_3–M_6. This allows the output to swing to within 2 $V_{d\text{sat}}$ of the positive and negative supplies while maintaining a high incremental voltage gain. At the bias currents used in this circuit, the value of $V_{d\text{sat}}$ is approximately 0.2 V. The differential-to-single-ended conversion is performed by the current source M_{13}–M_{14}–M_{17} in conjunction with the follower M_{20}. Since the gate of M_{13} is returned to ground, the resultant V_{gs} of M_{20} is the correct value needed to give a nominal zero output voltage and hence a negligible input-referred systematic offset. However, the low transconductance of M_{20} in combination with the gate capacitances of M_{15} and M_{18} result in poor frequency response in the differential-to-single-ended converter. To alleviate this problem, the feed-forward capacitor C_2 is included to bypass this signal path at high frequencies. Because high-frequency power supply rejection is a critical requirement, capacitors C_3 and C_1 are also included to balance the displacement current flowing through C_2 when signals are present on the positive power supply.

Compensation of the circuit is achieved with a single capacitor from the output node to ground. The actual

value of this capacitance depends on the bandwidth desired in the particular application.

One of the most challenging problems in achieving the performance objectives of the overall chip was the realization of the output stage so as to drive low impedance resistive and capacitive loads to voltages near the power supply while achieving good linearity, good supply rejection, low noise, and low quiescent power dissipation.

The approach taken in the class AB buffer is illustrated in Fig. 3(a). Here, the output swing is limited by the on-resistance of the output transistors which enter the triode region at the extremes of the output swing. Bipolar transistors cannot be used reliably because of the occurence of load faults such as short circuits and the possibility of chip latchup due to the substrate currents. In this configuration, the error amplifiers must meet a number of constraints. First, since the amplifiers will have input-referred offset voltages on the order of several millivolts, relatively modest values of open loop voltage gain on the order of 10 must be used. This ensures that the input offset voltages, when referred t o the output, will not cause gross variations in the quiescent current in the amplifier. This problem can also be attacked by including a crossover circuit [4]. A second constraint is that the amplifiers have an output swing near the supplies so as to provide good drive capability for the output transistor gate and an input common-mode range that extends over the same range of voltages as the output swing of the buffer. Finally, when

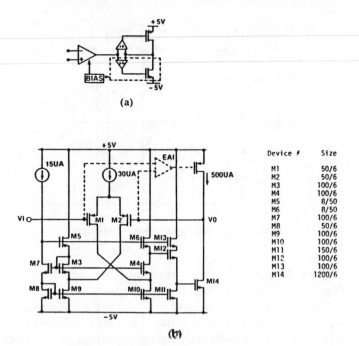

Fig. 3. (a) Complimentary class *AB* output buffer. (b) Circuit schematic of the class *AB* output buffer.

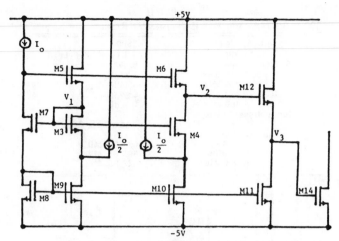

Fig. 4. Equivalent circuit of negative error amplifier for balanced quiescent condition.

the differential dc input voltage is zero, the error amplifiers must provide a quiescent dc output voltage which is the correct value to give the desired quiescent bias current in the output transistors.

The implementation used to satisfy these requirements is illustrated in Fig. 3(b). The device sizes shown are those used for the 300 Ω transformer driver output buffer. Only the error amplifier driving the n-channel output transistor is shown; a dual of this circuit is used to drive the p-channel output transistor. Transistors $M_1 - M_2$ form a source-coupled pair driving enhancement load transistors $M_5 - M_6$ through common-gate transistors $M_3 - M_4$. This combination produces a voltage gain in the error amplifier of about 8. Source follower M_{12} drives the output transistor gate. Transistors M_7, M_8, M_9, M_{10}, and M_{11} provide differential-to-single-ended conversion and produce the desired quiescent voltage at the gate of the output transistor.

The quiescent current in the output transistor can be calculated using the equivalent circuit shown in Fig. 4. Here, it has been assumed that the differential input voltage to the $M_1 - M_2$ pair is zero, so that the drain currents of M_1 and M_2 are each equal to $I_o/2$. From the symmetry of the circuit it is clear that voltages V_1 and V_2 at the drains of M_3 and M_4 are equal. Thus, the voltage V_2 is given by

$$V_1 = V_2 = 2V_T + \sqrt{\frac{2I_o}{\mu C_o \left(\frac{W}{L}\right)_8}} + \sqrt{\frac{2I_o}{\mu C_o \left(\frac{W}{L}\right)_7}} \quad (1)$$

where μ = carrier mobility; C_o = gate oxide capacitance density; V_T = threshold voltage; and $(W/L)_n$ = aspect ratio of device n.

The voltage applied to the gate of the output transistor is this voltage minus the gate-source drop of transistor M_{12}. The drain current of M_{12} is equal to the drain current of M_{11}, which is given by

$$I_{D11} = I_{D12} = I_o \frac{(W/L)_{11}}{(W/L)_8}. \quad (2)$$

Thus, the gate-source voltage of the output transistor is given by

$$V_{GS14} = V_T + \sqrt{\frac{2I_o}{\mu C_o}} \left(\frac{1}{\sqrt{(W/L)_8}} + \frac{1}{\sqrt{(W/L)_7}} - \sqrt{\frac{(W/L)_{11}}{(W/L)_{12}(W/L)_8}} \right). \quad (3)$$

Here it has been assumed that body effect in the various transistors can be neglected, and the threshold voltages are equal. Because the sources of M_7, M_4, and M_{12} are at almost the same potential, this assumption produces little error. Finally, using the preceding two equations, the quiescent current in the output transistor is given by

$$I_{\text{out}} = I_o \left(\frac{W}{L}\right)_{14} \left[\frac{1}{\sqrt{(W/L)_8}} + \frac{1}{\sqrt{(W/L)_7}} - \sqrt{\frac{(W/L)_{11}}{(W/L)_{12}(W/L)_8}} \right]. \quad (4)$$

The circuit produces a quiescent current which is directly proportional to the bias current. Furthermore, it is possible to adjust the W/L ratio of M_{11} and M_{12} so as to reduce the quiescent current in the output transistor to any desired value. In the particular case of the circuit described here, the bias current in the output transistor is set at about $10I_o$. This value of approximately 300 μA is a compromise between power dissipation and bandwidth. This bias point corresponds to a typical $V_{d\text{sat}}$ of 150 mV in the output transistor. Thus, with a gain of 8 in the error amplifiers, an

TABLE II
300 Ω Buffer Amplifier Experimental Results @ + 5 V, 40°C

Open Loop Gain	5500
Voltage Swing	± 4.1V @ 300u Load
Quiscent Power	6mW
VCC PSRR @ 1 kHz	−74dB
@ 100 kHz	−44dB
VBB PSRR @ 1 kHz	−62dB
@ 100 kHz	−55dB
Input Noise Density	
@ 1kHz	95 nV/√Hz
Active Area For	
300u Buffer	628 Sq. Mil.
1 k Buffer	563 Sq. Mil.
10 k Buffer	372 Sq. Mil.

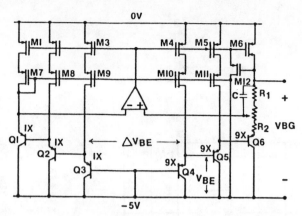

Fig. 6. Improved CMOS bandgap reference voltage circuit schematic.

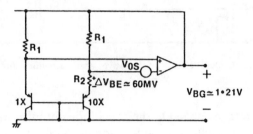

Fig. 5. Basic bandgap reference circuit implementation.

input-referred offset voltage of 3 mV would result in an output-referred offset of 24 mV, which would give shifts in the quiescent current on the order of 30 percent.

The maximum gate drive to the negative output transistor results when enough differential drive is applied to the error amplifier to turn off transistor M_2. The combination of the bias current I_o and the W/L of M_5 are chosen such that when this occurs the gate of M_5 assumes a potential near the positive supply. In this mode, the gate voltage of M_{14} is approximately equal to the supply voltage minus two threshold voltages. Alternatively, by choosing a smaller value of quiescent V_{gs} for M_5, the circuit can be operated in a current-limiting mode in which the maximum gate drive on the output transistor is programmed by the value of I_o in conjunction with the W/L of M_5.

Table II summarizes the measured performance of the 300 Ω power amplifier.

IV. Precision Bandgap Reference Circuit

A basic CMOS bandgap reference circuit is shown in Fig. 5 [5], [6]. Here the output voltage is

$$V_{BG} = V_{BE} + (\Delta V_{BE} + V_{OS}) \cdot \left(1 + \frac{R_1}{R_2}\right) \quad (5)$$

where V_{OS} is the amplifier offset voltage. However, this approach has two basic disadvantages. First, the amplifier offset voltage adds directly to the difference in base–emitter voltages ΔV_{BE} of the bipolar transistors. Offset voltages of typical CMOS operational amplifiers range between ± 15 mV and, when amplified by the resistor ratio gain factor, lead to a large variation in the reference voltage. This increases the reference voltage trimming requirements for a desired precision. Second, the offset voltage of a CMOS op amp drifts with time and has a temperature coefficient of

around 20 μV/°C. These variations are amplified and degrade the reference stability and performance.

One possible solution is to use chopper stabilization to null out the op amp offset voltage [7]. This is effective for a reference voltage valid only during a portion of the clock period. However, the analog signal paths of this VLSI processor require a continuous and stable reference to generate various bias voltages rendering this solution unsuitable.

Another approach is to reduce the relative effect of the offset voltage by increasing the contribution of the bipolar transistor base–emitter voltages. By using an area-ratioed stack of three closely matched bipolar transistors, the circuit, shown in Fig. 6, produces a basic reference voltage which is three times the silicon bandgap voltage. This reduces the effect of the offset by a factor of 3. The bandgap voltage is given by

$$V_{BG} = 3V_{BE} + (3\Delta V_{BE} + V_{OS}) \cdot \left(1 + \frac{R_1}{R_2}\right). \quad (6)$$

Transistors $M_1 - M_6$ are matched current sources, each of which forces a current equal to $3\Delta V_{BE}/R_2$ into each bipolar transistor. The transistors $M_7 - M_{11}$ drop the necessary voltage required to match the currents in $M_1 - M_6$ to within 0.5 percent. The resulting output voltage V_{BG} is 3.8 V.

In this approach, there are both positive and negative feedback paths around the amplifier. The negative feedback path must remain dominant at all frequencies to avoid instability. To ensure this, a capacitor C shunts the resistor R_1, increasing negative feedback at high frequency with feedforward action. The resistors R_1 and R_2 are made of p diffusion and $Q_1 - Q_6$ are vertical p-n-p transistors with a common collector.

Besides the normal desired operating state, this circuit also has a degenerate state where all the devices are off. Transistors M_{12} is used to start up the feedback loop and turns off in normal operation.

The bandgap voltage is referenced to the negative supply. The primary reason for this is to avoid collector-base reverse bias modulation of the emitter currents, which would degrade the reference power supply rejection. How-

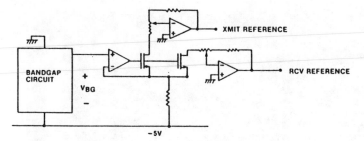

Fig. 7. Reference level shifting and gain adjust circuitry.

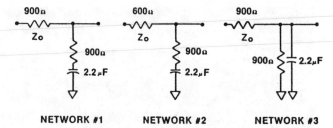

Fig. 8. Equivalent circuits of the three internal balance networks.

TABLE III
PERFORMANCE OF THE REFERENCE VOLTAGE CIRCUIT

Parameter Measured Data	
Temp. Coeff. (0–70°C)	< 100 ppm/°C
Power Supply Rejection	
+5 V @ 1 kHz	−75 dB
+5 V @ 100 kHz	−36 dB
−5 V @ 1 kHz	−70 dB
−5 V @ 100 kHz	−37 dB
Power Dissipation	8 mW
Active Area	3000 mils^2

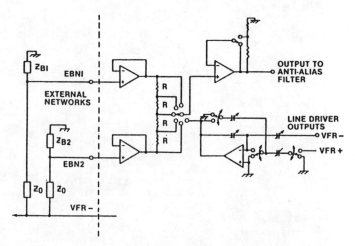

Fig. 9. Programmable balance network.

ever, the A/D and D/A converters need their reference voltages with respect to ground. Fig. 7 shows the circuits which provide the level shifting and gain adjust.

The reference voltage must be trimmed to have a zero temperature coefficient at room temperature. Because the relative offset voltage has been reduced by a factor of 3, only 4 bits are required to trim the bandgap voltage under all process variations. Also, to account for manufacturing variations, 3 bits in the level shifter and gain adjust circuit trim the transmit and receive gains to within ±1 percent accuracy. The complete trim procedure involves: 1) measurement of the amplifier offset voltage; 2) measurement and adjustment of the bandgap voltage V_{BG} and trimming it to three times the silicon bandgap voltage plus the amplifier offset voltage with the 4 bits of trim provided; and 3) adjustment of the transmit and receive reference voltages to nominally 3.2 V with the additional 3 bits of trim.

Table III presents the measured performance of the bandgap reference circuit.

V. LINE BALANCING FUNCTION

In a typical telecommunications line card design, the subscriber interface uses 2-wire transmission while the switching and transmission equipment employs 4-wire. The circuit which performs the 2- to 4-wire conversion requires a balance network to improve the return loss of the receive channel into the transmit channel. This line card function usually is implemented with discrete components and only recently has it been integrated into low-voltage MOS technology [8], [9]. This analog VLSI processor provides the system designer with two different options for implementing the 2- to 4-wire conversion and line balancing on the same chip. The first option provides three of the most commonly used balance networks, which

are shown in Fig. 8. These three networks use bilinear z-transform techniques and are implemented with one amplifier and a programmable switched-capacitor array clocked at 32 kHz. The user can select any of these three networks under software control.

However, there may be different system requirements calling for a special balance network. The second option allows the user to connect two different off-chip balance networks and select either of these or provide an interpolation between these two frequency characteristics under software control [10]. By connecting these networks to the analog VLSI, the user can design a compromise balance network transfer characteristic for different loop lengths by programming the interpolation coefficient a to be 0, 0.25, 0.5, 0.75, or 1.0. The resulting frequency characteristic will be

$$H(f) = aH_1(f) + (1-a)H_2(f) \qquad (7)$$

where

$H_1 =$ the transfer function of external Network 1

$H_2 =$ the transfer function of external Network 2.

This will cover many possible loop lengths, improving and easing the line balancing function in the line card. Also, because of different interface requirements of either transformer based hybrids or the new generation of integrated SLIC's, a programmable gain of 0 or 6 dB is provided in the path of the return signal.

Fig. 9 shows the programmable internal balance networks and how the two external balance networks are

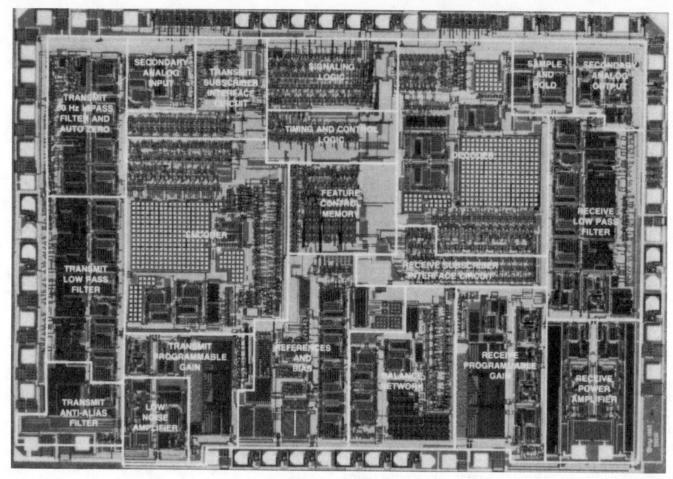

Fig. 10. Photomicrograph of the chip.

TABLE IV
SUMMARY OF THE CHIP PERFORMANCE

Parameter	Transmit Channel	Receive Channel
Gain Tracking @ 1020 Hz		
+3--40 dBmO	> -0.12, < +0.05 dB	< +0.10 dB
-40--55 dBmO	> -0.30, < +0.50 dB	< +0.20 dB
Sig./Dist. (C-msg) @ 1020 Hz		
+3--30 dBmO	> 37 dB	> 37 dB
-30--40 dBmO	> 31 dB	> 33 dB
-45 dBmO	28 dB	28 dB
Idle Channel Noise	15 dBrnCo	4 dBrnCo
Filter Frequency Response		
300-3000 Hz	> -0.01, < +0.08 dB	< +0.05 dB
Above 4600 Hz	< -35 dB	< -48 dB
Power Supply Rejection		
+5 V @ 1 kHz	-69 dB	-50 dB
+5 V @ 50 kHz	-60 dB	-28 dB
-5 V @ 1 kHz	-45 dB	-51 dB
-5 V @ 50 kHz	-40 dB	-28 dB

within 10 percent of these typical numbers. The programmable features are controlled by 30 bits. The chip is fabricated in a 4 μm n-well CHMOS process and is packaged in either a 28 or 22 pin DIP. Table IV summarizes the overall chip performance.

VII. CONCLUSIONS

A CMOS VLSI analog/digital interface circuit has been described. Its unique system architecture provides much higher level integration of the line card functions on a single chip than previously reported. The analog VLSI employs several novel circuit design techniques to achieve superior performance.

buffered by op amps. A resistive network and switches provide the selection and interpolation features.

VI. SYSTEM ATTRIBUTES/PERFORMANCE

The photomicrograph of the chip is shown in Fig. 10. The die size is approximately 50 000 mils^2, and it dissipates 80 mW active power with a 1 mW standby mode. A power trim circuit on the chip can adjust the power dissipation to

REFERENCES

[1] D. Senderowicz, S. F. Dreyer, J. M. Huggins, C. F. Rahim, and C. A. Laber, "A family of differential NMOS analog circuits for a PCM CODEC," *IEEE J. Solid-State Circuits*, vol. SC-17, pp. 1014–1023, Dec. 1982.
[2] B. K. Ahuja, M. R. Dwarkanath, T. E. Seidel, D. G. Marsh, "A single chip CMOS CODEC with filters," in *Proc. Int. Solid-State Circuits Conf.*, Feb. 1980, pp. 242–243.
[3] B. K. Ahuja, "An improved frequency compensation technique for CMOS operational amplifiers," *IEEE J. Solid-State Circuits*, vol. SC-18, pp. 629–633, Dec. 1983.
[4] K. E. Brehmer and J. B. Weiser, "Large swing CMOS power amplifier," *IEEE J. Solid-State Circuits*, vol. SC-18, pp. 624–629, Dec. 1983.

[5] K. E. Kujik, "A precision reference voltage source," *IEEE J. Solid-State Circuits*, vol. SC-8, pp. 222–226, June 1973.

[6] R. J. Widlar, "New developments in IC voltage regulators," *IEEE J. Solid-State Circuits*, vol. SC-6, pp. 2–7, Feb. 1971.

[7] B. S. Song and P. R. Gray, "A precision curvature compensated CMOS bandgap reference," *IEEE J. Solid-State Circuits*, vol. SC-18, pp. 634–643, Dec. 1983.

[8] M. Foster, H. El Sissi, V. Korsky, K. Silmens, R. Wallace, and W. Sin, "A monolithic NMOS filter and line balancing chip," in *Proc. Int. Solid-State Circuits Conf.*, Feb. 1980, pp. 182–183.

[9] R. Apfel, H. Ibrahim, and R. Ruebush, "Signal-processing chips enrich telephone line-card architecture," *Electronics*, pp. 113–118, May 5, 1982.

[10] A. De la Plaza, "Hybrid with automatic selection of balance networks," in *Proc. Int. Symp. Circuits Syst.*, Apr. 1981, pp. 725–728.

A Precision CMOS Bandgap Reference

JOHN MICHEJDA AND SUK K. KIM

Abstract —This paper describes the design of a precision on-chip bandgap voltage reference for applications with CMOS analog circuits. The circuit uses naturally occurring vertical n-p-n bipolar transistors as reference diodes. P-tub diffusions are used as temperature-dependent resistors to provide current bias, and an op-amp is used for voltage gain. The circuit is simple. Only two reference diodes, three p-tub resistors, and one op-amp are necessary to produce a reference with fixed voltage of −1.3 V. An additional op-amp with two p-tub resistors will adjust the output to any desired value.

The criteria for temperature compensation are presented and show that the properly compensated circuit can *in principle* produce thermal drift which is less than 10 ppm/°C. Process sensitivity analysis shows that in practical applications it is possible to control the output to better than 2 percent, while keeping thermal drift below 40 ppm/°C. Test circuits have been designed and fabricated. The output voltage produced was −1.30 ± 0.025 V with thermal drift less than 7 mV from 0°C to 125°C. Significant improvements in performance, at modest cost in circuit complexity, can be achieved if the op-amp offset contribution to the output voltage is reduced or eliminated.

I. INTRODUCTION

IN a large and complex LSI-CMOS analog circuit, the voltage reference is often a potentially most troublesome component since it must produce a temperature stable, process invariant, and precisely controlled output. In the past, most of efforts in voltage reference design have emphasized temperature compensation at the expense of output precision. The commonly used references based on the difference between gate/source voltages of enhancement and depletion mode MOS transistors realize low thermal drift; however, the absolute magnitude of output is poorly controlled because it depends on the accuracy of depletion and enhancement implants [1]. In the bandgap references, where the output is derived from the voltage difference of two diodes forward biased by ratioed currents, both the thermal drift and the absolute value of the output can be controlled with precision [2]–[4].

A CMOS bandgap voltage reference which uses bipolar-like source-to-drain transfer characteristics of MOS transistor in weak inversion was reported [5], [6]. The output voltage exhibited relatively low thermal drift and tight voltage spread from sample to sample. Another approach [7] used precision curvature-compensated switched capacitor CMOS bandgap reference. It required trimming and used a complex circuitry for generation of bias currents, thus consuming a large area.

This paper describes the design of another simple bandgap circuit that can be conveniently implemented in CMOS technology. The output of this circuit is both temperature stable and precise. The circuit configuration which follows that given by Kuijk [8] uses temperature dependent p-tub resistors to provide bias currents to the reference diodes, which are the emitter–base junctions of the bipolar transistors formed by the n^+ diffusion inside the p-tub. First, the basic circuit and the criteria for temperature compensation are presented, followed by discussion of the characteristics of the reference diodes and biasing resistors. The sensitivity of the output voltage to the most common process variation, and the power supply fluctuations will then be discussed in detail. Finally, the experimental results to verify circuit performance will be presented to illustrate the precision and the stability of the circuit.

II. THE BANDGAP CIRCUIT

In the bandgap circuit the output voltage is derived from the voltage difference across two identical diodes forward-biased by two unequal, precisely ratioed currents. The positive temperature coefficient of this difference is then cancelled by the negative temperature coefficient of voltage across one of the diodes. If the voltage across diode 1 is $V_1(T)$ and across diode 2 is $V_2(T)$, then the output can be expressed as

$$V_{out} = A(V_1(T) - V_2(T)) + B(V_2(T))$$
$$= aV_1(T) - bV_2(T) \tag{1}$$

where constants a and b are chosen to obtain a voltage V_{out} that has a minimum variation over the temperature range of interest.

The principles of bandgap references and the criteria for derivation of constants a and b are given in detail in [4]. These are derived for devices biased by either temperature independent constant current I, or currents which vary with temperature as T^α, where α is a constant. Neither one of these temperature variations of currents can be easily implemented in a CMOS analog circuit.

The bandgap circuit configuration used to derive the function in (1) is illustrated in Fig. 1. The first op-amp with transistors $Q1$ and $Q2$, and resistors $R1$, $R2$, and $R3$ is the bandgap circuit which produces a fixed voltage $V_{out} = -1.3$ V. The second op-amp with resistors $R4$ and $R5$ is a gain stage to adjust the output V_{ref} to a desired value. The discrete version of this bandgap circuit was first

Manuscript received August 12, 1983; revised July 13, 1984.
J. Michejda is with AT&T Bell Laboratories, Murray Hill, NJ 07974.
S. K. Kim is with Solid State Electronics Division, Honeywell, MN 55441.

Reprinted from *IEEE J. Solid-State Circuits*, vol. SC-19, no. 6, pp. 1014–1021, Dec. 1984.

34

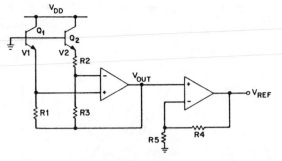

Fig. 1. Schematic of the CMOS bandgap circuit to produce negative output voltage. A bandgap source developing positive output with respect to ground is illustrated in [9].

proposed by Kuijk [8], who used integrated diode pairs and thin film resistors. More recently, Ye and Tsividis [9] have demonstrated this configuration, and a configuration producing a positive output voltage, using vertical n-p-n bipolar transistors and discrete external resistors. They also suggested using diffusion or polysilicon resistors in a fully integrated version of this circuit.

The gain constants a and b of circuit in Fig. 1 are given by

$$a = R_3/R_2 + 1 \qquad (2)$$

$$b = R_3/R_2. \qquad (3)$$

In addition to the gain of the circuit, which is determined by the ratio of resistors R_3 and R_2, the magnitude and the ratio of the bias currents is determined by the ratio of resistors R_1 and R_3:

$$I_1(T) = \frac{V_{\text{out}} - V_1(T)}{R_1} \qquad (4)$$

$$I_2(T) = \frac{V_{\text{out}} - V_1(T)}{R_3} = \frac{V_{\text{out}} - V_2(T)}{R_3 + R_2}. \qquad (5)$$

In this implementation, however, the biasing currents $I_1(T)$ and $I_2(T)$ are temperature dependent because of the variation of $V_1(T)$ and $V_2(T)$, and to a lesser extent because of the variation of V_{out} with respect to temperature. The conditions for the temperature compensation of this circuit with temperature independent R_1, R_2, and R_3 are given in [8].

In CMOS technology the situation is even further complicated because the only on-chip conductors that have large enough resistance values for proper biasing of the reference diodes are also temperature-dependent. The p-tub resistors approximately double their resistance for a temperature increase from 0 to 100°C.

The detailed derivation of the temperature compensation for the circuit illustrated in Fig. 1 is algebraically tedius and is briefly summarized in this section.

For a reference diode whose current I is given by Shockley's equation, for $qV \gg nkT$,

$$I = I_0 e^{qV/nkT}. \qquad (6)$$

The voltage drop at temperature T is given by

$$V(T) = n\left[V_G(T) + \left(\frac{kT}{q}\right)\ln\left(\frac{I}{AT^\beta}\right)\right] \qquad (7)$$

where n is the nonideality factor, $V_G(T)$ is the bandgap voltage at temperature T, k is the Boltzmann constant, q is the electron charge, A is a normalizing constant related to the geometry of the device, and β is a constant related to the fabrication process.

$V_G(T)$ is itself a function of temperature. Reference [10] gives the empirical expression for the bandgap value extrapolated from physical measurement over temperature range from 300 to 400 K.

$$V_G(T) = V_{G0} + \frac{dV_G}{dT}T \qquad (8)$$

where $V_{G0} = 1.20595$ V, and $dV_G/dT = -2.7325 \times 10^{-4}$ V/K.

For a circuit whose function is given by (1), subject to the condition that the temperature coefficient at temperature $T = T_0$ is zero,

$$\left.\frac{dV_{\text{out}}}{dT}\right|_{T=T_0} = 0. \qquad (9)$$

The value of the output voltage $V_{\text{out}}(T_0)$ is unique and given by

$$V_{\text{out}}(T_0) = n\left[V_{G0} + \left(\frac{kT_0}{q}\right)\right.$$
$$\left.\cdot\left(\beta - 1 + T_0\left(\frac{1}{R(T_0)}\right)\left(\left.\frac{dR}{dT}\right|_{T=T_0}\right)\right)\right] \qquad (10)$$

where $dR/dT|_{T=T_0}$ is the derivative of resistance of biasing resistors with respect to temperature at T_0.

The temperature response to the bandgap equation can be described by the following differential equation:

$$T\frac{dV_{\text{out}}}{dT} - V_{\text{out}} + \frac{nkT^2}{q}\left(\frac{1}{R}\right)\left(\frac{dR}{dT}\right)$$
$$+ \frac{nkT}{q}(\beta - 1) + nV_{g0} = 0. \qquad (11)$$

It is important to notice that the value of $V_{\text{out}}(T_0)$ as given in (10), and the bandgap temperature response as given in (11) depends only on physical diode parameters n and β, and resistor temperature coefficient $(1/R)(dR/dT)$.

III. The Reference Diode in a CMOS Bandgap Circuit

It is well known that nearly ideal diode characteristics can be obtained from the base–emitter voltage V_{be} of a bipolar transistor. In the twin tub CMOS technology the vertical n-p-n bipolar devices are readily available with n⁻ substrate collector, p-tub base, and n⁺ emitter.

In the layout of the bipolar devices five unit transistors are connected in parallel to make one reference diode. In this approach, similar to the one given in [6], the reference device can operate at larger biasing current, and in addition better matching of references can be achieved.

These devices, each unit transistor with 20 μm $\times$ 20 μm emitter, manufactured in the 3.5 μm linear twin tub CMOS process, were characterized to obtain the value of parameters n and β necessary to predict the voltage drop across the reference device as given in (7).

The value of n was determined by measuring I–V characteristics of the reference devices at room temperature between currents of 0.5 and 500 μA, and then fitting the measured voltage using (7), with n being the adjustable parameter. The voltage drop of the diode in the fit was normalized to the value at the lowest current. The best fit was obtained for $n = 1.01$ over a range of currents from 0.5 to 25 μA. At currents above 25 μA, the differences between measured voltage and the fit were greater than 0.5 mV.

To measure the β parameter directly, the precise I–V characteristics of reference devices over a wide range of temperatures are needed. Such a measurement is tedious and difficult to do precisely, because a minor inaccuracy (as little as 0.5°C) in temperature measurement of the references, can lead to large errors in estimate of β. When these measurements are made on devices placed on a wafer prober, an uncertainty in temperature between the thermocouple in the wafer chuck, and the wafer itself can also cause significant errors in the value of β.

A new indirect method of measurement was used in determining the value β. A precision op-amp with low input offset and high open-loop gain, and a set of precision discrete resistors whose values were individually measured, were externally connected to the bipolar devices on the wafer. The connections were identical to those of the bandgap circuit, and the values of discrete resistors were selected to minimize the output variations with temperature. The voltage produced by the circuit was measured as a function of temperature of the reference diodes. A fit to the data as a function of β was made using the computer simulation of the bandgap circuit. Fig. 2 illustrates the measured and predicted responses of the circuit. The best overall agreement is for $\beta = 1.775$.

The simulator used to obtain the best fit was written in Fortran. The program contains appropriate diode models given in (7), and (8), to generate the I–V–T characteristics of the reference diodes. Given diode characteristics, and the resistance values R_1, R_2, and R_3, the program solves iteratively for bias points V_1 and V_2, and output voltage V_{out} until,

$$\frac{V_{\text{out}} - V_1(R_1(T), T)}{R_1(T)} = \frac{V_{\text{out}} - V_2(R_3(T), T)}{R_2(T) + R_3(T)} \cdot \frac{R_3(T)}{R_1(T)}.$$

(12)

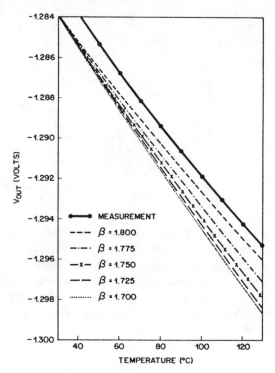

Fig. 2. Comparison of temperature response between simulated and measured fixed resistor bandgap circuit to determine the value of β.

This method of measurement is much less sensitive to vagaries in temperature measurement since the discrete resistors are chosen to minimize temperature dependence of the output voltage. This decreased output voltage sensitivity to temperature enables easier, and more precise determination of β.

IV. BIASING RESISTORS IN THE CMOS BANDGAP CIRCUIT

The results of the diode characterizations described in the previous section illustrate that the proper operation of reference devices requires small biasing currents in the microampere range. To provide these small currents, resistance values of the order of 10^5 Ω or higher are needed. The only on-chip conductor available in CMOS technology that has sufficiently high sheet resistance to render these resistors practical is the p-tub diffusion. The sheet resistance of the p-tub diffusion in the 3.5 μm twin tub CMOS technology is ~ 3 kΩ/$\square$. Thus, resistor values up to 0.5 MΩ can be readily realized.

P-tub resistors exhibit temperature dependent behavior due to mobility changes over the temperature range of interest. The measurements of temperature effect on mobility variation of p-type silicon samples [11] yielded mobility dependence of $T^{-2.2}$. The measurements of temperature dependence of p-tub resistance, illustrated in Fig. 3, yielded essentially the same result.

Therefore, for a p-tub resistor,

$$R(T) = R_0 T^{2.2}$$

(13)

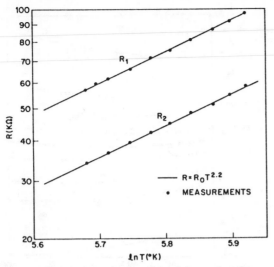

Fig. 3. Temperature characteristic of the p-tub resistors.

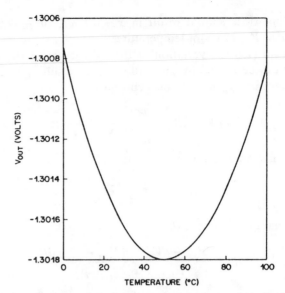

Fig. 4. Predicted temperature response of the bandgap circuit.

where R_0 is the normalizing constant. Therefore,

$$\left(\frac{1}{R}\right)\left(\frac{dR}{dT}\right) = \frac{2.2}{T} \qquad (14)$$

and the proper compensation, according to formula (10), occurs when V_{out} at T_0 is

$$V_{out}(T_0) = n\left[V_{G0} + \left(\frac{kT_0}{q}\right)(\beta + 1.2)\right]. \qquad (15)$$

Fig. 4 illustrates the predicted temperature response of the bandgap circuit from 0 to 100°C, where the resistance of biasing resistors varies as $T^{2.2}$, the value of $\beta = 1.775$, and the value of T_0 is 50°C. The value of $V_{out}(T_0)$ is -1.3018 V, and the temperature variation of V_{out} over 100°C is approximately 1 mV. The temperature coefficient that can be obtained using this approach is less than 8 ppm/°C.

V. PROCESS SENSITIVITY OF THE OUTPUT VOLTAGE

One of the advantages of the bandgap references over the threshold differencing scheme is that both magnitude and temperature compensation of the output voltage are relatively tolerant of processing variations. In this section an approximate analysis of the sensitivity of the output voltage to the processing variants will be presented. The current density ratio between reference diodes is $25:1$ yielding voltage difference $(V_1 - V_2) \sim 80$ mV at 25°C. The value of constants a and b in (1) is 10 and 9, respectively.

The main processing parameters that affect the output of the bandgap circuit are: p-tub doping, resistor mismatch, reference diode mismatch, and the threshold mismatch of the op-amp input devices. The threshold mismatch, which results in the op-amp offset error, mainly affects the magnitude of the output voltage if the offset itself is not a function of temperature. All other processing variations affect magnitude, as well as the temperature compensation of the output.

The doping of the p-tub affects the resistance of the biasing resistors and the V_{be} drop across the reference diodes. Both the value of the biasing resistors, and the V_{be} drop across the reference devices affect the voltages V_1 and V_2 of the circuit.

The resistance of the p-tub resistor uniformly doped by ion implantation is inversely proportional to the implant dose of boron N_s.

$$R \sim \frac{1}{N_s}. \qquad (16)$$

Therefore, for a fractional error in implant dose dN_s/N_s, the fractional error on the p-tub resistance dR/R is;

$$\frac{dR}{R} \sim \frac{-dN_s}{N_s}. \qquad (17)$$

For the n-p-n bipolar device biased with fixed V_{be}, the collector current is proportional to the number of impurities/unit area in the base [12] (also known as Gummel number). For the device with ion implanted p-tub base, this number is equal to the ion implant dose N_s. Therefore,

$$I \sim \frac{1}{N_s} e^{qV_{be}/kT} \qquad (18)$$

and the V_{be} drop across the reference device is

$$V_{be} \sim \frac{kT}{q} \ln(IN_s). \qquad (19)$$

The change in V_{be} of the reference device due to the p-tub ion implant error is

$$dV_{be} \approx \frac{kT}{q} \frac{dN_s}{N_s}. \qquad (20)$$

The doping of the p-tub in the twin tub CMOS process can be controlled to ± 10 percent. This variation in the ion implant dose should result in a $\pm 0.1\ kT$, or ± 2.5 mV error in V_{be} at room temperature.

From (19) the variation of the voltage across the reference device biased by the p-tub resistor resulting from the changes in V_{be} and the bias current is

$$dV_1 \approx dV_2 \approx \frac{kT}{q}\left[\frac{dN_s}{N_s} + \frac{dI}{I}\right]. \qquad (21)$$

For bias current I in the circuit,

$$I = \frac{V_{\text{out}} - V_{be}}{R}, \qquad (22)$$

the change in current dI is

$$dI = \frac{dV_{\text{out}} - dV_{be}}{R} - \frac{(V_{\text{out}} - V_{be})\,dR}{R^2} \qquad (23)$$

for 10% resistance variation, the first term in (23) is small. Therefore,

$$\frac{dI}{I} \approx -\frac{dR}{R} \qquad (24)$$

and using (17), (21), and (24)

$$dV_1 \approx dV_2 \approx \frac{2kT}{q}\left(\frac{dN_s}{N_s}\right). \qquad (25)$$

The bandgap output voltage change is then

$$dV_{\text{out1}} = a\,dV_1 - b\,dV_2 \approx (a - b)\,dV_1 \approx dV_1. \qquad (26)$$

The total variation of the output voltage due to variation in the p-tub doping is

$$dV_{\text{out1}} \approx \frac{2kT}{q}\left(\frac{dN_s}{N_s}\right). \qquad (27)$$

The ± 10 percent error in the p-tub implant will result in ± 5 mV error in the output voltage at room temperature. This is significantly better than the accuracy of the output voltage produced by the threshold differencing circuit due to the ion implant variation.

The V_{be} voltage produced by the reference devices is very uniform, and the mismatch of the voltage across different devices of the same size on the same chip is small. The measurements of the reference devices biased with constant current on the different chip sites of the same wafer yielded the maximum mismatch of less than 0.2 mV. Such a mismatch would result in a output voltage change of

$$dV_{\text{out2}} \approx a\,dV_1 \approx b\,dV_2 \approx 2 \text{ mV} \qquad (28)$$

where the value of $a = 10$ is used.

The mismatch of the resistor ratio R_3 and R_2 affects the gain of the circuit. The mismatch of the ratio R_1 and R_3 influences the current ratio supplied to the references, and thus the difference between V_1 and V_2.

The resistances R_1 and R_3 can be matched accurately because they can be ratioed by an exact integer factor, and

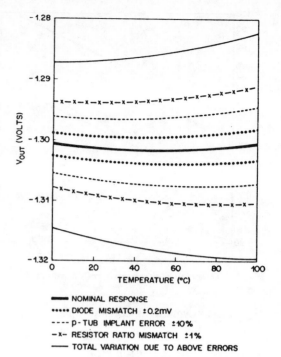

Fig. 5. Sensitivity of output voltage to processing variations.

also because R_1 and R_3 have an identical voltage across them thus eliminating problems due to any nonlinearities of p-tub resistance. The mismatch of R_2 and R_3 is more likely to occur since these resistors are ratioed by a noninteger number, and because R_2 is biased at a different potential from the substrate than R_1 and R_2. Therefore, only R_3/R_2 mismatch will be considered here.

The variation of the output voltage of the bandgap circuit due to mismatch of R_3, and R_2 is

$$dV_{\text{out3}} = d\left(\frac{R_3}{R_2} + 1\right)V_1 - d\left(\frac{R_3}{R_2}\right)V_2 = d\left(\frac{R_3}{R_2}\right)(V_1 - V_2).$$

$$(29)$$

In a careful layout the resistance values of the p-tub resistors can be matched better than 1 percent. Therefore, for 1 percent resistance mismatching and 80 mV difference between V_1 and V_2, with $R_3/R_2 = 10$ the output voltage error is

$$dV_{\text{out3}} = 0.01\left(\frac{R_3}{R_2}\right)(V_1 - V_2) \approx 8 \text{ mV}. \qquad (30)$$

The above discussion illustrated that excluding the offset error of the op-amp the various processing variations can affect the output voltage of the bandgap circuit by about ± 15 mV at room temperature in the worst-case analysis. This is only about 1.2 percent of the total voltage produced by the bandgap circuit. The estimate of the effect of these parameters on temperature compensation is considerably more difficult and was done using the numerical bandgap simulator discussed in Section III. Fig. 5 illustrates the predicted worst-case behavior of the bandgap circuit and how each processing variation contributes to errors in the output voltage temperature compensation.

In this simulation the output of each reference diode voltage was varied by ± 2.5 mV, reference device mismatch was varied by ± 0.2 mV, the sheet resistivity of biasing resistors was varied by ± 10 percent and the ratio of resistors was mismatched by ± 1 percent. The worst-case analysis yields the temperature compensation of the output voltage of about 5 mV over the temperature range from 0 to 100°C.

The input offset error of the summing op-amp can have significant and detrimental effect on the control of the magnitude of the output voltage. An analysis of the bandgap circuit shown in Fig. 1, which includes the offset error of the op-amp, gives the following relation for V_{out}.

$$V_{\text{out}} = aV_1 - bV_2 + cV_{os} \tag{31}$$

where a and b are given in (2) and (3), V_{os} is the offset, and

$$c = -\left(1 + \frac{R_3}{R_2}\right) = -a. \tag{32}$$

The offset of the op-amp is therefore multiplied by the factor c. In a typical bandgap circuit, $c \approx 10$; thus, small offset value of the op-amp can contribute a large error to the output voltage.

VI. Sensitivity of Output Voltage to Power Supply Variations

The primary effect of the power supply variation on the output voltage of the bandgap circuit comes from change of the reverse bias on the biasing p-tub resistors R_1, R_2, and R_3. This, in turn changes their resistance values resulting in modified gain factors and bias currents to reference diodes.

In the circuit, resistors R_1 and R_2 are identically biased with respect to the substrate, although they operate at different current densities. R_2 is biased approximately 80 mV more positive than the other two. Therefore, neglecting the influence of bias current, the voltage coefficients of R_1 and R_3 should be identical

$$\frac{1}{R_1}\frac{dR_1}{dV_{\text{sup}}} = \frac{1}{R_3}\frac{dR_3}{dV_{\text{sup}}}. \tag{33}$$

After tedious algebra, using (1), (2), (3), (4), (5), (7), and (33) one can compute the total variation of the output voltage to be

$$\frac{dV_{\text{out}}}{dV_{\text{sup}}} = \frac{R_3}{R_2}\left[\frac{1}{R_3}\frac{dR_3}{dV_{\text{sup}}} - \frac{1}{R_2}\frac{dR_2}{dV_{\text{sup}}}\right](V_1 - V_2)$$

$$- \frac{nkT}{q}\left[\frac{1}{1 + \dfrac{nkT}{q}\left(\dfrac{1}{V_{\text{out}} - V_1}\right)}\right]\left(\frac{1}{R_1}\right)\left(\frac{dR_1}{dV_{\text{sup}}}\right). \tag{34}$$

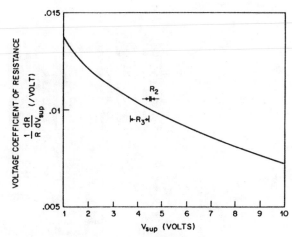

Fig. 6. Voltage coefficient of p-tub resistance as a function of substrate bias (V_{sup}).

The first term in brackets depends on differences between voltage coefficients of resistance between R_1 (or R_3) and R_2 due to different reverse biasing conditions. Fig. 6 illustrates the normalized voltage coefficient of the p-tub resistor as a function of reverse bias. (This curve is shown for an arbitrary current approximating operating point of R_3 resistor. The shape of the curve varies slightly at different currents.) On this curve the bias voltages are marked for resistors R_3 and R_2. Since the difference between the two coefficients of those points is small ($< 10^{-3}$), the first term in (34) is less than 0.8 mV/V.

The second term in (34) results from modified current through reference diodes. Again, a quick computation shows this value to be ~ 0.2 mV/V. The total expected variation of the output is therefore expected to be about 0.6 mV/V.

VII. Preliminary Results

Fig. 7 shows the photomicrograph of the circuit designed to test the performance of the resistor bandgap circuit. The operational amplifiers used in the circuit are described in [13]. The tester uses a second op-amp with p-tub resistors R_4 and R_5 to adjust the output voltage to a desired value. Because the gain of this stage is determined only by resistance ratio and not by resistance values themselves, it is reasonably precise and independent of temperature. The total circuit size, including the second op-amp and resistors R_4 and R_5, is 0.4 mm². The power consumption is 2 mW. In this paper only the results relating to the output of the first op-amp will be discussed. The data are based on measurements obtained from three device lots fabricated in the 3.5 μm linear twin tub CMOS process.

Fig. 8 shows the measured temperature response of one sample circuit along with the predicted response obtained from computer simulation. The output voltage shown has been compensated for the input offset error of the op-amp by measuring the offset contribution and subtracting it from the measured output of the circuit. The amplified

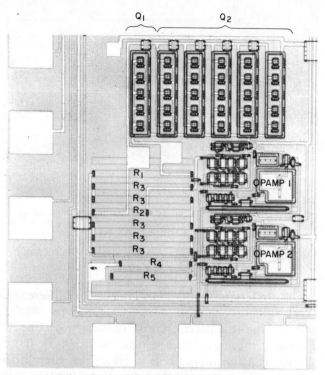

Fig. 7.   Photomicrograph of the bandgap test circuit.

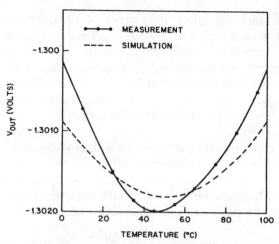

Fig. 8.   Measured and predicted temperature responses of the bandgap circuit after compensation for the op-amp offset.

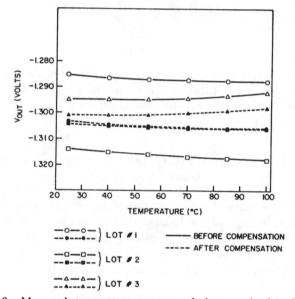

Fig. 9.   Measured temperature response of three randomly selected bandgap circuits from three separate wafer lots.

offset contribution was measured by grounding emitters of Q_1 and Q_2 through internal pads and measuring the output. The predicted temperature response and the value of the output voltage agree well with the compensated measured values. The temperature stability of the output is better than 2 mV from 0 to 125°C.

More complete measurements of circuits on different chip sites of a single wafer show that the offset compensated output voltage variation is 3 mV at room temperature. The wafer to wafer variation of the offset compensated output is 7 mV while the temperature stability of the output of most the circuits is better than 5 mV from 25 to 125°C.

As expected, the largest contribution to the output error comes from the input offset of the CMOS summing op-amp. Fig. 9 shows the typical temperature responses of three randomly selected circuits from three separate wafer slices from three wafer lots. For comparison, the output of these circuits with op-amp offset subtracted are also shown. The bulk of the output voltage variation in these samples is due to the offset of the op-amp. More extensive measurements indicate that the output voltage of individual circuits may vary by ±15 mV due to the offset error alone. In the extreme cases of large offset errors, the offset itself may be temperature-dependent, and may add 2 mV to the temperature instability of the bandgap circuit.

The sensitivity of the output to power supply variations was measured at room temperature with power supply voltage varied from 4.75 to 5.25 V. The output of most of the circuits changed by less than 0.3 mV, the mean was 0.4 mV. The maximum variation, observed on small percentage of samples, was 1.1 mV. The output variation is larger than predicted in Section VI. The differences are likely due to omission in analysis of the influence of bias current on the voltage coefficient of resistance of R_1 and R_3.

The preliminary data suggest that without any offset cancelling technique, the output voltage of the bandgap circuit is 1.30 ± .025 V, while the worst-case temperature drift is 7 mV from 0 to 125°C. Dramatic improvement in performance at modest cost in circuit complexity can be achieved if input error contribution is reduced either by cascading reference devices, or by offset cancelling techniques [14].

VIII.   Conclusions

The design of a simple and practical precision CMOS bandgap reference circuit which uses p-tub temperature dependent resistors and naturally occurring n-p-n bi-polar transistors is described. The criteria for the proper temperature compensation of the output voltage are derived and are shown to be independent of the design parameters such as current values and their ratios, resistor values, or diode bias points. The diodes manufactured in the 3.5 μm twin tub linear CMOS process are shown to be acceptable for the references in the bandgap circuit. The magnitude and temperature stability of the output voltage is shown to be tolerant to the most common variations of the CMOS process. The performance of the test circuits matches well the predictions of the bandgap response made by the bandgap computer simulations. The output voltage of the circuit is $-1.30 \pm .025$ V with temperature stability better than 7 mV from 0 to 125°C. A version of this circuit, which produces positive output is shown in [9]. A dramatic improvement in the performance can be achieved if op-amp offset error contribution is reduced by using offset correcting techniques.

ACKNOWLEDGMENT

The authors wish to acknowledge the support offered by H. J. Boll and J. G. Ruch, and to Y. P. Tsividis for many fruitful discussions. They are thankful to P. B. Smalley for performing the testing.

REFERENCES

[1] R. A. Blauschild, P. A. Tucci, R. S. Muller, and R. G. Meyer, "A new NMOS temperature-stable voltage reference," *IEEE J. Solid-State Circuits*, vol. SC-13, pp. 767–774, Dec. 1978.

[2] R. J. Widlar, "New developments in IC voltage regulators," *IEEE J. Solid-State Circuits*, vol. SC-6, pp. 2–7, Feb. 1971.

[3] A. P. Brokaw, "A simple three-terminal IC bandgap reference," *IEEE J. Solid-State Circuits*, vol. SC-9, pp. 388–393, Dec. 1974.

[4] P. R. Gray and R. G. Meyer, Analysis and Design of Analog Integrated Circuits. New York: Wiley, 1977.

[5] E. Vittoz and O. Neyrund, "A low voltage CMOS bandgap reference," *IEEE J. Solid-State Circuits*, vol. SC-14, June 1979.

[6] E. Vittoz, "MOS transistors operated in the lateral bipolar mode and their application in CMOS technology," *IEEE J. Solid-State Circuits*, vol. SC-18, June 1983.

[7] B. S. Song and P. R. Gray, "A precision curvature-compensated CMOS bandgap reference," in *ISSCC Dig. Tech. Papers*, vol. 26, Feb. 1983, pp. 240–241.

[8] K. Kuijk, "A precision reference voltage source," *IEEE J. Solid-State Circuits*, vol. SC-8, pp. 222–226, June 1973.

[9] R. Ye and Y. Tsividis, "Bandgap voltage reference sources in CMOS technology," *Electron. Lett.*, vol. 18, no. 1, pp. 24–25, Jan. 1982.

[10] Y. P. Tsividis, "Accurate analysis of temperature effects in I_c–V_{be} characteristics with applications to bandgap reference devices," *IEEE J. Solid-State Circuits*, vol. SC-15, pp. 1076–1084, Dec. 1980.

[11] C. Jacobini *et al.*, "A review of some charge transport properties of silicon," *Solid-State Electron.*, vol. 20, p. 77, 1977.

[12] S. M. Sze, *Physics of Semiconductor Devices*, 2nd ed. New York: Wiley, 1981.

[13] V. R. Saari, "Low-power high-drive CMOS operational amplifiers," *IEEE J. Solid-State Circuits*, vol. SC-18, pp. 121–127, Feb. 1983.

[14] K. C. Hsiech and P. R. Gray, "A low-noise chopper-stabilized differential switched-capacitor filtering technique," in *ISSCC Dig. Tech. Papers*, vol. 24, Feb. 1981, pp. 128–129.

A Precision Curvature-Compensated CMOS Bandgap Reference

BANG-SUP SONG, STUDENT MEMBER, IEEE, AND PAUL R. GRAY, FELLOW, IEEE

Abstract —A precision curvature-compensated switched-capacitor bandgap reference is described which employs a standard digital CMOS process and achieves temperature stability significantly lower than has previously been reported for CMOS circuits. The theoretically achievable temperature coefficient approaches 10 ppm/°C over the commercial temperature range utilizing a straightforward room temperature trim procedure. Experimental data from monolithic prototype samples are presented which are consistent with theoretical predictions. The experimental prototype circuit occupies 3500 mils^2 and dissipates 12 mW with ±5 V power supplies. The proposed reference is believed to be suited for use in monolithic data acquisition systems with resolutions of 10 to 12 bits.

I. INTRODUCTION

AN essential element of the analog and digital interface function is a voltage reference to control the scale factor of conversion. The temperature stability of a reference source is a key factor in the accuracy of the overall data acquisition function. Therefore, the ability to integrate an entire data acquisition system within a single CMOS VLSI chip is contingent upon the ability to realize a CMOS compatible voltage reference with a very low temperature drift. Since its introduction by Widlar [1], the bandgap referencing (BGR) technique has been widely employed for implementing a voltage reference source in bipolar integrated circuits. The temperature stability of the bandgap reference has been continuously improved via new circuit and technology innovations such as curvature compensation and laser trim [2]–[5]. In CMOS technology, the BGR technique has been directly applied [6]–[8]. However, the development of a high-performance CMOS bandgap reference has been hindered by several limiting factors attributable to the peculiarities of the bipolar devices available in a standard CMOS process, the high offset and drift of CMOS op amps that make up the circuit and the inherent curvature problem in the bandgap reference.

This paper will describe one circuit implementation of a precision CMOS bandgap reference which overcomes some of the drawbacks of a standard CMOS process, and embodies curvature compensation and differential offset cancellation to achieve experimental typical temperature drifts of 13.1 and 25.6 ppm/°C over the commercial and military temperature ranges, respectively. In the proposed reference, a temperature-stable voltage is developed by adding linear and quadratic temperature correction voltages to the forward-biased diode voltage which is obtained from the substrate p-n-p transistor available in CMOS processes. The linear temperature correction voltage is proportional to the absolute temperature (commonly called PTAT) while the quadratic temperature correction voltage is proportional to the absolute temperature squared (PTAT^2). They are independently adjustable to set the reference output voltage for a minimum temperature drift.

The offset voltage of the CMOS op amp is eliminated using the correlated-double sampling (CDS) technique [9]. The base current and base spreading resistance of the native substrate p-n-p transistor are cancelled to the first order and the amplification ratio is set by a capacitor ratio rather than a resistor ratio. Due to the cyclic behavior of the offset cancellation in this technique, the output reference voltage is not available at all times. However, the reference can be operated synchronously with other elements of the systems. Since the periodic offset sample and subtraction cycle effectively removes the low-frequency $1/f$ noise component of a CMOS op amp along with its offset, the dominant noise source is the thermal noise of a CMOS op amp which is designed to be on the order of 100 μV (rms) at the output in 500 kHz bandwidth.

In Sections II and III, the primary limitations in a conventional CMOS BGR implementation and the temperature curvature in bandgap references are discussed. In Section IV, a curvature-compensated switched-capacitor CMOS bandgap reference is introduced. In Section V, experimental results measured from monolithic prototype samples are presented and the problems related to the design of p-n-p transistors are discussed. The theoretical analysis of BGR temperature compensation techniques is included in the Appendix.

II. CONVENTIONAL CMOS BGR IMPLEMENTATION

One example of a conventional CMOS BGR implementation in an n$^-$-well CMOS process is shown in Fig. 1. Transistors Q_1 and Q_2 are substrate p-n-p transistors whose collectors are always tied to the most negative power

Manuscript received April 6, 1983; revised August 3, 1983. This work was supported by the National Science Foundation under Grants ECS-8023872 and ECS-8120012, IBM Corporation, and the MICRO Project.

B.-S. Song is with Bell Laboratories, Murray Hill, NJ 07974.

P. R. Gray is with the Department of Electrical Engineering and Computer Science, Electronics Research Laboratory, University of California, Berkeley, CA 94720.

Reprinted from *IEEE J. Solid-State Circuits*, vol. SC-18, no. 6, pp. 634–643, Dec. 1983.

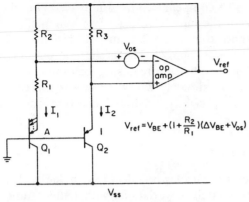

Fig. 1.   Example of a conventional CMOS bandgap reference.

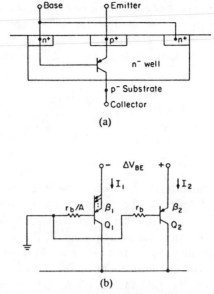

Fig. 2.   (a) Substrate p-n-p transistor profile. (b) Nonideal parameters in the PTAT correction voltage generation circuit.

supply because, in an n^--well CMOS process, the p^+ diffusion in the n^--well, the n^--well itself and the p^- substrate form a vertical p-n-p structure as shown in Fig. 2(a). Therefore, it is not possible to sense the collector current directly as in the bipolar bandgap reference so as to reduce the error due to the finite current gain [3], [4]. In a p^--well process, a dual circuit incorporating n-p-n transistors would be used. While many other circuit implementations are possible, this circuit appears to be as good as any. Therefore, individual error sources will be described one by one for this circuit in the rest of this section. All resistors are the p^+-diffusion resistor in the n^--well and the CMOS op amp is assumed to have an infinite gain with the offset voltage of V_{os}. This assumption is justified because CMOS op amps usually have enough gains such that the error due to finite-gain effects is negligible for this application.

Assuming that transistor Q_1 in Fig. 1 has an area that is larger by a factor A than transistor Q_2, and both are in the forward active region, the output voltage of the reference is given by

$$V_{ref} = V_{BE} + \left(1 + \frac{R_2}{R_1}\right)(\Delta V_{BE} + V_{os}) \qquad (1)$$

where V_{BE} is the emitter–base voltage of transistor Q_1, ΔV_{BE} is the difference between the emitter–base voltages of transistors Q_1 and Q_2, and V_{os} is the input offset voltage of the operational amplifier. The value of this expression is influenced by the nonidealities of the bipolar transistors as illustrated in Fig. 2(b). If these are taken into account, the transistor emitter–base voltage is given by

$$V_{BE} = V_T \ln \frac{I_1}{I_{s1}} + V_T \ln \frac{1}{1 + \frac{1}{\beta_1}} + \frac{r_b I_1}{A\beta_1} \qquad (2)$$

where V_T is the thermal voltage kT/q, I_1 is the emitter current of transistor Q_1, I_{s1} is the saturation current of transistor Q_1, β_1 is the current gain of transistor Q_1 and r_b is the effective series base resistance of Q_2. The second term in this expression results from the fact that while the collector current is a well-defined function of the emitter–base voltage, the current sensed and controlled by

this circuit is the emitter current. The third term results from the voltage drop in the finite series base resistance. The difference between the two emitter–base voltages is given by

$$\Delta V_{BE} = V_T \ln A + V_T \ln \frac{I_2}{I_1} + V_T \ln \frac{1 + \frac{1}{\beta_1}}{1 + \frac{1}{\beta_2}} + r_b \left(\frac{I_2}{\beta_2} - \frac{I_1}{A\beta_1} \right)$$

$$(3)$$

where I_2 is the emitter current of transistor Q_2 and β_2 is the current gain of transistor Q_2. If the bipolar transistors used to implement the reference are ideal in the sense that they have infinite current gain and zero base resistance, and if the emitter currents of the transistors are in fact equal, then only the first terms in (2) and (3) are nonzero. However, because of the relatively poor performance of CMOS-compatible devices, these terms can strongly influence the performance of the reference. The presence of the operational amplifier offset voltage in the output, multiplied by the gain factor $(1 + R_2/R_1)$, which is typically on the order of 10, is also an important degradation. Finally, the variation of the bias currents I_1 and I_2 with temperature must be carefully considered. In the following subsections, the effects of these nonidealities are examined in more detail.

A. Operational Amplifier Offset

The operational amplifier offset is the biggest error source that causes the nonreproducibility in the output voltage temperature coefficient. Normally, a bandgap reference is trimmed to an output voltage which is predetermined to give a near-zero temperature coefficient of the output. Large, non-PTAT components in the output due to

43

the op amp offset cause the trimming operation to give an erroneous result. If we assume the offset voltage V_{os} is independent of temperature, the resulting temperature coefficient error due to a 5 mV V_{os}, for example, is approximately

$$\text{TC error} = \frac{\left(1 + \frac{R_2}{R_1}\right)V_{os}}{V_{ref}T_o} \approx \frac{10 \times 5 \text{ mV}}{1.26 \text{ V} \times 300 \text{ K}}$$

$$\approx 132 \text{ ppm/}^\circ\text{C}. \qquad (4)$$

That is, a temperature coefficient on the order of 132 ppm/°C will result in the reference output temperature coefficient from a 5 mV temperature-independent offset voltage in the operational amplifier if the reference is trimmed assuming the offset is zero. This offset error contribution can be reduced by making ΔV_{BE} bigger, in effect decreasing the gain factor $(1 + R_2/R_1)$ as implied by (1). One way to achieve this is to obtain ΔV_{BE} by taking the difference of two cascaded transistor strings discussed later in Section IV. In this work, however, offset cancellation technique is employed to further reduce this error while the cascading scheme is used for the PTAT current generation.

B. Bias Current Variation

If the resistors R_1, R_2, and R_3 have a zero temperature coefficient, then the bias currents in transistors Q_1 and Q_2 must be PTAT since the voltage across R_1 is PTAT. The finite temperature coefficient of actual resistors formed from the source-drain diffusion or from polysilicon layers results in non-PTAT variation of the bias current. This in turn causes an additional component in the temperature variation of the V_{BE} term in the output. If only the first two terms of (2) and (3) are taken, the V_{BE} is given by

$$V_{BE} = V_T \ln \frac{I_1}{I_{s1}} = V_T \ln \frac{V_T \ln A}{R_1 I_{s1}}$$

$$= V_T \ln \frac{V_T \ln A}{R_1(T_o)I_{s1}} + V_T \ln \frac{R_1(T_o)}{R_1(T)} \qquad (5)$$

where T_o is the reference temperature, usually room temperature. Note that the first term is the V_{BE} variation that results when there is no resistor temperature coefficient, and the second is that which results when a temperature coefficient is present. Further insight can be obtained by expanding this second term as a Taylor series in temperature about T_o, and neglecting higher order terms:

$$V_{BE} = V_{BE}|_{\text{ideal}} - V_T$$

$$\cdot \frac{1}{R} \frac{dR}{dT}\bigg|_{T_o}(T - T_o) - V_T \frac{1}{2R} \frac{d^2R}{dT^2}\bigg|_{T_o}(T - T_o)^2 - \cdots. \quad (6)$$

From this relation it can be seen that even a purely linear variation in resistor value with temperature results in an output temperature variation with both PTAT and PTAT^2 temperature variation components. Assuming the resistor temperature behavior is known and reproducible, the PTAT

portion can be compensated by simply changing the target trim value of the output voltage. For example, (6) can be used to show that a 1000 ppm/°C resistor TC would result in a -21 ppm/°C reference output TC, which could be removed by simply raising the output voltage trim target by approximately 8 mV. However, cancellation of PTAT^2 term precisely requires curvature compensation discussed later.

C. Other Effects

The effects of various nonidealities, including base resistance, β mismatch, β variations with temperature, and β variations with collect current can be evaluated with the use of (1), (2), and (3). Which of these effects is the most important in a given circuit application is strongly dependent on the nature of the bipolar transistors in the particular technology used. If the well (base) doping is particularly light, as is often the case, then the intrinsic base resistance effect, represented by the last term in (3), may well be the most important. The temperature coefficient in the output due to this term is given by

$$\text{TC error} = \left(1 + \frac{R_2}{R_1}\right)\frac{r_b I_2}{V_{ref}\beta_2}\left(\frac{1}{r_b}\frac{dr_b}{dT} + \frac{1}{I_2}\frac{dI_2}{dT} - \frac{1}{\beta_2}\frac{d\beta_2}{dT}\right).$$

$$(7)$$

For example, assuming a 2 kΩ base resistance with 1000 ppm/°C TC, a 30 μA PTAT bias current level, a β of 150, and a β TC of 7000 ppm/°C, an output TC of -8.6 ppm/°C results. As in the case of the bias current variation, this can be partly compensated by modifying the trim target voltage if these parameters are reproducible. The other errors mentioned above are negligible if transistor performance is reasonably good. Another error source results from the temperature coefficient of the ratio of the diffused resistors R_1 and R_2. Data presented later show that for resistors used in this experimental work a differential ratio TC of 1 to 2 percent is achieved. Fortunately, this error is negligible compared to those already discussed.

III. CURVATURE IN BANDGAP REFERENCES

For a bandgap reference which is ideal in the sense that the operational amplifier is ideal, the bipolar transistors have infinite current gain, zero intrinsic base resistance, and have perfect exponential junction relationships, and the bandgap of silicon varies linearly with temperature, the output voltage is typically given by [10]

$$V_{ref} = V_{go} + V_T(4 - n - \alpha)\left(1 + \ln \frac{T}{T_o}\right) \qquad (8)$$

where V_{go} is the extrapolated silicon bandgap at 0 K, n is the exponent of the mobility variation in the base of the bipolar transistor (typically about 0.8), α is the exponent of the temperature variation of the bias current (1 for PTAT bias current, for example), and T_o is the temperature at which the reference output temperature coefficient is zero, usually chosen to be near room temperature. Equation (21) derived in the Appendix is the more general form of (8). This relation illustrates the well-known fact that even for an ideal bandgap with an optimally chosen T_o, the output voltage as a function of temperature displays a curvature which causes it to decrease both for temperatures higher or lower than T_o. Usually, the practical performance aspect of interest is the maximum total variation of the output voltage over the range of temperatures. A further complication is the fact that the bandgap of silicon in fact does vary with T^2, as well as linearly with temperature [11], [12]. This higher order temperature variation adds a curvature term, increasing the effective TC even for optimally adjusted references. These effects together combine to give a best achievable TC of about 25 ppm/°C for a temperature range of -55°C to 125°C.

Several approaches have been suggested for curvature correction. If the quantity $(4 - n - \alpha)$ in (8) could be made zero, then there would be no curvature other than the curvature of the silicon bandgap. This could be achieved for example by using a very strongly temperature-dependent bias current or by linearizing V_{BE} directly [5]. Several authors have proposed simply adding in higher order temperature-dependent terms in the output to cancel the PTAT^2 term of the output voltage variation [4]. This is the approach used in this paper to be described in the next section.

IV. Precision Switched-Capacitor CMOS BGR Technique

In order to implement voltage references in CMOS technology which have performance approaching that achievable in bipolar technology, special steps must be taken to counteract the relatively poor performance of CMOS op amps and CMOS compatible bipolar transistors. In addition, curvature compensation, already widely applied in bipolar references, must be incorporated. In this section, the implementation of the techniques in CMOS technology is described.

A. Curvature Compensation

The overall concept of BGR temperature compensation is illustrated in Fig. 3 and general BGR temperature compensation techniques are described in the Appendix. The first step is to add a PTAT correction voltage KV_T to V_{BE} to cancel out the linear temperature variation of V_{BE}. After the PTAT correction voltage is added, the reference output V_{ref} will exhibit mostly the quadratic temperature variation as shown in Fig. 3. If a PTAT^2 correction voltage FV_T^2 is added to that to cancel out the quadratic temperature variation of V_{BE}, the final reference output V_{ref} should drift only due to higher order temperature variations and a zero temperature coefficient is achieved at T_o. One implementation of a switched-capacitor bandgap reference which embodies curvature compensation as well as offset-cancelled amplification is illustrated in Fig. 4. The gain G of the gain

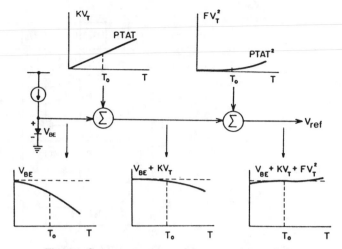

Fig. 3. Curvature-compensation concept (not scaled).

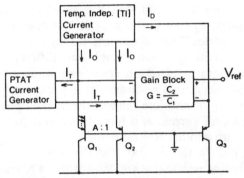

Fig. 4. Overall schematic of the curvature-compensated switched-capacitor CMOS bandgap reference.

block is determined by the capacitor ratio C_2/C_1. The current I_o is temperature-independent (TI) while I_T is PTAT. The bias current I_D is also TI.

If the effects of the base current and the base spreading resistance are neglected, the reference output V_{ref} is given by

$$V_{ref} = V_{BE} + \frac{C_2}{C_1} \Delta V_{BE} \qquad (9)$$

where

$$\Delta V_{BE} = V_T \ln A \frac{I_o + I_T}{I_o - I_T}$$

$$= V_T \ln A + 2 V_T \left(\frac{I_T}{I_o} \right) + \frac{2}{3} V_T \left(\frac{I_T}{I_o} \right)^3 + \cdots. \qquad (10)$$

The inclusion of the PTAT^2 voltage means that the trim procedure for the reference consists of two steps, one to give the correct output voltage for the uncompensated reference, and one to trim the value of the PTAT^2 voltage that is added in. A key advantage of the circuit configuration chosen is that it allows the PTAT^2 component to be adjusted independently from the basic reference. The first part of the trim procedure is to disconnect the PTAT^2 component (set I_T to zero), and trim the absolute output voltage. In the particular experimental device described

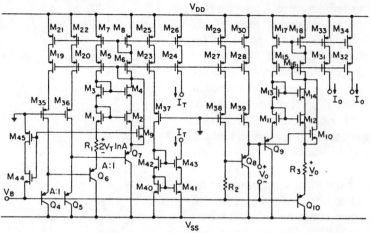

Fig. 5.   PTAT and TI current generators for the bias currents I_T and I_o.

here, it is most convenient to do this with a combination of a capacitor array to adjust the C_2/C_1 ratio, and a resistor string to provide fine adjustment through adjustment of the bias current I_d. Even though two different physical trim arrays are involved, in effect one trim operation is performed and a total resolution of approximately 12 bits is achieved in the absolute value of the final output voltage. Next the PTAT current I_T is turned on and its ratio to I_o is adjusted with another resistor string to give a change in the output equal to the desired PTAT^2 compensation value. Both trimming operations are done at room temperature.

The currents I_T and I_o are generated using the circuit shown in Fig. 5. The bias current I_D is generated in the same manner as the TI current I_o. The stacked-cascode connection formed by transistors M_1 to M_8 causes the emitter currents of transistors Q_6 and Q_7 to be equal. The same scheme is also employed for the matching of the emitter currents of transistors Q_9 and Q_{10}. Transistors M_9 and M_{10} form a start-up circuit for this self-biased circuit. As indicated, transistors Q_4 and Q_6 have emitter areas larger by a factor of A than the remaining transistors. Therefore, the voltage developed across the resistor R_1 is PTAT and the current I_T through R_1 is also PTAT. The voltage V_o formed by the transistor Q_8 and the resistor R_2 is approximately temperature independent. The temperature stability of V_o is not critical because it affects only the PTAT^2 component of the output voltage (not PTAT). The TI voltage V_o is developed across R_3 and the current I_o through R_3 is also TI. Therefore, the current ratio I_T/I_o is easily trimmed by R_3.

In the presence of the mismatches of $M_1 - M_2$ and $M_{11} - M_{12}$ transistor pairs, the voltages across R_1 and R_3 deviate from the ideal values. If the gate–source voltage mismatches of MOS transistor pairs inclusive in the bias circuit of Fig. 5 are assumed to be V_{m1}, V_{m2}, and V_{m3}, respectively, and current mirrors and p-n-p transistors are assumed to be ideal, the three bias currents I_T, I_o, and I_D are given by

$$I_T = \frac{1}{R_1}(2V_T \ln A + V_{m1}), \tag{11}$$

$$I_o = \frac{1}{R_3}(V_o + V_{m2}) \tag{12}$$

and

$$I_D = \frac{1}{R_4}(V_o + V_{m3}). \tag{13}$$

The factor 2 in (11) results from cascading transistors $Q_4 - Q_6$ and $Q_5 - Q_7$ to reduce the error contribution of the mismatch voltage V_{m1}. For this application, cascading of two devices is enough for the PTAT current generation. Depending on applications, several transistors can be cascaded to reduce the offset error contribution in such a BGR implementation, discussed in Section II. Substituting (11), (12), and (13) into (9) and (10), and neglecting higher orders, we obtain

$$V_{\text{ref}} = V_{BE}|_{\text{ideal}} + \frac{C_2}{C_1}V_T \ln A + 4\frac{C_2}{C_1}\frac{R_3}{R_1}\frac{\ln A}{V_o}V_T^2 + V_\epsilon$$

$$= V_{BE}|_{\text{ideal}} + KV_T + FV_T^2 + V_\epsilon \tag{14}$$

where

$$V_\epsilon = \frac{\sigma}{100}\frac{1}{R}\frac{dR}{dT}\Big|_{T_o}(V_{T_o} - V_T)T + 2\frac{C_2 R_3 V_T V_{BE}}{C_1 R_1 V_o^2}V_{m1}$$

$$- 4\frac{C_2 R_3 V_T^2 \ln A}{C_1 R_1 V_o^2}V_{m2} + \frac{V_T}{V_o}V_{m3}. \tag{15}$$

The constant σ represents the standard deviation (%) of the temperature coefficient of R_4. The error voltage V_ϵ is less than 1 mV at room temperature. Therefore, the temperature coefficient uncertainty resulting from the mismatch voltages is 50 times smaller than the uncertainty caused by the op amp offset of existing designs given by (4).

B. Offset-Cancelled Amplification

In order to remove dc offsets from the amplifier, it is divided into two stages as shown in Fig. 6. In the first offset-storage mode, all the MOS switches are closed to

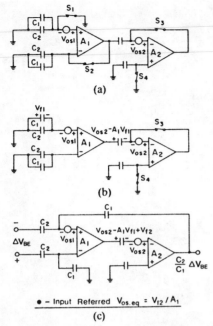

Fig. 6. Gain block which embodies offset-cancelled amplification. (a) First offset-storage mode. (b) Second offset-storage mode. (c) Amplification mode.

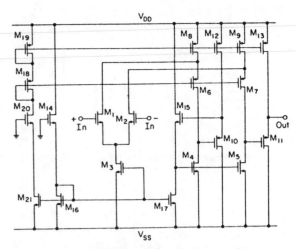

Fig. 7. Amplifier configuration for A_1 and A_2.

sample the offset voltages of the individual op amps. In the process of opening the MOS switches S_1 and S_2, the channel charges are injected into the op amp summing nodes to load the capacitors C_1 and C_2. The charge injection differential voltage V_{f1} due to the mismatch of switches S_1 and S_2 is sampled across C_1 and C_2 along with the offset voltage V_{os1}. In the second offset-storage mode, the first gain stage charges the coupling capacitor to compensate for the input differential voltage V_{f1}. After the switches S_3 and S_4 are opened, two stages are connected in a feedback amplification mode, and the amplification of ΔV_{BE} takes place by the capacitor ratio C_2/C_1. When referred to input, the feedthrough difference of the switches S_3 and S_4 is reduced by the open-loop gain of the first stage. Note the bottom plate of one capacitor C_1 should be connected to the diode voltage in the actual reference as shown in Fig. 4.

A single-pole folded-cascode CMOS op amp configuration for A_1 and A_2 is shown in Fig. 7. Two amplifiers are identical and designed to meet the following requirements:
1) moderate gain for each stage (100 to 300);
2) single dominant pole per stage;
3) inherent zero systematic offset voltage; and
4) capacitor driving capability (15 pF for one stage and 100 pF for two stages).

Transistors $M_{19}-M_{21}$ form a bias string for the amplifier. The replica bias circuit formed by transistors $M_{14}-M_{17}$ performs a level shift and a differential to single-ended conversion, and reduces the inherent systematic offset. In order to limit the gain of each stage, the lower part composed of transistors M_4-M_5 is not cascoded while the upper part M_6-M_9 is. The class-A source follower stages formed by transistors $M_{10}-M_{13}$ are added to meet the capacitor driving capability. In the offset-storage modes, each op amp is stabilized by connecting a frequency com-

pensation capacitor between the high impedance node and the ac ground V_{ss}. When the two op amps are cascaded and the feedback loop closed around the composite amplifier, a Miller capacitance is switched in from the high impedance node of the second stage to the same node of the first stage to achieve a pole-splitting compensation.

C. Base Current Cancellation

In the CMOS process used, the current gain of substrate p-n-p transistors is often limited and highly variable. Therefore, to compensate for the difference between the collector current and the emitter current, the base current has to be returned to the emitter as shown in Fig. 8(a). However, a simpler approach is to replicate the base current and allow it to flow into the emitter as shown in Fig. 8(b). The base current is cancelled with an accuracy of about 90 percent because the base currents typically match each other within 10 percent for the adjacent transistors on a single chip.

D. Base Resistance Cancellation

The effective series base resistance of the bipolar transistors consists of that produced by lateral flow in the base region under the emitter, and the extrinsic base resistance between the base contact and the active base. The former is bias dependent and difficult to predict, while the latter is more straightforward to predict given device geometry. The approach taken here is to include a lumped resistor R_{comp} made of the same n^--well diffusion material so as to achieve approximate tracking with temperature and process variations. The value of R_{comp} in Fig. 9 should be

$$R_{comp} = \left(\frac{1+\beta_1}{1+\beta_2} - \frac{1}{A} \right) r_{b1}. \tag{16}$$

For the process used here, the magnitude of the extrinsic compensation resistance is about one quarter of the measured intrinsic base resistance of p-n-p transistors with the same geometry.

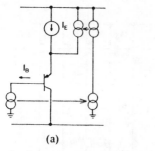
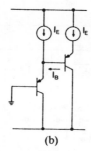

Fig. 8.   Base current cancellation schemes. (a) I_B returning. (b) I_B replication.

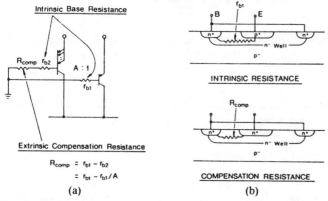

Fig. 9.   Base resistance cancellation. (a) Extrinsic compensation resistance R_{comp}. (b) Difference between intrinsic base resistance and extrinsic compensation resistance.

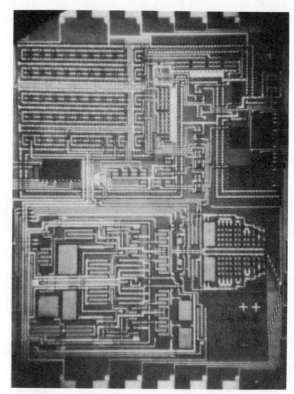

Fig. 10.   Chip photo of the experimental prototype.

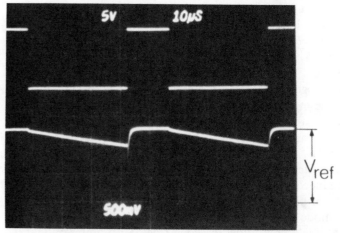

Fig. 11.   Waveforms. (a) Output sync clock. (b) V_{ref}.

V. EXPERIMENT AND DISCUSSIONS

The experimental prototype circuit implementing the proposed reference was fabricated employing a self-aligned single-poly Si-gate CMOS process on a 20–30 $\Omega \cdot$cm boron-doped p-type $\langle 100 \rangle$ substrate. The gate oxide is 0.07 μm thick and the drawn minimum feature is 6 μm. Fig. 10 shows the microphotograph of the prototype chip and Fig. 11 shows its output waveform as well as output sync pulse.

Experimental data were gathered from seven representative samples from one wafer. Every experimental chip contains three types of reference voltages. The Type I reference has no curvature compensation, no base current cancellation, no base resistance cancellation, and no offset cancellation, and amplification is performed by a resistor ratio. The Type II reference which uses a capacitor-ratio amplification has the cancellations of offset, base current and base resistance, but no curvature compensation. However, the Type III reference which also uses a capacitor ratio amplification has all components of the Type II reference plus curvature compensation.

Following the procedure described in Section IV and the Appendix, one sample was adjusted to give a minimum temperature drift and the other six samples were trimmed at room temperature to an output voltage predetermined from the first sample. Statistical data from seven samples are summarized in Table I for three types of reference voltages. The optimum values of the PTAT^2 correction voltage, the first-order corrected and second-order corrected V_{ref}'s at 25 °C were found to be 61 mV, 1.256 and 1.192 V, respectively. Estimating from the measured data, the parameters V_{go1}, V_{go2}, and $4-n-\alpha$ necessary for specifying the prototype bandgap reference were 1.181, 1.158, and 2.623 V, respectively.

Note that the offset-cancelled amplification and compensation of r_b and β effects give a factor of 5 improvement in temperature stability and its deviation over the approach without any compensation. By curvature com-

TABLE I
STATISTICS OF MEASURED TEMPERATURE COEFFICIENTS
OF 7 SAMPLES (ppm°C)

0 to 70°C:		Mean	Standard Deviation	Minimum	Maximum
	Type I	105	43	42	167
	Type II	22.3	10.8	11.1	42
	Type III	13.1	7.1	5.6	25.7
−55 to 125°C:					
	Type I	185	56.5	107	273
	Type II	35.1	18.8	17.6	66.7
	Type III	25.6	10.5	12.1	39.9

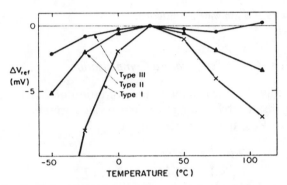

Fig. 12. Typical measured temperature variations of three types of references when they are optimally compensated.

TABLE II
PERFORMANCE SUMMARY

V_{ref}	1.192 V ± 1 mV at 25°C
TC	See Table I
Power	12 mW with ± 5 V supply
Load	100 pF capacitor
Cycle time	5 μs
+ PSRR	50 dB (dc)
− PSRR	60 dB (dc)
Clock RR	75 dB
Output noise	400 μV (500 kHz)

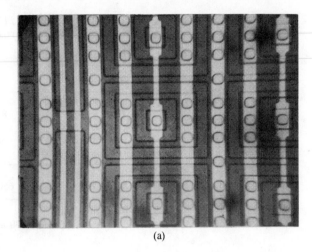

(a)

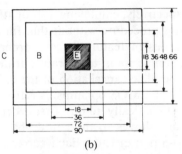

(b)

Fig. 13. Substrate p-n-p transistor. (a) A unit cell. (b) Drawn dimensions of a unit cell (μm).

pensation, a factor of 2 further improvement was obtained. Fig. 12 compares graphically those three types of optimally-compensated bandgap references. The experimental results are summarized in Table II. To improve the power supply rejection ratio (PSRR), the base of all p-n-p transistors should be biased at a constant voltage relative to the negative supply line. Otherwise, the base width modulation (Early effect) will limit the PSRR of the reference.

The critical aspect of the design is the substrate p-n-p transistor because the n⁻-well in CMOS processes usually has a relatively high resistivity. In order to minimize the intrinsic base spreading resistance, the base contact surrounds the emitter junction as shown in Fig. 13. The emitter junction and the base contact plug are separated by 9 μm. By the surrounding base, the intrinsic base resistance is reduced by a factor of 6 when compared to the parallel contact of emitter and base. For the process used here, the estimated r_b of this geometry is about 1.5 kΩ. With this geometry, the observed emitter area reduction due to the base crowding is insignificant over the 1 to 100 μA emitter current range. The current gain β is also relatively constant within this range. However, in the emitter current range over 100 μA, the base crowding effect is severe. Also, in the emitter current range below 1 μA, the current gain β decreases due to space charge recombination in the emitter–base junction. The unit cells of Fig. 13 were connected in parallel to obtain a multiple-emitter device. The current gain β is minimum at −55°C and the average value at room temperature is 175. Temperature data for the p-n-p transistors and p⁺ diffused resistors in the n⁻-well in the particular technology used here are listed in Table III. The temperature coefficients of diffused resistors match each other within 1.2 percent.

One potential problem in the use of the substrate p-n-p transistor is the fact that the dc collector current flows into the substrate. If this current gives rise to a large enough ohmic drop in the substrate, it could initiate latchup. To minimize the likelihood of this, each n⁻-well (base) was surrounded by the p⁺ diffusion (collector) as shown in Fig. 13. No latchup was observed during experimental measurements even under transient conditions.

VI. CONCLUSION

A precision curvature-compensated switched-capacitor CMOS bandgap reference is reported whose monolithic prototype exhibits average temperature drifts of 13.1 and 25.6 ppm/°C over the commercial and military tempera-

TABLE III
TEMPERATURE DATA OF DIFFUSED RESISTORS AND p-n-p TRANSISTORS
(−55 to 125 °C)

Diffused Resistors:	Mean	Standard Deviation	Minimum	Maximum
Sheet resistance (Ω/square)	68.9	3.1	64.6	72.1
TC or R^*	886	27.9	849	917
Ratio of $6R/R^{**}$	5.952	0.02	5.9304	5.9794
TC of $6R/R^*$	10.6	2.2	8.9	13.6
p-n-p Transistors:				
Current gain β at $I_E = 30\,\mu A$	175	83	91	273
TC of β^*	6503	872	5471	7792

*TC unit is ppm/°C.

**R and $6R$ are composed of 40 and 240 squares of 6 μm-wide p$^+$ diffusion.

ture ranges, respectively, employing a straightforward room temperature trim procedure without thin-film resistor and laser trim. The design features second-order temperature compensation, simple room temperature trim up to 12-bit accuracy and complete cancellation of the offset and long-term offset drift of CMOS op amps. A reference voltage is obtained by systematically adding first-order and second-order correction voltages to the emitter-base potential of the substrate p-n-p transistor. Each correction voltage is individually trimmable to minimize temperature drift. The proposed reference is compatible with a standard digital CMOS process and is applicable to high-resolution monolithic CMOS data acquisition systems.

APPENDIX
BGR TEMPERATURE COMPENSATION TECHNIQUES

Employing (6) and neglecting higher orders, the forward-biased diode voltage is given by [10]

$$V_{BE} = V_g - V_T[(4-n-\alpha)\ln T - \ln EG] + HV_T - LV_T^2 \tag{17}$$

where V_g is the bandgap of silicon which is a function of base doping [13] and temperature [11], [12], n and α are parameters illustrated in (8), E and G are the parameters whose magnitude are insignificant in the temperature analysis [10], and H and L are defined from (6)

$$H = T_o \frac{1}{R} \frac{dR}{dT}\bigg|_{T_o} \quad \text{and} \tag{18}$$

$$L = \frac{T_o}{V_{T_o}} \frac{1}{R} \frac{dR}{dT}\bigg|_{T_o}. \tag{19}$$

That is, if the bias current variation and the silicon bandgap curvature discussed in Sections II and III are included, the V_{BE} in (17) is the actual diode voltage whose temperature variation is to be compensated to give a temperature stable reference voltage. In the next two subsections, first- and second-order temperature compensation techniques are discussed in more detail.

A. First-Order Temperature Compensation

If only the linear temperature variation of V_{BE} in (17) is compensated by adding the first-order correction voltage KV_T to V_{BE}, the reference output V_{ref} is

$$
\begin{aligned}
V_{ref} &= V_{BE} + KV_T \\
&= V_g - V_T(4-n-\alpha)\ln T \\
&\quad + (K + H + \ln EG)V_T - LV_T^2.
\end{aligned}
\tag{20}
$$

By equating the derivative of V_{ref} at T_o to zero and eliminating the unknown constants, we obtain

$$V_{ref} = V_g - \frac{dV_g}{dT}\bigg|_{T_o} T + V_T(4-n-\alpha)\left(1 + \ln\frac{T_o}{T}\right) - LV_T^2\left(1 - 2\frac{T_o}{T}\right). \tag{21}$$

The first two terms and the last term of (21) result from the nonlinearity of the Si bandgap over temperature and the bias current variation, respectively, and will not disappear until higher order temperature variations are compensated. Without these, (21) is identical to (8). Tsividis [12] recently explained the Si bandgap temperature dependence employing equations from Bludau et al.'s [11]. From (21), the nominal voltage at T_o is therefore

$$V_{ref}|_{T_o} = V_{go1} + V_{T_o}(4-n-\alpha) + LV_{T_o}^2 \tag{22}$$

where

$$V_{go1} = V_g(T_o) - \frac{dV_g}{dT}\bigg|_{T_o} T_o \approx 1.205 \text{ V}. \tag{23}$$

As commonly called, V_{go1} is the linearly-extrapolated Si bandgap voltage at $T = 0$ K. Equation (22) indicates that the bias variation resulting from the temperature coefficient of diffused resistors causes the nominal voltage different from a theoretical value.

B. Second-Order Temperature Compensation

If the linear and the quadratic temperature variations of

V_{BE} in (17) are compensated as illustrated in Fig. 3 by adding both the first-order correction voltage KV_T and the second-order correction voltage FV_T^2 to V_{BE}, the reference output V_{ref} is

$$
\begin{aligned}
V_{\text{ref}} &= V_{BE} + KV_T + FV_T^2 \\
&= V_g - V_T(4 - n - \alpha)\ln T \\
&\quad + (K + H + \ln EG)V_T + (F - L)V_T^2. \quad (24)
\end{aligned}
$$

By equating the first-order and the second-order derivatives of V_{ref} at T_o to zero and eliminating the unknown constants, we obtain

$$
\begin{aligned}
V_{\text{ref}} = V_g &- \left.\frac{dV_g}{dT}\right|_{T_o} T - \frac{1}{2}\left.\frac{d^2V_g}{dT^2}\right|_{T_o} T^2\left(1 - 2\frac{T_o}{T}\right) \\
&+ V_T(4 - n - \alpha)\left(\ln\frac{T_o}{T} + \frac{1}{2}\frac{T}{T_o}\right). \quad (25)
\end{aligned}
$$

Therefore, the nominal voltage at T_o is

$$
\left.V_{\text{ref}}\right|_{T_o} = V_{go2} + \frac{1}{2}V_{T_o}(4 - n - \alpha) \quad (26)
$$

where

$$
V_{go2} = V_g(T_o) - \left.\frac{dV_g}{dT}\right|_{T_o} T_o + \frac{1}{2}\left.\frac{d^2V_g}{dT^2}\right|_{T_o} T_o^2 \approx 1.179 \text{ V}. \quad (27)
$$

Now V_{go2} is the quadratically-extrapolated Si bandgap voltage at $T = 0$ K. The voltage V_{go2} is closer to the Si bandgap at $T = 0$ K than V_{go1} of (23). The theoretical value of $V_g(0\text{ K})$ is approximately 1.179 V [12]. The bias current variation no longer affects (26) because it is compensated by the PTAT^2 correction voltage.

Only after the PTAT voltage is added, the intermediate voltage at T_o is

$$
\left.(V_{BE} + KV_T)\right|_{T_o} = V_{go2} + \frac{1}{2}\left.\frac{d^2V_g}{dT^2}\right|_{T_o} T_o^2 - LV_{T_o}^2. \quad (28)
$$

Correspondingly, the magnitude of the PTAT^2 voltage FV_T^2 at T_o is obtained by subtracting (28) from (26):

$$
\left.FV_T^2\right|_{T_o} = \frac{1}{2}V_{T_o}(4 - n - \alpha) - \frac{1}{2}\left.\frac{d^2V_g}{dT^2}\right|_{T_o} T_o^2 + LV_{T_o}^2. \quad (29)
$$

This PTAT^2 correction voltage includes the inherent Si bandgap curvature as well as the bias current variation. Other error sources neglected can be included in (29) easily in the same manner as the bias current variation. No matter how many error sources are included, the final V_{ref} given by (25) is independent of these instabilities as far as their temperature variations are compensated properly.

REFERENCES

[1] R. J. Widlar, "New developments in IC voltage regulators," *IEEE J. Solid-State Circuits*, vol. SC-6, pp. 2–7, Feb. 1971.
[2] K. E. Kujik, "A precision reference voltage source," *IEEE J. Solid-State Circuits*, vol. SC-8, pp. 222–226, June 1973.
[3] A. P. Brokaw, "A simple three-terminal bandgap reference," *IEEE J. Solid-State Circuits*, vol. SC-9, pp. 288–393, Dec. 1974.
[4] C. R. Palmer and R. C. Dobkin, "A curvature corrected micropower voltage reference," in *Proc. Int. Solid-State Circuits Conf.*, Feb. 1981, pp. 58–59.
[5] G. C. M. Meijer, P. C. Schmale, and K. van Zalinge, "A new curvature-corrected bandgap reference," *IEEE J. Solid-State Circuits*, vol. SC-17, pp. 1139–1143, Dec. 1982.
[6] Y. P. Tsividis and R. W. Ulmer, "A CMOS voltage reference," *IEEE J. Solid-State Circuits*, vol. SC-13, pp. 774–778, Dec. 1978.
[7] E. A. Vittoz and O. Neyroud, "A low-voltage CMOS bandgap reference," *IEEE J. Solid-State Circuits*, vol. SC-14, pp. 573–577, June 1979.
[8] R. Gregorian, G. A. Wegner, and W. E. Nicholson, Jr., "An integrated single-chip PCM voice codec with filters," *IEEE J. Solid-State Circuits*, vol. SC-16, pp. 322–333, Aug. 1981.
[9] W. H. White, D. R. Lampe, F. C. Blaha, and I. A. Mack, "Characterization of surface channel CCD image arrays at low light levels," *IEEE J. Solid-State Circuits*, vol. SC-9, pp. 1–14, Feb. 1974.
[10] P. R. Gray and R. G. Meyer, *Analysis and Design of Analog Integrated Circuits.* New York: Wiley, 1977, pp. 256.
[11] W. Bludau, A. Onton, and W. Heinke, "Temperature dependence of the bandgap in Si," *J. Appl. Phys.*, vol. 45, pp. 1846–1848, 1974.
[12] Y. P. Tsividis, "Accurate analysis of temperature effects in $I_C - V_{BE}$ characteristics with application to bandgap reference source," *IEEE J. Solid-State Circuits*, vol. SC-15, pp. 1076–1084, Dec. 1980.
[13] J. W. Slotboom and H. C. DeGraaf, "Measurements of bandgap narrowing in Si bipolar transistors," *Solid-State Electron.*, vol. 19, pp. 857–862, 1976.

CMOS Voltage References Using Lateral Bipolar Transistors

MARC G. R. DEGRAUWE, OSKAR N. LEUTHOLD, ERIC A. VITTOZ, MEMBER, IEEE, HENRI J. OGUEY, MEMBER, IEEE, AND ARTHUR DESCOMBES, MEMBER, IEEE

Abstract —Two bandgap references are presented which make use of CMOS compatible lateral bipolar transistors. The circuits are designed to be insensitive to the low beta and alfa current gains of these devices. Their accuracy is not degraded by any amplifier offset.

The first reference has an intrinsic low output impedance. Experimental results yield an output voltage which is constant within 2 mV, over the commercial temperature range (0–70°C), when all the circuits of the same batch are trimmed at a single temperature. The load regulation is 3.5 μV/μA and the Power Supply Rejection Ratio (PSRR) at 100 Hz is 60 dB.

Measurements on a second reference yield a PSRR of minimum 77 dB at 100 Hz. Temperature behavior is identical to the first circuit presented. This circuit requires a supply voltage of only 1.7 V.

I. Introduction

DURING THE LAST FEW YEARS, there has been an increasing trend to realize analog and digital circuits on the same chip. Bipolar technologies are more adequate to implement analog functions but CMOS technologies are more interesting if large digital parts have to be realized.

Voltage references are a key element of analog-to-digital converters. In the past, several CMOS compatible voltage references [1]–[5] have already been proposed but none of them achieves the precision of purely bipolar bandgap references. The circuits suffer from the weaknesses of CMOS technologies (large amplifier offset, poor matching of transistors, etc.) or are quite complex and do not deliver a continuous reference voltage [4].

In this paper, a family of more accurate voltage references is presented. The circuits which make use of lateral bipolar transistors eliminate the problems of the previously published CMOS voltage references.

CMOS compatible lateral bipolar transistors [6] are first briefly discussed and an alternative way for their realization is presented. In the next section, the basic principle of the references is studied in detail. Limitations of tempera-

ture stability due to technology variations are explained. A low output impedance reference is then presented with exhaustive experimental results. Finally a voltage reference with a very high Power Supply Rejection Ratio (PSRR) is discussed.

II. Lateral Bipolar Transistors

It has been shown that a bipolar transistor, with a collector which is not tied to the substrate, is available in any CMOS technology [6]. This device is obtained by operating a MOS transistor in the lateral bipolar mode.

A cross section of a concentric n-channel transistor realized in a p-well CMOS process is shown in Fig. 1. By biasing the gate of the MOS transistor far below its threshold voltage, an accumulation layer is created below the gate. This prevents MOS transistor operation between the two concentric n^+ diffusions. By properly biasing the p-well (B) and the drains (E, C), a bipolar operating mode is obtained. In addition a second, unwanted, vertical bipolar transistor is also activated. In order to favor the lateral bipolar transistor, the gate length should be as small as possible and the perimeter-to-surface ratio of the emitter should be maximized. A symbolic representation of this five-terminal device is shown in Fig. 2.

An alternative way to realize this device is shown in Fig. 3. The normal thin oxide MOS transistor has been replaced by a parasitic field oxide transistor with an aluminum gate. If the threshold voltage of this transistor is high enough, this device operates already in the bipolar mode when gate and emitter are tied together. This structure thus has the advantage that one terminal can be eliminated and that no negative gate voltage need be created on chip.

III. Basic Principle and Theory

A classical approach to realize CMOS bandgap references is shown in Fig. 4 [9], [10]. Assuming an ideal amplifier, the output voltage of the circuit is given by

$$V_{out} = V_{BE_1} + \frac{R_2}{R_1}(V_{BE_1} - V_{BE_2}). \quad (1)$$

Since V_{BE_1} decreases approximately linearly with absolute temperature T and $(V_{BE_1} - V_{BE_2})$ increases with T, the

Manuscript received April 22, 1985; revised July 22, 1985. This work was supported by the "Fonds National Suisse pour la Recherche Scientifique, PN13."

M. G. R. Degrauwe, E. A. Vittoz, and H. J. Oguey are with Centre Suisse d'Electronique et de Microtechnique S.A. (CSEM), Recherche et Développement (formerly CEH) Maladière 71, 2000 Neuchâtel 7, Switzerland.

O. N. Leuthold was with Ebauches Electroniques S.A., MEM, 2074 Marin, Switzerland. He is now with Hughes Aircraft Co., Newport Beach, CA.

A. Descombes is with Ebauches Electroniques S.A., MEM, 2074 Marin, Switzerland.

Reprinted from *IEEE J. Solid-State Circuits*, vol. SC-20, no. 6, pp. 1151–1157, Dec. 1985.

52

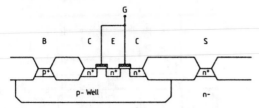

Fig. 1. Cross section of a lateral bipolar transistor.

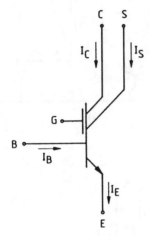

Fig. 2. Symbol for a lateral bipolar transistor.

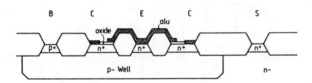

Fig. 3. Cross section of a lateral bipolar transistor formed with a parasitic MOS transistor.

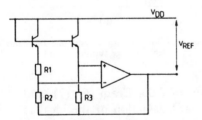

Fig. 4. Classical bandgap reference.

circuit can be made almost independent of temperature by choosing an appropriate ratio of R_2/R_1. The resulting output voltage at the reference temperature will be called the ideal reference voltage ($V_{\text{ref}}(T_r)$).

However the MOS amplifier is not ideal. It has a typical offset of a few millivolts. The output voltage will be given by (1) plus the additional term

$$\left(1 + \frac{R_2}{R_1}\right) \cdot V_{os} \tag{2}$$

Since the ratio R_2/R_1 is about 10, the output voltage will have an unwanted component of about 20–50 mV. Since the offset voltage is not proportional to absolute temperature (PTAT), it cannot be fully compensated. Furthermore, it decorrelates the relationship which exists be-

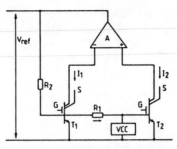

Fig. 5. Principle of voltage reference A-current comparator-V_{CC}-voltage controlled current source which draws a current proportional to I_1 and much larger than the base current of T_1.

tween output voltage and temperature behavior. Obtaining a very accurate voltage reference by trimming the output voltage at a single temperature becomes impossible.

To improve the accuracy of the voltage reference, the influence of the amplifier offset must be decreased. This can be achieved by using a chopper stabilized amplifier [4] or a stack of bipolar transistors [5]. These techniques lead, however, to circuits quite complex that are and/or require relatively large supply voltages.

A better solution is to use lateral bipolar transistors as shown in Fig. 5. The circuit consists of two bipolar transistors operated at different current densities, two resistances R_1 and R_2, a current comparator, and a voltage controlled current source. This current source draws a current much larger than the maximum base current of T_1 to achieve (1) independently of the current gain T_1. By realizing the current comparator by MOS transistors operating deep in strong inversion, the effect of the current offset can be neglected. Furthermore, any mismatch of bipolar pair T_1, T_2 from their nominal area ratio results in an output voltage component which is PTAT and can thus be compensated for.

The temperature behavior of the circuit will thus be as described by Tsividis [7] and is shortly repeated hereafter.

If the effective mobility for the minority carriers in the base can be represented with sufficient accuracy by

$$\mu(T) = C \cdot T^{-p} \tag{3}$$

with C and p appropriate constants, if the current through the bipolar transistors is given by

$$I_C = I_C(T_r)\left(\frac{T}{T_r}\right)^m \tag{4}$$

where

$$T_r = \text{reference temperature}$$

and if

$$\left.\frac{R_2}{R_1}\left(V_{BE_1} - V_{BE_2}\right)\right|_{T_r} = V_{G0_r} + (n-m)kT_r/q - V_{BE_1}(T_r)$$

$$\tag{5}$$

where

$$V_{G0_r} = \text{extrapolated bandgap voltage from } T_r \text{ to 0 K}$$

$$n = 4 - p$$

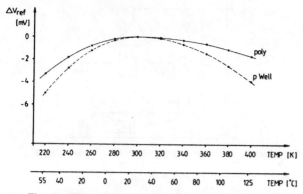

Fig. 6. Theoretical variation of the reference voltage.

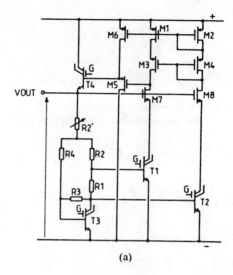

(a)

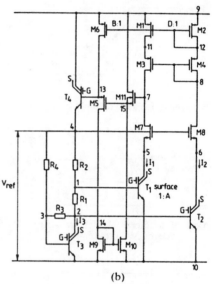

(b)

Fig. 7. (a) Circuit of low output impedance bandgap reference. (b) High-voltage supply variant.

then the output voltage of the circuit will be equal to

$$V_{\text{ref}}(T) = V_{\text{ref}}(T_r) + f_r(T) \qquad (6)$$

where

$$V_{\text{ref}}(T_r) = V_{G0_r} + (n - m) \cdot \frac{k \cdot T_r}{q} \qquad (7)$$

$$\begin{aligned}
f_r(T) = {} & V_G(T) - V_{G0_r} \\
& + \frac{T}{T_r}\left(V_{G0_r} - V_G(T_r)\right) \\
& + (n - m) \cdot \frac{k}{q} \cdot (T - T_r - T \ln(T/T_r)) \quad (8)
\end{aligned}$$

"$f_r(T)$" in (7) expresses the nonideal behavior of the voltage reference. The only way to decrease this nonideality is to choose an appropriate m.

Since the current in the bipolar transistors is fixed by PTAT voltage and a resistance, the coefficient m is determined by the temperature coefficient of this resistance. If the temperature behavior of the resistance is modeled by

$$R(T) = R(T_r) \cdot \left(\frac{T}{T_r}\right)^a \qquad (9)$$

then

$$m = 1 - a. \qquad (10)$$

In Fig. 6 the formula (8) is evaluated for standard poly resistances ($a = 0.1$) and p-well resistances ($a = 2$). For the bandgap, the model given in [7] was used and from measurements the coefficient n was found to be 2.1. A bandgap reference realized with poly resistances will thus be more accurate than one realized with p-well resistors. For the commercial temperature range, the reference with poly resistance is stable within about 0.4 mV and with p-well resistances within 0.8 mV.

In practice all the circuits of a same batch will be trimmed at a reference temperature T_r to the same reference voltage given by (7). However, due to technology variations the ideal reference voltages of the circuits differ from each other. Measurements have shown that the temperature coefficient a of the poly resistances has a stan-

dard deviation of 0.05. From (6), (7), and (10) it is seen that the standard deviation of the ideal reference voltage $V_{\text{ref}}(T_r)$ will thus be about 1.25 mV at T_r. This means that at worst (3σ) a circuit will be trimmed 3.75 mV above or below its ideal reference voltage. This results, with respect to (8), in an additional PTAT temperature variation of about 0.9 mV (plus or minus) for the commercial temperature range.

IV. LOW OUTPUT IMPEDANCE VOLTAGE REFERENCE

A straightforward implementation of the principle described above is shown in Fig. 7(a). The current comparator is realized as a cascode current mirror ($M_1 - M_4$) followed by two source followers (M_5, T_4). Transistors M_1, M_2, and M_6 operate deep in strong inversion in order to minimize the sensitivity to threshold mismatch [11]. MOS follower M_5 and cascode transistors M_7, M_8 (biased at V_{ref}) avoid any variation of the current density ratio of T_1

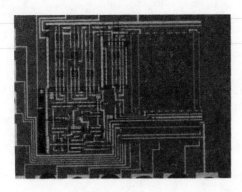

Fig. 8. Chip photograph of low output impedance bandgap reference (area = 0.42 mm²).

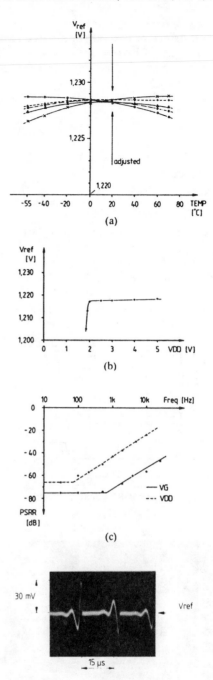

Fig. 9. (a) Temperature behavior of samples of different wafers. Output voltage is trimmed at a single temperature by short circuiting (8 bits) parts of a resistive divider; 90 percent of the measured circuits (30) were as precise as the samples shown. (b) Reference voltage as a function of supply voltage. (c) PSRR of voltage reference. (d) Dynamic behavior of voltage reference (pulsed load capacitance).

and T_2. The bipolar source follower provides a low output impedance. Output conductance is the product of the transconductance of T_4 and the loop gain.

The circuit has been integrated (Fig. 8) in a low threshold 4-μm-Si gate p-well technology with standard polysilicon resistors of 50 $\Omega/\square$. In order to satisfy the condition given by (5), several taps are placed on the resistor. These taps can be short-circuited outside the chip. In the final version, fixed resistors will be used and the adjustment of the reference voltage will be done by adjusting the current mirror ratio of $M_1 - M_2$ [8]. The lateral bipolar transistors are realized with active transistors (polygate). Their gate voltage is biased at about -2 V.

The most important measurement results are shown in Fig. 9 and summarized in Table I.

The standard deviation of the reference voltage is 5.3 mV before trimming and 150 μV after trimming. The output voltage is constant within 2mV, over the commercial temperature range (0–70°C), when all the circuits of the same batch are trimmed at the same voltage at a single temperature. It can be noted that, at room temperature, some circuits have a positive temperature coefficient while others have a negative one. This means that their ideal reference voltage is, respectively, lower or higher than the voltage on which they are trimmed.

The load regulation is 3.6 μV/μA. The circuit can thus be loaded with approximatively 4 K without loss of temperature accuracy (reference voltage changes only 1 mV).

The PSRR (Fig. 9(b), (c)) of the positive rail is 60 dB at 100 Hz and the PSRR of the bipolar gate voltage is 75 dB at 100 Hz. Long but straightforward calculations show that the PSRR of the positive rail can be further improved by increasing the gain of the folded amplifier formed by T_1, T_2, $M_1 - M_4$, M_7, and M_8. In the actual design, this gain is about 4000. A gain of 8000 would increase the PSRR by 6 dB.

The $1/f$ noise is essentially due to transistors M_1 and M_2. In this design they are only 20 μm by 10 μm and can eventually be made larger to reduce the $1/f$ noise.

The dynamic behavior of the voltage reference is shown in Fig. 9(d). The settling time is smaller than 15 μs with a load of 250 pF. Further reductions can be obtained by increasing the current (use smaller resistances), which is now only 70 μA for the whole reference circuit.

For high supply voltages, the threshold of transistor M_5 becomes large and can eventually push transistor M_7 out of saturation, this degrades circuit performances drastically. The variant scheme shown in Fig. 7(b) is less sensitive to this effect. This circuit has also been integrated and has the same characteristics as the circuit of Fig. 7(a). Only the minimal supply voltage has increased to 2.8 V.

TABLE I
MEASUREMENT RESULTS OF LOW OUTPUT IMPEDANCE VOLTAGE
REFERENCE

Output voltage $\bar{x}$	1.2285 V	
σ	150 μV	
Minimal supply voltage	2.2 V	
Supply current	79 μA	
Noise spectra		
white	316 nV/$\sqrt{\text{Hz}}$	
1/f (at 1 kHz)	560 nV/$\sqrt{\text{Hz}}$	
RMS noise voltage (0.01 - 250 kHz)	162 μV	
PSRR at 100 Hz	60 dB	
Load regulation ($\Delta V_{out}/I_{out}$)	3.6 μV/μA	
Chip area	0.42 mm^2	

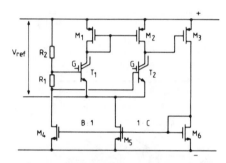

Fig. 10. High PSRR bandgap reference.

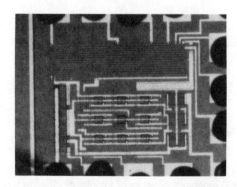

Fig. 11. Chip photograph of high PSRR bandgap reference.

V. HIGH PSRR VOLTAGE REFERENCE

The drawback of the above described voltage references is that polyresistances have to be used to obtain the mentioned PSRR. Indeed since resistors are fixed with respect to V_{SS}, the ratio of well resistors will be affected by bulk modulation of V_{DD} which will result in a degraded PSRR. Despite their poor temperature behavior, p-well resistances remain interesting to realize, on a small die area, micropower voltage references.

Fig. 10 presents a voltage reference which accepts p-well resistors without degrading PSRR. This circuit, which dif-

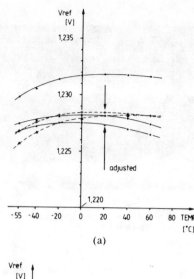

(a)

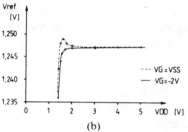

(b)

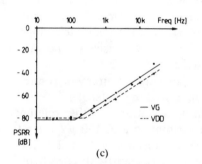

(c)

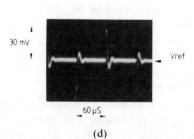

(d)

Fig. 12. (a) Temperature behavior of samples of different wafers. Output voltage is trimmed at a single temperature by short-circuiting (5 bits) parts of a resistive divider. (b) Reference voltage as a function of supply voltage. (c) PSRR of voltage reference. (d) Dynamic behavior of voltage reference (pulsed load capacitance).

fers slightly from the principle shown in Fig. 4, is also insensitive to low alfa and beta of the bipolar transistors.

The circuit consists of a two-stage amplifier, with lateral bipolar input transistors, and two feedback resistors. The current through R_1 is designed to be much larger than the maximal base current of T_1. The current density of the transistors T_1 and T_2 is different which causes a PTAT offset of about 54 mV at room temperature. Due to the feedback mechanism, this voltage is multiplied with R_2/R_1.

TABLE II
MEASUREMENT RESULTS OF HIGH PSRR VOLTAGE REFERENCE

Output voltage $\bar{x}$	1.2281	V
σ	350	μV
Minimal supply voltage	1,7	V
Supply current	20	μA
Noise spectra		
white	500	nV/$\sqrt{\text{Hz}}$
1/f (at 1 kHz)	1	μV/$\sqrt{\text{Hz}}$
PSRR at 100 Hz	77	dB
Load regulation ($\Delta V_{out}/I_{out}$)	4.1	mV/μA
Chip area	0.18	mm^2

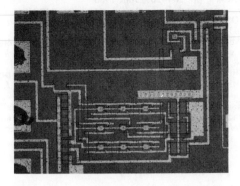

Fig. 13. Chip photograph of high PSRR bandgap reference which makes use of p-well resistors.

The reference voltage appears between the positive rail and the emitter of T_1. Better PSRR is achieved since the potentials of all critical nodes are constant with respect to the substrate. Changes of the supply voltage do not affect the value of resistances R_1 and R_2 and thus do not degrade the PSRR. They only affect the current mirror ratios B and C (see Fig. 10). Straightforward calculations show that the PSRR is inversely proportional to the length of transistors $M_4 - M_6$.

The circuit has been integrated in two different technologies. The most important measurement results of the integration (Fig. 11) in the aforementioned technology are shown in Fig. 12 and summarized in Table II.

The standard deviation of the reference voltage is 7.2 mV before trimming and 350 μV after trimming. The temperature behavior is identical to that of the first circuit presented. In Fig. 12(a) the reference voltage is shown as a function of temperature. The temperature behavior of an untrimmed circuit is also given.

Fig. 12(b) shows the reference voltage as a function of the supply voltage. The full line is for an externally applied gate voltage (V_G) of -2 V and the dotted line is in case the gate of the bipolar transistors is connected to the negative rail.

The PSRR of the negative rail and that of the bipolar gate voltage are almost identical. Measurement yields a PSRR of 77 dB at 100 Hz.

The load regulation ($\Delta V_{ref}/I_{out}$) is 4.1 mV/μA, which is the inverse of the sum of the transconductances of transistors T_1 and T_2.

The dynamic behavior of the voltage reference is shown in Fig. 12(d). The settling time is smaller than 40 μs with a load of 250 pF.

The current consumption of the circuit is 20 μA and is independent of the α of the bipolar transistors. Since the current through T_1 depends on α, transconductance of T_1 as well as the settling time depend on α. In case this is not acceptable, one can insert a lateral bipolar transistor, with base and collector tied together, between resistor R_1 and the drain of M_4; however, power consumption will then depend on α.

The circuit has also been integrated in a 4-μm p-well technology with threshold values of ± 1 V. In order to be able to bias the transistors M_1 and M_2 deep in strong inversion, a lateral bipolar transistor was inserted between resistor R_2 and the positive supply voltage (base and collector connected to V_{DD}). Reference voltage appears between V_{DD} and the base of T_1.

The circuit (Fig. 13) was realized with p-well resistors and lateral bipolar transistors formed with parasitic transistors. The PSRR of this circuit is 87 dB (dc). Current consumption is 4.2 μA and the minimal required supply voltage is 2.2 V.

VI. CONCLUSIONS

In this paper, two new bandgap references are presented which make use of CMOS compatible lateral bipolar transistors. The circuits are designed to be insensitive to the possible low beta and alfa current gains of these devices. Their accuracy is not degraded by any amplifier offset.

The circuits have been integrated in three consecutive batches. The experimental results yield a reference voltage which is stable within 2 mV after trimming at a single temperature. The temperature accuracy is limited mainly by technology fluctuations, especially those of the temperature coefficient of resistors.

The first circuit can be used as a voltage regulator since it has a low output impedance. The second circuit permits the use of p-well resistances without any degradation of the PSRR. Both circuits use only a few components, which results in a moderate die area, and deliver a continuous reference voltage.

REFERENCES

[1] Y. Tsividis and R. Ulmer, "A CMOS voltage reference," *IEEE J. Solid-State Circuits*, vol. SC-13, pp. 774–778, Dec. 1978.
[2] E. Vittoz and O. Neyround, "A low-voltage CMOS bandgap reference," *IEEE J. Solid-State Circuits*, vol. SC-14, pp. 573–577, June 1979.
[3] H. Oguey and B. Gerber, "MOS voltage reference based on poly-silicon gate work function difference," *IEEE J. Solid-State Circuits*,

vol. SC-15, pp. 264–269, June 1980.

[4] B. S. Song and P. Gray, "A precision curvature compensated CMOS bandgap reference," *IEEE J. Solid-State Circuits*, vol. SC-18, pp. 634–643, Dec. 1983.

[5] B. Ahuja, W. Baxter, and P. R. Gray, "A programmable CMOS dual channel interface processor," in *ISSCC Dig. Tech. Pap.*, Feb. 1984, pp. 232–233.

[6] E. Vittoz, "MOS transistors operated in the lateral bipolar mode and their application in CMOS technology," *IEEE J. Solid-State Circuits*, vol. SC-18, pp. 273–279, June 1983.

[7] Y. Tsividis, "Accurate analysis of temperature effects in $I_C - V_{BE}$ characteristics with application to bandgap reference sources," *IEEE J. Solid-State Circuits*, vol. SC-15, pp. 1076–1084, Dec. 1980.

[8] H. Oguey, "Référence de tension et détecteur de tension de pile à faible consommation," in *Proc. 57th Swiss Congress of Chronometry* (Montreux, Switzerland), Oct. 1982, pp. 59–63.

[9] R. Ye and Y. Tsividis, "Bandgap voltage reference sources in CMOS technology," *Electron. Lett.*, vol. 18, no. 1, pp. 24–25, Jan. 1982.

[10] J. Michejda and S. K. Kim, "A precision CMOS bandgap reference," *IEEE J. Solid-State Circuits*, vol. SC-19, pp. 1014–1021, Dec. 1984.

[11] E. Vittoz, "The design of high performance analog circuits on digital CMOS chips," *IEEE J. Solid-State Circuits*, vol. SC-20, pp. 657–665, June 1985.

A Low Drift Fully Integrated MOSFET Operational Amplifier

ROBERT POUJOIS AND JOSEPH BOREL, MEMBER, IEEE

Abstract—A fully integrated MOSFET amplifier with very low drift has been built using standard technology. Input offset voltages as low as 5 μV and drift values of this offset voltage less than 0.05 μV/°C are measured.

INTRODUCTION

ANALOG amplifiers using standard MOS technology could be thought to have dc performances very different from the bipolar ones mainly as far as drift and offset voltages are concerned. Specific advantages of MOS devices can be used to avoid these limitations [1], [2], and results similar to those obtained in chopper amplifiers can be measured [3], [4]. The basic principle is to use memory capacitances to store the existing defects (either drift, noise in a given bandwidth, or offset voltages) before amplification and refresh these analog data as soon as possible.

The main MOSFET's limitations when these are used in analog amplifiers are the following.

1) Offset voltages as high as 50 mV are observed and drift voltages in the range of 50 μV/°C are measured.

2) Noise voltage levels of 250 μV peak are typical.

These values are related to the Si–SiO$_2$ interface properties and can be decreased slightly by improved SiO$_2$ technology.

Other limitations are in the electrical properties of analog amplifiers built in standard MOS technologies.

1) High values of voltage gain per stage are difficult to obtain and a value of 3 is typical. It follows that a large number of stages (five or six) must be used, giving additional phase shift.

2) Maximum gain–bandwidth products of 10 MHz can be achieved, and increased values can only be obtained with new

Manuscript received January 13, 1978.
The authors are with the Laboratoire de Microelectronique Appliquee, C.E.A.-C.E.N.G., Grenoble, France.

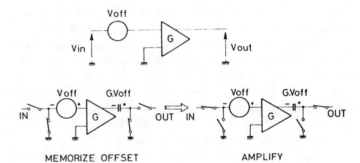

Fig. 1. Offset memorization principle.

technologies (self-registered gates, depletion enhancement technology, SOS technology, etc.).

On the other hand, MOSFET devices have specific advantages that can be used in analog amplifiers.

1) On chip compatibility with LSI logic circuits leads to complex arrays where digital and analog signals are processed.

2) Refresh memory capabilities allow one to store analog values of defects like offset voltages.

3) No offset current is seen due to the very high input impedance of the SiO$_2$ gate layer.

4) Fully integrated amplifiers without external components can be built.

5) Low cost of the MOS technology is a very attractive argument for standard products.

We present some results obtained on analog MOSFET amplifiers (chopper amplifiers) using the general considerations given above.

AMPLIFIERS WITH OFFSET VOLTAGE MEMORIZATION

The basic principle is to store the analog offset voltage in an MOS capacitance and to use this offset voltage to compensate for the amplifier input offset voltage. The circuit configuration is given Fig. 1 where we have represented the offset mem-

Reprinted from *IEEE J. Solid-State Circuits*, vol. SC-13, no. 4, pp. 499–503, Aug. 1978.

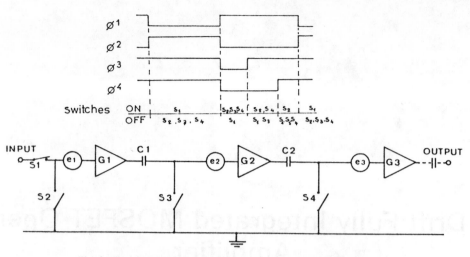

Fig. 2. Residual voltage successive memorization (RSM) amplifier.

orization principle. An actual amplifier has an offset voltage V_{off} due to the mismatch of the electrical parameters of the input devices.

For one elementary amplification stage, we use four switches and a storage capacitance as seen in the bottom part of Fig. 1.

During memorization time, input node and output node through the storage capacitance are grounded. In this capacitance a value $(G \cdot V_{off})$ is memorized. We must avoid saturation of the amplifier by the $(G \cdot V_{off})$ value at the output. So a limited value of the voltage gain per elementary stage is allowed $(G \cdot \max \leqslant 100)$. Several elementary stages in series are needed for a high voltage gain amplifier.

During the next sequence, the amplifier configuration is changed and the input and output nodes are connected in the signal path for amplification with a low residual input offset voltage. The sequence is repeated at the clock rate (16 kHz in our case), allowing us to use dc amplifiers (no large area connecting capacitances).

Storage capacitances must have high values compared to switch capacitances to avoid parasitic offset signals due to switching. A MOSFET switch presents a mean parasitic capacitance C_p of 0.1 pF between gate and channel and a memory capacitance C_m of 10 nF will be necessary if offset voltages ϵ as low as 100 μV are wanted. For a $V = 10$ V clock voltage on the gate of the switch transistor, we have

$$\epsilon = V \cdot \frac{C_p}{C_m} \simeq 10 \text{ V} \times \frac{0.1}{10^4} = 100 \ \mu\text{V}.$$

A differential configuration allows us to compensate for the switching parasitics and leakage currents of the switches and of memory capacitances. Only the mismatch between these elements is to be considered, allowing the use of smaller values for the memory capacitances (compatible with integration).

RESIDUAL VOLTAGE SUCCESSIVE MEMORIZATION (RSM) AMPLIFIER

This technique uses integrated storage capacitances as low as 4 pF and memorizes parasitic pulses as well as circuit defects. The elementary stages are sequentially compensated

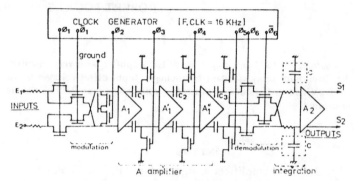

Fig. 3. Block diagram of the RSM amplifier.

starting from the input of the amplifier. Fig. 2 is a schematic view of the circuit with clock voltage waveforms: if switches 2, 3, and 4 are closed and switch 1 is opened, C_1 and C_2 are, respectively, charged with $G_1 \cdot e_1$ and $G_2 \cdot e_2$. These values must not saturate the amplifiers (relatively low values of G_1 and G_2).

If switch 3 is then opened, giving an extra contribution ϵ_2 to the offset voltage e_2, this can be compensated for in capacitance C_2. So the net input offset voltage comes only from the input offset voltage (and the parasitic input clock pulse) of the last amplifier stage. The input offset voltage is then given by

$$V_{off} = (e_n + \epsilon_n)/(G_1 \times G_2 \cdots \times G_{n-1})$$

and can be decreased by increasing the overall voltage gain. e_n and ϵ_n are, respectively, the input offset voltages of the nth amplifier stage and parasitic clock pulse at the input of this amplifier stage. $G_1 - G_{n-1}$ are the voltage gains of the various stages. Using such considerations, an MOS integrated amplifier has been built with integrated clock generator and modulation-demodulation stages to compensate for the parasitic signals coming from switches 1 and 2 at the input. Standard p channel aluminum gate technology is used.

Fig. 3 is a block diagram of the RSM amplifier. It has five basic blocks.

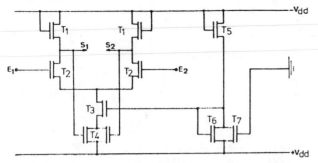

Fig. 4. Basic differential amplifier with compensated bias.

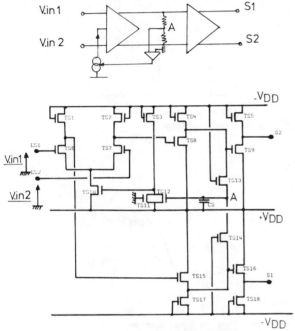

Fig. 5. Output amplifier stage diagram.

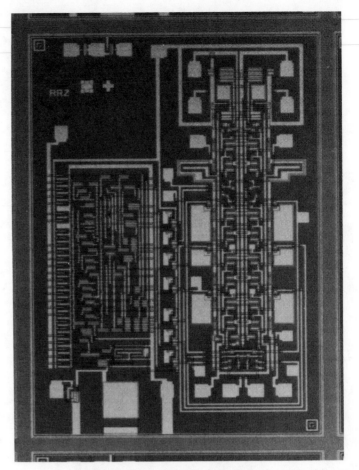

Fig. 6. Amplifier chip with clock generation.

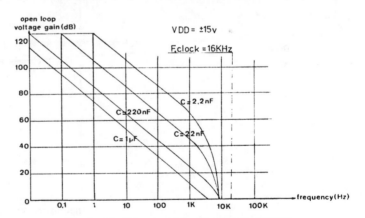

Fig. 7. Amplifier open loop frequency response.

1) The amplifier A: it is a three-stage differential RSM amplifier. All signals are amplified differentially to compensate for the parasitics coming from the memory switches.

2) The modulation-demodulation system: it is built to eliminate very low frequency drift input signals (mainly from thermal origin) and parasitic pulses from ϕ_1, $\overline{\phi_1}$, and ϕ_2.

3) The clock generator working at a clock frequency of 16 kHz.

4) The integration circuit working with two external capacitances C.

5) The output stage A_2 with a voltage gain of 200.

The crossed modulation-demodulation system is able to amplify very low frequency signals without any drift (input signals are sequentially switched toward the two inputs of A). Moreover, amplifier A is working at the clock frequency and the $1/F$ noise spectrum is lowered. The chopper amplifier behaves like a high-pass filter for the noise (see Fig. 8, for example).

Fig. 4 is a diagram of one elementary stage of the amplifier (A_1, A_1', or A_1''). The main features are

1) common mode rejection through $T4$,

2) output level stabilization ($S1$ and $S2$) versus supply voltage variations ($T7$), threshold shifts ($T6$), and geometrical spreads of devices ($T5$).

Fig. 5 shows a block diagram and a complete configuration of amplifier A_2. A feedback loop is used to obtain a virtual ground in A, resulting in symmetrical dc output voltages.

Fig. 6 is a view of the RSM amplifier including clock generation and amplifier stages—chip size is 3.1 mm × 2.2 mm.

The next figures give the measured performances of the amplifier.

1) Fig. 7 is a plot of open loop voltage gain versus frequency. Gain-bandwidth products of 4 MHz are achieved.

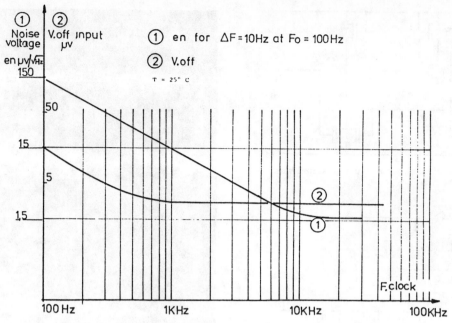

Fig. 8. Input offset voltage and input noise voltage versus frequency.

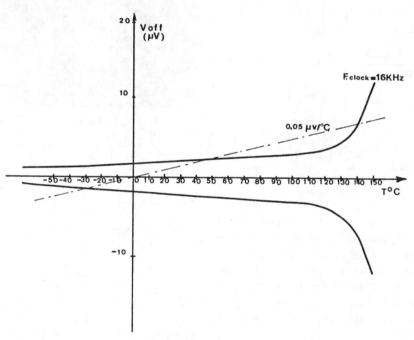

Fig. 9. Input offset voltage versus temperature.

Values of integration capacitances C are given as a parameter. Cutoff frequencies are limited below F clock/2 (Nyquist limit).

2) Fig. 8 is a plot of input offset voltage V_{off} and input noise voltage versus clock frequency. An offset voltage of 5 μV is reached and is independent of clock frequency in a wide range. Narrow-band noise voltage (ΔF = 10 Hz bandwidth) at central frequency F_0 = 100 Hz is a function of clock frequency showing evidence of $1/f$ noise contribution from the input devices.

3) Fig. 9 gives an example of the behavior of offset voltage versus temperature for a 16 kHz clock frequency. It is clearly seen that at higher temperatures leakage current compensation is not good, and higher clock frequencies should be necessary.

4) Fig. 10 is a typical spread of input offset voltages measured on the amplifier of Fig. 6. This corresponds to input offset voltages spread either for chips on the same wafer or from different wafers. Most of the input offset voltages are within 5 μV.

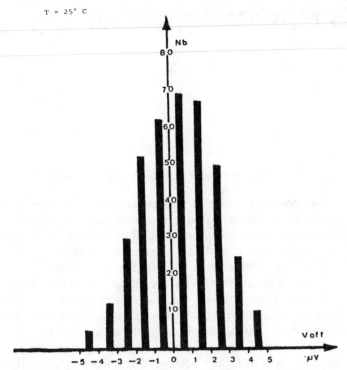

Fig. 10. Typical spread of input offset voltages on various chips coming from different wafers.

TABLE I
RSM Amplifier Characteristics (Ambient Temperature)

voltage gain	2 000 000
input offset voltage	<5 μV
drift voltage	<0.05 μV/°C
input noise voltage	2.5 μV/$\sqrt{Hz}$
supply voltage	±10 V to ±18 V
power consumption	240 mW
	(V_a = ±15 V)
common mode rejection ratio	120 dB
supply voltage rejection ratio	>120 dB
external components	2 capacitors

Conclusion

We have presented a new technique to compensate for offset voltages and drift voltages in analog amplifiers using standard MOSFET technology. Amplifiers with improved performances are obtained, and their association with digital MOSFET circuits is a powerful tool in custom integrated circuit design. These amplifiers can be used as elementary blocks in more complex analog or digital arrays or as standard products.

Some performance still needs to be improved for specific applications such as input noise level and maximum operating frequency.

In summary, we present in Table I measured results that have been obtained with the integrated amplifier of Fig. 6.

Let us only point out the very high value of supply voltage rejection ratio observed in the case of RSM amplifiers. This is due to the internal memorization of supply voltage shifts.

Best frequency responses are expected with depletion enhancement self-registered technologies or with SOS technology, and lower noise levels can be obtained when optimizing input stages.

Acknowledgment

Thanks are due to J. M. Ittel for making the measurements, to D. Barbier for valuable discussions, and to E. Mackowiak for a critical reading of the manuscript.

References

[1] R. Poujois, B. Baylac, D. Barbier, and J. M. Ittel, "Low level MOS transistor amplifier using storage techniques," in *ISSCC 73*, pp. 152–153.
[2] R. Poujois, J. M. Ittel, and J. Borel, "A low drift MOSFET operational amplifier: A.R.Z.," presented at ESSCIRC, Toulouse, France, Sept. 1976.
[3] J. L. Villevieille, "Amplificateur opérationnel stabilisé par chopper," in *Texas Instruments Seminar Dig.*, 1974, p. 177.
[4] I. Hackel and H. Hagemann, "Construction of chopper amplifiers," *Elektron. Reinschan* (Germany), vol. 16, pp. 509–512, Nov. 1962.

Large Swing CMOS Power Amplifier

KEVIN E. BREHMER, MEMBER, IEEE, AND JAMES B. WIESER

Abstract—A CMOS class AB power amplifier is presented wherein supply-to-supply voltage swings across low impedance loads are efficiently and readily handled. The amplifier consists of a high gain input stage and a push–pull unity gain amplifier output stage. The amplifier dissipates only 7 mW of dc power and delivers 36 mW of ac power to a 300 Ω load, using standard power supplies of ±5.0 V. Lower impedance loads can be driven to higher power levels, providing the internal current limiting level is not exceeded.

I. INTRODUCTION

DURING the past few years, CMOS has emerged as an industry standard because of its low power dissipation. However, the implementation of analog functions in MOS has presented a challenge to many circuit designers. One area of MOS analog design where considerable work is taking place is the design of an efficient, large dynamic range power amplifier [1], [2].

Prior art MOS power amplifiers used output stage configurations that were subject to various limitations. Such limitations included the size of the output driver devices and the control of the dc bias current in these output drivers. In order to control the dc bias current in the output driver devices, previous designs used a source follower device and a controlled current sink as an output stage [1], [2]. This type of design has dynamic range limitations and requires large output driver device sizes. By replacing the source follower output driver with a bipolar emitter follower, transistor die area and dynamic range limitations can be reduced considerably. However, instabilities arise when using a bipolar emitter follower to drive high capacitive loads, thus limiting its application. Also, bipolar transistors are parasitic devices in a standard CMOS process which are not necessarily well controlled.

In order to obtain an efficient power amplifier with a large dynamic range capable of driving a low impedance load, a push–pull class AB, fully CMOS power amplifier is presented. This amplifier is designed to operate at voice-band frequencies for telecommunication and audio applications, and to be used in an inverting configuration so that users can design gain or attenuation into their systems.

Manuscript received July 11, 1983; revised July 26, 1983.
The authors are with the Telecom Group, National Semiconductor, Santa Clara, CA 95051.

II. DESCRIPTION OF THE POWER AMPLIFIER

The complete power amplifier consists of a high gain input stage driving a unity gain push–pull output stage. The input stage is comprised of a differential amplifier and a common source amplifier. Compensation of this stage is achieved by a Miller multiplied feedback capacitor.

The output stage includes two unity gain amplifiers in push–pull configuration. Each amplifier contains a differential input stage whose output controls the gate of the output driver device. The drain of the output driver device is directly fed back to the noninverting input of the differential stage to form a noninverting unity gain amplifier, and one half of the push–pull configuration.

In the event of an offset between the two push–pull amplifiers, a feedback circuit controls the dc bias current in the output drivers. Fig. 1 shows a pseudo block diagram form of the power amplifier and the associated feedback circuitry required to stabilize the dc bias current in the output driver devices.

A. Output Stage Analysis

From Fig. 1 we can see that amplifier $A1$ and transistor $M6$ form the unity gain amplifier for the positive half of the output voltage swing, and conversely amplifier $A2$ and transistor $M6A$ form the negative half cycle circuit. For the sake of simplicity, only the circuit referring to the positive half output swing of the output stage will be discussed. The operation of the negative half circuit is an inverted mirror image of that of the positive half swing circuit. Components performing similar functions in each circuit are designated with an additional letter "A" for the negative half circuit.

Shown in Fig. 2 is a detailed schematic of the positive unity gain amplifier. The differential amplifier input stage of the unity gain amplifier has a large positive common mode range (CMR), which allows transistor $M6$ to source large amounts of current to the load, while still being of a reasonable physical size for incorporation into a monolithic circuit. Large current sourcing is provided by producing the highest possible gate drive on $M6$. The maximum gate-to-source voltage $M6$ can have while still keeping $M1$ and $M2$ in the saturation region is given by

$$V_{GS6_{max}} = -(V_{CC} - (V_{IN} - V_{GS1} + V_{DSAT1})) \qquad (1)$$

Reprinted from *IEEE J. Solid-State Circuits*, vol. SC-18, no. 6, pp. 624–629, Dec. 1983.

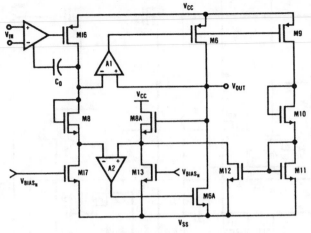

Fig. 1. Block diagram of power amplifier.

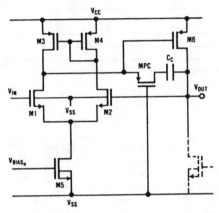

Fig. 2. Positive output stage unity gain amplifier.

It first appears that increasing V_{GS1} by increasing the overdrive voltage of $M1$ and $M2$ will increase the gate drive of $M6$. However, after substitution of the saturation equation (2) into the expression for V_{GS6max}, we see that this is not the case:

$$V_{DSAT} = V_{GS} - V_T \tag{2}$$

$$V_{GS6_{max}} = -(V_{CC} - V_{IN} + V_{T1}). \tag{3}$$

From (3), maximum gate drive on $M6$ is provided by increasing the threshold voltage (4) of $M1$ and $M2$. One method of increasing the threshold voltage of $M1$ and $M2$ is to modify V_{T0} by a threshold implant. Implanting devices $M1$ and $M2$ to produce a higher V_{T0} pushes the common source voltage lower, allowing the gate of $M6$ to drop more while still keeping $M1$ and $M2$ in saturation.

$$V_T = V_{T0} + \gamma\sqrt{V_{BS} - 2\varphi_F}. \tag{4}$$

Further enhancement of the positive CMR is obtained by connecting the substrate of transistors $M1$ and $M2$ to

V_{SS}, thereby modulating the source–substrate voltage of these transistors. The effect of this substrate modulation on the threshold voltage of $M1$ and $M2$ can be seen in (4). As the output swing increases, the common source voltage also increases, however, not by the same amount, since the threshold voltage of $M1$ and $M2$ has increased due to substrate modulation. The increased threshold voltage of $M1$ and $M2$ tends to reduce the common source mode voltage of the differential pair thus driving the gate of $M6$ more negative while still keeping $M1$ and $M2$ in saturation.

The current in the output driver device $M6$ is typically controlled by the current mirror developed in the differential amplifier of the positive unity gain amplifier and matches the current set in the negative output driver device $M6A$ by the negative unity gain amplifier. If an offset occurs between amplifiers $A1$ and $A2$, the current balance between output drivers $M6$ and $M6A$ no longer exists and either massive amounts of current or no current at all will flow through these devices. The feedback loop, shown in Fig. 1, consisting of transistors $M8A-M13$, stabilizes the current through output drivers $M6$ and $M6A$ in the event of an offset between amplifiers $A1$ and $A2$. The feedback loop operates as follows. Assume that amplifier $A1$ has an offset such that transistor $M6$ begins to source excessive amounts of current. The excessive current is sensed by transistor $M9$ and is fed back to the source follower $M8A-M13$. The increase of current provided to transistor $M8A$ produces a greater voltage drop across the source follower $M8A-M13$ and more differential signal on the input of amplifier $A2$. The larger differential signal on amplifier $A2$ results in lower gate drive on output driver $M6A$, thereby reducing the current in the output drivers $M6-M6A$. The output voltage now has increased due to the fact that the positive swing amplifier $A1$ attempts to keep both of its inputs at the same potential. The complete power amplifier is in a feedback loop, wherein amplifier feedback drops the voltage of the negative input of amplifier $A1$ in attempting to keep the output of the complete power amplifier at 0 V in the dc bias condition. Transistor $M8$ transfers this voltage drop to the negative input of amplifier $A2$, thus balancing offset of amplifier $A2$. The offset that was initially introduced by amplifier $A1$ is absorbed by the source follower transistor $M8A$.

Because the output stage current feedback is not unity gain, some current variation in transistors $M6$ and $M6A$ occurs. Offsets between amplifiers $A1$ and $A2$ can produce a 2:1 variation in dc current over temperature and process variations. Equation (5) predicts the change in the output driver current assuming that V_{out} is at ground, and any offset between amplifiers $A1$ and $A2$ can be reflected as a difference between the inputs of amplifier $A1$. From this equation, V_{off1} must be zero for ΔI_0 to be zero.

$$\Delta I_0 = -gm_{6A}A_2\left[V_{off1} - \sqrt{\frac{2\beta_9\beta_{12}}{\beta_{8A}\beta_6\beta_{11}}}\left[\sqrt{I_{B1}\left(\frac{\beta_6\beta_{11}}{\beta_9\beta_{12}} + \frac{1}{2}\frac{\beta_5\beta_6}{\beta_7\beta_3}\right) + \Delta I_0} - \sqrt{I_{B1}\left(\frac{\beta_6\beta_{11}}{\beta_9\beta_{12}} + \frac{1}{2}\frac{\beta_5\beta_6}{\beta_7\beta_3}\right)}\right]\right] \tag{5}$$

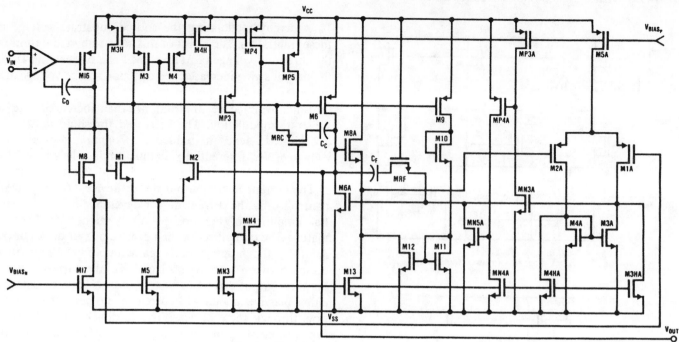

Fig. 3.   Complete power amplifier schematic.

where

$$\beta = \frac{\mu_{\text{eff}} C_{\text{ox}}}{2} \left(\frac{W}{L} \right)_{\text{eff}} \text{ and } I_{B1} = I_{MI7}.$$

Since transistor $M6$ can supply large amounts of current, care must be taken to ensure that this transistor is off during the negative half cycle of the output voltage swing. For large negative swings, the drain of transistor $M5$ pulls to V_{SS}, turning off the current source that biases the differential amplifier $A1$. As the bias is turned off, the gate of transistor $M6$ floats and tends to pull towards V_{SS}, turning on transistor $M6$.

Shown in Fig. 3 is circuitry which ensures that transistor $M6$ remains off for large negative voltage swings. As transistor $M5$ turns off, transistors $M3H$ and $M4H$ pull up the drains of transistors $M3$ and $M4$, respectively. As a result, transistor $M6$ is turned off and any floating nodes in the differential amplifier are eliminated. Positive swing protection is provided for the negative half cycle circuit by transistors $M3HA$ and $M4HA$, which operate in a manner similar to that described above for the negative swing protection circuit. The swing protection circuit does, however, degrade the step response of the power amplifier since the unity gain amplifier not in operation is completely turned off.

Short circuit protection is also included in the design of the amplifier. From Fig. 3, we can see that transistor $MP3$ senses the output current through transistor $M6$, and in the event of excessively large output currents, the biased inverter formed by transistors $MP3$ and $MN3$ trips, thus enabling transistor $MP5$. Once transistor $MP5$ is enabled, the gate of transistor $M6$ is pulled up towards the positive

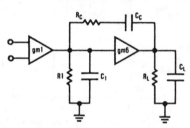

Fig. 4. RC compensation for the output stage unity gain amplifier.

supply V_{CC}, thus limiting the current transistor $M6$ source to approximately 60 mA.

B. AC Compensation of the Power Amplifier

AC stability of the complete power amplifier is achieved by providing a wide-band output stage and by using compensation at the input stage to produce the dominant pole. The dominant pole in the input stage is produced by the Miller multiplied capacitor C_D. The compensation of each unity gain amplifier in the output stage is achieved by a Miller multiplied capacitor and feedforward resistor. From Figs. 2 and 3, we can see that transistor MRC and capacitor C_C comprise the RC network for the positive unity gain amplifier and transistor MRF and C_F comprise the RC network for the negative unity gain amplifier. To simplify the calculation of the open loop transfer function of Fig. 2, a model as shown in Fig. 4 has been developed [1]. In this figure, the effective output impedance of the first stage is represented by resistor R_1 and capacitor C_1, while the effective output impedance of the second stage is represented by resistor R_L and capacitor C_L. From this model, it can be shown that the poles and zero of each unity gain amplifier before closing the feedback loop are

$$P_1 \simeq \frac{-1}{gm_2 R_L R_1 C_C} \; ; \text{ for high } R_L$$

$$\simeq \frac{-1}{gm_2 R_L R_1 C_C + R_1(C_1 + C_C)} \; ; \text{ for low } R_L$$

$$P_2 \simeq \frac{-gm_2 C_C}{C_L(C_1 + C_C) + C_1 C_C} \; ; \text{ for high impedance load}$$

$$\simeq \frac{-(gm_2 + g_2)C_C}{C_C(C_1 + C_C) + C_1 C_C} \; ; \text{ for low impedance load}$$

$$P_3 \simeq \frac{-1}{R_C C_1}$$

$$Z \simeq \frac{-1}{R_C C_C - \dfrac{C_C}{gm_2}} .$$

Note that the pole splitting between $P1$ and $P2$ is a function of the load resistance and load capacitance. For low resistive loads the open-loop gain of the positive swing amplifier drops, and $P1$ and $P2$ both move out in frequency. Since both $P1$ and $P2$ move out in frequency, phase degradation could occur depending on the location of $P3$; however, by bringing the zero shown above, into the left half plane, some phase shift that occurs can be cancelled. The zero placement also helps to cancel phase shift that will occur as a result of capacitive loading on the output. As the output capacitance increases, $P2$ decreases and phase margin degradation occurs; however, by careful placement of the zero, the reduction in the phase margin can be minimized.

The ac stability of the total output stage can be modeled, to a first order, as two independent amplifiers in parallel. Since the negative unity gain amplifier is an inverted mirror image of the positive unity gain amplifier with equal drive requirements, the dominant poles and zeros of each amplifier are approximately the same, and the complete output amplifier transfer function simplifies to that of a half cycle stage. The number and location of the poles and zeros of the complete output stage are identical to those in each of unity gain amplifiers; therefore, no additional compensation is required to stabilize the complete output stage.

The total amplifier circuit is capable of driving 300 Ω and 1000 pF to ground. The gain–bandwidth product is approximately 500 kHz and is limited by the output stage 1000 pF load requirement. The output stage bandwidth is approximately 1.0 MHz.

III. Experimental Results

The power amplifier presented was fabricated using National Semiconductor's proprietary P^2CMOS process. A die photo of the power amplifier is shown in Fig. 5. The total die area of the power amplifier is 1500 mils².

Table I shows a comparison between the simulated and measured results of the power amplifier and Table II shows the device sizes that correspond to Fig. 3. Since the ampli-

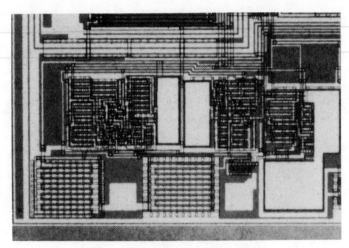

Fig. 5. Die photo of power amplifier.

TABLE I
POWER AMPLIFIER PERFORMANCE

Parameter	Simulation	Measured Results
Power dissipation	7.0 mW	5.0 mW
A_{vol}	82 dB	83 dB
F_u	500 kHz	420 kHz
V_{offset}	0.4 mV	1 mV
PSRR + (dc)	85 dB	86 dB
(1 kHz)	81 dB	80 dB
PSRR − (dc)	104 dB	106 dB
(1 kHz)	98 dB	98 dB
THD $V_{IN} = 3.3\ V_p\ R_L = 300\ \Omega$	0.03%	0.13%(1 kHz)
$C_L = 1000$ pF	0.08%	0.32%(4 kHz)
$V_{IN} = 4.0\ V_p\ R_L = 15\text{K}\ \Omega$	0.05%	0.13%(1 kHz)
$C_L = 200$ pF	0.16%	0.20%(4 kHz)
T settling (0.1%)	3.0 μs	< 5.0 μs
Slew rate	0.8 V/μs	0.6 V/μs
$1/f$ noise at 1 kHz	N/A	130 nV/Hz
Broad-band noise	N/A	49 nV/Hz
Die area		1500 mils²

TABLE II
COMPONENT SIZES
(μm,pF)

MI6	184/9	M8A	481/6
MI7	66/12	M13	66/12
M8	184/6	M9	27/6
M1, M2	36/10	M10	6/22
M3, M4	194/6	M11	14/6
M3H, M4H	16/12	M12	140/6
M5	145/12	MP3	8/6
M6	2647/6	MN3	244/6
MRC	48/10	MP4	43/12
CC	11.0	MN4	12/6
M1A, M2A	88/12	MP5	6/6
M3A, M4A	196/6	MN3A	6/6
M3HA, M4HA	10/12	MP3A	337/6
M5A	229/12	MN4A	24/12
M6A	2420/6	MP4A	20/12
MRF	25/12	MN5A	6/6
CF	10.0		

67

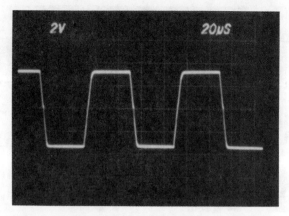

Fig. 6.   Output time response for a $\pm 3.3\ V_P$ input pulse.

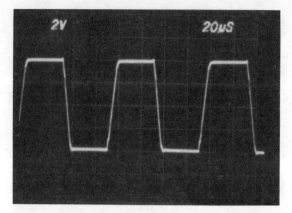

Fig. 7.   Output time response for a $\pm 4.0\ V_P$ input pulse.

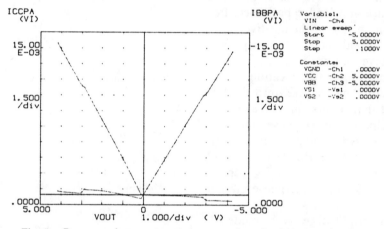

Fig. 8.   Power supply current versus output voltage level for $R_L = 300\ \Omega$.

fier was designed to be used in an inverting configuration, the common-mode range and CMRR are not applicable op amp parameters.

Shown in Figs. 6 and 7 are step responses of $\pm 3.3\ V_p$ and $\pm 4.0\ V_p$ for loads of 300 Ω/1000 pF and 15K Ω/200 pF, respectively, using ± 4.75 V supplies. The slight cross-over distortion, as seen in Fig. 6, while slewing negatively, is a result of the delay caused by the offset feedback circuit when amplifier $A2$ is required to drive large amounts of current to the load. This distortion is negligible in THD measurements for telecommunication and audio applications.

Fig. 8 shows the efficiency of the power amplifier as the output voltage swings from rail-to-rail. For this case the power supplies used were ± 5.0 V. The slight discontinuity in the supply current curves at approximately ± 3.0 V is a result of the opposite unity gain amplifier turning off in the output stage. This discontinuity in supply current has no effect on the distortion seen in the amplifier. The dc operating current at 0 V is the operating current for two power amplifiers since the application in which the power

amplifiers is to be used requires a differential signal. The dc operating current per amplifier is 500 μA. The ICCPA current in the positive supply is greater due to the fact that the high current $M6$ source is mirrored around to the source follower $M8A$ by the output driver current offset feedback circuit.

Amplifier offset voltages of typically 1 mV were measured. These low offsets were a result of a good process and careful layout. From these data and the fact that amplifier feedback will reduce any offset between the output unity gain amplifiers, the dc current variation in the output driver devices should be very small.

IV.   CONCLUSION

Presented was a fully CMOS class AB power amplifier wherein supply-to-supply voltage swings across low impedance loads are efficiently and readily handled. The amplifier dissipates only 7 mW of dc power and can deliver 36 mW of ac power to a 300 Ω load using ± 5.0 V power

IEEE JOURNAL OF SOLID-STATE CIRCUITS, VOL. SC-18, NO. 6, DECEMBER 1983

supplies. Other features of the design include typical offset voltages of 1 mV and THD of less than 0.4 percent.

ACKNOWLEDGMENT

The authors gratefully acknowledge the contributions of C. Laber to the design of the presented power amplifier.

REFERENCES

[1] W. C. Black, Jr., D. J. Allstot, and R. A. Reed, "A high performance low power CMOS channel filter," *IEEE J. Solid-State Circuits*, vol. SC-15, pp. 921–929, Dec. 1980.
[2] D. Senderowicz, D. Hodges, and P. Gray, "High-performance NMOS operational amplifier," *IEEE J. Solid-State Circuits*, vol. SC-13, pp. 760–766, Dec. 1978.

A High-Performance CMOS Power Amplifier

JOHN A. FISHER, MEMBER, IEEE

Abstract — A high-performance CMOS power amplifier consisting of a new input stage especially suited to power amplifier applications and a variation on a class *A B* output stage is presented which has been fabricated using a conventional silicon gate p-well process. The configuration results in several performance improvements over previously reported high-output current amplifiers without requiring process enhancements. Design details and experimental results are described.

I. INTRODUCTION

THE CONTINUING SEARCH for a better CMOS power amplifier seems to be leading to configurations which take advantage of a common source type output stage in order to achieve higher load current capability along with a higher output swing [6], [7]. By introducing a local feedback network around the common source transistor in the form of a simplified operational amplifier, a pseudo source follower is formed which offers substantially better voltage swing than that available from a conventional enhancement source follower. Unfortunately, the bandwidth of the amplifier must often be substantially reduced to maintain a stable frequency response due to an excessive amount of phase shift through this pseudo source-follower circuit.

Other recent developments include variations in amplifier compensation techniques which improve the amplifier's ability to reject noise on the power supplies. Usually this improvement comes at the expense of other performance parameters such as common-mode range, offset voltage, or output swing.

This paper presents a new input stage which exhibits excellent supply rejection properties without the previously mentioned disadvantages and allows a potentially large increase in gain over the classic two stage design. Also, an output stage variation is presented which exhibits an improved frequency response over previously reported large swing stages. Although the output swing of this stage is comparatively limited, it still provides up to ± 3 V into 200 Ω without requiring process enhancements.

Manuscript received January 4, 1985; revised July 25, 1985.
The author is with Siemens AG, WIS TE PE 23, Balanstrasse 73, 8000 Munich 80, West Germany.

II. CIRCUIT DESCRIPTION

A. Core Amplifier

In modern CMOS system design, many analog and digital circuits are often included on the same die. This situation can result in system degradation in the form of power supply noise contamination of the analog signals unless there is sufficient rejection of this noise within the amplifiers in the circuit.

Unfortunately, the classic *RC* compensation technique exhibits very poor supply rejection at high frequencies.

One solution to this problem is to return the compensation capacitor (*Cc*) to a virtual ground, such as the source of a cascode transistor in the input stage [1], [2]. This type of connection works well for inverting gain configurations but the cascode transistors limit the common-mode range for voltage follower or positive gain applications. A separate bias string connected to the output of the first stage can also be used to generate a virtual ground [3], [4]; however, inexact current cancellation in this type of connection can degrade the offset voltage of the amplifier.

Another solution is to use a single gain stage arrangement where the load capacitance is used to compensate the amplifier [4]. This approach leads to a degraded output swing due to the necessity of using cascode transistors to increase the gain of the amplifier to a reasonable level.

The core amplifier used in this design is shown in Fig. 1.[1] It was felt that a three-stage configuration, consisting of a wide bandwidth input stage and two high gain stages, offered much more flexibility in optimizing the performance requirements. As such, the principle of returning *Cc* to a virtual ground can be used for improving the supply rejection of the amplifier without introducing the previously mentioned disadvantages. Also, the three-stage architecture allows for a significant increase in gain without sacrificing the stability of the amplifier.

[1]Since the submission of this paper, a similar transconductance amplifier has been reported by D. B. Ribner and M. A. Copeland, "Design techniques for a cascoded CMOS op-amps with improved PSRR and common-mode input range," *IEEE J. Solid State Circuits*, vol. SC-19, pp. 919–925, Dec. 1984.

Reprinted from *IEEE J. Solid-State Circuits*, vol. SC-20, no. 6, pp. 1200–1205, Dec. 1985.

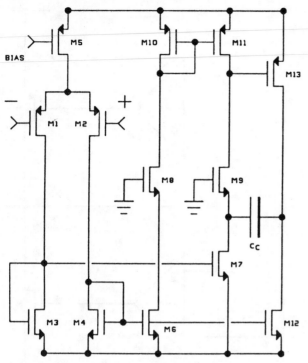

Fig. 1. Schematic diagram of the input stage.

the zero of the amplifier slightly, the configuration shown generally results in less phase shift at high frequencies.

The poles in (1) are not all widely spaced and, although $P1$ is easily extracted at

$$P1 \approx \frac{-g_{ds}10\ g0}{gm13\ Cc} \qquad (4)$$

$P2$ and $P3$ normally form complex conjugates at

$$P2,\ P3 \approx \frac{-gm8(Cc + C_0)}{2C_0Cc}$$

$$\pm j\left[\frac{gm8\ gm13}{C_0C_1} - \left(\frac{gm8\ (C_0 + Cc)}{2\ C_0Cc}\right)^2\right]^{1/2} \qquad (5)$$

Generally, this type of structure does not lend itself to hand calculation of the appropriate values for stability. Suggested guidelines are to make $gm8$ large and $gm13 \gg gm6$. The unity gain bandwidth occurs at

$$BW \approx \frac{gm2\ gm6}{gm4\ Cc}. \qquad (6)$$

The pole-zero structure of this type of compensation is different than that of the normal RC type. The transfer function of Fig. 1 is given in (1), where it should be understood that $gm1 = gm2$, $gm3 = gm4$, etc.:

$$A_V \approx \frac{a(1 + Sb)}{1 + Sc + S^2d + S^3e}$$

$$a \approx \frac{gm2\ gm6\ gm13}{gm4\ g_{ds}10\ g0}$$

$$b \approx \frac{Cc\ gm6\ gm13 + C1\ gm8\ gm12}{2\ gm6\ gm8\ gm13}$$

$$c \approx \frac{gm13\ Cc}{gds10\ g0}$$

$$d \approx \frac{C_1(Cc + C_0)}{g_{ds}10\ g0}$$

$$e \approx \frac{C_1CcC_0}{gm8\ g_{ds}10\ g0} \qquad (1)$$

where

$$g0 = g_{ds}12 + g_{ds}13$$

$$C_0 = C_L + C_{db}12 + C_{db}13$$

$$C_1 = C_{gs}13 + C_{db}11 + C_{db}9 + C_{gd}9. \qquad (2)$$

The zero of this transfer function occurs at

$$Z \approx \frac{-2\ gm6\ gm8\ gm13}{Cc\ gm6\ gm13 + C_1\ gm8\ gm12}. \qquad (3)$$

While reducing $M12$ to a simple current source changes

B. Output Stage

Fig. 2 shows two previously reported output stages [4], [5] which have been merged to form the output stage in this design. The configuration shown in Fig. 2(a) exhibits very desirable frequency characteristics in the form of one pole and one zero at very high frequencies for normal loads. The voltage swing for the stage is limited to slightly more than a V_{gs} from either supply, which becomes quite significant for large output currents into low impedance loads.

The configuration shown in Fig. 2(b) has several problems associated with it. First, connecting the inputs of Amp1 and Amp2 together results in a voltage between the gates of $M1$ and $M2$ that is offset dependent. This means that unless some method is devised to control the quiescent current through $M1$ and $M2$, the current through these transistors will vary widely with variations in V_{os_1} and V_{0s_2}. Second, the common-mode range of Amp1 and Amp2 must be equal to the desired swing in V_0 if the stage is to work properly. Third, careful consideration must be given to the frequency characteristics of Amp1 and Amp2 if the overall amplifier is to be stable. This type of output stage exhibits a large amount of phase shift at high frequencies which typically has required limiting the bandwidth of the overall amplifier in order to insure stability.

Many of these problems are easily solved within the merged output stage shown in Fig. 3. By building a small offset voltage into Amp1 and Amp2 as shown, transistors $M1$ and $M2$ are turned off in the quiescent state. The quiescent output current is therefore controlled by transistors $M3-M6$. The quiescent output current will be proportional to the current through $M3$ and $M4$ and is a function of the size ratio of $M5$ to $M3$ and $M6$ to $M4$. Under full load conditions in the negative direction, $M2$

71

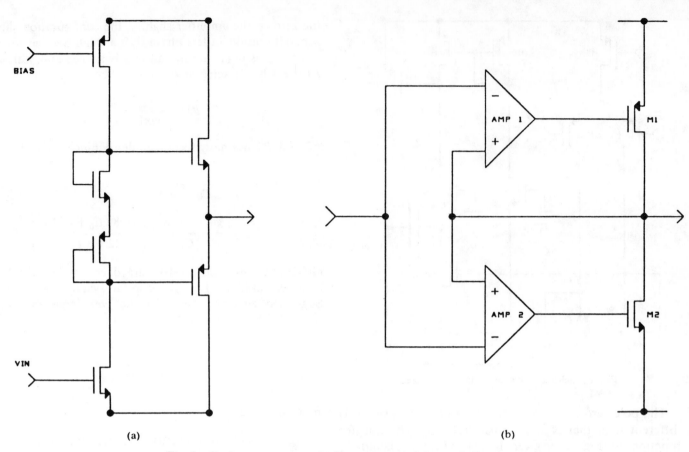

Fig. 2. Previous output stages. (a) Class AB source follower. (b) Pseudo source follower.

sinks approximately 95 percent of the required load current and $M6$ the remaining 5 percent. The actual current control mechanism through $M6$ takes place as Amp2 attempts to equalize the source voltages of $M4$ and $M6$, producing a similar V_{gs} across both transistors, and thus, ratioing their currents. A similar action takes place during the positive swing. This configuration results in a output swing which is still limited to within a V_{gs} of either supply. However, it is a relatively well controlled V_{gs}, dependent on the current through $M3$ and $M4$ rather than on a large output current through $M5$ or $M6$. The limiting factor on output swing in this design is the large threshold voltage of $M4$ and $M6$ due to back bias effects. Naturally, if a low threshold p-channel were available, the output swing could be improved considerably.

Although transistors $M5$ and $M6$ supply a fraction of the load current, their real usefulness lies in quiescent current control and in reducing the excess phase shift introduced by Amp1 and Amp2 by providing a feed-forward path to the output at high frequencies. Amp1 and Amp2 still require some minimal phase compensation in order to make them stable entities in the closed-loop mode, but the overall amplifier tends to adopt the frequency characteristics of $M5$ and $M6$ rather than that of the composite source followers. For example, during a fast input transient, it would be expected that the slow pseudo source-follower circuits would have a large amount of error associated with them. Observation, however, showed no

visually discernible crossover point nor the threshold point of the pseudo source followers. Apparently these errors are corrected by the much faster $M3$–$M6$ source-follower combination. This combination also insures stable operation into a pure capacitive load.

More specifically about Amp1, Fig. 4 shows the structure used in this design. The dc requirements for Amp1 are to be able to operate with its inputs near the positive supply while driving the gate of $M9$ to near the negative supply. Therefore, n-channel inputs are used which, with back bias effects, provide a common-mode range exceeding the positive supply. A second stage is used to maximize the gate drive to $M9$. During negative excursions of the output voltage, the gate–source voltage of $M1$ and $M2$ will tend to decrease the current through $M5$ as the negative common-mode range is exceeded. The effect of this current reduction is decreased gate drive to $M6$, which in turn drives $M9$ on. $M9$ corresponds to $M1$ in Fig. 3. As can be seen in Fig. 3, $M1$ should be prevented from turning on during the negative swing to maximize the amplifier's efficiency. Therefore, transistors $M13$–$M15$ have been added in Fig. 4 to force $M9$ off for output voltages more negative than the threshold of $M13$. As was also shown in Fig. 3, $M1$ and $M2$ are held off in the quiescent state by a small offset voltage built into Amp1 and Amp2. Note that when the output of the amplifier is at ground, the source of $M3$ and $M4$ are likewise near ground. Thus, connecting the inputs of Amp1 and Amp2 together as shown in Fig. 3 will

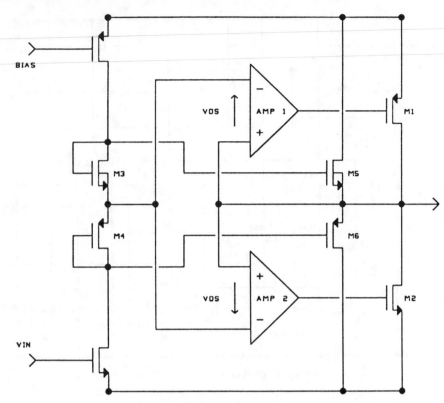

Fig. 3. Combined output stage.

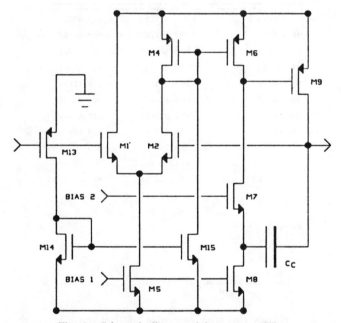

Fig. 4. Schematic diagram of the error amplifier.

standard RC type when used in the output stage because the zero in the transfer function is completely independent of $gm9$, which varies widely with output swing into a resistive load. This independence generally simplifies stabilizing the amplifier for varying load conditions. The transfer function of Amp1 with $M9$ is given in (7):

$$Av \approx \frac{a(1+Sb)}{1+Sc+S^2d+S^3e}$$

$$a \approx \frac{gm2\ gm6\ gm9}{2\ gm4\ g_{ds}6\ gL}$$

$$b \approx \frac{Cc+C_{gs}7}{gm7+gm_{bs}7}$$

$$c \approx \frac{CL+Cc\ \dfrac{gm9}{g_{ds}6}}{g1}$$

$$d \approx \frac{C_1(CL+Cc)}{g_{ds}6\ gL}$$

$$e \approx \frac{C_1\ Cc\ CL}{gm7\ g_{ds}6\ gL} \tag{7}$$

where

$$C_1 = C_{gs}9 + C_{db}6 + C_{db}7 + C_{gd}7. \tag{8}$$

As was the case for the input stage, $P2$ and $P3$ are not widely spaced and again normally form complex con-

result in positive gate drive to $M1$ and negative gate drive to $M2$ when the offsets of the error amplifiers are positive as labeled. These offsets do not affect the operation of the circuit for output voltages other than ground. In Fig. 4, the offset has been introduced by making $M2$ slightly wider than $M1$. A mirror image of this circuitry has been used within Amp2. Amp1 in conjunction with $M9$ can be viewed as a three-stage amplifier in unity gain with pole splitting compensation between the source of $M7$ and the output. This type of compensation has a large advantage over the

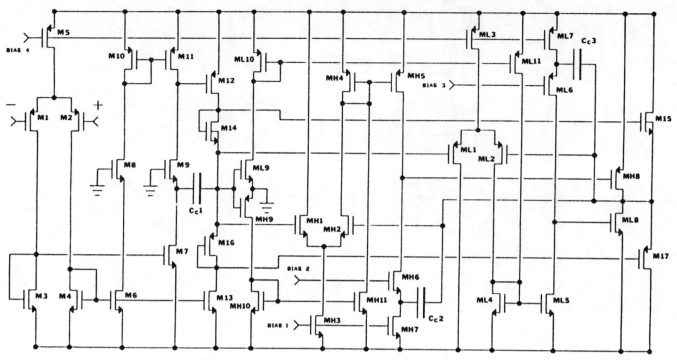

Fig. 5.   Schematic diagram of the complete power amplifier.

jugates. The open-loop poles and zero of Amp1 with $M9$ are given by

$$Z \approx -\frac{(gm7 + gm_{bs}7)}{Cc + C_{gs}7}$$

$$P1 \approx \frac{-gL}{CL + Cc\dfrac{gm9}{g_{ds}6}}$$

$$P2, P3 \approx \frac{-gm7\,(Cc + CL)}{2\,Cc\,CL}$$

$$\pm j\left[\frac{gm7\,g_{ds}6\left(CL + Cc\dfrac{gm9}{g_{ds}6}\right)}{Cc\,CL\,C_1}\right.$$

$$\left. - \left(\frac{gm7(Cc + CL)}{2\,CcCL}\right)^2\right]^{1/2}. \qquad (9)$$

Fig. 5 shows a complete schematic of the power amplifier circuit and a device size listing is given in Table I.

III.   Experimental Results

A die photograph of this amplifier is shown in Fig. 6. The amplifier was fabricated using a polysilicon gate CMOS process with an n+-implant to generate the capacitor bottom plate. The minimum geometry used in this circuit is 5 μm and the die area of the amplifier, excluding bonding pads, is 1000 mil². A summary of the amplifier characteristics is presented in Table II.

TABLE I
COMPONENT SIZES

M1 400/15	MH1 48/10	ML1 48/6
M2 400/15	MH2 50/10	ML2 50/6
M3 150/10	MH3 500/15	ML3 300/15
M4 150/10	MH4 300/6	ML4 150/5
M5 100/15	MH5 300/6	ML5 100/5
M6 150/10	MH6 200/5	ML6 300/6
M7 150/10	MH7 250/15	ML7 100/15
M8 300/5	MH8 700/6	ML8 400/5
M9 300/5	MH9 15/6	ML9 5/5
M10 300/10	MH10 10/15	ML10 5/15
M11 300/10	MH11 20/15	ML11 15/15
M12 1200/10	Cc1 20 pf	
M13 600/10	Cc2 4 pf	
M14 200/5	Cc3 4pf	
M15 200/5		
M16 600/6		
M17 600/6		

Fig. 7 shows the step response of the amplifier with a load of 200 Ω, 1000 pf using supplies of $\pm$ 5 V. Fig. 7(a) shows a large signal response of ±3.1 V and Fig. 7(b) shows a small signal response of ± 20 mV.

A power amplifier has been described which provides a high degree of performance from a standard polysilicon gate CMOS process. Several new circuit configurations have been incorporated.

Acknowledgment

The author wishes to thank W. C. Black, Jr. for many useful discussions on amplifier design.

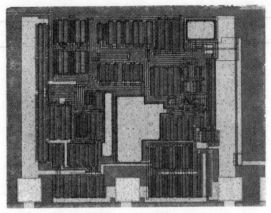

Fig. 6.   Die photograph of the power amplifier.

TABLE II
POWER AMPLIFIER PERFORMANCE SUMMARY
(First Revision)

Parameter	Measured Results
Supplies	±5 V
Open-Loop Gain	93 dB
Bandwidth	1.2 MHz
Power Dissipation $\bar{x}$	12.7 mW
σ	1.76 mW
Output Swing (R_L=200Ω)	±3.1 V
PSRR+ at DC	93 dB
1 kHz	91 dB
10 kHz	76 dB
100 kHz	60 dB
PSRR- at DC	102 dB
1 kHz	89 dP
10 kHz	75 dB
100 kHz	53 dB
Slew Rate	1.5 V/µs
Input Common Mode Range	+3.3 V
	-5.5 V
Die Area	1000 mils²
Harmonic Distortion (3 kHz) V_{in} = 3 V_p R_L = 200Ω	
HD2	-73 dB
HD3	-78 dB

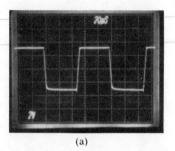

(a)

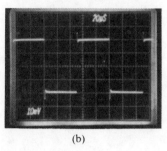

(b)

Fig. 7.   Step response. (a) Large signal. (b) Small signal.

REFERENCES

[1] R. D. Jolly and R. H. McCharles, "A low-noise amplifier for switched capacitor filters," *IEEE J. Solid-State Circuits*, vol. SC-17, pp. 1192–1194, Dec. 1982.

[2] D. J. Allstot and W. C. Black, Jr., "Technological design considerations for monolithic MOS switched-capacitor filtering systems," *Proc. IEEE*, vol. 71, pp. 967–986, Aug. 1983.

[3] B. K. Ahuja, "An improved frequency compensation technique for CMOS operational amplifiers," *IEEE J. Solid-State Circuits*, vol. SC-18, pp. 629–633, Dec. 1983.

[4] P. R. Gray and R. G. Meyer, "MOS operational amplifier design—A tutorial overview," *IEEE J. Solid-State Circuits*, vol. SC-17, pp. 969–982, Dec. 1982.

[5] W. C. Black, Jr., D. J. Allstot, and R. A. Reed, "A high performance low power CMOS channel filter," *IEEE J. Solid State Circuits*, vol. SC-15, pp. 921–929, Dec. 1980.

[6] K. E. Brehmer and J. B. Wieser, "Large swing CMOS power amplifier," *IEEE J. Solid-State Circuits*, vol. SC-18, pp. 624–629, Dec. 1983.

[7] B. K. Ahuja, W. M. Baxter, and P. R. Gray, "A programmable CMOS dual channel interface processor," in *Dig. Tech. Pap. Int. Solid-State Circuits Conf.*, Feb. 1984, pp. 232–233.

An Improved Frequency Compensation Technique for CMOS Operational Amplifiers

BHUPENDRA K. AHUJA

Abstract—The commonly used two-stage CMOS operational amplifier suffers from two basic performance limitations due to the RC compensation network around the second gain stage. First, this frequency compensation technique provides stable operation for limited range of capacitive loads, and second, the power supply rejection shows severe degradation above the open-loop pole frequency. The technique described here provides stable operation for a much larger range of capacitive loads, as well as much improved V_{BB} power supply rejection over very wide bandwidths for the same basic op amp circuit. This paper presents mathematical analysis of this new technique in terms of its frequency and noise characteristics followed by its implementation in all n-well CMOS process. Experimental results show 70 dB negative power supply rejection at 100 kHz and an input noise density of 58 nV/$\sqrt{\text{Hz}}$ at 1 kHz.

I. Introduction

LINEAR CMOS techniques have achieved significant progress over the last five years to provide high-performance low-power analog building blocks like opera-

Manuscript received July 11, 1983; revised August 23, 1983.
The author is with the Intel Corporation, Chandler, AZ 85224.

tional amplifiers (op amp), comparators, buffers, etc. These circuits have demonstrated comparable performance to their bipolar counterparts at much less silicon area and power dissipation, thus enabling single chip implementations of complex filtering functions, A/D and D/A conversions with quite stringent specification. Due to relatively simple circuit configurations and flexibility of design, CMOS technology has an edge over NMOS technology and is gaining rapid acceptance as the future technology for linear analog integrated circuits, especially in the telecommunication field [1], [2]. The most important building block in any analog IC is the op amp of which numerous implementations have been reported in both the technologies [3], [6].

The most commonly used op amp configuration in CMOS has two gain stages, the first one being the differential input stage with single-ended output, and the second one being either class A or class AB inverting output stage. Each stage typically is designed to have gain in the range of 40 to 100. Fig. 1(a) shows the circuit configuration while

Reprinted from *IEEE J. Solid-State Circuits*, vol. SC-18, no. 6, pp. 629–633, Dec. 1983.

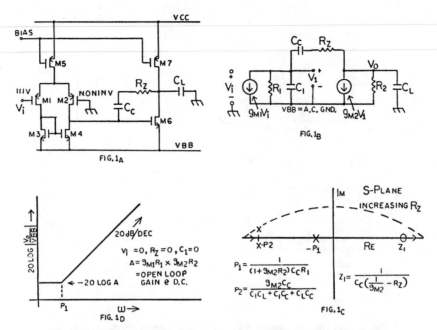

Fig. 1. (a) Commonly used two-stage operational amplifier. (b) Small signal equivalent model for the two-stage amplifier. (c) Pole-zero diagram of Fig. 1(b). (d) V_{BB} PSRR in unity gain configuration.

its first-order ac equivalent model is shown in Fig. 1(b). This configuration is most suitable for internal usage in the IC for driving capacitive loads only. Briefly, transistors $M1$ to $M5$ form the input differential stage and $M6$ and $M7$ form the output inverting gain stage. The series RC network across the second gain stage provides the frequency compensation for the op amp. This circuit, previously analyzed by many authors [5], [7], displays a dominant pole, two complex high frequency poles, and a zero which can be moved from the right half plane to the left half plane by increasing the compensating resistor value R_Z. This is pictorially shown in Fig. 1(c). Due to feedforward path with no inversion from the first stage output to the op amp output provided by the compensation capacitor at high frequencies, the op amp performance shows the following degradations:

1) The op amp stability is severely degraded for capacitive loads of the same order as compensation capacitor (C_L must be less than $g_{m2}C_c/g_{m1}$ to avoid second pole crossover of the unity gain frequency).

2) In case of p-channel MOS transistors for the input differential stage, the negative power supply displays a zero at the dominant pole frequency of the op amp in unity gain configuration. This results in serious performance degradation for sampled data systems which use high-frequency switching regulators to generate their power supplies. (In the case of n-channel MOS transistors for the input differential pair, it is the positive supply which shows similar

degradation.) This is illustrated in Fig. 1(d).

The circuit technique described in this paper overcomes both of these limitations. This technique has been referenced earlier [7] as a private communication by Read and Weiser [8]. This paper provides analysis, implementation, and experimental results on the realization in an n-well CMOS process.

II. IMPROVED FREQUENCY COMPENSATION TECHNIQUE

The technique is based on removing the feed forward-path from the first stage output to the op amp output. The circuit shown in Fig. 1 has a current $C_c d(V_0 - V_1)/dt$ flowing into the first-stage output. If one can devise a circuit where only $C_c dV_0/dt$ current flows into the first-stage output, one would have eliminated the feedforward path while still producing a dominant pole due to the Miller effect. The only difference is that Miller capacitance is now $A_2 C_c$ rather than $(1 + A_2)C_c$ where A_2 is the second-stage voltage gain. Thus, the conceptual ac equivalent of such a circuit is shown in Fig. 2(a). Here the compensation capacitor is shown to be connected between the output node and a virtual ground (or ac ground), while the controlled current source having the same value as $C_c dV_0/dt$ charges the first-stage output. It can be shown that for such an arrangement, the open-loop gain of the op amp is given by

$$A = \frac{-A_1 A_2}{1 + s(R_1 C_1 + R_2 C_L + R_2 C_c + A_2 R_1 C_c) + s^2 R_1 R_2 C_1 (C_c + C_L)}$$

where $A_1 = g_{m1} R_1 =$ dc gain of the first stage and

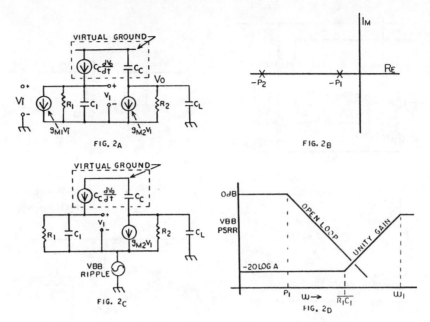

Fig. 2. (a) The new frequency compensation concept. (b) Resultant pole locations in s-plane. (c) Small signal model for V_{BB} PSRR analysis. (d) Expected V_{BB} PSRR frequency response of Fig. 1(a).

$$A_2 = g_{m2}R_2 = \text{dc gain of the second stage.} \quad (1)$$

Fig. 2(b) shows its pole-zero location. Notice that there is no finite zero in this circuit and that both the poles are real and are widely spaced.

$$P_1 \cong \frac{1}{(g_{m2}R_2)C_cR_1} \quad (2)$$

$$P_2 \cong \frac{g_{m2}C_c}{C_1(C_c+C_L)}. \quad (3)$$

Assuming the internal node capacitance C_1 being much smaller than the compensation capacitor C_c or the load capacitance C_L, the unity gain frequency W_1 is still given by g_{m1}/C_c. This results in

$$\frac{P_2}{W_1} = \frac{g_{m2}}{g_{m1}} \cdot \frac{C_c}{C_1} \cdot \frac{C_c}{(C_c+C_L)}. \quad (4)$$

Taking some typical design values of a two-stage amplifier as given by

$$g_{m2}/g_{m1} = 10, C_c = 5 \text{ pF}, C_1 = 0.5 \text{ pF, and } P2/W_1 \geqslant 5,$$

the new compensation technique can drive up to 100 pF capacitive load as compared to 10 pF capability of the commonly used RC technique as shown in Fig. 1. Thus, the new technique offers an order of magnitude improvement in capacitive load capability for the same performance. The improvement factor is given by C_c/C_1, where C_1 can be reduced by careful layout and design of the first stage.

Another major performance improvement is found in the negative power supply rejection characteristics. Fig. 2(c) shows the model for computing the open-loop negative power supply rejection with grounded inputs. It can be shown that open-loop V_{BB} PSRR is given by

$$\frac{V_0}{V_{BB}} =$$

$$\frac{1+sC_1R_1}{1+s[R_1C_1+R_2(C_c+C_L)+A_2R_1C_c]+s^2R_1R_2C_1(C_c+C_L)}$$

$$\cong \frac{1+sC_1R_1}{(1+s/P_1)(1+s/P_2)} \quad (5)$$

which indicates that is has the same poles as the open-loop gain and a zero which is created by the parasitic capacitance at the first-stage output. Thus, in a unity gain configuration, the V_{BB} PSRR is given by

$$\frac{V_0}{V_{BB}} =$$

$$\frac{1+sC_1R_1}{(1+s/P_1)(1+s/P_2)} \cdot \frac{1}{1+A_1A_2/(1+s/P_1)(1+s/P_2)}$$

$$\cong \frac{(1+sC_1R_1)}{A_1A_2(1+s/W_1)}. \quad (6)$$

This implies a flat response at $-20\log A_1A_2$, until the parasitic zero frequency of the first stage where it starts to degrade at 6 dB/octave rate and becomes flat again at unity gain frequency W_1. This is illustrated in Fig. 2(d).

III. A Circuit Implementation and Experimental Results

Although the above described scheme can be applied to any MOS amplifier design, it lends a relatively simple

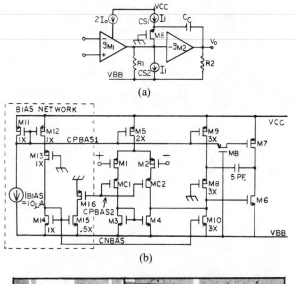

(a)

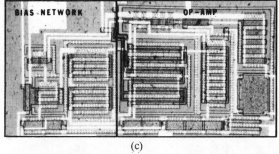

(b)

(c)

Fig. 3. (a) Implementation of the current transformer providing virtual ground. (b) Circuit schematic of the implemented amplifier. (c) Photomicrograph of the amplifier.

TABLE I
$V_{CC} = +5 \text{ V}$, $V_{BB} = -5 \text{ V}$, AND $T = 27\degree\text{C}$

Parameter	Measured Value
Open-Loop Gain	80 dB
Unity Gain Frequency	3.8 MHz
Phase Margin with $C_L = 15$ pF	70°
Input Common Mode Range	+4 to −2.5 V
CMRR at 1 kHz	−74 dB
Input noise density at	
1 kHz	58 nV/$\sqrt{\text{Hz}}$
100 kHz	8 nV/$\sqrt{\text{Hz}}$
C-msg Input Noise	−21 dBrnc
V_{CC} PSRR at	
1 kHz	−84.5 dB
10 kHZ	−84.5 dB
100 kHz	−73 dB
V_{BB} PSRR at	
1 kHz	−84 dB
10 kHz	−84 dB
100 kHz	−70 dB

implementation in CMOS technology. Fig. 3(a) shows an implementation of the current transformer providing a virtual or ac ground to the compensation capacitor, while still able to dump $C_c dV_0/dt$ current into the second-stage input. The current source $CS1$ biases the source of $M8$ at a fixed dc potential above ground, thus providing the ac ground for the compensation capacitor. By matching the $CS2$ value to $CS1$, all displacement current $C_c dV_0/dt$ flows into to or out of the first stage output.

Under large differential input conditions, the output can slew at a rate determined by the total input differential bias current $2I_0$, i.e.,

$$C_c \cdot \frac{dV_0}{dt} = \pm 2I_0. \qquad (7)$$

In order to keep the current transformer biased during the slewing intervals, one must make I_1 greater than $2I_0$. Also, the size of M_8 and the value of I_1 should be large enough to keep V_{GS} of M_8 relatively constant under worst-case slewing conditions.

Fig. 3(b) shows a circuit schematic of the implemented amplifier. The input differential stage, formed by $M1$ to $M5$ transistors, uses cascode devices $MC1$ and $MC2$ to reduce supply capacitance from the negative power supply for switched-capacitor applications [5]. The current transformer is being realized by $M8$, $M9$, and $M10$. Due to its

cascode configuration, this technique has been referred to as the "grounded gate cascode compensation" in [7]. The output stage is formed by $M6$ and $M7$. The transistor MB and the gate capacitance of the $M7$ transistor provide RC low-pass filtering of the high-frequency noise on the bias line $CPBAS1$. The associated bias circuit shown in the dotted box is shared among several such amplifiers, thus reducing power and area overhead cost due to this compensation technique. Fig. 3(c) shows the die photo of the amplifier. The amplifier has been designed in a 4 μm n-well CMOS process and occupies about a 165 mil² die area.

The input referred noise of this amplifier is slightly worse than the one shown in Fig. 1(a) due to the noise contributions from transistors $M9$, $M10$, $M12$, and $M14$. However, these contributions can be reduced significantly by choosing large values of channel lengths of these devices with respect to the channel lengths of input transistors $M1$ and $M2$ [3], [7].

Some of the measured performance parameters are listed in Table I. The op amp exhibits open-loop gain of 80 dB, unity gain frequency of 3.8 MHz, and a phase margin of 70° with 15 pF load capacitance. The V_{CC} and V_{BB} PSRR at low frequencies are better than −80 dB due to the bias circuit design and the cascode transistors $MC1$ and $MC2$, respectively. The V_{BB} PSRR shows zero at about 60 kHz, which closely matches the simulated value of the parasitic zero frequency. The op amp displays an input referred noise density of 58 and 8 nV/$\sqrt{\text{Hz}}$ at 1 and 100 kHz frequencies, respectively.

CONCLUSIONS

An improved frequency compensation technique has been described with a brief review of the existing techniques. A CMOS implementation of the technique has also been presented with experimental results which show considerable high-frequency power supply rejection improvement over the existing techniques which would result in approximately −30 to −35 dB V_{BB} PSRR at 100 kHz.

Furthermore, the technique provides extended capacitive drive capability for the same size of the compensation capacitor.

ACKNOWLEDGMENT

The author would like to thank Dr. P. Gray for technical discussions on the noise analysis of this compensation technique. Also, the technical assistance in the performance evaluation by T. Barnes is greatly appreciated.

REFERENCES

[1] R. Gregorian and G. Amir, "A single chip speech synthesizer using a switched-capacitor multiplier," *IEEE J. Solid-State Circuits*, vol. SC-18, pp. 65–75, Feb. 1983.

[2] B. K. Ahuja *et al.*, "A single chip CMOS PCM codec with filters," in *ISSCC Dig. Tech. Papers*, pp. 242–243, Feb. 1981.

[3] P. R. Gray, "Basic MOS operational amplifier design—An overview," in *Analog MOS Integrated Circuits*. New York: IEEE Press, 1980, pp. 28–49.

[4] D. Senderowicz, D. A. Hodges, and P. R. Gray, "A high performance NMOS operational amplifier," *IEEE J. Solid-State Circuits*, vol. SC-13, pp. 760–768, Dec. 1978.

[5] W. C. Black *et al.*, "A high performance low power CMOS channel filter," *IEEE J. Solid-State Circuits*, vol. SC-15, pp. 929–938, Dec. 1980.

[6] V. R. Saari, "Low power high drive CMOS operational amplifiers," *IEEE J. Solid-State Circuits*, vol. SC-18, pp. 121–127, Feb. 1983.

[7] P. R. Gray and R. G. Meyer, "MOS operational amplifier design—A tutorial overview," *IEEE J. Solid-State Circuits*, vol. SC-17, pp. 969–982, Dec. 1982.

[8] R. Read and J. Wieser, as referred in [7].

Design Considerations for a High-Performance 3-μm CMOS Analog Standard-Cell Library

CARLOS A. LABER, MEMBER, IEEE, CHOWDHURY F. RAHIM, MEMBER, IEEE, STEPHEN F. DREYER,
GREGORY T. UEHARA, MEMBER, IEEE, PETER T. KWOK, MEMBER, IEEE,
AND PAUL R. GRAY, FELLOW, IEEE

Abstract —Several design aspects of a high-performance analog cell library implemented in 3-μm CMOS are described, including an improved central biasing scheme, a new circuit for high-swing cascode biasing, an impact ionization shielding technique, and a family of operational transconductance amplifiers (OTA's) including a precision low offset-voltage amplifier utilizing lateral bipolar transistors.

I. INTRODUCTION

THE application of standard-cell methodology to the design of digital integrated circuits has greatly reduced their design time and associated engineering costs. While the same potential advantages are available in the mixed analog–digital domain, the application of standard-cell-based design methodology in the analog domain is more difficult because of the much wider range of applications encountered, the wider variety of types of cell functions required, and the tendency of analog blocks to interact with one another in a variety of ways, some difficult to predict *a priori*. This variety of different performance levels and types of functions required makes the systematic application of a standard-cell-based methodology to high-performance mixed analog and digital designs a challenging task.

This paper will describe several circuit design approaches intended to alleviate some of the problems mentioned above, and to improve circuit robustness in the presence of wide process parameter variations found in different silicon foundries. The techniques have been applied to a general-purpose CMOS analog cell library, described in general terms at the end of the paper. This library consists of a variety of analog and digital standard cells intended for application in IC's for high-performance telecommunications systems, precision data-acquisition and instrumentation systems, and general analog processing. In Section II, the overall characteristics of the cell

library are discussed. In Section III, an impact ionization shielding concept which allows the realization of 10-V circuitry using technologies which display significant impact ionization in the 10-V range of V_{ds} without encountering the degrading effects of impact ionization will be described. In Section IV, the approach taken to the optimum biasing of the active circuitry under the control of a central bias source is discussed, including an optimized central biasing scheme that incorporates circuitry to allow the accurate control of the bias points of the active circuitry in the presence of wide excursions in process parameters such as resistor sheet resistance. Also, an approach for optimum biasing high-swing cascode current sources so as to allow maximum possible signal swing in the active circuitry will be discussed.

In Section V, several basic analog cells are described including a low-input offset-voltage operational transconductance amplifier which utilizes lateral n-p-n bipolar transistors. Finally, a summary of the standard-cell library is given, as well as examples of actual integrated circuit implementations which make use of this library.

II. LIBRARY OBJECTIVES

The circuit design techniques described in this paper were dictated largely by the specific objectives the library was intended to address. A key objective of the cell library was to allow integration of telecommunication and data-acquisition subsystems with performance compatible with 12-bit linearity and 14-bit resolution in A/D interfaces and 90-dB dynamic range in analog signal paths, including switched-capacitor filters. This consideration dictated relatively high levels of performance in areas such as operational-amplifier input noise and input offset voltage, power supply rejection, and voltage reference drift. The requirement for high-linearity A/D interfaces was realized through the use of self-calibration techniques. The library also makes extensive use of differential circuitry in filters and amplifiers, so as to optimize power supply rejection and dynamic range. Single-ended structures are also utilized for less demanding applications.

A second objective was that the functions to be implemented include operating voltages from 4.5 to 11 V, split or single supplies, and temperature ranges of from -55 to

Manuscript received September 9, 1986; revised January 5, 1987.

C. A. Laber, C. F. Rahim, and S. F. Dreyer are with Micro Linear Corporation, San Jose, CA 95131.

G. T. Uehara was with Micro Linear Corporation, San Jose, CA 95131. He is now with the Department of Electrical Engineering and Computer Sciences, University of California, Berkeley, CA 94720.

P. T. Kwok was with Micro Linear Corporation, San Jose, CA 95131. He is now with Exar Corporation, Sunnyvale, CA 94088.

P. R. Gray is with the Department of Electrical Engineering and Computer Sciences, University of California, Berkeley, CA 94720.

IEEE Log Number 8613379.

Reprinted from *IEEE J. Solid-State Circuits*, vol. SC-22, no. 2, pp. 181–189, Apr. 1987.

+125C. They must be sufficiently insensitive to process variations such that the same circuit can be fabricated on different foundry fabrication lines with similar but not equal process parameters utilized. As discussed later, an important aspect in achieving this goal is the provision for accurate control of the active-circuitry bias currents in the presence of variations in supplies, temperature, and process. Good control of the bias points is also important in order to achieve an optimum speed, power, and swing trade-off for a given application, particularly when extensive use is made of folded cascode amplifiers as is the case here.

Finally, due to the wide spectrum of requirements on speed, power dissipation, dynamic range, complexity, and cost which are encountered in practice, a set of cells is required for each function which encompasses a spectrum of area/performance and power/performance trade-offs so as to economically address a wide range of applications. This dictates a design approach in which cells can be scaled through bias current modification and other means to easily achieve different levels of speed and dynamic range with near-optimum power dissipation and area.

III. IMPACT IONIZATION SHIELDING

Impact ionization is a severe problem in scaled MOS technologies operating at supplies voltages above 5 V. In a typical n-channel transistor, illustrated for reference in Fig. 1, as the drain–source voltage is increased the electric field strength at the drain end of the channel eventually becomes high enough to induce significant impact ionization currents originating in the drain depletion region. The magnitude of the peak field for a given bias point is a function of gate oxide thickness, drain junction depth, doping concentration in the substrate, the voltage between the drain terminal and the drain end of the channel region, and the gate–drain voltage. It is not a strong function of channel length and the magnitude of the impact ionization current is not dramatically reduced by simply making the channel length longer. For technologies with feature sizes in the 2–3-μm range which use a nongraded implanted arsenic source–drain region for the n-channel device, the condition at which the substrate current equals 1 percent of the drain current typically occurs at voltages between the drain and the drain end of the channel of between 4 and 9 V, assuming the device is biased in saturation with a V_{gs}-V_t of a few hundred millivolts. In p-channel devices the effect occurs at substantially higher field strengths.

Operation in the impact ionization mode has several undesirable effects from a circuit standpoint. One potentially catastrophic consequence is the inducement of latch-up due to the ohmic drops induced by the ionization current. Assuming this can be controlled by proper strapping, another negative consequence is the degradation of impedances at the output of cascode current sources, as used, for example, in folded cascode amplifiers which rely on high output impedances to function properly. Also, the presence of a significant amount of substrate current in-

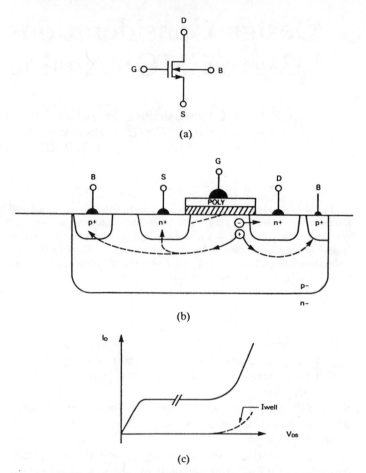

Fig. 1. (a) N-channel device. (b) N-channel device cross section showing impact ionization current flow. (c) I–V characteristic observed as a result of impact ionization.

creases the noise of the device. A third serious potential consequence of continuous operation in the impact ionization mode is threshold shift. If the electric field is high enough, high-energy carriers can be created which can be trapped in the gate oxide, and the resulting long-term threshold shift can have the effect of debiasing the active circuitry if the threshold shift occurs in a device whose threshold voltage is important in setting up bias currents [2].

A number of process modifications have been proposed and implemented to alleviate this problem by lowering the impurity gradient in the drain junction using lightly doped drain (LDD) structures. These structures are effective at raising the voltage at which impact ionization becomes a problem. However, major modifications are required for a typical 2–3-μm digital CMOS technology to completely eliminate impact ionization from the n-channel transistor for values of V_{ds}-$V_{d\,sat}$ of 11 V, as required in 10-V $\pm$ 10-percent supply systems. Since in this case the objective was to realize A/D mixed functions operating on 10 V with near-standard foundry digital CMOS technology, a circuit solution to the impact ionization problem was preferable to extensive technology modification. A design philosophy was adopted in which all 10-V circuitry would be designed in such a way that no n-channel transistor would experience a drain–source voltage larger than one-half the power

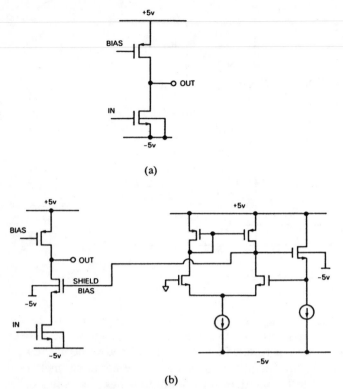

Fig. 2. (a) Conventional CMOS gain stage. (b) Same as (a) but with impact ionization shielding device and associated bias circuit.

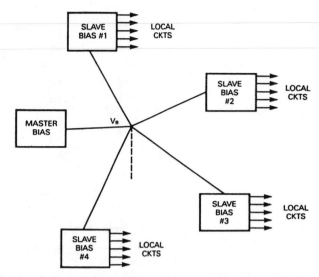

Fig. 3. Central biasing configuration with master bias supplying bias reference to a number of slave bias circuits placed around the chip.

supply voltage, except for the case of transmission gates carrying only transient displacement currents. This was achieved by inserting shielding devices in series with each common-source n-channel transistor.

In a conventional CMOS gain stage, illustrated in Fig. 2, the n-channel active device experiences impact ionization for positive output swings, as the output gets closer to the positive rail. One way to deal with this problem is to place a shield n-channel device in series with the driver, as shown in Fig. 2(b). For positive swings, the shield transistor goes into saturation, preventing either n-channel device from having a V_{ds} greater than approximately one-half of the supply. For negative swings, the shield transistor enters the deep triode region, without significantly affecting the performance of the circuit.

The shield transistor must be biased in such a way that its source resides approximately at ground potential. Grounding the gate of this transistor would be simplest, but because of the large body effect in some CMOS technologies, this would result in large V_{ds} drops across the shield transistor itself. Instead, a dedicated bias circuit is used to more precisely set the source of the shield transistor at ground. This is accomplished by the circuit on the right of Fig. 2(b), which contains a shield mirror transistor, whose source is forced by the feedback loop to a voltage which is approximately ground (or the midpoint potential for single supply systems). This bias scheme is independent of the body effect of the technology, because of the feedback action around the mirror device.

The shield device was incorporated in all 10-V circuitry in the library at an area cost of about 10 percent. The

shield bias voltage is contained in the standard bias bus and is routed to all analog cells. In a total of ten products designed for 10-V operation and fabricated with the library, no performance or reliability problems attributable to impact ionization have been encountered.

IV. CELL BIAS TECHNIQUE

One critical element in achieving high-performance CMOS analog circuits is the accurate control of the bias currents of the different transistors employed in a given circuit topology. Excessive variation of bias currents with temperature and process tends to sacrifice speed at the low extreme of bias current, and power dissipation and output voltage swing at the high end. Furthermore, the variation of bias currents with supply voltage results in poor power supply rejection (PSR). Achieving the goal of minimizing the dependence of bias points on supply voltage, temperature, and process variations requires a bias circuit of some complexity, and one which is therefore uneconomical to implement on a per-cell basis. Therefore a central biasing scheme, as illustrated in Fig. 3, was adopted for this cell library. A central master bias circuit, described below, produces a voltage which is distributed around the chip. This voltage is used by slave bias cells, which generate the locally required bias voltages, in order to power the nearby circuitry.

The main advantage of this approach is the flexibility of the architecture, which allows different slave bias cells to work at different current levels, as required in certain applications. This scheme also helps in preventing cross-talk between critical circuits through the bias lines, as can occur when a bias line sharing approach is employed. The device which generates the output control master bias voltage is designed with a large $V_{d\,sat}$, so that small voltage gradients across the chip in the supply lines do not significantly affect the resulting current.

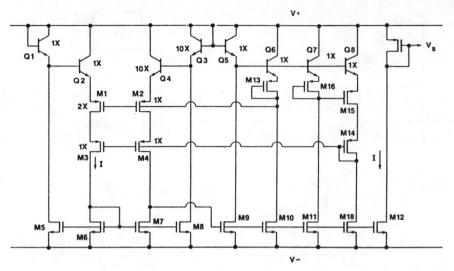

Fig. 4. Simplified schematic of the master bias circuit.

In critical circuits where tight limits on power dissipation or speed require a tighter control on bias current, the master bias cell allows the use of polysilicon fuse trimming, giving the desired result at the cost of trim pad area. Another advantage of central biasing is that full-chip power down can be accommodated, assuming that the slave bias cells and active circuits are properly designed to avoid floating-gate-induced parasitic paths in the power-down state.

A. Master Bias Implementation

A simplified schematic of the master bias cell is shown in Fig. 4. This is a self-biased circuit whose output current has been optimized for best temperature coefficient, absolute tolerance, and power supply rejection. The core of the circuit is the vertical bipolar transistors $Q1–Q4$. Since they are forced to conduct the same amount of current, by the bottom n-channel current mirrors $M5–M8$, a voltage difference is developed between the emitters of $Q2$ and $Q4$, which is approximately equal to $2(kT/q)\ln(10)$, or about 120 mV at room temperature. When this voltage is applied across p-channel devices $M1$ and $M2$, which are biased by the same gate voltage, it is easy to show that a current is generated whose value is given to a first order by

$$I = \frac{2\mu C_{ox}\left(\frac{W}{L}\right)_2\left[\left(\frac{kT}{q}\right)\ln 10\right]^2}{\left[1 - \sqrt{\frac{(W/L)_2}{(W/L)_1}}\right]^2}.$$

Taking into account the effects of tolerance on oxide thickness and mobility, as well as the effects of offset voltages in the MOS and bipolar transistors, this results in an absolute room-temperature tolerance on bias current which is somewhat better than for the more typical $(kT/q)/R$ based current reference for most technologies,

where R is a poly, source–drain diffusion, or well resistor. More importantly, the process dependence of the derived bias current results in a very well-defined value of the $V_{d\,sat}$ of the MOS transistors in the active circuitry, making it possible to maintain high-voltage swings over wide variations in process parameters. Further, since the channel mobility of the MOS device has an approximate $T^{-3/2}$ temperature dependence, where T is the absolute temperature, the result is a bias current with a net residual $T^{+1/2}$ temperature coefficient (TC). This is a very desirable feature, since in amplifiers biased by this current the transconductance of saturated MOS device is to a first order, inversely proportional to the square root of absolute temperature. The net consequence is that the duration of the slewing portion of the transient response of those active analog circuits exhibits a small positive TC, whereas the duration of the small-signal settling transient portion has a small negative TC. As a result the overall settling time behavior of the active circuitry does not display severe degradation at temperature extremes.

The right-hand part of the circuit shown in Fig. 4 is responsible for the high-swing cascode biasing of transistors $M1–M4$, and also the negative feedback which insures proper setting of the voltage at the drain of $M7$, which is the only high-impedance node in the circuit. The output current is finally forced to flow through diode-connected p-channel transistor $M17$, which produces the output master bias voltage V_b. As mentioned above, $M17$ has a large $V_{d\,sat}$ to absorb any difference in the threshold voltage between $M17$ and the slave bias mirror device, as well as V_{DD} drops across the chip. The high-swing cascode biasing scheme illustrated in Fig. 4 is the same as the one used in the slave bias cell, and is discussed below. Not shown in Fig. 4 are the start-up circuit which prevents the zero-current state at power up, the impact ionization shield devices and shield bias generator for 10-V operation, and cascode transistors on the n-channel current sources. Table I summarizes the main performance parameters of the master bias cell.

TABLE I
TYPICAL MASTER BIAS CIRCUIT
PERFORMANCE (ROOM TEMPERATURE)

Parameter	value
Output bias current	13.1uA
Output current standard deviation	1.3UA
Output current temperature coefficient	+1700ppm/deg C
PSRR	1500ppm/volt
Supply range	4.5-11 volts
Power dissipation	2mW@10V

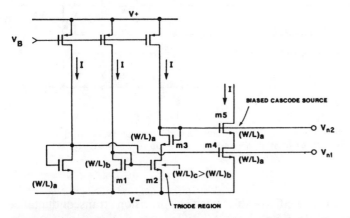

Fig. 5.   Simplified schematic of the slave bias circuit. The bias lines V_{n1} and V_{n2} are used to supply the high-swing cascode current sources in the local active circuitry. A similar circuit is used to supply the p-channel current courses.

B. Slave Bias Implementation

As mentioned earlier, local slave bias cells are used in conjunction with each local group of circuits to produce the much smaller $V_{d\,\text{sat}}$'s required to bias the gates of transistor current sources within each active cell. The slave bias cell produces bias voltages for p- and n-channel current sources, and the proper voltages to optimally bias the cascode devices in series with these current sources as used in, for example, folded cascode operational amplifiers.

Within the various cells of the library, cascode current sources are used extensively to generate high-impedance current source loads. Because of the requirement that the cells operate on a single 5-V supply, it is important that these cascodes be biased for optimum voltage swing [3]. The generation of high-swing bias voltages can be accomplished in many ways. However, previous reported techniques [4] have suffered from a strong body-effect sensitivity and poor control of the absolute tolerance on the current-source current value.

A simplified schematic of the slave bias cell is shown in Fig. 5. The basic concept utilized in this circuit is to force device $M2$ into the triode region by making its aspect ratio greater than that of transistor $M1$, since both $M1$ and $M2$ conduct the same amount of current. The V_{ds} of $M2$ then

takes a value, neglecting second-order effects, of

$$V_{ds2} = \left\{ \sqrt{\frac{2I}{\mu C_{ox}(W/L)_1}} \right\} \left[1 - \sqrt{1 - \frac{(W/L)_1}{(W/L)_2}} \right]$$

or, in terms of the $V_{d\text{sat}}$ of $M1$

$$V_{ds2} = V_{d\,\text{sat}1} \left[1 - \sqrt{1 - \frac{(W/L)_1}{(W/L)_2}} \right].$$

Assuming that devices $M3$ and $M5$ are sized so that they have equal gate–source voltages, the drain–source voltage of $M4$ will be equal to the drain–source voltage of $M2$, given above. Thus by choosing device ratios, the V_{ds} of $M4$ can be chosen to be an arbitrary multiple, usually on the order of 1.5, of its $V_{d\,\text{sat}}$, independent of process parameters, by means of transistors $M1$ and $M2$. The margin of drain–source voltage over $V_{d\,\text{sat}}$ required to insure operation fully in the saturation region with accompanying output conductance is in the 200-mV range. Another feature of this circuit is that when the master bias described earlier is used to generate the input current I in Fig. 5, then $V_{d\,\text{sat}}$ and thus the V_{ds} of $M2$ become approximately independent of process variations, and only proportional to absolute temperature. The body-effect sensitivity is avoided because, to a first order, the body effect of $M3$ cancels that of $M5$, resulting in a voltage across $M4$ which is approximately independent of gamma.

V.   BASIC ANALOG ACTIVE CELLS

Most low- and mid-frequency active circuitry is implemented using a family of folded cascode operational transconductance amplifiers (OTA's) which can drive internal capacitive loads and a family of class AB unity-gain buffers for driving on-chip resistive loads and for driving signals off-chip. The OTA's include an n-channel input single-ended amplifier, a p-channel input single-ended amplifier, a bipolar input single-ended cell, and a differential output amplifier. Simplified circuits for the NMOS and PMOS input amplifiers are shown in Fig. 6. The NMOS input OTA includes cascode devices in the input stage and has the input devices in a well so as to optimize power supply rejection in single-ended switched-capacitor filter applications. The bipolar input amplifier is discussed below.

The OTA configuration was chosen as the basic active element because of the flexibility of these circuits and the ease with which they can be combined into more complex blocks. For example, in applications requiring very high voltage gain, two OTA's can be cascaded and compensated with a single pole-splitting capacitor and nulling resistor between the two high-impedance nodes. For applications requiring higher transconductance or lower $1/f$ noise and thermal noise, OTA's can be connected directly in parallel, unlike two-stage operational amplifiers. This capability

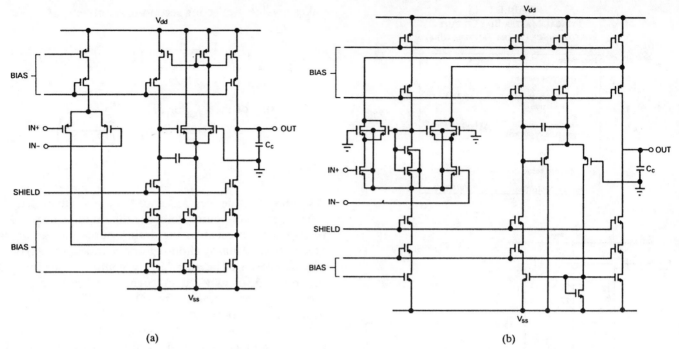

(a)   (b)

Fig. 6.   Schematic diagram of the (a) PMOS and (b) NMOS input OTA's. The auxiliary differential pairs in the differential-to-single-ended converter balance the output stage voltages so as to achieve a small value of systematic offset voltage.

also permits the realization of a rail-to-rail common-mode capability by paralleling an NMOS input OTA with a PMOS input OTA. The large phase margin of the structure allows it to be used as the front-end gain stage in composite off-chip driver operational amplifiers in which a unity feedback path is closed from output to input. Another important advantage for the wide spectrum of requirements encountered in the custom environment is the fact that the compensation capacitor can be connected to whatever potential is being used as the signal reference, giving good high-frequency power supply rejection. Depending on the application, this can be either supply or a separate ground.

These elements also have great flexibility for use in conjunction with nonlinear feedback elements to perform such functions as peak detectors and comparators. In such applications the compensation capacitor can often be removed completely to maximize the speed of operation. In all applications, the bandwidth and transient response of the block can be improved at the cost of power dissipation and voltage swing by scaling up the bias current using the slave bias approach described earlier.

The MOS amplifiers described above display the broad input offset-voltage distribution typical of MOS differential amplifiers. While switched-capacitor offset cancellation techniques can be used to remove system offsets in many cases, the need often arises for continuous amplifiers with input dc offset voltages which are smaller than those achievable in MOS amplifiers, and also which display the PTAT drift characteristic found in bipolar amplifiers. The latter characteristic is important in bandgap references, for example. This is best satisfied with a bipolar input stage,

and to satisfy this need a precision transconductance amplifier using a combination of lateral and vertical transistors was designed. This cell is then used as an element of more complex blocks such as references and instrumentation amplifiers that require low offset or PTAT drift.

This circuit, which is illustrated in simplified form in Fig. 7, uses the lateral transistor structure previously described by Degrauwe et al. [5]. In a p-well CMOS technology, this device is the lateral n-p-n transistor formed by the source and drain diffusions of an NMOS transistor in a p^- well. The well terminal becomes the base and the source and drain diffusions the emitter and the collector.

The drawback of this device is the inevitable presence of the parasitic n-p-n vertical transistor, which has a saturation current from 4 to 20 times larger than the lateral device. To eliminate the effects of this excess current, a biasing scheme was utilized as shown in Fig. 7, which insures that the lateral transistor maintains a well-controlled collector current, and hence transconductance, even in the presence of large variations in lateral-to-vertical current ratio. This is achieved by placing a dummy transistor Q_1 in a feedback loop, which forces an emitter current I_e, such that the lateral collector current of Q_1 is equal to the desired value I. The current I_e becomes the tail current of the input stage which biases devices Q_3-Q_5. Because of this and assuming that Q_1 and Q_3-Q_5 are matched, the lateral current of transistors Q_3-Q_5 is approximately given by I. Notice that this is independent of the lateral-to-vertical current ratio, and also that this causes the actual tail current of the input stage to vary over a 4- or 5-to-1 ratio while keeping the collector currents constant.

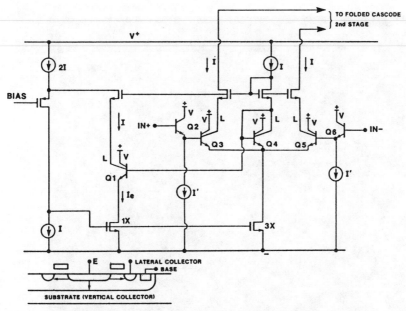

Fig. 7. Simplified schematic of the lateral bipolar input stage. The lateral transistor is shown in cross section at the bottom of the drawing. The gate overlying the lateral base region is tied to the negative power supply.

TABLE II
OTA TYPICAL PERFORMANCE (ROOM TEMPERATURE, ±5-V SUPPLIES)

Parameter	PMOS OTA	NMOS OTA	Bipolar OTA	units
Small-signal transconductance	80	90	200	uA/V
Max output current	10	10	10	uA
Open loop gain	40K	40K	50K	V/V
Unity-gain bandwidth C_L=3pF	2.0	1.8	2.0	Mhz
Output swing	0.6	0.6	0.6	Volts from supply
Input offset volt. (std deviation)	3	3	1	mV
Input bias current	neg	neg	2	na
Equiv. input noise (1khz)	75	75	20	$\frac{nV}{\sqrt{hz}}$

Another important factor in the design is that the vertical-to-lateral current ratio is a strong function of the collector–emitter voltage V_{ce} of the vertical and lateral devices because of the high Early effect of these transistors. For this reason, the dummy device Q_1 is biased off the common-mode point of the input stage, so as to keep the V_{ce}'s approximately the same as that of Q_3–Q_5. This is particularly important, for example, in voltage followers where the input common-mode voltage can be quite high. Transistor Q_4 is used to derive the bias voltages for the cascode n-channel devices, which are used to avoid the degradation in common-mode rejection ratio (CMRR) caused by the poor Early voltage of these bipolar devices. Transistors Q_2 and Q_6 are used to decrease the input bias current due to the low beta of the laterals (typically 50). The remainder of the amplifier is a folded-cascode second stage in which the p- and n-channel current source devices

have been made large in geometry, interdigitated, and operated at relatively large $V_{d\,sat}$ in order to reduce their contribution to input offset voltage. This is also true of the NMOS current sources which bias the input transistors Q_2 and Q_6.

Over a large sample of units, the input offset voltage of this amplifier displays a mean value of 0.25 mV, and a standard deviation of 1 mV in a sample taken from many wafers. The typical input bias current is 2 nA. The performance parameters of all three OTA's are summarized in Table II.

For driving off-chip loads and resistive on-chip loads, four class AB unity-gain output stages are provided which are capable of driving from 10 kΩ down to 300 Ω with full-swing signals, and lower resistive loads with corresponding lower output swing. A complete power operational amplifier can be assembled by combining one of these output stages with one of the amplifiers mentioned above. The configuration used is the conventional composite-device common-source class B [4].

Additional building blocks include a one-pin crystal oscillator, a voltage control oscillator (VCO) capable of operating with a center frequency of up to 5 MHz, and two bandgap references with different area–performance trade-offs. The precision reference is oriented toward high-performance applications and is polysilicon fuse trimmable to 0.1 percent with a typical tempco of 20 ppm/C at room temperature. This cell is described in more detail elsewhere [7]. In addition, the library contains a large number of utility functions such as 5–10-V logic level shifters, analog switches, analog multiplexers, clock generators for switched capacitor filters, and so forth. Finally, these analog cells are supplemented with a conventional family of 75 digital cells which implement most common logic functions.

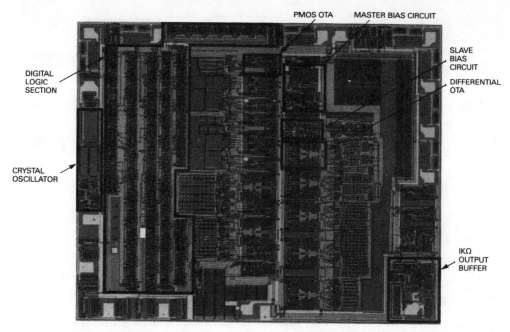

PMOS OTA   MASTER BIAS CIRCUIT

SLAVE BIAS CIRCUIT

DIFFERENTIAL OTA

DIGITAL LOGIC SECTION

CRYSTAL OSCILLATOR

IKΩ OUTPUT BUFFER

Fig. 8.   Tone signaling chip example. The chip contains 20 operational amplifiers and 5000 gates of digital logic. The die size is 25 200 mils^2 (16.3 mm^2).

From a geometric viewpoint, internal cells containing active circuitry are laid out at constant height of 500 μm with standard placement of power supply, ground, and bias lines running along the top and bottom interior of the cells so that these lines are automatically routed when the cells are placed abutting each other. I/O cells such as analog output buffers are configured so that they can lie directly on the chip periphery and incorporate pads. Passive components and random transistor level circuitry are hand placed and interconnected in the areas above and below the assemblages of active cells, although the cell layouts are compatible with the ultimate use of analog place and route packages and block compilers. In a typical application the cells are combined into blocks as just described, and these blocks together with automatically placed and routed digital blocks are assembled and interconnected to compose the complete chip.

VI.   HIGHER LEVEL ANALOG FUNCTIONAL BLOCKS

More complex functions such as filters and A/D converters are made up from these basic cells together with additional passive and active circuitry. In order to address a wide spectrum of applications, a very large number of the higher level cells is required in order to adequately cover the spectrum of performance at economical die area and cost. Thus the number of cells in a practical library will continually expand over time as the spectrum of applications expand. Initially, the D/A conversion function is provided in this library by a conventional untrimmed current switched segmented 300-ns DAC cell at the 8-bit level. At the 12-bit level a self-calibrated cyclic D/A cell is utilized. The details of operation of this cell have been described elsewhere [7]. For the A/D function,

the same algorithmic self-calibrating cell is used at the 12-bit level, giving a conversion time of 25 μs. This cell incorporates both the sample/hold function as well as a programmable gain function. For the 8-bit level, a two-step flash 1.5-μs cell is used with no trimming. These are not fixed-height internal cells but are placed as large blocks in the chip layout and manually interconnected.

Switched-capacitor filters are implemented using internal fixed-height operational amplifier, bias, and clock generator cells together with manual placement and interconnect of capacitor elements. Both single-ended and fully differential filter architectures can be accommodated with this approach.

A.   Example

The application of the cell library described here is illustrated by the chip shown in Fig. 8. This integrated circuit is used for tone signaling on trunk circuits and contains switched-capacitor filters, peak detectors, timers, and so forth for performing precision voice-band tone detection. In addition it provides an audio speech-path filtering function. Various of the cells mentioned above are used on this chip and are indicated in the photograph. Most of the digital cells are concentrated on one side of the die and were manually placed and routed in this case, although automatic place and route has been subsequently used. The analog blocks implemented in the rest of this chip include continuous active *RC* filters, differential switched-capacitor filters, discrete-time gain blocks, tone detectors, and so forth. Cells included in this die are the master bias, slave bias, differential OTA, single-ended operational amplifiers with resistive drive capability, comparators, one-pin crystal oscillator, and a number of smaller cells.

A second example of the application of the cell library is the telephone trunk equalizer, discussed elsewhere in this issue [6], which illustrates the dynamic range capability of the library.

REFERENCES

[1] C. Laber, C. Rahim, S. Dreyer, G. Uehara, P. Kwok, and P. R. Gray, "A high-performance 3μ CMOS analog standard cell library," in *Proc. IEEE 1986 Custom Integrated Circuits Conf.* (Rochester, NY), May 1986, pp. 21–24.

[2] C. Hu, "Hot electron effects in MOSFETS," in *IEDM Tech. Dig.*, 1983.

[3] T. Choi *et al.*, "High-frequency CMOS switched-capacitor filters for communications applications," *IEEE J. Solid-State Circuits*, vol. SC-18, pp. 652–665, Dec. 1983.

[4] B. K. Ahuja *et al.*, "A programmable CMOS dual interface processor for telecommunications applications," *IEEE J. Solid-State Circuits*, vol. SC-19, pp. 892–899, Dec. 1984.

[5] M. Degrauwe, E. Vittoz, and H. Oguey, "A family of CMOS compatible bandgap references," in *Dig. Tech. Papers, IEEE Int. Solid-State Circuits Conf.* (New York, NY), Feb. 1985.

[6] C. F. Rahim, C. A. Laber, B. L. Pickett, and F. J. Baechtold, "A high-performance custom standard-cell CMOS equalizer for telecommunications applications," *IEEE J. Solid-State Circuits*, vol. SC-22, no. 2, pp. 174–180, Apr. 1987.

[7] M. Armstrong, H. Ohara, H. Ngho, and P. R. Gray, "A self-calibrating CMOS 13-bit self-calibrating analog interface processor," in *Dig. Tech. Papers, IEEE Int. Solid-State Circuits Conf.* (New York, NY), Feb. 1987.

A MOS Switched-Capacitor Instrumentation Amplifier

ROBERT C. YEN AND PAUL R. GRAY, FELLOW, IEEE

Abstract—This paper describes a precision switched-capacitor sampled-data instrumentation amplifier using NMOS polysilicon gate technology. It is intended for use as a sample-and-hold amplifier for low level signals in data acquisition systems. The use of double correlated sampling technique achieves high power supply rejection, low dc offset, and low $1/f$ noise voltage. Matched circuit components in a differential configuration minimize errors from switch channel charge injection. Very high common mode rejection (120 dB) is obtained by a new sampling technique which prevents the common mode signal from entering the amplifier. This amplifier achieves 1 mV typical input cffset voltage, greater than 95 dB PSRR, 0.15 percent gain accuracy, 0.01 percent gain linearity, and an rms input referred noise voltage of 30 μV/input sample.

Manuscript received April 27, 1982; revised June 7, 1982.

This work was sponsored by the Xerox Corporation and the National Science Foundation under Grant ENG78-11397.

R. C. Yen was with the Department of Electrical Engineering and Computer Sciences and the Electronics Research Laboratory, University of California, Berkeley, CA 94720. He is now with the Hewlett-Packard Laboratory, Hewlett-Packard, Inc., Palo Alto, CA 94304.

P. R. Gray is with the Department of Electrical Engineering and Computer Sciences and the Electronic Research Laboratory, University of California, Berkeley, CA 94720.

I. INTRODUCTION

ANALOG data acquisition systems often need to perform analog-to-digital conversion on signals of very small amplitude or signals superimposed on large common mode components. This problem is traditionally solved by using fixed gain differential amplifier implemented as a stand-alone component in bipolar technology [1], [2]. However, MOS technology is increasingly being utilized to implement monolithic data acquisition systems, either as a stand-alone component or as part of a control-oriented microcomputer or signal processor [3].

A typical data acquisition system generally consists of an input amplifier, sample-and-hold stage, and an A-to-D converter. The amplifier serves to increase the signal level prior to analog-to-digital conversion. Input offset voltage is a key aspect of amplifier performance since it can limit dc system accuracy. In some systems, it is possible to measure and subtract the dc offset, but the equivalent input noise voltage represents a fundamental limit on the resolution of the system.

Reprinted from *IEEE J. Solid-State Circuits*, vol. SC-17, no. 6, pp. 1008–1013, Dec. 1982.

Also, gain accuracy and gain linearity are critical parameters for instrumentation applications. In some cases, a low-amplitude signal input is superimposed on large common mode components due to electrostatic or electromagnetic induction. This adds a requirement for high common mode rejection. The data acquisition circuit may reside on the same chip as the digital LSI processor; therefore, the ability to reject power supply noise is also very important.

Compared to bipolar devices, MOS transistors display smaller transconductance at a given drain current level. This makes it difficult to achieve large values of voltage gain in a single MOS amplifier stage, and also results in high dc offsets in source coupled pairs. Also, the MOS devices inherently displays much larger $1/f$ noise than bipolar devices.

This paper describes a switched-capacitor circuit technique for the implementation of the instrumentation amplifier and sample/hold function in MOS technology. Double correlated sampling [4] is used to reduce the circuit dc offset and low-frequency noise, and a balanced circuit configuration is used to achieve first-order cancellation of switch channel charge injection. A charge redistribution scheme is described in this paper which allows the circuit CMRR to be independent of the op amp CMRR, thus resulting in a very high overall common mode rejection ratio.

In Section II, a circuit approach to implement the sample-and-hold instrumentation amplifier is described. The prototype implementation of this circuit using NMOS technology is depicted in Section III. The switch channel charge injection problems are addressed in Section IV, and the fundamental noise limitation of kT/C noise is also discussed in Section V. Finally, in Section VI, the experimental results of this circuit fabricated using local oxidation NMOS polysilicon gate technology are presented.

II. Circuit Description

The MOS implementation of the differential double-correlated sampling amplifier is shown in Fig. 1. This circuit consists of a pair of sampling capacitors $C1$, $C2$; gain setting capacitors $C3$, $C4$; offset cancellation capacitors $C5$, $C6$; and two differential amplifiers $A1$ and $A2$ where amplifier $A1$ is a broad-band low-gain differential preamplifier and $A2$ is a high-gain differential operational amplifier. The input signal is sampled on to the sampling capacitors $C1$, $C2$, and subsequently transferred to the gain setting capacitors $C3$, $C4$ through a sequence of switching operations. The output voltage will be a replica of the input differential signal with a voltage gain defined by the capacitor ratio $C1/C3$ if capacitor $C1$ matches $C2$ and $C3$ matches $C4$. The circuit is fully differential so that all the switch charge injection and power supply variations are cancelled to the first order.

Operation of the circuit takes place in two phases as illustrated in Fig. 2. In the sample mode, the switches are closed as shown in Fig. 2(a). In this mode, the differential and common mode input voltages appear across both $C1$ and $C2$. The difference between the offset voltages $A1$ and $A2$ is impressed across $C5$ and $C6$. The instantaneous value of the $1/f$ noise of both amplifiers is also stored. A requirement on this amplifier, $A1$, is that its gain be low enough so that its output does not saturate on its own offset when the inputs are shorted. The input signal is sampled and a transition to the hold

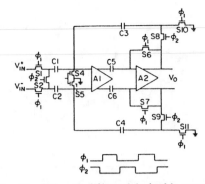

Fig. 1. Circuit schematic of differential double-correlated sampling instrumentation amplifier.

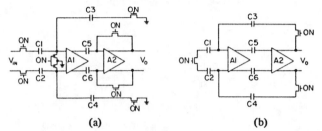

(a) (b)

Fig. 2. Illustration of the switching sequence of the instrumentation amplifier.

mode is made when clock one goes negative, turning off the input sampling switches and feedback switches. Subsequently, the switches are closed as shown in Fig. 2(b), and the voltage difference between the two inputs is forced to zero. This causes a charge redistribution in capacitors $C1$, $C2$, $C3$, and $C4$, which results in an output voltage which is only proportional to the input difference voltage. Any common mode input voltage will not cause charge redistribution error, even if the capacitors do not match each other exactly. Another requirement of the amplifier $A1$ is that it must have a high enough bandwidth such that the loop stability is assured in this mode. Differential amplifiers $A1$ and $A2$ together must provide enough loop gain to achieve the desired closed-loop gain accuracy.

This circuit has several advantages compared to other techniques. Because the amplifier does not experience any common mode shift, the overall common mode rejection of the circuit is independent of the common mode rejection of the operational amplifier. Because of the balanced nature of this circuit, switch charge injection and clock feedthrough are cancelled to the first order. Because of the equal and opposite voltage excursion on the capacitors, the capacitor nonlinearity is also cancelled to the first order. The sampling bandwidth of this circuit is determined by the RC time constant of the input switch and capacitor, which is usually much faster than the settling time of an operational amplifier. The gain is set by capacitor ratios, which has good initial accuracy [5], very good temperature stability, and is trimmable. Both the $1/f$ noise and the dc offsets are reduced by the use of double correlated sampling; and as a result of this fact and the balanced nature of the circuit, the power supply rejection is also very high.

The overall performance of the circuit is limited by the mismatch of charge injection from the input switches. In this switched-capacitor instrumentation amplifier circuit, cancellation of switch channel charge injection is guaranteed by the

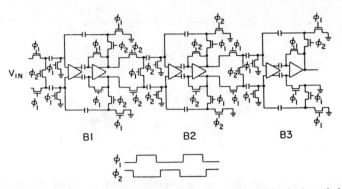

Fig. 3. Experimental implementation of a programmable single-ended output instrumentation amplifier.

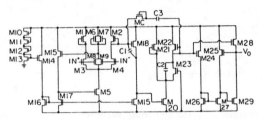

Fig. 4. Circuit schematic of single-ended NMOS operational amplifier.

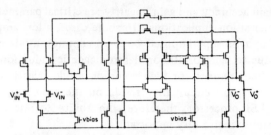

Fig. 5. Circuit schematic of differential NMOS operational amplifier.

symmetry of the circuit, and the offset becomes limited primarily by the mismatches of the switch charge injection. The mismatch in the switch channel charge is determined by mismatches in device parameters such as threshold voltage, channel geometry, and so forth. Experimental results to be presented later indicate that for the particular technology used here, the channel charge mismatch in an 8 μm MOS device is typically on the order of one percent.

III. EXPERIMENTAL IMPLEMENTATION OF SWITCHED-CAPACITOR INSTRUMENTATION AMPLIFIER

Precision preamplifiers may be required to provide voltage gains from less than 10 to over 1000. The use of a single stage to obtain very large values of voltage gain requires operational amplifiers with very large open-loop gain, and also very large capacitor ratios. In NMOS technology, the voltage gain achievable in operational amplifiers is often limited, and as a result, it is more desirable to use a relatively small value of closed-loop gain. High values of overall closed-loop gain can be achieved by either cascading multiple stages, as shown in Fig. 3, or by using a single stage in a recirculating mode [6]. In the example described here, a fixed gain of ten is used. Another problem is the fact that many A-to-D converters require a single-ended input voltage. Thus, a single-ended output referenced to ground must be produced. This can be achieved as shown in Fig. 3 where the first stage of amplification is realized in a fully differential mode, and the last stage uses a single-ended output operational amplifier to generate a single-ended output voltage.

Operational Amplifier Design

The broad-band low-gain differential preamplifier used in this system is a single differential pair with enhancement load devices. The single-ended output operational amplifier shown in Fig. 4 is a conventional NMOS operational amplifier design [7]. Transistors $M1-M5$, $M20-M23$, and $M24-M29$ are the

input, gain, and output stages, respectively. $C1$ and $C2$ are feedthrough capacitors, and $C3$ is the Miller compensation capacitor. Transistor MC is a depletion mode resistor for right half-plane zero compensation. The dc bias points of this op amp are set by the replica bias circuit composed of transistors $M6-M17$ [8]. The output stage has the ability to drive a large capacitive load without severely degrading the loop phase margin. This op amp realizes a voltage gain of about 1500 with output voltage swing of about 6.5 V (±7.5 V supply).

A schematic of the fully differential operational amplifier is shown in Fig. 5. Two differential stages are used to achieve a voltage gain of 1500. Common mode feedback is used in these two stages to stabilize the dc bias condition. The output stages are simple source followers, and the capacitors and depletion device resistors are utilized for frequency compensation. Since this op amp is to be used only in the early stages of amplification where the signal swing is small, the output of this amplifier is required to develop a differential voltage of less than 1 V.

IV. MOS SWITCH-INDUCED ERRORS

MOS switches introduce a significant amount of error due to clock voltage feedthrough through the gate–source, gate–drain overlap capacitance and the channel charge stored in the MOSFET device. As shown in Fig. 6, the MOS switch is connected to a sampling capacitor which is charged to the input voltage level. When the MOS switch is turned off, the amount of channel charge injected into the sampling capacitor represents an error source as a result of the sudden release of the charge under the MOS gate. The amount of channel charge that flows into the sampling capacitor as opposed to the amount that flows back to the input terminal is a complex function of the gate voltage fall time, input impedance level, and the size of the sampling capacitor [9]. For a typical switch size of 8×8 μm and a sampling capacitor of 5 pF, for example, a 5 V gate overdrive will introduce an error on the order of 20 mV if half of the channel charge flows into the sampling capacitor.

One approach to the reduction of this type of error is the use of large external capacitors [10], [11]. However, the added complexity and the decreased circuit operating speed due to the large external capacitors would limit the usefulness of the circuit as a subsystem of a VLSI processor. A second approach is to use an on-chip capacitor with a dummy switch, as shown in Fig. 7(a), to cancel the charge from the main switch. Unfortunately, all of the channel charge in the dummy switch flows onto the sampling capacitor, while only a fraction of that from the main switch does. As mentioned above, this fraction is a function of gate waveform and source impedance. Another approach is to use dummy switches in combination

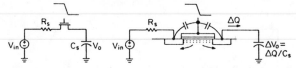

Fig. 6. Clock feedthrough and channel charge injection of an MOS switch.

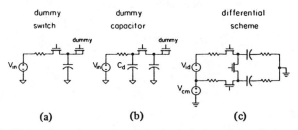

(a)   (b)   (c)

Fig. 7. Illustration of several switch channel charge cancellation techniques.

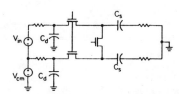

Fig. 8. Differential switch charge cancellation with dummy capacitors.

with dummy capacitors [12] as shown in Fig. 7(b). This configuration assures by symmetry that exactly one half of the channel charge will flow into the sampling capacitor. Thus, the dummy switch with one half the size of the input switch can guarantee the exact cancellation of the channel charge injection and the clock feedthrough problems. This technique works well when the source impedance, clock frequency, and fall time are all well controlled. However, when the signal is driven from an external source whose impedance level is low, the effect of the dummy capacitor on the channel charge cancellation is reduced and becomes a function of the source impedance.

The differential sampling configuration shown in Fig. 7(c) uses two matched switches and capacitors to sample the differential signal. The channel charge injection will introduce the same error voltage on these two sampling capacitors, thus giving no differential error in the sampled values. Although the differential input voltage introduces a difference of channel charge in the two matched switches, this error term is proportional to the input differential voltage, and can be considered as a gain error. One drawback in the configuration shown in Fig. 7(c) is that the channel charge cancellation relies on the matching of the two differential input impedances. However, the effects of source impedance mismatch can be reduced by the inclusion of additional dummy capacitors as shown in Fig. 8 [9].

V. NOISE PERFORMANCE

A key aspect of the performance of the circuit is the noise introduced into the signal path by the circuit, which in this case results from two principal sources, the $1/f$ noise in the operational amplifiers and the thermal noise in the channel resistance of the MOS switches making up the circuit.

An inherent aspect of the offset cancellation technique used is the fact that the noise contribution from the operational amplifier on any sample is the difference between the instantaneous input-referred noise voltage at the time when the input voltage is sampled and when the op amp output is sampled. For the case of the $1/f$ noise, since the noise energy is concentrated at low frequency, the successive samples of the input-referred noise are highly correlated. The noise spectrum which results from this correlated sampling process has been treated analytically elsewhere [13]. Assuming that all of the $1/f$ noise energy is concentrated below the sampling frequency, the spectrum of the noise added to the signal path has the form

$$\bar{\sigma}^2_{eq}(w)$$

$$= S(w) * \{1 + |\text{sinc}(wT/2)|^2 - 2 * \text{sinc}(wT/2) * \cos(wT)\}$$

where $S(w)$ is the spectral density of the original input-referred $1/f$ noise and T is the sampling period. In the case of the amplifier described here where the sampling rate is in the 200 kHz range and the $1/f$ noise corner frequency is in the 10 kHz range, this reduction in noise contribution at low frequencies is enough such that the other noise source, kT/C noise, is dominant.

The fundamental limitation on the noise performance of the circuit results from thermal noise in the MOS switches. When the switches are on in the sampling mode, a low-pass filter is formed by the channel resistance of the switch and the sampling capacitor which band limits this noise. The resulting mean-square noise voltage appearing across the capacitor is kT/C. When the switch is opened, this noise is sampled, with the result that each sample of the incoming signal has a noise sample added to it. The amplitude distribution of these noise samples is Gaussian with a standard deviation of the square root of kT/C V. For example, for a 10 pF sampling capacitor, this standard deviation is 25 μV at room temperature.

A frequent application of an amplifier of this type would be in a signal processing system in which it samples an input signal periodically, after which the samples are processed to form an output sequence which is subsequently converted to analog form with a DAC and sample/hold and smoothed with a reconstruction filter which passes output spectral components up to one half the sampling rate. For this case, it can be shown that the contribution of the kT/C noise is equivalent to that of a continuous time white noise source at the input with a spectral density of $2\,kT/fC$ where f is the sampling frequency. Again, assuming a 10 pF capacitor and a 200 kHz sampling rate, the equivalent input noise spectral density would be 65 nV/$\sqrt{\text{Hz}}$ or equivalent to the noise in a 250 kΩ resistor. The noise behavior of the amplifier is very similar to that of a switched-capacitor integrator with a 10 pF sampling capacitor. Increasing the sampling rate decreases the input noise density, but the total noise energy below the sampling frequency remains constant.

VI. EXPERIMENTAL RESULTS

Experimental circuits for the fully differential stage and single-ended output stage were designed and fabricated using

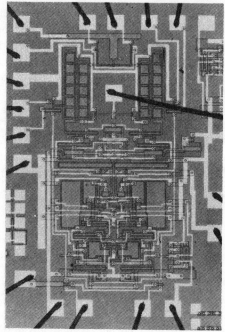

Fig. 9. Die photo of a fully differential instrumentation amplifier gain block.

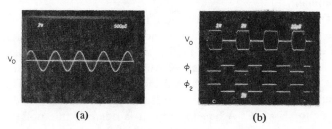

(a)   (b)

Fig. 10. Measured circuit output waveform.

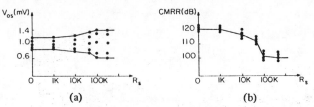

(a)   (b)

Fig. 11. Experimentally measured V_{os} and CMRR as a function of source resistance value.

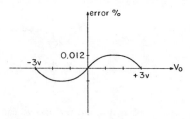

Fig. 12. Experimentally measured gain nonlinearity.

TABLE I
SUMMARY OF EXPERIMENTAL RESULTS; 25°C (15 SAMPLES)

Gain accuracy ($G = 10$)	
average value	0.15 percent
standard deviation	0.03 percent
Input offset voltage	
average value	1 mV
standard deviation	0.5 mV
Input offset drift	2 μV/C
Gain linearity	0.01 percent
Common mode rejection	
dc	120 dB
10 kHz	95 dB
Power supply rejection	
dc	>95 dB
1 kHz	>95 dB
Equivalent input noise	
(200 kHz sampling freq.)	65 nV/$\sqrt{\text{Hz}}$
Input voltage range	+4 V, −6 V
Power supply	+7.5 V, −7.5 V
Power dissipation	10 mW

a local oxidation polysilicon gate NMOS process with depletion load. The minimum geometry used in these circuits was 7 μm. The input sampling capacitors were 10 pF each, feedback capacitors were 1 pF, and the DCS capacitors were 3 pF.

A die photo of the switched-capacitor amplifier with differential output stage is shown in Fig. 9. The top part of the die photo contains the matched capacitor arrays; the bottom part contains the differential amplifier. The symmetrical layout of the circuit is crucial to the matching of circuit components. The die area of this circuit is 2500 mils^2.

In Fig. 10(a), the output waveform is shown with a 1 kHz sinusoidal input signal. The staircase-shaped waveform is the result of the sample-and-hold operation. Shown in Fig. 10(b) is the output waveform with a square-wave input signal. The output is reset to its own offset value during the sample mode when clock 1 is high, and generates an amplified input sample during the hold mode when clock 2 goes high.

The experimentally observed input offset voltage and common mode rejection ratio are shown in Fig. 11 for five typical samples of the circuit as a function of source resistance. For small values of source resistance, the average value of the input

offset voltage is 1 mV; this actually results from a layout-induced systematic offset in the amplifier offset cancellation circuit and not from charge injection in the input switches. The spread of this offset is about 500 μV, which increases at high values of source resistance because the percentage of the channel charge injected into the sampling capacitors increases [9]. The CMRR at low values of source resistance is about 120 dB. This is also degraded at high values of source resistance for the same reason as the offset voltage. The CMRR and PSRR of this circuit were measured using a spectrum analyzer. A large single frequency sinusoidal signal was applied to the common mode input and connected in series with the power supply, respectively, and the magnitude of the output component was measured.

Fig. 12 illustrates the typical gain nonlinearity observed for the device. This nonlinearity results primarily from the non-linear open-loop characteristic of the operational amplifier. The peak deviation from linear behavior is about 0.012 percent of full scale at plus and minus 3 V output swing. Table I shows a summary of the data measured at room temperature for power supply voltages of plus and minus 7.5 V.

VII. CONCLUSION

This paper has described one approach to the implementation of the instrumentation amplifier/sample–hold function in MOS technology. The performance levels achieved are generally somewhat inferior to recently reported bipolar instrumentation amplifiers. The significance of the results achieved is that the compatibility with MOS technology allows a higher level of integration in the analog data acquisition interface with the digital interfacing and/or processing circuitry without adding to the complexity of basic digital MOS technologies. The experimental amplifier described in this paper was implemented in NMOS technology, but it appears that a CMOS implementation could achieve significantly higher operating speed because of the higher available gain per stage and the resulting improvement in operational amplifier gain and bandwidth.

ACKNOWLEDGMENT

The authors would like to thank Prof. D. A. Hodges for his support on this project. The discussion with many graduate students in the Berkeley IC Group on the development of the IC process is also acknowledged.

REFERENCES

[1] M. Timko and A. P. Brokaw, "An improved monolithic instrumentation amplifier," in *Dig. 1975 ISSCC*, Feb. 1975.
[2] C. T. Nelson, "A 0.01% linear instrumentation amplifier," in *Dig. 1980 ISSCC*, Feb. 1980.
[3] M. Townsend, M. E. Hoff, Jr., and R. E. Holm, "An NMOS microprocessor for analog signal processing," *IEEE J. Solid-state Circuits*, vol. SC-15, Feb. 1980.
[4] J. Stremler, *Introduction to Communication Systems*.
[5] J. L. McCreary, "Matching properties and voltage and temperature dependence of MOS capacitors," *IEEE J. Solid-State Circuits*, vol. SC-16, Dec. 1981.
[6] R. H. McCharles and D. A. Hodges, "Charge circuits for analog LSI," *IEEE Trans. Circuits Syst.*, vol. CAS-25, July 1978.
[7] Y. P. Tsividis and D. L. Fraser, Jr., "A process insensitive NMOS operational amplifier," in *Dig. 1979 ISSCC*, Feb. 1979.
[8] P. R. Gray, D. A. Hodges, and R. W. Brodersen, Eds., *Analog MOS Integrated Circuits*. New York: IEEE Press, 1980.
[9] R. C. Yen, "High performance MOS circuits," Ph.D. dissertation, Univ. California, Berkeley, Dec. 1981.
[10] R. Poujois and J. Borel, "A low drift fully integrated MOSFET operational amplifier," *IEEE J. Solid-State Circuits*, vol. SC-13, Aug. 1978.
[11] M. Coln, "Chopper stabilization of MOS operational amplifiers using feed-forward techniques," *IEEE J. Solid-State Circuits*, vol. SC-16, Dec. 1981.
[12] L. A. Bienstman and H. DeMan, "An 8-channel 8b μP compatible NMOS converter with programmable ranges," in *Dig. 1980 ISSCC*, Feb. 1980.
[13] R. W. Brodersen and S. P. Emmons, "Noise in buried channel charge coupled devices," *IEEE J. Solid-State Circuits*, vol. SC-11, Feb. 1976.

A Micropower CMOS-Instrumentation Amplifier

M. DEGRAUWE, E. VITTOZ, MEMBER, IEEE,
AND I. VERBAUWHEDE

Abstract —A CMOS switched capacitor instrumentation amplifier is presented. Offset is reduced by an auto-zero technique and effects due to charge injection are attenuated by a special amplifier configuration. The circuit which is realized in a 4-μm double poly process has an offset (σ) of 370 μV, an rms input referred integrated noise $(0.5 - f_c/2)$ of 79 μV, and consumes only 21 μW ($f_c = 8$ kHz, $V_{DD} = 3$ V).

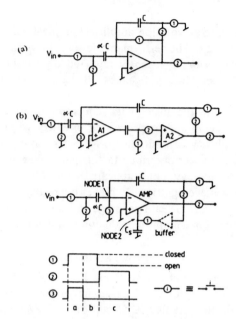

Fig. 1. Auto-zero techniques. (a) Information stored at the input. (b) Information stored after a first gain stage. (c) Information stored at a low sensitive auxiliary input.

I. INTRODUCTION

For the realization of intelligent sensors an instrumentation amplifier is very often required in order to detect small differential signals in the presence of a large common mode signal. Such amplifiers should have very low offset and $1/f$ noise, large CMRR, and especially for biomedical applications consume as less as possible current and be able to operate at a low supply voltage.

First an offset cancellation technique will be presented. Further the amplifier realization will be discussed and experimental results will be given.

II. AUTO-ZERO REALIZATIONS

Offset cancellation by means of auto-zero techniques can be implemented in different ways [1]. Up to now basically two approaches are used.

The first method (see Fig. 1(a)) consists of storing the offset information at the input of the amplifier. Ideally the stored information will be equal to the offset value. However due to charge injection of the switch S_1, this value will be significantly changed resulting in a degraded offset cancellation.

A second method (see Fig. 1(b)) [2] consists of storing the offset information at the output of the first amplifier stage. In practice the stored offset information will be about 100 times the offset value. Therefore, the charge injection of the switches $S_1 - S_2$ will not significantly degrade the offset cancellation. However the need of a two-stage amplifier gives rise to potential stability problems, degraded noise, and PSRR performances [3].

In this correspondence an alternative auto-zero topology is presented (Fig. 1(c)) [5], [8].

AMP is an ordinary amplifier with gain A_1 (between node 1 and output) and offset V_{off1}. An auxiliary input (node 2) of reduced sensitivity (gain A_2) is added to the main amplifier. It is controlled by a compensation voltage V_2 stored in capacitor C_s. The buffer is added to speed up the compensation phase.

The compensation works as follows. During time slot "a", the input signal is sampled. In the same time the input terminals of the main amplifier are short-circuited. This amplifier will thus amplify its own offset. However, due to the feedback path through the auxiliary input, the output voltage will be stabilized and across the store capacitor C_s there

Manuscript received October 29, 1984; revised December 20, 1985. This work was partially supported by the "Fonds National Suisse pour la Recherche Scientifique, PN13."
M. Degrauwe and E. Vittoz are with Centre Suisse d'Electronique et de Microtechnique S.A. CSEM—Recherche & Développement—(formerly CEH) Maladière 71, 2000 Neuchâtel 7, Switzerland.
I. Verbauwhede was with CEH in 1983 on leave from Katholieke Universiteit Leuven, Kardinaal Mercierlaan 94, B-3030 Heverlee, Belgium.

will be a voltage equal to

$$V_2 = \frac{-A_1 V_{off1}}{1 + A_2}.\tag{1}$$

At the beginning of time slot "b", the switches "3" are opened and a charge is injected at node 1. This causes an additional offset ΔV_1. The feedback mechanism is however still active and the voltage across C_s is now given by

$$V_2 = \frac{-A_1(V_{off1} + \Delta V_1)}{1 + A_2}.\tag{2}$$

At the end of time slot "b" the switch "1" opens and causes charge injection (ΔV_2) at the node 2. The voltage at this node is now given by

$$V_2 = \frac{-A_1(V_{off1} + \Delta V_1)}{1 + A_2} + \Delta V_2.\tag{3}$$

At the output node appears then a voltage equal to

$$V_{out} = -A_1 \cdot \left(\frac{(V_{off1} + \Delta V_1)}{1 + A_2} + \Delta V_2 \cdot \frac{A_2}{A_1} \right)\tag{4}$$

which corresponds to an equivalent input which is A_1 times smaller.

Finally during the time slot "c" the charge stored on the capacitor αC (sample of the input signal) is transferred to the integration capacitor.

The equivalent offset of the whole amplifier contains thus two residual parts

the sum of the initial offset and the charge injection at node 1 which both are attenuated by a factor $(1 + A_2)$; and

the charge injection at node 2 which is attenuated by a factor A_1/A_2.

The optimal value of the gain of the auxiliary input is obtained by differentiation of (4)

$$A_2 = \left(\frac{(V_{off1} + \Delta V_1) A_1}{\Delta V_2} \right)^{1/2} - 1\tag{5}$$

Reprinted from *IEEE J. Solid-State Circuits*, vol. SC-20, no. 3, pp. 805–807, June 1985.

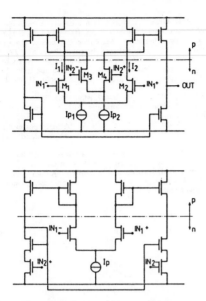

Fig. 2. Amplifier with auxiliary input. (a) Two parallel input stages. (b) Degenerated current mirror.

and results in a residual offset of

$$V_{\text{off}} = 2 \cdot \left(\frac{(V_{\text{off1}} + \Delta V_1) \cdot \Delta V_2}{A_1} \right)^{1/2} \qquad (6)$$

The choice of A_1 and A_2 has to be so that the amplifier never saturate. Therefore the method will result in a smaller achievable residual offset for larger supply voltages. Further, it can be shown that the offset of the auxiliary amplifier can be neglected, provided it is of the same order of magnitude as the input offset.

From (5) it is seen that the optimum value of the auxiliary input depends on the initial offset of the main amplifier and on the clock injection on nodes 1 and 2. For small values of $(V_{\text{off1}} + \Delta V_1)$, the gain A_2 should be small and for large $(V_{\text{off1}} + \Delta V_1)$, the gain A_2 should be larger. For optimum compensation, the gain A_2 should thus not be constant. Recently it has been shown that a quadratic auxiliary input results in a better offset reduction [4].

If a fixed gain A_2 is used, this gain should be optimized for the largest expected values of the initial offset and of the clock feedthrough.

III. CIRCUIT CONFIGURATION AND EXPERIMENTAL RESULTS

There are several ways to add an auxiliary input at a conventional amplifier. A first method consists of using two parallel input stages (Fig. 2(a)). The ratio of the gain A_1/A_2 will be determined by the ratio of the transconductances of the two input stages. The bias current I_{p2} can however not be chosen arbitrary small. Due to the offset voltage V_{off1}, the current I_1 and I_2 can deviate as much as 10 percent of their ideal value. Therefore, the current I_{p2} should be at least 10 percent of I_{p1}. Large A_1/A_2 can thus only be achieved by operating transistors $M_3 - M_4$ much deeper in strong inversion than transistors $M_1 - M_2$. This will however result in a large voltage drop across $M_3 - M_4$ which can be unacceptable for low-voltage battery operation.

An alternative method consists of degrading a current mirror ratio by inserting transistors operating in the linear region (Fig. 2(b)) [6], [7]. In this case the second input is realized with no additional current consumption. However, the gain A_1/A_2 will now depend on the supply voltage which will result in a reduced PSRR. For battery operation, the specifications for the PSRR can however be somewhat relaxed.

An SC instrumentation amplifier was developed according to the principles of Figs. 1(c) and 2(b). The circuit was realized in a differential way in order to further improve the performances [2].

The amplifier realization is shown in Fig. 3 (sampling and integrating capacitors are not shown). The main amplifier $M_1 - M_{15}$ is a differential transconductance amplifier whose input stage has a low gain. The secondary inputs are realized by adding four transistors to the main amplifier

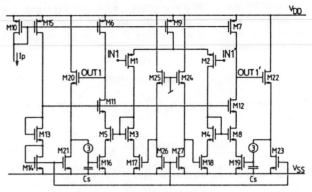

Fig. 3. Realized amplifier.

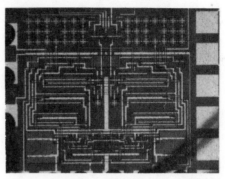

Fig. 4. Chip photograph.

TABLE I
MEASUREMENT RESULTS

Gain $(= \propto C/C)$	20 dB
Max. clock frequency	8 kHz
$(C_L = 22$ pF$)$	
Offset (at 8 kHz) $\bar{x}$	90 µV
σ	370 µV
Equivalent input noise (0.5 Hz-4 kHz)	79 µV
No 1/f noise was observed above 0.5 Hz	
(= under limit of measurement equipment)	
CMRR	>95 dB
Current consumption	7 µA
PSRR$^-$ (at DC)	54 dB
PSRR$^+$ (at DC)	66 dB

$(M_{16} - M_{19})$. Those transistors who operate in the linear region can modulate the current ratio of two current mirrors. The buffer stages are simple source followers $(M_{20} - M_{23})$.

The circuit has been realized in a 4-µm double poly CMOS process. A chip photograph is shown in Fig. 4. The total chip area is 640 µm × 700 µm ($\simeq 0.45$ mm^2).

The most important measurement results are given in Table I. The standard deviation of the offset is 370 µV. Further reduction of the offset can be obtained by combining the offset cancellation technique of Fig. 1(a) and (c).

Recently the presented circuit has been redesigned in a 3-µm technology [9]. Measurement results of this circuit will be reported very soon [10].

IV. CONCLUSIONS

A micropower instrumentation amplifier has been presented which has a typical offset of 370 µV and consumes only 21 µW. The performances are achieved by the use of a new offset cancellation technique in which the effects of charge injection are attenuated. For larger supply voltages ($\cong 10$ V) residual offsets of less than 50 μV are obtainable with the presented technique.

ACKNOWLEDGMENT

The authors wish to thank Dr. H. Oguey for useful discussions.

REFERENCES

[1] R. Poujois and J. Borel "Low-level MOS transistor amplifier using storage techniques," in *ISSCC Dig. Tech. Papers*, 1973, pp. 152–153.

[2] R. C. Yen and P. R. Gray, "A MOS switched capacitor instrumentation amplifier," *IEEE J. Solid-State Circuits*, vol. SC-17, pp. 1008–1013, Dec. 1982.

[3] B. J. Hosticka, W. Brockherde, and M. Wrede, "Effects of the architecture on noise performance of CMOS operational amplifiers," in *Proc. ECCTD*, 1983.

[4] E. Vittoz, "Dynamic analog techniques," "Advanced Summer Course on 'Design of MOS-VLSI Circuits for Telecommunications,'" L'Aquila, Italy, June 18–29, 1984.

[5] E. Vittoz and H. Oguey, Swiss Patent Application No. 2179/84, filed on April 5, 1984.

[6] H. Oguey, CEH Rep. HO/81, pp. 56–57, Mar. 1981.

[7] A. Acovic, "Amplificateur à offset compensé," Masters work of the EPFL, Lausanne, July 1983, (in French).

[8] M. Degrauwe, E. Vittoz, and I. Verbauwhede, "A Micropower CMOS instrumentation amplifier," in *Proc. ESSCIRC '84* (Edinburgh, Scotland), September 1984, pp. 31–34.

[9] I. Verbauwhede, "Design and integration of a low-power CMOS SC-instrumentation amplifier," M.S. thesis, K.U. Leuven, June 1984 (in Dutch).

[10] P. Van Peteghem, I. Verbauwhede, and W. Sansen, "A micropower high performance S.C. building block for integrated low level signal processing," submitted for publication in the *J. Solid-State Circuits*.

A Precision Variable-Supply CMOS Comparator

DAVID J. ALLSTOT, MEMBER, IEEE

Abstract—Several new techniques are presented for the design of precision CMOS voltage comparator circuits which operate over a wide range of supply voltages. Since most monolithic A/D converter systems contain an on-chip voltage reference, techniques have been developed to replicate the reference voltage in order to provide stable supply-independent dc bias voltages, and controlled internal voltage swings for the comparator. These techniques are necessary in order to eliminate harmful bootstrapping effects which can potentially occur in all ac-coupled MOS analog circuits. An actively controlled biasing scheme has been developed to allow for differentially autozeroing the comparator for applications in differential A/D converter systems. A general approach for selecting the gain in ac-coupled gain stages is also presented. The comparator circuit has been implemented in a standard metal-gate CMOS process. The measured comparator resolution is less than 1 mV, and the allowable supply voltages range from 3.5 to 10 V.

I. INTRODUCTION

MOS switched capacitor analog circuits [1] have evolved rapidly during the past several years, with particular emphasis on industrial and telephony applications. As the power supplies in these applications are both fixed and relatively large (typically 10 V), depletion-load NMOS and CMOS have emerged as viable technologies, although CMOS is gaining wider acceptance because of its superior analog performance capabilities. This technology trend will continue as these techniques are used in new applications such as consumer and medical electronics where operation with both small supply voltages and large supply voltage compliances are required for battery-powered systems. Additionally, more process- and supply-in-

Manuscript received July 16, 1982; revised August 9, 1982.
The author is with Nova Monolithics, Inc., Carrollton, TX 75006.

dependent design techniques will be needed for the analog–digital interfaces in the lower voltage digital VLSI technologies.

In this paper, some of these issues are addressed with new design techniques for a precision, variable-supply CMOS comparator. This particular comparator is used in the switched capacitor A/D section [2]–[3] of a through-the-lens autofocusing system for SLR cameras [4]. It is designed to operate with total supply voltages ranging from 3.5 to 10 V, with a nominal quiescent current of 500 μA. It is capable of resolving a 1 mV peak input signal in about 12 μs, although for this application, it was only required to resolve a 4 mV peak signal with a typical switching time of about 3 μs with a 10 V power supply. Differential structures are used extensively to minimize the effects of power supply coupling and to reduce the input-referred offset voltage due to switch feedthrough effects. Replication of the on-chip reference voltage as well as controlled biasing techniques are used to generate stable, supply-independent dc bias voltages for the comparator.

II. SELECTION OF COMPARATOR TOPOLOGY

The most basic issue in selecting a comparator topology is to determine the minimum required gain. In this design, in order to obtain a 10 V logic swing with a 500 μV peak input signal, a minimum gain of 20 000 is necessary, which obviously requires multiple gain stages. There are several key issues involved in selecting both the number of gain stages, as well as the partitioning of the total gain relative to the placement of the inter-stage coupling capacitor(s). These considerations are illustrated in the following examples.

An obvious comparator topology is a conventional two-stage differential amplifier as shown in Fig. 1(a). It is initially configured in unity gain through a p-channel reset switch with the

Reprinted from *IEEE J. Solid-State Circuits*, vol. SC-17, no. 6, pp. 1080–1087, Dec. 1982.

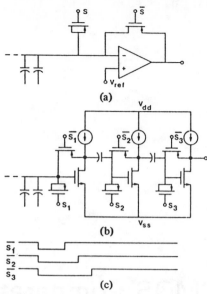

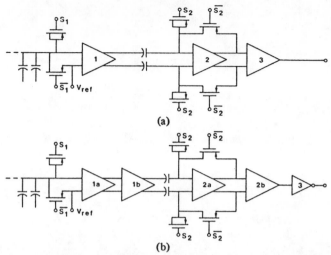

Fig. 1. Possible comparator topologies with offset cancellation. (a) A conventional two-stage differential amplifier which is periodically connected in unity gain. (b) A cascade of n self-biased inverters which are periodically reset using (c) n overlapping clock phases.

Fig. 2. Differential comparator topology. (a) A simplified block diagram of a three-stage design using only a single set of coupling capacitors as a result of the differentially reset second stage. (b) A more detailed block diagram of the actual implementation showing the partitioning of the five amplifiers relative to the coupling/reset network.

intent of storing the input-referred offset voltage on the common top plate of the A/D capacitor array where it is effectively subtracted during the A/D conversion process. This technique works well for removing the input offset voltage contribution of the amplifier itself. However, the clock-feedthrough [2]–[3] and channel charge-pumping [5] effects of the reset switch contribute a *residual* offset charge which is not completely eliminated, even with the use of a charge cancellation device as shown in the figure. Furthermore, the accuracy of the charge cancellation is strongly dependent on the supply voltage (clock swing) which is not constant in this application. Thus, in attempting to reduce the input offset voltage due to the residual feedthrough charge $V_{os} = Q_{ft}/C_{array}$, the array capacitance may be increased. Unfortunately, to maintain an equivalent time constant for charging the larger capacitor array, the size (W/L) of the reset and cancellation switches must also be increased, which results in a larger residual feedthrough offset charge, and hence, little or no effective decrease in the input-referred feed-through offset voltage. Furthermore, the two-stage amplifier is usually pole-split frequency compensated for unity gain, which results in both limited switching speed because of slewing the compensation capacitance and poor high-frequency PSRR due to V_{dd} variations coupling directly to the amplifier output through the compensation capacitor.

Another commonly used comparator configuration is a cascade of ac coupled self-biased inverters [6] as shown in Fig. 1(b) where, for n gain stages, n overlapping reset signals are required [Fig. 1(c)]. By staggering the reset signals in time, only the feedthrough offset associated with the last clock phase contributes directly to the input offset voltage of the comparator as

$$V_{os} = \frac{V_{ft_3}}{A_1 A_2}. \tag{1}$$

All other offset terms are stored on the coupling capacitors.

Obviously, the input-referred offset voltage can be made very small if the dc gain preceding the last stage, $A_1 A_2$, is made large. Unfortunately, V_{ft_1} and V_{ft_2} are indirectly very important since the dc output voltages of the first two stages are $A_1 V_{ft_1}$ and $A_2 V_{ft_2}$ (A_i is the dc gain of the ith stage), respectively, which can result in saturated gain stages for even modest dc gains considering the relatively large changes in the V_{ft}'s due to processing and power supply variations. Hence, the gain per stage must be kept small to avoid saturation, which results in a large number of gain stages for precision applications. Due to the use of single-ended amplifiers, another major disadvantage of the inverter cascade is its poor power-supply rejection which can limit the minimum resolvable signal to several millivolts in systems where a large amount of digital circuitry is contributing to power supply noise.

In order to circumvent these problems, as well as new problems which arise as a result of the variable-supply requirement (to be described later), a fully differential topology of the type shown in Fig. 2(a) is desirable. As in the single-ended case, the value of A_1 must be chosen sufficiently small to ensure that V_{ft_1} does not saturate the first stage. However, because of the differential outputs, A_1 can be chosen twice as large as in the single-ended case for the same degree of output saturation. The input-referred offset voltage for this circuit is given by

$$V_{os} = \frac{\Delta V_{ft_2}}{A_1} + \frac{V_{os_3}}{A_1 A_2} \tag{2}$$

where ΔV_{ft_2} is the *difference* in clock feedthrough voltages at the second stage inputs. In other words, the common-mode feedthrough charge terms are eliminated, with only the differential feedthrough referred to the input of the comparator. Although V_{ft_2} is large and varies widely with supply voltage and process changes, ΔV_{ft_2} is much smaller due to the matching and tracking between the V_{ft}'s for the two sides. Furthermore, the degree of saturation at the output of the second stage depends on $A_2 \Delta V_{ft_2}$, implying that for the same degree of output saturation, the second-stage gain can be significantly increased relative to the single-ended design. Hence, by using the differential resetting and charge cancellation techniques, a precision design (small V_{os}) is obtained with only a single stage of ac coupling. Because the common-mode feedthrough terms are eliminated, the coupling capacitance can be reduced, which is beneficial in higher speed designs. (It should be noted that although the common-mode feedthrough terms do not contribute to the input offset voltage of the comparator, charge cancellation devices should still be included to ensure that the changes in the common-mode dc bias voltages as a result of the switching transients are small.) A more detailed block diagram showing that the three stages actually consist of five amplifiers is shown in Fig. 2(b).

The next section focuses on dc-biasing ac-coupled comparator circuits which is, in general, a very critical issue with variable power supplies.

III. DC BIASING CONSIDERATIONS

DC biasing of the second gain stage will first be considered in order to explain a rather subtle bootstrapping effect which can occur in *any* comparator with ac coupling, but which is much more likely to occur when the power supplies vary widely as in this application. Fig. 3 shows a simplified schematic of the second gain stage including the resetting and ac coupling networks. For reasons to be explained later, this stage is designed so that the dc bias voltage between V_{dd} and the n-channel inputs during the reset time approximately equals the reference voltage V_{ref}. (Note that this could be achieved by simply shorting the gates of the n-channel inputs directly to V_{ref}. However, in order to remove the second-stage offset, this approach requires another set of coupling capacitors with yet another relatively low gain stage.) When the switches are open, C_{in} represents the input capacitance of the second gain stage with C_s as the ac coupling capacitance. Thus, for a voltage change of ΔV_x at node $\textcircled{X}$, the voltage change at node $\textcircled{Y}$ is

$$\Delta V_y = \Delta V_x \, C_s/(C_s + C_{in}). \tag{3}$$

Usually, by design, $C_s \gg C_{in}$, and therefore, $\Delta V_y \cong \Delta V_x$. Assuming p-channel reset switches, there exists a parasitic diode (Fig. 3) from node $\textcircled{Y}$ to V_{dd}, namely, the drain-bulk diode of the p-channel switch. For a large positive voltage swing $\Delta V_y > V_{ref}$, node $\textcircled{Y}$ is bootstrapped above V_{dd}, and the diode becomes forward biased, resulting in a parasitic substrate current I_{sub} which has the deleterious effect of removing charge from C_s. Since this effect occurs for a range of intermediate to maximum input signals, it results in a gain change (nonlinearity) in the A/D converter transfer characteristic. Hence, another basic requirement in dc biasing the comparator is to ensure that the bias points and positive internal signal swings are well controlled to prevent this bootstrapping effect from occurring.

In biasing the comparator (as well as other analog circuitry), significant advantage can be taken of the on-chip voltage reference. For example, Fig. 4 shows a simplified schematic of the ΔV_t reference [7]–[8] with resistive trim circuitry as used in this design. By forming the reference voltage relative to V_{dd},

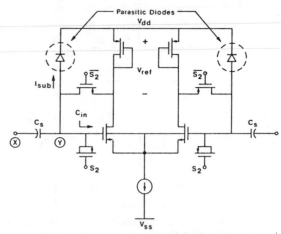

Fig. 3. A simplified schematic of the capacitive network coupling into the second gain stage (a_{2a}) showing the critical parasitic diodes which can conduct current if node $\textcircled{Y}$ is bootstrapped above V_{dd}.

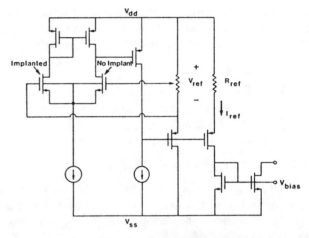

Fig. 4. A simplified schematic of the delta-V_t voltage reference with resistor trim network. An additional p-channel source follower replicates V_{ref} across R_{ref} to generate the bias voltage for other analog circuitry including the comparator.

with careful layout, and by also referencing the input signal(s) to V_{dd}, the substrate noise appears as a common-mode signal and is substantially rejected. In this design, the reference voltage was trimmed to $V_{ref} = 2.048$ V. The measured temperature coefficient of V_{ref} was less than 100 ppm/°C, and the measured dc PSRR was better than 55 dB [9]. In order to exploit these characteristics, circuit techniques were developed which replicated V_{ref} to produce stable supply-independent dc bias voltages for the comparator. Thus, a p-channel source follower was added to the basic reference to replicate V_{ref} across the p-well resistor R_{ref}, resulting in a current $I_{ref} = V_{ref}/R_{ref}$, which subsequently flows into the n-channel current mirror to generate the basic bias voltage V_{bias}. (Note that for biasing analog circuitry which is remote from the reference, it is preferable to supply a current bias since it is less susceptible to the substrate noise coupled through the interconnect stray capacitance.) Although V_{ref} is constant, I_{ref} varies by ±50 percent due to changes in the p-well resistivity over processing and temperature (6400 ppm/°C).

A schematic of the two differential gain stages [labeled 1a and 1b in Fig. 2(b)] preceding the coupling capacitors is shown in

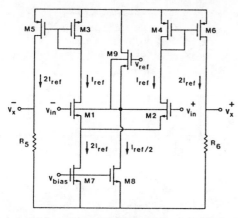

Fig. 5. Comparator input stage consisting of a cascade of two differential amplifiers with the resistive loads used in the second gain stage to replicate the reference voltage. This technique produces supply-independent dc bias voltages and controlled supply-independent output swings for the critical stages of the comparator.

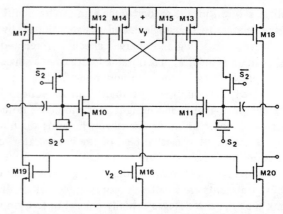

Fig. 6. Comparator second stage consisting of the ac coupling network and a cascade of two differential amplifiers. Positive feedback is used to provide increased gain. The bias voltage $V_2 \neq V_{\text{bias}}$ is derived from the bias control amplifier.

Fig. 5 with all bias currents indicated relative to I_{ref}. Rather than conventional active loads, p-well resistive loads are used in this design to achieve replication of V_{ref}.[1] In particular,

$$V_{R_5} = V_{R_6} = 2I_{\text{ref}}R_5 = 2 V_{\text{ref}}(R_5/R_{\text{ref}}) \qquad (4)$$

which generates a very reproducible fraction of V_{ref} since it is determined by a resistor *ratio*. In addition to providing a known bias voltage, the maximum *positive* output swing of the first stage is now well controlled at a value of

$$\Delta V_x(\text{max}) = 2I_{\text{ref}}R_5 = 2 V_{\text{ref}}(R_5/R_{\text{ref}}) \qquad (5)$$

where (R_5/R_{ref}) is chosen to eliminate bootstrapping, i.e., ΔV_x (max) = V_{ref} in this design. (Negative output swings are of no direct concern since bootstrapping can occur only on large positive output swings.)

Note that to a first order, the first-stage bias voltage and output swing are independent of V_{dd}. However, the voltage coefficient of the p-well resistors introduces a second-order V_{dd} dependence since the voltage across R_{ref} is constant (V_{ref}) with respect to the substrate V_{dd}, while the voltage across R_5 relative to the substrate increases directly with V_{dd}. Experimentally, V_{R_5} varied by about 100 mV as V_{dd} was changed from 4 to 8 V.

The source follower $M8$–$M9$ produces a replicated dc bias voltage ($V_{GS_1} = V_{GS_2} = V_{GS_9}$) for the p-well of the input pair which is independent of the common source voltage. This simple technique provides a speed improvement for single-ended comparator applications since the large p-well-to-substrate capacitance is not charged or discharged as a function of the common-mode input signal component.

Fig. 6 shows a schematic of the second gain stage with V_2 as the bias voltage for the current source device $M16$. p-channel reset switches are used to differentially store the offset voltages onto the coupling capacitors, with the load voltages assumed equal to V_y during the reset time. Several important observations can be made regarding the ac-coupled differential stage.

[1] A potential disadvantage of this technique is the poor rejection of noise on the negative power supply. However, because of the differential topology used in this design, the V_{ss} noise appears as a common-mode signal to the second stage, and is substantially eliminated.

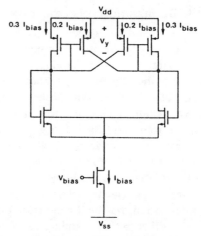

Fig. 7. A model of the second stage (a_{2a}) of the comparator during the reset interval with $V_2 = V_{\text{bias}}$.

Normally, in order to minimize the attenuation through the capacitive coupling network, n-channel cascode devices would be used in series with the input transistors $M10$–$M11$ to reduce the input capacitance due to the Miller-multiplied C_{gd}. Unfortunately, with small supply voltages, cascodes cannot be used. Next, assume that $V_2 = V_{\text{bias}}$, and that the p-channel reset switches are turned on and modeled as short circuits, as shown in Fig. 7. I_{bias} is nominally 150 μA with ± 50 percent variations. Thus, over the worst case processing variations (Table I), the p-channel load bias voltage given by

$$V_y = |V_{tp}| + \left[\frac{2(0.3 I_{\text{bias}})}{\mu_p C_{\text{ox}}(W/L)}\right]^{1/2} \qquad (6)$$

ranges from 1.3 to 2.9 V, as shown in the graph of Fig. 8. In the minimum case, V_y (min) = 1.3 V, substrate injection can occur since the first stage positive output swing is ΔV_x (max) = V_{ref} = 2.048 V. In the maximum case, V_y (max) = 2.9 V, the p-channel reset switches may not be sufficiently turned on due to the small amount of available gate drive V_{GS} (min) = V_{dd} (min) - V_y (max). If necessary, full CMOS transmission gates can be used as reset switches to eliminate this problem at the expense of greater uncertainty in the clock feedthrough charge cancellation. The approach taken in this design overcomes

TABLE I
WORST CASE PARAMETER VARIATIONS FOR DEVICES WITH DRAWN
CHANNEL LENGTHS OF 12 μm; $T = 25°$C

	N-Channel		P-Channel	
	min.	max.	min.	max.
V_t (Volts)	0.5	1.0	-0.5	-1.0
uCox (μA/volts^2)	14.4	24.0	5.9	9.7
T_{ox} (Angstroms)	800	1000	800	1000
Gamma (volts)$^{\frac{1}{2}}$	1.6	2.0	0.6	0.8
Lambda (V^{-1}) L=12μ	0.0031	0.0063	0.0104	0.0209

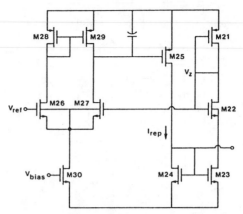

Fig. 9. Bias control amplifier. Negative feedback adjusts I_{rep} so that $V_z \cong V_{ref}$. By replication, the second-stage reset voltage V_y is approximately equal to V_{ref}.

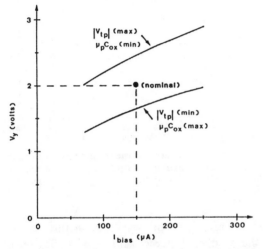

Fig. 8. Variations in the second-stage reset voltage as a function of processing and bias current variations for $V_2 = V_{bias}$.

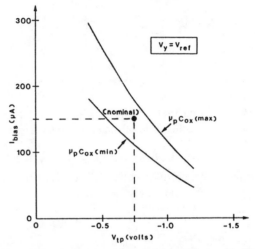

Fig. 10. Variations in the second-stage bias current as a function of processing variations. Controlled biasing is used so that the second-stage reset voltage $V_y \cong V_{ref}$.

both problems by combining the V_{ref} replica biasing techniques within an active feedback network to provide a solution whereby $V_y \approx V_{ref}$ independently of processing and operating variations. This so-called bias control amplifier shown in Fig. 9 consists of two major parts: 1) the replica bias string $M21$–$M23$ which nominally operates at the same current densities (same V_{GS}'s) as $M12$–$M10$–$M16$ in the second stage (Fig. 6), with $I_{rep} = 0.4 I_{bias}$ to conserve power; and 2) negative feedback around the differential amplifier forces $V_z \approx V_{ref}$ by varying the current I_{rep}, which results in a similar variation in I_{bias}, and hence, by replication, $V_y \approx V_z \approx V_{ref}$.

In simple terms, the controlled biasing of the gain stage(s) merely trades off the variation in the reset voltage V_y that existed previously (Fig. 8) for an increased variation in the bias current as shown in Fig. 10. With $V_2 = V_{bias}$, the bias current varied from 75 to 250 μA, while with the new controlled biasing, the current varies from 60 to 300 μA. This also results in a slight increase in the gain variance of the comparator (Table II).

Four nonideal effects contribute to a deviation of V_y from V_{ref}: 1) the common-mode reset/cancellation switch feed-

TABLE II
CALCULATED DC GAIN VARIATIONS FOR EACH OF THE COMPARATOR STAGES

	a_{1a}	a_{1b}	a_{2a}	a_{2b}	a_3	a_{tot}
min	1.52	2.62	7.74	40.26	20.13	2.50×10^4
nom	1.96	3.68	9.98	87.13	43.57	2.73×10^5
max	2.52	5.81	12.83	231.04	115.52	5.01×10^6

through (in metal-gate CMOS technology, alignment-insensitive switch layouts should be used to provide the best possible matching between the V_{ft}'s [see Fig. 14 below stage 2]); 2) the input-referred offset voltage of the bias control amplifier (typically 20–50 mV); 3) the closed-loop gain error of the control amplifier (typically 0.05 V_{ref}); and 4) the matching of the replicated transistor strings (typically 1 percent mismatch). The worst case deviation in V_y from V_{ref} is thus about 200 mV. The amount of bias offset that can be tolerated is dependent on the attenuation through the capacitive coupling network $\Delta V_y/\Delta V_x = C_s/(C_s + C_{in})$ where, by design, $\Delta V_x \approx V_{ref}$. Assuming a gain of 0.9 through the capacitive divider and $V_{ref} = 2.048$ V, then ΔV_y is approximately 1.8 V, which partially compensates for the 200 mV maximum offset in the bias

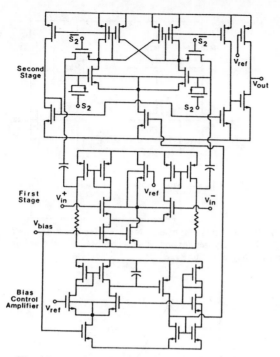

Fig. 11. A complete schematic of the comparator.

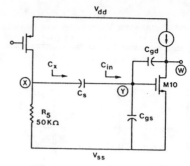

Fig. 12. A differential half circuit of the first-stage output and the series capacitor coupling into the input of the second gain stage.

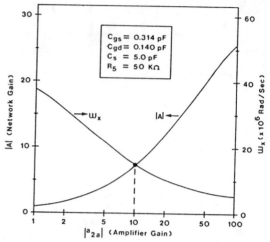

Fig. 13. The effective gain (A) of the second stage including the capacitive divider and the small-signal bandwidth (ω_x) of the first-stage output, both as a function of the gain (a_{2a}) of the second-stage input amplifier.

voltage V_y ensuring that no bootstrapping can occur. Fig. 11 shows a complete schematic of the comparator circuit.

IV. Gain/Bandwidth Considerations

The low frequency differential gain of the first comparator stage (Fig. 5) is approximately given by

$$a_1 = a_{1a}a_{1b} \approx \frac{g_{m1}}{g_{m3}} \cdot g_{m5}R_5 \tag{7}$$

which can be written as

$$a_1 \approx \sqrt{\frac{\mu_n}{\mu_p}} \sqrt{\frac{(W/L)_1}{(W/L)_3}} \cdot \frac{2I_5 R_5}{(V_{GS} - V_{tp})_5} \tag{8}$$

where I_5 ($2I_{ref}$) is the dc bias current through $M5$. The first two terms in (8) represent the gain of an n-channel common-source amplifier with a p-channel diode load. It is similar in form to the gain of a single-channel enhancement inverter [10] with increased gain based on the square root of the n-channel to p-channel mobility ratio, but with more gain variance due to the independence of μ_n and μ_p. The third term represents the gain of the resistively loaded common-source amplifier where $I_5 R_5 = V_{ref}$ is the dc bias voltage across the load resistor and $(V_{GS} - V_{tp})_5$ is the effective turn-on voltage of $M5$. (It is interesting to note that this gain term is similar to that of a resistively loaded common-emitter bipolar amplifier with kT/q replaced by $(V_{GS} - V_{tp})_5/2$.) The bias current levels must be chosen to provide an adequate large signal slew rate (note that there is no slewing in resistively loaded stages) and small-signal bandwidth. These are standard calculations which will not be considered further with respect to the first comparator stage.

Having completed the design of the first stage, the next issue is to design the ac coupling network. The size of the coupling capacitor(s) is usually selected based on feedthrough matching

considerations. $C_s = 5$ pF was used in this design. At this point, the designer is faced with a conflicting set of requirements. Fig. 12 shows a differential half circuit of the coupling network between the first-stage output and the second-stage input. On the one hand, it is desirable to make C_x as small as possible to maximize the small signal bandwidth at node $\widehat{X}$ which is given by

$$\omega_x = 1/(R_5 C_x) = (1 + C_s/C_{in})/R_5 C_s \tag{9}$$

where $C_{in} = C_{stray} + C_{gs_{10}} + C_{gb_{10}} + (1 - a_{2a}) C_{gd_{10}}$. Assuming that $C_s \gg C_{in}$, this implies that a_{2a} should be small to minimize the Miller-multiplied $C_{gd_{10}}$ term. On the other hand, the gain from node $\widehat{X}$ to node $\widehat{W}$ is given by

$$A = a_{2a} \cdot C_s/(C_s + C_{in}) \tag{10}$$

which implies that a_{2a} should be large. Fortunately, a near-optimum solution to this problem can be obtained graphically as shown in Fig. 13 where (9) and (10) are plotted versus the low frequency gain of the stage being ac coupled into a_{2a}. For low gain values, the bandwidth approaches $(R_5 C_{in})^{-1}$, and for high gain values, it approaches $(R_5 C_s)^{-1}$. For high a_{2a} values, A approaches $-C_s/C_{gd_{10}}$ since the input becomes a virtual ground with $C_{gd_{10}}$ as the feedback capacitor. The crossover point where $a_{2a} = 10$ was chosen for this design.

One approach for obtaining the gain of 10 is to again use an n-channel differential pair with p-channel diode loads as used in the first stage. However, based on (8), it is difficult to achieve this relatively large gain with a reasonable aspect ratio. Therefore, it was decided to use a controlled amount of positive feedback to effectively increase the driver device transconductance [11]. The gain of the positive feedback cell of Fig. 6 is given by

$$a_{2a} = \frac{g_{m10}}{g_{m12}} (1 - \alpha)^{-1} \tag{11}$$

where $\alpha = (W/L)_{14}/(W/L)_{12}$. In this design, $\alpha = \frac{2}{3}$, giving a factor of three gain increase. Care must be exercised in selecting the value of α. If $\alpha = 1$, the stage becomes a positive feedback latch. If $\alpha < 1$, linear amplification with increased gain is obtained as in this design. For $\alpha > 1$, the stage becomes a Schmitt trigger circuit, with the amount of hysteresis dependent on the value of α. For linear amplification, $\alpha_{nom} = 0.9$ is a practical maximum because with mismatches and processing variations, the effective α can approach or even exceed one which leads to undesirable comparator hysteresis. The remaining comparator stages shown in Fig. 11 are conventional and will not be considered further.

V. EXPERIMENTAL RESULTS

Fig. 14 shows a die microphotograph of the comparator circuit which was fabricated using a p-well metal-gate CMOS process with 8 μm minimum feature sizes. The die size is 560 μm (22.0 mils) by 950 μm (37.4 mils).

The most basic, and often the first, test for a comparator circuit is the qualitative "good or bad" test based on the performance of the complete A/D converter system. For example, Fig. 15(a) shows the measured dc transfer characteristic of an A/D converter which was judged to be bad due to the rather gross nonlinearity as indicated. This problem was traced to the comparator circuit which was different from the design described in this paper. More specifically, internal comparator nodes were being bootstrapped above V_{dd} as described earlier. This resulted in the gain change in the transfer characteristic (nonlinearity) due to the flow of parasitic substrate current which removed charge from the interstage coupling capacitors for large internal positive swings. Fig. 15(b) shows the measured transfer characteristic for the same A/D converter system using the comparator design described herein. The use of the controlled supply-independent biasing techniques has obviously eliminated the bootstrapping effect since the performance is now good. Furthermore, while the old design exhibited increased nonlinearity for larger supply voltages (larger positive voltage swings), the present transfer characteristic maintained linearity over the entire range of supply voltages.

Due to probe and instrument parasitics, direct comparator measurements are difficult to obtain without interfering with the performance of the circuit. Fortunately, many of the important parameters can be inferred quantitatively from the A/D transfer characteristic. For example, by increasing the conversion frequency, the available response time for the comparator is decreased; eventually, there will be insufficient time for the comparator to respond to an LSB input. As a result, the A/D digital output code will remain at all zeros for small input levels. The conversion frequency at the onset of the nonlinear-

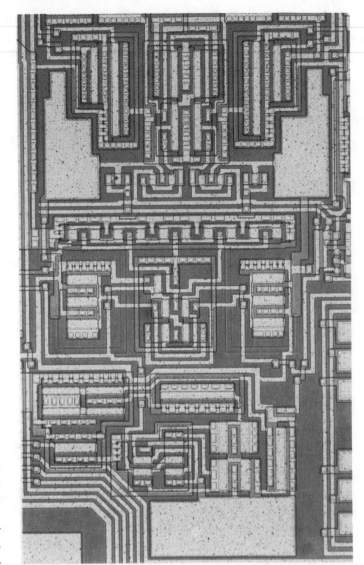

Fig. 14. Die microphotograph of the metal-gate CMOS comparator circuit. The layout orientation is identical to the schematic orientation of Fig. 11.

ity in the transfer characteristic indicates the switching speed. Using this method, the comparator's switching time was determined to be about 3 μs for a 4 mV LSB and about 12 μs for a 1 mV LSB.

Although not performed here, this technique can also be used to infer the equivalent input noise of the comparator. As the LSB input level is decreased, the probability of error in the A/D digital code is increased. By taking a large number of transfer characteristics, the error probability versus input level can be determined, and from these data, assuming a Gaussian distribution, the equivalent rms input noise can be determined. The calculated rms input noise for this comparator is nominally about 100 μV.

VI. CONCLUSIONS

Several new design techniques for variable-supply MOS analog circuits have been described in terms of a precision CMOS comparator design. In A/D conversion systems that contain an on-chip voltage reference, techniques have been developed to replicate the reference voltage to produce stable supply-independent dc bias voltages for other analog circuitry including

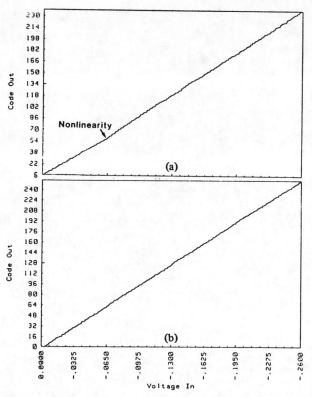

Fig. 15. Measured A/D dc transfer characteristics. (a) Using a previous comparator design, the A/D characteristic exhibited nonlinearity due to internal nodes of the comparator being bootstrapped above the supply V_{dd}. (b) Using the comparator described herein, the same A/D converter system maintained linearity over the entire range of supply voltages. V_{ref} was untrimmed in both cases.

the comparator. Controlled biasing techniques have also been developed which allow differential resetting in the comparator, which is beneficial in minimizing the input-referred offset voltage due to clock feedthrough and channel charge-pumping

effects. These controlled biasing techniques eliminate the possibility of internal comparator nodes being bootstrapped above the power supply, which results in A/D nonlinearity if it occurs. A generalized technique for designing ac-coupled gain stages was also presented.

ACKNOWLEDGMENT

The author gratefully acknowledges the essential contributions and suggestions of Dr. W. C. Black, Jr., and J. R. Hellums, J. R. Ireland, and M. B. Terry. J. Baechler and V. Allstot are also acknowledged for helping prepare the final manuscript.

REFERENCES

[1] P. R. Gray, D. A. Hodges, and R. W. Brodersen, *Analog MOS Integrated Circuits.* New York: IEEE Press, 1980.
[2] J. L. McCreary and P. R. Gray, "All-MOS charge redistribution analog-to-digital conversion techniques—Part I," *IEEE J. Solid-State Circuits*, vol. SC-10, pp. 371–379, Dec. 1975.
[3] R. E. Suarez, P. R. Gray, and D. A. Hodges, "All-MOS charge redistribution analog-to-digital conversion techniques—Part II," *IEEE J. Solid-State Circuits*, vol. SC-10, pp. 379–385, Dec. 1975.
[4] F. Guteri, "Cameras that 'think,'" *IEEE Spectrum*, pp. 32–37, June 1982.
[5] J. S. Brugler and P. G. A. Jespers, "Charge pumping in MOS devices," *IEEE Trans. Electron Devices*, vol. ED-16, pp. 297–302, Mar. 1969.
[6] A. R. Hamade, "A single chip all-MOS A/D converter," *IEEE J. Solid-State Circuits*, vol. SC-13, pp. 785–791, Dec. 1978.
[7] M. E. Hoff, Jr., J. Huggins, and B. M. Warren, "An NMOS telephone codec for transmission and switching applications," *IEEE J. Solid-State Circuits*, vol. SC-14, Feb. 1979.
[8] R. A. Blauschild, P. A. Tucci, R. S. Muller, and R. G. Meyer, "A new NMOS temperature-stable voltage reference," *IEEE J. Solid-State Circuits*, vol. SC-13, pp. 767–774, Dec. 1978.
[9] M. B. Terry, private communication.
[10] Y. P. Tsividis, "Design considerations in single-channel MOS analog integrated circuits—A tutorial," *IEEE J. Solid-State Circuits*, vol. SC-13, pp. 383–391, June 1978.
[11] K. B. Ohri and M. J. Callahan, Jr., "Integrated PCM codec," *IEEE J. Solid-State Circuits*, vol. SC-14, pp. 38–46, Feb. 1979.

A 750MS/s NMOS Latched Comparator

David C. Soo, Alexander M. Voshchenkov, Gen M. Chin,

Vance D. Archer, Maureen Lau, Mark Morris

AT&T Bell Laboratories

Holmdel, NJ

Ping K. Ko*, Robert G. Meyer**

University of California

Berkeley, CA

Bruce A. Wooley*

Stanford University

Stanford, CA

THIS PAPER WILL DESCRIBE THE DESIGN of analog circuits integrated in a polysilicon gate NMOS technology with 1μ effective channel length devices[1]. In particular, a latched comparator with 4b input resolution at 750MS/s and a wideband amplifier with 10dB of voltage gain over a bandwidth of 1.25GHz, when driving 130fF of on-chip capacitance, will be reported. The comparator was designed primarily for application to high-speed flash A/D conversion, but is is also suitable for use in integrated broadband fiber optic receivers. Its circuit configuration differs from previous designs[2,3] in that negative feedback is used in the preamplifier section to trade gain for bandwidth.

Preamplification in the comparator has been accomplished by differential implementation of the wideband amplifier topology shown in single-ended form in Figure 1. This amplifier design is suitable for a wide variety of both small and large-signal on-chip applications within larger NMOS circuits. For driving off-chip loads, a simple broadband output buffer, also shown in Figure 1, was added.

In Figure 1, transistor M_1 forms a transconductance input stage, while M_2 and the feedback transistor M_3 form a transresistance stage. M_4 and M_5 are source followers used to buffer the transconductance stage output, and M_6-M_{10} are depletion mode current sources that provide dc biasing for the amplifier. The compensation capacitor, C_c, is used to optimize the complex pole response of the amplifier to suit the particular application.

Active, rather than resistive, feedback is used in the circuit in Figure 1, because it provides higher loop gain, and the amplifier gain is, therefore, less sensitive to variations in device transconductance and output resistance. If the effects of M_4 and M_5 are neglected, the dc gain and the per-stage gain-bandwidth product of the amplifier are approximated by

---

*Bruce A. Wooley and Ping K. Ko were formerly with AT&T Bell Laboratories.

**Research supported in part by the U.S. Army Research Office under Grant DAAG29-84-K-0043.

[1] Ko, P.K., et. al., "SIGMOS-A Silicon Gigabit/sec NMOS Technology", *IEDM Technical Digest*, p. 751-753; 1983.

[2] Suzuki, M., et. al., "A Bipolar Monolithic Multigigabit/s Decision Circuit", *IEEE J. Solid State Circuit*, Vol. SC-19, p. 462-467; Aug., 1984.

[3] Meignant, D., et. al., "A High Performance 1.8GHz Strobed Comparator for A/D Converter", *IEEE GaAs IC Symposium Technical Digest*, p. 66-69; 1983.

$$G = \frac{g_{m1}}{g_{m2}}\left[1 + \frac{1}{g_{m2}r_{o2}}\left(1 + \frac{1}{g_{m2}r_{o1}}\right)\right] - 1 \qquad (1)$$

$$\sqrt{G} \times BW = \left[\frac{g_{m1}g_{m2}}{C_1 C_2}\left(\frac{1}{1 + \frac{C_c}{C_1} + \frac{C_c}{C_2}}\right)\right]^{1/2} \qquad (2)$$

As is apparent from (1), the dc gain is proportional to the ratio of two well-matched transconductances and is therefore relatively insensitive to variations in temperature and processing. As seen from (2), the per-stage gain bandwidth product of the amplifier approaches the transistor ω_T ($=g_m/C_s$) as C_c approaches zero.

To test the small-signal response of the preamplifier, the single-ended version was fabricated together with a broadband output buffer capable of driving a 50Ω load. The effects of the output buffer were subtracted from the test system measurements to determine the preamplifier response. The buffer has an input capacitance of 130fF and a voltage gain of 0.75 when driving a 50Ω load. The frequency response measured for the preamplifier is shown in Figure 2.

The amplifier configuration described has been used for preamplification in a 750MS/s NMOS latched comparator. This comparator differs from conventional designs in that wideband controlled gain, rather than open-loop high-gain, preamplification, is combined with a sensitive latch in a 2-phase clock architecture. By this means the large signal digital outputs are obtained via regeneration in the latch rather than gain in the preamplifier, and an increase in comparator speed is thus achieved.

Figure 3 is the schematic of the latched comparator. It is essentially a differential wideband amplifier (M_1-M_{23}) driving a positive feedback latch (M_{22}-M_{30}), M_1 and M_2, together with the input capacitance of M_3 and M_4, form an analog sample-and-hold circuit with a -3dB bandwidth above 1GHz. Common mode charge injection and clock coupling from M_1 and M_2 are rejected by the differential pair M_3 and M_4. ϕ_1 and ϕ_2 are complementary overlapping clocks. At the falling edge of ϕ_1, the input signal is sampled onto the amplifier input capacitance and held for the entire time clock ϕ_2 is high. While ϕ_2 is high, M_{21} enables the negative feedback loop and the wideband amplifier establishes a differential voltage at its output which represents the initial latch voltage, V_i. When ϕ_1 goes high again (and ϕ_2 falls) M_{30} enables the latch. Regenerative feedback then drives the latch output to one of the two stable states as governed by the initial imbalance V_i. The large signal voltage developed across the outputs during ϕ_1 will be referred to as the final latch voltage, V_f.

The length of time ϕ_1 must be kept high (denoted as τ_1) is determined by the regeneration in the latch. Neglecting effects

Reprinted from *1985 IEEE Int. Solid-State Circuits Conf.*, pp. 146–147, 327, Feb. 1985.

of source followers M_{28} and M_{29}, and assuming linear operation, τ_1 is approximated by

$$\tau_1 \approx \tau_T \left(\frac{A_l}{A_l - 1} \right) \ln \frac{V_f}{V_i} \qquad (3)$$

where $\tau_T = C_l/g_{m21,22}$, A_l^2 is the positive feedback loop gain and C_l is the capacitive loading at the output nodes. Depending on the technology used, τ_T can be made to approach the transistor transit time, C_g/g_m. Regeneration thus offers the fastest possible generation of a large signal from a small one, assuming $A_l \gg 1$.

The length of time ϕ_2 is high, τ_2, is governed by the acquisi-for the output voltage to change from V_f to the minimum initial latch voltage, $V_i(min)$, when the input of the amplifier is driven only by a voltage, V_{lsb}, equal to the minimum resolvable voltage expected of the comparator; that is, $V_i(min) = A_p V_{lsb}$, where A_p is the preamplifier gain. Qualitatively, τ_2 increases with increases in both A_p and, to a lesser extent, the size of the compensation capacitor, C_c.

Circuit simulation was used to determine the optimum choices for A_p and C_c. Since τ_2 decreases and τ_1 increases with decreasing A_p, A_p is chosen so that $\tau_1 = \tau_2$. This choice also simplifies the design requirements for generating two clock phases. After A_p is chosen, C_c is then adjusted so that overshoot at the output is always less than the minimum initial latch voltage, $V_i(min)$. In the present design A_p is about 3.

The comparator was integrated together with clock drivers and output buffers that facilitate monitoring the outputs with 50Ω test equipment.

Measured dc offset voltage is on the order of 25mV and the comparator input common mode range is +1V to −2V. For a full-scale input voltage of 2.5V, the comparator has an equivalent input resolution of 4b at a sampling rate of 750MS/s. Figure 4 shows the output waveform at 750MS/s when a periodic pattern (101100111000...) is applied to the input. Figure 5 is the eye diagram obtained when a 1/2b density pseudorandom pattern is applied to the input. In Figure 6, the measured input resolution of the comparator for an error rate of 10^{-9} is plotted as a function of the sampling rate.

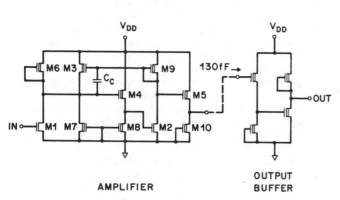

FIGURE 1—Wideband feedback amplifier.

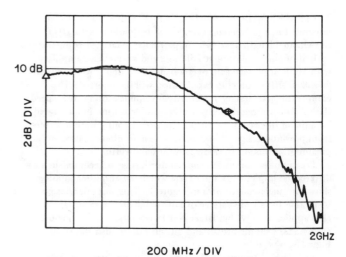

200 MHz / DIV

FIGURE 2—Frequency response of wideband amplifier.

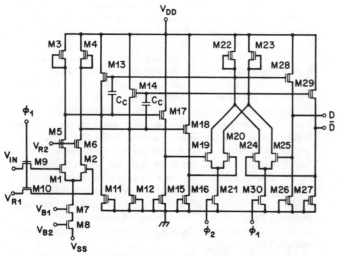

FIGURE 3—Latched comparator.

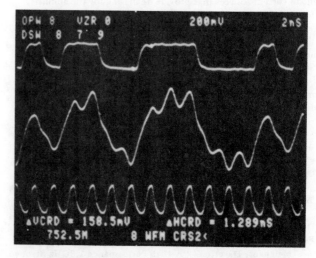

FIGURE 4—Output waveform of comparator.

108

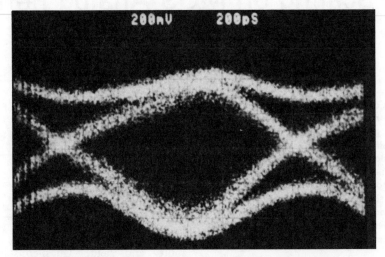

FIGURE 5—Eye diagram of comparator output (return to zero).

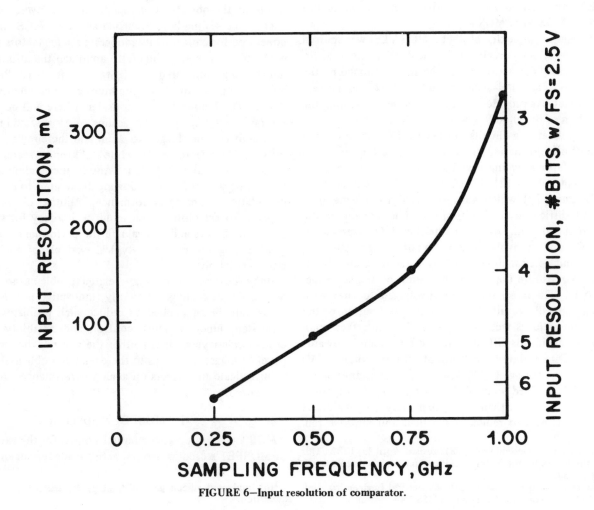

FIGURE 6—Input resolution of comparator.

Impact of Scaling on MOS Analog Performance

STEPHEN WONG AND C. ANDRE T. SALAMA, MEMBER, IEEE

Abstract—A first-order analysis of the impact of scaling on MOS analog performance under moderate scaling conditions is presented in this paper. Assuming a polysilicon gate ion-implanted MOS technology, quasi-constant voltage (QCV) scaling is shown to be the optimal scaling law, offering the best overall analog performance and resulting in an increase in functional density, gain–bandwidth product with a moderate degradation in gain, and signal-to-noise ratio. The first-order analysis agrees fairly well with computer simulation. A typical case study shows that under moderate scaling conditions, CMOS can generally offer a higher voltage gain when compared to depletion load NMOS and is the preferred technology for scaled analog implementations.

TABLE I
SCALING LAWS

	CE	QCV	CV
Voltages	λ^{-1}	$\lambda^{-1/2}$	1
Lateral dimensions	λ^{-1}	λ^{-1}	λ^{-1}
Vertical dimensions	λ^{-1}	λ^{-1}	$\lambda^{-1/2}$
Doping concentration	λ	λ	λ

I. INTRODUCTION

WITH the ever increasing complexity of very large scale integrated circuits, it has become highly desirable to integrate analog and digital circuits on a single silicon chip. The primary VLSI focus has been on those technologies which permit high density integration of analog and digital circuits on a single chip. The most useful MOS technologies in this context are NMOS and CMOS [1].

The increasing complexity of digital MOS VLSI was spurred by the concept of device scaling [2]. The density of digital circuits has increased continuously and the area consumed by the analog portion of a digital–analog circuit has become relatively larger. This fact has provided a strong impetus for scaling the analog portion of the circuit as well as the digital one.

The scaling of analog circuits can benefit from the considerable amount of information available on the scaling of digital MOS IC's. Various scaling laws have been proposed and are summarized in Table I [2], [3]. The first of these is constant field (CE) scaling [2] which was introduced as a means of alleviating the difficulties arising from two-dimensional parasitic effects associated with short channels in MOS transistors. Basically, CE scaling involves a reduction of voltages, currents, and lateral and vertical dimensions by a factor λ and an increase in the substrate doping concentration by the same factor λ ($\lambda \geq 1$). The linear reduction of voltage associated with CE scaling generally results in a sizeable deterioration of the signal-to-noise ratio as well as a lack of TTL interface compatibility. To avoid this difficulty, the current trend is towards nonconstant field scaling in the form of constant voltage (CV) or quasi-constant voltage (QCV) scaling [3]. Constant voltage scaling implies a fixed nonscaled power supply (compatible with TTL) and a slower than λ (approximately $\sqrt{\lambda}$) scaling of gate–oxide thickness to reduce oxide field and improve reli-

ability and yield. Quasi-constant voltage scaling implies a slower than λ (taken as $\sqrt{\lambda}$ for convenience sake) decrease in voltage while the other factors scale as in the CE case.

QCV scaling was found to provide optimum drive current capability in digital VLSI MOS circuits [3]. Although this factor is an important criterion in digital applications, parameters such as gain, bandwidth, and signal-to-noise ratio are more suitable criteria to evaluate the performance of analog circuits.

One of the objectives of this paper is to investigate the impact of the scaling laws discussed above on MOS analog component performance. Only moderate scaling factors are considered and no attempt is made to approach the ultimate scaling limits being considered for digital circuits [4]. Present day MOS analog circuits have minimum channel lengths of about 8 μm. The scaling factors considered here will be limited to the range $\lambda = 1$ to $\lambda = 4$. A scaling of $\lambda = 4$ would reduce the minimum channel length to 2 μm and the circuit area by approximately a factor of 16 while still keeping second-order effects within bounds [5], [6]. Any further scaling would lead (as will be seen from the following discussion) to unacceptable degradation in analog performance. Another objective of this paper is to establish the optimum technology for scaled analog MOS circuits and to evaluate the performance of a scaled MOS analog op amp as a typical example of analog circuit implementation.

Many intricate technology related problems are expected during device scaling. Presently, innovative processing techniques are being applied to reduce such problems as device mismatch, junction depth control, and reliability of scaled down oxide layers. It is neither the aim nor the scope of this paper to suggest solutions to these related problems, but merely to investigate the effects on analog performance when scaling is achieved.

II. MOSFET MODELING

MOS analog integrated circuits[1] consist for the vast majority of MOSFET's used as drivers, active loads for analog switches,

Manuscript received December 4, 1981; revised April 23, 1982. This work was supported by the Natural Sciences and Engineering Research Council of Canada.

The authors are with the Department of Electrical Engineering, University of Toronto, Toronto, Ont., Canada M5S 1A4.

[1]Ion-implanted silicon gate technology is assumed throughout the discussion.

Reprinted from *IEEE J. Solid-State Circuits*, vol. SC-18, no. 1, pp. 106–114, Feb. 1983.

110

as well as MOS capacitors, resistors (diffused or polysilicon), and interconnections.

Since the MOSFET's are the most critical components in any analog implementation, the discussion will focus mainly on those MOSFET parameters which have a direct influence on analog performance.

A. I–V Characteristics

In order to account for second-order effects which may arise due to scaling, a set of analytical equations must be used. Modeling of small geometry (short and narrow channel) devices have been treated extensively by many authors [7]–[10]. These models generally include the following phenomena: 1) threshold dependence on two-dimensional charge sharing effects [11], [12], 2) mobility-degradation due to increased normal and lateral fields [13], and 3) velocity saturation of carriers causing premature saturation of current and lower output conductance [14], [15].

When moderate scaling effects are considered, two additional factors must be considered. These are 1) mobility degradation due to impurity scattering when the effective substrate doping exceeds 5×10^{16} cm^{-3} [16], and 2) gain reduction via feedback through the parasitic drain and source resistances.

The saturation region of the MOSFET plays an important role in determining analog performance parameters such as transconductance, output resistance, and voltage gain. The set of equations selected for the analysis are based on a semi-empirical model of the MOSFET operation in this particular region of operation.

The threshold voltage V_T of the device can be expressed as

$$V_T = V_{TO} - \Delta V_{TS} + \Delta V_{TN} \tag{1}$$

where V_{TO} is the threshold voltage for a large geometry device and ΔV_{TS} and ΔV_{TN} are the corrections for short and narrow channel effects, respectively [7]. V_{TO} is given by

$$V_{TO} = \phi_{MS} + 2\phi_F - \frac{Q_{SS} + Q_B}{C_0} \pm \frac{qN_iD}{C_0} \tag{2}$$

where ϕ_{MS} is the work function difference between gate and bulk material, ϕ_F is the Fermi potential given by

$$\phi_F = \frac{kT}{q} \ln \frac{N_B}{n_i} \tag{3}$$

where N_B is the bulk doping concentration under the channel and Q_{SS}, Q_B are the oxide and bulk charges, respectively. The last term in (2) represents the effect of the ion implant used to adjust the threshold voltage. N_iD is the effective implant dose and D is the implant depth which is assumed to be very shallow.

Mobile carriers in the channel of a scaled MOST normally experience scattering effects due to the electric fields and increased impurity levels. This leads to a significant degradation of the channel mobility μ which is usually expressed as

$$\mu = \frac{\mu_0}{\delta_1 \delta_2} \tag{4}$$

where μ_0 is the zero field mobility and δ_1, δ_2 are second-order parameters describing the degradating effects of high field and

high doping concentration on mobility, respectively. Above saturation, the parameter δ_1 is defined as

$$\delta_1 = 1 + \frac{\theta}{t_0}(V_G - V_T) + \frac{V_{D,\text{sat}}}{E_c L} \tag{5}$$

where θ is a constant (typically 3×10^{-7} cm/V), E_c is the critical field for velocity saturation (2×10^4 V/cm for NMOS and 2×10^5 V/cm for PMOS), $V_{D,\text{sat}}$ is the saturation voltage and is taken as $(V_G - V_T)$ for a first-order analysis. The parameter δ_2 accounts for impurity scattering effects and is defined as

$$\delta_2 = \left(1 + \frac{N_B}{5 \times 10^{16}}\right)^{1/2}. \tag{6}$$

The gain constant β of the transistor is defined as

$$\beta = \frac{\mu C_0}{\delta_3} \frac{Z}{L} \tag{7}$$

where Z is the channel width and δ_3 accounts for the feedback degradation due to the source resistance R_s

$$\delta_3 = 1 + \beta R_s(V_G - V_T). \tag{8}$$

R_s is proportional to the length but inversely proportional to the width, depth, and doping of the source junction. The effects of the inversion layer capacitance and contact resistance on β are neglected here. These effects are only relevant for channel lengths below 1.5 μm [6].

The saturation current of the device can be expressed as

$$I_{D,\text{sat}} = \frac{\mu C_0 Z}{2L} \frac{(V_G - V_T)^2}{\delta_3 \delta_4} \tag{9}$$

where δ_4 is a second-order parameter which takes into account the body effect on the channel. δ_4 is defined as

$$\delta_4 = 1 + \frac{\gamma}{2(2\phi_F - V_B)^{1/2}} \tag{10}$$

where

$$\gamma = \frac{(2\epsilon_s q N_B)^{1/2}}{C_0} \tag{11}$$

where ϵ_s is the dielectric constant of silicon.

The transconductance in the saturation region can be expressed as

$$g_m = \frac{\mu C_0 Z}{L} \frac{(V_G - V_T)}{\delta_3 \delta_4}. \tag{12}$$

B. Output Conductance

The output conductance of the device in the saturation region is critical in determining the voltage gain of inverters as well as the output impedance of current source. Three important phenomena affect the channel conductance of the saturated enhancement mode MOSFET. These are 1) classical pinchoff [17], 2) velocity saturation [14], [15], and 3) feedback caused by drain induced barrier lowering [18], [19].

For moderate channel lengths, the first two phenomena dominate. The current in the device beyond saturation is increased by channel length modulation and is given by

$$I_D = I_{D,\text{sat}} \frac{L}{L - \Delta L}. \tag{13}$$

For long channel lengths, ΔL is the reverse bias depletion width formed between the drain and the channel and is defined as

$$\Delta L = K_1 [V_D - V_{D,\text{sat}}]^{1/2} \tag{14}$$

where

$$K_1 = [2\epsilon_s/qN_B]^{1/2}. \tag{15}$$

As the internal field in a scaled device increases as a result of nonconstant field scaling, velocity saturation of mobile carriers, and the deterioration of the gradual channel approximation make it necessary to modify the drain depletion length ΔL as follows [14], [15]:

$$\Delta L = \left[\left(\frac{E_p K_1^2}{2} \right)^2 + K_1^2 (V_D - V_{D,\text{sat}}) \right]^{1/2} - \frac{E_p K_1^2}{2} \tag{16}$$

where E_p is the lateral field at the drain during channel pinch-off. Out of convenience, E_p is usually equated to the critical field for velocity saturation E_c [20]. A more realistic value of E_p has been derived by Rossel [15] to ensure the continuity of current and conductance in the triode and saturation regions. His expression for E_p can be approximated by

$$E_p \simeq \left[\frac{qN_B(V_G - V_T)^2}{2\epsilon_s L \delta_3 \delta_4} \right]^{1/3}. \tag{17}$$

For moderate E_p, (16) can be rewritten as (see Appendix I)

$$\Delta L = K_1 (V_D - V_{D,\text{sat}})^{1/2} \delta_5 \tag{18}$$

where δ_5 is a correction factor for channel shortening given by

$$\delta_5 = 1 - \frac{E_p K_1}{2(V_D - V_{D,\text{sat}})^{1/2}}. \tag{19}$$

Differentiating (13), using the value of ΔL given by (18), yields the output conductance

$$g_{ds} = \frac{dI_D}{dV_D} = \frac{I_{D,\text{sat}} K_1}{2L(V_D - V_{D,\text{sat}})^{1/2} \delta_6} \tag{20}$$

where δ_6 is defined as

$$\delta_6 = \left(1 - \frac{\Delta L}{L} \right)^2. \tag{21}$$

From (20), g_{ds} is proportional to $I_{D,\text{sat}}$, as observed experimentally in moderately scaled devices.

For very short channel lengths, extensive drain induced barrier lowering (DIBL) occurs resulting in a large dependence of threshold voltage on V_D as observed by Masuda [21]. As a direct consequence of this effect, g_{ds} will be directly proportional to $(V_G - V_T)$ and inversely proportional to L^3 (as shown in Appendix II). This strong dependence of g_{ds} on L cannot be tolerated in analog applications and in general the DIBL regime must be avoided.

C. Subthreshold Current

The exponential dependence of the subthreshold current plays an important role in analog applications. For $V_G < V_T$ and $V_D \gg kT/q$, the weak inversion current can be expressed as [2]

$$I_{D,\text{sat}} \simeq \frac{\mu C_0 Z}{L} \left(\frac{kT}{q} \right)^2 \exp \left[\frac{q(V_G - V_{\text{on}})}{nkT} \right] \tag{22}$$

where V_{on} is the turn on voltage at a surface potential $\phi_s = 1.5\phi_F$, C_d is the surface depletion region capacitance defined as

$$C_d = \left[\frac{\epsilon_s q N_B}{2 \left(\phi_s + \frac{kT}{q} \right) - V_B} \right]^{1/2} \tag{23}$$

and n is given by

$$n = 1 + \frac{C_d}{C_0}. \tag{24}$$

Due to the exponential nature of the subthreshold current, it does not scale linearly with λ. A commonly used figure of merit to describe subthreshold behavior is the slope S_s of the $\log I_D$ versus V_G curve. S_s can be expressed as

$$S_s = \frac{d \log I_{D,\text{sat}}}{dV_G} = \frac{q}{2.3 \, nkT}. \tag{25}$$

In an analog switch, for instance, S_s is a significant factor in determining the on–off current ratio of the device.

D. Noise

The noise in MOS devices, working at low frequencies, is high due to the dominant contribution of $1/f$ noise [22]. The rms equivalent gate noise voltage, in this case, can be expressed as

$$V_{ng} = \left(\frac{a_n}{ZL} \frac{q^2}{C_0^2} \frac{\Delta f}{f} \right)^{1/2} \tag{26}$$

where a_n is a constant dependent on the interface trap density at the Si-SiO$_2$ interface, Δf is the bandwidth and f is the frequency.

III. EFFECT OF SCALING ON MOS PARAMETERS

In this section, the effect of scaling on the MOS device parameters previously discussed in first investigated. The effect of scaling on MOS circuit components is then discussed. The fact that scaling will limit the accuracy and the ability to match components is not considered in detail here since it is technology dependent.

A. Subthreshold Current

In order to avoid large subthreshold currents (and the onset of substantial DIBL), the long channel index M suggested by

[2] Assuming the surface state density at the Si-SiO$_2$ interface to be negligible.

Brews *et al.* [23] can be used. This index is defined as

$$M = \frac{A\left[(x_j t_0)(W_S + W_D)^2\right]^{1/3}}{L} \qquad (27)$$

where A is a constant and W_S, W_D are the widths of source and drain depletion regions, respectively, and are given by

$$W_S = \left[\frac{2\epsilon_s}{qN_B}(2\phi_F - V_B)\right]^{1/2} \qquad (28)$$

$$W_D = \left[\frac{2\epsilon_s}{qN_B}(2\phi_F - V_B + V_D)\right]^{1/2}. \qquad (29)$$

For x_j, W_S, W_D, and L in microns and t_0 in Å, the value of A is $0.41(\text{Å})^{1/3}$ for n-channel devices. Upon scaling, a value of $M \leqslant 1$ will guarantee that long channel subthreshold behavior is maintained. When $M > 1$ undesirable short channel effects are expected.

B. Threshold Voltage

For successful scaling, the threshold voltage must scale with the other voltages. However, due to the nonscalable term $(\phi_{MS} + 2\phi_F)$ in (2), it is unlikely that V_T can be scaled properly without compensation. In general, scaling of this term produces a V_T which is too large for CE scaling and too small for QCV scaling. The problem is more critical in the PMOS case (assuming an n$^+$ polysilicon gate technology), where $(\phi_{MS} + 2\phi_F)$ is more significant. While some adjustment can be made by varying V_B in NMOS technology, this is not possible in CMOS.

Fortunately, ion implantation can be used in both NMOS and CMOS to adjust the threshold voltage. In a typical device which uses a shallow implant as a threshold adjust, it is common to obtain cancellation between the implant term and $(\phi_{MS} + 2\phi_F)$, whenever necessary[3], so that V_T becomes approximately

$$V_T \simeq \frac{-Q_B - Q_{SS}}{C_0} = \frac{\pm\sqrt{4\epsilon_s q N_B \phi_F} - Q_{SS}}{C_0} \qquad (30)$$

(positive sign applies for NMOS while the negative sign applies for PMOS). If $Q_{SS} \ll Q_B$, the above equation will generally yield the desired threshold voltages.

With (30), proper scaling for V_T is easily achieved under the CV and QCV laws, resulting in a threshold voltage scaled by ≈ 1 and $\approx \lambda^{-1/2}$, respectively. To ensure that (30) remains valid, one requires that the condition

$$\left|\frac{qN_iD}{C_0}\right| \approx \left|\phi_{MS} + 2\phi_F\right| \qquad (31)$$

holds under scaling. Since it is desirable that N_i scales with λ, this implies that the implant depth D must scale with $\lambda^{-1/2}$ in the CV case, and remain fixed for QCV and CE scaling. However, adjustment of D may be necessary to scale V_T with $1/\lambda$ (instead of $1/\sqrt{\lambda}$) under the CE law, and to compensate for the second-order short channel and narrow width effects.

[3]For instance, in an n$^+$ polysilicon gate p-well CMOS process, the PMOS threshold voltage is adjusted in this manner.

TABLE II
DEVICE SCALING

	CE	QCV	CV	Equation
$I_{D,\text{sat}}$	λ^{-1}	1	$\lambda^{1/2}$	9
g_m	1	$\lambda^{1/2}$	$\lambda^{1/2}$	12
g_{ds}(enh)	1	$\lambda^{3/4}$	λ	20
g_{ds}(dep)	1	$\lambda^{1/4}$	$\lambda^{1/2}$	34
$\frac{C_d}{C_0}$	1	$\lambda^{-1/4}$	1	23
S_s	1	$\geqslant 1$	1	25
V_{ng}	1	1	$\lambda^{1/2}$	26
M	$\lambda^{-1/3}$	$\lambda^{-1/6}$	$\lambda^{1/3}$	27

Nevertheless, proper scaling of V_T seems feasible using ion implantation and has been assumed in the following discussion.

C. First-Order Parameter Scaling

In the model already described, the δ's represent second-order effects which do not scale linearly with λ. Their magnitudes are strong functions of the initial unscaled device and the scaling laws. To obtain a first-order estimate of the effect of scaling on device parameters as well as to keep the analysis independent of the unscaled conditions, the parameters δ are first assumed to be unity. This implies an ideal long channel MOSFET as the unscaled device. Justification for this assumption, and a comparison of first-order scaling estimates with computer simulated results on analog performance, taking into consideration the effect of δ's, are given in Section V.

The first-order scaling for some relevant device parameters are listed in Table II. Limitations can be expected from the parameters S_s and V_{ng}, which do not scale-down as desired.

D. Capacitor Scaling

In MOS technology, capacitors are realized between metal and heavily doped silicon or between two layers of heavily doped polysilicon. Thermally grown silicon dioxide is used as the dielectric. Scaling of capacitors is straightforward. A direct scale down of the surface area and dielectric thickness lead to a reduction in total capacitance. However, difficulties arise in absolute accuracy and matching of component values as the effects of misalignment and fringing become relatively more important. Electron quantization noise may also become a problem as capacitors become smaller.

E. Resistor Scaling

In MOS technology, resistors are fabricated using either diffused layers or polysilicon layers deposited on oxide. The latter alternative is preferred since it produces minimal parasitic capacitance. Scaling, again, is relatively simple; however accuracy and matching as well as conduction mechanisms [24] in the polysilicon become serious limits as the resistors become smaller.

F. Interconnection Scaling

Three items are of prime concern in interconnection scaling. These are electromigration, in the very thin, narrow aluminum conductors, contact resistance in the very shallow junctions, and polysilicon interconnection sheet resistance. For the scaling conditions under consideration here, the first two items do not present serious constraints. As far as the third item is concerned, refractory metal silicides, which offer low sheet resistance, are a good alternative to polysilicon. In addition, it is reasonable to expect that, at least some of the longest interconnections will not scale-down at all, resulting in adverse effects on parasitic capacitance and line driving capabilities.

IV. EFFECT OF SCALING ON ANALOG BUILDING BLOCK PERFORMANCE

This section investigates the scaling tendencies of some basic analog building blocks with the objective of determining the optimal scaling law for analog applications. Two technologies are considered: CMOS and NMOS with depletion load.

A. Gain Stage Performance

The most useful figure of merit in evaluating performance of gain stages is the voltage gain. In a CMOS stage, this gain is given by

$$A_{V,\text{CMOS}} = -\frac{g_m}{(g_{ds,n} + g_{ds,p})} \qquad (32)$$

where $g_{ds,n}$ and $g_{ds,p}$ are the n- and p-channel device output conductances respectively, as defined by (20). In NMOS technology, where a depletion device is used as load, the bulk threshold feedback effect normally dominates over channel length modulation effects in determining $g_{ds,\text{dep}}$ [20]. The voltage gain, in this case, is given by

$$A_{V,\text{NMOS}} = -\frac{g_m}{g_{ds,\text{dep}}} \qquad (33)$$

where

$$g_{ds,\text{dep}} = \frac{\gamma I_{D,\text{sat}}}{|V_{T,\text{dep}}| (V_{DD} - V_0 + 2\phi_F)} \qquad (34)$$

where $V_{T,\text{dep}}$ is the threshold voltage of the depletion mode device, V_0 is the quiescent output voltage, and V_{DD} is the supply voltage.

B. Op Amp Performance

Since the operational amplifier is one of the basic building blocks in analog circuitry, an evaluation of its performance subject to scaling would be useful. For such an analysis, the following performance indices must be considered.

Voltage Gain A_v—In a typical two-stage op amp which uses a Miller capacitor for compensation, the voltage gain is

$$A_V \propto \left(\frac{g_m}{g_{ds}}\right)^2. \qquad (35)$$

Unity Gain Bandwidth (GBW)—The unity bandwidth is mainly determined by the transconductance of the first stage g_{m1}, and the compensation capacitor C_c and can be expressed as

$$\text{GBW} = \frac{g_{m1}}{C_c}. \qquad (36)$$

TABLE III
SCALING OF ANALOG CIRCUITS

	CE	QCV	CV	Equation
Gain Stages				
$A_{V,\text{CMOS}}$	1	$\lambda^{-1/4}$	$\lambda^{-1/2}$	32
$A_{V,\text{NMOS}}$	1	$\lambda^{1/4}$	1	33
Op Amp				
$A_{V\text{tot},\text{CMOS}}$	1	$\lambda^{-1/2}$	λ^{-1}	35
$A_{V\text{tot},\text{NMOS}}$	1	$\lambda^{1/2}$	1	35
Power/Area	1	$\lambda^{3/2}$	$\lambda^{5/2}$	39
C_c	λ^{-1}	λ^{-1}	$\lambda^{-3/2}$	—
GBW	λ	$\lambda^{3/2}$	λ^2	36
S_R/V_{DD}	λ	$\lambda^{3/2}$	λ^2	38
S/N	λ^{-1}	$\lambda^{-1/2}$	$\lambda^{-1/2}$	37

The GBW is normally determined by the second parasitic pole of the amplifier which can be approximately expressed as

$$\text{GBW} = \frac{g_{m2}}{C_1 + C_2} \qquad (37)$$

where g_{m2} is the effective transconductance of the second stage, and C_1 and C_2 are the input (gate) and load capacitances associated with the second stage [25]. From (36) and (37), it appears that C_c will be proportional to $(C_1 + C_2)$. If one assumes C_2 to be the input (gate) capacitance of the subsequent stage, and that the overlap gate capacitance is small, then $(C_1 + C_2)$, and therefore C_c, will be directly proportional to AC_0, where A is the active gate area of an individual transistor. This implies that the area of C_c will scale with A and that the ratio of capacitor to op amp areas will remain constant under scaling.

Signal-to-Noise Ratio (S/N)—From (26), the signal to noise ratio for low-frequency analog applications can be expressed as

$$S/N = \frac{V_{sg}}{V_{ng}} \propto \frac{V_{sg}\sqrt{ZL}}{t_0} \qquad (38)$$

where the input signal amplitude V_{sg} is assumed to scale with other voltages. This implies that the signal-to-noise ratio is dependent on device geometry.

Slew Rate S_R—The slew rate normalized to the supply voltage V_{DD} can be used to provide an indication of large signal op amp response

$$\frac{S_R}{V_{DD}} = \frac{I_{D,\text{sat}}}{V_{DD}C_c}. \qquad (39)$$

Power Density—The power density is a useful figure of merit in determining the maximum packing density of op amps per unit area.

$$\text{power density} \propto \frac{I_{D,\text{sat}}V_{DD}}{ZL}. \qquad (40)$$

C. Optimum Scaling Law

The three scaling laws presented in the introduction were applied to the first-order scaling of the basic gain stages and the op amp parameters. The results are listed in Table III.

Other than an increase in speed resulting from reduction of C_c, the majority of the analog performance parameters remain

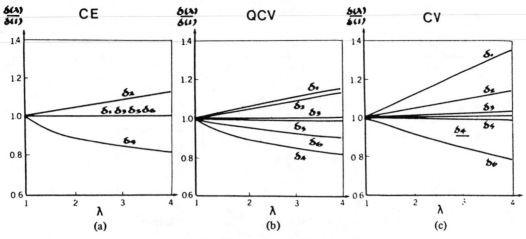

Fig. 1. Effect of scaling on δ's. (a) CE. (b) QCV. (c) CV.

invariant to the application of CE scaling. The exception being that the signal-to-noise ratio is severely degraded.

Although both nonconstant field scaling laws offer improvements in speed and frequency response, CV scaling appears unacceptable for analog applications because it results in the largest voltage gain degradation, the highest power dissipation per unit area without a significant gain in signal-to-noise performance. In addition (referring to (27) and Table II), the parameter M increases with CV scaling, resulting in short channel subthreshold effects becoming dominant at low values of λ.

QCV scaling offers an acceptable compromise, yielding improved speed and signal-to-noise ratio over the CE case, while exhibiting a higher gain and lower power dissipation than CV scaling.

V. CASE STUDY

The scaling analysis performed in the previous sections neglected all second-order effects by assuming an ideal long channel MOSFET as the unscaled device. This allows simple first-order scaling factors to be computed without involving the nonlinear δ terms. In this section, computer simulation involving the δ's is performed using 7 μm channel length MOSFET's as the initial unscaled devices. The simulation focuses mainly on the QCV law in light of the results of the previous section. These results are compared with the first-order scaling theory of Tables II and III. Scaling simulation is achieved by manually scaling all voltages, dimensions, and doping concentrations in accordance with the QCV law and then inputing them into a simulation program which accommodates our model. A representative case study of such a simulation is presented in the following paragraphs.

A. Effect of Scaling on the δ Parameters

As an example, consider a typical n-channel device having the following initial unscaled characteristics: $Z = 70 \,\mu$m, $L = 7 \,\mu$m, $t_0 = 800$ Å, $x_j = 1.2 \,\mu$m, $N_A = 5 \times 10^{15}$ cm^{-3}, $E_p = 1.1 \times 10^4$ V/cm, $V_D = 5$ V, $(V_G - V_T) = 1.5$ V, $V_B = 0$, $R_S = 20 \,\Omega$. In this case the value of M is 0.788, implying that the device is still in its long channel mode of operation. From Table II and (27), it is seen that M does not increase for CE or QCV scaling, but becomes greater than 1 for $\lambda \geqslant 2$ in the case of CV scaling. Using the unscaled device characteristics defined above, the δ values for $\lambda = 1$ are $\delta_1(1) = 1.163$, $\delta_2(1) = 1.049$, $\delta_3(1) = 1.008$, $\delta_4(1) = 1.590$, $\delta_5(1) = 0.855$, and $\delta_6(1) = 0.785$. The effect of scaling on the $\delta(\lambda)$ parameters were computed and are plotted in Fig. 1(a), (b), and (c) for CE, QCV, and CV scalings. The graphs show that in most cases the $\delta(\lambda)/\delta(1)$ ratio remains very near unity (i.e., the δ's do not change drastically with scaling). For δ_3 and δ_5, this is true for all three scaling laws, implying that parasitic resistance and velocity saturation effects are still negligible in the scaling range under consideration. Under worst case conditions, the $\delta(\lambda)/\delta(1)$ factors do not deviate by more than 20 percent from unity for CE and QCV scaling and by more than 40 percent for CV scaling. If CV scaling is restricted to $\lambda \leqslant 2$ (which as discussed above is required to prevent the onset of DIBL), the maximum deviation is 20 percent. These results imply that the second-order effects described by the δ terms can be ignored in establishing the general trend of scaling on device parameters as was done in Sections III and IV. The error caused by ignoring the δ's may sometimes be significant but not dominant.

B. Gain Stage Simulation

The performance of the NMOS and CMOS gain stages, shown in Fig. 2, were investigated using computer simulation. To provide a fair comparison, the two stages were designed to have identical operating conditions and Z/L ratios. To achieve a zero quiescent output voltage in the NMOS stage with quiescent input voltage $V_{in} = V_{DD}/2$ (typical operating conditions in a gain stage), one requires the ratio $(Z/L)_{driver} : (Z/L)_{load}$ to be 2.8. In the CMOS stage, the same current and aspect ratios are maintained if V_{bias} is set at $-0.9 \, V_{DD}$.

Scaling was carried out starting with a set of unscaled devices with characteristics listed in Table IV. The simulated gain for both gain stages as a function of the scaling factor λ, is shown in Fig. 3. Also shown in the figure are the corresponding gains observed from first-order scaling theory (Table IV). The maximum discrepancy between simulation and first-order theory is of the order of 20 percent. The unscaled gain of the CMOS stage remains higher than that of the NMOS stage. However, the gain in the NMOS case increases with increasing λ while the opposite is true in the CMOS case. The two gains approach each other at $\lambda \simeq 4$. The relative increase

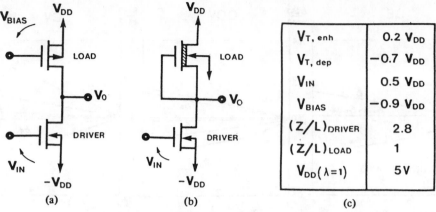

$V_{T, enh}$	$0.2\ V_{DD}$
$V_{T, dep}$	$-0.7\ V_{DD}$
V_{IN}	$0.5\ V_{DD}$
V_{BIAS}	$-0.9\ V_{DD}$
$(Z/L)_{DRIVER}$	2.8
$(Z/L)_{LOAD}$	1
$V_{DD}(\lambda=1)$	5 V

Fig. 2. Gain stages. (a) CMOS. (b) NMOS with depletion load. (c) Specifications.

TABLE IV
NMOS AND CMOS UNSCALED DEVICE CHARACTERISTICS

	NMOS Stage		CMOS Stage	
	Driver	Load	Driver	Load
Type	Enhancement	Depletion	n-channel	p-channel
$V_T(V)$	1	-3.5^a	1	-1^a
$N_B(cm^{-3})$	5×10^{15}	5×10^{15}	5×10^{15}	1.5×10^{15}
$\mu_0(cm^2/V \cdot s)$	600	600	600	300
$E_c(V/cm)$	2×10^4	2×10^4	2×10^4	2×10^5
$X_j(\mu m)$	1.2	1.2	2.2	1.2
$L(\mu m)$	7	7	7	7

^{a}Threshold voltages adjusted by means of shallow implants.

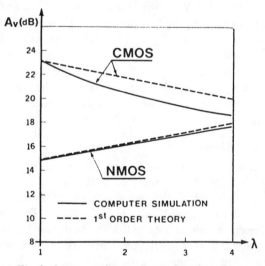

Fig. 3. Inverter voltage gain as a function of λ.

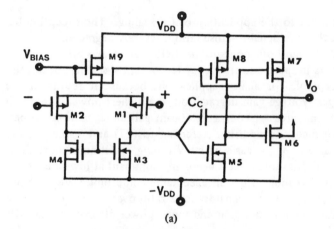

(a)

	Z (μm)	L (μm)
M 1	100	7
M 2	100	7
M 3	100	7
M 4	100	7
M 5	400	7
M 6	400	5
M 7	100	7
M 8	200	7
M 9	100	7
V_{DD}	5 V	
V_{BIAS}	2 V	
C_c	10 pF	

(b)

Fig. 4. CMOS operational amplifier. (a) Circuit diagram. (b) Unscaled specifications.

in the gain of the NMOS stage can be attributed to the fact that $g_{ds, dep}$ due to body effect increases at a slower rate ($\lambda^{1/4}$) than that due to channel length modulation ($\lambda^{3/4}$). However, this cannot be regarded as an advantage, since the latter mechanism present in the depletion NMOS would eventually overtake that due to body effect. From the above results, it appears that in the low λ ($\lambda < 4$) range, the CMOS stage can offer a higher gain than that of its NMOS equivalent. This result is not specific to the particular set of device parameters chosen, but was found to be true in the majority of the cases investigated.

C. Op Amp Simulation

In light of the conclusion that, in general, scaled CMOS offers a gain advantage over scaled NMOS gain stages, a CMOS op amp was selected for simulation. The unscaled op amp and its important layout characteristics are shown in Fig. 4. The

TABLE V
SCALING RESULTS OF OP AMP

	Voltage Gain Av		GBW (MHz)		Power Dissipation (mW)		R_{out} (kΩ)	
	Simulation	First-Order Theory	Simulation	First-Order Theory	Simulation	First-Order Theory	Simulation	First-Order Theory
1	903	903	5	5	16.1	16.1	28	28
2	589	639	12.6	14.1	11.3	11.4	15.7	16.6
4	357	452	31.6	40	8.3	8.2	8.7	9.9
Scaling Factor	$\lambda^{-1/2}$		$\lambda^{3/2}$		$\lambda^{-1/2}$		$\lambda^{3/4}$	

circuit configuration is typical of op amps presently used in telecommunication applications. It consists of two gain stages with a total gain of about 60 dB. A buffer stage is used in the feedback compensation loop to eliminate an undesirable zero at g_m/C_c [25]. The op amp has a useful feature associated with the fact that all the transistor channel lengths are nearly equal which guarantees equal V_T for all the transistors, even under scaled conditions.

Table V lists the results of the simulation as a function of λ. Also shown in the table are the results obtained from first-order scaling theory. Agreement between the simulation and first-order theory appears reasonable, confirming the validity of using the first-order theory to estimate trends in analog scaling.

VI. CONCLUSION

A theoretical analysis of the impact of scaling on analog component and circuit performance has been presented. Among the three scaling laws considered, QCV appears to be the optimum for analog scaling. Its application results in small area, high speed and moderate degradation in gain, power density, and signal-to-noise ratio. The selection of QCV is compatible with Chatterjee's [3] choice of the same scaling law for digital applications. Thus it appears feasible to scale both the analog and digital portions of a circuit using the same scaling law.

A typical case study comparing the performance of NMOS and CMOS gain stages under moderate scaling conditions shows that CMOS offers the optimum gain configuration for scaled analog implementations. A comparison between computer simulation (of gain stages and a CMOS op amp) and first-order scaling theory shows that second-order effects produce significant but nondominant errors in evaluating the performance of analog components under scaled conditions. Thus, first-order theory can be used to estimate scaling tendencies in analog applications.

APPENDIX I

Considering (16), if one lets

$$x = \frac{E_p K_1}{2(V_D - V_{D\,sat})^{1/2}},$$ (A1)

this equation becomes

$$\Delta L = K_1(V_D - V_{D\,sat})^{1/2} \{(1 + x^2)^{1/2} - x\}.$$ (A2)

If x is small, implying a moderate E_p, (A2) can be simplified to

$$\Delta L \simeq K_1(V_D - V_{D\,sat})^{1/2} \left(1 - x + \frac{x^2}{2}\right).$$ (A3)

Differentiating (A3) yields

$$\frac{\delta \Delta L}{\delta V_D} = \frac{1}{2} \frac{K_1}{(V_D - V_{D\,sat})^{1/2}} \left(1 - x + \frac{x^2}{2}\right)$$

$$+ K_1(V_D - V_{D\,sat})^{1/2} (x - 1) \frac{\delta x}{\delta V_D}$$

$$= \frac{1}{2} \frac{K_1}{(V_D - V_{D\,sat})^{1/2}} \left\{1 - x + \frac{x^2}{2} + x\right\}$$

$$= \frac{1}{2} \frac{K_1}{(V_D - V_{D\,sat})^{1/2}} \left\{1 - \frac{x^2}{2}\right\}.$$ (A4)

Substituting (A3) and (A4) into

$$g_{ds} = \frac{I_{D,\,sat}}{L\left(1 - \frac{\Delta L}{L}\right)^2} \frac{\delta \Delta L}{\delta V_D}$$ (A5)

and neglecting all x^2 terms yields

$$g_{ds} = \frac{I_{D,\,sat} K_1}{2L(V_D - V_{D\,sat})^{1/2} [1 - K_1(V_D - V_{D\,sat})^{1/2} (1 - x)]^2}$$ (A6)

which is (20). Note also that by neglecting x^2, (A3) becomes equivalent to (18).

APPENDIX II

Consider the ideal current equation

$$I_{D\,sat} = \frac{\beta}{2}(V_G - V_T)^2.$$ (B1)

When DIBL is present, the threshold voltage V_T is a function of V_{DS}, according to Masuda [21], the dependence is

$$V_T = V_{T_0} - \eta(V_{DS} - \phi)$$ (B2)

where

$$\eta = \frac{\eta_0(x_j, N_B)}{L^3}.$$ (B3)

V_{T_0}, η_0, and ϕ are constants dependent on the technology. By substituting V_T into (B1) and differentiating with respect to V_{DS}, one obtains

$$g_{ds} = \frac{\delta I_{D\,sat}}{\delta V_{DS}} \simeq \frac{\beta\eta}{L^3}(V_G - V_T) \simeq \frac{g_m\eta}{L^3}. \tag{B4}$$

Therefore, as L becomes smaller, this effect will become a dominant factor in determining g_{ds}.

REFERENCES

[1] C. A. T. Salama, "VLSI technology for telecommunication IC's," *IEEE J. Solid-State Circuits*, vol. SC-16, pp. 253-250, 1981.

[2] R. H. Dennard *et al.*, "Design of ion implanted MOSFET's with very small physical dimensions," *IEEE J. Solid-State Circuits*, vol. SC-9, pp. 256-266, 1974.

[3] P. K. Chatterjee, "The impact of scaling laws on the choice of n-channel or p-channel for MOS VLSI," *Electron. Dev. Lett.*, vol. EDL-1, pp. 220-223, Oct. 1980.

[4] B. Hoeneisen and C. A. Mead, "Fundamental limitations in microelectronics—I–MOS technology," *Solid-State Electron.*, vol. 15, pp. 819-829, 1972.

[5] E. Demoulin, "Process statistics of submicron MOSFET's," in *Tech. Dig., IEDM Conf.*, Washington, DC, 1979, pp. 34-37.

[6] Y. A. El-Mansy, "On scaling MOS devices for VLSI," in *Proc. IEEE Int. Conf. Circuits Comput.*, 1980, pp. 457-460.

[7] G. Merckel, "A simple model of the threshold voltage of short channel and narrow channel MOSFET's," *Solid-State Electron.*, vol. 23, pp. 1207-1213, 1980.

[8] P. P. Wang, "Device characteristics of short channel and narrow width MOSFET's," *IEEE Trans. Electron. Devices*, vol. ED-25, pp. 779-786, July 1978.

[9] Y. A. El-Mansy *et al.*, "A simple 2-dimensional model for IGFET operation in the saturation region," *IEEE Trans. Electron. Devices*, vol. ED-24, pp. 254-262, Mar. 1977.

[10] M. H. White *et al.*, "High-accuracy MOS models for computer-aided design," *IEEE Trans. Electron. Devices*, vol. ED-27, pp. 899-906, 1980.

[11] H. S. Lee, "An analysis of the threshold voltage for short channel IGFET's," *Solid-State Electron.*, vol. 16, pp. 1407-1417, 1973.

[12] L. D. Yau, "A simple theory to predict the threshold voltage of short channel IGFET's," *Solid-State Electron.*, vol. 17, pp. 1059-1063, 1974.

[13] O. Leistiko, "Electron and hole mobilities in inversion layers on thermally oxidized silicon surfaces," *IEEE Trans. Electron. Devices*, vol. ED-12, pp. 248-254, 1965.

[14] F. M. Klaassen and W. C. J. de Groot, "Modeling of scaled down MOS transistors," *Solid-State Electron.*, vol. 23, pp. 237-242, 1980.

[15] G. Baum and H. Beneking, "Drift velocity saturation in MOS transistors," *IEEE Trans. Electron. Devices*, vol. ED-17, pp. 481-482, 1970.

[16] C. Hilsum, "Simple empirical relationship between mobility and carrier concentration," *Electron. Lett.*, vol. 10, no. 13, pp. 259-260, Jan. 1974.

[17] V. G. K. Reddi and C. T. Sah, "Source to drain resistance beyond pinchoff in MOS transistors," *IEEE Trans. Electron. Devices*, vol. ED-12, pp. 139-141, 1965.

[18] T. Poorter and J. H. Satter, "A dc model for an MOS transistor in the saturation region," *Solid-State Electron.*, vol. 23, pp. 765-772, 1979.

[19] R. R. Troutman, "VLSI limitations from drain induced barrier lowering," *IEEE Trans. Electron. Devices*, vol. ED-26, pp. 461-468, Apr. 1979.

[20] B. Hoefflinger *et al.*, "Model and performance of hot electron MOS transistors for VLSI," *IEEE J. Solid-State Circuits*, vol. 14, pp. 435-442, 1979.

[21] H. Masuda *et al.*, "Characteristics and limitation of scaled down MOSFET's due to two-dimensional field effect," *IEEE Trans. Electron Devices*, vol. ED-26, pp. 980-986, 1979.

[22] H. Katto *et al.*, "MOSFET's with reduced low frequency $1/f$ noise," *Oyo Buturi* (Japan), vol. 44, pp. 243-248, 1975.

[23] J. R. Brews, "Generalized guide for MOSFET miniaturization," in *Tech. Dig., IEDM Conf.*, 1979, pp. 10-13.

[24] N. C. C. Lu *et al.*, "A new conduction model for polycrystalline silicon films," *Electron. Dev. Lett.*, vol. EDL-2, pp. 95-98, 1981.

[25] P. R. Gray, "Basic MOS operational amplifier design—An overview," in *Analog MOS Integrated Circuits*. New York: IEEE Press, 1980, pp. 28-49.

[26] Y. P. Tsividis, "Design consideration in single channel MOS analog integrated circuits—A tutorial," *IEEE J. Solid-State Circuits*, vol. SC-13, pp. 383-391, June 1978.

Matching Properties, and Voltage and Temperature Dependence of MOS Capacitors

JAMES L. McCREARY, MEMBER, IEEE

Abstract—The matching properties of MOS capacitors are modeled and compared with measured data. A weighted-capacitor array design approach is described. Voltage and temperature dependence of MOS capacitors are analyzed, modeled, and compared with measured data. It is shown that to a first-order heavily doped polysilicon accumulates and depletes similar to crystallyine silicon.

LIST OF SYMBOLS

A/P	Area-to-perimeter ratio
BWC	Binary weighted capacitor
C_{ox}	Silicon dioxide capacitance per unit area, k_{ox}/t_{ox} (F/cm^2)
C_s	Semiconductor space charge capacitance per unit area (F/cm^2)
C_t	Total MOS capacitance per unit area (F/cm^2)
E_f	Fermi level (eV)
ϵ_0	Permittivity of free space
ϵ_{ox}	Dielectric constant of silicon dioxide
ϵ_{si}	Dielectric constant of silicon
k	Boltzmann's constant
L_{si}	Coefficient of linear thermal expansion of silicon, 2.8 ppm/C
M[WCD]	Mean value of the WCD
N_d	Donor impurity concentration
Phis	Surface electrostatic potential relative to semiconductor bulk
q	Electronic charge
Q_s	Semiconductor space charge per unit area
SD	Standard deviation
SD[x]	Standard deviation of the measured values of x
SD%[x]	Percent of full scale standard deviation of x
sgn[x]	The sign of x; or x divided by the absolute value of x
t_{ox}	Silicon dioxide thickness
T	Absolute temperature
T_{CC}	Temperature coefficient of capacitance (ppm/°C)
T_{CCj}	T_{CC} for a junction capacitor (ppm/°C)
U_f	Dimensionless Fermi potential, E_f/kT
U_s	Dimensionless surface potential, $q\,Phis/kT$
V_{CC}	Voltage coefficient of capacitance (ppm/V)
V	Voltage applied at the top plate of MOS capacitor with substrate or lower power held at ground
V_b	Barrier potential
V_{fb}	Flat-band voltage
V_j	Reverse-bias junction voltage
WCD	Worst-case deviation
%FS	Percent of full scale

Manuscript received May 5, 1981; revised June 6, 1981.
The author is with the Intel Corporation, Santa Clara, CA 95051.

I. INTRODUCTION

IN MOS capacitor circuits that perform A/D conversion [1]–[3], precision analog gain and attenuation [4], [5], and filtering [6], [7], the performance and also the cost effectiveness depend upon the accuracy of the capacitor ratio matching. This paper describes a systematic approach to capacitor array design that uses a statistical model for capacitor ratio errors. The voltage coefficient and temperature coefficient of MOS capacitors are also discussed. Measured data are compared with first-order calculations. The overall objective of the paper is not only to report this investigation, but also to establish a comprehensive design approach for MOS capacitor array circuits.

In the past, capacitor ratio errors have been modeled using a statistical analysis which treats these errors as random variables [8], [9]. More recently, Yee *et al.* [10] have done an analysis of capacitor errors in a two-stage capacitor array. All of these models have assumed that the errors were random, uncorrelated, and normally distributed. These assumptions are supported by data shown in this paper. However, little has been published regarding a systematic design approach. In addition, all previously reported capacitor matching data have been based upon limited data gathered using measurement techniques of limited accuracy. The analysis which follows is supported by data from more than 32 000 capacitor arrays taken from a total of 16 groups of wafers processed in five different technologies. Measurements were made using a 16-bit accurate, self-calibrating technique previously described [11]. Since the database is large enough, useful empirically deduced relationships are identified for aid in design.

II. MATCHING PROPERTIES OF MOS CAPACITORS

Array Designs with Constant Area-to-Perimeter Ratio

Attention is now focused upon arrays with constant A/P ratio. Here, the ratio of the area of the capacitor to the total perimeter of the plate defining the capacitor is constant for each capacitor in the array. For purposes of this discussion, a capacitor array is defined as a set of capacitors with one common node connection. Consider an array of two capacitors C_m and C_k, made of m and k identical unit capacitors $C1$, respectively. Let SD[$C1$] be the SD (the standard deviation) in

Reprinted from *IEEE J. Solid-State Circuits*, vol. SC-16, no. 6, pp. 608–616, Dec. 1981.

119

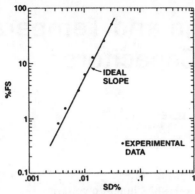

Fig. 1. Capacitor ratio in %FS versus the standard deviation of ratio errors for a 128-unit BWC array.

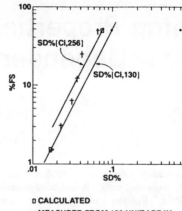

□ CALCULATED
• MEASURED FROM 130 UNIT ARRAY
† MEASURED FROM 256 UNIT ARRAY

Fig. 2. Measured and calculated values of capacitor ratio versus ratio error for a 130-unit array and a 256-unit array.

the distribution of errors in $C1$ from its ideal value for a particular process. Then, assuming that the errors are uncorrelated, random and normally distributed, the following equations can be derived:

$$SD[C_m] = [m]^{1/2} SD[C1], \qquad (1)$$

and

$$SD[C_k] = [k/m]^{1/2} SD[C_m]. \qquad (2)$$

A more useful parameter, however, is SD% which is the SD of the error in capacitor ratio (rather than capacitor value) expressed as percent of full scale, %FS:

$$SD\%[C_k, k+m] = 100\, SD[C_k]/(C_k + C_m)$$
$$= [k/m]^{1/2} SD\%[C_m, k+m] \qquad (3)$$

where $k + m$ is the total number of units in the array. For a BWC array, this equation predicts that the slope of log(capacitor ratio) versus log(SD%) is 2 as shown by the line on the graph of Fig. 1 where the capacitor ratio is expressed as %FS. This is true since capacitor ratio is directly proportional to k while SD% is proportional to the square root of k. This figure also shows measured data points for a certain BWC array design. These closely follow the predicted slope. For this array the unit plate size was approximately 72 μm × 72 μm and consisted of 128 total units and was fabricated in a technology denoted as CMOS-1. The A/P ratio was 18 μm.

Dependence upon Total Area for a Constant A/P Ratio

Consider now a 130-unit capacitor array having four capacitors: $C64A$, $C64B$, $C1A$, and $C1B$. Both $C64A$ and $C64B$ are each composed of 64 unit capacitors $C1$. $C1A$ and $C1B$ are each composed of a 1 unit capacitor. The unit capacitor size under study was 25 μm × 25 μm corresponding to an A/P ratio of 6.25 μm. When taken in parallel $C1A + C1B$ can also be considered as a single two-unit capacitor $C2$. From measurements on this array SD%[$C64A$, 130] was found to be 0.10%FS. Then the following equation can be written for $C2$:

$$SD\%[C2, 130] = [2/64]^{1/2} SD\%[C64A, 130] = 0.017\,\%FS$$
$$(4)$$

which agreed closely with the measured value of 0.018 %FS.

This is illustrated graphically in Fig. 2 along with a line labeled SD%[C_i, 130] representing the ideal behavior.

We can also calculate the matching of $C1A$ to $C1B$ by assuming that these are the only capacitors in a two-unit array. Using (2) and (3), it can be shown that in the general case the SD% of capacitor C_i composed of i units of $C1$ in an array of j units of $C1$ can be expressed in terms of the SD% of capacitor C_m having m units of $C1$ in a different array of k units of $C1$:

$$SD\%[C_i, j] = (k/j)\,[i/m]^{1/2} SD\%[C_m, k]. \qquad (5)$$

Using this equation, the SD% can be calculated for each capacitor in the two-unit array based upon the SD% value measured for capacitor $C64A$:

$$SD\%[C1, 2] = (130/2)\,[1/64]^{1/2}$$
$$\cdot SD\%[C64A, 130] = 0.80\,\%FS. \qquad (6)$$

The measured value of the same parameter was found to be 0.64 %FS and is shown in Fig. 2 along with the calculated value.

Now assume that it is desirable to reduce the SD% and that this is done by increasing the total area such that the entire array now has 256 units of the same capacitor $C1$. The new SD% of the largest capacitor $C128$ is given by

$$SD\%[C128, 256] = (130/256)\,[128/64]^{1/2}$$
$$\cdot SD\%[C64A, 130] = 0.072\,\%FS. \qquad (7)$$

The measured value of the above parameter for the 256-unit array was found to be 0.068 %FS. Only limited data (provided by C. Laber) were available for this particular array. The measured data points for all capacitors in this array are also shown in Fig. 2 along with a line labeled SD%[C_i, 256] which denotes the ideal behavior of the array as predicted by the single data point for SD%[$C64A$, 130].

Capacitor Ratio Error Dependence upon A/P Ratio

Now consider the effect of different A/P ratios upon capacitor matching errors. One would expect that capacitor plate edge resolution limitations would cause larger matching errors for smaller plate sizes. It is obvious that at some point, photo-

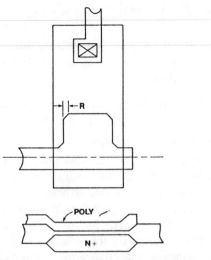

Fig. 3. Unit capacitor layout type-*A* for poly-to-Si capacitor showing slanted corners and poly interconnect.

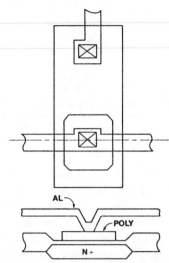

Fig. 4. Unit capacitor layout type-*B* for poly-to-Si capacitor using slanted corners and Al interconnect.

lithography limitations result in edge location uncertainties which cause significant capacitor area variations. Hence a larger A/P ratio would seem to be desirable. In fact, a single square or even a single circle would be a preferable geometry over a multiple-plate capacitor. However the largest capacitor cannot be a single plate since then the capacitor ratios would not be process insensitive (see [8] for a detailed discussion). Previous approaches have involved unit plate sizes ranging from the smallest capacitor [12] to intermediate values [8]. Other approaches have used voltage division between the main BWC array and an additional BWC array [13], [10]. These techniques have the advantage of keeping the A/P ratio larger than would otherwise occur for a single array and also reduce the total area required. The obvious extension of this approach would be a *C*–2*C* ladder array in which two is the largest ratio between capacitors. However, this and the previous approach introduces an additional matching constraint upon the large voltage dependent parasitic capacitors in the array in order to maintain the required voltage divider precision. In this case absolute linearity may be difficult to achieve. Furthermore, it can be shown that although a second array or resistor string improves the overall resolution when added to a larger array, it does not substantially affect the integral linearity or WCD expressed as %FS. However, the linearity and WCD expressed in LSB (least significant bits) are degraded by approximately the same factor that the resolution is increased. This is a consequence of increasing the resolution without a corresponding increase in component matching accuracy.

How small the unit plate size may be before the perimeter effects become important is a function of the particular process technology. For very large unit plate sizes, the perimeter effects may be neglected. Assume that this is the case for the 128-unit array with A/P = 18 μm, previously shown in Fig. 1. Using this assumption, we now calculate the error SD%[*C*64, 130] for the 130-unit array having A/P = 6.25 μm. For square geometries, a general equation may now be written which normalizes for differences in the unit plate A/P ratio

$$SD\%[C_i, j] = (A_k/A_j)^{1/2} \frac{k}{j} (i/m)^{1/2} SD\%[C_m, k]. \qquad (8)$$

Here A_k and A_j represent the unit plate areas of the *k*-unit and *j*-unit arrays, respectively. Using this equation, the error for *C*64 in the smaller 130-unit array may be calculated from measured data on the larger array. We calculate SD%[*C*64, 130] to be 0.085 %FS. This is slightly less than the 0.1 %FS measured value shown in Fig. 2. The difference is probably attributable to the nonnegligible perimeter effects in the 130-unit array.

Ratio Error Dependence upon Layout Design

Two experimental arrays, *A* and *B*, were identical in die area except that *A* used the layout shown in Fig. 3 while *B* used that of Fig. 4. The ratio error SD%[*C*64, 128] for the largest capacitor in each array was measured. For type *A*, this value was 0.026 %FS, while for type *B*, the value was 0.037 %FS. The reason that type *A* provides better matching is not obvious, however, this may be due to the larger A/P ratio (for the same die area) that is possible with type *A* layout. Also the process of etching the metal is not well controlled leading to variations in the capacitance of the plate-to-plate metal interconnect.

Other layout techniques involve the use of common centroid geometry [8] which offers some improvement for long-range oxide gradients. Short-range gradients often encountered with poly-to-poly capacitors are not cancelled by this technique.

Ratio Error Dependence upon Process Technology and Capacitor Structures

For a given array design, capacitor ratio errors are a function of the capacitor structure and may also depend upon the process technology. A particular 10-bit BWC array using the unit plate layout of Fig. 3 was fabricated in five different technologies. This particular design used a 72 μm $\times$ 72 μm unit plate size for the large array capacitors *C*8, *C*16, *C*32, $\cdots$ *C*512. However, the smaller capacitors *C*4, *C*2, *C*1*A*, and *C*1*B* were approximately square geometries. The goal of this technique was to find an optimal compromise between a small unit plate size to minimize process sensitivity and a large unit plate size to increase A/P (and hence improve matching). Each design used identical geometries. The first technology was a metal-gate NMOS process denoted as NMOS-MG. This design has been

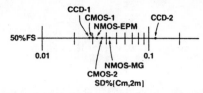

Fig. 5. Measured standard deviation of ratio error for a 10-bit BWC array fabricated in five different technologies. Only the error for the largest capacitor in each array is shown (the 50 %FS point).

previously reported [8]. The unit capacitor structure was an Al top plate with an n$^+$ silicon lower plate. The second technology was a two-level polysilicon CCD process used to fabricate capacitor arrays having two different structures. The first structure CCD-1 used a poly2 (upper level polysilicon) to n$^+$ silicon unit capacitor structure. The second CCD-2 was a poly2-to-poly1 capacitor structure similar to Fig. 3 except that the n$^+$ lower plate was replaced by poly1. The third technology, was an older poly gate CMOS process CMOS-1. Here the unit capacitor structure was identical to CCD-1. The fourth technology was an advanced, high-speed CMOS technology CMOS-2, which also used the same poly-to-silicon capacitor structure as CCD-1. The fifth technology was a modified NMOS EPROM process denoted as NMOS-EPM. This used a poly2-to-poly1 capacitor structure similar to CCD-2.

Fig. 5 shows the measured ratio-error data for each of the five arrays. Only the data point SD%[C_m, $2m$] at 50 %FS for the largest capacitor in the array C_m is shown in the figure. It is evident that the poly-to-silicon capacitor structures CMOS-1, CMOS-2, and CCD-1 have the best matching properties in the technologies examined. However, the matching properties of the poly2-to-poly1 structure is good for NMOS-EPM but extremely poor for CCD-2. The reason for this is not known, however, the CCD-2 structure was observed to have an extremely rough poly1 surface that may have contributed to the high ratio-error and also to an abnormally high oxide defect density. On the other hand, an EPROM process requires a high-quality, uniform poly-to-poly oxide. Lastly, the NMOS-MG structure matches worse than do any of the other except for CCD-2.

Total Yield Analysis

An important factor in capacitor array design is yield to a given linearity. For purposes of this discussion, the total capacitor array yield can be expressed as $Y_t = Y_f Y_r$. The functional yield Y_f is dependent primarily upon oxide defect density D and total capacitor oxide area A such that $Y_f = f(A, D)$. Several functional yield models have appeared in the literature [14], [15]. In a well-monitored process, the parameter D is known or else can be determined from test patterns.

The ratio-matching yield Y_r can be calculated from the required linearity and measured data. Let z be the maximum allowable nonlinearity in %FS. Let M[WCD] and SD%[WCD] be the mean and SD% of the WCD expressed as %FS. Y_r can be calculated directly by integrating the area under the Gaussian probability density function (having the standard deviation SD%[WCD]), centered at M[WCD], from $-z$ to $+z$. The following empirical relationship has been observed from measured data:

$$SD\%[WCD]/SD\%[C_m, 2m] = 1.5 \text{ to } 1.0. \quad (9)$$

For the 10-bit BWC array previously discussed, typical values of Y_f and Y_r were 95 and 80 percent (for WCD less than 0.05 %FS), respectively.

III. VOLTAGE DEPENDENCE OF MOS CAPACITORS WITH DEGENERATELY DOPED SILICON PLATES

Voltage Coefficient of Capacitance

It is evident from the literature [16], [17] that the equations which describe the *CV* behavior of an MOS capacitor may be cumbersome, especially for structures involving degenerately doped surfaces. Usually, however, simplicity of description is desired. For those cases, which are often circuit design oriented, it is sufficient to specify the nominal capacitance value and the rate of change of capacitance over some voltage interval. For this reason, the voltage coefficient of capacitance V_{CC} is used: $V_{CC} = 1/C (dC/dV)$. This is the rate of fractional change in C per unit voltage at some dc voltage V. For large N_d and small V, C tends to become a linear function of applied voltage. The total MOS capacitance is given by the series combination of oxide and space charge capacitance:

$$C_t = \frac{1}{(1/C_{ox} + 1/C_s)}. \quad (10)$$

Equation (10) is valid for an MOS capacitor with one semiconductor surface, while the other surface is assumed to have a negligible space charge region (a metal). Using Maxwell–Boltzmann (MB) occupation statistics, the space charge capacitance is given by the well-known equation [18]

$$C_s = q/kT \, dQ_s/dU_s$$
$$= \text{sgn}[U_s] \left[\frac{\epsilon_{si} q^2 N_d}{2kT} \right]^{1/2} [e^{U_s} - 1] [e^{U_s} - 1 - U_s]^{-1/2}. \quad (11)$$

The *CV* model using this equation for C_s will be referred to as the MB *CV* model. V_{CC} may be expressed as

$$V_{CC} = \frac{C_{ox}^2}{\epsilon_{si} q N_d} \frac{e^{2U_s} - 2U_s e^{U_s} - 1}{(e^{U_s} - 1)^3}. \quad (12)$$

Taking the limit as U_s approaches zero (V approaches V_{fb}),

$$V_{CC}(V_{fb}) = \frac{C_{ox}^2}{3q\epsilon_{si} N_d}. \quad (13)$$

Since V_{fb} is small, (13) is nearly identical to the value at $V = 0$ for large values of N_d.

The values of V_{CC} as a function of doping are plotted in Fig. 6 [curve labeled $V_{CC}(0 \text{ V})$]. It can be seen from (13) that V_{CC} can be reduced by increasing N_d, and increasing t_{ox}. To first order, V_{CC} is independent of temperature in this analysis. In addition $V_{CC}(-10)$ and $V_{CC}(+10)$, the voltage coefficients at -10 and $+10$ V, respectively, were also computed from (12). The V_{CC} curves for these two voltages are also plotted in Fig. 6.

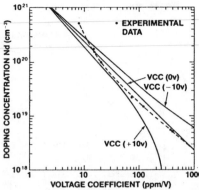

Fig. 6. Plot of calculated surface doping concentration versus voltage coefficient for 0, -10, and +10 V. A dashed line is an approximate fit to the experimental data points.

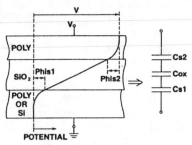

Fig. 7. Model for poly-to-Si and poly-to-poly capacitor showing two space charge capacitances.

Experimental Procedure and Results For the Al-to-Si Capacitor

Several n-type Si wafers received a range of phosphorus predepositions from medium to heavy concentrations. Dry 1000 Å thermal oxides were then grown. Half of each wafer received a poly deposition followed by a second phosphorus doping. Then, Al was evaporated and capacitor top plates were patterned in the Al. This was followed by a poly etch and final anneal in hydrogen at 450°C. It has been assumed that this process results in negligible interface charge and fast surface state densities. However, some studies have suggested that this may not be true if the surface concentration becomes too large [19]. In this manner, both poly-to-Si and Al-to-Si capacitors were fabricated on the same Si substrate.

C-V measurements were performed at 1.5 MHz on all samples using a modified PAR model 410 C-V plotting system. Capacitance bridge measurements were also made as a function of dc voltage to confirm the plotted data. The samples were then grooved and spreading resistance data gathered from which the doping profiles in the poly and in the Si were determined [20]. Using the measured values of oxide thickness together with doping information, the CV curves were then computed for the MB CV model described in this paper. Allowing for the uncertainty in the exact position of the origin, the measured curve was then translated onto the same axes as the calculated curve to allow for comparison of slopes (voltage coefficients) rather than absolute values.

The measured values for V_{CC} obtained near the origin for all of the samples are plotted in Fig. 6. The experimental value for $N_d = 5 \times 10^{20}$/cm^3 was obtained from the literature [9]. One surprising result is the relatively good agreement between the MB model and the measured data. This is unexpected since it is well known that the MB model is valid only for the nondegenerate case (N_d less than 10^{19}). For the case of degenerately doped Si, the formal analysis becomes extremely complex [21].

Analysis of Poly-to-Si and Poly-to-Poly MOS Capacitor Structures

It will be shown in this paper that heavily doped poly accumulates and depletes in the same manner as heavily doped single crystalline Si. Consequently, the CV model for crystal-line Si can be applied to the poly surface. In the case of a poly-to-poly or poly-to-Si capacitor, one expects that since one surface accumulates while the other depletes and vice versa that the voltage dependence tends to cancel if both surfaces are equally doped. A direct solution for CV dependence of this type of capacitor involves a simultaneous solution for both surface potentials and spaces charges such that the electric field in the oxide is identical at both interfaces. Referring to Fig. 7, an assumption is now made that simplifies the analysis. The electrostatic surface potential $Phis_2$ is considered negligible when computing $Phis_1$ and vice versa. This is a reasonable approximation when at least one surface is heavily doped. The device model can then be reduced to two space charge capacitances C_{s1} and C_{s2} in series with C_{ox} as shown in Fig. 7. Now the CV curves for a poly-to-Si device having different surface dopings can be computed in a direct manner:

$$1/C_t = 1/C_{ox} + 1/C_{s1} + 1/C_{s2}. \tag{14}$$

The analytical CV curves are obtained using (11) and (14). Based on this simple device model, the voltage coefficient for poly-to-Si and poly-to-poly capacitors is also approximately given by $V_{CC} = V_{CC1} - V_{CC2}$, where the plate numbering convention used is that shown in Fig. 7. The individual plate voltage coefficients are determined in the usual manner using Fig. 6.

Experimental Results for the Poly-to-Si Capacitor

Comparisons of some analytical and experimental results are shown in Figs. 8 and 9. The devices selected for display in these figures include one having both plates heavily doped, device $D1$ (Fig. 8). Fig. 9 illustrates a device $D2$ having a much smaller N_d in the poly than in the Si, while the opposite situation exists for the device $D3$. Agreement between the calculated and experimental data is not as good for lightly doped poly surfaces (doping less than 10^{19}/cm^3). This may be associated with the error in spreading resistance measurements on thin lightly doped films or may be due to the onset of grain boundary effects.

It is interesting to note that for the device $D2$ having a poly plate that is more lightly doped than the crystalline silicon (Fig. 9) the poly plate provides the dominant space charge capacitance and hence the voltage dependence. In this case, the total capacitance decreases with increasing positive voltage since the poly surface is depleting and the Si surface is undergoing negligible accumulation. Hence the fundamental assertion that a heavily doped poly surface accumulates and depletes similar to that to crystalline Si is confirmed.

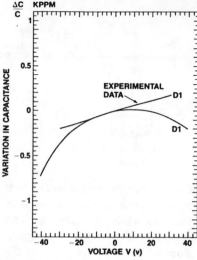

Fig. 8. Calculated and measured CV curves for a poly-to-Si capacitor $D1$ with $N_d(\text{poly}) - 1.5 \times 10^{20}/\text{cm}^3$ and $N_d(\text{Si}) = 1.1 \times 10^{20}/\text{cm}^3$.

If both the poly and Si were identically doped, the total capacitance variation would be symmetrical about the flat-band voltage, resulting in a nearly zero voltage coefficient. The MOS capacitor $D1$ in Fig. 8 exhibits a voltage coefficient of approximately 7 ppm/V.

Nonlinearities in Capacitor Arrays Used for A/D and D/A Conversion

When capacitor arrays are used in A/D and D/A conversion, a charge transfer error may be caused by the capacitor voltage dependence. This error may be expressed as a coding nonlinearity in terms of V_{CC}. Fig. 10 illustrates the final charge distribution between two capacitors $C1$ and $C2$. Here the V_{CC} polarities shown are chosen to minimize parasitic capacitance and leakage current at node X. The figure shows that one plate of $C1$ is raised to the full-scale reference voltage V_r. The nonlinearity in the voltage V_x due to V_{CC} can now be calculated by setting the charge on $C1$ equal to that on $C2$ and integrating the capacitance over the voltage transition. By using simplifying approximations, we get

$$V_x = V_r \frac{C1}{C1 + C2} - V_r^2 \frac{V_{CC}}{2} \frac{C1C2}{(C1 + C2)^2}. \tag{15}$$

The error voltage at node X can now be expressed as a nonlinearity in %FS:

$$\text{nonlinearity due to } V_{CC} = -50 \, V_{CC} V_r N (1 - N) \, \%\text{FS} \tag{16}$$

where $N = C1/(C1 + C2)$. N corresponds to the fraction of full-scale signal being coded. It is seen that the maximum nonlinearity occurs at one-half full scale ($N = 1/2$) and has the value

$$\text{maximum nonlinearity due to } V_{CC} = -50 \, V_{CC} V_r/4 \, \%\text{FS} \tag{17}$$

This is identical to that calculated for a capacitor A/D converter in which a comparator is connected to node X and a successive approximation algorithm is used to drive node X from $-V_{in}$ back to zero.

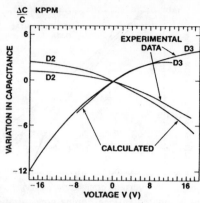

Fig. 9. Calculated and measured CV curves for poly-to-Si capacitor $D2$ with $N_d(\text{poly}) = 9 \times 10^{18}/\text{cm}^3$ and $N_d(\text{Si}) = 1.65 \times 10^{20}/\text{cm}^3$, and curves for poly-to-Si capacitor $D3$ with $N_d(\text{poly}) = 1.2 \times 10^{20}/\text{cm}^3$ and $N_d(\text{Si}) = 5.5 \times 10^{18}/\text{cm}^3$.

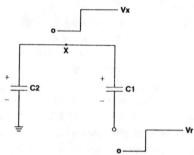

Fig. 10. Simplified representation of capacitor array charge redistribution during A/D or D/A conversion.

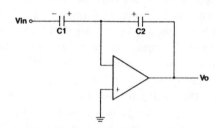

Fig. 11. Simplified schematic of precision analog amplifier, attenuator, or D/A converter.

Harmonic Distortion in Capacitor Array Circuits

Voltage coefficient of capacitance can also lead to harmonic distortion. Consider the precision analog amplifier, attenuator, or D/A converter shown in Fig. 11. Here again the polarity of the capacitor V_{CC} is chosen as shown due to considerations for leakage current and parasitic capacitance. By equating the charges transferred on each capacitor, we get an approximate equation:

$$V_0 = -\frac{C1}{C2} V_{in} + \frac{C1}{C2} \frac{V_{CC}}{2} \left(1 + \frac{C1}{C2}\right) V_{in}^2. \tag{18}$$

Letting $V_{in} = A \sin wt$, the second-harmonic distortion HD2 may be calculated:

$$\text{HD2} = (1 + G) \, V_{CC} \frac{A}{4} \tag{19}$$

in which $G = C1/C2$ is the nominal gain of the circuit and is greater than zero. As seen from the equation above, distortion increases with gain and voltage amplitude.

IV. Temperature Coefficient of MOS Capacitors

The temperature coefficient of capacitance is defined as

$$T_{CC} = \frac{1}{AC_t} \frac{d}{dT}(AC_t) \qquad (20)$$

in which C_t is the total MOS capacitance per unit area and A is the plate area of the device. For this study the MOS capacitor was the same Al-to-Si structure discussed in Section III.

The temperature coefficient of capacitance T_{CC} represents the fractional rate of change of total capacitance per unit temperature. Using the fact that C_s is much greater than C_{ox} for these structures, the value of T_{CC} may be computed and reduced to the following:

$$T_{CC} = \left[\frac{1}{A}\frac{dA}{dT} - \frac{1}{t_{ox}}\frac{dt_{ox}}{dT}\right] + \frac{C_{ox}}{C_s^2}\frac{dC_s}{dT} + \frac{1}{\epsilon_{ox}} + \frac{d\epsilon_{ox}}{dT}$$

$$= T_{CC}(th) + T_{CC}(sc) + T_{CC}(\epsilon_{ox}). \qquad (21)$$

In this form, T_{CC} is resolved into three components. The first term represents the change in capacitance for a plate area A and dielectric thickness t_{ox} due to thermal expansion. The second term corresponds to the temperature dependence of space charge capacitance. The third term represents the temperature dependence of the dielectric constant of the oxide k_{ox}. In this section, the first two components of T_{CC} will be discussed resulting in an expression which allows experimental evaluation of the third term.

Evaluation of the thermal expansion component $T_{CC}(th)$ requires an analysis of stresses and strains. This requires knowledge of coefficients of linear thermal expansion, values of elastic moduli and Poisson's ratios of the materials [22]. An exact solution has been performed by Thurston [23]; however, a reasonably good approximation can be made which simplifies the analysis and results in less than 5 percent error. The approximation involves neglecting the term $(1/t_{ox})\, dt_{ox}/dT$ and also the assumption that the thickness of the substrate (200 μm) is so large compared with that of the dielectric (0.1 μm) and the top plate (0.6 μm) that the Si expands freely carrying both thin films along with it. Hence for a plate of any geometry:

$$\frac{1}{A}\frac{dA}{dT} = 2L_{si} = 5.6 \text{ ppm/}°\text{C (from Appendix B)}. \qquad (22)$$

The space charge component $T_{CC}(sc)$ requires evaluation of the first derivative of C_s with respect to T which is difficult to evaluate for an arbitrary applied voltage since U_s is an intricate function of T. However the analysis becomes simplified if $U_s = 0$ (the flat-band condition). This is approximately achieved if the applied voltage is held at zero during the temperature excursion. This is valid for heavily doped Si substrates where U_s is small for applied voltages within several volts of flat band. Equation (11) for C_s in Section III may be simplified by using the series expansion of e^{U_s} for small U_s:

$$\frac{e^{U_s} - 1}{(e^{U_s} - 1 - U_s)^{1/2}} = (2)^{1/2}. \qquad (23)$$

Therefore, at flat band, the space charge capacitance is

$$C_s(fb) = \left[\frac{\epsilon_{si} N_d q^2}{kT}\right]^{1/2} \qquad (24)$$

Differentiating $C_s(fb)$ with respect to T and allowing k_{si} to be a function of T, it follows that

$$T_{CC}(sc) = \frac{C_{ox}}{C_s^2}\frac{dC_s}{dT} = -\frac{C_{ox}k}{2q(\epsilon_{si}N_d kT)^{1/2}}$$

$$\cdot \left[1 - T\left(\frac{1}{\epsilon_{si}}\frac{d\epsilon_{si}}{dT}\right)\right] \qquad (25)$$

which is valid in the vicinity of flat band. The term $(1/\epsilon_{si})\, d\epsilon_{si}/dT$ has been experimentally evaluated in Appendix A and its value was found to be 252 ppm/°C.

Calculated values of $T_{CC}(sc)$ are plotted in Fig. 12 as the curve labeled "0V." Since $T_{CC}(sc)$ is actually negative, the horizontal axis is the absolute value of $T_{CC}(sc)$. As illustrated, for large donor concentration, the value of $T_{CC}(sc)$ becomes small. $T_{CC}(sc)$ is also plotted for different applied voltages (with the Si substrate at ground). These are labeled in Fig. 12. For negative voltages and small N_d, $T_{CC}(sc)$ becomes a strong function of voltage, changes sign when strong surface depletion occurs, and then becomes very large.

It is suggested by Fig. 12 that $T_{CC}(\epsilon_{ox})$ could be evaluated for a device having N_d as large as possible in order to minimize the temperature dependence associated with the space charge relative to that for the oxide dielectric constant. The literature contains little information regarding measured values of $T_{CC}(\epsilon_{ox})$. Furthermore the value appears dependent upon the process associated with the oxide. For this reason, the value of $T_{CC}(\epsilon_{ox})$ is determined by the experimental technique just mentioned. This is described in Appendix B and the value of $T_{CC}(\epsilon_{ox})$ was found to be 21 ppm/°C, which is similar to (although not equal to) a previously reported value of 15 ppm/°C [24] for deposited oxides. Using (B2) and previous assumptions, it is now possible to compute T_{CC} for values of N_d from 10^{18} to 10^{21}/cm^3 for $V = 0$. This was done and the results are plotted in Fig. 13. Measured data points are also plotted along with a dashed line fitted to these points. As seen in this figure, there is generally good agreement between the measured and calculated data. This is better than would have been expected from the nondegenerate MB CV theory. The intersection of measured data and calculated data at $N_d = 1.6 \times 10^{20}$/cm^3 is of course due to the evaluation of $T_{CC}(\epsilon_{ox})$ at that point.

V. Conclusion

A comprehensive technique for designing MOS capacitor arrays has been discussed. This included a method of calculating capacitor ratio errors and subsequent total yield. Data illustrating the sensitivity of ratio matching to capacitor layout, structures, and technology were also presented.

This paper has compared measured and calculated voltage coefficients of MOS capacitors as a function of surface concen-

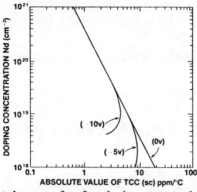

Fig. 12. Calculated curves of surface doping concentration versus absolute value of space charge capacitance temperature coefficient of and Al-to-Si capacitor for different applied voltages (0, −5, and −10 V). For +10 V, the calculated curve is close to that for 0 V.

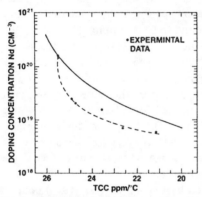

Fig. 13. Calculated and measured curves for surface doping concentration versus temperature coefficient of total capacitance of an Al-to-Si capacitor. The dashed line represents an approximate fit to the measured data.

tration. It has been shown that heavily doped polysilicon accumulates and depletes in a manner similar to that for single crystalline silicon.

It has also been demonstrated that the temperature dependence of space charge capacitance, thermal expansion, and temperature dependence of dielectric constant are the major components of T_{CC}. Of these, it has been found that the dielectric constant term dominates for the case of high substrate doping and low applied voltages.

For poly-to-Si capacitors with heavily doped plates, additional experiments have shown that the T_{CC} of these structures is also predicted by the same model. This is expected since the T_{CC} is due largely to expansion and to the dielectric properties as has been demonstrated.

APPENDIX A

The graded-junction capacitance per unit area is given by

$$C_j = K_j \epsilon_{si}^{2/3} (V_b - V_j)^{-1/3} \quad (A1)$$

in which K_j is a device constant and V_b is the barrier potential. Hence for a junction capacitor, of total capacitance AC_j, the temperature coefficient is

$$T_{CCj} = \frac{1}{A}\frac{dA}{dT} + \frac{2}{3}\frac{1}{\epsilon_{si}}\frac{d\epsilon_{si}}{dT} - \frac{1}{3}\frac{1}{(V_b - V_j)}\frac{dV_b}{dT}. \quad (A2)$$

The junction potential barrier is given by

$$V_b = \frac{KT}{q}\ln\left[\frac{N_a N_d}{n_i^2}\right] \quad (A3)$$

where N_a is the acceptor impurity concentration. The intrinsic carrier concentration n_i and silicon bandgap energy E_g are approximately given by [21]

$$n_i = 3.8 \times 10^{16} T^{3/2} \exp[-E_g/2kT] \quad (A4)$$

and

$$E_g = 1.2 - (2.83 \times 10^{-4})T. \quad (A5)$$

Finally, the desired result is obtained after the appropriate substitutions:

$$\frac{1}{\epsilon_{si}}\frac{d\epsilon_{si}}{dT} = \frac{3}{2}\left[T_{CCj} - 2L_{si} + \frac{1}{3}\frac{1}{(V_b - V_j)}\frac{dV_b}{dT}\right] \quad (A6)$$

Hence, to evaluate $(1/\epsilon_{si})\,d\epsilon_{si}/dT$, it is only necessary to measure the temperature coefficient of the junction capacitor and calculate those terms containing N_a, N_d, and L_{si}. This experiment has been performed and appears in the literature [8]. From that experiment, a value of T_{CCj} = 230 ppm/°C was measured. Using this value and data from that experiment we get

$$\frac{1}{\epsilon_{si}}\frac{d\epsilon_{si}}{dT} = 252 \text{ ppm/°C.} \quad (A7)$$

APPENDIX B

The following experiment was performed on an Al-to-Si capacitor. The surface concentration N_d was measured to be 1.65×10^{20}/cm³ using a spreading resistance technique. C_{ox} for the 1000 Å thermal oxide was 3.4×10^{-4} F/cm².

The thermal expansion term $T_{CC}(th)$ becomes

$$2L_{si} = 5.6 \text{ ppm/°C.} \quad (B1)$$

From (25), Section IV, $T_{CC}(sc)$ is calculated to be −1.5 ppm/°C. The temperature coefficient T_{CC} was then measured. For ten devices, the average value of T_{CC} was 25.5 ppm/°C with a standard deviation less than 0.4 ppm/°C. Using these data in (21), Section IV, the effective value of $T_{CC}(\epsilon_{ox})$ becomes

$$T_{CC}(\epsilon_{ox}) = \frac{1}{\epsilon_{ox}}\frac{d\epsilon_{ox}}{dT} = 21 \text{ ppm/°C.} \quad (B2)$$

REFERENCES

[1] R. Suarez and D. A. Hodges, "All-MOS charge redistribution A/D conversion techniques: Part II," IEEE J. Solid-State Circuits, vol. SC-10, pp. 379–383, Dec. 1975.
[2] J. L. McCreary and P. R. Gray, "All-MOS charge redistribution A/D conversion techniques: Part I," IEEE J. Solid-State Circuits, vol. SC-10, pp. 371–379, Dec. 1975.
[3] Y. P. Tsividis et al., "A segmented mu-255 law PCM encoder utilizing NMOS technology," IEEE J. Solid-State Circuits, vol. SC-11, pp. 740–753, Dec. 1976.
[4] R. McCharles and D. A. Hodges, "Charge transfer circuits for analog LSI," IEEE Trans. Circuits Syst., vol. CAS-25, pp. 490–497, July 1978.
[5] G. L. Baldwin and J. L. McCreary, "A CMOS digitally-controlled analog attenuator for voice band signals," in Proc. IEEE Int. Symp. Circuits Syst., Phoenix, AZ, Apr. 1977, p. 519.
[6] B. J. Hosticka, R. W. Brodersen, and P. R. Gray, "MOS sampled

data recursive filters using state variable techniques," in *Proc. Int. Symp. Circuits Syst.*, Phoenix, AZ, Apr. 1977, pp. 525–529.

[7] J. T. Caves *et al.*, "Sampled analog filtering using switched capacitors as resistor equivalents," *IEEE J. Solid-State Circuits*, vol. SC-12, pp. 592–599, Dec. 1977.

[8] J. L. McCreary, Ph.D. dissertation, Univ. California, Berkeley, 1975.

[9] R. Suarez, Ph.D. dissertation, Univ. California, Berkeley, 1975.

[10] Y. S. Lee, L. M. Terman, and L. G. Heller, "A two-stage weighted capacitor network for D/A and A/D conversion," *IEEE J. Solid-State Circuits*, vol. SC-14, pp. 778–780, Aug. 1979.

[11] J. L. McCreary and D. A. Sealer, "Precision capacitor ratio measurement technique for integrated circuit capacitor arrays," *IEEE Trans. Instrum. Meas.*, vol. IM-28, pp. 11–17, Mar. 1979.

[12] G. Smarandouiu and G. F. Landsberg, "A two-chip CMOS codec," in *Dig. Int. Solid-State Circuits Conf.*, Feb. 1978, pp. 180–181.

[13] K. Chri and M. Callahan, "Integrated PCM codec," *IEEE J. Solid-State Circuits*, vol. SC-14, pp. 38–46, Feb. 1979.

[14] R. M. Warner, *IEEE J. Solid-State Circuits*, p. 86, June 1957.

[15] R. Stapper, *IBM J. Res. Dev.*, vol. 20, no. 3, p. 228, 1976.

[16] A. S. Grove, *Physics and Technology of Semiconductor Devices.* New York: Wiley 1967.

[17] R. Seiwatz and M. Green, "Space charge calculations for semiconductors," *J. Appl. Phys.*, vol. 29, p. 1034, July 1958.

[18] A. Many *et al.*, *Semiconductor Surfaces.* Oxford, England: Pergamon, 1962.

[19] J. Snel, "Insulating films on semiconductors," in *Proc. Inst. Phys. Conf.*, Series 50, 1979, p. 119.

[20] *Spreading Resistance Symposium*, Nat. Bureau Standards Spec. Publ. 400-10, Dec. 1974.

[21] S. M. Sze, *Physics of Semiconductor Devices.* New York: Van Nostrand Reinhold, 1972.

[22] L. D. Landau and E. M. Lifshitz, *Theory of Elasticity.* Reading, MA: Addison-Wesley, 1959.

[23] R. N. Thurston, *Physical Acoustics*, vol. 1A. New York: Academic, 1964.

[24] A. B. Grebene, *Analog Integrated Circuit Design.* New York: Van Nostrand Reinhold, 1972, p. 96.

Random Error Effects in Matched MOS Capacitors and Current Sources

JYN-BANG SHYU, GABOR C. TEMES, FELLOW, IEEE, AND FRANCOIS KRUMMENACHER

Abstract — Explicit formulas are derived using statistical methods for the random errors affecting capacitance and current ratios in MOS integrated circuits. They give the dependence of each error source on the physical dimensions, the standard deviations of the fabrication parameters, the bias conditions, etc. Experimental results, obtained for both matched capacitors and matched current sources using a 3.5 µm NMOS technology, confirmed the theoretical predictions. Random effects represent the ultimate limitation on the achievable accuracy of switched-capacitor filters, D/A converters, and other MOS analog integrated circuits. The results indicate that a 9-bit matching accuracy can be obtained for capacitors and an 8-bit accuracy for MOS current sources without difficulty if the systematic error sources are reduced using proper design and layout techniques.

Manuscript received June 8, 1984; revised July 30, 1984. This work was supported by the National Science Foundation under Grants ECS 81-05166 and ECS 83-15221 and by the Xerox Corporation and the University of California under MICRO Grant 104.

J.-B. Shyu was with the Department of Electrical Engineering, University of California, Los Angeles, CA 90024. He is now with American Microsystems, Inc., Santa Clara, CA 95051.

G. C. Temes is with the Department of Electrical Engineering, University of California, Los Angeles, CA 90024.

F. Krummenacher is with the Department d'Electricité, Laboratorie d'Électronique Générale (LEG), École Polytechnique Fedérále De Lausanne, CH-1007 Lausanne, Switzerland.

I. INTRODUCTION

THE metal–oxide–semiconductor (MOS) technology employed in the large-scale integrated (LSI) fabrication of digital circuits has been also used recently to realize

Reprinted from *IEEE J. Solid-State Circuits*, vol. SC-19, no. 6, pp. 948–955, Dec. 1984.

analog circuits [1]. Unlike for digital integrated circuits, the performance of analog MOS integrated circuits depends heavily upon the element matching accuracy [2], [3]. The key elements are usually the capacitors, but in some applications [4], [5], the matching accuracy of MOS transistors used as current sources is also critical. Therefore, the matching properties of both MOS capacitors and transistors are investigated in this paper. It is a continuation of our earlier work [6] in which we examined the local random errors of MOS capacitors only.

The elements of MOS integrated circuits are inherently subject to errors from two sources. One is the *systematic error*, which affects adjacent elements with identical geometries similarly. It can thus be reduced by proper matching techniques. The other is the *random error*, which differs from element to element, and therefore cannot be corrected by improved matching techniques. It hence represents the ultimate limitation on the achievable accuracy.

A statistical model proposed recently [6] is used to analyze the random errors which are due to *both* the *local* and *global* variations of the linear dimensions and parameters of the circuit elements. These random variations are modeled as random processes, with a zero mean and with stationary characteristics. This is more relevant to the realistic properties of MOS capacitor and transistor matching than the earlier model [6].

Two random capacitance error mechanisms are examined first. One is the random *edge effect* which is due to the local and global random variations of the ideally straight edges of the capacitor, and the other is the random *oxide effect* which is due to the local and global random fluctuations of the oxide thickness and permittivity in the capacitor. Multiple-plate capacitor effects are also considered. Then, four random error mechanisms for the transistor current are examined. First, the *random edge effect* is considered. Second, the *random surface-charge effect* due to both local and global variations of surface-state and ion-implanted charges is analyzed. Third, the *random oxide effect* is studied. Finally, the *random mobility effect* due to both local and global variations of the carrier channel mobility is examined.

In addition to the theoretical predictions, experimental results are presented, based on tests performed on both types of devices, to confirm the theoretical results. A large number of experimental test chips with "on-chip" measurement capability were fabricated for both devices, using 3.5 μm silicon-gate NMOS technology. The experimental results indicate that a 9-bit capacitor matching accuracy and an 8-bit transistor current matching accuracy can be easily achieved. The test results are in good agreement with the theoretical predictions.

II. RANDOM ERRORS IN MOS CAPACITORS

A. Single-Plate Capacitors

In the fabrication of an integrated circuit pattern, the edges of the lines and devices cannot be exactly located due to the uncertainty in the locations of the particle beams and mask dimensions [7]. The position of an edge is thus affected by a certain amount of "noise" so that an ideally straight line appears wavy. The edge variation includes a local jagged edge variation and a global distorted edge variation. The *local* edge variation is usually caused by the granular nature of the aluminum (or polysilicon) edge which is due to the evaporation onto a heated wafer, and also by the jagged edges of the developed photoresist which are caused, e.g., by light interference patterns [8]. The *global* edge variation may be caused by large-scale edge distortion which is due to the fact that the etchant solution may become saturated with the etched material in some areas of the chip [8] or by quantization effects if digital methods are used in the fabrication process, etc. The *local* jagged edge variation is similar to a wide-band thermal noise, containing high-frequency components in its spectrum. It has a very narrow autocorrelation range in terms of displacement [6]. The *global* distorted edge variation is like a narrow-band flicker noise containing only low-frequency components in its spectrum; it has a rather wide autocorrelation range in terms of displacement. Both of these variations introduce a random deviation of the area, and hence the capacitance, of the device.

In addition to the randomly varying edges, the uncertainty in the oxide thickness t permittivity ϵ also causes random errors in a capacitor. Therefore, e.g., two capacitors which have the same area will not have the same capacitance. As for the edge variations, the oxide variations also include *local oxide variations* and *global oxide variations*.

The local oxide variations may be due to the granularity of polysilicon, surface defects of crystalline silicon, etc. The global variations may be caused by slow variation of surface flatness, wafer warping, variations of oxide growth rate, etc.

In [6], the random capacitance errors due to *local* edge and oxide effects were derived. Following somewhat similar arguments, it is possible to find the corresponding errors due to *global* effects as well. Assuming that all of these effects are independent, the combined relative capacitance error can be expressed as

$$\frac{\Delta C}{C} = \sqrt{K_{le}C^{-3/2} + K_{ge}C^{-1} + K_{lo}C^{-1} + K_{go}} \qquad (1)$$

where K_{le} is the local edge effect factor, K_{ge} is the global edge effect factor, K_{lo} is the local oxide effect factor, and K_{go} is the global oxide effect factor. Somewhat pessimistic estimates can be derived for these factors [11], as follows:

$$K_{le} \simeq 8 d_e \sigma_{le}^2 \left(\frac{\epsilon}{t}\right)^{3/2}, \quad K_{ge} \simeq \frac{7\epsilon}{t} \sigma_{ge}^2$$

$$K_{lo} \simeq 4 d_o^2 \frac{\epsilon}{t} \left(\frac{\sigma_{l\epsilon}^2}{\epsilon^2} + \frac{\sigma_{lt}^2}{t^2}\right)$$

$$K_{go} \simeq \frac{\sigma_{ge}^2}{\epsilon^2} + \frac{\sigma_{gt}^2}{t^2}. \qquad (2)$$

In (2), d_e is the correlation radius and σ_{le} is the standard deviation of the local edge variation [6], [11], ϵ is the

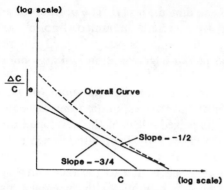

Fig. 1. The relative capacitance error versus capacitance due to local and global edge effects.

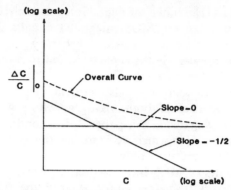

Fig. 2. The relative capacitance error versus capacitance due to local and global oxide effects.

permittivity and t is the thickness of the oxide layer, σ_{ge} is the standard deviation of the global edge variation, σ_{le} and σ_{lt} are the standard deviations of ϵ and t for local effects, and d_o is the local oxide correlation radius. Finally, σ_{gt} and σ_{gt} are the standard deviations of ϵ and t for global effects.

Figs. 1 and 2 illustrate on a log–log scale the relations between the relative capacitance error and the value of C due to the various terms in (1). For very small C, the local edge effect ($\Delta C/C \propto C^{-3/4}$) dominates; for very large C, the global oxide error dominates. The other two error terms can be appreciable for intermediate values of C. The factors given in (2), and hence the relative importance of the four effects, are technology dependent.

B. Multiple-Plate Capacitors

In order to avoid process-caused systematic errors, MOS capacitors are often realized as the parallel combinations of several smaller "unit capacitors." Therefore, the random errors affecting such multiple-plate capacitors also are discussed. For an n-plate capacitor $C = nC_i$, the random capacitance error due to the local edge effect can be obtained from [6, eq. (25)] as $\sigma_{nC_i} = n^{1/4}\sigma_C = n^{1/2}\sigma_{C_i}$. Similarly, the random capacitance error due to the global edge effect can be found. The result is $\sigma_{nC_i} = n^{1/2}\sigma_C = n\sigma_{C_i}$. Here, the errors of the unit capacitors were assumed to be fully correlated since the *global* edge variation is considered. As far as both of the local and global oxide effects are concerned, there is no physical difference between the random errors of the single- and multiple-plate realizations. If both edge and oxide effects are considered for the multiple-plate capacitor, the overall relative capacitance error is

$$\frac{\sigma_{nC_i}}{C} = \sqrt{\frac{n^{1/2}K_{le}}{C^{3/2}} + \frac{K_{lo}}{C} + \frac{nK_{ge}}{C} + K_{go}} \qquad (3a)$$

or

$$\frac{\sigma_{nC_i}}{nC_i} = \sqrt{\frac{K_{le}}{nC_i^{3/2}} + \frac{K_{lo}}{nC_i} + \frac{K_{ge}}{C_i} + K_{go}}. \qquad (3b)$$

Equation (3a) shows that, *for a given C, the more parti-*

tions we make, the worse the matching becomes. Equation (3b) shows, however, that *for a given unit capacitance C_i,* the matching can be improved by paralleling more unit capacitors since this increases C.

Consider now a realistic example of a single-plate capacitor. Assume that the edge length L of the unit capacitor is 50 μm, $t_o = 700$ Å, and d_e and d_o are about 1 μm. From (1), assuming also that $\sigma_{lt} = \sigma_{gt} = 0$, $\sigma_{le} = \sigma_{ge} = 0.2$ μm, and $\sigma_{lt} = \sigma_{gt} = 10$ Å, the various terms in (1) turn out to be $K_{le}^{1/2}C^{-3/4} \simeq 0.16$ percent, $K_{ge}^{1/2}C^{-1/2} \simeq 1.06$ percent, $K_{lo}^{1/2}C^{-1/2} \simeq 0.069$ percent, and $K_{go}^{1/2} \simeq 1.74$ percent. $\Delta C/C$ is thus dominated by the terms containing K_{ge} and K_{go}.

As this example shows, for a typical MOS process, the *global edge and oxide variations represent the crucial limitations on the achievable matching accuracy.* Of course, the relative random errors in MOS capacitors which are made with a different process could be significantly different. However, the parameters in (4) give us an insight into these random error sources and their effects for a typical situation. The dominance of the global effects shows why steps aimed at their elimination (common centroid geometries [8], guard rings, etc.) are particularly important for high-accuracy matching.

In most applications, it is the *ratio* α of two capacitances C_1 and C_2, rather than their individual values, which is of interest. Assuming that $C_1 = nC_i$ and $C_2 = mC_i$ are both multiplate capacitors, it can readily be shown [11] that the relative rms error of α is

$$\frac{\sigma_\alpha}{\alpha} = \sqrt{\left(\frac{\sigma_{nC_i}}{C_1}\right)^2 + \left(\frac{\sigma_{mC_i}}{C_2}\right)^2}$$

$$= \sqrt{\left(\frac{1}{n} + \frac{1}{m}\right)\left(\frac{K_{le}}{C_i^{3/2}} + \frac{K_{lo}}{C_i}\right) + \frac{K_{ge}}{C_i} + 4K_{go}}. \qquad (4)$$

III. Experimental Results on Capacitance Matching

A large number of experimental capacitor test chips with on-chip capacitance measurement capability was fabricated using the 3.5 μm silicon gate NMOS technology of the Xerox Microelectronics Center. A total of 42 test struc-

TABLE I
LAYOUT STRATEGIES OF MATCHED MOS CAPACITORS IN CAPACITOR TEST CHIPS

	Capacitor Ratio	Unit Capacitor Size (μm^2)	Layout Strategy	Number of Test Structure
Group 1	1:1	25×25	Precision unit capacitor layout technique: multiple-plate realization, common-centroid geometry, guard ring, corner cutting, mask alignment tabs.	25
	2:1	38×38		
	4:1	50×50		
	8:1	75×75		
	16:1	100×100		
Group 2	1:1	25×25	Multiple-plate realization used only.	12
	2:1	50×50		
	4:1	100×100		
	8:1			
Group 3	1:1	25×25	Same as Group 1, except that corner cutting was not used.	5
	4:1	50×50		
	2:2	100×100		

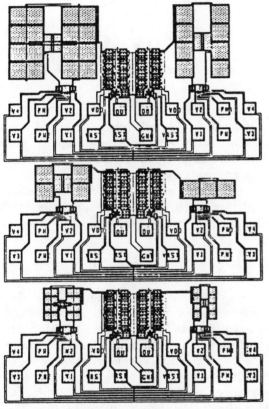

Fig. 3.   Parts of the layout of Group 1 in a capacitor test chip.

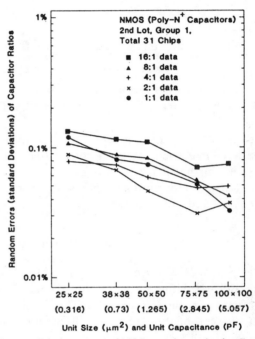

Fig. 4.   Random errors of MOS capacitor ratios for Group 1.

tures, which included different capacitor ratios with different unit capacitor sizes, were designed using three different layout strategies and were implemented on 6×6 mm² monolithic chips. The properties of the three groups, according to their layout strategies, are summarized in Table I. Part of the layout of a capacitor test chip (Group I capacitors) is shown in Fig. 3. The test circuit was a stray-insensitive buffered switched-capacitor circuit which transforms the ratio of two capacitors to a voltage ratio. After the circuit is zeroed, the variable voltage gives the capacitance ratio [11]. Since the circuit is highly sensitive and immune to parasitics, its accuracy is much better than 0.1 percent. The experimental test data are plotted in two charts. One (Fig. 4) shows the random errors for Group 1 of the test structures, and the other (Fig. 5) shows the random errors for Groups 2 and 3. Each data point gives the measured standard deviation of the capacitance ratio for pairs of capacitors, with nominal ratios 16:1, 8:1, etc.

The conclusions which can be drawn from the experimental results are as follows.

1) Since the slopes of the relative random error versus C plots on a log–log scale are close to $-1/4$ and the capacitance ratio errors are slowly varying functions of the capacitance ratios, the dominant random error sources for this process are clearly the global edge effect and the global oxide effects. This means that in (10b), K_{ge} and K_{go} are the dominant factors. The theoretical predictions made in connection with (12) have thus been verified.

2) Group 1 provided major improvements over Group 2 for large capacitor ratios and small unit capacitor sizes. Hence, a more careful layout strategy indeed gives a better capacitor matching.

3) The optimum geometries for this process are between 50×50 μm^2 and 75×75 μm^2 because the crossovers between systematic edge and oxide effects for all three groups of test data (not shown) are within this range. For area

131

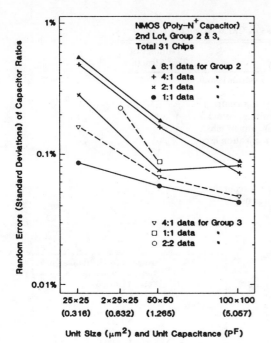

Fig. 5.   Random errors of capacitor ratios for Groups 2 and 3.

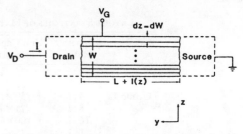

Fig. 6.   Top view of an MOS transistor with random length variations.

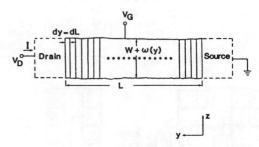

Fig. 7.   Top view of an MOS transistor with random width variations.

efficiency, the $50 \times 50 \ \mu\text{m}^2$ unit capacitor size can be used to achieve a 9-bit capacitor ratio accuracy since the random errors at this geometry are around 0.1 percent. This accuracy is adequate for most precision switched-capacitor circuits or A/D and D/A converters.

IV.   RANDOM ERRORS IN MOS TRANSISTORS

Consider next a $W \times L \ \mu\text{m}^2$ polysilicon gate n-channel MOS transistor where the width W and the length L are so large that first-order models are valid. The simple square-law drain current relation of a MOSFET operating in saturation is then [9]

$$I_o = \frac{\overline{\mu C_{ox}}}{2} \frac{W}{L} (V_{GS} - \overline{V}_T)^2 \qquad (5)$$

where I_o is the nominal drain current, $\bar{\mu}$ is the nominal effective electron mobility, $\overline{C}_{ox} \triangleq \epsilon / \bar{i}$ is the nominal gate capacitance per unit area, W is the nominal channel width, and L is the nominal channel length, while $\overline{V}_T$ is the nominal threshold voltage (including the body effect).[1]

A.   Random Errors Due to Edge Effects

Because the gate area of a MOSFET is determined by several different masks (i.e., the poly mask determines the channel length, while the diffusion mask determines the channel width), we should analyze the edge effects due to random length variation and random width variation independently. As for MOS capacitors, the edge effects also include the local and global random variations. First we

[1]Note that the random variations of substrate doping can also have significant effects on the matching, e.g., for PMOS transistors in a CMOS process. Our simple drain–current relation is unsuitable for an analysis of this effect, and a more elaborate device model is required.

consider random length variations only. A MOSFET can then be modeled as a parallel combination of many differential-width "strip transistors" with equal widths dW and slightly different lengths $L + l(z)$ as shown in Fig. 6. Here, $l(z)$ represents the random length variation and is assumed to be a zero-mean stationary random process. On the basis of this model, the relative current error, due to both local and global length variations, can be found as

$$\frac{\Delta I}{I_o} = \frac{1}{L} \sqrt{B_{gL} + \frac{B_{lL}}{W}} \ . \qquad (6)$$

Here, $B_{gL} = \sigma_{gl}^2$ where σ_{gL} is the standard deviation of L due to global variations and $B_{lL} = 2d_e\sigma_{lL}^2$ where d_e is the correlation radius of the local length variation and σ_{lL} is the standard deviation of L due to local variations.

Next, the channel width variations are considered. The transistor can then physically be modeled as a series combination of many wide "strip transistors" with equal differential lengths dL and slightly different widths $W + \omega(y)$ as shown in Fig. 7. On the basis of this model [9], the relative current error due to both local and global width variations can be obtained as

$$\frac{\Delta I}{I_o} = \frac{1}{W} \sqrt{B_{gW} + \frac{B_{lW}}{L}} \qquad (7)$$

where $B_{gW} \triangleq \sigma_{gW}^2$ is the global-width-effect factor and $B_{lW} \triangleq 2d_e\sigma_{lW}^2$ is the local-width-effect factor.

If *both* length and width random variations are taken into account, the relative current error due to edge effect is found

$$\left.\frac{\Delta I}{I_o}\right|_e = \sqrt{\left(\frac{B_{gL}}{L^2} + \frac{B_{gW}}{W^2}\right) + \left(\frac{B_{lL}}{L} + \frac{B_{lW}}{W}\right)\frac{1}{LW}} \ . \qquad (8)$$

Equation (8) shows that for a given current I_o, the current matching can be improved by increasing the device size. The matching is not affected by the body effect if the edge effect is the dominant factor.

B. Random Errors Due to Surface-State-Charge and Ion-Implanted-Charge Effects

In a typical MOS process, ion-implanted charges are usually employed to shift the threshold voltage [10]; also, some inherent interface charges always exist at and inside the interface of the silicon oxide and substrate. The surface-state charge density Q_{ss} and the ion-implanted charge density Q_{ii} are randomly distributed over the whole channel area. Hence, the deviations of these charge densities from their mean values cause a deviation on the nominal threshold voltage $\overline{V}_T$ and thus also a current error. The nominal threshold voltage in the presence of body effect is given by [9], [14]

$$\overline{V}_T = \phi_{MS} + 2|\phi_F| + \frac{\sqrt{2q\epsilon_{Si}N_a(2|\phi_F| + V_{BS})}}{\overline{C}_{ox}} - \frac{\overline{Q}_{ss}}{\overline{C}_{ox}} + \frac{\overline{Q}_{ii}}{\overline{C}_{ox}} \tag{9}$$

where the meaning of the symbols is as follows.

ϕ_{MS}: Potential difference of the work functions of the gate and substrate.

ϕ_F: Built-in potential in the bulk of the substrate.

$Q_D \triangleq \sqrt{2q\epsilon_{Si}N_a(2|\phi_F| + |V_{BS}|)}$: Space-charge density per unit area.

Q_{ss}: Surface-state charge density per unit area.

Q_{ii}: Ion-implanted charge density per unit area.

$C_{ox} \triangleq \epsilon/t$: Gate capacitance per unit area.

The bar above a symbol indicates the nominal value. Hence, the nominal drain current, from (5), becomes

$$I_o = \beta \left[K_1 + \frac{1}{\overline{C}_{ox}} (\overline{Q}_{ss} - \overline{Q}_{ii}) \right]^2 \tag{10}$$

where $\beta \triangleq (\overline{\mu}\overline{C}_{ox}/2)(W/L)$ is the gain factor and $K_1 \triangleq V_{GS} - \phi_{MS} - 2|\phi_F| - (Q_D/\overline{C}_{ox})$ is a constant.

We then describe the random variations of Q_{ss} and Q_{ii} as functions of y and z as follows:

$$Q_{ss}(y, z) = \overline{Q}_{ss} + q_{ss}(y, z)$$
$$Q_{ii}(y, z) = \overline{Q}_{ii} + q_{ii}(y, z) \tag{11}$$

where $q_{ss}(y, z)$ and $q_{ii}(y, z)$ are assumed to be zero-mean stationary two-dimensional random processes. We can express $\overline{Q}_{ii}$ and $\overline{Q}_{ss}$ in (10) using (11) and then integrate it over the whole channel area. After some fairly complicated calculations [11], the relative current error due to *both* local and global variations of Q_{ss} and Q_{ii} is found to be

$$\left. \frac{\Delta I}{I_o} \right|_q = \frac{1}{\overline{C}_{ox}(V_{GS} - \overline{V}_T)} \sqrt{(B_{gss} + B_{gii}) + \frac{1}{WL}(B_{lss} + B_{lii})} \tag{12}$$

where $B_{gss} \triangleq 4\sigma_{gq}^2$ and $B_{gii} \triangleq 4\sigma_{gq_{ii}}^2$ represent the global surface-state charge and ion-implanted charge-effect factors, while $B_{lss} \triangleq 16d_q^2\sigma_{lq_{ss}}^2$ and $B_{lii} \triangleq 16d_q^2\sigma_{lq_{ii}}^2$ are the

local surface-state charge and ion-implanted charge effect factors, respectively. The subscript "q" here and in the following means the random surface-charge effect. Equation (12) shows that a larger $\overline{C}_{ox}$ and V_{GS} and smaller body effect in V_T results in a better current matching for a given transistor size, but at the cost of a poorer dynamic range due to the larger V_{GS}.

A practical application of our results is to the random input offset voltage of a differential input amplifier. This is found to be [12]

$$\Delta V_{GS} = \frac{\Delta I}{g_{mo}} \tag{13}$$

where $g_{mo} \triangleq \overline{\mu}\overline{C}_{ox}(W/L)(V_{GS} - \overline{V}_T)$ is the nominal transconductance, while ΔI represents the random current error. Hence, from (12) and (13), the random input offset voltage due to random surface-charge effects is

$$\Delta V_{GS}|_q = \frac{1}{2\overline{C}_{ox}} \sqrt{(B_{gss} + B_{gii}) + \frac{1}{WL}(B_{lss} + B_{lii})} . \tag{14}$$

C. Random Errors Due to Oxide Effects

Consider now the random fluctuations of the oxide thickness t. The gain factor as well as the threshold voltage in (5) are affected by the oxide thickness variation, and hence so is the drain current of the device. From (5) and (9), the nominal drain current can be expressed as

$$I_o = \frac{K_2}{\overline{t}}(K_3 - \overline{t}K_4)^2 \tag{15}$$

where $K_2 \triangleq \beta\overline{t}$ and $K_3 \triangleq V_{GS} - \phi_{MS} - 2|\phi_F|$ and $K_4 \triangleq (Q_D - \overline{Q}_{ss} + \overline{Q}_{ii})/\epsilon$. Next, t is considered to be composed of a mean $\overline{t}$ and a random variation $\Delta t(y, z)$ across the gate area, i.e., $t(y, z) = \overline{t} + \Delta t(y, z)$.

Performing a statistical analysis, the relative current error due to both local and global variations of oxide thickness is found to be

$$\left. \frac{\Delta I}{I_o} \right|_o = \frac{1}{\overline{t}} \left[1 + \frac{2Q_T}{\overline{C}_{ox}(V_{GS} - \overline{V}_T)} \right] \sqrt{B_{go} + \frac{B_{lo}}{WL}} \tag{16}$$

where $Q_T \triangleq Q_D - \overline{Q}_{ss} + \overline{Q}_{ii}$ represents the surface charge, and $B_{go} \triangleq \sigma_{g\Delta t}^2$ and $B_{lo} = 4d_o^2\sigma_{l\Delta t}^2$ are the global oxide and local oxide-effect factors, respectively. (The subscript "o" refers to the oxide effect.) An interesting aspect of (16) is that the first term in the square brackets represents the oxide effect acting through the gain factor of the device, while the second term represents the oxide effects due to the threshold voltage of the device. If the first term is dominant because, e.g., a large $V_{GS} - \overline{V}_T$ is used, the matching accuracy will be improved by increasing $\overline{t}$; however, it should be recalled that as (12) showed, the relative error due to random surface-charge effects increases with increasing $\overline{t}$. Therefore, the relation between $\overline{t}$ and the matching accuracy due to both oxide and surface-charge effects is not obvious; it depends upon the process, i.e., the actual values of B_{gss}, B_{go}, $\cdots$, etc.

Our results can be applied again to a differential input amplifier. The random input offset voltage variation due to oxide effects acting through the threshold voltage is found to be

$$\Delta V_{GS}|_o = \frac{Q_T}{\epsilon} \sqrt{B_{go} + \frac{B_{lo}}{WL}} . \tag{17}$$

If both oxide and surface-charge effects are taken into account using (14) and (17), then the random input offset voltage due mainly to the threshold voltage mismatch can be expressed as

$$\Delta V_{GS}|_{V_T}$$

$$= \sqrt{ \frac{B_{gss} + B_{gii}}{4\overline{C}_{ox}^2} + \frac{Q_T^2 B_{go}}{\epsilon^2} + \frac{1}{WL}\left(\frac{B_{lss} + B_{lii}}{4\overline{C}_{ox}^2} + \frac{Q_T^2 B_{lo}}{\epsilon^2} \right) } . \tag{18}$$

Hence, ΔV_{GS} is *not* a function of $V_{GS} - \overline{V}_T$. This is in agreement with the argument in [12].

D. Random Errors Due to Channel Mobility Effects

Due to the random impurity scattering and lattice scattering mechanisms [13], the effective channel mobility of the carriers varies randomly over the entire channel area of the device. The effective channel mobility μ can be regarded as composed of a mean value $\bar{\mu}$ and a random variation $\Delta\mu(y, z)$, i.e., $\mu(y, z) \triangleq \bar{\mu} + \Delta\mu(y, z)$. Hence, the deviation of I can be found to be

$$\Delta I = K_9 \int_o^W \int_o^L \Delta\mu(y, z)\, dy\, dz . \tag{19}$$

Performing a statistical analysis [11], the relative current error due to both local and global variations of μ turns out to be

$$\frac{\Delta I}{I_o}\bigg|_\mu = \frac{1}{\bar{\mu}} \sqrt{B_{g\mu} + \frac{B_{l\mu}}{WL}} \tag{20}$$

where $B_{g\mu} \triangleq \sigma_{g\Delta\mu}^2$ and $B_{l\mu} = 4d_\mu^2\sigma_{l\Delta\mu}^2$ represent the global and local channel mobility-effect factors, respectively.

In conclusion, combining the relative current errors due to all four random effects, the overall relative current error from (8), (12), (16), and (20) can be expressed as

$$\frac{\Delta I}{I_o} = \Bigg\{ \Bigg[\left(\frac{B_{gL}}{L^2} + \frac{B_{gW}}{W^2} \right) + \frac{B_{gss} + B_{gii}}{\overline{C}_{ox}^2 (V_{GS} - \overline{V}_T)^2} $$

$$+ \frac{B_{go}}{\bar{t}^2}\left(1 + \frac{2Q_T}{\overline{C}_{ox}(V_{GS} - \overline{V}_T)} \right)^2 + \frac{B_{g\mu}}{\bar{\mu}^2} \Bigg]$$

$$+ \frac{1}{WL}\Bigg[\left(\frac{B_{lL}}{L} + \frac{B_{lW}}{W} \right) + \frac{B_{lss} + B_{lii}}{\overline{C}_{ox}^2 (V_{GS} - V_T)^2}$$

$$+ \frac{B_{lo}}{\bar{t}^2}\left(1 + \frac{2Q_T}{\overline{C}_{ox}(V_{GS} - \overline{V}_T)} \right)^2 + \frac{B_{l\mu}}{\bar{\mu}^2} \Bigg] \Bigg\}^{1/2} . \tag{21}$$

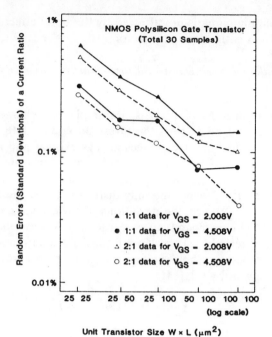

Fig. 8.   Random errors of the current matching.

Here, the terms in the first pair of square brackets in (21) represent the *global* random effects, while those in the second pair of square brackets represent the *local* random effects. Since WL enters in the denominators of all terms in the second brackets, all local random effects can be improved by using a larger transistor size.

V. EXPERIMENTAL RESULTS ON TRANSISTOR CURRENT MATCHING

An experiment aimed at verifying the MOSFET's predicted current matching properties was carried out using the same 3.5 μm NMOS technology as for capacitor matching at the Xerox Microelectronics Center. Hence, the results obtained previously for capacitance matching regarding edge effects, oxide effects, and optimum geometry could be utilized in the analysis of the transistor current matching. A total of 300 test transistors with five unit transistor sizes ($W \times L = 25 \times 25$, 25×50, 25×100, 50×100, and 100×100 μm^2) were fabricated and tested using off-chip current meters on five test chips. These large transistor sizes guaranteed that the short-channel and narrow-width effects were not significant. The nominal threshold voltage for this technology was $\overline{V}_T = 0.9$ V. Two different gate-to-source voltages ($V_{GS} = 2$ and 4.5 V) were used in order to be able to distinguish the current errors due to the threshold voltage effects. The nominal current levels for $V_{GS} = 2$ V were 5.06, 10.13, and 20.25 μA, while for $V_{GS} = 4.5$ V, they were 53.69, 107.39, and 214.79 μA. Matched pairs of MOSFET's were designed for two current ratios (1:1 and 2:1) using careful layout techniques. The experimental data obtained are shown in Fig. 8.

The following conclusion can be drawn from our experimental results.

1) The global random edge variations were the dominant effects for smaller unit transistor sizes (25×25 to 25×100

μm^2), while the global random oxide effects dominated for larger unit transistor sizes (50×100 to 100×100 μm^2). Furthermore, the *global* random surface-charge effects were also important over the entire range of unit transistor sizes. This follows because the relative current errors for $V_{GS} = 2$ V were always larger than for $V_{GS} = 4.5$ V. Also, the *local* random surface-charge effects were influential for the smaller unit transistor sizes. This follows because the improvement of random errors in (21) due to a larger V_{GS} is seen to be reduced by increasing the unit transistor size. This means that the random surface-charge effects are functions of W and L. From (21), we realize that the *local* random surface-charge variations also appear to have a large effect for smaller unit transistor sizes.

2) The optimum geometry for this process is about 50×100 μm^2 ($W \times L$). To reduce power dissipation, the smaller gate-to-source voltage (i.e., $V_{GS} = 2$ V) can be used to obtain an 8-bit current ratio accuracy. This is adequate for many precision applications.

3) As our data show, a careful layout technique incorporating such features as guard rings, common-centroid geometry, etc., is important for good current matching.

VI. CONCLUSIONS

Theoretical formulas have been derived for random effects, both local (granular) and global (large-scale), affecting the ratios of matching capacitances and currents. For capacitors, the effects considered were edge variations and oxide variations. Both single-plate and multiple-plate capacitances were analyzed. For MOSFET current sources, the error sources considered included edge variations, surface-state and ion-implanted charge variations, oxide-variation effects, and carrier mobility variations.

Using a 3.5 μm NMOS technology, a large number of matched capacitors and MOSFET current sources were fabricated and tested using on-chip measurement circuits. The results confirmed the general conclusions drawn from the theoretical relations, and also gave the optimal dimensions and bias conditions for the process.

The importance of random effects is that they represent the ultimate limitation on the achievable accuracy of MOS analog circuits in the absence of systematic errors. Our results indicate that random errors permit a 9-bit accuracy for capacitance matching and an 8-bit accuracy for MOSFET current-source matching without excessive area requirements.

ACKNOWLEDGMENT

The authors are grateful to S. Law and S. Eckert of the Xerox Corporation and to Prof. K. Yao of UCLA for useful discussions, and to the Xerox Microelectronics Center for fabricating the test circuits described.

REFERENCES

[1] P. R. Gray, D. A. Hodges, and R. W. Brodersen, *Analog MOS Integrated Circuits.* New York: IEEE Press, 1980.
[2] R. Gregorian, K. W. Martin, and G. C. Temes, "Switched-capacitor circuit design," *Proc. IEEE*, vol. 71, pp. 941–966, Aug. 1983.
[3] J. L. McCreary and P. R. Gray, "All-MOS charge redistribution analog-to-digital conversion techniques—Part I," *IEEE J. Solid-State Circuits*, vol. SC-10, pp. 371–379, Dec. 1975.
[4] H. U. Post and K. Waldschmidt, "A high-speed NMOS A/D converter with a current source array," *IEEE J. Solid-State Circuits*, vol. SC-15, pp. 295–300, June 1980.
[5] S. Kelly and D. Ulmer, "A single-chip CMOS PCM codec," *IEEE J. Solid-State Circuits*, vol. SC-14, pp. 54–59, Feb. 1979.
[6] J.-B. Shyu, G. C. Temes, and K. Yao, "Random errors in MOS capacitors," *IEEE J. Solid-State Circuits*, vol. SC-17, pp. 1070–1076, Dec. 1982.
[7] J. T. Wallmark, "A statistical model for determining the minimum size in integrated circuits," *IEEE Trans. Electron Devices*, vol. ED-26, pp. 135–142, Feb. 1979.
[8] J. L. McCreary, "Successive approximation analog-to-digital conversion in MOS integrated circuits," Ph.D. dissertation, Univ. California, Berkeley, 1975.
[9] W. N. Carr and J. P. Mize, *MOS/LSI Design and Application.* New York: McGraw-Hill, 1972, p. 46.
[10] B. S. Song and P. R. Gray, "Threshold-voltage temperature drift in ion-implanted MOS transistors," *IEEE J. Solid State Circuits*, vol. SC-17, pp. 291–298, Apr. 1982.
[11] J.-B. Shyu, "The obtainable accuracy of analog MOS integrated circuit elements," Ph.D. dissertation, Univ. California, Los Angeles, 1984.
[12] P. R. Gray and R. G. Meyer, "MOS operational amplifier design—A tutorial overview," *IEEE J. Solid-State Circuits*, vol. SC-17, pp. 969–982, Dec. 1982.
[13] A. S. Grove, *Physics and Technology of Semiconductor Devices.* New York: Wiley, 1967.
[14] R. S. Muller and T. I. Kamins, *Device Electronics for Integrated Circuits.* New York: Wiley, 1977.
[15] J. L. McCreary, "Matching properties, and voltage and temperature dependence of MOS capacitors," *IEEE J. Solid-State Circuits*, vol. SC-16, pp. 608–616, Dec. 1981.

Characterization and Modeling of Mismatch in MOS Transistors for Precision Analog Design

KADABA R. LAKSHMIKUMAR, MEMBER, IEEE, ROBERT A. HADAWAY,
AND MILES A. COPELAND, SENIOR MEMBER, IEEE

Abstract —This paper is concerned with the design of precision MOS analog circuits. Section II of the paper discusses the characterization and modeling of mismatch in MOS transistors. A characterization methodology is presented that accurately predicts the mismatch in drain current over a wide operating range using a minimum set of measured data. The physical causes of mismatch are discussed in detail for both p- and n-channel devices. Statistical methods are used to develop analytical models that relate the mismatch to the device dimensions. It is shown that these models are valid for small-geometry devices also. Extensive experimental data from a 3-μm CMOS process are used to verify these models.

Section III of the paper demonstrates the application of the transistor matching studies to the design of a high-performance digital-to-analog converter (DAC). A circuit design methodology is presented that highlights the close interaction between the circuit yield and the matching accuracy of devices. It has been possible to achieve a circuit yield of greater than 97 percent as a result of the knowledge generated regarding the matching behavior of transistors and due to the systematic design approach.

I. INTRODUCTION

THE DESIGN of precision analog circuits requires a thorough understanding of the matching behavior of components available in any given technology. In MOS technology, capacitors are being widely used for designing precision analog circuits such as data converters [1] and switched-capacitor filters [2], [3] because of their excellent matching characteristics [4]. The matching behavior of MOS capacitors has been discussed in detail [5]–[7]. However, all precision analog circuits cannot be designed using capacitors alone. For applications such as high-speed data conversion, capacitive techniques tend to be too slow. Further a digital VLSI process may not offer linear capacitors. These factors motivated us to study the matching behavior of MOS transistors.

Section II of this paper discusses the characterization and modeling of mismatch in MOS transistors. The interest in such a study is evidenced by recent publications [7], [8]. In [8] experimental results of matching of MOS current mirrors are discussed without any reference to the physical causes of mismatch. The work reported in [7] attempts to break down the causes of mismatch but the experimental results are limited to large-area n-channel devices only. The work reported here is aimed at providing a more comprehensive understanding of the causes of mismatch in both p- and n-channel devices of large and small geometry. As the circuit designer has freedom to choose only the device dimensions, analytical models have been developed that relate the electrical mismatch to the dimensions. Extensive data to verify these models are obtained from a 5-V, 3 μm, p-well CMOS process that is in use at Northern Telecom Electronics Limited, Ottawa, Canada.

Section III of the paper demonstrates the application of the knowledge of matching behavior for the design of a high-speed digital-to-analog converter (DAC). Of late, the area of high-speed data converters in MOS technology is gaining importance (for example, [23] and [24]). However, these designs do not indicate the circuit yield obtainable for a given resolution, or the possibility of extension of these techniques to higher resolution converters. Therefore a circuit design methodology is presented here that relates the achievable linearity and yield to the matching accuracy of the components. A high-performance DAC with a circuit yield of greater than 97 percent has been realized without using any post-process trimming and yet occupying a small chip area using this design methodology [9].

II. TRANSISTOR MATCHING STUDIES

In general, there are two variations to consider in an integrated circuit process. *Global variation* accounts for the total variation in the value of a component over a wafer or a batch. *Local variation* or *mismatch* reflects the variation in a component value with reference to an adjacent component on the same chip. As the design of precision analog circuits is based on component ratios rather than their

Manuscript received April 21, 1986; revised July 14, 1986. The Carleton University part of this work was supported by the Canadian Commonwealth Scholarship and Fellowship Committee and the Natural Sciences and Engineering Research Council of Canada.

K. R. Lakshmikumar was with the Department of Electronics, Carleton University, Ottawa, Ont. K1S 5B6, Canada. He is now with AT&T Bell Laboratories, Murray Hill, NJ 07974.

R. A. Hadaway is with Northern Telecom Electronics Ltd., Ottawa, Ont., Canada.

M. A. Copeland is with the Department of Electronics, Carleton University, Ottawa, Ont. K1S 5B6, Canada.

IEEE Log Number 8610606.

Reprinted from *IEEE J. Solid-State Circuits*, vol. SC-21, no. 6, pp. 1057–1066, Dec. 1986.

absolute values, we have concentrated our study on the mismatch behavior.

The characterization of mismatch in MOS transistors is more complex than that in the case of capacitors. The drain current matching not only depends on the device dimensions but also on the operating point. In Section II-A, a characterization methodology is developed that accurately determines the mismatch in drain current over a wide operating region, using a minimum set of measurement data. The physical causes of mismatch are discussed in Section II-B, and analytical expressions to relate the mismatch to device dimensions are developed. We call these quantitative relationships *mismatch models*.

A. *Characterization Methodology*

Our aim is to predict the mismatch in the drain current over a wide range of operating conditions using a minimum set of measured data, and simultaneously to throw light on the detailed causes of mismatch. This problem can be best approached by measuring the mismatch in various parameters of a suitable circuit model [10]. The model chosen should be such that it gives an adequate description of the electrical behavior of the device, and at the same time should have readily measurable parameters that are amenable to statistical description. As an elaborate circuit model may greatly exceed the accuracy of measurable data or may hamper the extraction of statistically significant model parameters, we chose the simple square-law model. The current–voltage relationship in the triode region is given by

$$I = K(V_{GS} - V_T - V_{DS}/2)V_{DS} \qquad (1)$$

where I is the drain current, K is the conductance constant, V_T is the threshold voltage, and V_{DS} is the drain-to-source voltage. The statistically significant parameters of this model are V_T and K. The mismatch in V_T accounts for the variations in the different charge quantities, and in the gate oxide capacitance per unit area. The variations in the dimensions, channel mobility, and gate oxide capacitance per unit area are measured as the mismatch in K. As both V_T and K are dependent on the gate oxide capacitance per unit area, we need to measure the correlation between the mismatches in V_T and K also.

The square-law model (1) is not an accurate description of the current–voltage relationship. It should be noted that we are only looking for local variations and are not so much concerned about the estimation of the absolute value of the parameters. Therefore any small model error would cancel out to a first order while estimating the mismatch, and hence the square-law model should suffice for our application. Several assumptions are made while deriving this one-dimensional model. As some of these are not strictly applicable, a further discussion to justify the use of this model is in order.

The gradual channel approximation and the assumption that the substrate is uniformly doped do not necessarily hold for small-geometry devices. An accurate analysis calls for a two-dimensional solution of Poisson's equation. However, in order to develop analytical expressions for the mismatch behavior, we use a one-dimensional circuit model and apply appropriate corrections to account for the effects of dimensional dependence on threshold voltage and nonuniform doping of the substrate. Also we have assumed a simplified picture of the oxide–silicon interface, i.e., that oxide fixed charge and interface trap charges are considered to be smeared-out uniform charge sheets. In fact the oxide fixed charge and charged interface traps are localized, and not sheets [11]. The accurate calculation of the surface potential would then be a two-dimensional problem. We are simplifying the problem by estimating the aggregate of the localized nonuniformity over the entire area of the channel by using the simplified one-dimensional circuit model. With these approximations, we develop analytical expressions for the mismatch behavior that compare remarkably well with experimental data.

Generally, MOS transistors will be operating in the saturation region in analog circuits. Therefore we should relate the measured mismatches in V_T and K to the saturation region, where the drain current is given by

$$I = \frac{K}{2}(V_{GS} - V_T)^2. \qquad (2)$$

Then the variance in the drain current may be written as

$$\frac{\sigma_I^2}{\bar{I}^2} = \frac{\sigma_K^2}{\bar{K}^2} + 4\frac{\sigma_{VT}^2}{(V_{GS} - \bar{V}_T)^2} - 4r\frac{\sigma_{VT}}{V_{GS} - \bar{V}_T} \cdot \frac{\sigma_K}{\bar{K}} \qquad (3)$$

following the derivation in [12] concerning the variance of a function of two random variables. Here r is the correlation coefficient between the mismatches in V_T and K, $\bar{I}$ is the expected value of the random variable I, σ_I is the standard deviation of I, and so on. Thus the mismatch in drain current at any operating point may be estimated if σ_K, σ_{VT}, and r are known.

Experimentally, V_T and K are determined by measuring the drain current versus gate voltage for a small value of V_{DS}. The maximum slope of the I versus V_{GS} curve provides the value of K. V_T is obtained from the intercept of the maximum slope on to the V_{GS} axis. If ΔK_i is the difference in the value of K for the ith matched pair of devices, the standard deviation of K is given by

$$\sigma_K = \left[\frac{1}{N-1}\left\{\sum_{i=1}^{N}(\Delta K_i)^2 - \frac{1}{N}\left(\sum_{i=1}^{N}\Delta K_i\right)^2\right\}\right]^{1/2} \qquad (4)$$

where N is the number of matched pairs measured on each wafer. The second term in (4) is close to zero as the matched pairs are laid out in such a way as to minimize systematic mismatch. σ_{VT} is also computed in a similar way.

B. Factors Causing Mismatch

1. Threshold Voltage Mismatch: The threshold voltage of a transistor may be expressed as

$$V_T = \phi_{MS} + 2\phi_B + \frac{Q_B}{C} - \frac{Q_f}{C} + \frac{qD_I}{C} \qquad (5)$$

where ϕ_{MS} is the gate–semiconductor work function difference, ϕ_B is the Fermi potential in the bulk, Q_B is the depletion charge density, Q_f is the fixed oxide charge density, D_I is the threshold adjust implant dose, and C is the gate oxide capacitance per unit area. The last term in (5) accounts for the threshold adjust implant where the implanted ions are assumed to have a delta function profile at the silicon–silicon dioxide interface [14]. The standard deviation of V_T may be determined if we can find the standard deviations of the various terms on the right-hand side of (5). The Fermi potential ϕ_B has a logarithmic dependence on the substrate doping, and ϕ_{MS} has a similar dependence on the doping in the substrate and in the polysilicon gate. Hence these terms may be regarded as constants not contributing to any mismatch.

Next we consider oxide fixed charge which is reported to have a Poisson distribution [11, p. 242]. Then its variance is given by

$$\sigma_{Qf}^2 = \frac{q\overline{Q}_f}{\overline{LW}} \qquad (6)$$

where L is the effective length and W is the effective width of the channel.

The depletion charge per unit area Q_B is also a random variable dependent on the distribution of the dopant atoms. This is an important difference between the treatment given here and the one reported in [7], where Q_B is treated as a constant. In fact, it is reported in [11, p. 237] that the dopant ions are nonuniformly distributed in MOS devices. No theoretical treatment of fluctuations in dopant ion density is available. However, we shall show that the physical conditions in the substrate favor a Poisson distribution [13]. The number of atoms per unit volume in silicon is 5×10^{22} cm^{-3}. Only a very small fraction of these sites are occupied by the dopant atoms. The number of dopant atoms in nonoverlapping volumes is independent. Further, for domains of sufficiently small volume, the probability of finding exactly one dopant atom in a domain is proportional to the volume, and the probability of finding more than one atom is negligible. Hence the dopant ions may be considered to have Poisson distribution. Then the variance in Q_B may be shown to be [15]

$$\frac{\sigma_{QB}^2}{\overline{Q}_B^2} = \frac{1}{4\overline{LW}W_d N_A} \qquad (7)$$

where W_d is the depletion layer width and N_A is the substrate doping.

Similarly, assuming the implanted ions to follow a Poisson distribution, the variance of D_I is given by

$$\sigma_{DI}^2 = \frac{\overline{D}_I}{\overline{LW}}. \qquad (8)$$

The variance in C may be determined by estimating the variances in oxide thickness and permittivity [7]. It can be shown that [15]

$$\frac{\sigma_C^2}{\overline{C}^2} = \frac{1}{\overline{LW}}A_{ox} \qquad (9)$$

where A_{ox} is a parameter to be determined from measurements.

The random variables Q_f, Q_B, D_I, and C are all independent. Hence the variance in V_T may be written using (5) as

$$\sigma_{VT}^2 = \frac{1}{\overline{C}^2}\left(\sigma_{QB}^2 + \sigma_{QF}^2 + q^2\sigma_{DI}^2\right) + \frac{\sigma_C^2}{\overline{C}^2}\left(\frac{\overline{Q}_B^2}{\overline{C}^2} + \frac{\overline{Q}_f^2}{\overline{C}^2} + \frac{q^2\overline{D}_I^2}{\overline{C}^2}\right). \qquad (10)$$

Substituting (6)–(9) into (10) we have

$$\sigma_{VT}^2 = \frac{1}{\overline{LWC}^2}\left[q(\overline{Q}_B + \overline{Q}_f + q\overline{D}_I) + A_{ox}(\overline{Q}_B^2 + \overline{Q}_f^2 + q^2\overline{D}_I^2)\right]. \qquad (11)$$

Now let us examine the importance of the various terms on the right-hand side of (11), for both p- and n-channel devices. Consider n-channel devices first. In our process, the threshold adjust implant is carried out for p-channel transistors only. Therefore $qD_I = 0$ for n-channel devices. The depletion charge density per unit area is

$$Q_B = 7.7 \times 10^{-8} \text{ C/cm}^2. \qquad (12)$$

In a well-controlled process the number of fixed oxide charges can be reduced to about 2×10^{10}/cm^2, and hence

$$Q_f = 3.2 \times 10^{-9} \text{ C/cm}^2. \qquad (13)$$

Comparing (12) and (13) we may infer that the contribution of the variability of the fixed oxide charges to threshold voltage mismatch (11) may be neglected.

The measured relative standard deviation of the threshold voltage ($\sigma_{VT}/\overline{V}_T$) is plotted against the reciprocal of the square root of the effective channel area in Fig. 1 for n-channel devices with six different W/L (drawn) values. $\sigma_{VT}/\overline{V}_T$ is chosen as the ordinate so as to express the variation as a percentage, independent of the operating point. However, if one is interested in current mismatch only, then $\sigma_{VT}/(V_{GS} - \overline{V}_T)$ should be plotted so that it may be directly used in (3).

For collecting statistics, 128 device pairs of each size were measured on every wafer. The vertical error bars reflect the spread in measured values over four wafers. Although the error bars appear large in this figure, in reality they represent relatively small deviations. For ex-

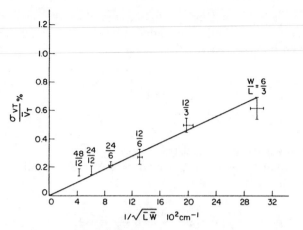

Fig. 1. Threshold voltage mismatch versus dimensions for n-channel devices.

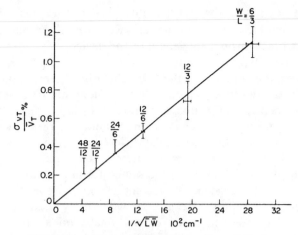

Fig. 2. Threshold voltage mismatch versus dimensions for p-channel devices.

ample, devices with $W/L = 6$ μm/3 μm have a spread in the standard deviation of matching of threshold voltage of 3.5–4.9 mV. This spread is partly due to the nominal process variations from wafer to wafer, and partly due to the dependence of the matching on the electrical dimensions of the device. The effective channel length and width of devices were measured electrically at different places on a wafer and also on different wafers. The spread in the values is indicated by the horizontal error bars. The measured data fit well with the theoretical straight line relationship given by (11), which may be approximated for n-channel devices as

$$\sigma_{VT}/\overline{V}_T = \frac{1}{\sqrt{\overline{LW}}} \left(2.5875 \times 10^{-12} + 1.2421 A_{ox} \right)^{1/2}/\overline{V}_T. \tag{14}$$

Comparing the slope of the line in Fig. 1 with that in (14), it is found that

$$A_{ox} = 6.4631 \times 10^{-14} \text{ cm}^2. \tag{15}$$

Then from (9), $\sigma_C/\overline{C} = 0.02$ percent for a 24×6-μm^2 gate. This low value agrees with the extremely uniform nature of the gate oxide thickness observed in other measurements [16].

Now we consider p-channel devices. As the threshold adjust implant is a very shallow one, a considerable portion of the implanted ions is retained in the gate oxide. Although this results in charged states, they are readily annealed during subsequent processing [17]. However, the presence of these impurity atoms in the oxide may cause a degradation in the capacitance matching of p-channel devices as compared to n-channel ones. For our process $Q_B = 4.810 \times 10^{-8}$ C/cm^2 and $qD_I = 8.0 \times 10^{-8}$ C/cm^2. Hence the contribution of Q_f to threshold voltage mismatch may be ignored and (11) may be written as

$$\sigma_{VT}/\overline{V}_T = \frac{1}{\sqrt{\overline{LW}}} \left(4.2945 \times 10^{-12} + 1.8463 A_{ox} \right)^{1/2}/\overline{V}_T. \tag{16}$$

The numerical coefficients in (16) are larger than the corresponding ones in (14) indicating a larger mismatch in

p-channel devices, owing to the additional threshold adjust implant. This may be physically interpreted as a larger variation in the surface concentration due to the differential doping occurring at the surface.

The mismatch in threshold voltage of p-channel devices is plotted in Fig. 2. The data fit very well into the theoretical straight line relationship, and it is found that

$$A_{ox} = 3.0369 \times 10^{-12} \text{ cm}^2. \tag{17}$$

Comparing (15) and (17) we may infer that the gate oxide capacitance matching is poorer for p-channel devices than that for n-channel ones.

We will now summarize the threshold voltage mismatch behavior. These findings are particular to this work.

a) The standard deviation of mismatch is inversely proportional to the square root of the effective channel area.

b) In a well-controlled process the nonuniform distribution of the fixed oxide charges has negligible effect on threshold voltage mismatch.

c) The nonuniform distribution of the dopant atoms in the bulk is a major contributor to the threshold voltage mismatch. The assumption that these atoms follow a Poisson distribution has resulted in excellent agreement with measurements.

d) Devices which use a compensating threshold adjust implant have a higher mismatch in threshold voltage due to the differential doping occurring at the surface. This is the major reason for the significantly larger mismatch noticed in p-channel devices as compared to n-channel transistors.

e) The gate oxide capacitance is quite uniform and hence has little influence on the threshold voltage mismatch. However, between n- and p-channel devices, the gate oxide capacitance of the latter has slightly poorer matching characteristic. This could be due to the nonuniform distribution of the threshold adjust implant atoms in the gate oxide.

2. Conductance Constant Mismatch: The conductance constant is given by

$$K = \mu C W/L \tag{18}$$

where μ is the channel mobility. We can express the variance in K in terms of the variances in μ, C, W, and L. Let us first consider the length of the device.

Electrically, the channel length is the average distance between the source and the drain diffusions. Any raggedness in the definition of the polysilicon may not be exactly reproduced in the source and drain diffusion edges. Further, we are not so much concerned about this raggedness; rather we are interested in the difference in the electrical length from one device to the next. Such a mismatch in length may not be due to the nonuniformity of the edge alone. In the absence of a complete knowledge of the causes of variation in length, we will simply indicate the variance of the length by σ_L^2 and make no attempt to derive any expression for it. In [7] the nonuniformity of the edges is the only cause considered for the mismatch and an expression for σ_L^2 is derived which is inversely proportional to the width of the device. This would mean that the mismatch in length would tend to zero for very wide devices. We have observed results that contradict this. For example, the conductance constant matching of devices with $W = 100$ μm and $L = 2$ μm is not all that different from devices with $W = 200$ μm and $L = 2$ μm. In fact we have noticed that σ_L is more or less independent of the device width.

The width of the device may be treated similarly and thus we let the variance in W be σ_W^2. The definitions of the length and width occur during different stages of processing and under different conditions. Hence L and W may be treated as independent random variables.

To determine the variance in mobility, a knowledge of the factors that affect it is required. It is reported in [18] that at room temperature and moderate gate bias the electron mobility is mainly governed by scattering due to interface charge centers and phonons. An empirical relationship for μ is [18]

$$\mu = \frac{\mu_0(N_A)}{1 + \alpha(N_A)N_f} \qquad (19)$$

where $\mu_0(N_A)$ and $\alpha(N_A)$ are empirical constants with very little dependence on the dopant concentration. Thus the mismatch in μ may be approximated to be entirely due to the nonuniformity of Q_f. As the fixed oxide charges have a Poisson distribution, we may write

$$\sigma_\mu^2 = \frac{\mu_0^2 \alpha^2}{(1 + \alpha \overline{Q}_f)^4} \cdot \frac{\overline{N}_f}{L W}. \qquad (20)$$

The discussion given above for the mobility mismatch is for electrons only. We are not aware of any model that relates the mobility of holes to the doping concentration in the bulk and the fixed oxide charge density. The situation in the case of p-channel devices is further complicated by the threshold adjust implant. This could cause some damage in the substrate which may not be completely annealed, resulting in a poorer mobility matching than in the case of n-channel devices. In spite of these uncertain-

ties, it is still reasonable to assume that the mobility mismatch has a similar dependence on channel dimensions as given by (20). Then

$$\frac{\sigma_\mu}{\overline{\mu}} = \left(\frac{A_\mu}{\overline{L}\overline{W}} \right)^{1/2} \qquad (21)$$

where $\sqrt{A_\mu} = 4.95 \times 10^{-7}$ cm for n-channel devices and is not known for p-channel transistors.

The factors on the right-hand side of (18) are all independent. Thus

$$\frac{\sigma_K^2}{\overline{K}^2} = \frac{\sigma_L^2}{\overline{L}^2} + \frac{\sigma_W^2}{\overline{W}^2} + \frac{\sigma_\mu^2}{\overline{\mu}^2} + \frac{\sigma_C^2}{\overline{C}^2}. \qquad (22)$$

From (9) and (21)

$$\frac{\sigma_K^2}{\overline{K}^2} = \frac{1}{\overline{L}\overline{W}}(A_\mu + A_{ox}) + \frac{\sigma_L^2}{\overline{L}^2} + \frac{\sigma_W^2}{\overline{W}^2}. \qquad (23)$$

After substituting the values of A_μ and A_{ox} for n-channel devices, (23) may be solved for σ_L and σ_W using the measured values of σ_K of different sized devices. It is found that σ_L and σ_W are approximately the same and in the range of 0.01–0.03 μm. To provide a feel for the relative importance of the factors causing mismatch in K, we may substitute $\sigma_L = \sigma_W = 0.02$ μm in (23). Then

$$\frac{\sigma_K^2}{\overline{K}^2} = (2.46 \times 10^{-13} + 0.646 \times 10^{-13}) \cdot \frac{1}{\overline{L}\overline{W}}$$
$$+ 4 \times 10^{-12} \left(\frac{1}{\overline{L}^2} + \frac{1}{\overline{W}^2} \right) \qquad (24)$$

where the effective dimensions $\overline{L}$ and $\overline{W}$ have the units of centimeters. $\sigma_K / \overline{K}$ is plotted against $(1/\overline{L}^2 + 1/\overline{W}^2)^{1/2}$ in Fig. 3. The plotted relationship is not linear as shown by (24), with the curvature increasing for smaller geometry devices due to the increasing contribution of the $1/\overline{L}\overline{W}$ term. A similar plot for p-channel devices is shown in Fig. 4. The p-channel devices have a larger mismatch in conductance constant. One reason for this is the poorer gate oxide capacitance matching as has already been pointed out in connection with threshold voltage mismatch. Another factor could be a larger mobility variation.

We will now summarize the mismatch behavior in the conductance constant. These results are particular to this work.

a) The mismatch in K due to edge variations is proportional to $(1/\overline{L}^2 + 1/\overline{W}^2)^{1/2}$. The standard deviation of mismatch in length and width is in the range 0.01–0.03 μm. For n-channel devices, this is the dominant source of mismatch in K.

b) The larger gate oxide capacitance variation observed in p-channel devices in connection with V_T mismatch agrees with the larger mismatch in K.

c) For n-channel devices, the variation in mobility has little effect on the mismatch in K. The corresponding quantity for p-channel transistors, however, could be larger

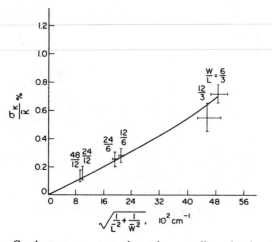

Fig. 3. Conductance constant mismatch versus dimension for n-channel devices.

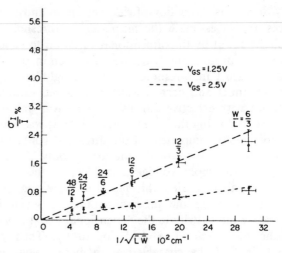

Fig. 5. Drain current mismatch versus dimension for n-channel devices. The dots are the estimated values using (25).

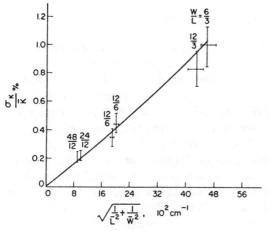

Fig. 4. Conductance constant mismatch versus dimension for p-channel devices.

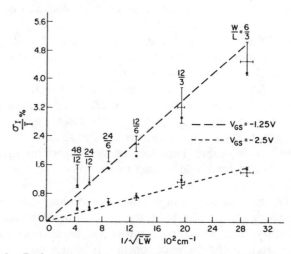

Fig. 6. Drain current mismatch versus dimension for p-channel devices. The dots are the estimated values using (25).

due to any damage in the substrate caused by the threshold adjust implant.

3. Correlation Between Mismatches in V_T and K: A common contributing factor to the mismatches in V_T and K is the variation in the gate oxide capacitance. Hence we can expect a dependence between the mismatches in V_T and K. A theoretical expression for the correlation coefficient is derived in [15]. Also the value has been experimentally measured. The agreement is excellent for n-channel devices and fair for p-channel ones. However, both the theoretical and experimental values are close to zero indicating that the mismatches in V_T and K are almost independent.

4. Mismatch in Drain Current: The drain current mismatch in the saturation region is given by (3). As the correlation coefficient is nearly equal to zero, we have

$$\frac{\sigma_I^2}{\bar{I}^2} = \frac{\sigma_K^2}{\bar{K}^2} + 4 \frac{\sigma_{VT}^2}{(V_{GS} - \bar{V}_T)^2}. \qquad (25)$$

At low values of gate-to-source voltage the dominant factor causing the mismatch in drain current is the threshold voltage variation. For bias levels approaching the mid-rail,

the conductance constant and the threshold voltage mismatches have almost equal contributions to the drain current mismatch. From (25) σ_I may be estimated from the measured values of σ_K and σ_{VT}. Also we have actually measured σ_I at different gate biases to compare with the estimated values. Fig. 5 is a plot of $\sigma_I/\bar{I}$ versus $1/\sqrt{LW}$ for n-channel devices for two gate voltages. The estimated values obtained from (25) using the measured average values of σ_{VT} and σ_K are indicated by the dots. A similar result is shown for the p-channel devices in Fig. 6. The excellent agreement between the measured values and the estimated ones for both p- and n-channel devices validates the characterization methodology and also verifies the mismatch models.

5. Range of Applicability: It is important to consider the dimensional range over which the mismatch models we have developed in the preceding sections are accurate. In our analysis we have assumed that the dimensional variations are accounted for entirely by the mismatch in conductance constant and have no influence on the threshold voltage mismatch. As the threshold voltage of small-geometry devices is a function of channel length and width, it is necessary to estimate the mismatch in threshold voltage brought about by the dimensional variation to validate the above assumption and hence the mismatch models. To this end, the shift in threshold voltage brought about due to short-channel and narrow-width effects was estimated for a device with effective dimensions 2×2 μm^2 fabricated in our process, using the expressions in [14]. It was found that the mismatch component of the threshold voltage brought about by the dimensional variations is only 10 percent of the total threshold voltage mismatch, in the worst case [15]. We also verified this fact for even smaller device geometries using the process parameters given in [19]. Hence we may attribute the dimensional variations entirely to the mismatch in K and not to V_T. Thus we may conclude that the characterization methodology and the mismatch models are valid for small-geometry devices also. However, as new processes emerge permitting smaller geometry devices, further experimental work is needed to characterize the mismatch.

6. Effect of Temperature: As the threshold voltage and conductance constant vary with temperature, it is interesting to know their matching behavior as a function of temperature. In the case of the threshold voltage, as expressed by (5), the only terms that are dependent on temperature are ϕ_{MS} and ϕ_B. We have seen that the contribution of these terms to the threshold voltage mismatch is negligible. Therefore we may expect the matching behavior of threshold voltage to be almost independent of temperature.

The only factor through which the conductance constant matching can be affected is the temperature dependence of mobility. For n-channel devices we have seen that the mismatch in conductance constant is largely due to photolithographic edge variations, and mobility variations have the least effect. Thus temperature variations should have very little effect on the conductance constant matching of n-channel devices. Since the mismatch in drain current is due to mismatches in threshold voltage and conductance constant, we can expect the current mismatch in n-channel devices to be almost unaffected over a wide temperature range. Limited experimental results seem to agree with this prediction. As far as the p-channel devices are concerned, since the mobility behavior of holes is not clearly understood, no theoretical explanation of the temperature effect is possible.

III. Design Methodology

To demonstrate the usefulness of the study of the matching behavior of transistors and the related models, we took up the task of designing an 8-bit current-steering CMOS DAC. Circuit details of the DAC have already been presented [9]. Here we only indicate the design methodology. In Section III-A, a brief description of the DAC configuration is presented. Section III-B discusses statistical error analysis. The close interaction between the DAC configuration, the matching accuracy of devices, and the circuit

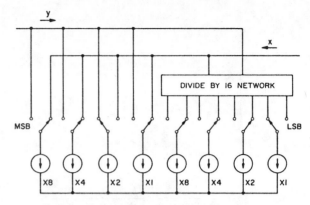

Fig. 7.   Schematic of a multiple current-source DAC.

yield is brought out here. Finally, yield results are presented in Section III-C.

A. DAC Configuration

A multiple current-source approach was chosen to realize the binary-weighted currents. This was done primarily to overcome the problems of nonlinear relationship between the drain current and aspect ratio of small-geometry MOS devices [14], [15], [20], and the voltage coefficient of resistance of R-$2R$ networks. The configuration is shown in Fig. 7 and is similar to that reported in [21]. The least significant bit (LSB) has one *unit* current source, the next significant bit has two unit current sources connected in parallel, and so on. The exponential growth in the number of unit current sources is overcome by having an interstage $16:1$ resistive current divider.

B. Statistical Error Analysis

In general, the errors generated by a DAC consist of linearity, offset, and gain errors. Usually, a DAC may be calibrated for zero gain and offset errors. Linearity error, however, occurs due to the random mismatch in the conversion elements. Hence, the circuit yield is a function of the matching accuracy of the unit current sources.

Integral nonlinearity of a DAC is generally defined as the difference between the actual output to the desired output normalized to the full-scale output of the DAC. This enables the nonlinearity to be expressed in terms of fractions of LSB or as a percentage.

Let x be the output of the DAC for a given input word and y be the analog complement of the output. The full-scale output of the converter is the sum $x + y$. To express the error as a fraction of the full scale, first we normalize the output to the full scale as follows:

$$z(x, y) = \frac{x}{x + y} \tag{26}$$

where z is the normalized output, and x and y are dependent on the input digital word and the DAC configuration.

For the 8-bit interstage divider DAC shown in Fig. 7

$$x = \left[8D_1 + 4D_2 + 2D_3 + D_4 \right.$$
$$\left. + \frac{1}{16}(8D_5 + 4D_6 + 2D_7 + D_8) \right] I_{unit} \quad (27)$$

and

$$y = \left[8(1 - D_1) + 4(1 - D_2) + 2(1 - D_3) + (1 - D_4) \right.$$
$$\left. + \frac{1}{16} \{ 8(1 - D_5) + 4(1 - D_6) + 2(1 - D_7) + (1 - D_8) \} \right] I_{unit}$$
$$(28)$$

where

$$D_i = 0 \text{ or } 1$$
$$i = 1, 2, \cdots, 8$$

D_1 is the most significant bit (MSB), D_8 is the LSB, and I_{unit} is the unit current source with a mean value $\bar{I}$ and standard deviation of matching σ. As the unit current sources are random and uncorrelated, we may treat x and y to be independent random variables with standard deviations σ_x and σ_y, respectively. Now we may determine the standard deviation of z in terms of σ_x and σ_y [12]

$$\sigma_z^2 = \frac{\bar{y}^2 \sigma_x^2 + \bar{x}^2 \sigma_y^2}{(\bar{x} + \bar{y})^4} \quad (29)$$

where σ_x and σ_y are evaluated as follows:

$$\bar{x} = \left[8D_1 + 4D_2 + 2D_3 + D_4 \right.$$
$$\left. + \frac{1}{16}(8D_5 + 4D_6 + 2D_7 + D_8) \right] \bar{I}$$

and

$$\sigma_x^2 = \left[8D_1 + 4D_2 + 2D_3 + D_4 \right.$$
$$\left. + \frac{1}{16}(8D_5 + 4D_6 + 2D_7 + D_8) \right] \sigma^2$$
$$= \frac{\bar{x}}{\bar{I}} \cdot \sigma^2. \quad (30)$$

Similarly

$$\sigma_y^2 = \frac{\bar{y}}{\bar{I}} \cdot \sigma^2. \quad (31)$$

Substituting (30) and (31) into (29)

$$\sigma_z^2 = \frac{1}{\bar{I}} \cdot \frac{\bar{x}\bar{y}}{(\bar{x} + \bar{y})^3} \cdot \sigma^2. \quad (32)$$

The expected value of z may be shown to be [12]

$$\bar{z} = \frac{\bar{x}}{\bar{x} + \bar{y}}. \quad (33)$$

Using (33) in (32), we have

$$\sigma_z^2 = \frac{\bar{z}(1 - \bar{z})}{\bar{I}(\bar{x} + \bar{y})} \cdot \sigma^2. \quad (34)$$

The above result may also be derived by determining the joint probability density function of z as shown in [22]. Equation (34) expresses the variances of the D/A output for different digital words. The function $\bar{z}(1 - \bar{z})$ will have a maximum value when $\bar{z} = 1/2$, i.e., when the output is halfway through the full scale, and falls off towards zero for minimum and maximum input word combinations. This observation suggests that the MSB current could be the most critical and should have the highest accuracy. Error contributions of the bits taper off towards the LSB, and hence the relative error contributions of all the bit current sources need not be the same. Such an error distribution is indeed a natural consequence in the multiple current-source approach where the relative accuracy of the bits improves towards the MSB. This may be shown as follows. If the unit current source has a mean value $\bar{I}$ and standard deviation of matching σ, connecting n such sources in parallel would produce an equivalent current source with mean value $n\bar{I}$ and standard deviation $\sqrt{n}\,\sigma$, as the current sources are uncorrelated. Thus there is an improvement in accuracy by a factor $\sqrt{n}$.

The analysis so far has shown that maximum error occurs halfway through the full scale and the error contributions of the individual bits taper off towards the LSB. Now we proceed further to relate the circuit yield to the standard deviation of the unit current sources. Here we define circuit yield as the percentage of functional devices that have integral nonlinearity less than 1/2 LSB. In other words, we are eliminating catastrophic device failures due to defects, etc. With this definition, a theoretical estimate of the circuit yield of the DAC is obtained by multiplying the probabilities that each of the 256 outputs of the DAC have less than 1/2 LSB error. For normal distribution with variances given by (34), the yield is

$$G = \prod_{i=2}^{255} \frac{1}{\sqrt{2\pi}\,\sigma_z} \int_{\bar{z}-1/512}^{\bar{z}+1/512} \exp\left\{ -\frac{(z - \bar{z})^2}{2\sigma_z^2} \right\} \cdot dz$$
$$= \prod_{i=2}^{255} \text{erf}(Q/\sqrt{2}) \quad (35)$$

where 1/512 is the normalized 1/2 LSB value and

$$Q = \frac{1}{512 \left[\dfrac{\bar{z}(1 - \bar{z})}{15 + 15/16} \right]^{1/2}} \cdot \frac{\sigma}{\bar{I}}. \quad (36)$$

The method used to derive (35) is quite general and may be easily extended to converters of different resolutions and accuracies. The circuit yield as given by (35) is plotted as a function of $\sigma/\bar{I}$ in Fig. 8.

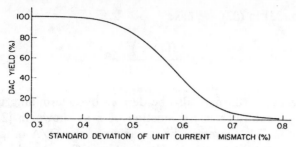

Fig. 8. DAC yield versus current-source mismatch.

TABLE I
DAC YIELD

Wafer Number	Devices Tested	Functional Devices	Good Devices	Circuit Yield %
1	32*	32	31	97
2	55*	55	55	100
3	64	60	60	100
4	64	52	52	100
5	64	60	60	100

*Broken Wafers. Hence the number of devices tested are less than 64.

Because of the error function nature of the relation between yield and current-source matching, there is an almost flat region close to the 100-percent yield level, followed by a very steep region and finally the yield asymptotically approaches zero. To avoid the possibility of any process variations from batch to batch affecting the yield adversely, the design should be such as to avoid the region where the yield is very sensitive to the matching accuracy. On the other hand, a very conservative matching tolerance is also not desirable because the improvement in yield is marginal with improvement in matching accuracy beyond a certain point. Therefore we chose 95-percent yield level as an optimum value that would not appreciably shift due to process variations from batch to batch. From the theoretical relationship shown in Fig. 8, the standard deviation of matching of the current sources should be about 0.45 percent to achieve this yield level. It may be noted that for an 8-bit DAC, an integral nonlinearity of $\pm 1/2$ LSB is equivalent to 0.2 percent of full scale. Thus, with this configuration, it is possible to provide good integral linearity associated with a high circuit yield without requiring an equivalent degree of component matching. Further it is shown in [15] that the divide-by-16 network is highly tolerant to component mismatch and hence is not a potential source of yield loss.

The analysis given above provides a systematic design approach for data converters. A similar approach may be used to design any precision analog circuit in general.

C. Results

Based on the understanding of the matching behavior of MOS devices and the systematic design methodology, an 8-bit high-speed DAC has been designed. The electrical performance results are reported in [9]. The unit current source used in the DAC is made up of a 24-μm-wide and 6-μm-long n-channel transistor in combination with a 4.7-kΩ source degradation resistor. This configuration has a better matching accuracy than a transistor alone for the same current value, owing to the better matching of resistors and the local negative feedback offered by the resistor [9]. The combination has a standard deviation of current matching of 0.45 percent when biased at a current of 128 μA. This should result in a circuit yield of approximately 95 percent. It may be noted that the same degree of matching may be obtained without using the source de-

gradation resistor by choosing larger area devices, and/or operating the devices with a larger gate-to-source voltage. Knowledge of the mismatches in V_T and K in conjunction with (25) may be used to obtain curves such as those in Figs. 5 and 6, for any process and operating condition. This information when used with (35) or Fig. 8 completes the design cycle.

Circuit yield statistics of the DAC are presented in Table I. Column 2 indicates the number of devices that are tested on each wafer. The next column shows the number of devices that are functional. In other words, we are eliminating catastrophic failures here. The fourth column shows the number of devices with integral nonlinearity less than $\pm 1/2$ LSB. Finally, the circuit yield is shown in the last column. In most cases the circuit yield is 100 percent, demonstrating the accuracy of the device characterization and the circuit design methodology. We have been able to achieve this high level of yield using relatively small devices and without using any trimming because of the knowledge we generated regarding the matching behavior of the devices as a function of dimensions, and the systematic design methodology we have followed.

IV. CONCLUSION

The design of precision analog circuits presents challenges in the areas of device matching characterization and circuit design. Novel methodologies relevant to both aspects of the design are presented in this paper. Section II is devoted to the study of transistor matching behavior. The overall objective has been not only to provide a clear understanding of the random mismatch, but also to develop a comprehensive design approach for precision analog circuits. The parameters a circuit designer will have freedom to choose are the dimensions of the devices. Therefore analytical models have been developed that relate the mismatch to device dimensions.

Section III of the paper discusses the application of the matching characterization in precision analog design. Design methodology for a high-performance DAC is illustrated. This is presently important because of the need for high-speed data converters in MOS technology. The close interaction between device matching and circuit yield is discussed. Experimental results of circuit yield are also presented.

IEEE JOURNAL OF SOLID-STATE CIRCUITS, VOL. SC-21, NO. 6, DECEMBER 1986

ACKNOWLEDGMENT

The authors are indebted to the management of Northern Telecom Electronics Ltd., for extending their facilities to carry out this research. They would like to thank Dr. M. Simard-Normandin for her help with device characterization. Thanks are also due to M. King for many suggestions and discussions. The assistance of N. Prasad concerning statistical analysis is gratefully acknowledged.

REFERENCES

[1] J. L. McCreary and P. R. Gary, "All MOS charge redistribution analog-to-digital conversion techniques—Part I," *IEEE J. Solid-State Circuits*, vol. SC-10, pp. 371–379, Dec. 1975.

[2] J. T. Caves, M. A. Copeland, C. F. Rahim, and S. D. Rosenbaum, "Sampled analog filtering using switched capacitors as resistor equivalents," *IEEE J. Solid-State Circuits*, vol. SC-12, pp. 592–599, Dec. 1977.

[3] B. J. Hosticka, R. W. Brodersen, and P. R. Gray, "MOS sampled data recursive filters using switched capacitor integrators," *IEEE J. Solid-State Circuits*, vol. SC-12, pp. 600–608, Dec. 1977.

[4] D. A. Hodges, P. R. Gray, and R. W. Brodersen "Potential of MOS technologies for analog integrated circuits," *IEEE J. Solid-State Circuits*, vol. SC-13, pp. 285–294, June 1978.

[5] J. L. McCreary, "Matching properties and voltage and temperature dependence of MOS capacitors," *IEEE J. Solid-State Circuits*, vol. SC-16, pp. 608–616, Dec. 1981.

[6] J. B. Shyu, G. C. Temes, and K. Yao, "Random errors in MOS capacitors," *IEEE J. Solid-State Circuits*, vol. SC-17, pp. 1070–1076, Dec. 1982.

[7] J. B. Shyu, G. C. Temes, and F. Krummenacher, "Random error effects in matched MOS capacitors and current sources," *IEEE J. Solid-State Circuits*, vol. SC-19, pp. 948–955, Dec. 1984.

[8] M. Akyia and S. Nakashima, "High-precision MOS current mirror," *Proc. Inst. Elec. Eng.*, vol. 131, pt. I, pp. 170–175, Oct. 1984.

[9] K. R. Lakshmi Kumar, R. A. Hadaway, M. A. Copeland, and M. I. H. King, "A high-speed 8-bit current steering CMOS DAC," in *Proc. IEEE 1985 Custom Integrated Circuits Conf.*, pp. 156–159.

[10] J. Logan, "Characterization and modelling for statistical design," *Bell Syst. Tech. J.*, vol. 50, pp. 1105–1147, Apr. 1971.

[11] E. H. Nicollian and J. R. Brews, *MOS Physics and Technology*. New York: Wiley, 1982.

[12] A. Papoulis, *Probability, Random Variables and Stochastic Processes*. Tokyo: McGraw-Hill, Kogakusha, 1965.

[13] W. Feller, *An Introduction to Probability Theory and Its Applications*, vol. I. New York, Wiley, 1957, p. 146.

[14] S. M. Sze, *Physics of Semiconductor Devices*, 2nd ed. New York: Wiley, 1981.

[15] K. R. Lakshmi Kumar, "Characterization and modelling of mismatch in MOS devices and application to precision analog design," Ph.D. dissertation, Carleton Univ., Ottawa, Ont., Canada, 1985.

[16] M. Simard-Normandin, private communication, 1985.

[17] S. K. Ghandhi, *VLSI Fabrication Principles Silicon and Gallium Arsenide*. New York: Wiley, 1983, p. 353.

[18] S. C. Sun and J. D. Plummer, "Electron mobility in inversion and accumulation layers on thermally oxidized silicon surfaces," *IEEE J. Solid-State Circuits*, vol. SC-15, pp. 562–573, Aug. 1980.

[19] J. R. Brews, W. Fitchner, E. H. Nicollian, and S. M. Sze, "Generalized guide for MOSFET miniaturization," *IEEE Electron Device Lett.*, vol. ED.-1, pp. 2–4, Jan. 1980.

[20] P. Yang and P. K. Chatterjee, "SPICE modelling for small geometry MOSFET circuits," *IEEE Trans. Computer-Aided Des.* vol. CAD-1, pp. 169–182, Oct. 1982.

[21] P. H. Saul *et al.*, "An 8-bit, 5-ns monolithic D/A converter subsystem," *IEEE J. Solid-State Circuits*, vol. SC-15, pp. 1033–1039, Dec. 1980.

[22] S. Kuboki *et al.*, "Nonlinearity analysis of resistor string A/D converters," *IEEE Trans. Circuits Syst.*, vol. CAS-29, pp. 383–390, June 1982.

[23] P. H. Saul *et al.*, "An 8b CMOS video DAC," in *ISSCC Dig. Tech. Papers*, 1985, pp. 32–33.

[24] T. Miki *et al.*, "An 80 MHz 8b CMOS D/A converter," in *ISSCC Dig. Tech. Papers*, Feb. 1986, pp. 132–133.

Measurement and Modeling of Charge Feedthrough in n-Channel MOS Analog Switches

WILLIAM B. WILSON, STUDENT MEMBER, IEEE, HISHAM Z. MASSOUD, MEMBER, IEEE,
ERIC J. SWANSON, MEMBER, IEEE, RHETT T. GEORGE, JR., MEMBER, IEEE, AND
RICHARD B. FAIR, SENIOR MEMBER, IEEE

Abstract —Charge feedthrough in analog MOS switches has been measured. The dependence of the feedthrough voltage on the input and tub voltages, device dimensions, and load capacitances was characterized. Most importantly, it was observed that the feedthrough voltage decreases linearly with the input voltage. The significance of this observation when considering harmonic distortion in sample-and-hold circuits is discussed. A first-order computer simulation based on the quasi-static small-signal MOSFET capacitances shows good agreement with experimental results.

I. INTRODUCTION

CMOS CIRCUITS have historically been used in digital logic and memory applications as a result of their high packing density and low power consumption. Recently, analog CMOS has combined the benefits of both analog and digital circuits on the same die. This combination offers many advantages in signal processing and analog-to-digital conversion. A key component in interfacing digital and analog circuitry is the analog switch, or transmission gate, where a signal is fed from the input side of the device to a load connected at the output. The on–off switching capability is provided by the gate voltage governing the presence of an inversion channel.

MOSFET's are not ideal switches. Consider a typical sample-and-hold circuit shown in Fig. 1. Ideally, when the switch is turned off, the output voltage on the load capacitance should remain at its value at the time of switching. However, the MOSFET actually couples some of the charge in the inversion layer onto the load capacitor. Thus the final capacitor voltage consists of two components, one directly related to the input voltage and an error component arising from the charge stored in the switch. This error component is the feedthrough voltage. In this paper, this feedthrough voltage is measured in MOS switches implemented in a 3.5-μm CMOS technology. In addition, first-order simulations based on quasi-static small-signal

models of MOSFET capacitances have been made to predict the magnitude of the feedthrough voltage. Experimental results and computer simulations are presented and the conclusions drawn lead to a first-order understanding of the phenomenon of charge feedthrough.

II. MEASUREMENT SYSTEM

When an n-channel MOSFET is turned on, an inversion channel of electrons is formed underneath the gate. When the device is turned off, these electrons are dispersed, flowing either into the substrate of the device, or to the loads at the source and drain. The effect produced by electrons flowing into the substrate is called charge pumping [1]. The flow of electrons to the source and drain nodes is called charge or clock feedthrough. It was first discussed by Stafford *et al.* in 1974 [2]. This study is concerned with measuring charge feedthrough in an n-channel MOSFET in a sample-and-hold circuit under a variety of bias and loading configurations. The circuit used is shown in Fig. 2, where it can be seen that both the source (connected to nC) and the drain (connected to $2C$) nodes are floating. This is in contrast with recent studies of charge feedthrough where only one node was floating [3], [4]. The value of C is 1.15 pF and the index n in the load capacitance nC can be varied from $n = 1$ to $n = 96$. For $n = 96$, the source load capacitance simulates a voltage source. This permits the measurement of the feedthrough voltage in a situation where only one node is allowed to float, and makes possible comparisons with other studies [3], [4]. The feedthrough voltages at the source and drain are measured as V_{OUT1} and V_{OUT2} after source followers M_3 and M_4, respectively. The distinction between source and drain becomes irrelevant once steady-state conditions are attained and the labels V_{OUT1} and V_{OUT2} are used to identify one side of the device from the other. The source followers supply the power to drive the inherent capacitances of the package and the cable connections. Thus these capacitances do not need to be considered. Bias circuitry (not shown in Fig. 2) maintains the transistors M_1 and M_2 of the source followers biased in saturation with a 25-μA current. The gate length

Manuscript received March 4, 1985; revised July 8, 1985. W. B. Wilson was supported by a Microelectronics Center of North Carolina Fellowship.
W. B. Wilson, H. Z. Massoud, R. T. George, Jr., and R. B. Fair are with the Department of Electrical Engineering, Duke University, Durham, NC 27706.
E. J. Swanson was with AT&T Bell Laboratories, Reading, PA 19604. He is now with Crystal Semiconductor, Austin, TX 78744.

Reprinted from *IEEE J. Solid-State Circuits*, vol. SC-20, no. 6, pp. 1206–1213, Dec. 1985.

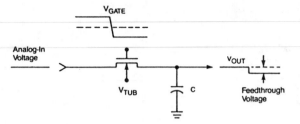

Fig. 1. Charge-feedthrough definition in a sample-and-hold circuit.

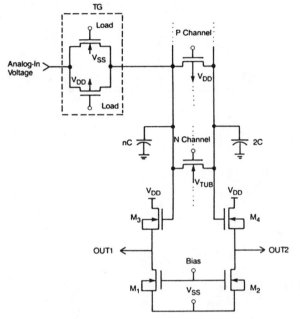

Fig. 2. Basic circuit used in measuring charge feedthrough.

in both the n- and p-channel test devices varies between 4 and 20 μm.

III. MEASUREMENT PROCEDURE

In this investigation, only the charge feedthrough on n-channel devices was characterized. The parallel p-channel device was used to initialize the system ensuring that the prefeedthrough voltage on each load capacitor is identical. This is necessary when n-channel devices are body-effect limited.

The basic measurement procedure is started by applying a gate voltage of $+5$ V to the n-channel test transistor, which turns the switch on. Then the two load capacitors ($2C$ and nC) are charged to their initial value called the analog-in voltage V_{AI}. At this point, the gate voltage is quickly driven to -5 V turning the switch off. As this occurs, the stored charge within the device is dumped onto the now-floating load capacitors shifting the voltage across each of them. The difference between the measured voltages at each node (V_{OUT1} and V_{OUT2}) before and after the gate transition yields the feedthrough voltage at the corresponding node. This is shown graphically in Fig. 1.

The following variables affect charge feedthrough and will be studied.

1) The initial voltage on each of the load capacitors is called the analog-in voltage V_{AI}. It is placed on the two load capacitors by closing the CMOS transmission gate, labeled TG in Fig. 2, and by turning on both the n-channel and p-channel test devices. The magnitude of V_{AI} affects the amount of charge stored in the n-channel device in two ways. First, it directly affects the amount of charge on the gate through its effect on the gate-to-source voltage V_{gs}. Secondly, it affects the amount of charge stored in the bulk through the body effect. It should be noted that this latter effect is nonlinear.

2) The tub voltage V_{TUB} also has an important effect on the amount of charge stored in the device. However, unlike V_{AI}, V_{TUB} only contributes via the nonlinear body effect. Thus the change in the feedthrough voltage with V_{TUB} is less dramatic than with V_{AI}.

3) The gate length L of the device has an obvious effect. By increasing the area of the gate, the total amount of charge stored in the device is increased, as is the feedthrough voltage.

4) Finally, the value of the load capacitance is expected to affect the feedthrough voltage.

Several other parameters, which also have a pronounced effect on the amount of charge feedthrough, were held constant in this study. The fall time of the gate signal, which directly affects the amount of charge pumping [1], was held constant at 10 ns. Changing the fall time of the gate signal has been shown both theoretically and experimentally to affect the measured feedthrough voltage [3]–[5]. Such effects are currently being investigated. Also, the gate voltage was always pulsed from $+5$ to -5 V. The device width was the same (20 μm) for all devices, with the overlap capacitance estimated to be 11.5 fF, based on an assumed overlap distance of 1 μm. Accurate measurements of the overlap distances at the source and drain are presently being made. The p-well was doped with an acceptor concentration of 1.1×10^{16} cm^{-3} resulting in a zero-bias threshold voltage (V_{T0}) of 0.7 V.

A large matrix of feedthrough voltage measurements was obtained where each data point was the average of ten measurements. The accuracy of the measurements was 5 mV, and the standard deviation between the ten measurements was consistently less than one least significant bit (5 mV).

Due to the layout of the circuit, there are several additional parasitic capacitances in parallel with the source and drain load capacitances. These capacitances have different values and, as a consequence, for the symmetrical load case of $n = 2$, the feedthrough voltages measured at the source and drain are not identical.

IV. EXPERIMENTAL RESULTS

In this section, the measured dependence of the feedthrough voltage at the source and drain is presented as a function of the analog-in voltage, the tub voltage, the device dimensions, and the value of the load capacitances.

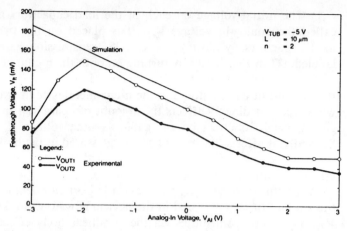

Fig. 3. Dependence of the feedthrough voltage V_{ft} on the analog-in voltage V_{AI}. Circles (experiment), solid lines (simulation).

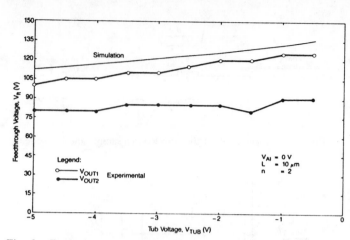

Fig. 5. Dependence of the feedthrough voltage V_{ft} on the tub voltage V_{TUB}. Circles (experiment), solid lines (simulation).

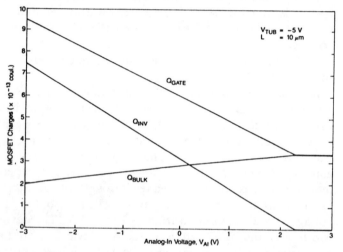

Fig. 4. Dependence of the gate charge Q_{GATE}, the inversion charge Q_{INV}, and the bulk charge Q_{BULK} on the analog-in voltage V_{AI}.

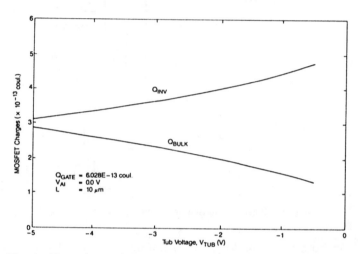

Fig. 6. Dependence of the inversion charge Q_{INV}, and the bulk charge Q_{BULK} on the tub voltage V_{TUB}.

The results of the simulations, which will be described later, are also shown in Figs. 3, 5, 7, and 8.

A. Effect of the Analog-In Voltage

Before discussing the measurement results, it is important to mention that the gain of the source-follower circuit is nearly unity (0.974) over the entire range of values of the input voltage V_{AI}. A typical plot of the dependence of the feedthrough voltage on V_{AI} is shown in Fig. 3. The important observation here is the linearity of the feedthrough voltage from -2 to $+2$ V. In this range, the nonlinear body effect, dependent on the bulk-to-source voltage, does not significantly affect the feedthrough voltage. This suggests that the amount of inversion channel charge initially stored in the device is the predominant source of the feedthrough voltage. This is particularly important as it implies that the inversion charge varies linearly with the channel potential. This dependence is demonstrated in Fig. 4 where the calculated inversion charge is plotted as a function of the analog-in voltage V_{AI}. This linear dependence of Q_{INV} on V_{AI} confirms the inver-

sion channel charge as the predominant source of feedthrough charge.

There are two portions of the plots that deviate from linearity. The first is for low values of V_{AI}. It is suspected that this dropoff is a function of the fall time of the gate signal, and that the curve would be extended if the fall time was increased. The second deviation from linearity is for large values of V_{AI}. In this case, there is no inversion charge stored in the device in its initial state. The difference in feedthrough voltage at V_{OUT1} and V_{OUT2}, in this case, can be attributed to different overlap capacitances at the source and drain. The cause of the deviation is shown clearly in Fig. 4 where stored charge is plotted against the initial channel voltage V_{AI}. This graph represents the initial amount of stored charge in the intrinsic device only. The equations describing the stored charge are listed in the Appendix. The gate voltage is $+5$ V, and as can be seen, the inversion charge goes to zero as V_{AI} reaches 2.3 V. Because of the dependence of the feedthrough voltage on the inversion channel charge, the remaining discussion of the dependence of V_{ft} on V_{AI} will be limited to $V_{AI} \leq 2$ V where the inversion charge goes to zero.

B. Effect of the Tub Voltage

Fig. 5 shows the feedthrough voltage dependence on the tub voltage V_{TUB}. Here, the effect of the backgate bias is demonstrated. Again, it is remarkable to note that the curve is reasonably linear given that a nonlinear effect is involved. The cause of this linearity can be shown by examining a plot of the initial conditions as shown in Fig. 6. It is evident here that the body effect is nonlinear. However, the nonlinearity is quite small and becomes even less significant in the feedthrough process.

It is also interesting to note that increasing the backgate bias decreases the feedthrough voltage. As V_{TUB} is increased, the stored depletion charge increases and the inversion charge decreases. This is significant because V_{TUB} can be fixed at a large negative value, minimizing V_{ft}. It is also desirable to have a large negative V_{TUB} because it increases the allowed range of input voltages to the switch. Consequently, all remaining discussion and data will be for a switch with $V_{TUB} = -5$ V, minimizing the feedthrough voltage and maximizing the allowable input range.

C. Effect of the Channel Length

The width of the experimental devices has been fixed at 20 μm, while the length ranged from 4 to 20 μm. In Fig. 7, the systematic alternation of measured values of V_{ft} at V_{OUT1} and V_{OUT2} is a direct result of the different overlap capacitances, as discussed earlier, at the diffusions used as source and drain. Due to the special layout of this circuit, these diffusions were alternately used as source or drain with increasing channel length. The feedthrough voltage appears to have a linear dependence on device dimensions as expected. Theoretically, this can be attributed to the linear dependence of the inversion channel charge on the gate area. From this, it follows that to minimize the feedthrough voltage, a minimum-sized switch should be used. If such devices are not available, then feedthrough cancellation schemes should be used [6].

D. Effect of the Load Capacitance

The feedthrough voltage is plotted against the inverse of the load capacitance in Fig. 8. It can be seen that the amount of charge fed out of the switch is independent of the value of the load capacitance. This is demonstrated in two ways. First, the feedthrough voltage is constant on the drain side, where the load capacitance is fixed (2C). Thus, changing the load capacitance on one side (nC) of the device does not appear to affect the amount of charge feedthrough on the other. The second observation comes from the feedthrough voltage at the source, where the load capacitance nC is varying. If V_{ft} is plotted versus the inverse load capacitance then the amount of feedthrough charge is independent of the load capacitance if the curve is linear. Thus the feedthrough voltage is minimized by increasing the load capacitance which would, however, increase the data capture time in a sample-and-hold circuit.

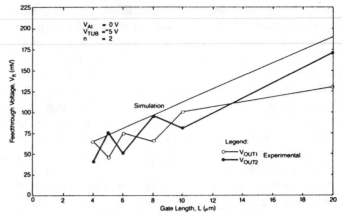

Fig. 7.   Dependence of the feedthrough voltage $\bar{V}_{ft}$ on the gate length L. Circles (experiment), solid lines (simulation).

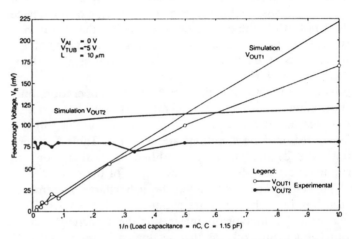

Fig. 8.   Dependence of the feedthrough voltage V_{ft} on $1/n$, the inverse of the load capacitance index number. Circles (experiment), solid lines (simulation).

Consequently, a compromise must be reached based on requirements on the circuit performance.

V. Modeling Charge Feedthrough

A. Model Description

It is important to be able to predict the amount of charge feedthrough in MOSFET circuits, especially when simulating the behavior of sample-and-hold circuits. The following is a discussion of a simple program which simulates charge feedthrough in a transmission gate where both the source and drain are floating when the switch is turned off.

The simulation is based on the MOSFET quasi-static small-signal capacitance model developed by Liu and Nagel [7]. The simulation is divided into four sections corresponding to four regions of the model. Each region is simulated in turn yielding a contribution to the feedthrough voltage for that region. Finally, the voltages from each of the four regions are summed together yielding the total feedthrough voltage. In spite of the fact that the original model has been considerably simplified, the simulations show qualitative agreement with the measured results.

The first assumption in the simulation is that charge pumping is ignored. This assumption manifests itself in two ways. First, it is not necessary to include the gate-to-bulk capacitance, since this element cannot contribute charge to the load capacitors at the source and drain. Second, an ideal 50–50 split of the inversion channel charge is assumed between the source and the drain. This assumption will cause the simulation to predict values slightly larger than will actually occur.

The second assumption in the simulation is that the drain-to-source voltage is zero when the device is biased in the linear region. In principle, this is valid because the initial voltages at the source and drain are equal, producing a uniform inversion channel. Also, from the experimental results it is observed that the load on one side of the device does not affect the other side. This implies that the device is never biased in the saturation region. As the gate voltage is decreased, the device will move from a uniformly distributed inversion layer directly into depletion. It is then possible to ignore an entire region of device operation, thus shortening the program considerably. This assumption also requires modification of the capacitance equations in the linear region. The equations are shown in Table I, where both the original and the modified expressions for the capacitances are shown for each region of the model.

The equations for the feedthrough charge increments resulting in each region are obtained in closed form. This was done by analytically integrating the capacitance equations. Each integral was then evaluated at the initial and final voltages to obtain the charge increment. This integration made it necessary to accept a discontinuity in the bulk depletion capacitance. This should not cause, however, any significant errors.

B. Simulation Results

In Fig. 9, a plot of the feedthrough voltage predicted by the simulation for each region is shown. It should be noted that the backgate bias is -5 V, to minimize the feedthrough voltage. This constrains the device from ever actually entering the accumulation region, because the gate voltage reaches -5 V while the device is biased in the depletion-accumulation region. The linearity of the feedthrough voltage versus V_{AI} for each individual region confirms the dominance of the gate-to-source and gate-to-drain capacitances on the feedthrough process. Also, the feedthrough voltage from the inversion region dominates over the feedthrough voltage from other regions, which are not a strong function of V_{gs}. From this observation, one can reduce the problem of minimizing the feedthrough voltage to minimizing the stored inversion charge, or the use of feedthrough cancellation schemes [6].

Comparisons between the simulation and measurements are shown in Figs. 3, 5, 7, and 8, where the plots of the measured and calculated values of the feedthrough voltage are in qualitative agreement. The simulation overestimates

TABLE I(a)
MODEL EQUATIONS

Original Equations [7]	This Work
Inversion Region (Region 1)	
$C_{GS} = C_{OV} + \frac{2}{3} C_{OX} W L \dfrac{V_{SAT}[3V_{SAT} - 2V_{DS}]}{[2V_{SAT} - V_{DS}]^2}$	$C_{GS} = C_{OV} + \frac{1}{2} C_{OX} W L$
$C_{GD} = C_{OV} + \frac{2}{3} C_{OX} W L \dfrac{[V_{SAT} - V_{DS}][3V_{SAT} - V_{DS}]}{[2V_{SAT} - V_{DS}]^2}$	$C_{GD} = C_{OV} + \frac{1}{2} C_{OX} W L$
$C_{BS} = \dfrac{C_{JS}}{\left[1 - \frac{V_{BS}}{\phi_B}\right]^{m_B}} \left[1 + \frac{2}{3} \frac{C_{BG}}{C_{JS}} W L \dfrac{V_{SAT}[3V_{SAT} - 2V_{DS}]}{[2V_{SAT} - V_{DS}]^2}\right]$	$C_{BS} = \dfrac{C_{JS}}{\left[1 - \frac{V_{BS}}{\phi_B}\right]^{m_B}} \left[1 + \frac{1}{2} \frac{C_{BG}}{C_{JS}} W L\right]$
	$C_{BD} = \dfrac{C_{JD}}{\left[1 - \frac{V_{BD}}{\phi_B}\right]^{m_B}} \left[1 + \frac{1}{2} \frac{C_{BG}}{C_{JD}} W L\right]$
$C_{BD} = \dfrac{C_{JD}}{\left[1 - \frac{V_{BD}}{\phi_B}\right]^{m_B}} \left[1 + \frac{2}{3} \frac{C_{BG}}{C_{JD}} W L \dfrac{[V_{SAT} - V_{DS}][3V_{SAT} - V_{DS}]}{[2V_{SAT} - V_{DS}]^2}\right]$	

TABLE I(b)
MODEL EQUATIONS

Original Equations [7]	This Work
Depletion-Inversion Region (Region 2)	
$C_{GS} = C_{OV} + \frac{2}{3} C_{OX} W L \left[1 - 4[V_{GS} - V_{TH}]^2\right]$	$C_{GS} = C_{OV} + \frac{1}{2} C_{OX} W L \left[1 - 4[V_{GS} - V_{TH}]^2\right]$
$C_{GD} = C_{OV}$	$C_{GD} = C_{OV} + \frac{1}{2} C_{OX} W L \left[1 - 4[V_{GS} - V_{TH}]^2\right]$
$C_{BS} = \dfrac{C_{JS}}{\left[1 - \frac{V_{BS}}{\phi_B}\right]^{m_B}} \left[1 + \frac{2}{3} \frac{C_{BG}}{C_{JS}} W L \left[1 - 4[V_{GS} - V_{TH}]^2\right]\right]$	$C_{BS} = \dfrac{C_{JS}}{\left[1 - \frac{V_{BS}}{\phi_B}\right]^{m_B}}$
$C_{BD} = \dfrac{C_{JD}}{\left[1 - \frac{V_{BD}}{\phi_B}\right]^{m_B}}$	$C_{BD} = \dfrac{C_{JD}}{\left[1 - \frac{V_{BD}}{\phi_B}\right]^{m_B}}$

the value of the feedthrough voltage. This is possibly due to ignoring the effects of charge pumping. The errors involved range from 15 to 40 percent in some cases. It is also interesting to note that the simulation results in a nearly linear dependence of the feedthrough voltage on each of the parameters studied. This further supports the conclusion that the feedthrough process is dominated by the inversion channel charge, as discussed earlier.

C. Discussion

Despite the discrepancy between measured results and simulations, charge-feedthrough modeling in this investiga-

TABLE I(c)
MODEL EQUATIONS

Original Equations [7]	This Work
Accumulation-Depletion Region (Region 3)	
$C_{GS} = C_{OV} \qquad C_{GD} = C_{OV}$	$C_{GS} = C_{OV} \qquad C_{GD} = C_{OV}$
$C_{BS} = \dfrac{C_{JS}}{\left[1 - \dfrac{V_{BS}}{\phi_B}\right]^{m_B}}$	$C_{BS} = \dfrac{C_{JS}}{\left[1 - \dfrac{V_{BS}}{\phi_B}\right]^{m_B}}$
$C_{BD} = \dfrac{C_{JD}}{\left[1 - \dfrac{V_{BD}}{\phi_B}\right]^{m_B}}$	$C_{BD} = \dfrac{C_{JD}}{\left[1 - \dfrac{V_{BD}}{\phi_B}\right]^{m_B}}$
Accumulation Region (Region 4)	
$C_{GS} = C_{OV} \qquad C_{GD} = C_{OV}$	$C_{GS} = C_{OV} \qquad C_{GD} = C_{OV}$
$C_{BS} = \dfrac{C_{JS}}{\left[1 - \dfrac{V_{BS}}{\phi_B}\right]^{m_B}}$	$C_{BS} = \dfrac{C_{JS}}{\left[1 - \dfrac{V_{BS}}{\phi_B}\right]^{m_B}}$
$C_{BD} = \dfrac{C_{JD}}{\left[1 - \dfrac{V_{BD}}{\phi_B}\right]^{m_B}}$	$C_{BD} = \dfrac{C_{JD}}{\left[1 - \dfrac{V_{BD}}{\phi_B}\right]^{m_B}}$

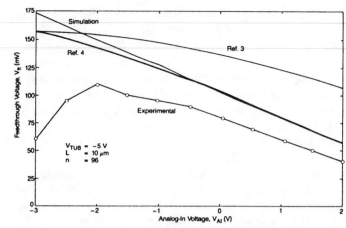

Fig. 10. Comparison of the simulation and experimental results obtained in this study with results from [3] and [4].

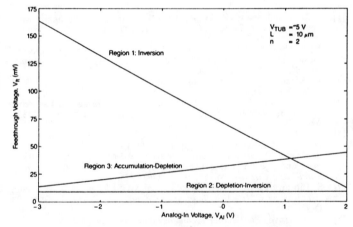

Fig. 9. Dependence of the regional feedthrough voltage V_{ft} on the analog-in voltage V_{AI}. Circles (experiment), solid lines (simulated).

tion offers the following enhancements over previous studies [3]–[5].

1) It allows for both the source and drain nodes to be floating after the switch is turned off. This allows measurement of all transient charges flowing from the source and drain, where previous efforts cannot account for the charge exiting at the node held at a fixed potential. The special case of $V_{TUB} = 0$ has been treated by Vittoz [5].

2) It is based on the quasi-static small-signal MOSFET capacitances and within the limits of this study offers good agreement with the measured results. This indicates the capability of quasi-static small-signal models to accurately simulate transient effects in a MOSFET under various doping and bias conditions.

3) It results in the linear dependence of V_{ft} on V_{AI}, which was observed experimentally but not predicted by MacQuigg [3] (see Fig. 10).

4) It includes the influence of V_{TUB} on V_{ft}, which was measured experimentally. This was not accounted for in the lumped model of Sheu and Hu [4]. It should be remembered that although simulation results in this work and those obtained using the lumped model in [4] agree in Fig. 10, these two calculations of V_{ft} were made for different values of V_{TUB}.

5) It accounts for the junction capacitances associated with the source and drain diffusions.

It should be again noted, however, that charge-feedthrough modeling in this study did not include the influence of the fall time of the gate voltage, currently under investigation, or that of charge pumping on the feedthrough voltage.

The observed linearity of the feedthrough voltage implies that the MOS switch can be used effectively in analog design. If the transfer function of a switch is considered, it will have two terms. First, the unity gain of an ideal switch will yield the input voltage at the output. Second, the linear feedthrough term is added. Thus, the transfer function will remain linear, and the analog switch should introduce little harmonic distortion into the signal passing through the switch over a limited range of input voltages. These conclusions could only be made because the source–follower circuit has a constant gain of nearly unity over the active range of values for the input voltage and all measurements are made after the transients have settled as in a typical sample-and-hold application.

VI. CONCLUSION

This paper presents measurements and first-order modeling of charge feedthrough. The feedthrough voltage in an n-channel MOSFET was measured with respect to several parameters, including analog-in voltage, tub voltage, device dimensions, and load capacitances. It was found that the feedthrough voltage is nearly linear with respect to the analog-in voltage. Thus a typical analog switch appears to introduce practically no harmonic distortion in the output signal over a limited input range. Also, the feedthrough voltage is minimized for a large back bias, concurrently allowing for a wide input voltage swing. Minimum device dimensions are desirable to reduce the feedthrough voltage. Also, it was observed for the first time that the magnitude of the load capacitance on one side of the device has virtually no effect on the amount of charge feedthrough at the other side. In other words, the amount of charge feedthrough is independent of the load capacitance. A simulation program that provides first-order qualitative agreement with the measured data, using a quasi-static MOSFET capacitance model, was developed.

APPENDIX

This appendix lists the relationships of the different charge components stored in an MOS switch [8]. The charge stored on the gate Q_{GATE} is given by

$$Q_{\text{GATE}} = C_{\text{OX}} \left[V_{\text{GATE}} - V_{f_b} - 2\phi_F - V_{AI} \right] \tag{1}$$

where C_{OX} is the gate oxide capacitance, V_{GATE} the gate voltage, V_{fb} the flatband voltage, ϕ_F the bulk potential difference between the Fermi and intrinsic levels, and V_{AI} the analog-in voltage.

The bulk charge Q_{BULK} is given by

$$Q_{\text{BULK}} = -\sqrt{2\epsilon_{\text{si}} q N_A [\phi_s]} \tag{2}$$

where N_A is the substrate doping, and ϕ_s the band bending in the silicon, which is given by

$$\phi_s = 2\phi_F + V_{AI} - V_{\text{TUB}}. \tag{3}$$

The inversion charge is the difference of these two charge components and is given by

$$Q_{\text{INV}} = -Q_{\text{GATE}} - Q_{\text{BULK}}. \tag{4}$$

As V_{AI} increases, the amount of inversion charge Q_{INV} decreases as the bulk charge increases, and the charge on the gate decreases. This continues until the inversion channel disappears, and the bulk charge mirrors the gate charge. Now the source and drain do not affect the surface potential and, therefore, do not influence the amount of charge stored underneath the gate of the device. Consequently, Q_{BULK} and Q_{GATE} will not change with continued increase in V_{AI}, where the gate and substrate voltages are held constant. This is shown graphically in Fig. 4, and is the cause of the nonlinearity in Fig. 3 for large values of V_{AI}. Equations (1)–(3) and (4) are used in Fig. 6 where the tub voltage is varied. Here, the effect is to shift the relative distribution of charge between the inversion channel and the bulk charge, where the total amount of charge stored in the silicon remains unchanged as the tub voltage is varied.

ACKNOWLEDGMENT

The authors would like to thank Dr. J. J. Paulos of North Carolina State University for many valuable discussions, and the reviewers for their comments.

REFERENCES

[1] J. S. Brugler and P. G. A. Jespers, "Charge pumping in MOS devices," *IEEE Trans. Electron Devices*, vol. ED-16, pp. 297–302, Mar. 1969.
[2] K. R. Stafford, P. R. Gray, and R. A. Blanchard, "A complete monolithic sample/hold amplifier," *IEEE J. Solid-State Circuits*, vol. SC-9, pp. 381–387, Dec. 1974.
[3] D. MacQuigg, "Residual charge on a switched capacitor," *IEEE J. Solid-State Circuits*, vol. SC-18, pp. 811–813, Dec. 1983.
[4] B. J. Sheu and C. Hu, "Switched-induced error voltage on a switched capacitor," *IEEE J. Solid-State Circuits*, vol. SC-19, pp. 519–525, Aug. 1984.
[5] E. Vittoz, "Microwatt switched capacitor circuit design," *Electrocomponent Sci. Technol.*, vol. 9, pp. 263–273, 1981.
[6] E. Suarez, P. Gray, and D. A. Hodges, "All-MOS charge redistribution analog-to-digital conversion techniques—Part II," *IEEE J. Solid-State Circuits*, vol. SC-10, p. 379, 1975.
[7] S. Liu and L. W. Nagel, "Small-signal MOSFET models for analog circuit design," *IEEE J. Solid-State Circuits*, vol. SC-17, pp. 983–998, Dec. 1982.
[8] A. S. Grove, *Physics and Technology of Semiconductor Devices*. New York: Wiley, 1969.

Measurement and Analysis of Charge Injection in MOS Analog Switches

JE-HURN SHIEH, STUDENT MEMBER, IEEE, MAHESH PATIL, STUDENT MEMBER, IEEE,
AND BING J. SHEU, MEMBER, IEEE

Abstract —Charge injection in MOS switches has been analyzed. The analysis has been extended to the general case including signal-source resistance and capacitance. Universal plots of percentage channel charge injected are presented. Normalized variables are used to facilitate usage of the plots. The effects of gate voltage falling rate, signal-source level, substrate doping, substrate bias, switch dimensions, as well as the source and holding capacitances are all included in the plots. A small-geometry switch, slow switching rate, and small source resistance can reduce the charge injection effect. On-chip test circuitry with a unity-gain operational amplifier, which reduces the disturbance imposed by measurement equipment to a minimum, is found to be an excellent monitor of the switch charge injection. The theoretical results agree with the experimental data.

I. INTRODUCTION

IN A monolithic sample and hold, a signal is stored on a capacitor. The accuracy of sample-and-hold circuits is disturbed by charge injected when the sampling switch turns off. The majority of sample-and-hold circuits are implemented using MOS technologies because the high input impedance of MOS devices performs excellent holding function. When the switch connecting the signal-source node and the data-storage node is turned on, the sampling function is performed. When the switch is turned off, the data stored in the storage node will be held until the next operation step occurs. However, an MOS switch is not an ideal switch. A finite amount of mobile carriers are stored in the channel when an MOS transistor conducts. When the transistor turns off, the channel charge exits through the source, the drain, and the substrate electrodes. The charge transferred to the data node during the switch turning-off period superposes an error component to the sampled voltage. In addition to the charge from the intrinsic channel, the charge associated with the feedthrough effect of the gate-to-diffusion overlap capacitance also enlarges the error voltage after the switch turns off [1]. This charge injection problem was identified in the early stage of switched-capacitor circuit development. Various compensation schemes [2],[3] have been used to reduce the switch-induced error voltage. As the design of higher preci-

Manuscript received August 18, 1986; revised October 13, 1986. This work was supported by the Defense Advanced Research Projects Agency under Contract MDA903-81-C-0335.

The authors are with the Department of Electrical Engineering and Information Sciences Institute, University of Southern California, Los Angeles, CA 90089.

IEEE Log Number 8613218.

sion sample-and-hold circuits progresses, the need for effective test patterns to accurately monitor the switch charge injection becomes increasingly important. In 5-V technologies, the resolution in a 10-bit analog-to-digital (A/D) converter is 4.88 mV, while the resolution in a 16-bit A/D converter is only 76 μV. The error voltage caused by the switch charge injection is usually in the millivolt range. The fully differential circuit approach [3] cancels the switch charge injection to the first order. Precise characterization and detailed analysis of the switch charge injection is of prime interest in designing high-performance integrated circuits.

There have been some attempts to model the switch charge injection. MacQuigg [4] made a qualitative observation and did SPICE simulation on a simplified case. Sheu and Hu [1] developed an analytical model corresponding to infinite source capacitance. A two-transistor source follower was used by Wilson *et al*. [5] with an attempt to improve the measurement accuracy. In this paper, analysis on the general case of switch charge injection is described in Section II. Analytical models of special cases are also presented. A better test structure for monitoring switch charge injection is proposed in Section III. Experimental results are presented in Section IV. A conclusion is given in Section V.

II. ANALYSIS

We assume that the charge pumping phenomenon due to the capture of channel charge by the interface traps is insignificant and all the channel charge exits through the source and drain electrodes when the transistor turns off. The turn-off of an MOS switch consists of two distinct phases. During the first phase, the gate voltage is higher than the transistor threshold voltage. There is a conduction channel that extends from the source to the drain of the transistor. As the gate voltage decreases, mobile carriers exit through both the drain end and the source end and the channel conduction decreases. During the second phase, the gate voltage is below the transistor threshold voltage and the conduction channel does not exist any more. The coupling between the gate and the data-holding node is merely through the gate-to-diffusion overlap capacitance. In our analysis, attention is focused on the switch charge

Reprinted from *IEEE J. Solid-State Circuits*, vol. SC-22, no. 2, pp. 277–281, Apr. 1987.

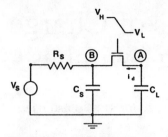

Fig. 1.   Circuit for analysis of switch charge injection.

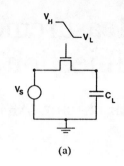

(a)

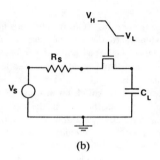

(b)

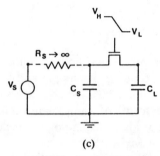

(c)

Fig. 2.   Special cases of switch charge injection. (a) No source resistance and capacitance. (b) No source capacitance. (c) Infinitely large source resistance.

injection due to the first phase of the switch turn-off. The circuit schematic corresponding to the general case of switch charge injection is shown in Fig. 1. Capacitance C_L is the lumped capacitance at the data-holding node. Resistance R_S could be the output resistance of an operational amplifier, while capacitance C_S could be the lumped capacitance associated with the amplifier output node.

Let C_G represent the total gate capacitance, including both the channel capacitance and gate-to-source/gate-to-drain overlap capacitances:

$$C_G = WLC_0 + C_{0vs} + C_{0vd}. \qquad (1)$$

By following the derivation presented in [1], Kirchkoff's current law at node A and node B requires

$$C_L \frac{dv_L}{dt} = -i_d + \frac{C_G}{2} \frac{d(V_G - v_L)}{dt} \qquad (2)$$

and

$$\frac{v_S}{R_S} + C_S \frac{dv_S}{dt} = i_d + \frac{C_G}{2} \frac{d(V_G - v_S)}{dt} \qquad (3)$$

where v_L and v_S are the error voltages at the data-holding node and the signal-source node, respectively. Gate voltage is assumed to decrease linearly with time from the ON value V_H:

$$V_G = V_H - Ut \qquad (4)$$

where U is the falling rate. When the transistor is operated in the strong inversion region

$$i_d = \beta(V_{HT} - U)(v_L - v_S) \qquad (5)$$

where

$$\beta = \mu C_0 \frac{W}{L}$$

and

$$V_{HT} = V_H - V_S - V_{TE}. \qquad (6)$$

Here V_{TE} is the transistor effective threshold voltage including the body effect. For small-geometry transistors, narrow- and short-channel effects should be considered in determining the V_{TE} value. Under the condition $|dV_G/dt| \gg |dv_L/dt|$ and $|dv_S/dt|$, (2) and (3) simplify to

$$C_L \frac{dv_L}{dt} = -\beta(V_{HT} - Ut)(v_L - v_S) - \frac{C_G}{2} U \qquad (7)$$

and

$$\frac{v_S}{R_S} + C_S \frac{dv_S}{dt} = \beta(V_{HT} - Ut)(v_L - v_S) + \frac{C_G}{2} U. \qquad (8)$$

No closed-form solution to this set of equations can be found. Numerical integration can be employed to find the results. Analytical solutions to special cases are given below.

Fig. 2(a) shows the case with only a voltage source at the signal-source node. Since $C_S \gg C_L$, v_S can be approximated as zero and the governing equation reduces to

$$C_L \frac{dv_L}{dt} = -\beta(V_{HT} - Ut)v_L - \frac{C_G}{2} U. \qquad (9)$$

When the gate voltage reaches the threshold condition, the error voltage at the data-holding node is

$$v_L = -\sqrt{\frac{\pi U C_L}{2\beta}} \left(\frac{C_G}{2C_L} \right) \mathrm{erf} \left(\sqrt{\frac{\beta}{2U C_L}} V_{HT} \right). \qquad (10)$$

Another special case is when the source capacitance is negligibly small, as is shown in Fig. 2(b). The governing equations reduce to

$$C_L \frac{dv_L}{dt} = -\beta(V_{HT} - Ut)(v_L - v_S) - \frac{C_G}{2}U \quad (11)$$

and

$$\frac{v_S}{R_S} = \beta(V_{HT} - Ut)(v_L - v_S) + \frac{C_G}{2}U. \quad (12)$$

When the gate voltage reaches the threshold condition, (5) breaks down and the error voltage at the data-holding node is

$$v_L = -\frac{UC_G}{2C_L} \exp\left(-\frac{V_{HT}}{UC_L R_S}\right)$$

$$\cdot \int_0^{V_{HT}/U} [\beta R_S(V_{HT} - U\xi) + 1]^{1/C_L \beta R_S^2 U}$$

$$\cdot \exp\left(\frac{\xi}{C_L R_S}\right)\left(2 - \frac{1}{1 + \beta R_S(V_{HT} - U\xi)}\right) d\xi. \quad (13)$$

If the time constant $R_S C_S$ is much larger than the switch turn-off time, then the channel charge will be shared between C_S and C_L, as is shown in Fig. 2(c). For the case of a symmetrical transistor and $C_S = C_L$, half of the channel charge will be deposited to each capacitor. Otherwise the following equations can be used to find out the results:

$$C_L \frac{dv_L}{dt} = -\beta(V_{HT} - Ut)(v_L - v_S) - \frac{C_G}{2}U \quad (14)$$

and

$$C_S \frac{dv_S}{dt} = \beta(V_{HT} - Ut)(v_L - v_S) + \frac{C_G}{2}U. \quad (15)$$

We now multiply (15) by the ratio C_L/C_S, and then subtract the result from (14), to obtain

$$C_L \frac{d(v_L - v_S)}{dt} = -\beta(V_{HT} - Ut)\left[1 + \frac{C_L}{C_S}\right](v_L - v_S)$$

$$- \frac{UC_G}{2}\left(1 - \frac{C_L}{C_S}\right). \quad (16)$$

When the gate voltage reaches the threshold condition, the amount of voltage difference between the data-holding node and the signal-source node is

$$v_L - v_S = -\sqrt{\frac{\pi UC_L}{2\beta(1 + C_L/C_S)}}\left(\frac{C_G(1 - C_L/C_S)}{2C_L}\right)$$

$$\cdot \text{erf}\left(\sqrt{\frac{\beta(1 + C_L/C_S)}{2UC_L}}\, V_{HT}\right). \quad (17)$$

Fig. 3 shows the calculated percentage of channel charge injected to the data-holding node when the source resistance is infinitely large. Similar plots were obtained by

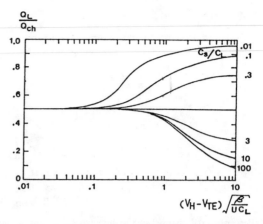

Fig. 3. Percentage of channel charge injected to the data-holding node. Source resistance is assumed to be infinitely large. A family of curves corresponding to various C_S/C_L ratios has been plotted.

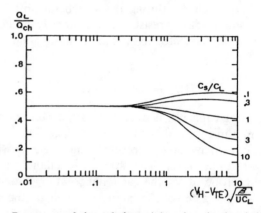

Fig. 4. Percentage of channel charge injected to the data-holding node with $V_{HT}/UR_S C_S = 1$.

numerical integration after some special transformation of the problem [6]. The dimensionless quantity $V_{HT}\sqrt{\beta/UC_L}$ has been identified as the driving force of the switch charge injection effect. It has the same functional dependence as the argument of the error function in (10). A family of curves corresponding to various C_S/C_L ratios have been plotted. When the switch turns off, the channel charge exits to the signal-source node and the data-holding node under capacitive coupling and resistive conduction. In the fast switching-off conditions, the transistor conduction channel disappears very quickly. There is not enough time for the charge at the signal-source side and the charge at the data-holding side to communicate. Hence, the percentage of charge injected into the data-holding node approaches 50 percent independent of the C_S/C_L ratio. In the slow switching-off conditions, the communication between the charge at the signal-source side and the charge at the data-holding side is so strong that it tends to make the final voltages at both sides equal. This allows the majority of channel charge to go to the node with larger capacitance.

Another important factor in switch charge injection is the relative magnitude of the falling rate compared with the signal time constant $R_S C_S$. The curves corresponding to two different $V_{HT}/UR_S C_S$ values are shown in Figs. 4 and 5. Source resistance effectively offers a leakage path

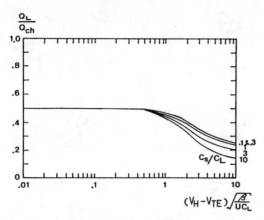

Fig. 5. Percentage of channel charge injected to the data-holding node with $V_{HT}/UR_SC_S = 5$.

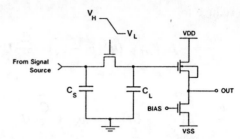

Fig. 6. A two-stage source-follower measurement approach.

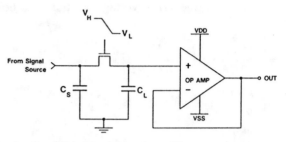

Fig. 7. A unity-gain operational-amplifier measurement approach.

for the channel charge during the switch turn-off period. Hence, a small source resistance will greatly reduce the amount of charge injected to the data-holding node.

III. MEASUREMENT

The holding capacitor is usually chosen around or above 1 pF to minimize the thermal noise voltage. Direct measurement of switch charge injection using single transistors has severe limitations. The stray capacitance of the equipment probe alters the capacitance at the interested node. When the gate voltage falling rate is high, the probe capacitance and inductance greatly perturbs measurement accuracy. On-chip circuitry can be used to circumvent the problem. It offers good buffering between the interested node and the measurement equipment. The insertion of a two-transistor source follower between the interested node and the external probe, as used by Wilson et al. [5], achieves the buffering function to the first order. However, a two-transistor source follower has nonlinear voltage characteristics and limited driving capability. Fig. 6 shows the two-transistor source-follower test configuration.

The unity-gain operational amplifier is found to be a better monitor of the switch charge injection. The output of the amplifier precisely tracks the input voltage. The amplifier possesses an excellent driving capability to interface with the measurement equipment. The unity-gain op-amp test configuration is shown in Fig. 7.

The circuit schematic of the operational amplifier used in the studies is shown in Fig. 8. The operational amplifier is a conventional two-stage design with a source-follower output stage [7]. It is similar to the amplifier used in the on-chip capacitance measurement of MOS transistors in some respects [8]. This circuit configuration provides good common-mode range, output swing, voltage gain, and common-mode rejection ratio. Transistors $M1-M3$ are p-channel current sources. The input stage consists of $M5-M8$. They are a p-channel differential pair with n-channel active loads and double-to-single-ended conversion. Transistors $M11$ and $M12$ are the dummy biasing string for the tracking compensation scheme and also offer dc bias to the output stage. Transistors $M4$ and $M10$ form

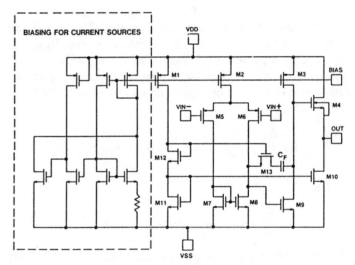

Fig. 8. Schematic of an operational amplifier suitable for charge injection monitoring.

a class-A output stage. A pole-splitting capacitor is used to compensate for the frequency response. If a fabrication process is primarily used for digital circuits and high-quality capacitors are not readily available, a thin-gate transistor can be connected to supply the necessary capacitance. Since the input-referred noise is inversely proportional to the size of the input devices, large-geometry transistors with $W/L = 99~\mu m/6~\mu m$ are used to keep the input-referred noise small. Notice that the substrate and source terminals of the output transistor $M4$ are connected together to eliminate the body effect. This configuration improves the amplifier output range. If an n-well process is used instead of a p-well process, then the output transistors would be changed to p-channel transistors because the transistor inside a well can have its substrate and source tied together. The bias of the current sources can be derived in two ways. A dedicated biasing circuit can be used. The other alternative is to apply an external bias to the pad BIAS. The latter approach turns out to be a good

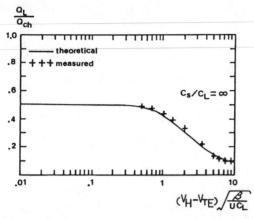

Fig. 9. Comparison of measured and theoretical charge injection results for the special case of Fig. 2(a). The gate voltage falling rate was varied in the experiments.

choice in the application because it reduces the size of the whole test pattern without sacrificing any measurement accuracy. The dotted portion in Fig. 8 denotes the optional biasing block for the current sources.

IV. EXPERIMENTAL RESULTS

The transistors used in the experiments were fabricated using a 3-μm CMOS process. The transistor gate-oxide thickness is 50.0 nm, substrate doping is 10^{16} cm^{-3}, and zero-bias threshold voltage is 0.9 V. Percentage charge injection was measured against the gate voltage falling rate ranging from 1.25×10^6 to 5×10^8 V/s. Fig. 9 shows the measured data and theoretical results. Good agreement between the theoretical results and experimental data is found.

V. CONCLUSION

Charge injection in MOS switches has been analyzed. The analysis has been extended to the general case which includes signal-source resistance and capacitance. This extension makes the results useful for the various conditions encountered in integrated-circuit applications. Plots of the percentage charge injection corresponding to various normalized parameters are presented. The source resistance effectively offers a leakage path for the channel charge during the switch turn-off period. On-chip test circuitry with a unity-gain operational amplifier, which reduces the disturbance imposed by the measurement equipment to a minimum, is found to be an excellent monitor of switch charge injection.

ACKNOWLEDGMENT

The authors wish to thank Prof. P. R. Gray of the University of California, Berkeley for his suggestion of the test patterns and the anonymous reviewers for their valuable suggestions. Generous support from G. Lewicki, V. Tyree, and the MOSIS group is highly appreciated. Discussions with J. Tzeng and K.-Y. Toh were beneficial.

REFERENCES

[1] B. J. Sheu and C. Hu, "Switched-induced error voltage on a switched capacitor," *IEEE J. Solid-State Circuits*, vol. SC-19, pp. 519–525, Aug. 1984.
[2] R. E. Suarez, P. R. Gray, and D. A. Hodges, "All-MOS charge redistribution analog-to-digital conversion techniques: Part II," *IEEE J. Solid-State Circuits*, vol. SC-10, pp. 379–385, Dec. 1975.
[3] R. C. Yen and P. R. Gray, "An MOS switched-capacitor instrumentation amplifier," *IEEE J. Solid-State Circuits*, vol. SC-17, pp. 1008–1013, Dec. 1982.
[4] D. MacQuigg, "Residual charge on a switched capacitor," *IEEE J. Solid-State Circuits*, vol. SC-18, pp. 811–813, Dec. 1983.
[5] W. B. Wilson, H. Z. Massoud, E. J. Swanson, R. T. George, Jr., and R. B. Fair, "Measurement and modeling of charge feedthrough in n-channel MOS analog switches," *IEEE J. Solid-State Circuits*, vol. SC-20, no. 6, pp. 1206–1213, Dec. 1985.
[6] E. Vittoz, "Microwatt switched capacitor circuit design," *Electrocomponent Sci. and Technol.*, vol. 9, no. 4, pp. 263–273, 1982.
[7] P. R. Gray and R. G. Meyer, *Analysis and Design of Digital Integrated Circuits*, 2nd ed. New York: Wiley, 1984.
[8] J. J. Paulos and D. A. Antoniadis, "Measurement of minimum-geometry MOS transistor capacitances," *IEEE Trans. Electron Devices*, vol. ED-32, no. 2, pp. 357–363, Feb. 1985.

Design Techniques for MOS Switched Capacitor Ladder Filters

GORDON M. JACOBS, DAVID J. ALLSTOT, MEMBER, IEEE, ROBERT W. BRODERSEN, MEMBER, IEEE, AND PAUL R. GRAY, SENIOR MEMBER, IEEE

Abstract—Design techniques for monolithic, high-precision, MOS sampled-data active-ladder filters are described. Switched capacitor integrators are used to implement the "leapfrog" configuration for simulating doubly terminated *LC* ladder networks. Techniques are presented for designing all-pole low-pass filters, as well as methods for including transmission zeros. An approach for implementing bandpass filters is described which is derived from the conventional low-pass-to-bandpass transformation. Monolithic realizations for two different low-pass filters are briefly described which show excellent agreement with theory.

Manuscript received April 24, 1978; revised July 10, 1978. This research was supported by the Joint Services Electronics Program under Contract F44620−76−C−0100, and by the Army Research Office under Grant DAAG29−76−G−0244. This paper was presented in part at the International Circuits and Systems Symposium, New York, May 1978.

G. M. Jacobs was with the Department of Electrical Engineering and Computer Sciences and the Electronics Research Laboratory, University of California, Berkeley, CA. He is now with Silicon Systems, Inc., Irvine, CA 92714.

D. J. Allstot, R. W. Brodersen, and P. R. Gray are with the Department of Electrical Engineering and Computer Sciences and the Electronics Research Laboratory, University of California, Berkeley, CA 94720.

I. INTRODUCTION

IN THE EARLY 1970's, filters using periodically operated switches were investigated [15]. Recent work has demonstrated that by using metal-oxide-semiconductor (MOS) integrated circuit technology, it is possible to implement fully integrated second-order filters using analog sampled-data techniques [1]–[4]. The basic element of most of these filters is a switched capacitor integrator which is used to implement a state variable or biquad active filter circuit [2]–[4]. Monolithic MOS realization of these circuits results in two-pole filters with excellent performance [3]. However, if higher order filters are implemented by cascading these second-order sections, the frequency response of the resulting filter will have a relatively high sensitivity to component variations.

It is well known from modern filter theory that for high-order filters, a passive doubly terminated *RLC* ladder achieves very low sensitivity to component varia-

Reprinted from *IEEE Trans. Circuits Syst.*, vol. CAS-25, no. 12, pp. 1014–1021, Dec. 1978.

tions in the passband response, and in fact, has zero sensitivity when the power transfer is matched between source and load [5]. This low sensitivity can be maintained in active filters [6],[16]. "Leapfrog" or "active-ladder synthesis" is one approach which was developed to simulate *RLC* ladder networks exactly using active filter building blocks [7]. In order to obtain minimum sensitivity high-order filters, a similar approach to the "leapfrog" design was taken in this paper which makes use of new switched capacitor techniques. The precision elements in these filters are monolithic MOS capacitors whose ratios determine the frequency response. The inherent temperature stability (<25 ppm/°C) and high matching accuracy (0.1 percent) of MOS capacitor ratios [8] make it possible to implement monolithic high-order filters with very precise frequency characteristics [9], [10].

An interesting aspect of the switched capacitor filters is that they simultaneously have properties of both discrete sampled-data and continuous-time filters. For example, their frequency response scales with the frequency of the clocks which are used to drive the switched capacitors. This is a property which would appear to classify them as analog discrete-time filters. However, they also have continuous-time characteristics, such as delay-free loops, that can contain arithmetic operations, as well as unsampled paths that extend completely through the filter.

The approach to synthesis taken in this paper will be to configure the circuits to simulate the conventional continuous-time ladder circuits as much as practicable, so that the extensive design tables that are available for these filters can be used [11]. It will be shown that it is possible to minimize most of the discrete-time effects (particularly phase shifts due to time delays) which will allow design using classical continuous-time theory directly. However, one discrete-time characteristic which must be considered is the necessity of providing a continuous-time antialiasing prefilter preceding the switched capacitor filter. The requirements on this prefilter can be very relaxed (a single-pole *RC* is often adequate) since the filter structures described in this paper can be designed to have a sample rate which is a large multiple of the highest passband frequency without increasing sensitivity or requiring large amounts of integrated circuit area for implementation compared to other monolithic approaches.

II. SWITCHED CAPACITOR INTEGRATORS

The conventional building block of the "active-ladder" filter circuits is an operational amplifier (op amp) integrator which is shown in Fig. 1(a). The transfer function of this integrator is given by

$$H(\omega) = -\frac{\omega_0}{j\omega} \tag{1}$$

where the integrator bandwidth is $\omega_0 = 1/R_1 C_2$. A switched capacitor version of this integrator is shown in Fig. 1(b), in which the resistor R_1 has been replaced by the capacitor C_1 which is switched at the clock rate f_c. In each clock period, $T_c = 1/f_c$, C_1 is charged to the voltage

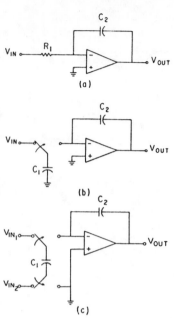

Fig. 1. (a) A conventional *RC* integrator. (b) A single input switched capacitor sampled-data integrator. (c) A switched capacitor differential integrator.

V_{in}, and is then discharged by the op amp. The charge which flows onto the capacitor is $Q_c = C_1 V_{in}$, and the average current flow i can be calculated to be $i = C_1 V_{in} / T_c$. Thus an effective resistance "R_{eff}" of the switched capacitor can be defined

$$R_{eff} = \frac{1}{f_c C_1}. \tag{2}$$

The bandwidth of the switched capacitor integrator in Fig. 1(b) is therefore

$$\omega_0 = \frac{1}{R_{eff} C_2} = f_c \left[\frac{C_1}{C_2} \right]. \tag{3}$$

Since the integrator bandwidth is now determined by a ratio of capacitors, high accuracy and stability can be obtained in a monolithic implementation [2]–[4].

A differential input integrator can easily be obtained using switched capacitors as shown in Fig. 1(c). C_1 is charged to the difference between the two input voltages on the first half clock cycle. When C_1 is then switched into the op amp, its bottom plate is referenced to ground and the charge transferred is $Q_t = C_1(V_{in_1} - V_{in_2})$. If V_{in_1} is grounded, the circuit performs exactly as the integrator in Fig. 1(b) without the sign inversion in the resulting transfer function. Note that for the noninverting integrator, the time constant is still defined by a single capacitance ratio.

An important difference between the conventional and switched capacitor integrators is that the switched capacitor integrator only samples the input signal once each clock cycle. Hence, there can be a time delay between the input signal and the integrator output of up to one full clock period. This time delay results in an excess phase shift through the integrator which results in *Q* enhancement if it is used to directly replace a conventional *RC* integrator in an active-ladder circuit. This phase shift can, therefore, considerably complicate the design, and more importantly, it can break down the analogy of the design with the *RLC* passive ladder circuit which is being used as a prototype. If this happens, the extremely low sensitivities possible with a doubly terminated ladder can be lost. Fortunately, by properly phasing the switches of adjacent integrators, the effect of this deleterious phase shift can be almost completely eliminated.

In order to analyze the effect of the time delay, it is useful to develop a z-transform model of the integrator of Fig. 1(b). In order to be specific, it will be assumed that the switched capacitor C_1 will be charged to the input voltage $V_{in}(t)$ at the beginning of each clock period, and will be switched to the op amp halfway through each cycle. Therefore, at the beginning of the nth clock cycle (i.e., $t = nT_c$), C_1 has a charge $Q_{C_1} = C_1 V_{in}(nT_c)$. The value of the output signal at this time is stored as the charge Q_{C_2} on capacitor C_2. Its value was determined a *half cycle before* at $t = [(n-1/2)T_c]$ and, therefore, equals $Q_{C_2} = C_2 V_{out}[(n-1/2)T_c]$. At the next half cycle time $[t = (n+1/2)T_c]$, the capacitor C_1 is discharged by the operational amplifier thus transferring the charge from C_1 to C_2. The charge on C_2 is now

$$C_2 V_{out}[(n+1/2)T_c] = C_2 V_{out}[(n-1/2)T_c] - C_1 V_{in}(nT_c).$$
(4)

The transfer function for the output taken at this *half cycle time* $H_{1/2}(z)$ is therefore

$$H_{1/2}(z) = -\left[\frac{C_1}{C_2}\right]\left[\frac{z^{-1/2}}{1-z^{-1}}\right]$$
(5)

where the subscript 1/2 indicates there is only 1/2 clock cycle delay in the forward path of this integrator.

For the next half cycle from $(n+1/2)T_c$ to $(n+1)T_c$, the output does not change in value, which yields another half cycle of delay. Therefore, the transfer function of the output taken at the end of this interval (at $t = (n+1)T_c$) has a full cycle of delay in the forward path and is given by

$$H_1(z) = -\left[\frac{C_1}{C_2}\right]\left[\frac{z^{-1}}{1-z^{-1}}\right].$$
(6)

An important result which was first obtained by Bruton [12] in an investigation of digital ladder filters is that a discrete-time integrator which has only a half cycle of delay (such as $H_{1/2}(z)$ in (5)) has exactly the same phase shift as a continuous-time integrator; on the other hand, the integrator represented by (6) has significant phase shift errors. To demonstrate this very interesting result, it is only necessary to evaluate the frequency responses of $H_{1/2}(z)$ and $H_1(z)$ by setting $z = e^{j\omega T_c}$ to obtain

$$H_{1/2}(\omega) = -\frac{\omega_0}{j\omega}\left[\frac{\omega T_c}{2 \sin (\omega T_c/2)}\right]$$
(7)

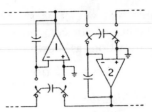

Fig. 2. A two-integrator loop with proper switch phasing.

and

$$H_1(\omega) = \frac{\omega_0}{j\omega}\left[\frac{\omega T_c \exp (-j\omega T_c/2)}{2 \sin (\omega T_c/2)}\right].$$
(8)

These expressions are factored so that the term in brackets is the deviation from the response of a continuous-time integrator given in (1). The error in the magnitude of the integrator bandwidths in (7) and (8) is relatively unimportant because it is equivalent to a small error in the component values, and the doubly terminated LC ladders which are being simulated were chosen because of their extreme insensitivity to errors of this type. Also, since the magnitude error decreases as the sample period is decreased, the sample rate can be increased (at the expense of increased capacitor ratios) to reduce the error to any desired level.

Unfortunately, this low sensitivity does not apply for the extra phase shift of $\omega T_c/2$ radians in the transfer function for $H_1(\omega)$ in (8). When integrators with this response are used in an active-ladder configuration, this phase shift is equivalent to introducing loss (finite element Q) in the inductors and capacitors which are being simulated. However, the sign of this loss is *opposite* to that which is obtained in real circuits elements due to parasitic resistances and conductances, so that instead of a droop in the frequency response (which is usually associated with finite-element Q) the response exhibits peaking. It is interesting to note that this peaked response appears to retain the low sensitivity to magnitude errors that is characteristic of the active-ladder configuration [12]. However, if predistortion techniques are used to compensate for the peaking, computer simulations show that the sensitivity is considerably increased. In most cases, this peaking is not desirable so that the integrators should be operated to obtain the $H_{1/2}(\omega)$ of (7) which does not contain any phase shift errors.

The phasing of the switched capacitors of integrators which are connected together determines whether the responses of the integrators are given by $H_{1/2}(\omega)$ or $H_1(\omega)$. A ladder section containing a two integrator loop is shown in Fig. 2 to demonstrate the proper phasing of the switches needed to obtain the $H_{1/2}(\omega)$ transfer function. Assuming ideal components, the signal at the output of integrator 1 is available as soon as the capacitor is switched to the op amp, so in order to avoid an extra half cycle of delay (and the resulting $H_1(\omega)$ response), the switches of the second integrator must be phased to immediately sample that output as shown in the figure. Therefore, the switches of adjacent integrators should be thrown in *opposite* directions as shown in Fig. 2.

III. DESIGN OF ACTIVE-LADDER NETWORKS

A. The Signal Flow Diagram

A convenient technique for designing an active-ladder network is to transform the differential equations describing the network into a pictorial representation called a flow diagram [13]. The flow diagram, unlike a circuit schematic, contains nodes for both voltage and current variables in the circuit. The branches which interconnect these nodes represent the transfer functions of each circuit element. There are usually several valid flow diagram representations of a given network which require different circuit realizations. The objective here is to manipulate the signal flow diagrams in order to obtain a representation that can be realized with the switched capacitor techniques. To construct a diagram for a canonical network, one simply creates a node for each voltage and current in the circuit, and then interconnects them with the proper impedance or admittance using Kirchhoff's nodal and loop equations. There are also many rules for the proper reduction of redundant branches in the flow diagram [13]. Once the proper flow diagram is constructed, the transformation into a switched capacitor circuit easily follows. Additional techniques for simplifying specific switched capacitor networks will be described in later sections.

The complete design procedure for a low-pass ladder will be presented beginning with the construction of the flow diagram. In subsequent sections, the modifications necessary to implement bandpass filters will be given, as well as an efficient technique for implementing finite transmission zeros.

B. Low-Pass Ladder Design

A passive doubly terminated low-pass ladder which will be used as the prototype for the switched capacitor implementation is shown is Fig. 3. In this figure, the voltages and currents of each circuit element are labeled. A complete set of loop and node equations which involves only integrations is shown below

$$V_0 = V_{\text{in}} - V_1 \qquad \text{9(a)}$$

$$I_0 = \frac{V_0}{R_1} \qquad \text{9(b)}$$

$$I_1 = I_0 - I_2 \qquad \text{9(c)}$$

$$V_1 = (1/sC_1)I_1 \qquad \text{9(d)}$$

$$V_2 = V_1 - V_3 \qquad \text{9(e)}$$

$$I_2 = (1/sL_2)V_2 \qquad \text{9(f)}$$

$$I_3 = I_2 - I_4 \qquad \text{9(g)}$$

$$V_3 = (1/sC_3)I_3 \qquad \text{9(h)}$$

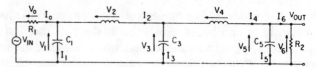

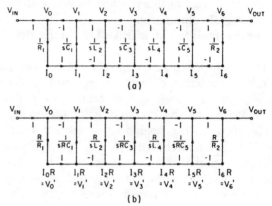

Fig. 3. Doubly terminated *LC* fifth-order all-pole low-pass filter.

Fig. 4. Flow diagram for fifth-order all-pole low-pass ladder.

$$V_4 = V_3 - V_5 \qquad \text{9(i)}$$

$$I_4 = (1/sL_4)V_4 \qquad \text{9(j)}$$

$$I_5 = I_4 - I_6 \qquad \text{9(k)}$$

$$V_5 = (1/sC_5)I_5 \qquad \text{9(l)}$$

$$V_6 = V_5 \qquad \text{9(m)}$$

$$I_6 = V_6/R_2 \qquad \text{9(n)}$$

$$V_{\text{out}} = V_6. \qquad \text{9(o)}$$

The flow diagram which represents these equations is shown in Fig. 4. Each node (voltage or current) is defined by the signal paths flowing into it. The factor written next to each arrow, denoting the direction of the path, is the gain for that path. Multiple inputs into a single node are to be considered summed. In this flow diagram, the nodes which represent currents produce integrations that are bordered by both voltages and currents. Since the actual implementation will use voltage-controlled voltage sources (operational amplifiers) as integrators, it is necessary to transform current nodes to voltage nodes. This is performed by multiplying all current nodes by a scaling resistance R so that the currents I_i are now represented as the voltages $V_i' = RI_i$. In order to maintain the proper relationships between the voltage and current nodes, the gain factors must also be scaled by R.

There are tradeoffs in capacitor circuit area and filter dynamic range which are involved in choosing a value for R. In general, a value of $R = 1 \ \Omega$ is a good compromise, and for this case, the integrator time constants are the original L or C values [14]. After scaling, the terminations R_1 and R_2 are simulated by connecting the inputs of the terminating integrators to their outputs multiplied by the gains R/R_1 and R/R_2. The optimum choice for the values of R_1 and R_2 also depends on many factors, and

162

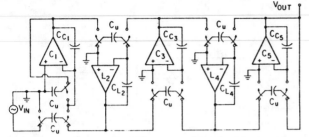

Fig. 5. Switched capacitor version of a fifth-order all-pole low-pass filter.

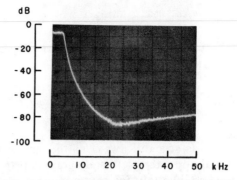

Fig. 6. Measured frequency response for a monolithic MOS fifth-order low-pass filter.

they will also be assumed to equal 1 Ω in the designs in this paper.

It should be emphasized that the flow diagram in Fig. 4 is one of many that describe the ladder of Fig. 3 (for example, differentiations could have been used), but this one was chosen because of its ease of implementation when using switched capacitor techniques.

In the flow diagram, it is apparent that the basic element is a differential integrator of the form shown in Fig. 1(c). If five of these integrators are interconnected as shown in the flow diagram, then the result is the complete switched capacitor circuit shown in Fig. 5. In this figure, the phasing of the switches is alternated as required by the rule developed in Section II in order to minimize the effect of delays.

The final step in the design is to determine the capacitor ratios required in the circuit. It will be assumed that $R_1 = R_2 = R = 1$ Ω and that the element values of the passive prototype (C_1, L_2, C_3, L_4, and C_5) were obtained from standard design tables so that they correspond to a cutoff frequency of 1 rad/s [11]. The ratio of the integrating capacitors (C_{C_1}, C_{L_2}, C_{C_3}, C_{L_4}, and C_{C_5} in Fig. 5) to the switched capacitors C_u (all switched capacitors were set equal for convenience) can be found from (3), and the flow diagram to be given by

$$\frac{C_{C_1}}{C_u} = \frac{f_c C_1}{\omega_{c0}} \qquad \text{10(a)}$$

$$\frac{C_{L_2}}{C_u} = \frac{f_c L_2}{\omega_{c0}} \qquad \text{10(b)}$$

$$\frac{C_{C_3}}{C_u} = \frac{f_c C_3}{\omega_{c0}} \qquad \text{10(c)}$$

$$\frac{C_{L_4}}{C_u} = \frac{f_c L_4}{\omega_{c0}} \qquad \text{10(d)}$$

$$\frac{C_{C_5}}{C_u} = \frac{f_c C_5}{\omega_{c0}} \qquad \text{10(e)}$$

where ω_{c0} is the desired cutoff frequency of the filter, and f_c is the sampling frequency.

The terminations are paths using unit sized capacitors from the output to the input of the first and last integrators. Unfortunately, there is an extra half cycle delay in the integrator termination loops. This extra delay causes

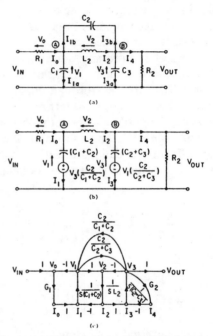

Fig. 7. (a) Doubly terminated third-order elliptic low-pass filter. (b) An equivalent form. (c) The corresponding flow diagram.

an error in the simulated termination resistance. For practical high-order filters, the error in the response due to this incorrect termination is often negligible [9], [10], [12], [14].

A monolithic version of a fifth-order low-pass filter of Fig. 5 has been fabricated using standard MOS technology. The filter was designed for a Chebyshev low-pass response with 0.1-dB passband ripple, and a cutoff frequency of 3400 Hz when clocked at 128 kHz. The measured frequency response shown in Fig. 6 is in very close agreement with the predicted design values [9], [10].

IV. THE ADDITION OF TRANSMISSION ZEROES TO FILTER RESPONSE

The addition of finite transmission zeros to a low-pass ladder filter response has great importance in many filter applications. The zero addition is easily accomplished on the RLC low-pass prototype by adding feedthrough capacitors across the series arm of the ladder network such as C_2 in Fig. 7(a). Imaginary axis zero locations are the resonant frequencies of the LC tank circuit, i.e., $\omega_{zero} = [C_2 L_2]^{-1/2}$. The flow diagram for this noncanonical

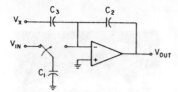

Fig. 8. Switched capacitor integrator/summer.

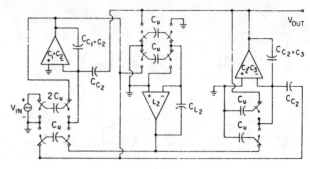

Fig. 9. Switched capacitor third-order elliptic low-pass filter.

network is not as straightforward as the simple low-pass case. The usual approach to flow diagram construction of the circuit is suitable only for continuous-time active *RC* implementations as it contains voltage attenuators (multiplications separate from op amp integrators) [13]. This is not desirable in a switched capacitor implementation since additional op amps would be required.

In order to design a switched capacitor network with zeros which does not require any additional operational amplifiers, it is useful to examine in detail the operations that are performed by the feedthrough capacitors added to the low-pass ladder structure. Referring to Fig. 7(a), a three-pole two-zero *RLC* filter is shown with voltages and currents defined. Using Kirchhoff's current law at nodes *A* and *B*, the following equations are derived to explain the function of C_2: (This is shown symbolically in Fig. 7(b).)

$$V_1 = \frac{(I_0 - I_2)}{s(C_1 + C_2)} + V_3 \left[\frac{C_2}{C_1 + C_2} \right] \qquad (11)$$

and

$$V_3 = \frac{(I_2 - I_4)}{s(C_2 + C_3)} + V_1 \left[\frac{C_2}{C_2 + C_3} \right]. \qquad (12)$$

Thus C_2 has been identified as an element that feeds some of the voltage V_3 to node V_1 and vice-versa. As illustrated in Fig 7(b), in order to implement a transmission zero pair, it is necessary to change the integrator time constants that represent shunt capacitors in the low-pass case to account for the feedthrough capacitor. This action, along with creating the feedforward and feedback paths, completely simulates the added series capacitance.

In a switched capacitor implementation, the integrator time constants are easily changed by adjusting their capacitor ratios. However, the new paths linking V_1 and V_3 of Fig. 7(c) present a problem because both voltages are op amp outputs, and are not suitable for adding without using an additional amplifier. Fig. 8 shows a circuit that achieves the required addition without extra op amp stages. The circuit performs a standard sampled data integration on V_{in}, and in addition, continuously multiplies V_x by a constant and sums it to the output. Since C_2 and C_3 are held to a virtual ground on one side by the op amp, C_3 charges to $Q_3 = C_3 V_x$ and the output due to V_x is given by

$$V_{out} = -\frac{Q_3}{C_2} = -\left[\frac{C_3}{C_2} \right] V_x. \qquad (13)$$

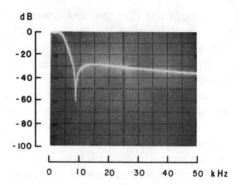

Fig. 10. Measured frequency response of a monolithic MOS third-order elliptic low-pass filter.

Although the summation is continuous, in the filter, V_x will be derived from another integrator whose output changes only once every clock cycle. Using the integrator/summer in place of the conventional integrators for C_1 and C_3 (now $(C_1 + C_2)$ and $(C_2 + C_3)$) in Fig. 7(b) allows the necessary additions at nodes V_1 and V_3. Since the summations can only be brought in with a sign inversion, some minor modifications must be made to the flow diagram. For example, if node V_1, the output of an op amp, must contain a fraction of another node voltage V_3, the two voltages must be of opposite sign on the flow diagram.

The method described previously for obtaining transmission zeros requires very little additional hardware over the all-pole filter circuit. The example chosen to demonstrate the design methods of this section is a third-order elliptic filter. The *RLC* network contains four energy storing devices, while the final switched capacitor circuit shown in Fig. 9 requires only three operational amplifiers. In addition, only two switches and four small capacitors are required over the simple low-pass structure. A monolithic implementation of this version of an elliptic filter has been integrated using standard MOS technology. The filter was designed for an elliptic low-pass response with 0.1-dB passband ripple, a cutoff frequency of 3400 Hz, and a transmission zero at 8.8 kHz when clocked at 128 kHz. The insertion loss was set at 0 dB by simply doubling the size of the switched capacitor connected to the input node. Again, the measured performance shown in Fig. 10 agrees very closely with the design goals [9], [10].

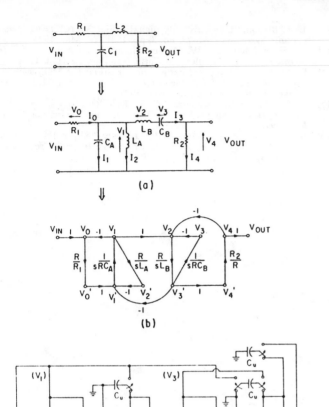

Fig. 11. (a) A two-pole *RLC* low-pass prototype filter, and a four-pole *RLC* bandpass filter. (b) The all-integrator signal flowgraph for the bandpass filter of Fig. 11(a). (c) A switched capacitor bandpass ladder filter.

V. BANDPASS LADDER FILTERS

The bandpass ladder is obtained by performing the standard low-pass-to-bandpass transform on the low-pass prototype [5]. This is done by letting

$$s \rightarrow \frac{\omega_0}{B}\left[\frac{s}{\omega_0} + \frac{\omega_0}{s}\right] \qquad (14)$$

where *B* is the desired 3-dB bandwidth of the passband and ω_0 is the center frequency. Fig. 11(a) shows a two-pole low-pass prototype network and its corresponding four-pole bandpass network after transformation. An all-integrator signal flow diagram for the bandpass ladder is given in Fig. 11(b).

From (14), the elements in the bandpass network of Fig. 11(a) are found to be

$$C_A = \frac{C_1}{B} \qquad 15(a)$$

$$L_A = \frac{B}{C_1 \omega_0^2} \qquad 15(b)$$

$$L_B = \frac{L_2}{B} \qquad 15(c)$$

$$C_B = \frac{B}{L_2 \omega_0^2}. \qquad 15(d)$$

The switched capacitor circuit for the flow diagram of Fig. 11(b) is shown in Fig. 11(c). From the element values above, the capacitor ratios are calculated

$$\frac{C_{C_A}}{C_u} = \frac{C_1 f_c}{B} \qquad 16(a)$$

$$\frac{C_{L_A}}{C_u} = \frac{B f_c}{C_1 \omega_0^2} \qquad 16(b)$$

$$\frac{C_{L_B}}{C_u} = \frac{L_2 f_c}{B} \qquad 16(c)$$

$$\frac{C_{C_B}}{C_u} = \frac{B f_c}{L_2 \omega_0^2}. \qquad 16(d)$$

Note that the bandpass circuit has the same form as a low-pass ladder circuit with two-integrator loops substituted for single integrators. This corresponds to the transformations of single low-pass elements into *LC* tank circuits as given by (14).

VI. CONCLUSIONS

Design techniques for MOS switched capacitor ladder filters have been described. The major advantage of this approach is that precision monolithic high-order filters can be efficiently realized with transfer functions that are insensitive to component variations. Monolithic MOS filters have been designed and tested, and the results confirm the theoretical predictions.

ACKNOWLEDGMENT

The authors would like to acknowledge helpful discussions with K. S. Tan, I. A. Young, and Prof. D. A. Hodges who are involved in other monolithic filtering approaches.

REFERENCES

[1] I. A. Young, P. R. Gray, and D. A. Hodges, "Analog NMOS sampled data recursive filters," in *Dig. Int. Solid-State Circuits Conf.* (Philadelphia, PA), pp. 156–157, Feb. 1977.
[2] B. J. Hosticka, R. W. Brodersen, and P. R. Gray, "MOS sampled data recursive filters using state variable techniques," in *Proc. Int. Symp. on Circuits and Systems* (Phoenix, AZ), pp. 525–529, Apr. 1977.
[3] ——, "MOS sampled data recursive filters using switched capacitor integrators," *IEEE J. Solid-State Circuits*, vol. SC-12, pp. 600–608, Dec. 1977.
[4] J. T. Caves, M. A. Copeland, C. F. Rahim, and S. D. Rosenbaum, "Sampled analog filtering using switched capacitors as resistor equivalents," *IEEE J. Solid-State Circuits*, vol. SC-12, pp. 592–599, Dec. 1977.
[5] E. A. Guillemin, *Synthesis of Passive Networks.* New York: Wiley, 1957.

[6] H. J. Orchard, "Inductorless filters," *Electron. Lett.*, vol. 2, pp. 224–225, June 1966.

[7] F. E. J. Girling and E. F. Good, "The leapfrog or active ladder synthesis," *Wireless World*, pp. 341–345, July 1970.

[8] J. L. McCreary and P. R. Gray, "All-MOS charge-redistribution analog-to-digital conversion techniques—Part I," *IEEE J. Solid-State Circuits*, vol. SC-10, pp. 371–379, Dec. 1975.

[9] D. J. Allstot, R. W. Brodersen, and P. R. Gray, "Fully-integrated high order NMOS sampled data ladder filters," in *Dig. Int. Solid-State Circuits Conf.* (San Francisco, CA), Feb. 1978.

[10] ——, "MOS switched capacitor ladder filters," *IEEE J. Solid-State Circuits*, vol. SC-13, pp. 000–000, Dec. 1978.

[11] A. I. Zverev, *Handbook of Filter Synthesis*. New York: Wiley, 1967.

[12] L. T. Bruton, "Low sensitivity digital ladder filters," *IEEE Trans. Circuits Syst.*, vol. CAS-22, no. 3, pp. 168–176, Mar. 1975.

[13] W. E. Heinlein and W. Harvey Holmes, *Active Filters for Integrated Circuits*. Englewood Cliffs, NJ: Prentice-Hall, 1974.

[14] G. M. Jacobs, "Practical design considerations for MOS switched capacitor ladder filters," Memorandum No. UCB/ERL-M77/69, University of California, Berkeley, 1977.

[15] K. Hirano and S. Nishimura, "Active RC filters containing periodically operated switches," *IEEE Trans. Circuit Theory*, vol. CT-19, May 1972.

[16] K. R. Laker, M. S. Ghausi, and J. J. Kelly, "Minimum sensitivity active (leapfrog) and passive ladder bandpass filters," *IEEE Trans. Circuits Syst.*, vol. CAS-22, pp. 670–677, Aug. 1975.

Technological Design Considerations for Monolithic MOS Switched-Capacitor Filtering Systems

DAVID J. ALLSTOT, SENIOR MEMBER, IEEE, AND WILLIAM C. BLACK, JR., MEMBER, IEEE

Invited Paper

Abstract—In this paper, various technological and topological considerations for the design of monolithic MOS switched-capacitor (SC) filtering systems are described. The properties of the passive and active devices typically available in depletion-load NMOS and CMOS technologies are presented as they relate to various SC performance parameters. Layout techniques for improving specific performance parameters are also given. An overview of MOS operational-amplifier design emphasizes technological considerations in the design of high-performance SC systems. Finally, several important techniques are reviewed for maximizing dynamic range in SC circuits.

I. INTRODUCTION

METAL–OXIDE–SEMICONDUCTOR (MOS) integrated circuit technology is unique in its abilities to store signal-carrying charge packets for relatively long periods of time, to move the packets under clock control, and/or to continuously sense the charge without destroying the information contained therein. This inherent analog memory capability has traditionally been used in the design of dynamic logic circuits and dynamic random-access memories. In each of these applications, the charge packets, although analog in the sense that they may assume a continuous range of values, are eventually quantized and interpreted as either digital ones or zeros. As a substantial noise margin is usually assigned to the quantizing operation, the charge packets may be stored on small, relatively imprecise nodal capacitances within the circuit. By slightly modifying the MOS technology to include a *precision* MOS capacitor, these same inherent technological capabilities can be used to realize *analog* functions compatibly with dense, low-cost digital LSI circuitry [1].

In the mid 1970's, all-MOS charge-redistribution analog–digital (A/D) conversion techniques were investigated [2], [3]. It was initially demonstrated that with proper process design, the monolithic MOS capacitor possessed remarkably stable characteristics in terms of its temperature and voltage coefficients (typically 20–50 ppm/°C, and 10–200 ppm/V, respectively). The absolute value of the capacitance, however, exhibited random processing variations on the order of 10–20 percent. This limitation was soon overcome by developing circuit techniques wherein the *ratio* of two MOS capacitors defined the fundamental precision analog quantity. With proper layout techniques, it was shown that the monolithic MOS capacitor *ratio* was reproducibly accurate to within 0.1 percent.

Manuscript received November 16, 1982; revised April 19, 1983.
D. J. Allstot is with Nova Monolithics, Inc., Carrollton, TX 75006.
W. C. Black, Jr. is a private consultant, Cedar Rapids, IA 52403.

During the same time period, another key advance was the development of an internally compensated MOS operational amplifier [4]. Although having modest gain in comparison to its bipolar counterparts, the economies of die area and power consumption indicated the potential for integrating many operational amplifiers on a single chip.

Motivated by the desire to economically integrate complete systems containing dense digital circuitry as well as high-accuracy A/D converters and precision frequency-selective filters, MOS sampled-data recursive filtering techniques, derived from digital filter prototypes, were developed [5]. In addition to MOS operational amplifiers for performing arithmetic functions, the technique also used precision-ratioed MOS capacitors to realize the filter coefficients, and a precision external clock to define accurate time delays. Later, these techniques were improved by the use of the switched-capacitor resistor concept [6] which lead to the development of precision state-variable MOS switched-capacitor (SC) filters [7], [8] based on the more familiar active *RC* prototypes, and to high-order SC ladder filters [9], [10] derived from passive *RLC* prototypes. An important characteristic of these filters is the low passband sensitivity, which allows the sampling frequency of the SC filter to be made large relative to the baseband frequencies, thus making on-chip antialiasing viable, even with the imprecise *RC* products realizable in MOS technology. (It should be noted that interest in the concurrent alternative approach to monolithic analog filtering, the charge-coupled-device (CCD) transversal filter [11], soon dwindled due to the fundamental advantages that SC filters have over CCD transversal filters for implementing highly selective frequency responses [12].)

The first commercially available monolithic MOS SC filtering systems (for telephony applications) were introduced in 1979 [13]–[15]. Although exhibiting superior accuracy to any previously existing monolithic filtering approach, they generally suffered from higher noise output, lower power supply rejection, and higher power dissipation than desired. It was subsequently shown that the performance of conventional SC filters was significantly enhanced by using improved circuit and layout techniques [16]–[19]. The purpose of this paper is to review these techniques, as well as other technological considerations in the design of SC *systems*.

The characteristics of the passive and active devices commonly available in complementary-MOS (CMOS) and depletion-load n-channel MOS (NMOS) technologies are described in Section II. Considerations for the design of operational amplifiers for use in

Reprinted from *Proc. IEEE*, vol. 71, no. 8, pp. 967–986, Aug. 1983.

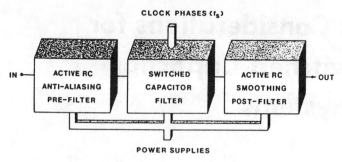

Fig. 1. Block diagram of a typical SC filter system.

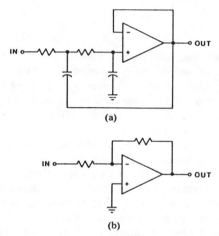

Fig. 2. Commonly used circuits requiring resistors include (a) the Sallen-Key second-order antialiasing/smoothing filter and (b) the inverting gain-setting amplifier.

active-*RC* and SC filters are reviewed in Section III, and in Section IV, various design considerations for SC systems are described. Section V summarizes this paper.

II. PROPERTIES OF RESISTORS, CAPACITORS, AND MOSFET SWITCHES

The performance of a monolithic SC filtering system (Fig. 1) is critically dependent upon the properties of the basic components available in MOS technology. In this section, these properties are reviewed.

A. Monolithic Resistors in MOS Technology

In MOS SC filtering systems, monolithic resistors are commonly used in realizing continuous-time *RC* antialiasing and smoothing filters (Fig. 2(a)) [20], [21], and occasionally as feedback elements in gain-setting amplifier configurations (Fig. 2(b)). The performance of these circuits depends on a number of the properties of the various resistor types within a specific MOS process; thermally diffused crystalline and/or chemically deposited polycrystalline silicon (polysilicon) resistors are usually available. Fig. 3 shows cross sections of the five resistor types existing in a typical depletion-load NMOS process, and Fig. 4 shows the six different resistor types in a typical silicon-gate CMOS process.

Resistor Absolute Accuracy: Large resistance values (typically several hundred kilohms) are often required in realizing on-chip continuous-time *RC* filters. Square or rectangular "serpentine" layouts of the type shown in Fig. 5(a) are commonly used where the absolute resistance value is given by

$$R = \frac{\bar{\rho}L}{WX_t} = \frac{L}{q\bar{\mu}\bar{N}\,WX_t}. \tag{1}$$

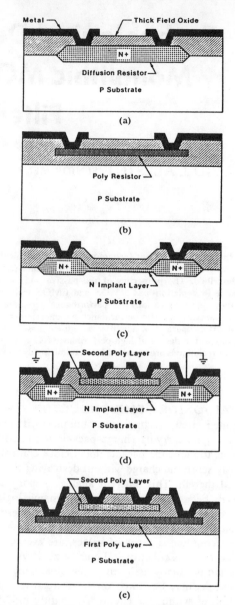

Fig. 3. The five resistor types commonly available in a depletion-load NMOS process. (a) An n^+ diffusion resistor. (b) A polysilicon (poly-I) resistor. (c) A depletion-implant resistor. (d) A poly-II resistor with substrate shield. (e) A double-polysilicon distributed *RC* structure with the resistor formed in the poly-II layer.

If these parameters are assumed to be statistically independent, then the standard deviation of the resistance value is

$$\sigma_R = \left[\left(\frac{\delta\bar{\rho}}{\bar{\rho}}\right)^2 + \left(\frac{\delta L}{L}\right)^2 + \left(\frac{\delta W}{W}\right)^2 + \left(\frac{\delta X_t}{X_t}\right)^2 \right]^{1/2} \tag{2}$$

where it is generally true that $(\delta L/L) \ll (\delta W/W)$ since $L \gg W$ in most monolithic resistors. The standard deviation is typically about ± 25 percent for diffusion resistors (with ion-implanted dopants), and about ± 50 percent for polysilicon resistors. One reason for the difference is that the polysilicon resistors are composed of a conglomerate of many independently oriented small *grains of crystalline silicon* as opposed to the *uniformly crystalline* diffusion resistors. The variations in the size and structure of these grain boundaries in polysilicon resistors result in an increased variation in the average mobility value, thus increasing σ_R [22].

Resistor Ratio Accuracy: Somewhat different layout considerations are required for ratio matching a pair of resistors. For example, long-range first-order gradients in the average resistivity $\bar{\rho}$ and thickness X_t are compensated by using an interdigitated

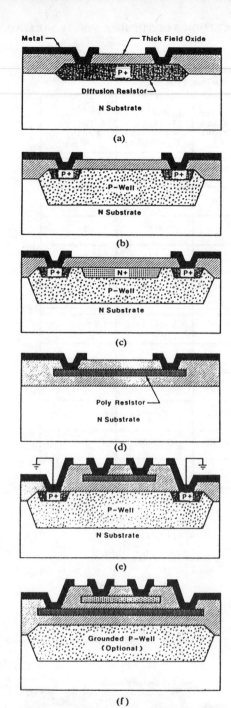

Fig. 4. The six resistor types commonly available in a p-well CMOS process. (a) A p$^+$ (or n$^+$) diffusion resistor. (b) A diffused p-well resistor. (c) A diffused p-well "pinch" resistor. (d) A polysilicon resistor. (e) A polysilicon resistor with substrate shield. (f) A double-polysilicon distributed *RC* structure with the resistor formed in the poly-II layer (p-well substrate shield is optional).

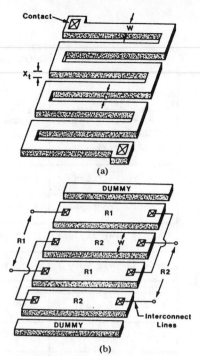

Fig. 5. Typical topological mask layouts for (a) a large-value resistor (Sallen–Key applications), and (b) a precision matched pair of resistors (gain-setting amplifier applications).

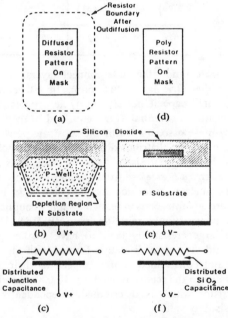

Fig. 6. Diffused versus deposited resistor features. For a p-well resistor, (a) shows the mask opening and subsequent enlargement due to outdiffusion; (b) cross section showing the depletion region, and (c) a distributed *RC* model. For a polysilicon resistor, (d) shows the mask opening; (e) a cross section, and (f) an *RC* distributed model.

resistor layout as shown in Fig. 5(b). Furthermore, in accordance with (2), it is often necessary to increase W from its minimum value (typically 5 μm) in order to reduce the error term due to random width variations ($\delta W/W$) to an acceptably small amount. Using relatively large widths of 20–50 μm, precision resistor pairs are typically matched to within about ± 0.25 percent [23]. The "dummy" resistor segments shown in Fig. 5(b) are included to provide a uniform processing environment for the edges of all resistor segments being matched.

Temperature and Voltage Coefficients of Resistance: Resistor types with large average resistivities are advantageous in reducing the die area required to implement a given resistance value. Unfortunately, there are basic tradeoffs between the resistivity

and the temperature and voltage coefficients of resistance [24], [25]. As the average doping densities are reduced in order to increase the resistivity, the mobility variations with temperature are increased [25, p. 41], resulting in a substantial increase in the *temperature coefficient* of resistance. This property is characteristic of both diffusion and polysilicon resistors. A similar tradeoff exists between the resistivity of diffusion resistors and the *voltage coefficient* of resistance [26], [27].

Resistor Parasitic Capacitances: All of the diffused or ion-implanted resistor types are contained within a reverse-biased p-n junction (Fig. 6(b)), while the polysilicon resistors are isolated

TABLE I
PROPERTIES OF MONOLITHIC RESISTORS IN DEPLETION-LOAD
NMOS (TOP) AND CMOS (BOTTOM) TECHNOLOGIES

Resistor Type (See Figs. 2 and 3)	Nominal Sheet Resistivity (Ohms/□)	Absolute Accuracy σ_R (%)	Approximate Temperature Coefficient (ppm/°C)	Approximate Voltage Coefficient (ppm/volt)	Relative Merit
N$^+$ Diffusion	20-80	25-50	200-2000	50-500	4
N$^+$ Polysilicon	50-150	50	500-1500	20-200	3
N$^-$ Depletion Implant	10K	25	20K	25K	5
N$^+$ Poly over Implant	50-150	50	500-1500	20-200	2
Poly II over Poly I	50-150	50	500-1500	20-200	1
P$^+$ Diffusion	50-200	25-50	200-2000	50-500	6
P$^-$ Well Diffusion	3-5K	25	5K	10K	1
P-Well Pinch Resistor	5-10K	50	10K	20K	5
N$^+$ Polysilicon	50-150	50	500-1500	20-200	4
N$^+$ Poly over P$^-$ Well	50-150	50	500-1500	20-200	3
Poly II over Poly I	50-150	50	500-1500	20-200	2

from the substrate (and other interconnect layers) by a thick silicon dioxide dielectric layer (Fig. 6(e)). In either case, the parasitic distributed capacitance effectively ac couples the resistor to the substrate potential (Fig. 6(c) and (f)); therefore, high-frequency noise associated with the substrate power supply can couple through the distributed parasitic capacitance to the signal flowing through the resistor. In many cases, resistor types which can be *electrically shielded* from the substrate by an intermediate layer connected to analog ground (such as in Fig. 4(e), for example), are advantageous in reducing supply coupling which improves the power-supply rejection ratio (PSRR). Diffusion resistors with larger resistivity tend to have smaller parasitic capacitances, and vice versa. Resistors formed with the polysilicon layer are not constrained in this respect, as their parasitic capacitance is due to the relatively thick (~ 10 000-Å) field oxide under the resistor which is determined independently of the polysilicon resistivity value.

Table I summarizes the properties of commonly available resistor types, and gives a relative figure-of-merit for their use in SC systems.

B. Monolithic Capacitors in MOS Technology

In SC filtering systems, monolithic MOS capacitors are used in realizing continuous-time *RC* filters (Fig. 2(a)), and as the precision elements in the sampled-data integrator/summer (Fig. 7) which is the basic building block of most SC filters.

MOS capacitors can be classified into either of two types depending on the host material for the *thermally grown* silicon dioxide dielectric layer: 1) capacitors formed on heavily doped crystalline silicon (Fig. 8(a)), or 2) capacitors formed on polycrystalline silicon (Fig. 8(b)). Note that a *chemically deposited* silicon dioxide dielectric layer is generally not suitable for preci-

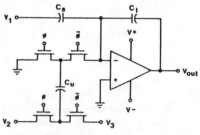

Fig. 7. The basic SC integrator/summer building block.

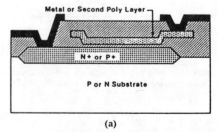

(a)

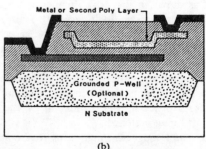

(b)

Fig. 8. Two basic precision capacitor types. (a) Poly-II (or metal) over a heavily doped diffused layer. (b) Poly-II (or metal) over poly-I.

sion capacitors since it exhibits large variations in thickness (poor matching), and undesirable charge–voltage hysteresis effects. Many variations of the two basic capacitor types are available, and their properties may be substantially different in terms of capacitance per unit area, matching accuracies, parasitic capacitances, and temperature and voltage coefficients [28], [29].

Capacitor Absolute Accuracy: Ideally, the value of a monolithic MOS capacitor is given by

$$C = \frac{\epsilon_o \epsilon_{ox} A}{t_{ox}} = \frac{\epsilon_o \epsilon_{ox} WL}{t_{ox}}. \tag{3}$$

If the variables are statistically independent, the standard deviation in the value of C is

$$\sigma_C = \left[\left(\frac{\delta \epsilon_{ox}}{\epsilon_{ox}} \right)^2 + \left(\frac{\delta t_{ox}}{t_{ox}} \right)^2 + \left(\frac{\delta W}{W} \right)^2 + \left(\frac{\delta L}{L} \right)^2 \right]^{1/2} \tag{4}$$

where it is generally true that $(\delta W/W) \cong (\delta L/L)$ since $W \cong L$ for most precision monolithic capacitors.

Since a large capacitance per unit area is desirable for reducing total chip area, the capacitors with oxides thermally grown on *crystalline* silicon are advantageous since the high-quality silicon dioxide layer can be made relatively thin (~ 500 Å with up to 50-V breakdown voltage) [17] as compared to the medium-quality oxides thermally grown on *polysilicon* (polyoxides). The presence of the polysilicon grain boundaries tends to cause additional defects in the polyoxide layer which lowers the breakdown voltage; therefore, polyoxide capacitor types usually require a minimum oxide thickness of greater than 1000 Å to achieve a minimum breakdown voltage of 40 V. Some technologies use a silicon dioxide (1000 Å)/silicon nitride (200 Å) sandwich to form the capacitor's dielectric layer. Although increasing the process complexity, this technique has two advantages: 1) Due to the large relative dielectric constant of Si_3N_4 (7.50), the capacitance per unit area is increased over conventional polyoxide capacitors of the same total thickness, and 2) the thin Si_3N_4 layer tends to fill in many of the defects in the polyoxide, thus improving process yield as well as the breakdown (reliability) characteristics of the capacitor.

Almost any type of precision capacitor can be used in implementing SC filters if the capacitor plates are either metallic or degenerately doped such that the capacitance is relatively constant over the total operating range of signal voltages. The metal or polysilicon capacitor plates are deposited to a thickness of about 5000 Å which provides for good photolithographic edge definition simultaneously with good step coverage. Absolute monolithic capacitance values typically vary by ± 10–20 percent.

Capacitor Ratio Accuracy [28], [29]: The most basic issues involved in defining a precise capacitance ratio can be understood with reference to (4) wherein errors associated with the first two terms, $\delta \epsilon_{ox}/\epsilon_{ox}$ and $\delta t_{ox}/t_{ox}$, are called *oxide effects*, and variations associated with the third and fourth terms, $\delta W/W$ and $\delta L/L$, are called *edge effects*. For large capacitors, the oxide effects are dominant, while for small capacitors, the edge effects usually determine the ratio accuracy. Existing data indicate that the crossover point between the two regions occurs for values of W (and L) ranging from about 20 to 50 μm. The dominant error mechanism can be determined experimentally by plotting the standard deviation in the capacitance ratio versus the ratio for an *array* of capacitors as indicated in Fig. 9 which is a plot of McCreary's data [28]. If the slope of the resulting line is 2, as in the examples shown, oxide effects are dominant, and if the slope is equal to 4, edge effects are dominant [29], [30]. In McCreary's experiments, where the smallest capacitor size was $W = L = 25$ μm, oxide effects generally dominated. For a given capacitance

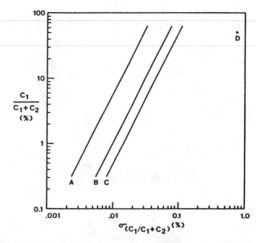

Fig. 9. Typical capacitor-matching data [28]. There were 128 unit capacitors in Array A, 256 in B, 130 in C, and 2 in D. The unit-capacitor sizes were 72 μm $\times$ 72 μm for A, and 25 μm $\times$ 25 μm for the others. The corresponding area/perimeter ratios were 18 μm for A, and 6.25 μm for the other arrays. Oxide effects were dominant in these experiments.

ratio, the magnitude of the random error can only be reduced by increasing the *total capacitance*. In Fig. 9, Array A was constructed with about 8.17 times more total capacitance than Array C, and as shown, the capacitor ratio errors in A were approximately $(8.17)^{1/2}$ times smaller than those in Array C.

Capacitor ratio accuracy is also a function of the capacitor type [28]; due to their better oxide quality, capacitors formed over crystalline silicon usually match better than those with polyoxide dielectrics. As mentioned earlier, deposited oxides are generally unacceptable for fabricating precision capacitors. Although no data are available in the literature regarding the matching characteristics of the polyoxide/silicon nitride sandwich capacitors, it can be conjectured that their matching accuracy would be relatively poor due to the fact that the silicon nitride layer is very thin (200 Å) and difficult to control, since it is chemically deposited rather than thermally grown. Thus although the nitride layer increases the capacitance per unit area, the apparent chip area advantage may be lost since larger unit–capacitor layouts may be required to achieve the same degree of ratio matching attainable with conventional capacitor types.

In addition to the purely *random* effects described above, ratio-matching errors due to oxide or edge effects can also be classified in terms of *systematic* errors. With regard to oxide effects, *common-centroid capacitor layouts* can be used to eliminate systematic errors due to *long-range* gradients in the oxide thickness and/or permittivity; for edge effects, *constant area/perimeter (A/P) layouts* can be used to eliminate systematic errors due to incorrect mask sizing, improper exposure of photoresist, undercutting, etc.

In SC filters, a *unit capacitor* is typically used as the switched capacitor, while the summation and integration capacitors are constructed using a number (usually noninteger) of unit capacitors, as illustrated in Fig. 10. This approximately constant A/P layout technique improves matching accuracy by minimizing systematic edge effects. Due to the low-sensitivity properties of most SC filters, common-centroid layouts (which would eliminate errors due to systematic oxide effects) are generally not required. (There is also evidence that errors due to short-range gradients which typically exist in double-poly capacitors are not canceled using common-centroid layout techniques [28].) For the same reason, dummy strips surrounding the capacitors to provide equal edge environments are also not commonly used. It is preferable to combine fractional unit capacitors with unit capacitors as

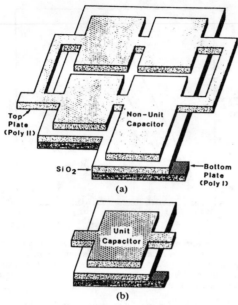

Fig. 10. Typical capacitor layouts for SC filter applications. (b) A unit capacitor is usually used for the sampling or switched capacitor while (a) a number of unit capacitors are interconnected to form the integration and summation capacitors.

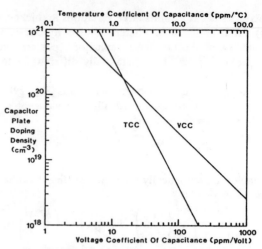

Fig. 11. Typical temperature and voltage coefficients of capacitance as functions of the capacitor-plate doping density [28].

shown in Fig. 10(a) in order to increase the A/P ratio of the noninteger section, thus reducing edge effects in comparison to a separate subunit layout.

Important Note on Ratio Matching: Many companies use a pattern generator (PG) to photographically expose the master reticles from which the photomasks are made. The PG operates by mechanically changing an aperture size to selectively expose rectangular regions of a photosensitive plate to ultraviolet light. Unfortunately, there is a certain amount of mechanical backlash or hysteresis associated with the changing of the aperture size. The ratio matching of both resistors and capacitors can be significantly improved by *sorting* the geometries being matched (e.g., resistor segments or unit capacitors). (This technique is also applicable to the design of input stages of operational amplifiers requiring low dc offset voltages.) The sorting can be performed in either of two ways: 1) All *unit* capacitor geometries on a given layer can be exposed *sequentially*, or 2) additional PG layers can be defined which include only unit capacitors. For example, a layer poly-IA can be defined as containing transistor gates, nonunit capacitor plates, and interconnect lines, while another PG layer, poly-IB, includes only unit capacitor plates. (These PG layers are subsequently combined on a single poly-I mask layer.) When the pattern generator exposes all of the critical geometries in sequence without changing the aperture size, all mechanical hysteresis effects are eliminated. It has been shown experimentally that resistor matching can be improved by as much as an order of magnitude by this simple technique [31]. Further study is required to quantize the improvement in capacitor ratio matching. The ratio accuracy achievable with electron-beam exposure systems is also a topic for further study.

It should again be emphasized that the capacitor-matching characteristics are strongly dependent on process details including the types of equipment used. Therefore, capacitor-matching data should be taken for each MOS process, as well as for each processing facility.

Voltage and Temperature Coefficients of Capacitance [28]: The equations describing the capacitance/voltage characteristics of a MOS capacitor [32], [33] are cumbersome, and will not be presented here. The polyoxide capacitors with a relatively thick oxide typically exhibit lower voltage coefficients than the capacitor types with a thin oxide layer. The double-poly capacitor exhibits a very low voltage coefficient when the doping densities of the two poly layers are approximately equal due to the fact that while one poly capacitor plate depletes charge, the other poly plate accumulates charge which effectively cancels the voltage dependence. Voltage coefficients of less than 200 ppm/V are typically required in precision SC filter applications. Approximate voltage coefficients as a function of the capacitor-plate doping densities are shown in Fig. 11 [28].

The temperature coefficient of capacitance is also dependent on the oxide thickness and on the doping densities of the capacitor plates. Capacitors with a voltage coefficient of 200 ppm/V typically exhibit a negligible temperature coefficient of less than 25 ppm/°C. Furthermore, if all capacitors are assumed to be at the same temperature, the first-order temperature variations in the capacitor *ratio* cancel. Approximate temperature coefficients of capacitance versus capacitor-plate doping densities are also shown in Fig. 11 [28].

Capacitor Parasitic Capacitances: Different capacitor types exhibit substantially different parasitic capacitances. The capacitors grown on crystalline silicon have a relatively large voltage-dependent reverse-biased p-n junction parasitic capacitance associated with the bottom plate, whereas the polyoxide capacitors are isolated from the substrate by a relatively small voltage-independent parasitic capacitance due to the thick field oxide. In many applications, capacitors which can be *electrically shielded* from the substrate in a manner similar to the resistor shielding described previously, are advantageous for improving the PSRR of the filter [16]. For example, Yamakido *et al.* [34] have shown experimentally that the PSSR performance of a CMOS PCM low-pass SC filter is improved by more than 10 dB by electrically shielding both plates of the double-poly integration capacitors from the substrate using a p-well connected to analog ground. Table II summarizes the properties of precision capacitor types commonly used in MOS SC filters.

C. MOS Transistors

The MOSFET is a nearly ideal switch in that it has virtually infinite dc input impedance at the controlling node (gate); in the OFF state, the dc isolation between the switch points (source and drain) is nearly infinite, and in the ON state, there is zero dc offset voltage between the switch points when driving capacitive loads (no dc current flow) as in SC filters.

TABLE II
PROPERTIES OF MONOLITHIC CAPACITORS IN DEPLETION-LOAD
NMOS (TOP) AND CMOS (BOTTOM) TECHNOLOGIES

Precision Capacitor Type	T_{ox} (Angstroms)	Approximate Absolute Accuracy σ_C (%)	Approximate Temperature Coefficient (ppm/°C)	Approximate Voltage Coefficient (ppm/volt)	Relative Merit
Metal-N$^+$	500-1000	10	10-50	20-200	3
Poly-N$^+$	500-1000	10	10-50	20-200	2
Poly II-Poly I	1000-1500	20	10-50	20-200	1
Metal-Poly	1000-1500	20	10-50	20-200	4
Metal-N$^+$	500-1000	10	10-50	20-200	3
Metal-P$^+$	500-1000	10	10-50	20-200	6
Poly-N$^+$	500-1000	10	10-50	20-200	2
Poly-P$^+$	500-1000	10	10-50	20-200	5
Metal-Poly	1000-1500	20	10-50	20-200	4
Poly II-Poly I	1000-1500	20	10-50	20-200	1

MOSFET DC Characteristics: There are two basic MOS device types: 1) *Enhancement-mode* or normally OFF transistors, and 2) *depletion-mode* or normally ON transistors. Enhancement MOSFET's are typically used as charge switches and amplification or driver devices, while depletion-mode devices are commonly used in NMOS technology as load devices for logic gates and operational amplifiers [8], [35], and in some CMOS technologies in the design of output stages for high-performance power amplifiers [16], [27].

Cross sections of typical n-channel enhancement MOSFET's in NMOS and CMOS technologies are shown in Fig. 12, and typical drain characteristics for n- and p-channel enhancement devices are shown in Fig. 13. The MOSFET is usually characterized for two regions of operation; for small drain–source voltages, the MOSFET operates in the *nonsaturation region*, and behaves as a voltage-controlled resistor with a first-order characteristic equation of [36]–[38]

$$I = \mu C_{ox}\left(\frac{W}{L}\right)\left[(V_{GS} - V_T)V_{DS} - \frac{1}{2}V_{DS}^2\right]. \quad (5)$$

For large drain–source voltages, the MOSFET operates in the *saturation region*, and approximates a voltage-controlled current source of value [36]–[38]

$$I = \frac{\mu C_{ox}}{2}\left(\frac{W}{L}\right)(V_{GS} - V_T)^2(1 + \lambda V_{DS}). \quad (6)$$

In this simple model, when the applied gate–source voltage is less than the threshold voltage, the enhancement-mode MOSFET is assumed to be turned OFF with no current flowing between the source and drain. (Actually, when the V_{GS} applied to the MOSFET is within about ± 300 mV of the threshold voltage, the drain current exhibits an exponential characteristic [39]. In standard SC applications, proper gate voltages should be used to avoid biasing any MOSFET's in this *subthreshold region* of operation.)

By differentiating (5), the large-signal ON resistance of a MOS

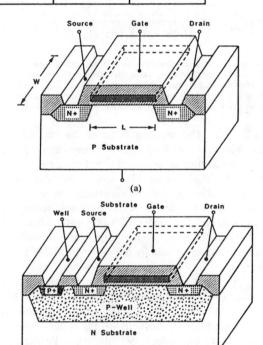

(a)

(b)

Fig. 12. Enhancement n-channel devices in (a) NMOS and (b) CMOS technologies.

transistor is derived as

$$R_{ON} = \left[\mu C_{ox}\left(\frac{W}{L}\right)(V_{GS} - V_T - V_{DS})\right]^{-1} \quad (7)$$

which is the reciprocal slope of the $I-V$ characteristics in the nonsaturation region of Fig. 13. There are two features of the MOSFET which limit its use as a precision resistor in most

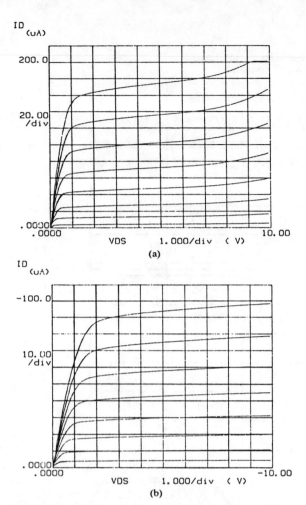

Fig. 13. Typical *I–V* characteristics for (a) n-channel and (b) p-channel MOSFET's. Note the increasing drain current (due to avalanche effects) for the n-channel device for drain–source voltages above 6 V. Both devices are 25 μm wide and 8 μm long, as drawn. For the n-channel, V_{GS} varies from 1.1 to 2.5 V in 0.2-V increments. For the p-channel, V_{GS} varies from −1.5 to −3.6 V in −0.3-V increments.

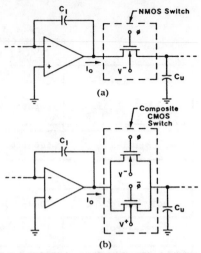

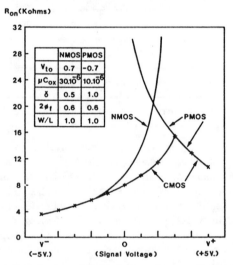

Fig. 14. For small-signal output voltages, the ON resistance of (a) the NMOS switch or (b) the CMOS transmission gate determines response time. For large signals, the charging current may be limited by the available slewing current from the operational amplifier.

Fig. 15. Variations in the small-signal ON resistance of several commonly used switch types versus changes in the steady-state voltage level. The turnoff of single-channel switches can limit filter dynamic range as indicated.

analog circuits: 1) The absolute value of the ON resistance exhibits large variations with changes in temperature and process parameters, and 2) the ON resistance is not constant with changes in the applied signal voltage. In some continuous-time filter applications, the desired absolute resistance can be obtained by using an additional frequency-locked feedback network to control the gate voltage of the MOSFET [40], [41]. Also, by using a balanced fully differential filter topology, the second-harmonic distortion term (which is usually dominant in MOSFET circuits) is canceled [41].

In SC applications where it is used to transfer charge to or between small (< 1-pF) capacitors, a minimum-sized MOSFET ($W = L = 5\ \mu$m) typically has an average ON resistance of a few kilohms resulting in time constants on the order of 10 ns. Thus with modern MOS technology, SC filters can be designed with sampling frequencies up to tens of megahertz. (SC filters with sampling frequencies as high as 18 MHz have recently been reported [42].)

The aspect ratios of the switch transistors are chosen based on transient response-time requirements which are determined from the sampling frequency (f_S) as K/f_S where K is the clock-phase duty-cycle factor. The transient time usually consists of a large-signal slewing component preceding a small-signal settling component. The large-signal interval occurs during the time when the switch transistor(s) is operating in the saturation region wherein the *ideal* charging current is determined from (6). However, in many cases, the *actual* charging current is not dependent on the device aspect ratio, but is, in fact, limited to the slewing current I_o available at the output of the previous operational amplifier as indicated in Fig. 14. This consideration may determine the required amplifier slew rate in high-frequency SC filters.

In terms of the small-signal settling time, it is important to consider the changes in switch resistance as a function of the steady-state signal level. Fig. 15 shows typical ON resistance variations for both polarities of single-channel MOSFET's and for a composite CMOS switch. Due to the body effect on the threshold voltage

$$V_T = V_{T0} + \gamma\left(\sqrt{V_{BS} + 2\phi_f} - \sqrt{2\phi_f}\right) \qquad (8)$$

a single-channel switch may actually turn OFF for large-signal swings (positive for NMOS, negative for PMOS) resulting in a sharp increase in harmonic distortion due to waveform clipping.

The small-signal settling time constant $R_{ON}C_u$ is partially determined by the device aspect ratio (7). Typically, R_{ON} is selected to allow small-signal settling to within 0.1 percent of the correct final value which requires 6.9 time constants. It should be emphasized that it is the *maximum* worst case ON resistance over the *full range* of signal swings which should be used in sizing the MOSFET switches.

174

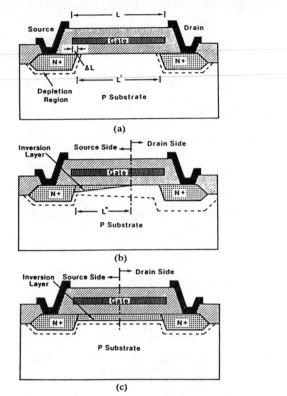

Fig. 16. Cross-sectional views of an NMOS device operated in the (a) cutoff, (b) saturation, and (c) nonsaturation regions.

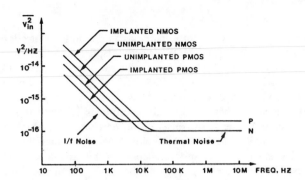

Fig. 17. Typical noise characteristics of various n- and p-channel MOSFET's.

MOSFET AC Characteristics: In general, parasitic capacitance exists between each pair of terminals in the MOSFET device; it is important to characterize these capacitances as they may have a profound effect on the performance of monolithic SC filtering systems.

The intrinsic components of the terminal capacitances of the MOSFET device are strongly dependent on its region of operation [36]–[38], [43], while the extrinsic components due to layout strays, overlapping regions, etc., are relatively constant.

In the first-order device model, the parasitic capacitances are calculated by making simple assumptions regarding the characteristics of the inversion layer. When the MOSFET is OFF (Fig. 16(a)), the region under the gate is assumed to be accumulated, and the terminal capacitances are given as

$$C_{GS} \simeq W\Delta L C_{ox}$$
$$C_{GD} \simeq W\Delta L C_{ox}$$
$$C_{GB} \simeq WL'C_{ox} = W(L - 2\Delta L)C_{ox}$$
$$C_{SB} \simeq A_S C_{pn}(V_{SB})$$
$$C_{DB} \simeq A_D C_{pn}(V_{DB}). \quad (9)$$

When the MOSFET is operating in the saturation region (Fig. 16(b)), it is commonly assumed that the channel begins at the source and extends two-thirds of the distance to the drain; in this region, the device capacitances are approximately

$$C_{GS} \simeq W\Delta L C_{ox} + WL''C_{ox} = WC_{ox}[\Delta L + 2/3(L - 2\Delta L)]$$
$$C_{GD} \simeq W\Delta L C_{ox}$$
$$C_{GB} \simeq 1/3W(L - 2\Delta L)C_{ox}C_{pn}(V_{DB})/(C_{ox} + C_{pn}(V_{DB}))$$
$$C_{SB} \simeq A_S C_{pn}(V_{SB}) + 2/3W(L - 2\Delta L)C_{pn}(V_{SB})$$
$$C_{DB} \simeq A_D C_{pn}(V_{DB}). \quad (10)$$

In nonsaturation, the inversion layer extends all the way from the source to the drain as indicated in Fig. 16(c). In this region, it is commonly assumed that half of the channel is connected to the source, and the other half to the drain, resulting in

$$C_{GS} \simeq W\Delta L C_{ox} + 1/2WL'C_{ox}$$
$$= WC_{ox}[\Delta L + 1/2(L - 2\Delta L)]$$
$$C_{GD} \simeq W\Delta L C_{ox} + 1/2WL'C_{ox}$$
$$= WC_{ox}[\Delta L + 1/2(L - 2\Delta L)]$$
$$C_{GB} \simeq 0 \text{ (neglecting layout strays)}$$
$$C_{SB} \simeq A_S C_{pn}(V_{SB}) + 1/2W(L - 2\Delta L)C_{pn}(V_{SB})$$
$$C_{DB} \simeq A_D C_{pn}(V_{DB}) + 1/2W(L - 2\Delta L)C_{pn}(V_{DB}). \quad (11)$$

In calculating the various parasitic capacitances, several often-neglected components must be considered: 1) All junction capacitances should be calculated using the sidewall component in addition to the planar component. Furthermore, the sidewall component can be significantly increased by the abutting field implant which forms a shallow p^+-n^+ junction at the surface. 2) In saturation, C_{GB} exists in the region between the end of the channel and the drain, and is composed of an oxide capacitance in series with a surface capacitance as illustrated in Fig. 16(b). 3) C_{DB} and C_{SB} must include the junction capacitance between the inversion layer and the bulk. 4) In CMOS technology, the transistors formed in the well or tub region have a large parasitic junction capacitance C_{well} from the back gate of the MOSFET to the substrate (Fig. 12(b)).

In typical monolithic SC filters, the parasitic MOSFET nodal capacitances usually range from 0.02 to 0.2 pF depending on the specifics of the technology. Hence, the *RC* time constants associated with the switches are usually negligible. However, as will be shown later, these parasitics can adversely affect the frequency response, PSRR, noise, and clock feedthrough in SC systems.

MOSFET Noise Characteristics: MOS transistors exhibit two different types of random noise generation as indicated in Fig. 17.

1) Flicker noise: The low-frequency component, which is referred to as the flicker or $1/f$ noise, is commonly modeled in terms of the device geometry and bias conditions as [44]–[46]

$$S_f = \overline{v_{eq}^2}/\Delta f \simeq \frac{K_f}{C_{ox}WL}\frac{1}{f}. \quad (12)$$

The "$1/f$" designation arises from the fact that the input-referred power-spectral density decreases with increasing frequency approximately as $1/f$, as shown in the figure.

Using a boron threshold-adjust implant, which is standard in most NMOS and CMOS technologies, the following qualitative relationships between the flicker-noise coefficients of different devices have been observed (Fig. 17): 1) Unimplanted NMOS devices exhibit less flicker noise than implanted NMOS devices;

furthermore, the flicker noise increases as the number of channel implants increases. 2) PMOS devices display significantly less flicker noise than NMOS devices with similar geometries. Contrary to what occurs in NMOS devices, however, implanted PMOS devices typically exhibit less flicker noise than unimplanted PMOS devices. Limited experiments indicate that this phenomenon is due to the onset of buried-channel conduction in the implanted PMOS devices. Note that for other types of implants, such as a phosphorus depletion implant in an NMOS process, these relationships may not be valid. The physical mechanisms responsible for the generation of $1/f$ noise are not well understood at the present time. Unfortunately, conservative design techniques, therefore, dictate that the K_f parameter in (12) be experimentally determined for each different device at each specific operating point.

2) Thermal noise: White noise is thermally generated in the channel resistance of the MOSFET with its spectral density given as

$$S_t = \overline{v_{eq}^2}/\Delta f = 4kT(2/3g_m).\qquad(13)$$

For a given device, the thermal-noise spectral density depends on the bias current through g_m. However, the total rms noise power obtained by integrating the spectral density over the device bandwidth g_m/C_{in} is equal to $8kT/3C_{in}$. This implies that the thermal-noise performance of n- and p-channel *devices* is identical, and that low thermal noise can only be achieved by using large devices. A similar conclusion can be drawn with regard to flicker noise, although in that case, various circuit techniques can be used to reduce the low-frequency flicker noise components [47], [48].

MOSFET Leakage Current Characteristics: At room temperature, reverse-biased p-n junctions associated with MOSFET's, diffusion resistors, or some types of capacitors exhibit leakage currents on the order of $1-100$ nA/cm^2; the magnitude of the leakage current changes by a factor of about two for each ten degrees of temperature change. In conventional SC filter topologies, this effect limits the minimum sampling frequency to a few kilohertz [49].

III. OPERATIONAL-AMPLIFIER DESIGN CONSIDERATIONS

Most monolithic MOS amplifiers are fabricated using either a single-channel depletion-load NMOS process, or CMOS. (Although single-channel all-enhancement amplifiers have been realized [4], [50], they are not widely used at the present time, and will not be considered further.) Many of the SC performance parameters are closely linked to the characteristics of the various operational amplifiers. As such, it is essential that the design of the amplifiers reflect a number of considerations, the most important of which are summarized in this section. The reader is also directed to [52] for a tutorial overview of MOS amplifier design techniques.

A. Amplifier Voltage Gain

Any real amplifier or buffer displays some nonlinearity in its transfer function which causes any waveform passing through the circuit to become harmonically distorted. In most applications, nonlinearity is reduced by employing negative feedback, wherein a portion of the amplifier gain is used to provide a more stable and more linear transfer characteristic. If the open-loop gain of an amplifier is too low, however, an unacceptable distortion level,

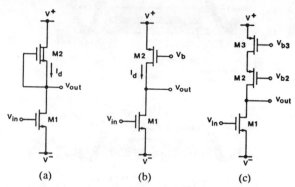

Fig. 18. Simple gain stages. (a) Depletion-load NMOS. (b) CMOS. (c) CMOS with p-channel cascode load.

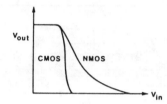

Fig. 19. Typical transfer characteristics for simple depletion-load NMOS and CMOS gain stages.

and an inaccurate filter response may still occur. Distortion products may assume unusual significance in sampled-data circuits since they may be aliased if they fall above the Nyquist frequency of a subsequent sampling stage. While this is most important in high-pass or all-pass networks where distortion terms may pass through without attenuation, it is also important in low-pass or bandpass circuits where the upper band-edge frequency is relatively near the Nyquist rate. As a rule of thumb, the amplifier in a single-stage SC circuit should maintain an open-loop gain of at least $A \times 10^{-D/20}$, where D is the total harmonic distortion specification in decibels, and A is the desired closed-loop gain.

In order to realize a high-gain amplifier, two, or at most three gain stages are usually used. While more stages would of course result in higher gain, amplifiers with more than three stages are nearly impossible to compensate. Thus it is important that each stage display as much gain as possible.

Several single-ended amplifier stages are shown in Fig. 18. Assuming both $M1$ and $M2$ are in saturation, the small-signal voltage gain of the depletion-load NMOS amplifier of Fig. 18(a) is given by [35]

$$a_v \simeq -\sqrt{\frac{(W/L)_1}{(W/L)_2}} \cdot \frac{1}{\eta}\qquad(14)$$

where for the load device

$$\eta = \frac{\gamma}{2\sqrt{V_{BS} + 2\phi_f}}\qquad(15)$$

and

$$\gamma = \frac{\sqrt{2\epsilon qN}}{C_{ox}}.\qquad(16)$$

Due to the body-bias term V_{BS} appearing in (15), the transfer characteristic of the NMOS amplifier (Fig. 19) exhibits significantly reduced gain (increased distortion) at low output voltages. Using this circuit, gains of 15–50 are achievable over a limited output range. Higher gains are difficult to realize, however, due to the large device ratios required. As modern scaled processes are moving towards more heavily doped channel regions and lower power supply voltages [51], modest voltage gains from depletion-load amplifiers will be even more difficult to achieve in the future.

A properly designed CMOS amplifier stage (Fig. 18(b)) provides higher gain than its NMOS counterpart, and maintains it over a wider output voltage range as shown in Fig. 19. In Fig. 18(b), maximum gain is obtained by using an NMOS driver device, and a PMOS load device, wherein the voltages V_{in} and V_b are adjusted so that each device is operating in the saturation region. The small-signal voltage gain of this simple CMOS inverter is given by [52]

$$a_v \simeq g_{m1}(r_{01}\|r_{02}) \simeq \frac{1}{\sqrt{I_d}}\left(\frac{1}{\lambda_1 + \lambda_2}\right)\sqrt{2\mu_1 C_{\text{ox}}(W/L)_1}. \quad (17)$$

As is evident, the gain is increased either by decreasing the bias current, and/or by increasing the channel lengths (which reduces λ). Unfortunately, both of these steps result in decreased bandwidth. Furthermore, if I_d is decreased sufficiently far that the driver device operates in the subthreshold region [39], the small-signal voltage gain becomes a constant [52]. As a general rule, gains of 100–200 are easily achieved in a simple CMOS stage, while gains of several thousand are possible from single stages employing *cascodes* as in Fig. 18(c).

Yet another technique for increasing gain is to increase the mobility of the driver device. Efforts to this effect have resulted in the "inverted" or "n-well" CMOS process [53]. By placing the NMOS device in a lightly doped p-type substrate, instead of in a heavily doped p-well, substantial mobility improvement occurs. However, care must be taken to avoid shallow threshold-adjust implants which may void much of the mobility improvement. This process modification also reduces avalanching effects in the NMOS device [54] (see Fig. 13(a)) which, in a conventional p-well CMOS process, substantially reduce the gain given by (17) at high output voltages (6–10 V). Although the PMOS device is now in a heavily doped n-well region, its avalanche effects are minimal since the avalanche multiplication coefficient is several orders of magnitude smaller for holes than for electrons [25]. Furthermore, channel-length modulation effects are also reduced due to the heavily doped n-well, and thus the p-well device becomes a nearly ideal current source load.

B. Transient Response Time

Since an amplifier in an SC circuit is used in a clocked or pulsed mode, its response time is of considerable importance. If the output of an amplifier is sampled prior to reaching steady state, the filter response will deviate from its design values, and may become noisy or distorted. In general, a stage in which total harmonic distortion cannot exceed a specified amount (such as 0.1 percent), is typically required to settle to approximately the same precision within a given clock period. Note that in some filters, where cascades of amplifiers are used, sufficient time must be allowed for all stages to settle consecutively. This may dramatically reduce the allowable settling time per amplifier, although usually not as severely as $1/n$ for an n-stage cascade.

Two performance parameters of the operational amplifier will determine transient response behavior: 1) AC small-signal settling, and 2) the large-signal nonlinear slewing. Both parameters are ultimately related to the type of phase compensation used in the amplifier. Perhaps the simplest compensation technique is represented in Fig. 20(a). In this circuit, the ac response is

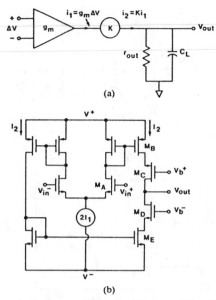

(a)

(b)

Fig. 20. Output-load phase compensation. (a) A small-signal equivalent circuit for a single high-gain stage employing transconductance multiplication. (b) A single high-gain CMOS stage employing output-load compensation.

dominated by the pole at the output node; r_{out} is the intrinsic output resistance of the amplifier, and C_L includes *both* the load and compensating capacitances. (Note that no low-impedance output stage is used in this circuit, as only capacitors are driven in the various filter stages.) The small-signal voltage gain of this amplifier is given by

$$a_v \simeq g_m \cdot K \cdot r_{\text{out}} \quad (18)$$

and the unity-gain frequency is approximately

$$\omega_u \simeq \frac{g_m \cdot K}{C_L}. \quad (19)$$

With proper design, the internal amplifier poles (and zeros) are placed well above the unity-gain frequency ω_u and the small-signal settling time constant is approximately $\tau \simeq 1/\omega_u$. For the typical implementation of this circuit [53], [55], [56] shown in Fig. 20(b), the overall transconductance is

$$G_m = 2g_{mA}(I_2/I_1) \quad (20)$$

with the output impedance given by

$$r_{\text{out}} \simeq r_{0\downarrow}\|r_{0\uparrow} \quad (21a)$$

where

$$r_{0\downarrow} \simeq r_{0D}(1 + g_{mD}r_{0E}) \quad (21b)$$

and

$$r_{0\uparrow} \simeq r_{0C}(1 + g_{mC}r_{0B}). \quad (21c)$$

It is important to note that this type of amplifier may only be implemented using CMOS technology, because of the requirement for cascodes in the high-gain second stage. Due to its large available transconductance, and the push–pull nature of the second stage, this circuit is especially useful in high-speed SC applications.

For large input excursions, wherein the amplifier of Fig. 20(b) operates in a nonlinear fashion, the maximum output slew rate is

$$SR \simeq 2I_2/C_L. \quad (22)$$

Thus the total transient response time, including n small-signal time constants, is given by

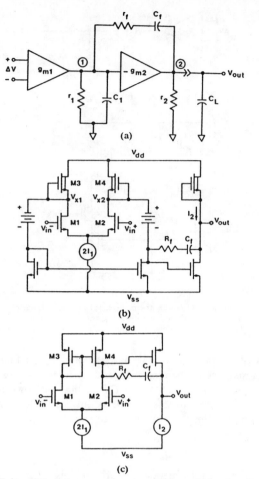

Fig. 21. Pole-splitting phase compensation. (a) A small-signal equivalent circuit. (b) Simplified schematic of a typical depletion-load NMOS amplifier. (c) A typical CMOS amplifier.

$$T_{tot} \simeq \frac{C_L}{2 I_2} [\Delta V + n I_1 / g_{mA}] \qquad (23)$$

where ΔV is the output slewing range.

Although more complicated, the most commonly used compensation technique is that represented in Fig. 21(a), wherein "pole splitting" is employed to create an approximate single-pole frequency response. This method is applicable to both NMOS and CMOS technologies, as it allows the use of two gain stages. In this technique, an on-chip capacitor is used as the feedback element in an inverting gain stage to realize a very large effective capacitance, For example, in Fig. 21(a), a dominant low-frequency pole is created at node ① by using the "Miller effect" to multiply the value of C_f by approximately the second-stage gain. Simultaneously, the pole at node ② is moved to a higher frequency. The transfer function of this circuit is given by [16], [61]

$$a(S) = \frac{a_v(1 + S/Z)}{(1 + S/P_1)(1 + S/P_2)(1 + S/P_3)} \qquad (24a)$$

where

$$Z \simeq \frac{1}{r_f C_f - C_f / g_{m2}} \qquad (24b)$$

$$P_1 \simeq \frac{-1}{g_{m2} r_1 r_2 C_f} \simeq \frac{-g_{m1}}{a_v C_f} \qquad (24c)$$

and

$$P_2 \simeq \frac{-g_{m2} C_f}{C_1 C_L + C_f (C_1 + C_L)} \simeq \frac{-g_{m2}}{C_L} \qquad (24d)$$

$$P_3 \simeq \frac{-1}{r_f C_1}. \qquad (24e)$$

In order for the amplifier to exhibit good closed-loop stability, only the dominant pole P_1 should appear below the unity-gain frequency ω_u. If any of the nondominant singularities P_2, P_3, or Z appear below ω_u, they should exist only as *doublets* (one left-half-plane (LHP) pole, and one LHP zero at very similar frequencies). For cases where the capacitive load C_L is small, the zero is commonly placed at infinity by setting $r_f = 1/g_{m2}$. Alternately, the zero may be eliminated (actually changed to a LHP doublet) by driving the compensation capacitor with a unity-gain voltage follower from the output [4]. If the compensation capacitance is made sufficiently large so that

$$C_f > C_L \cdot \frac{g_{m1}}{g_{m2}} \qquad (25)$$

and C_1 sufficiently small so that

$$C_1 \ll \frac{C_L}{g_{m2} r_f} \qquad (26)$$

then the second and third poles are placed well beyond the nominal unity-gain frequency. Assuming no other nondominant singularities (typically from level shifters or current mirrors), a unity-gain phase margin of at least 45° is obtained, resulting in an approximate single-pole frequency response with a unity-gain bandwidth of

$$\omega_u \simeq \frac{g_{m1}}{C_f}. \qquad (27)$$

When the load capacitance is large, it may be impractical to use a compensation capacitance of the size given by (25). In these cases, a doublet compression sequence may be used, which places Z of (24) at the same frequency as P_2. This technique both reduces the required compensation capacitance and increases the bandwidth of the amplifier. Unfortunately, if the pole and zero become separated in frequency over temperature and process variations, the amplifier may exhibit a slow-settling transient response. One method for compressing the doublet is with a "tracking" compensation circuit, wherein the pole–zero doublet frequencies are identically dependent on process parameters [16]. An idealized schematic of a two-stage pole-split NMOS amplifier is shown in Fig. 21(b), and a pole-split CMOS amplifier is shown in Fig. 21(c). Various two-stage NMOS amplifiers are described in [8], [35], [57]–[59], while several CMOS pole-split amplifiers are described in [13], [15], [16], [60].

The nonlinear response time of an amplifier employing pole-splitting compensation is usually determined by the time required for the output current of the first stage to charge the compensation capacitor (when $C_f/2I_1 > C_L/I_2$). A more complete treatment of the settling behavior of pole-split amplifiers is given in [62].

C. Noise Considerations

At the present time, the dynamic range of most SC circuits is limited by the noise generated within the operational amplifiers, and by their ability to reject noise generated externally. The equivalent input noise voltage for the simple gain stages of Fig. 18(a) and (b) is given by

$$e_n = \left[e_{n1}^2 + \left(\frac{g_{m2}}{g_{m1}} \right)^2 e_{n2}^2 \right]^{1/2} \qquad (28)$$

where e_{n1} and e_{n2} represent the equivalent input noise of each *device* (over any specified bandwidth), and e_n indicates the equivalent input noise of the *circuit* over the same bandwidth. Substituting the thermal noise from (13) into (28), the equivalent input noise for both simple inverter stages of Fig. 18 is given in terms of device geometries as

$$e_{nT} \simeq \left[\frac{8}{3} \frac{kT\Delta f}{g_{m1}} \left(1 + \frac{g_{m2}}{g_{m1}} \right) \right]^{1/2} \simeq \left[\frac{8}{3} \frac{kT\Delta f}{\sqrt{2\beta_1 I_d}} \left(1 + \sqrt{\frac{\beta_2}{\beta_1}} \right) \right]^{1/2}. \qquad (29)$$

Thus in order to minimize circuit noise, the input device should be made as wide as possible, and the load device should be as long as possible. It is relatively easy to make $\beta_2/\beta_1 \ll 1$ in CMOS designs, and in depletion-load NMOS circuits, this condition is necessary anyway in order to achieve modest voltage gain. For these cases, (29) reduces to

$$e_{nT} \simeq e_{n1T} \simeq \left[\frac{8}{3} \frac{kT\Delta f}{\sqrt{2\beta_1 I_d}} \right]^{1/2}. \qquad (30)$$

The flicker-related equivalent input noise for both circuits is obtained by applying the low-frequency noise model (12) to (28) [46]

$$e_{nf} \simeq \left[\frac{a_{n1}}{W_1 L_1} \left(1 + \frac{K_2' a_{n2}}{K_1' a_{n1}} \left(\frac{L_1}{L_2} \right)^2 \right) \right]^{1/2} \qquad (31)$$

where $a_n = K_f \Delta f / C_{ox}$ from (12). As in the thermal-noise case, the flicker noise of the circuit is reduced by making the input device wide, and the load device long. For the case where $L_2 \gg L_1$, the input device noise is dominant, and (31) reduces to

$$e_{nf} \simeq e_{n1f} \simeq \sqrt{\frac{a_{n1}}{W_1 L_1}}. \qquad (32)$$

Thus the total rms noise of a simple amplifier stage up to a frequency f is given by

$$e_{n\,\text{tot}} \simeq \left[e_{nf}^2 + e_{nT}^2 (f - f') \right]^{1/2} \qquad (33)$$

where the frequency f' is typically the $1/f$ noise corner frequency.

D. DC Offset Voltage

The residual dc output voltage of an SC filter is due to a combination of switch feedthroughs at the various integrating nodes, and the dc offset voltages of the various amplifiers. The switch feedthrough is usually caused by gate-overlap capacitances and charge-pumping effects [63], while amplifier offset is usually caused by a systematic or random asymmetry of some type in the input stage. Systematic offsets may occur because of design or mask errors, while random variations are usually due to process or environmentally induced mismatches in device thresholds or g_m's. Systematic offsets are eliminated by properly designing the circuit bias levels for zero *intrinsic* offset voltage, and by good quality control in both layout and mask generation. Random variations in device parameters, however, are impossible to eliminate, although the effects of these variations on amplifier offset may be minimized.

For equal input voltages, mismatches in device dc performance parameters result in unequal currents in the two sides of a differential input stage. These currents are equalized by skewing the input voltage by an amount which is defined as the dc offset voltage. For the amplifiers shown in Fig. 21(b) and (c), the offset voltage is given by

$$V_{os} \simeq \frac{1}{g_{m1,2}} \left[g_{m1,2} \Delta V_{T1,2} + \frac{I_1 \Delta \beta_{1,2}}{\bar{\beta}_{1,2}} + K_1 \Delta V_{T3,4} + K_2 \Delta \beta_{3,4} \right] \qquad (34)$$

where the first two terms represent error currents resulting from input-device mismatch, and the latter two represent load-device mismatch. The constants K_1 and K_2 are generally different for NMOS and CMOS amplifiers. It should be noted that level shifters and subsequent gain stages may also contribute to amplifier offset. These contributions are typically greater in NMOS designs due to the lower input-stage gain. However, in practice, these additions to the offset voltage may be kept small, and are not included here.

In NMOS amplifiers, offset due to load-device mismatch occurs because of a difference between bias voltages, V_{X1} and V_{X2}. When these two voltages are not equal, the input devices conduct slightly different currents due to channel-length modulation effects. The sensitivity to bias-point mismatch is given by

$$K_1 \simeq (I \cdot \lambda)_{1,2} \cdot \frac{\partial V_X}{\partial V_{T3,4}} \qquad (35a)$$

and

$$K_2 \simeq (I \cdot \lambda)_{1,2} \cdot \frac{\partial V_X}{\partial \beta_{3,4}}. \qquad (35b)$$

If the level-shift stages do not affect the bias-point calculations, the derivatives in (35a) and (35b) are easily obtained from (6) and (8). Unfortunately, this is not the case in NMOS amplifiers using low-impedance level shifters [35], [59], and thus K_1 and K_2 can be difficult to specify, in general. For CMOS amplifiers, these parameters are given by

$$K_1 \simeq g_{m3,4} \qquad (36a)$$

and

$$K_2 \simeq \frac{I_1}{\beta_{3,4}}. \qquad (36b)$$

As a general rule, input offsets are minimized by employing physically large and interdigitated input devices. Furthermore, when the input devices are designed to have a large transconductance, input-referred offset contributions from load-device mismatches are small. By using these techniques, offset voltage variations of 5 [58] to 15 mV [35] have been reported in silicon-gate NMOS amplifiers, while distributions of 2 mV have been observed using silicon-gate CMOS [64]. For reasons not well understood, metal-gate designs typically display offset variations which are 3–5 times higher.

E. Rejection of Power Supply Noise

One of the more severe problems afflicting SC filters is that of poor rejection of power-supply noise (poor PSRR). This limitation is due to various parasitics, and to certain undesirable amplifier characteristics. In configurations which use a pole-split amplifier (Fig. 22), the low-frequency supply coupling occurs

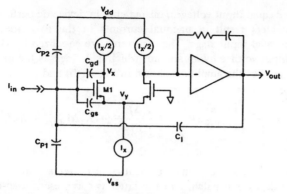

Fig. 22. Simplified integrator model for use in low-frequency PSRR calculations.

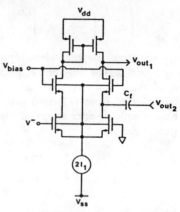

Fig. 23. Improved CMOS input stage employing cascodes for increased PSRR from the positive supply.

principally through the input stage, while high-frequency coupling occurs through the second gain stage.

Low-Frequency Power Supply Rejection: At low frequencies, variations in the power-supply voltages appear at the output of the integrator of Fig. 22 as [16]

$$\frac{\partial V_{\text{out}}}{\partial V_{dd}, V_{SS}} \simeq -\frac{1}{C_I}\left[C_{gd}\frac{\partial V_X}{\partial V_{dd}, V_{SS}} + C_{gs}\frac{\partial V_y}{\partial V_{dd}, V_{SS}} + C_{P2,1}\right].$$
(37)

While this equation is valid for both NMOS and CMOS amplifiers, the derivatives are modified for different types of input stages. Using the dependence of threshold voltage on back bias given in (8), for an NMOS amplifier with depletion current source loads

$$\frac{\partial V_X}{\partial V_{dd}} \simeq \frac{\partial V_X}{\partial I_X}\cdot\frac{\partial I_X}{\partial V_{dd}} \simeq \frac{-\sqrt{V_X - V_{SS} + 2\phi_{fL}}}{\gamma g_{mL}}\cdot\frac{\partial I_X}{\partial V_{dd}} \simeq \frac{-1}{2g_{mbSL}}\cdot\frac{\partial I_X}{\partial V_{dd}}$$
(38a)

$$\frac{\partial V_X}{\partial V_{SS}} \simeq 1 - \frac{\partial V_X}{\partial I_X}\cdot\frac{\partial I_X}{\partial V_{SS}} \simeq 1 - \frac{\sqrt{V_X - V_{SS} + 2\phi_{fL}}}{\gamma g_{mL}}\cdot\frac{\partial I_X}{\partial V_{SS}} \simeq 1 - \frac{1}{2g_{mbSL}}\cdot\frac{\partial I_X}{\partial V_{SS}}$$
(38b)

$$\frac{\partial V_y}{\partial V_{dd}} \simeq \frac{\partial V_y}{\partial I_X}\cdot\frac{\partial I_X}{\partial V_{dd}} \simeq -\frac{1}{2g_{m1}}\left[1 + \frac{\gamma}{2\sqrt{V_y - V_{SS} + 2\phi_{f1}}}\right]^{-1}\cdot\frac{\partial I_X}{\partial V_{dd}} \simeq \frac{1}{2g_{m1}}\left[\frac{\gamma}{2\sqrt{V_y - V_{SS} + 2\phi_{f1}}} - 1\right]\cdot\frac{\partial I_X}{\partial V_{dd}}$$
(38c)

$$\frac{\partial V_y}{\partial V_{SS}} \simeq \frac{\partial V_y}{\partial V_{T1}}\cdot\frac{\partial V_{T1}}{\partial V_{SS}} + \frac{\partial V_y}{\partial I_X}\cdot\frac{\partial I_X}{\partial V_{SS}} \simeq \frac{\gamma}{2\sqrt{V_y - V_{SS} + 2\phi_{f1}}} + \frac{1}{2g_{m1}}\left[\frac{\gamma}{2\sqrt{V_y - V_{SS} + 2\phi_{f1}}} - 1\right]\cdot\frac{\partial I_X}{\partial V_{SS}}$$
(38d)

For rejecting noise from the positive supply, these equations indicate that a supply-independent current source I_X should be used. Unfortunately, some residual coupling will still remain to the negative supply as given by

$$\frac{\partial V_{\text{out}}}{\partial V_{SS}} \simeq -\frac{1}{C_I}\left[C_{gd} + \frac{\gamma}{2\sqrt{V_y - V_{SS} + 2\phi_{f1}}}\cdot C_{gs} + C_{P1}\right].$$
(39)

The C_{gd} term is reduced or eliminated by using cascodes on the input stage [16]; however, the remaining term is usually dominant as typically $C_{gs} \gg C_{gd}$, and thus

$$\frac{\partial V_{\text{out}}}{\partial V_{SS}} \simeq -\frac{1}{C_I}\left[\frac{\gamma}{2\sqrt{V_y - V_{SS} + 2\phi_{f1}}}\cdot C_{gs} + C_{P1}\right].$$
(40)

The inability to cancel this first-order supply coupling is a *fundamental* problem with single-channel NMOS implementations. While the effect of this coupling may be reduced by using a large integrating capacitor (C_I), doing so will increase the circuit

die size and amplifier drive requirements (power dissipation). Alternatively, NMOS SC filters requiring better power supply rejection use more complicated circuitry in the form of on-chip substrate regulation [18], or differential filtering techniques [48].

For a CMOS input stage, the V_y dependence on the power supplies is identical to that given in (38). However, the V_X dependence is now given by

$$\frac{\partial V_X}{\partial V_{dd}} \simeq 1 - \frac{1}{2g_{mL}}\cdot\frac{\partial I_X}{\partial V_{dd}}$$
(41a)

$$\frac{\partial V_X}{\partial V_{SS}} \simeq -\frac{1}{2g_{mL}}\cdot\frac{\partial I_X}{\partial V_{SS}}.$$
(41b)

Note that if I_X is supply independent, as before, the V_{dd} coupling is not eliminated. As mentioned previously, this limitation is easily overcome by using cascode devices in the input stage, as shown in Fig. 23, wherein a supply-independent source is used to bias node V_X at a constant potential regardless of variations in V_{dd}.

It is also easy to eliminate V_{SS} noise coupling into the integrator by placing the amplifier's input devices in their own isolated well, as shown in the figure. This is only possible, however, if the input devices are NMOS in a p-well process, or PMOS in an n-well process. This configuration completely eliminates the input-device threshold variation terms of (40), making both the V_{dd} and V_{SS} low-frequency supply coupling solely dependent upon C_{P1} and C_{P2}. These "supply capacitances" are minimized or eliminated by using careful layout procedures as previously described. Note that (41) is equally valid for a PMOS input stage with NMOS loads if V_{dd} is simply switched with V_{SS}.

In summary, CMOS circuits can be made less sensitive to low-frequency supply coupling than NMOS. In both technologies, however, the use of larger integrating capacitors improves the low-frequency rejection performance, as will techniques employing differential filtering or on-chip supply regulation.

High-Frequency Power Supply Rejection: At high frequencies, supply coupling to the output occurs for different reasons than those presented previously. Specifically, as the gain of an amplifier decreases at higher frequencies, its ability to compensate for supply-induced (and other) errors decreases. In some systems, this coupled high-frequency noise is not important if it occurs outside of the filter passband. In other systems, however, such as those employing switching regulators, or those combining considerable digital logic and analog functions on the same power supplies, coupled supply noise may substantially reduce system performance. This is especially true if any of the coupled high-frequency noise becomes aliased into a valid signal passband.

In the pole-split amplifiers of Fig. 21, the principle path of high-frequency supply coupling is through the compensation capacitor. Since this capacitor presents a low impedance at high frequencies, the output and input of the second gain stage will tend to maintain constant relative potentials. This results in poor rejection of variations in the second-stage source supply voltage. Thus the circuit of Fig. 21(b) displays poor high-frequency rejection from the negative supply while that of Fig. 21(c) displays poor rejection of noise from the positive supply. In each case, the other power supply will display a small and frequency-independent coupling onto the second stage, as described in [52]. Three methods are commonly employed for improving the high-frequency supply rejection.

1) Compensation capacitor feedback from virtual ground: By removing the compensation capacitor from the input to the second stage and placing it instead at a virtual ground potential, the high-frequency supply coupling to the output may be greatly reduced. Of course, the essential signal feedback through the compensation capacitor must still be provided for. One means of accomplishing this is to feed the compensation capacitor back through the cascode circuit in the input stage as shown in Fig. 23 [65]–[67]. An alternative approach has been presented in [52] wherein a separate cascode device (not associated with the input stage) has been added.

2) Differential filtering techniques: If the amplifier uses differential outputs as well as inputs, then high-frequency supply coupling will (to a first order) appear as a common-mode signal. This then, allows cancellation of the supply coupling to the degree that both outputs match, and that common-mode signals can be rejected. As discussed previously, this technique is also useful in reducing low-frequency supply coupling.

3) Amplifier broadbanding techniques: As the rejection of high-frequency error signals is proportional to amplifier gain, this rejection may be improved by increasing the high-frequency performance of the amplifier. Specifically, if a single-pole amplifier response is maintained, the high-frequency supply rejection is proportional to amplifier bandwidth. Broadbanding often requires increased power and die size, however, and may adversely effect filter noise performance.

F. Common-Mode Range and Output Voltage Range

In most SC filters, there is no common-mode range (CMR) requirement for the internal amplifiers since they are always used with the noninverting input at ground. This is not the case, however, for amplifiers which are used as noninverting output drivers, or are used in a conventional pre- or post-filter. For these circuits, the CMR will directly affect the filter's maximum input or output level. In general, the CMR is determined by the voltage

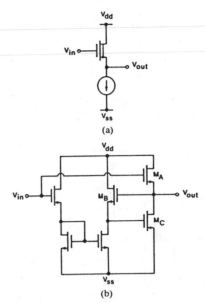

Fig. 24. Typical NMOS output stages. (a) Simple source-follower (enhancement or depletion device may be used). (b) An improved class-AB output stage.

at which the input pair or the input-stage current source enters the nonsaturation region. Large CMR is obtained by biasing all devices with relatively small $V_{GS} - V_T$ values. For depletion-load circuits, the bias current is adjusted so that the load devices are held at a voltage close to the supply. This is typically accomplished via "replica" biasing or feedback. Amplifiers which employ current injection [8] or attempt body-effect cancellation [57] exhibit reduced CMR, and should either be avoided or very carefully designed. For CMOS amplifiers, the cascode circuit of Fig. 23 also exhibits substantially reduced CMR.

Output voltage range is important both within the SC circuits and in any peripheral input/output circuits. It is perhaps more important internally, however, if the SC circuit is "unscaled," and peaking occurs in the SC frequency response. For high-level input signals, this peaking may cause internal filter stages to "clip," resulting in large amounts of distortion. (Various SC scaling techniques are detailed elsewhere [49].) As with CMR, output swings approaching the full supply range are obtained by maintaining small bias voltages.

G. Output Drivers

As discussed previously, output stages are rarely used in amplifiers which only drive on-chip capacitors. Driving off chip, however, an amplifier may see loads of very low effective resistance and/or very high capacitance. For these applications, an output stage is almost always employed.

The simplest output stage in either NMOS or CMOS is a source-follower, as shown in Fig. 24(a). This *class-A* circuit has considerable current-sourcing capability when the follower device is large; however, its current-sinking ability is limited to the value of the bias current source. Thus it is inefficient for driving low-value resistive and/or high-value capacitive loads. In CMOS circuits, the "parasitic" vertical substrate bipolar transistor may be used in the design of an output emitter–follower circuit. Unfortunately, the low (and poorly controlled) f_T of the bipolar transistor can lead to stability problems when driving capacitive loads [68], and can also result in reduced second-stage gain in the presence of low-value resistive loads.

An improvement of this circuit is the class-AB output stage shown in Fig. 24(b), wherein local feedback is obtained from the

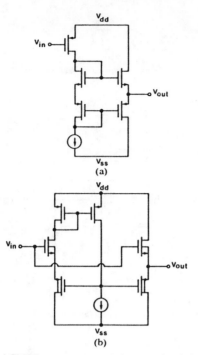

Fig. 25. Typical CMOS output stages. (a) A conventional class-AB gain/output stage. (b) An improved output stage with increased voltage swing due to the use of depletion devices.

output node via M_B, to modify the gate-drive on current–source device M_C [4], [35], [50].

Efficient output stages in CMOS are typically of the push–pull form shown in Fig. 25(a). In this class-AB circuit, bias voltages for the output drivers are established by the diode-connected MOSFET's placed within the second gain stage. However, while the threshold voltage of one of the devices may be held constant by placing it in its own well, the other device will experience the full body effect at high output levels, which limits the available output swing. This limitation often forces the use of a simple source-follower, or NMOS-type output stage described previously. An improved class-AB output stage is shown in Fig. 25(b), wherein a depletion PMOS device has been added (at the expense of an additional masking step) to increase the output swing. In this circuit, the PMOS device still sees the full body effect at output levels near V_{SS}, but due to the depletion implant, its maximum threshold voltage is close to 0 V. The additional bias circuitry shown maintains constant output quiescent current [16], [27].

An advantage that each of the circuits of Fig. 25 share over the NMOS output stage of Fig. 24, is the absence of local feedback. When local feedback is used in the presence of a large capacitive load, it often causes peaking in the amplifier gain–phase response which almost always adversely affects the closed-loop stability of the amplifier.

H. Power Dissipation

It is always desirable to minimize filter power dissipation. This is particularly true if the system employs many SC filters (as in a PCM switching system), or if the power consumed by circuitry within a chip approaches the maximum package dissipation limit.

For any amplifier within an SC filter, sufficient power must be dissipated so that each of the previously discussed design considerations are satisfied over all the potential process and operating conditions. While minimum power design is, in fact, very complicated due to the interrelation of many performance parameters, several methods for minimizing power dissipation in SC filters are available.

Reference-Current Trimming Techniques: As stated earlier, it is desirable from a power-supply rejection standpoint to use a bias source which is supply independent. Regretfully, most supply-independent current references, including a depletion-load device, exhibit very large changes in current over all process and temperature variations. Thus in order to assure proper filter operation over all conditions, it is necessary to substantially increase all current levels above those which are actually needed in the *nominal* case. By trimming the reference current after wafer fabrication, the process sensitivity and the need for high current levels is eliminated [16]. Note that this is only possible in CMOS amplifiers; in NMOS circuits, the current levels are determined by the depletion-load devices which are not usually accessible.

Current Scaling with Capacitive Load: It is generally the case that different capacitive loading occurs at different stages of the filter. If a single amplifier "cell" is used, it is often overdesigned for use in the stages with small capacitive loading. For the class-A transconductance amplifiers of Fig. 21, for example, this means that excess power is being dissipated in the second gain stage. By modifying the second-stage current in accordance with the load capacitance, substantial power savings are realized. This simple technique can reduce dissipated power by up to 50 percent in filters with widely varying integrating and feedforward capacitances [16].

Class-AB and Adaptively-Biased Gain Stages: In most of the gain stages considered thus far, the pull-down current is constant, regardless of the instantaneous need for current. More efficient designs were presented in Fig. 20 for the single-stage amplifier, and Fig. 25 for the class-AB output stage. In these circuits, the quiescent operating current is maintained at a low value; the pull-down (and -up) current is increased or decreased only when the circuit is not in equilibrium. For the push–pull amplifier of Fig. 20, the maximum charging current is equal to twice the value of the quiescent current, assuming one side of the input stage is completely turned off. This push–pull technique has also been applied to the second gain stage of a pole-split amplifier [13] with a similar increase in the maximum available charging current. It should be noted that although both of these circuits achieve better output efficiency, both require additional current paths when compared to simple class-A circuits, and hence, do not always result in reduced quiescent power dissipation. Further improvements in efficiency are realized with class-AB gain stages which can source or sink many times their typical quiescent current [52]. Circuits of this type, however, will usually display higher dc offset and equivalent input noise voltages than a simpler class-A amplifier.

Yet another alternative is that presented in [69], where the bias current of a class-AB amplifier, similar to that shown in Fig. 20(b), is made adaptive. That is, the bias current is made to be a function of the input signal, and is increased substantially during output transitions. This technique offers many advantages in micropower circuits, although its general usefulness at extremely low quiescent currents is questionable due to the attendant low amplifier bandwidth.

Synchronous Dynamic Biasing Techniques: Because the transition periods of amplifiers within an SC filter are synchronous with the clock signal, advantage can be taken of a pulsed biasing method [70]. In this technique, the bias current at the beginning of a clock cycle is increased to allow faster slewing in a class-A amplifier circuit. Subsequently, this bias current is decreased to a low quiescent value for the remainder of the clock cycle. While

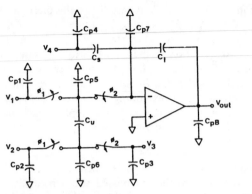

Fig. 26.   An SC integrator/summer with lumped nodal "parasitic" capacitances.

this technique is not as efficient as an adaptive technique, it is relatively simple to realize, and has proven to be useful in SC applications [71]. In practice, this technique allows the efficiency of a class-AB circuit to be obtained with a somewhat simpler class-A design.

Finally, it should be noted that most of the high-efficiency design techniques can only be implemented in CMOS technology.

IV. IMPORTANT SWITCHED-CAPACITOR SYSTEM DESIGN CONSIDERATIONS

In this section, several other important technological and topological considerations in designing monolithic SC filtering systems are presented.

A. SC Integrator Parasitic Capacitances

When designing or analyzing an SC circuit, it should be clearly understood that parasitic capacitances exist at each and every node within the circuit. Furthermore, every parasitic capacitance will affect some performance parameter within the system. One key to successful SC integration is to partition the parasitics as they relate to each level of the design. To illustrate this important point, consider the basic SC integrator/summer circuit of Fig. 26 in which lumped parasitic capacitances have been assigned to each node. In compensating the operational amplifiers, for example, parasitics C_{P1}, C_{P2}, C_{P3}, C_{P4}, and C_{P8} contribute to the loading capacitance of the various amplifiers, and should be considered at that level of the design. At other levels, these parasitics can be neglected. In designing the switch sizes, parasitics C_{P5} and C_{P6} must be included in calculating RC time constants. In terms of the high-level system design considerations, parasitics C_{P5} and C_{P7} are the most important and will be considered separately.

In some SC filters, wherein V_1 is used as an inverting input to the integrator, the parasitic C_{P5} is charged in parallel with C_u, thus perturbing the gain constant associated with that input [9]. Although this parasitic is small, it will adversely affect the frequency response of the SC filter by an amount which depends on the specific SC topology, as well as on the location of this parasitic within the given topology [72]. For extremely precise applications, V_1 should be connected to analog ground and not used as an integrator input. Since C_{P5} is now switched between analog ground and the virtual ground of the amplifier, its charge is constant, and therefore does not affect the frequency response of the filter (assuming large amplifier gain). By using V_2 as the noninverting integrator input and V_3 as the inverting input, the frequency response is, to a first order, *parasitic-insensitive*. The clock-phasing requirements associated with the various inputs are

described elsewhere in this issue, and in previous reports [10], [13], [15]–[17], [49], [73].

In determining the power-supply rejection performance of an SC filter, parasitic C_{P7} is of critical importance. In general, it consists of a component to the negative supply, and a component to the positive supply. To a first order, the ratio of these supply capacitances to the integrating capacitance C_I determines the low-frequency PSRR. Techniques for minimizing the contribution to supply capacitances from the operational amplifiers were described in Section III. For the capacitors, careful layout, including substrate shielding (if available), should be used to minimize supply capacitance. The capacitor plate, which is physically the top plate (see Fig. 8), should always be connected to the virtual ground node, as shown, to reduce parasitics (and possibly leakage current effects). In minimizing the contribution to C_{P7} from the switches, several procedures are suggested: 1) For each integrator, use only one set of top-plate switches; the number of bottom-plate switches is determined by the number of inputs (and their signs) from the other stages. 2) If possible, shield all interconnect lines which are connected to the inverting input of the amplifier. 3) Design the top-plate switches to be as small as possible. (Silicon-gate technologies are preferred due to the smaller overlap capacitances.) 4) In CMOS technologies, using parasitic-insensitive integrators, consider using only single-channel switches for the top plates, and full transmission gates, if necessary, for the bottom plates. Furthermore, in some applications, it may be possible to connect the p-well of the NMOS switches used for the top plates to a negative potential which is different from the negative supply. Assuming that this additional voltage source is free from power-supply noise, this technique eliminates coupling to the negative supply through the relatively large well-substrate capacitance [19].

In complete SC systems, it is important to consider the performance of the continuous-time filters, as well, as they suffer from supply-coupling mechanisms similar to those which exist in SC filters. Many of the shielding and layout considerations described above are applicable to these designs. Furthermore, depending on the specifics of the system, different resistor types may be used to optimize the area/performance tradeoff. For example, in a low-pass system, it is advantageous to use p-well resistors in the pre-filter, and shielded polysilicon resistors in the post-filter. The p-well resistors require less area, but are subject to supply coupling through the well-substrate parasitic capacitance, most of which occurs at higher frequencies. Fortunately, this high-frequency noise is subsequently removed by the low-pass SC filter which follows. The shielded poly resistors are free from supply coupling, and should be used in the post-filter since noise generated in that stage is not filtered by any other stages.

In designing differential SC integrators, in addition to minimizing parasitic capacitances using the techniques described above, it is also important to *match* the parasitics between the two channels.

B. Offset Voltage and Clock-Feedthrough Effects

In many SC systems, it is important to minimize residual dc voltages appearing in the output signal. As indicated in Fig. 27, amplifier dc offset voltages and clock-feed-through effects are usually the principal contributors, although in some systems wherein relatively slowly sampled decimation filters are used, leakage currents may also become significant. Based on the considerations given in Section III, and possibly using PG sorting techniques (Section II), amplifier dc offset voltages can be reduced to the millivolt range. If this is not adequate, *local offset*

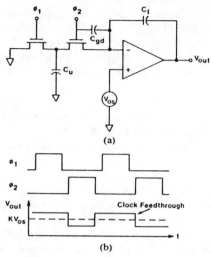

Fig. 27. (a) Circuit model for dc offset voltage and clock feedthrough due to capacitive coupling to the sampling clock. (b) Typical clock and output waveforms.

cancellation [74] may be attempted, although at this level of precision, the clock-feedthrough contributions from the required additional switches may void the apparent advantage of this technique. In many systems, *global* offset autozeroing using feedback techniques can be used to reduce effective dc offset voltages to the tenth-millivolt range, or less [19], [34].

Two different mechanisms are responsible for clock-feed-through offsets: 1) Capacitive coupling. As shown in Fig. 27(a), as the ϕ_2 clock voltage changes, a fraction of this transition, C_{gd}/C_I, is capacitively coupled to the integrator output. 2) Channel-charge pumping effects [63]. As the switches turn OFF and ON, a fraction of their *channel charge* is exchanged with C_u. This charge division is a complex function of many variables, including the rise and fall times of the clock waveforms, and is therefore difficult to accurately specify. Fortunately, there are four commonly used methods for reducing clock feedthrough: 1) Design the MOSFET switches to be as small as possible in order to minimize the overlap capacitance (silicon-gate technology is preferred); 2) design the capacitors to be as large as possible; 3) use charge-cancellation techniques [2], [3], i.e., add additional switches (and possibly additional clock phases) to inject a compensating charge onto the integration capacitor; and 4) use a continuous-time RC post-filter to remove clock feedthrough. (The post-filter may add additional amplifier offsets, however.)

In some systems, wherein the SC filter output is subsequently sampled by an asynchronously clocked stage, clock-feedthrough terms are potentially aliased into a valid signal passband (cf. [16, p. 931]), thus degrading system noise performance in a manner similar to the high-frequency supply coupling described earlier. In these situations, adequate post-filtering must be used to reduce the alias terms to an acceptably small level.

C. Noise Considerations

In addition to the noise sources existing in MOS operational amplifiers (Section III), it is also important to consider the thermal noise generated by the MOSFET's which are used as charge switches. When the MOSFET is turned ON, with an average on-resistance of R_{ON}, its thermal-noise power spectral density in a bandwidth Δf is given by

$$S_t = V_{eq}^2/\Delta f = 4kTR_{ON}. \qquad (42)$$

When charging a sampling capacitor C_u, a single-pole low-pass filter is formed with a -3-dB bandwidth of $\omega_{-3\,dB} = (R_{ON}C_u)^{-1}$.

This simple RC network low-pass filters the thermal noise generated by the MOSFET switch, resulting in a total rms noise voltage of $\sqrt{kT/C}$. Physically, this "kT on C" noise represents the standard deviation in the voltage which is sampled onto C_u, and it is the fundamental limit to the noise performance of SC filters. As always, the high-frequency noise components are aliased into a frequency band extending from dc to $f_S/2$. In analyzing kT/C noise, it is often convenient to replace each SC combination by an equivalent resistor of value

$$R_{EQ} \simeq 1/f_s C_u \qquad (43)$$

with a corresponding noise spectral density.

Noise generated within or passed through an SC filter stage may appear at the output as 1) noise at the original frequency, or 2) noise aliased to a lower than original frequency. The first of these is generally most important when the $1/f$ noise from the amplifier falls within the filter passband. While low-frequency noise from the amplifier may be minimized by appropriate input-stage design as described in Section III, or by sampling techniques, it may limit circuit dynamic range in many cases. Filter noise of this type is described as

$$V_0^2 \simeq \sum_{i=1}^{n} \int_0^{j\pi f_S} H_i^2(S)V_i^2(S)\,dS \qquad (44)$$

where $H_i(S)$ is the transfer function from the ith stage to the output, and $V_i(s)$ is the noise spectral density of the ith stage due to it alone. This latter term is an rms sum of the ith stage amplifier noise and the thermal noise of the switches which appears at the amplifier output. This expression is identical to that of a continuous filter's noise up to a frequency of $f_S/2$. Herein however, lies a fundamental difference between continuous and SC circuits. Because SC circuits are truly sampled-data systems, high-frequency signals (or noise) will be subject to aliasing. Thus high-frequency noise generated at one stage will be downshifted in frequency by any following stage which samples it, and it will thus appear to be at a frequency below the Nyquist rate. In fact, most filter output noise components at a frequency higher than the Nyquist rate will be those generated by the clocks, the last filter stage, or those passed forward to the last stage via a coupling capacitor from a previous stage. In most cases, it is only the noise reflected into the passband that is important, as other frequencies will be attenuated by the filter. The total filter noise including both aliased nd nonaliased terms is given by

$$V_{tot}^2 \simeq \sum_{i=1}^{n} \int_0^{j\pi f_S} H_i^2(S)\left[V_i^2(S) + \sum_{k=1}^{BW/f_s} V_i^2(kf_s \pm s)\right] dS \qquad (45)$$

where BW is the upper bandwidth of the sampling circuits, and the right V_i term indicates noise which is aliased. As is evident from (45), aliased noise is minimized if the ratio BW/f_S is kept as small as possible. Thus RC time constants in the various SC's should not be made excessively short relative to the clock rate, or conversely, the sampling rate should be as high as possible. In addition, by making the individual integrating capacitors as large as possible, the switch noise is reduced, thus providing a lower overall output noise at each stage.

V. Conclusions

Because of their inherent ease of manufacture and long-term stability, monolithic MOS SC filters offer many advantages over their continuous-time counterparts. In order for these advantages to be fully realized, however, considerable care must be exercised in every aspect of the design, and the relationships between the properties of the particular MOS technology and the parameters of the SC system must be clearly understood. In this paper, many of the fundamental implementation issues have been addressed and analyzed, with recommendations made regarding favored methods or solutions.

In view of the ongoing technology scaling efforts, the future role of analog signal processing techniques is a subject of considerable debate. It is true that the performance of conventional analog circuits is degraded when any of the popular technology scaling laws are applied [78]. Having first been implemented only about six years ago, however, and considering the rapid development of new circuit techniques which are overcoming many of the technology limitations, it appears that this debate may continue for many years to come.

ACKNOWLEDGMENT

The authors gratefully acknowledge the many helpful suggestions of Prof. P. R. Gray and Prof. Y. P. Tsividis in improving the organization of this paper. Special thanks are due V. G. Allstot for her meticulous efforts in preparing the figures.

REFERENCES

[1] P. R. Gray, D. A. Hodges, and R. W. Brodersen. *Analog MOS Integrated Circuits*. New York: IEEE Press, 1980.
[2] J. L. McCreary and P. R. Gray, "All-MOS charge redistribution analog-to-digital conversion techniques—Part I," *IEEE J. Solid-State Circuits*, vol. SC-10, pp. 371–379. Dec. 1975.
[3] R. E. Suarez, P. R. Gray, and D. A. Hodges, "All-MOS charge redistribution analog-to-digital conversion techniques—Part II," *IEEE J. Solid-State Circuits*, vol. SC-10, pp. 379–385, Dec. 1975.
[4] Y. P. Tsividis and P. R. Gray, "An integrated NMOS operational amplifier with internal compensation," *IEEE J. Solid-State Circuits*, vol. SC-11, pp. 748–754, Dec. 1976.
[5] I. A. Young, D. A. Hodges, and P. R. Gray, "Analog NMOS sampled-data recursive filter," in *Dig. IEEE Int. Solid-State Circuits Conf.*, pp. 156–157, 1977.
[6] D. L. Fried, "Analog sampled-data filters," *IEEE J. Solid-State Circuits*, vol. SC-7, pp. 302–304, Aug. 1972.
[7] J. T. Caves, M. A. Copeland, C. F. Rahim, and S. D. Rosenbaum, "Sampled analog filtering using switched capacitors as resistor equivalents," *IEEE J. Solid-State Circuits*, vol. SC-12, pp. 592–599, Dec. 1977.
[8] B. J. Hosticka, R. W. Brodersen, and P. R. Gray, "MOS sampled data recursive filters using switched capacitor integrators," *IEEE J. Solid-State Circuits*, vol. SC-12, pp. 600–608, Dec. 1977.
[9] D. J. Allstot, R. W. Brodersen, and P. R. Gray, "MOS switched capacitor ladder filters," *IEEE J. Solid-State Circuits*, vol. SC-13, pp. 806–814, Dec. 1978.
[10] G. M. Jacobs, D. J. Allstot, R. W. Brodersen, and P. R. Gray, "Design techniques for MOS switched capacitor ladder filters," *IEEE Trans. Circuits Syst.*, vol. CAS-25, pp. 1014–1021, Dec. 1978.
[11] D. D. Buss, D. R. Collins, W. H. Bailey, and C. R. Reeves, "Transversal filtering using charge transfer devices," *IEEE J. Solid-State Circuits*, vol. SC-8, pp. 138–146, Apr. 1973.
[12] R. W. Brodersen and T. C. Choi, "Comparison of switched capacitor ladder and CCD transversal filters," in *Proc. 5th Int. Conf. on Charge-Coupled Devices*, pp. 268–278, 1979.
[13] R. Gregorian and W. E. Nicholson, Jr., "CMOS switched-capacitor filters for a PCM voice CODEC," *IEEE J. Solid-State Circuits*, vol. SC-14, pp. 970–980, Dec. 1979.
[14] P. R. Gray, D. Senderowicz, H. Ohara, and B. M. Warren, "A single-chip NMOS dual channel filter for PCM telephony applications," *IEEE J. Solid-State Circuits*, vol. SC-14, pp. 980–991, Dec. 1979.
[15] B. J. White, G. M. Jacobs, and G. F. Landsburg, "A monolithic dual tone multifrequency receiver," *IEEE J. Solid-State Circuits*, vol. SC-14, pp. 991–997, Dec. 1979.
[16] W. C. Black, Jr., D. J. Allstot, and R. A. Reed, "A high performance low power CMOS channel filter," *IEEE J. Solid-State Circuits*, vol. SC-15, pp. 929–938, Dec. 1980.
[17] I. A. Young, "A low-power NMOS transmit/receive IC filter for PCM telephony," *IEEE J. Solid-State Circuits*, vol. SC-15, pp. 997–1005, Dec. 1980.
[18] H. Ohara, P. R. Gray, W. M. Baxter, C. F. Rahim, and J. L. McCreary, "A precision low-power PCM channel filter with on-chip power supply regulation," *IEEE J. Solid-State Circuits*, vol. SC-15, pp. 1005–1013, Dec. 1980.
[19] D. G. Marsh, B. K. Ahuja, T. Misawa, M. R. Dwarakanath, P. E. Fleisher, and V. R. Saari, "A single-chip CMOS PCM CODEC with filters," *IEEE J. Solid-State Circuits*, vol. SC-16, pp. 308–315, Aug. 1981.
[20] R. P. Sallen and E. L. Key, "A practical method of designing RC-active filters," *IRE Trans. Circuit Theory*, vol. CT-2, pp. 74–85, 1955.
[21] B. K. Ahuja, "Implementation of active distributed RC anti-aliasing/smoothing filters," *IEEE J. Solid-State Circuits*, vol. SC-17, pp. 1076–1080, Dec. 1982.
[22] N. C-C. Lu, L. Gerzberg, C-Y. Lu, and J. D. Meindl, "Modeling and optimization of monolithic polycrystalline silicon resistors," *IEEE Trans. Electron Devices*, vol. ED-28, pp. 818–830, July 1981.
[23] G. Kelson, H. H. Stellrecht, and D. S. Perloff, "A monolithic 10-bit digital-to-analog converter using ion implantation," *IEEE J. Solid-State Circuits*, vol. SC-6, pp. 396–403, Dec. 1973.
[24] W. W. Gartner, *Transistors, Principles, Design and Applications*. Princeton, NJ: Van Nostrand, 1960.
[25] S. M. Sze, *Physics of Semiconductor Devices*. New York: Wiley, 1969.
[26] H. L. Lawrence and R. M. Warner, Jr., "Diffused junction depletion layer calculations," *Bell Syst. Tech. J.*, vol. 39, pp. 389–403, 1960.
[27] W. C. Black, Jr., "High-speed CMOS A/D conversion techniques," Ph.D. dissertation, University of California, Berkeley, 1980.
[28] J. L. McCreary, "Matching properties, and voltage and temperature dependence of MOS capacitors," *IEEE J. Solid-State Circuits*, vol. SC-16, pp. 608–616, Dec. 1981.
[29] J-B. Shyu, G. C. Temes, and K. Yao, "Random errors in MOS capacitors," *IEEE J. Solid-State Circuits*, vol. SC-17, pp. 1070–1076, Dec. 1982.
[30] Y. S. Lee, L. M. Terman, and L. G. Heller, "A two-stage weighted capacitor network for D/A-A/D conversion," *IEEE J. Solid-State Circuits*, vol. SC-14, pp. 778–781, Aug. 1979.
[31] D. B. Hildrebrand, private communication.
[32] A. S. Grove, *Physics and Technology of Semiconductor Devices*. New York: Wiley, 1977.
[33] R. Siewatz and M. Green, "Space charge calculations for semiconductors," *J. Appl. Phys.*, vol. 29, p. 1034, July 1958.
[34] K. Yamakido, T. Suzuki, H. Shirasu, M. Tanaka, K. Yasunari, J. Sakaguchi, and S. Hagawara, "A single-chip CMOS filter/CODEC," *IEEE J. Solid-State Circuits*, vol. SC-16, pp. 302–307, Aug. 1981.
[35] D. Senderowicz, D. A. Hodges, and P. R. Gray, "High-performance NMOS operational amplifier," *IEEE J. Solid-State Circuits*, vol. SC-13, pp. 760–766, Dec. 1978.
[36] H. Shichman and D. A. Hodges, "Modeling and simulation of insulated-gate field-effect transistors," *IEEE J. Solid-State Circuits*, vol. SC-3, pp. 285–289, Sept. 1968.
[37] W. M. Penney and L. Lau, Ed., *MOS Integrated Circuits*. New York: Van Nostrand-Reinhold, 1972.
[38] R. S. C. Cobbold, *Theory and Applications of Field-Effect Transistors*. New York: Wiley, 1970.
[39] R. M. Swanson and J. D. Meindl, "Ion-implanted complementary CMOS transistors in low-voltage circuits," *IEEE J. Solid-State Circuits*, vol. SC-7, pp. 146–153, Apr. 1972.
[40] K. S. Tan and P. R. Gray, "Fully integrated analog filters using Bipolar-JFET technology," *IEEE J. Solid-State Circuits*, vol. SC-13, pp. 814–821, Dec. 1978.
[41] M. Banu and Y. P. Tsividis, "Fully integrated active RC filters in MOS technology," in *Dig. Int. Solid-State Circuits Conf.*, pp. 244, 245, 313, 1983.
[42] K. Matsui, T. Matsuura, and K. Iwasaki, "2 micron CMOS switched-capacitor circuits for analog video LSI," in *Dig. 1982 Int. Symp. on Circuits and Systems*, pp. 241–244, May 1982.
[43] Y. P. Tsividis, "Relation between incremental intrinsic capacitances and transconductances in MOS transistors," *IEEE Trans. Electron Devices*, vol. ED-27, pp. 946–948, May 1980.
[44] M. B. Das and J. M. Moore, "Measurement and interpretation of low-frequency noise in FETs," *IEEE Trans. Electron Devices*, vol. ED-21, pp. 247–257, Apr. 1974.
[45] F. M. Klassen, "Characterization of low-frequency 1/f noise in MOS transistors," *IEEE Trans. Electron Devices*, vol. ED-18, pp. 887–891, Oct. 1971.
[46] J. Bertails, "Low frequency noise considerations for MOS amplifier design," *IEEE J. Solid-State Circuits*, vol. SC-14, pp. 773–776, Aug. 1979.
[47] R. W. Brodersen and S. P. Emmons, "Noise in buried channel charge-coupled devices," *IEEE J. Solid-State Circuits*, vol. SC-11, pp. 147–155, Feb. 1976.
[48] K. C. Hsieh and P. R. Gray, "A low-noise chopper-stabilized differential switched-capacitor filtering technique," in *Dig. Solid-State Circuits Conf.* pp. 128, 129, 266, 1981.
[49] D. J. Allstot, "MOS switched capacitor ladder filters," Ph.D. dissertation, University of California, Berkeley, 1979.
[50] I. A. Young, "A high performance all-enhancement NMOS operational amplifier," *IEEE J. Solid-State Circuits*, vol. SC-14, pp. 1070–1077, Dec. 1979.
[51] R. H. Dennard, F. H. Gaenssler, H. N. Yu, V. L. Rideout, E. Bassouts, and A. R. Leblanc, "Design of ion implanted MOSFETs with very small physical dimensions," *IEEE J. Solid-State Circuits*, vol. SC-9, pp.

256–268, Oct. 1974.

[52] P. R. Gray and R. G. Meyer, "MOS operational amplifier design—A tutorial overview," *IEEE J. Solid-State Circuits*, vol. SC-17, pp. 969–982, Dec. 1982.

[53] W. C. Black, Jr., R. H. McCharles and D. A. Hodges, "CMOS process for high performance, analog LSI," in *Proc. IEDM*, pp. 331–339, 1976.

[54] R. Troutman, "Avalanche currents in MOS devices," in *Proc. IEDM*, pp. 43–46, 1974.

[55] R. H. McCharles and D. A. Hodges, "Charge circuits for analog LSI," *IEEE Trans. Circuits Syst.*, vol. CAS-25, pp. 490–497, July 1978.

[56] F. Krummenacher, "High voltage-gain CMOS OTA for micropower switched capacitor filters," *Electron. Lett.*, vol. 17, no. 4, pp. 160–164, Feb. 1981.

[57] E. Toy, "An NMOS operational amplifier," in *Dig. Int. Solid-State Circuits Conf.*, pp. 134–135, 1979.

[58] Y. P. Tsividis, D. L. Fraser, Jr., and J. E. Dziak, "A process-insensitive high performance NMOS operational amplifier," *IEEE J. Solid-State Circuits*, vol. SC-15, pp. 921–928, Dec. 1980.

[59] D. Senderowicz and J. H. Huggins, "A low noise NMOS operational amplifier," *IEEE J. Solid-State Circuits*, vol. SC-17, pp. 999–1008, Dec. 1982.

[60] G. Smarandoiu, D. A. Hodges, P. R. Gray, and G. F. Lansburg, "CMOS pulse-code-modulation voice CODEC," *IEEE J. Solid-State Circuits*, vol. SC-13, pp. 504–510, Aug. 1978.

[61] J. E. Solomon, "The monolithic operational amplifier—A tutorial study," *IEEE J. Solid-State Circuits*, vol. SC-9, pp. 314–322, Dec. 1974.

[62] C. T. Chuang, "Analysis of the settling behavior of an operational amplifier," *IEEE J. Solid-State Circuits*, vol. SC-17, pp. 74–80, Feb. 1982.

[63] J. S. Brugler and P. G. A. Jespers, "Charge pumping in MOS devices," *IEEE Trans. Electron Devices*, vol. ED-16, pp. 297–302, Mar. 1969.

[64] R. A. Reed, private communication.

[65] J. B. Weiser, private communication.

[66] M. B. Terry, private communication.

[67] R. D. Jolly and R. H. McCharles, "A low-noise amplifier for switched capacitor filters," *IEEE J. Solid-State Circuits*, vol. SC-17, pp. 1192–1194, Dec. 1982.

[68] P. R. Gray and R. G. Meyer, *Analysis and Design of Analog Integrated Circuits*. New York: Wiley, 1977.

[69] M. G. Degrauwe, J. Rigmenants, E. A. Vittoz, and H. J. DeMan, "Adaptive biasing CMOS amplifiers," *IEEE J. Solid-State Circuits*, vol. SC-17, pp. 522–528, June 1982.

[70] B. J. Hosticka, "Dynamic CMOS Amplifiers," *IEEE J. Solid-State Circuits*, vol. SC-15, pp. 887–894, Oct. 1980.

[71] B. J. Hosticka, D. Herbst, B. Hoefflinger, U. Kleine, J. Pandel, and R. Schweer, "Real-time programmable low-power SC bandpass filter," *IEEE J. Solid-State Circuits*, vol. SC-17, pp. 499–506, June 1982.

[72] D. J. Allstot and K. S. Tan, "Simplified MOS switched-capacitor ladder filter structures," *IEEE J. Solid-State Circuits*, vol. SC-16, pp. 724–729, Dec. 1981.

[73] K. Martin, "Improved circuits for the realization of switched-capacitor filters," *IEEE Trans. Circuits Syst.*, vol. CAS-27, pp. 237–244, Apr. 1980.

[74] R. Gregorian, "An offset-free switched-capacitor biquad," *Microelectron. J.*, vol. 13, no. 4, pp. 37–40, 1982.

[75] C. A. Gobet and A. Knob, "Noise analysis of switched-capacitor networks," in *Dig. Int. Symp. on Circuits and Systems*, pp. 856–859, 1981.

[76] B. Furrer and W. Guggenbuhl, "Noise analysis of sampled-data circuits, in *Dig. Int. Symp. on Circuits and Systems*, pp. 860–863, 1981.

[77] J. H. Fischer, "Noise sources and calculation techniques for switched-capacitor filters," *IEEE J. Solid-State Circuits*, vol. SC-17, pp. 742–752, Aug. 1982.

[78] S. Wong and C. A. T. Salama, "Impact of scaling on MOS analog performance," *IEEE J. Solid-State Circuits*, vol. SC-18, pp. 106–114, Feb. 1983.

Switched-Capacitor Circuit Design

ROUBIK GREGORIAN, MEMBER, IEEE, KENNETH W. MARTIN, MEMBER, IEEE,

AND GABOR C. TEMES, FELLOW, IEEE

Invited Paper

Abstract—Circuit design techniques are described for switched-capacitor filters, modulators, rectifiers, detectors, and oscillators. The applications of these circuits in telecommunications, speech processing, and other signal-processing systems are also briefly discussed.

I. INTRODUCTION

THE PRACTICAL USE of switched-capacitor integrated circuits began on a major scale only about five years ago. During this short interval, many commercial integrated circuits utilizing switched-capacitor (SC) techniques have been fabricated and marketed. Also, a very large number of theoretical papers (some with and some without influence on the practical developments) were published on the design and analysis of SC circuits. Because of this embarrassment of riches in design approaches, it was felt that a selective summary of some useful available design techniques may be warranted at this time. Rather than aim for completeness (which would have defeated the main purpose of the paper), we selected a few key design techniques, and have described them at an introductory level.

All practically useful (as opposed to merely clever, or ingenious) circuit design techniques for SC circuits are invariably characterized by a respectful observance of the inherent limitations and idiosyncrasies of the components that can be fabricated on a MOS chip. These limitations are discussed in detail by Allstot and Black in a companion paper in this issue. Hence, we mention them only briefly in this article. However, the selection of the circuit design techniques discussed was performed with these practical aspects in mind. In fact, only those techniques which have been successfully used in the design of commercial integrated circuits were included. Thus a conservative approach was strictly followed. This forced us to forego the inclusion of several novel and efficient design methods (e.g., those described in [9], [12], [15]–[17]) in favor of better tested techniques.

For some circuit types (such as SC biquads) several comparable design approaches exist. The choice then was dictated by the time limitation on the completion of this paper; thus we discussed the technique most familiar to us. No effort was made to select the best of several nearly equally good design procedures.

All the above explanations are intended, on the one hand, to encourage the industrial designer to use confidently the circuits discussed herein; and, on the other hand, soothe the ire of those researchers whose brilliant work is relegated to the list of references, or omitted altogether. We could not afford the time or space required to be either encyclopedic, or even meticulously discriminating in our choice of topics!

Most of the literature on SC circuits deals with the design of

filters and D/A converters only. Recently, however, SC techniques have been successfully applied to a number of other signal-processing functions. Hence, in addition to a detailed discussion of SC filter design, we have included a brief description of nonfiltering SC circuits. Finally, a survey of complete integrated systems which apply SC technology was carried out, and a valiant attempt made to predict some future applications—always a hazardous task in a rapidly changing field.

A large and representative list of recent literature was also compiled to enable the interested reader to go back to the source on some of the topics discussed.

II. INTEGRATED FILTERS

High-quality analog filters had been historically realized as passive *LCR* circuits. Since inductors are physically large, electrically lossy and noisy, and unsuitable for miniaturization [1], an effort to replace them by active elements had begun in the 1960's. The resulting circuits were the *active-RC filters* which gained wide acceptance over the past 20 years. To reduce their sizes, they are often realized in a hybrid construction, with monolithic op-amps and chip capacitors soldered on a board containing thick-film resistors. The next step in miniaturization was to realize *fully integrated filters*. Since the MOS technology offers high-quality capacitors, low-leakage charge storage, offset-free switches, and nondestructive charge sensing [2], it is usually preferred to bipolar technology for filtering applications. The straightforward integration of an active-*RC* filter, however, leads to difficulties. Since for an oxide thickness of 700 Å a 1-pF capacitor requires about 3-mil^2 (or about 2000-μm^2) chip area, MOS capacitors are seldom made larger than about 100 pF. Since integrated filters are commonly used in the voice-frequency (0- to 4-kHz) range, they require time constants of the order $RC \sim 10^{-4}$ s. Even for a large capacitor (say, $C = 10$ pF), this requires a resistor of order 10^7 Ω. Such a resistor, made by using a polysilicon line or diffusion region, occupies an area around 1600 mil$^2 \approx 10^6$ μm^2, or nearly 10 percent of the average chip area of an analog MOS integrated circuit [2]. In addition, MOS resistors tend to be nonlinear. Finally, since both the capacitors and resistors have absolute accuracies of only about 5–10 percent, and their errors are not correlated, the overall error of an *RC* time constant can be as high as 20 percent. This error will also vary with temperature and signal level.

To overcome these difficulties, *simulated resistors* can be constructed from capacitors and switches [3]. Two such circuits are shown in Fig. 1. In the circuit of Fig. 1(a), *C* is alternately charged to v_1 and v_2. Each time, a charge $\Delta q = C(v_1 - v_2)$ flows, with the polarity indicated. In the branch of Fig. 1(b), *C* is alternately discharged, and recharged to a voltage $v_1 - v_2$. During recharging, a charge $\Delta q = \Delta q_1 = \Delta q_2 = C(v_1 - v_2)$ flows with the polarity shown. The charge flows in sharp pulses, at the

Manuscript received March 29, 1983.

R. Gregorian is with American Microsystems Inc., Santa Clara, CA 95051.

K. W. Martin and G. C. Temes are with the Department of Electrical Engineering, University of California at Los Angeles, Los Angeles, CA 90024.

Reprinted from *Proc. IEEE*, vol. 71, no. 8, pp. 941–966, Aug. 1983.

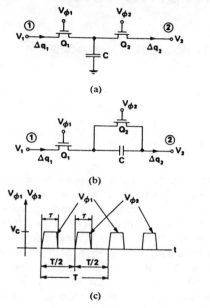

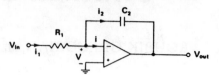

Fig. 2. Active-RC integrator.

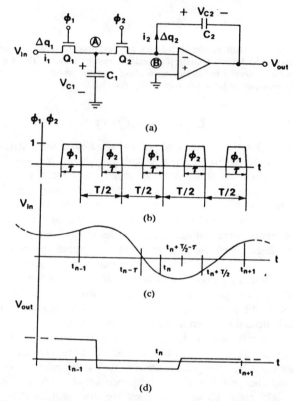

Fig. 1. Switched capacitor simulated resistors. (a) Shunt circuit. (b) Series circuit. (c) Clock waveforms.

outset of the clock pulses. However, we can define the *average current* i_{av} as the charge flow Δq in each clock period T, divided by T. Then

$$i_{av} = \frac{\Delta q}{T} = \frac{v_1 - v_2}{T/C}. \tag{1}$$

Thus both branches behave (on the average) as resistors of value

$$R = T/C \tag{2}$$

connected between nodes ① and ②. Physically, the switches thus transform the capacitor C, a nondissipative memoried element, into a dissipative memoryless (i.e., resistive) one.

It is now plausible that an active-RC filter can be modified by replacing all resistors by equivalent SC branches. A time constant, say $T_{12} = R_1C_2$, will then be transformed according to (2) into $T_{12}' = TC_2/C_1$. Thus it depends on the clock frequency $f_c = 1/T$, which can be accurately controlled using a crystal resonator in the clock oscillator, and on the *ratio* C_2/C_1 of two capacitances. Now, while the *value* of a MOS capacitor can only be controlled with an accuracy of 5-10 percent, the *ratio* of two capacitances can be made accurate to within a fraction of 1 percent. Hence, an accuracy of 0.1-0.5 percent is achievable for the time constant. This is because most error sources affect capacitors on the same chip (especially those located close to each other) the same way. This tracking behavior extends to variations with temperature and aging.

The area requirement for resistors is also drastically reduced by using the equivalent branches. For the value $R = 10^7$ Ω quoted above, assuming a 100-kHz clock frequency, by (2) a switched capacitor of value $C = 1$ pF is required. The area occupied is thus about 3 mil^2 (the area of the switches is negligible) rather than 1600 mil^2: a reduction by a factor of 500!

A. Switched-Capacitor Integrators

The basic building block of active-RC filters is the active-RC integrator (Fig. 2). Its input–output relation is

$$v_{out}(t) = -\frac{1}{R_1C_2} \int_{-\infty}^{t} v_{in}(\tau)\, d\tau \tag{3}$$

or, in the Laplace-transform domain,

Fig. 3. Active-SC integrator using shunt simulated resistor. (a) Circuit diagram. (b) Clock waveforms. (c) Input waveforms. (d) Output waveforms.

$$V_{out}(s) = \frac{-V_{in}(s)}{sR_1C_2}. \tag{4}$$

Replacing R_1 by the simulated resistor of Fig. 1(a), the SC integrator of Fig. 3(a) results. The waveforms of the clock signals, v_{in} and v_{out}, are shown in Figs. 3(b), (c), and (d), respectively.

The equivalence of the circuits of Figs. 2 and 3 is only approximate (even for ideal op-amps). To obtain the exact relations for the circuit of Fig. 3, we note that at $t = t_{n-1}$ the capacitor C_1 charges to $v_{in}(t_{n-1})$ and hence acquires $C_1 v_{in}(t_{n-1})$ charge. At $t = t_{n-1} + T/2$, this charge has been added to the charge already stored in C_2, and the new charge is held. Hence at $t = t_n$, the output voltage is

$$v_{out}(t_n) = -v_{C2}(t_n) = v_{out}(t_{n-1}) - \frac{C_1}{C_2} v_{in}(t_{n-1}). \tag{5}$$

Using z-transformation to solve this difference equation, the transfer function

$$H(z) \triangleq \frac{V_{out}(z)}{V_{in}(z)} = -\frac{C_1}{C_2} \frac{z^{-1}}{1 - z^{-1}} \tag{6}$$

results. It is instructive to compare the operation of the RC integrator of Fig. 2 with that of the SC integrator of Fig. 3 for sinewave signals. For $s = j\omega$, (4) becomes

$$H_{RC}(j\omega) \triangleq \frac{V_{out}(j\omega)}{V_{in}(j\omega)} = -\frac{1/(R_1C_2)}{j\omega} \tag{7}$$

while for $z = \exp(j\omega T)$, (6) leads to

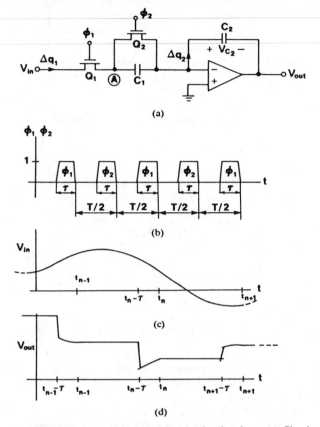

(a)

(b)

(c)

(d)

Fig. 4. Active-SC integrator using series simulated resistor. (a) Circuit diagram. (b) Clock waveforms. (c) Input waveforms. (d) Output waveforms.

$$H(e^{j\omega T}) = \frac{-C_1/C_2}{e^{j\omega T} - 1}. \tag{8}$$

Since $\exp(j\omega T) = 1 + j\omega T - (\omega T)^2/2 - + \cdots$ for $|\omega T| < 1$, we have

$$H(e^{j\omega T}) = \frac{-C_1/(C_2 T)}{j\omega - \omega^2 T/2 - + \cdots}. \tag{9}$$

Thus if $C_1 = T/R_1$ is chosen, $H(e^{j\omega T}) \approx H_{RC}(j\omega)$ for $\omega \ll 1/T = f_c$. The error is negligible if the signal frequency $f = \omega/2\pi < f_c/100$. For somewhat higher frequencies f, the error can be approximated by the second term in the denominator. The resulting transfer function

$$H(e^{j\omega T}) \approx \frac{-C_1/T}{j\omega C_2 - \omega^2 T C_2/2} \tag{10}$$

behaves as if C_2 had a finite negative Q-factor: $Q = \omega C_2/(-\omega^2 T C_2/2) = -2/(\omega T) = -f_c/(\pi f)$. For $|Q| \geqslant 100$, the restriction $f_c \geqslant 314f$ is thus necessary.

A direct comparison of (4) and (6) indicates also that the replacement of all *RC* integrators in an otherwise frequency-independent active-*RC* circuit by the SC circuit of Fig. 3 is equivalent to replacing s by $(z - 1)/T$. This is the well-known *forward-difference (or forward-Euler) mapping* [4], used in the design of digital filters from an analog model. It can easily be shown that it leads to Q-enhancement effects, in agreement with our earlier results.

A similar analysis can be performed for the SC integrator [5] obtained by using the resistor equivalent of Fig. 1(b). The resulting circuit is shown in Fig. 4; its transfer function turns out to be

$$H(z) = \frac{V_{\text{out}}(z)}{V_{\text{in}}(z)} = \frac{-C_1/C_2}{1 - z^{-1}}. \tag{11}$$

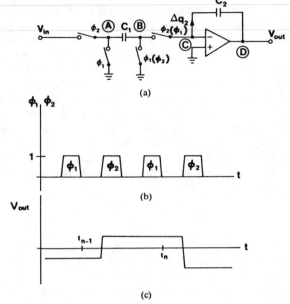

(a)

(b)

(c)

Fig. 5. Stray-insensitive SC integrator. (a) Circuit diagram. (b) Clock waveforms. (c) Output waveforms.

For $z = \exp(j\omega T)$ and $|\omega T| \ll 1$

$$H(e^{j\omega T}) \simeq \frac{-C_1/T}{j\omega C_2 + \omega^2 T C_2/2} \tag{12}$$

results. Hence, this circuit has a Q-factor equal to that of Fig. 3; however, Q is now positive. The mapping represented by the replacement of an *RC* integrator by this SC one is $s \to (1 - z^{-1})/T$. This is the *backward-difference (backward-Euler) mapping* [4], which is known to result in Q-degradation.

Another important difference between the SC integrators of Figs. 3 and 4 is that for the former, as (5) illustrates, $v_{\text{out}}(t_n)$ is affected only by $v_{\text{in}}(t_{n-1})$ and not by $v_{\text{in}}(t_n)$. Thus a delay of T exists between v_{in} and v_{out}. This is also evident from the z^{-1} factor in the numerator of $H(z)$ in (6). The corresponding equations for the circuit of Fig. 4 show that $v_{\text{out}}(t_n)$ depends on $v_{\text{in}}(t_n)$, and thus no delay is present.

Both SC integrators suffer from an important shortcoming: they are sensitive to the effects of stray capacitances between the various nodes (and lines) and ground. For the circuit of Fig. 3, e.g., node Ⓐ is connected to the source/drain diffusions of Q_1 and Q_2 which have appreciable capacitance to the substrate; also, the leads to Q_1, Q_2 and C_1 have parasitic capacitance to the substrate. The resulting stray capacitance C_A may be as large as 0.05 pF; its value is poorly controlled, and thus it makes the actual value of C_1 uncertain. If a 1-percent accuracy is needed for C_1, we must, therefore, choose $C_1 \geqslant 5$ pF. Normally, $C_2 \gg C_1$, and hence a large area is needed for the integrator. A similar argument concerning C_A in Fig. 4 results in similar conclusions.

The effects of stray capacitances can largely be eliminated by using the *stray-insensitive* integrator shown in Fig. 5, [6]–[8]. In this circuit, C_1 is again periodically charged by the input source to v_{in}, and it feeds the charge $C_1 v_{\text{in}}$ into C_2. Analysis of the stray capacitances C_A, C_B, C_C, and C_D loading nodes Ⓐ, Ⓑ, Ⓒ, and Ⓓ, respectively, reveals however that (for infinite op-amp gain) none of them contribute to the charge q_2 in C_2. The reason for this insensitivity is that every capacitor terminal is switched between low-impedance nodes (i.e., ground and an op-amp output) or is switched between ground and a virtual ground (which are both at the same potential). Hence, none of them affect $v_{\text{out}} = -q_2/C_2$.

If the clock phases shown *without* parentheses in Fig. 5 are used, then the operation is similar to that of the circuit of Fig. 4, with ϕ_1 and ϕ_2 interchanged. Thus C_1 discharges during $\phi_1 = 1$, and recharges through v_{in} and C_2 during $\phi_2 = 1$. Hence, (11) holds for its transfer function assuming that its output is sampled during $\phi_2 = "1."$ Note that the stray insensitivity is obtained at the cost of two extra switches. The equivalent Q is positive.

Consider next the clock phasing indicated *in* parentheses. Now, during $\phi_2 = "1,"$ capacitor C_1 charges to v_{in}, while during $\phi_1 = "1"$ it discharges into C_2. The output is again sampled during ϕ_2. Thus the operation is similar to that of Fig. 3. In particular, the equivalent Q is negative. However, the polarity of C_1 is inverted during the discharging, and hence the sign of $H(z)$ changes from that of (6); thus it is now

$$H(z) = \frac{C_1}{C_2} \frac{z^{-1}}{1 - z^{-1}}. \tag{13}$$

For a positive v_{in}, v_{out} thus increases. Hence, this circuit is a *noninverting* integrator, in contrast with all previous ones which were *inverting*. Both circuits described by (11) are *delay-free* integrators; the two circuits described by (6) and (13) are *delaying* ones.

At the cost of additional components and some effort in matching components in the layout, a stray-insensitive *inverting delaying integrator* can also be designed [9].

For a stray-insensitive integrator, the capacitance values C_1 and C_2 need not be much larger than the strays. They still should be much larger than the capacitance between the lines leading to their electrodes. This is typically 1–5 fF (1 fF = 10^{-15} F), and can be reduced by shielding. Thus minimum values as small as 0.1 pF can often be used. This represents a reduction of about 10–50 in size (and hence area), as compared to stray-sensitive integrators. In addition, higher accuracy can be attained for the capacitance ratios: errors as small as 0.1–0.5 percent are possible.

B. Cascaded SC Sections [9]

Using the stray-insensitive integrators of Fig. 5, it is possible to construct simple cascadable filter sections. Of special importance are sections with the biquadratic transfer function

$$H(z) = -\frac{a_2 z^2 + a_1 z + a_0}{b_2 z^2 + b_1 z + 1}. \tag{14}$$

These are often called *biquads*. Using the approximation

$$z = e^{sT} \simeq 1 + sT \tag{15}$$

(valid for $s = j\omega$, $|\omega| \ll 1/T$), $H(z)$ becomes a biquadratic transfer function $H_a(s)$ in s

$$H_a(s) = \frac{V_{out}(s)}{V_{in}(s)} = -\frac{K_2 s^2 + K_1 s + K_0}{s^2 + (\omega_0/Q)s + \omega_0^2}. \tag{16}$$

Here, ω_0 is the *pole frequency* and Q the *pole-Q*; if the pole is $s_p = \sigma_p + j\omega_p$ then

$$\omega_0 \triangleq |s_p| = \sqrt{\sigma_p^2 + \omega_p^2}$$

$$Q \triangleq \frac{|s_p|}{2|\sigma_p|} = \frac{1}{2}\sqrt{1 + (\omega_p/\sigma_p)^2}. \tag{17}$$

In constructing the SC biquads, we will make use of active-RC biquads developed to realize $H_a(s)$. The following steps will be used:

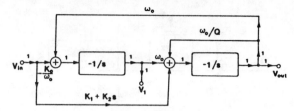

Fig. 6. Block diagram of a biquadratic system.

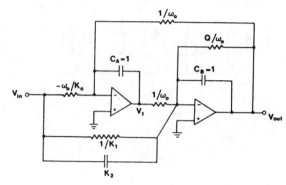

Fig. 7. Active-*RC* biquad.

1) The block diagram of a continuous-time system containing two cascaded integrators and coupling branches is constructed from $H_a(s)$.

2) An equivalent active-RC circuit is found.

3) Each resistor in the active-RC section is replaced by an equivalent SC branch containing a capacitor and four switches. By (2), the value of the switched capacitor C_i replacing a resistor R_i is given by $C_i \simeq T/R_i$.

4) If the approximations of (15) and (2) are not accurate enough, the exact transfer function of the SC circuit can be found and matched to the prescribed $H(z)$ given in (14) (which can be obtained using standard digital filter design techniques such as matched-z transform or the bilinear transform [4]). This gives a set of relations between the exact capacitance values C_i and the coefficients (a_j, b_k), from which the C_i values can be obtained.

To carry out the first step, (16) is rewritten in the form

$$s^2 V_{out} = -\left[K_2 s^2 + K_1 s + K_0\right] V_{in} - \left[\frac{\omega_0}{Q}s + \omega_0^2\right] V_{out}. \tag{18}$$

Dividing by s^2 and rearranging gives

$$V_{out} = -\frac{1}{s}\left\{[K_1 + K_2 s]V_{in} + \frac{\omega_0}{Q}V_{out} + \omega_0 V_1\right\} \tag{19}$$

where

$$V_1 \triangleq \frac{1}{s}\left[\frac{K_0}{\omega_0}V_{in} + \omega_0 V_{out}\right]. \tag{20}$$

The block diagram of the corresponding system is shown in Fig. 6. (Clearly, the rearrangement process, and the resulting block diagram are not unique!) This completes Step 1).

Using an op-amp with a unit-valued feedback capacitor to realize each integrator with a voltage/current transfer function $(-1/s)$, and RC admittances to realize the coupling branches, the active-RC filter section of Fig. 7 results. Note that two negative resistors are needed in the first integrator. This concludes Step 2).

Next, each resistor is replaced by an equivalent circuit similar

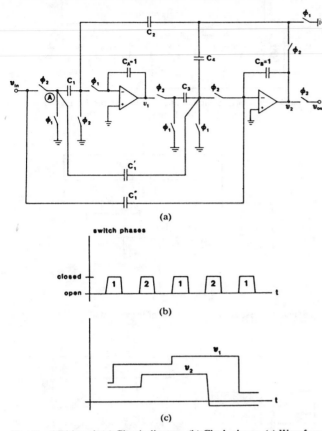

(a)

(b)

(c)

Fig. 8.   SC biquad. (a) Circuit diagram. (b) Clock phases. (c) Waveforms.

to the input branch of the integrator of Fig. 5. A positive resistor requires the clock phases shown *without* parentheses; a negative one those *in* parentheses. The resulting biquad[1] [10] is shown in Fig. 8. The element values can be found from the approximation $C_1 \simeq T/|R_i|$. Thus we get

$$C_1 \simeq TK_0/\omega_0 = \left(K_0/\omega_0^2\right)\omega_0 T = |A_{dc}|\omega_0 T$$

$$C_2 \simeq C_3 \simeq \omega_0 T$$

$$C_4 \simeq \omega_0 T/Q$$

$$C_1' \simeq K_1 T$$

$$C_1'' = K_2. \qquad (21)$$

Here, $A_{dc} = -K_0/\omega_0^2$ is the dc gain; usually, $|A_{dc}| \geqslant 1$. This completes Step 3).

If the accuracy represented by the above design process—which is based on the approximations (15) and (2)—is inadequate, the exact expression for the transfer function of the circuit must be obtained. This is efficiently done by way of the z-domain block diagram of the SC section shown in Fig. 9. In the diagram, each op-amp and its feedback capacitor (C_A or C_B) is replaced by its voltage-to-charge transfer function

$$\frac{V(z)}{\Delta Q(z)} = \frac{-1/C_F}{1 - z^{-1}}. \qquad (22)$$

Here C_F is the feedback capacitor, $V(z)$ the z-transform of the op-amp output voltage $v(t_n)$, and $\Delta Q(z)$ the z-transform of the input charge sequence $\Delta q(t_n)$. Similarly, each coupling branch is replaced by its charge/voltage transfer function $\Delta Q/V$. This

[1] In this circuit, and in subsequent ones, switches performing the same operations are combined to reduce the numbers of switches and lines.

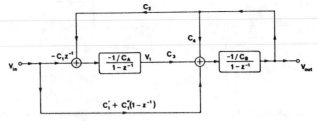

Fig. 9.   Z-domain block diagram representing the exact operation of the SC biquad of Fig. 8.

function equals

$C(1 - z^{-1})$ for an unswitched capacitor (e.g., C_1'')

C for a noninverting capacitor (C_1', C_3, C_4)

$-Cz^{-1}$ for an inverting capacitor (C_1, C_2) with the phasing indicated in Fig. 8(c).[2]

From the block diagram, the exact transfer function

$$\frac{V_{out}(z)}{V_{in}(z)} = -\frac{\left(C_1' + C_1''\right)z^2 + \left(C_1 C_3 - C_1' - 2C_1''\right)z + C_1''}{(1 + C_4)z^2 + (C_2 C_3 - C_4 - 2)z + 1}$$

$$(23)$$

can readily be obtained. Comparison with (14) then yields the relations between the capacitance values and the prescribed coefficients of $H(z)$

$$C_1'' = a_0$$

$$C_1' = a_2 - C_1'' = a_2 - a_0$$

$$C_1 = \frac{1}{C_3}\left(a_1 + C_1' + 2C_1''\right) = \frac{1}{C_3}(a_0 + a_1 + a_2)$$

$$C_4 = b_2 - 1$$

$$C_2 C_3 = b_1 + C_4 + 2 = b_1 + b_2 + 1. \qquad (24)$$

Note that that the circuit is realizable only if all capacitors are nonnegative; this requires $a_2 \geqslant a_0$ and $b_2 \geqslant 1$. Thus both zeros and poles must be inside (or on) the unit circle. For the poles, this is necessary anyway to insure the stability of the section. For the zeros, this is not always the case; e.g., for all-pass delay sections, the zeros must be outside the unit circle, and hence the circuit of Fig. 8 is not suitable.

The five relations in (24) do not fully determine the six capacitance values; hence, one more constraint may be added. We may note from (21) that for the usual case of $\omega_0 T \ll 1$ the capacitance spread is at least $C_A/C_2 \simeq 1/(\omega_0 T)$. If $Q > 1$, the spread is even larger. Thus we will use this section for $Q \leqslant 1$ only, and hence anticipate that C_2 (and/or C_3) will be the smallest capacitor(s) in the stage. Since (24) determines only the $C_2 C_3$ product, the smallest capacitance will be maximized, and thus the capacitance spread minimized, if $C_2 = C_3$ is chosen. This yields

$$C_2 = C_3 = \sqrt{1 + b_1 + b_2}$$

$$C_1 = \frac{a_0 + a_1 + a_2}{\sqrt{1 + b_1 + b_2}}. \qquad (25)$$

Note that all capacitance values are normalized, with $C_A = C_B = 1$ being the unit capacitance.

[2] In fact, C_3 and its switches provide only a half-period delay ($z^{-1/2}$). However, due to the sampled-and-held character of its input voltage v_1, a full period delay results.

191

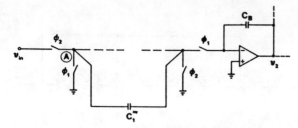

Fig. 10. Added SC branch needed to realize transfer functions with zeros outside the unit circle.

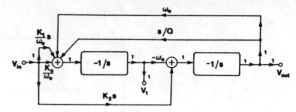

Fig. 11. Block diagram of a biquadratic system suitable for high-Q transfer functions.

The discussion up to now has ignored the nature of the filter characteristics. For a *low-pass* response, $K_1 = K_2 = 0$ can be set in (16); by (21), this corresponds to $C_1' = C_1'' = 0$ in Fig. 8. As (23) indicates, the resulting circuit realizes a transfer function $H(z)$ with $a_0 = a_2 = 0$.

For a *high-pass* characteristics, $K_0 = K_1 = 0$. Now, $C_1 = C_1' = 0$ and, by (23), $a_0 = a_2 = -a_1/2$. Hence, both zeros are at $z = 1$, corresponding to $s = 0$.

For a *notch* (*bandstop*) response with a pair of conjugate zeros on the $j\omega$-axis, $K_1 = 0$ can be set in (16). Then $C_1' = 0$ as (21) shows, and in (23), $a_2 = a_0$. Hence, a pair of conjugate complex zeros occurs on the unit circle of the z plane.

For a *bandpass* characteristics, let $K_0 = K_2 = 0$ in (16). This results in $C_1 = C_1'' = 0$ in (21), and hence $a_0 = 0$, $a_1 = -a_2$ in $H(z)$. Thus the zeros are at $z = 0$ and $z = 1$.

As mentioned earlier, the use of the circuit of Fig. 8 is restricted to transfer functions with zeros inside or on the unit circle. This restriction can be eliminated by adding an inverting SC branch (Fig. 10) between node A and the inverting input of the second op-amp in Fig. 8. As a result, the linear and constant coefficients in the numerator of $H(z)$ become

$$a_1 = C_1 C_3 - C_1' - 2C_1'' - C_1'''$$

$$a_0 = C_1'' + C_1'''. \tag{26}$$

For $a_0 > a_2$, we need $C_1''' > C_1'$. Thus we can use $C_1' = 0$ for $a_0 > a_2$, and $C_1''' = 0$ otherwise.

As mentioned earlier, the circuit of Fig. 8 is best suited for low-Q ($Q < 1$) applications. Then, the capacitance spread is close to $C_A/C_2 \approx 1/(\omega_0 T)$. To obtain a circuit better suited for the realization of high-Q ($Q > 1$) transfer function, (16) is reorganized in the form

$$V_{\text{out}} = -\frac{1}{s}[K_2 s V_{\text{in}} - \omega_0 V_1] \tag{27}$$

where

$$V_1 = -\frac{1}{s}\left[\left(\frac{K_0}{\omega_0} + \frac{K_1}{\omega_0}s\right)V_{\text{in}} + \left(\omega_0 + \frac{s}{Q}\right)V_{\text{out}}\right]. \tag{28}$$

The block diagram of the corresponding system is shown in Fig. 11; an active-RC realization is given in Fig. 12.

An SC circuit is obtained, as before, if all resistors are replaced by SC branches. The resulting section [11] is shown in Fig. 13. The element values can be found from the approximation $C_i \approx T/|R_i|$. This gives

$$C_1 \approx K_0 T/\omega_0 = \left(\frac{K_0}{\omega_0^2}\right)\omega_0 T = |A_{\text{dc}}|\omega_0 T$$

$$C_2 \approx C_3 \approx \omega_0 T$$

$$C_4 \approx 1/Q$$

$$C_1' \approx K_1/\omega_0$$

$$C_1'' \approx K_2. \tag{29}$$

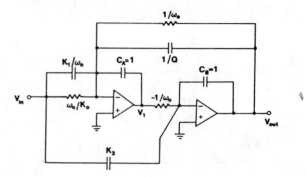

Fig. 12. Active-RC realization of the system of Fig. 11.

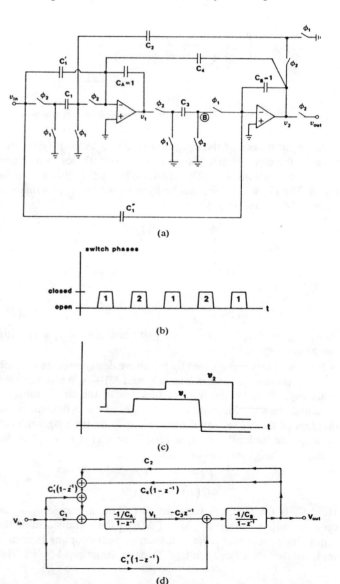

(a)

switch phases

(b)

(c)

(d)

Fig. 13. SC biquad. (a) Circuit diagram. (b) Clock waveforms. (c) Signal waveforms. (d) Block diagram corresponding to the exact Z-domain operation of the SC biquad.

192

Here A_{dc} is the dc gain; usually $|A_{dc}| \geq 1$. For $Q \geq 1$, the capacitance spread is then again $C_A/C_2 \approx 1/(\omega_0 T)$. Note that for the circuit of Fig. 8, the spread is $C_A/C_4 = Q/(\omega_0 T)$ if $Q > 1$.

The exact transfer function can be found from the block diagram of Fig. 13(d); the result is

$$H(z) = \frac{V_{out}}{V_{in}}$$

$$= -\frac{C_1''z^2 + (C_1 C_3 + C_1' C_3 - 2C_1'')z + (C_1'' - C_1' C_3)}{z^2 + (C_2 C_3 + C_3 C_4 - 2)z + (1 - C_3 C_4)}.$$

$$(30)$$

As indicated in (29), it is again a good strategy to choose $C_2 = C_3$. Then, equating the coefficients of the $H(z)$ functions given in (30) and in (14) yields

$$C_1'' = a_2/b_2$$

$$C_1' = (C_1'' - a_0/b_2)/C_3 = (a_2 - a_0)/(b_2 C_3)$$

$$C_1 = (a_1/b_2 - C_1' C_3 + 2C_1'')/C_3$$

$$\quad = (a_0 + a_1 + a_2)/(b_2 C_3)$$

$$C_4 = (1 - 1/b_2)/C_3$$

$$C_3^2 = (b_1/b_2 - C_3 C_4 + 2) = (b_1 + b_2 + 1)/b_2. \quad (31)$$

From the last equation, C_3 can be found. Then all other capacitances are defined.

For a *low-pass* response, $K_1 = K_2 = 0$ in (16); hence, by (29), $C_1' = C_1'' = 0$. This gives $a_0 = a_2 = 0$, as before. For a *high-pass* characteristics, $K_0 = K_1 = 0$. Then $C_1 = C_1' = 0$ and hence, by (30), $a_0 = a_2 = -a_1/2$. For a *notch* response, $K_1 = 0$ which leads to $C_1' = 0$. Then $a_2 = a_0$, and a pair of conjugate complex zeros occurs on the unit circle. For a *bandpass* response, $K_0 = K_2 = 0$. This gives $C_1 = C_1'' = 0$ in (29), and $a_2 = 0$, $a_1 = -a_0$ in (14). Hence, the zeros are at $z \to \infty$ and $z = 1$.

As (30) shows, the circuit of Fig. 13 is realizable only if $a_2 \geq a_0$ so that $C_1' C_3 = a_2 - a_0 \geq 0$. Hence, all zeros are restricted to be inside (or on) the unit circle.[3] If this constraint is not satisfied, the coupling branch shown in Fig. 10 must be included between the input node and node Ⓑ. This modifies the linear and constant coefficients in the numerator to

$$a_1 = C_1 C_2 + C_1' C_3 - 2C_1'' - C_1'''$$

$$a_0 = C_1'' - C_1' C_3 + C_1'''. \quad (32)$$

For $a_0 > a_2$, we must have $C_1''' > C_1' C_3$. Thus we can use $C_1' = 0$ for $a_0 > a_2$, and $C_1''' = 0$ otherwise.

For a *first-order section* with a transfer function

$$H(z) = -\frac{a_1 z + a_0}{b_1 z + 1}. \quad (33)$$

Using (15) results in the continuous-time approximation

$$H_a(s) = -\frac{K_1 s + K_0}{s + \omega_0}. \quad (34)$$

The block diagram of a system realizing $H_a(s)$ is shown in Fig. 14(a), an active-*RC* circuit realization in Fig. 14(b), and an SC equivalent in Fig. 14(c). If the clock phases without parentheses are used, the element values can be estimated from

$$C_1 \simeq K_1$$

$$C_1' \simeq TK_0$$

$$C_2 \simeq \omega_0 T. \quad (35)$$

[3]As before, the poles are similarly restricted.

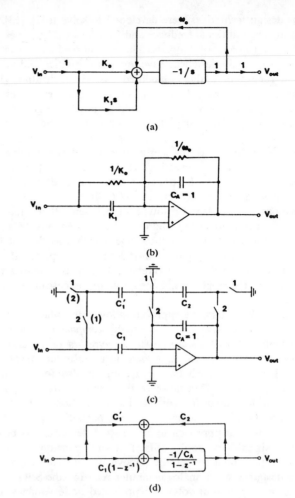

Fig. 14. First-order section. (a) Block diagram. (b) Active-*RC* realization. (c) Switched-capacitor realization. (d) Block diagram illustrating the exact Z-domain operation.

The exact transfer function can be found from the block diagram of Fig. 14(d). It is

$$H(z) = \frac{V_{out}}{V_{in}} = -\frac{(C_1 + C_1')z - C_1}{(1 + C_2)z - 1}. \quad (36)$$

Both the pole and the zero have nonnegative real values, and are *inside* (or on) the unit circle. If a zero *outside* the unit circle is required, the parenthesized clock phases should be used. The transfer function is then

$$H(z) = \frac{V_{out}}{V_{in}} = -\frac{C_1 z - (C_1 + C_1')}{(1 + C_2)z - 1}. \quad (37)$$

A *negative* zero (and/or pole) can be obtained, at the cost of losing the stray-insensitivity property, if the simulated resistor of Fig. 1(a) is used in the input (and/or feedback) branch.

III. LADDER FILTERS

As mentioned earlier, the conceptually simplest design technique for SC filters is to replace all resistors in an active-*RC* prototype circuit by switched capacitors. This approach was followed in the preceding section for integrators and first- and second-order sections. By cascading such sections, higher order filters can also be realized. However, for complicated filters (say, seventh or higher order), some of the required pole-Q's are usually very high, and the element-value sensitivities of the corresponding section become too high for reliable fabrication. Thus the yield tends to be low, and the circuit becomes uneconomical to fabricate.

This problem occurs, of course, for active-*RC* filters, and

193

various design techniques were developed to solve it [1], [13]. In one approach, additional feedback and feed-forward branches are added to a cascade of biquads. These multifeedback circuits, when appropriately designed, have lower sensitivities and wider dynamic ranges than the unmodified cascade realization [13], [14]. Typically, however, these circuits are complicated to design, and the improvement of the sensitivities not as large as that obtainable using some other design techniques to be discussed next.

The most widely used low-sensitivity circuit design strategy is based on Orchard's observation [18] that doubly terminated reactance two-ports can be designed with near-zero sensitivities in their passbands. Hence, an active-RC filter which is closely modeled on such a passive prototype—in the sense that it has the same transfer function $H(s)$, and that its $H(s)$ depends the same (or nearly the same) way on the element values as that of the passive circuit—will also have this desirable property. If the passive filter has a ladder configuration, then the sensitivities will be low also in the stopbands, comparable to the cascade realization [1].

For active-RC filters, the approach followed to obtain the final circuit from its passive model is usually *component simulation*. Thus simple active-RC stages, such as gyrators or generalized immittance converters, are used to replace the inductors of the model [1]. This approach has been adopted also for SC filter design (see, e.g. [15]–[17]); however, the resulting structures have not been widely used for commercial applications. The main reason is that they are usually sensitive to parasitic capacitances.

Another popular approach to active-RC filter design is based on the signal-flowgraph (SFG) representation of the current–voltage relations of the passive model. The active-RC filter is designed as an analog computer realizing the SFG, such that the op-amp output voltages represented node voltages and branch currents. This design technique[4] is not used extensively for high-order active-RC filters, since the number of op-amps required is high: at least equal to n, the order of the filter. Also, many resistors are needed in the circuit [1], [29]–[31].

For SCF's, the op-amps are integrated on the chip and (with proper design) require a relatively small area: 0.05 to 0.15 mm² per op-amp is typical. Thus as many as 40 op-amps can be placed on a single chip. Resistors, as realized by switched capacitors, are also easily integrated. Hence, the SFG design strategy *is* suitable for integrated SCF's. Since it can result in stray-insensitive circuits with very low element-value sensitivities, it has been a favorite industrial design strategy for high-order, high-accuracy SCF's (see, e.g., [7], [10], [2], [28]).

Historically, SFG-type SCF's (commonly, but inaccurately, called "SC ladder filters") have been designed using approximate techniques, valid only for very large clock frequency/signal frequency ratios. More recently, some exact design methods have been developed. Both approaches are briefly discussed in the next two subsections.

A. The Approximate Design of Ladder Filters

Rather than discussing the various design approaches in general and abstract terms, we shall use as an illustration the simple example of a third-order low-pass filter with a pair of finite transmission zeros. The corresponding doubly terminated LC two-port is shown in Fig. 15(a). Since in the SFG representation the parameters will be the state variables V_1, V_3, and I_2, the circuit is first transformed into the form shown in Fig. 15(b). From Fig. 15(b), the state variables can be expressed as follows:

[4]Called SFG design or (due to the form of the resulting configuration) leap-frog design.

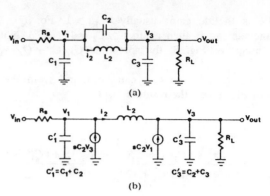

Fig. 15.   Doubly terminated LC two-port. (a) Circuit diagram. (b) Equivalent circuit.

$$-V_1 = \frac{-1}{s(C_1 + C_2)}\left[\frac{-V_1 + V_{in}}{R_S} + sC_2 \cdot V_3 - I_2\right]$$

$$-I_2 = \frac{+1}{sL_2}(V_3 - V_1)$$

$$V_3 = \frac{-1}{s(C_2 + C_3)}\left[-I_2 - sC_2 \cdot V_1 + \frac{V_3}{R_L}\right]. \qquad (38)$$

The block diagram of a system which realizes these relations and which contains only integrators and simple coupling branches is shown in Fig. 16(a). An active-RC realization is illustrated in Fig. 16(b). Note that in the second stage, all coupling branch impedances have been multiplied by the arbitrary scale factor R, and the feedback impedance by R^2. This leaves all currents to and from the stage (and hence all outside voltages) unchanged, but assures correct dimensions within the stage. The final SCF is obtained by replacing all resistors by SC branches: positive resistors by noninverting and negative by inverting. The resulting circuit is shown in Fig. 16(c). Using the approximation $C_i = T/R_i$ to determine the value C_i of the switched capacitor which replaces the resistor R_i, the element values

$$C_S = T/R_S$$
$$C_A = C_1 + C_2$$
$$C = T/R$$
$$C_B = L_2/R^2 = C^2 L_2/T^2$$
$$C_C = C_2 + C_3$$
$$C_L = T/R_L \qquad (39)$$

result.

As explained earlier in connection with (7)–(12), the approximation corresponding to the replacement of the continuous-time integrators of Fig. 16(a) and (b) by the SC ones of Fig. 16(c) is equivalent (to a first-order approximation) to introducing a finite Q for the feedback capacitors C_A, C_B, and C_C. Since C_A and C_C belong to integrators similar to those shown in Fig. 5 with the original clock phases, their Q's are positive: $Q = 2/(\omega T)$. By contrast, that of C_B is $-2/(\omega T)$ since the input branch is now a delaying and inverting one so that its transfer function is given by (6) (but with a positive sign) [10].

Since the feedback branches in Fig. 16 correspond to the reactive elements C_1', L_2, and C_3' of the prototype shown in Fig. 15(b), the same Q's can be assigned to these reactances in predicting the resulting distortion. To a first-order approximation, the loss distortion (in nepers) can be obtained from the classical formula (see, e.g., [20, eqs. (10)–(22)])

$$\Delta\alpha = \left(\frac{1}{Q_L} + \frac{1}{Q_C}\right)\frac{\omega\tau(\omega)}{2}. \qquad (40)$$

Here Q_L and Q_C are the Q factors of the L's and C's, respec-

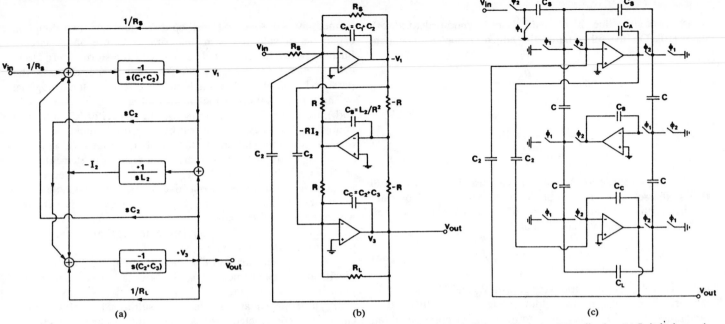

Fig. 16. Switched-capacitor filter equivalent to the doubly terminated LC two-port of Fig. 15. (a) Block diagram. (b) Active-RC realization. (c) Switched-capacitor realization.

tively, and $\tau(\omega)$ is the group delay. Since in our case $Q_L = -Q_C$, $\Delta\alpha \approx 0$. Thus by arranging the alternation of the integrator types as in Fig. 16(c), the loss distortion can be (to a first-order approximation) canceled. The approximation becomes poor, however, if the condition $f_c \gg f$ does not hold, or if $\tau(\omega)$ is very large which is the case for highly selective filters. Other relevant formulas are contained in [23]–[25].

Another way of looking at the approximately lossless behavior of the circuit of Fig. 16(c) is to consider the transfer function of two cascaded integrators, say those in the first two stages. Apart from a constant factor, the transfer function for this partial circuit is, by (6) and (11), $-z^{-1}/(1 - z^{-1})^2$. This is the same as for a cascade of two (hypothetical) *infinite-Q* integrators with transfer functions $\pm z^{-1/2}/(1 - z^{-1})$. Thus if all integrators could be so paired, the response would be ideal. In fact, of course, the pairing breaks down at the two terminations [26], [28]. The s-to-z mapping $s \rightarrow (z^{1/2} - z^{-1/2})/T$ is called *lossless digital integrator* (LDI) transformation [26], [27]. If carried out exactly (which is fortunately not possible), it leads to an unstable sampled-data filter. This is because any value of s maps into two values of z, one inside the unit circle and one outside.[5]

If necessary, the actual loss distortion $\Delta\alpha$ can be analyzed using one of the available SCF analysis computer programs [33]–[35]. However, a simple and general result can also be obtained, which also provides some insight into the nature of the distortion. If in the circuit of Fig. 16(c) each integrator and coupling branch is replaced by a block or branch with its actual z-domain transfer function, then the block diagram of Fig. 17 results. In the diagram, the notation

$$\hat{s} \triangleq \frac{z^{1/2} - z^{-1/2}}{T} \qquad (41)$$

is used. In comparing Figs. 17 and 16(a), we note that $\hat{s}$ replaced s in the integrator blocks. For $z = \exp(j\omega T)$ the new variable

$$\hat{s} = j\hat{\omega} = j\frac{\sin(\omega T/2)}{T/2} \qquad (42)$$

is purely imaginary; hence, the replacement of s by $\hat{s}$ is merely a

[5]See Section V-E of the companion paper by Y. Tsividis (this issue, pp. 926–940) for an interpretation of fractional powers of z.

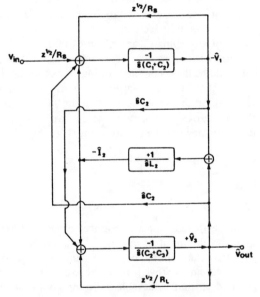

Fig. 17. Equivalent circuit corresponding to the switched-capacitor circuit of Fig. 16.

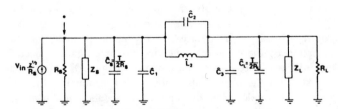

Fig. 18. Modified prototype circuit showing the effect of the imperfect modeling.

nonlinear rescaling of the frequency axis. Its effect can be minimized by prewarping the frequency limits of the prototype filter. More important is the change of the terminations from R_S and R_L to $R_S z^{-1/2}$ and $R_L z^{-1/2}$, respectively. This can be represented by the modified prototype circuit shown in Fig. 18. In the figure, the carets over the element symbols ($\hat{C}_1$, $\hat{L}_2$, etc.) indicate that the corresponding impedances depend on $\hat{s}$, rather than s. The effect of the added capacitors $\hat{C}_S$ and $\hat{C}_L$ on the loss is small;

in any case, the values of $\hat{C}_1$ and $\hat{C}_3$ can be modified to absorb them [10]. On the other hand, the added frequency-dependent resistors

$$Z_S = \frac{R_S}{\cos\left(\dfrac{\omega T}{2}\right) - 1}$$

$$Z_L = \frac{R_L}{\cos\left(\dfrac{\omega T}{2}\right) - 1} \qquad (43)$$

modify the response appreciably. It can readily be shown [36] that if the reflection factors at the input and output ports are much less than 1 both before and after Z_S and Z_L have been added, then the approximate formula

$$\frac{\hat{V}_{\text{out}}(\omega)}{V_{\text{out}}(\omega)} \cong \cos^{-4}(\omega T/4) \qquad (44)$$

holds. Here $\hat{V}_{\text{out}}$ is the output voltage with Z_S and Z_L included in the circuit, and V_{out} is the output voltage without them. It is remarkable that the distortion is, to a first-order approximation, independent of the order, structure, or any other characteristics of the SCF. This formula, therefore, enables the designer to get an *a priori* estimate of the distortion caused by the LDI design technique. Computer simulations indicate that this simple approximation is reasonably accurate [36].

The process of absorbing $\hat{C}_S$ and $\hat{C}_L$ in $\hat{C}_1$ and $\hat{C}_3$, respectively, makes the terminations in Fig. 18 purely real.[6] An alternative approach is to make them complex-conjugate quantities. This was suggested in [38], where a changed configuration was used to achieve it. It can be shown, however, that the same result can be obtained by changing only the values of the two switched capacitors in the output stage of the circuit of Fig. 16(c) as follows:

$$C_L \rightarrow \frac{C_L}{1 - C_L/C_C}$$

$$C \rightarrow \frac{C}{1 - C_L/C_C} \qquad (45)$$

where C is the switched capacitor connected between the output of op-amp 2 and the input of op-amp 3. Numerical experiments indicate [39] that the elimination of the imaginary parts of the terminations gives an improvement in the loss distortion which is at least as good as that achievable by using conjugate terminations.

Several comments should be made about the configuration shown in Fig. 16(c). Clearly, the signs of V_1, I_2, and V_3 used in (38) and hence the resulting phasing of the switches in Fig. 16(c) is quite arbitrary. The choice actually made is such that the alternation of the signs of Q factors, discussed earlier, occurs. This reduces the loss distortion inherent in the LDI approximation. Also, for the phasing used, the output terminal of the first op-amp is decoupled from the input of the second one during the interval $\phi_2 = $ "1," i.e., during its recharging transient. Similarly, the second op-amp is decoupled during its transient (when $\phi_1 = $ "1") from both other op-amps, and so is the third op-amp during $\phi_2 = $ "1." This is important for high clock rates, when the op-amp settling time represents a lower limit of the clock phase duration [21], [22]. If two (or more) op-amps are coupled during their transients, then their settling times are extended. This same consideration shows that the unswitched continuous-signal loop

---

[6]An alternative technique for this is described in [37]. The resulting circuit is, however, sensitive to stray capacitances.

formed by the first and last op-amps and the two coupling capacitors of value C_2 is an undesirable feature of the circuit. Such loops can be eliminated using circuit transformations. However, if the finite transmission zero is not too close to the passband edge, then C_2 is small and its effect is often negligible.

The transformation of the circuit of Fig. 15(a) to that of Fig. 15(b), resulting in the splitting of the capacitor C_2 into two equivalent branches, clearly destroys the one-to-one correspondence between the elements of the doubly terminated LC two-port and those of its SCF counterpart. As a result, the element-value sensitivities are somewhat higher for the latter. However, except in the case of exceptionally selective filters, the obtainable sensitivities are below 0.1 dB per 1 percent element-value change, which is usually acceptable.

In conclusion, the SFG design using the LDI approximation is performed in the following steps:

1) A doubly terminated LC two-port is designed from the SCF specifications, prewarped using the $\hat{\omega} \leftrightarrow \omega$ transformation implied in (42), if necessary.

2) The state equations and the resulting SFG of the LCR circuit are constructed. The signs of the voltage and current variables should be such that inverting and noninverting integrators alternate in the SFG.

3) The integrators and coupling branches of the SFG are replaced by their SCF equivalents. The integrators now should have Q's of alternating signs. Also, for high-frequency filters, each integrator should be decoupled during its transient, and unswitched loops should be eliminated if necessary.

4) The element values of the terminating sections should be modified to eliminate the imaginary parts of the simulated terminations.

5) The remaining loss distortion can be estimated using (44).

6) If necessary, for critical applications, the performance can be optimized using iterative computations [37]. For such optimization, it is faster and simpler to use the equivalent passive ladder (Fig. 18) rather than the SCF. This is because passive ladders are quickly and accurately analyzed using an algorithm which assumes their output voltage and works backwards toward their input terminals [40].

The design of high-pass and bandpass filters can be performed in similar steps.

The optimization step described above can be avoided if one of the exact design processes, described in the next subsection, is applied.

B. The Exact Design of Ladder Filters

The errors inherent in the LDI design techniques can be avoided by using a ladder synthesis based on the bilinear s-to-z transformation

$$s \leftrightarrow \frac{2}{T}\frac{z-1}{z+1}. \qquad (46)$$

This mapping has the same desirable property as the LDI transformation of (41): for $z = \exp(j\omega T)$, s is purely imaginary. In addition, unlike the LDI transformation, it maps a stable transfer function $H(s)$ into a stable function $\hat{H}(z)$ which is exactly achievable with an SCF [39], [41]–[43].

As before, the third-order RLC filter of Fig. 15(a) will be used to illustrate the design process. Referring to Fig. 19, the circuit is first transformed into the form shown in Fig. 19(b). Thus instead of leaving only L_2 in the series branch, the transformation introduces the negative capacitance

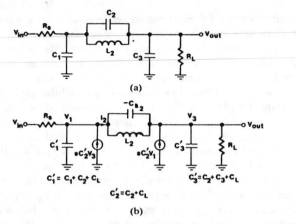

Fig. 19. Doubly terminated *LC* ladder filter. (a) Circuit diagram. (b) Equivalent circuit.

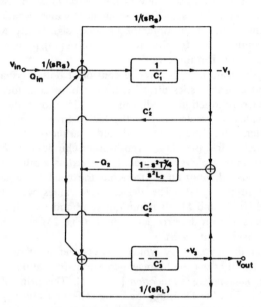

Fig. 20. Block diagram modeling the circuit of Fig. 19(b).

$$-C_{L_2} \triangleq -\frac{T^2}{4L_2} \qquad (47)$$

in parallel with L_2. As will be shown, this makes it possible to find a simple SCF realization for the the branch. Next, the state equations are written for the circuit, choosing as state variables now $V_1(s)$, $Q_2(s) \triangleq I_2(s)/s$, and $V_3(s)$

$$-V_1(s) = \frac{-1}{C_1'}\left[\frac{-V_1 + V_{in}}{sR_S} + C_2'V_3 - Q_2\right]$$

$$-Q_2(s) = \left(-C_{L_2} + \frac{1}{s^2L_2}\right)(V_3 - V_1)$$

$$= \frac{1 - s^2T^2/4}{s^2L_2}(V_3 - V_1)$$

$$V_3(s) = \frac{-1}{C_3'}\left[-Q_2 - C_2'V_1 + \frac{V_3}{sR_L}\right]. \qquad (48)$$

The corresponding block diagram is shown in Fig. 20. In the diagram, the first and third blocks have charge-to-voltage transfer functions of the form $V_{out}/Q_{in} = -1/C$. This relation can (as before) be realized by an op-amp with a feedback capacitor C. Similarly, the coupling branches with voltage-to-charge transmittances C_2' can simply be capacitors of value C_2'. The realization of

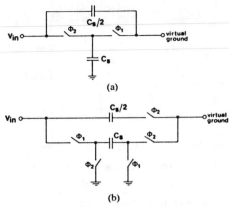

Fig. 21. Input branch realization. (a) Stray-sensitive circuit. (b) Stray-insensitive realization.

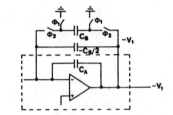

Fig. 22. Realization of the feedback branch.

the other building blocks is performed using the mapping of (46). Thus the input branch should satisfy the relation

$$Q_{in} = \frac{V_{in}(z)}{\left(\frac{2}{T}\frac{z-1}{z+1}\right)R_S}. \qquad (49)$$

Hence, the z-transform of the incremental charge sequence (i.e., of the charge flow per clock period) is given by

$$\Delta Q_{in}(z) = Q_{in}(z) - z^{-1}Q_{in}(z)$$

$$= \frac{T}{2R_S}(1 + z^{-1})V_{in}(z). \qquad (50)$$

Two realizations [41]–[43] for this branch are shown in Fig. 21, where $C_S \triangleq T/R_S$. Both circuits require a sampled-and-held input voltage that changes only when ϕ_1 is "1." The circuit of Fig. 21(a) is not stray-insensitive; however, the parasitic capacitance in parallel with C_S only introduces a constant gain change into the response. For the feedback branch with voltage-to-charge transmittance $1/(sR_S)$, the circuit of Fig. 22 can be used. The negative capacitance $-C_S/2$ can be absorbed in the feedback capacitance C_A which is usually much larger than $C_S/2$. The same circuit can be used to realize the $1/(sR_L)$ feedback branch (Fig. 20).

For the central building block, the prescribed z-domain transfer function is

$$\frac{\Delta Q(z)}{V_2 - V_1} = (1 - z^{-1})\frac{Q(z)}{V_3(z) - V_1(z)}$$

$$= (1 - z^{-1})\frac{1}{L_2}\left(s^{-2} - \frac{T^2}{4}\right)\bigg|_{s=(2/T)(z-1)/(z+1)}$$

$$= \frac{T^2}{L_2}\frac{z^{-1}}{1 - z^{-1}}. \qquad (51)$$

This transfer function can be realized by the circuit of Fig. 23, where the element values must satisfy $C_1C_3/C_2 = T^2/L_2$. It is also possible to obtain the same transfer function with different switching phases [42].

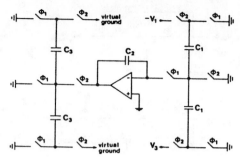

Fig. 23. Internal stage of the SC ladder filter.

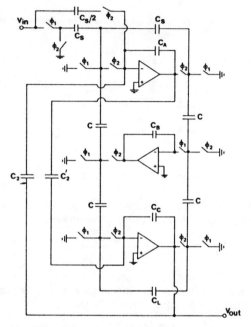

Fig. 24. The complete circuit of the SC bilinear ladder circuit.

Combining all branches, the overall "bilinear ladder" circuit of Fig. 24 is obtained. In the figure, C is arbitrary and the other element values are

$$C_S = \frac{T}{R_S}$$

$$C_{L_2} = \frac{T^2}{4L_2}$$

$$C_2' = C_2 + C_{L_2}$$

$$C_A = C_1 + C_2 + C_{L_2} - C_S/2$$

$$C_B = \frac{C^2}{4C_{L_2}} = \frac{C^2 L_2}{T^2}$$

$$C_C = C_2 + C_3 + C_{L_2} - C_L/2$$

$$C_L = \frac{T}{R_L}. \tag{52}$$

A comparison of the circuit of Fig. 24 with that of the "LDI ladder" reveals that they have nearly identical configurations: only the input branches are slightly different. Similarly, the element values are (for usual specifications) quite close. However, as a result of these seemingly minor changes, the loss distortion due to the LDI approximation is fully eliminated.

As the LDI ladder, the bilinear ladder can also be used to realize bandpass and high-pass filters. It is especially easily

applied when the LC filter model contains at most a parallel L and C in each branch. In such a circuit, inductive loops may exist, which, however (unlike for LDI filters), do not result in saturation of the op-amps, since now the currents modeled are the inductive cutset currents which become zero at dc.

A unique problem arises in the design of bilinear high-pass (and bandstop) ladders. For an input sampled-and-held signal of discrete frequency $f = f_c/2$, the input branch does not transmit any charge to the first op-amp, as can be verified from (50) for $z = -1$, or by inspection from Fig. 21. Hence, for a high-pass response which requires $V_{out}(z)$ to be nonzero at $z = -1$, the transfer impedance of the rest of the circuit must be infinite. This causes instability if the circuit is designed using the above-described process. The problem can be overcome by using an impedance scaling; the reader is referred to [42], [43] for the details of this process.

Another problem arises for very-narrow-band bandpass filters. For such circuits, the sensitivity to element-value variations and to the finite gain of the op-amps becomes significant. Methods for overcoming this difficulty are described in [43], [46] and in the references quoted in these works.

As an illustration of the precise control of both passband and stop-band characteristics afforded by the bilinear ladder design even under production conditions, Fig. 25 shows the typical measured loss characteristics of the S3528 Programmable Low-Pass Filter. This filter, produced and marketed by American Microsystems, Inc. [49], is a seventh-order elliptic low-pass filter with programmable clock frequency, realized as a bilinear ladder with a modified input stage (to avoid the need for a sampled-and-held input signal). The design loss values were 0.05-dB passband ripple and 51-dB minimum stopband loss; the typical values measured on the production units (Fig. 25) were the same within measurement accuracy.

Another design strategy which permits the simulation of an RLC ladder filter via the bilinear s-to-z transformation of (46) is based on *node-voltage simulation* [44], [11]. The principle is illustrated in Fig. 26. Defining the node self-admittance y_{ii} as the sum of the admittances connected to node i, and the mutual admittance $y_{ij} = y_{ji}$ as the total admittance connected directly between node i and j, the node equations of the circuit can be written in the form

$$V_1 = \frac{1}{y_{11}} \left(\frac{V_{in}}{R_S} + y_{13}V_3 \right)$$

$$V_3 = \frac{1}{y_{33}} (y_{13}V_1 + y_{35}V_5)$$

$$V_5 = \frac{1}{y_{55}} (y_{35}V_3 + y_{57}V_7)$$

$$\vdots$$

$$V_{N-2} = \frac{1}{y_{N-2,N-2}} (y_{N-2,N-4}V_{N-4} + y_{N-2,N}V_N)$$

$$V_N = \frac{y_{N-2,N}}{y_{NN}} V_{N-2}. \tag{53}$$

Let the LC ladder branches all contain at most an inductor and a capacitor in parallel. Then, for the internal nodes (nodes $3, 5, \cdots, N-2$) the transfer ratios $T_{ik} = \partial V_k/\partial V_i = y_{ik}/y_{kk}$ are all in the form

$$T_{ik} = \frac{y_{ik}}{y_{kk}} = a \frac{s^2 + b}{s^2 + c} \tag{54}$$

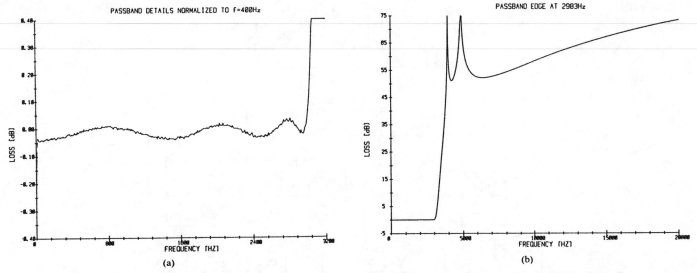

PASSBAND DETAILS NORMALIZED TO f=400Hz

(a)

PASSBAND EDGE AT 2983Hz

(b)

Fig. 25. The frequency responses of the AMI programmable low-pass filter. (a) Measured passband responses. (b) Measured overall responses.

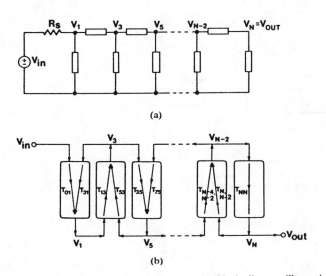

(a)

(b)

Fig. 26. Ladder circuit. (a) Circuit diagram. (b) Block diagram illustrating node-voltage simulation.

while those corresponding to the input and output nodes (nodes 1 and N) have the forms

$$T_{01} = \frac{1}{R_S y_{11}} = \frac{ds}{s^2 + es + f}$$

$$T_{N-2, N} = \frac{y_{N-2, N}}{y_{NN}} = \frac{s^2 + g}{s^2 + hs + i}. \qquad (55)$$

Here, $a, b, \cdots, i$ are constants determined by the element values of the ladder circuit.

Hence, after using the bilinear s-to-z transformation (46), all T_{ik}'s can be realized by the stray-insensitive biquads discussed earlier. Furthermore, any two T_{ik}'s with common second indices (T_{13} and T_{53}, T_{35} and T_{75}, etc.) have the same denominators. Hence, they can be pairwise realized by the *same* biquad, with two different sets of input branches. The block diagram of the SCF is hence that shown in Fig. 26(b). The number of biquads is equal to the number of nodes. Typically, for an nth-order filter, the number of op-amps is n (if n is even) or $n + 1$ (if n is odd). The configuration, element values, and sensitivities are similar to those obtainable by the "bilinear ladder" design method. As an

illustration, Fig. 27 shows the *LC* prototype and final SCF realization of an eighth-order bandpass filter [11] requiring eight op-amps in the form of four coupled biquads.

C. Scaling

The properties of a SCF obtained by one of the previously described design techniques can usually be significantly improved by using *scaling* [10], [45], [42]. This process is based on the following two statements, both evident from physical considerations:

1) If the values of all capacitors (including feedback capacitors) connected or switched to the inverting *input* terminal of an op-amp are multiplied by the same constant, then all voltages in the SCF remain unchanged. This is true since all voltages are affected only by the ratios of these capacitances.

2) If the values of all capacitors (including feedback capacitors) connected or switched to the *output* terminal of an op-amp in an SCF are multiplied by the same constant k, then the output voltage of this op-amp will be divided by k; all other op-amp output voltages remain unchanged. This follows since the described changes leave all charges flowing to and from the affected op-amp unchanged.

The second theorem can be used to achieve optimum scaling for maximum signal swing. Specifically, let the maximum voltage swing at the output of the ith op-amp be $V_{i,\max}$, and that of the whole SCF be $V_{\text{out},\max}$. Then, multiplying all capacitors connected to the output terminal of the ith op-amp by $V_{i,\max}/V_{\text{out},\max}$ will assure that this op-amp saturates for the same input voltage level as the output op-amp. Performing this scaling operation for all op-amps, the maximum possible signal level and hence dynamic range is obtained for the filter. Often, the maximum output voltage $V_{\text{out},\max}$ is simply chosen as V_{in}. The maximum signal levels $V_{i,\max}$ normally occurs just outside the passband of the filter.

Having achieved maximum dynamic range, the total capacitance in the SCF can be minimized in a second scaling step based on the first theorem. Now the values of all capacitors connected to the input terminal of each op-amp are scaled so that the smallest one becomes $C_{\min}$, the minimum value permitted by the technology used. Usually, $C_{\min}$ is 0.1–0.5 pF.

As an illustration of the improvement achievable, in the eighth-order bandpass filter of Fig. 27, scaling reduces the capaci-

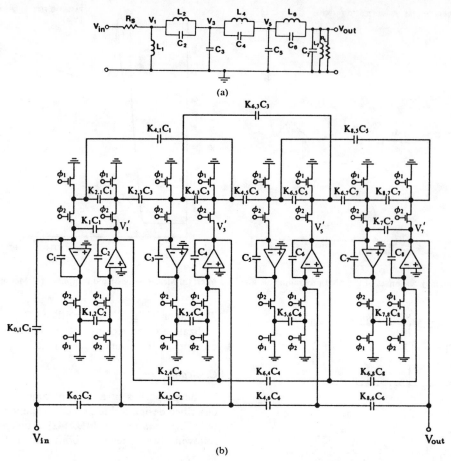

Fig. 27. An example of a bandpass filter system using node-voltage simulation. (a) *LCR* prototype circuit. (b) The resulting switched-capacitor filter.

tance spread from 7×10^7 to 35, and significantly improves the signal-to-noise ratio [11].

IV. Nonfiltering Applications of Switched-Capacitor Circuits

Although filters received the most attention recently among all SC circuits, other applications (such as the nonlinear ADC's and DAC's used in PCM telephone systems) resulted in the earliest fully integrated high-volume SC products. In these circuits, the MOS technology originally developed for memory applications was applied to charge-transfer ADC's [50]. In reviewing the current literature on commercial MOS analog integrated circuits, it becomes clear that most of them contain other SC circuits besides filters. In this section, a sample (intended to be fairly typical) of these circuits is described.

A. Switched-Capacitor Gain Stages

One of the most common building blocks in analog signal processing is the constant-gain voltage amplifier. Historically, the first SC amplifiers were used in charge-coupled device (CCD) filters; in fact, these circuits contained also some of the earliest integrated MOS op-amps. The general configuration of the commonly used voltage amplifier is shown in Fig. 28; the voltage gain is $V_{out}/V_{in} = -Z_2/Z_1$. For relaxed specifications, Z_1 and Z_2 can be realized by polysilicon resistors [47]. In principle, they can also be chosen simply as capacitors; however, due to charge buildup caused by leakage current flowing to its dc-insulated inverting input terminal, the op-amp would soon saturate and latch in such a circuit.

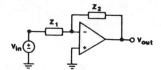

Fig. 28. General voltage amplifier circuit.

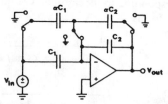

Fig. 29. A dc-stabilized switched-capacitor realization.

To prevent saturation, the dc-stabilized circuit of Fig. 29 can be used [51]. This is equivalent to using a parallel R and C in both the Z_1 and Z_2 branches. The gain is $-C_1/C_2$. In spite of the switched branches, the output is thus ideally just a scaled and inverted facsimile of the input, and no aliasing occurs. Clock feedthrough will be caused, however, by parasitic capacitances, and hence the clock frequency should be well above the signal frequency range. The circuit can easily be generalized to provide the linear combination of several input voltages (Fig. 30); the output voltage is then given by

$$V_{out} = -\sum_{i=1}^{n} \frac{C_i}{C_f} V_i. \qquad (56)$$

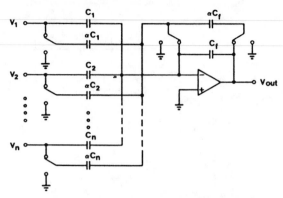

Fig. 30. Voltage amplifier which can provide the bilinear combination of several input voltages.

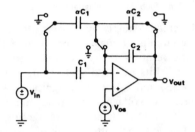

Fig. 31. Voltage amplifier including the dc offset voltage of the op-amp.

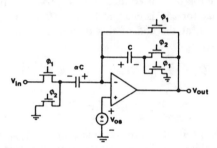

Fig. 32. Voltage amplifier with offset compensation.

An important disadvantage of the circuits of Figs. 29 and 30 is that the dc offset voltage V_{os} of the op-amp affects the output voltage V_{out} (Fig. 31). Assuming infinite gain and bandwidth for the op-amp, and zero input voltage V_{in}, the output voltage in the steady state is $V_{os}(1 + C_1/C_2)$. For typical values ($V_{os} = 15$ mV, $C_1/C_2 \cong 10$), the offset-caused output can be as high as 0.165 V. This voltage also varies with temperature and time.

To reduce the effects of the op-amp offset, the compensated circuit of Fig. 32 can be used [52]. When $\phi_1 = $ "1," the op-amp acts as a unit-gain voltage follower; its output is V_{os}. Hence, capacitor C charges to V_{os}, and αC to $V_{os} - v_{in}$, with the polarities indicated. When next $\phi_1 \to$ "0" and $\phi_2 \to$ "1," the voltage across C becomes $V_{os} - v_{out}$ and that across αC becomes V_{os}. From charge conservation at the inverting input terminal of the op-amp

$$CV_{os} - C[V_{os} - v_{out}(t)] + \alpha C[V_{os} - v_{in}(t - T/2)]$$
$$- \alpha C V_{os} = 0 \quad (57)$$

results. This gives $V_{out}/V_{in} = \alpha z^{-1/2}$, where the $z^{-1/2}$ factor is due to the half-period delay between the sampling of v_{in} and the appearance of the corresponding sample of v_{out}. Note that V_{os} does not enter v_{out}; it is canceled by the switching arrangement. This compensation principle can also be applied to the design of delay-free inverting amplifiers (Fig. 33) with a voltage gain ($-\alpha$).

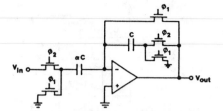

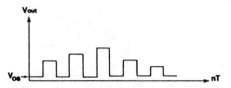

Fig. 33. Delay-free inverting amplifier using offset compensation.

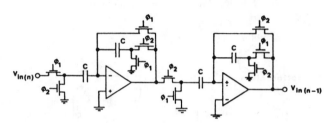

Fig. 34. The output voltage of the circuit of Fig. 33.

Fig. 35. A cascade of two voltage amplifiers forming a delay stage.

The output voltages of the offset-free amplifiers shown in Figs. 32 and 33 are useful only during the half-periods when $\phi_2 = $ "1;" when $\phi_1 = $ "1," the output is V_{os}, as illustrated in Fig. 34 for the circuit of Fig. 33. This waveform requires a fast slew rate and settling time of the op-amp, a disadvantage which can be significant for fast clock rates.

The offset compensation arrangement shown is equivalent to the correlated double sampling system proposed earlier [48], [53]. Hence, it reduces also the level of the low-frequency noise ($1/f$ noise) generated by the op-amp at the output of the amplifier.

By cascading two noninverting voltage amplifiers with complementary clocking phases (Fig. 35), a clock-controlled full-period delay stage results. Such delay stages, in conjunction with switches and digital control logic, can be used to perform a variety of sampled-data functions [54].

Some output offset voltage is generated also by asymmetric clock feedthrough. This effect can be greatly reduced by using fully differential circuitry [55] or, as a simpler alternative, differential-input/single-ended-output circuits [56].

If the periodic resetting of the output voltage cannot be tolerated, then it is also possible to use a first-order SC filter stage as an amplifier. By using a low-pass circuit with a small passband loss variation, the sample-and-hold function can be incorporated in the gain stage; by using a high-pass section, ac coupling can be achieved.

B. Programmable Capacitor Arrays

A useful feature of sampled-data analog circuits is that they can be made digitally programmable, by replacing some capacitors by programmable capacitor arrays (PCA's) [50]. Such a binary array is shown in Fig. 36. PCA's can be used to help realize programmable filters [58], speech synthesizers [59], and adaptive equalizers [60], as well as many other useful circuits.

Care must be taken in the design of PCA's to minimize the noise injection into the circuit. The bottom plate of a capacitor

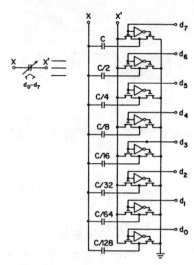

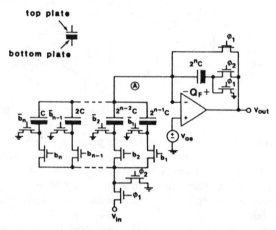

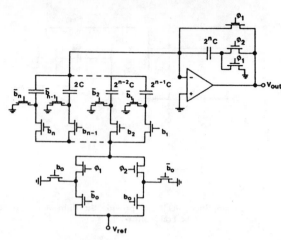

Fig. 36. Programmable capacitor array.

Fig. 37. Switched-capacitor digital-to-analog converter (DAC).

Fig. 38. A DAC in which the input signal may be positive or negative.

should never be connected to the inverting input (virtual ground) terminal of an op-amp; otherwise, the noise from the substrate power supply will be coupled into the op-amp input and amplified. If a PCA is programmed (i.e., switched) while signals propagate through the system, then the switched side (node x' in Fig. 36) of the array must be switched between ground and virtual ground only. This will reduce switching noise due to programming. If the array is programmed during a period when no signals are processed (e.g., while the op-amps are being reset), then the unswitched node (node x) of the PCA should be connected either to the virtual ground, or switched between ground and virtual ground. This keeps the feedback factor unchanged. Thus the effect of finite op-amp gain will be constant, independent of the programmed value of PCA. This can be of importance in such precision application as high-accuracy D/A conversion.

It often saves chip area to replace a large (say, 8-bit) PCA by two smaller (say, 4-bit) PCA's since the capacitance spread is reduced, even if this requires an extra op-amp [59].

C. Digital-to-Analog and Analog-to-Digital Converters

A very important application of programmable SC circuits is their use as digital-to-analog (DAC) or analog-to-digital (ADC) converters. Indeed, as mentioned earlier, some of the earliest SC circuits actually fabricated in integrated form were such converters [50], [61].

A DAC can be created simply by replacing the input capacitor αC in one of the circuits of Figs. 32 or 33 by a PCA. For the former circuit, the result is the n-bit DAC shown in Fig. 37, where the b_i are the bits of the digital input, with b_1 the most significant bit (MSB) and b_n the least significant bit (LSB). V_{in} may be a fixed reference voltage, or an analog input signal.

The operation of the DAC is as follows. During the $\phi_1 = $ "1" (sample-recharge) period, the bottom plate of capacitor $2^{n-i}C$ in the PCA is either connected to ground (if $b_i = $ "0") or to V_{in} (if $b_i = $ "1"). The top plate is connected to V_{os}. At the same time, the feedback capacitor $2^n C$ is charged to the offset voltage V_{os} of the op-amp. Hence, at the end of this period, the charge stored in the PCA is

$$Q_1 = \sum_{i=1}^{n} \left[b_i C V_{in} 2^{n-i} - C V_{os} 2^{n-i} \right]$$
$$= C V_{in} \sum_{i=1}^{n} b_i 2^{n-i} - (2^n - 1) C V_{os} \qquad (58)$$

while the charge in the feedback capacitor is $Q_{F1} = -2^n C V_{os}$. Next, during the $\phi_2 = $ "1" period, the bottom plates of all input capacitors are grounded, and hence most of their charges is transferred to the feedback capacitor. At the end of this period, the charge remaining in the PCA is $Q_2 = -(2^n - 1) C V_{os}$, while that on the feedback capacitor is $Q_{F2} = (V_{out} - V_{os})2^n C$. From charge conservation at node Ⓐ

$$(Q_{F2} - Q_{F1}) - (Q_2 - Q_1) = 0 \qquad (59)$$

hence

$$V_{out} = V_{in} \sum_{i=1}^{n} b_i 2^{-i}. \qquad (60)$$

Thus V_{out} is the product of the digital input $(b_1, b_2, \cdots, b_n)$ and V_{in}. The circuit can thus operate either as a simple DAC (if V_{in} is a temperature-stabilized reference voltage), or as a multiplying DAC (MDAC) if V_{in} is an analog input voltage.

If the binary input signal can have either a positive or a negative sign, as indicated by a sign bit b_0, then the circuit of Fig. 38 can be used as a DAC. For $b_0 = $ "1," the stage inverts V_{ref}; for $b_0 = 0$, it does not. For $n \geq 8$, the capacitance spread may become excessive, and the matching accuracy inadequate. In this case, as mentioned earlier, the capacitance spread can be greatly reduced (at the cost of an additional op-amp) by realizing the DAC in a circuit containing *two* programmable gain stages [52].

The majority of MOS sampled-data ADC's use a PCA and a comparator, controlled by digital logic, to perform a successive-approximation conversion. Modern designs often also include

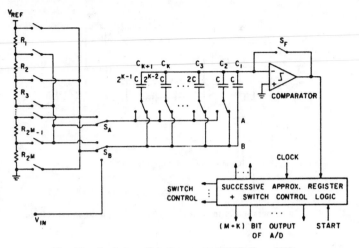

Fig. 39. Analog-to-digital converter (ADC) (from [63]).

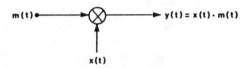

Fig. 40. Schematic diagram of a balanced modulator.

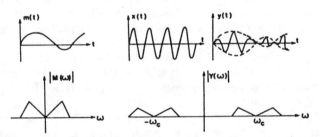

Fig. 41. The waveforms and the spectra of the signals in a balanced modulator.

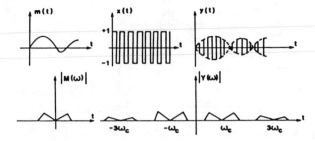

Fig. 42. Waveforms and spectra of the signals in a balanced modulator which uses a square-wave carrier signal.

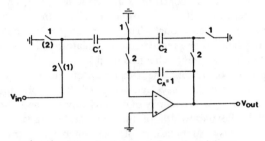

Fig. 43. SC realization of a balanced modulator.

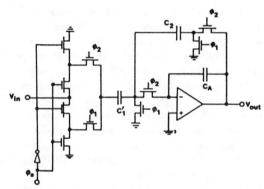

Fig. 44. A full-wave rectifier using an SC modulator.

multiple-tap resistive voltage-divider strings to assure monotonicity [62], [63]. A typical example, taken from [63], is shown in Fig. 39. Such ADC's are capable of an accuracy of 12-bits; i.e., the differential nonlinearity is less than 2^{-13}.

For faster operation, MOS ADC's can be realized in a parallel or "flash" configuration. Then, for an *n*-bit converter, each of the $(2^n - 1)$ outputs from a multiple-tap resistive divider is fed simultaneously into a separate comparator, and the output bits are computed digitally from the comparator outputs. In [64], a 6-bit converter operating at 20 MHz is described, based on using simple comparators with offset-voltage cancellation.

Alternative approaches, based on cascading 1-bit ADC's in a pipeline configuration, have also been suggested [65], [66], as viable alternatives that could achieve both high speed and high accuracy. However, to date these converters have not been fabricated, and hence their effectiveness is as yet untested.

D. Balanced Modulators

A balanced modulator, illustrated schematically in Fig. 40, is essentially an analog multiplier which provides the product of the carrier signal $x(t)$ and the modulating (baseband) signal $m(t)$. If the carrier is a sinusoid signal $x(t) = \cos \omega_c t$, then the spectrum of the output signal is given by

$$Y(\omega) = \tfrac{1}{2}M(\omega + \omega_c) + \tfrac{1}{2}M(\omega - \omega_c) \qquad (61)$$

(Fig. 41), where $M(\omega)$ is the spectrum of $m(t)$. Thus the carrier signal $x(t)$ is suppressed, and only the sidebands $M(\omega \pm \omega_c)$ are

produced. In general, if $x(t)$ is any periodic signal with a Fourier series representation

$$x(t) = \sum_{n=-\infty}^{\infty} a_n e^{jn\omega_c t} \qquad (62)$$

then the output spectrum is

$$Y(\omega) = \sum_{n=-\infty}^{\infty} a_n M(\omega + n\omega_c). \qquad (63)$$

A periodic carrier signal which is easy to generate using SC circuits is the square wave alternating between the values $+1$ and -1. This will, according to (63), generate the sidebands $M(\omega \pm \omega_c)$, along with the odd higher order products $M(\omega \pm 3\omega_c)$, $M(\omega \pm 5\omega_c)$, etc. Since the a_n for a square wave are zero if n is even, and decrease as $1/n$ with increasing n if n is odd, the higher order sidebands are smaller than the main ones (Fig. 42).

An easy way to achieve the square-wave carrier $x(t)$ is to switch the polarity of the input signal $m(t)$ periodically. Consider the circuit of Fig. 43 which is a simplified version (with $C_1 = 0$) of that shown in Fig. 14(c). For the clock phases shown without parentheses, the circuit realizes a damped integrator with a dc gain of $- C_1'/C_2$. If the parenthesized phases are used, the circuit still functions as a damped integrator; however, the dc gain is now $+ C_1'/C_2$, and there is also an additional half-period signal delay. Hence, if the carrier acts to interchange the clock phases

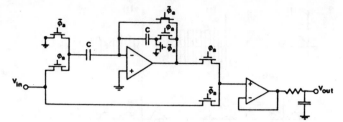

Fig. 45. A balanced modulator which uses the carrier as the clock signal.

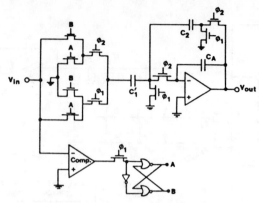

Fig. 46. The generation of the carrier clock signal from the input voltage.

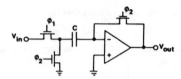

Fig. 47. An offset-canceling comparator.

periodically (Fig. 44), the desired modulation is achieved [60]. The half-period delay contributes an unwanted jitter, equivalent to a dc offset in the effective carrier $x(t)$. This, in turn, causes a baseband signal feedthrough. The jitter can be eliminated if a sample-and-hold stage precedes the modulator, or if a somewhat more elaborate switching scheme is used.

For proper operation, the frequency of the clock (ϕ_1, ϕ_2) should be much higher (say, by a factor of 30 or more) than that of ϕ_a. Moreover, the frequency of ϕ_a is usually chosen to be much larger than the bandwidth of the modulating signal $m(t)$, to permit the easy suppression of the sidebands centered at $\pm 3\omega_c, \pm 5\omega_c$, etc. (and thus avoid aliasing).

For optimum operation, the clock frequency (i.e., the frequency of ϕ_1 and ϕ_2) should be an integer multiple of the carrier frequency ω_c. This is necessary to avoid generating those intermodulation products, arising from the higher harmonics in $x(t)$, which overlap with the modulated spectrum $M(\omega + \omega_c)$. Since the amplitudes a_n decrease slowly with n, even very-high-order harmonics can cause significant intermodulation distortion otherwise.

Another way to avoid intermodulation distortion in the modulated spectrum is to use continuous-time band-limiting of the modulated signal *before* additional sampled-data filtering. As an example, a simple balanced modulator circuit (different, in principle, from that of Fig. 44) is shown in Fig. 45. In the modulator, the carrier acts also as the clock signal. When $\phi_a = 0$, the input signal is applied directly to the buffer, while the inverter stage (containing the op-amp and the two capacitors) is reset and its offset voltage stored for cancellation. When $\phi_a = 1$, v_{in} is inverted and connected to the buffer. The continuous-time RC filter following the buffer should have a large enough time constant so that an SC post filter processing v_{out} can extract the spectrum centered at $\pm \omega_c$ without aliasing effects. A single-sideband (SSB) modulator can also be realized, if a bandpass post filter is used.

E. Rectifiers and Peak Detectors

A *full-wave rectifier* converts an input signal into its absolute value $|v(t)|$. A simple way of implementing a full-wave SC rectifier is to add a comparator to one of the balanced modulators discussed in the previous section. Consider the modulator of Fig. 44. If the carrier clock signal ϕ_a is derived from v_{in} such that $\phi_a = $ "1" for $v_{in} > 0$ and $\phi_a = $ "0" for $v_{in} < 0$, then the circuit will invert negative input signals, but not positive ones. Thus the output voltage will be proportional to $|v_{in}|$. In fact, the proportionality is only approximate since the section consisting of C_1', C_2, C_4, and the op-amp forms a low-pass filter. However, if the 3-dB bandwidth of the filter is much larger than the bandwidth of v_{in}, then the approximation will be good. Otherwise, the circuit will calculate the average of $|v_{in}(t)|$.

The generation of ϕ_a is illustrated in Fig. 46, [60]. Here, two

new logic signals $A = [sign(v_{in}) + 1]/2$ and $B = \bar{A}$ are created by the comparator and the following latch each time $\phi_1 = $ "1." A and B then determine whether the main signal path from v_{in} to v_{out} inverts v_{in} or not.

For precision applications, the offset-canceling comparator shown in Fig. 47 can be used [50]. In this circuit, the input-referred offset voltage V_{os} is stored in C, and subtracted from v_{in} during comparison.

The circuits of Figs. 46 and 47 (as the modulators discussed earlier) require that v_{in} be a sampled-and-held signal which changes at the leading edge of the ϕ_1 clock signal. Otherwise, a polarity-dependent jitter, one-half-period large, will occur due to the change in the sampling time of v_{in}.

The output voltage v_{out} of the rectifier shown in Fig. 46 contains a term $(1 + C_1'/C_2)V_{os}$, where V_{os} is the (input-referred) offset voltage of the output op-amp. If this cannot be tolerated, then the offset canceling gain amplifier of Fig. 34 can be used, with the clock phases of the input switches modulated in a similar way. However, in this case the output voltage is valid only during the $\phi_2 = $ "1" interval.

A *peak detector* is a circuit whose output v_{out} holds the largest positive (or, if so specified, negative) value earlier attained by the input signal (Fig. 48). A possible realization is shown in Fig. 49. Here, the op-amp acts as a comparator of the current value of v_{in} and its largest earlier value $v_{max} = v_{out}$. If $v_{in} > v_{max}$, then the latch sets A to "1," and v_{out} is updated. Otherwise, $A = $ "0" and v_{out} remains unchanged. The operation assumes that v_{in} is a sampled-and-held signal which changes at the leading edge of ϕ_1; otherwise, the values of v_{in} used in the comparison and in the updating of v_{out} may differ.

To reset the circuit instantaneously, a switch may be connected in parallel with C. For slow reset, a small current source or a switched-capacitor resistor may be used across C. As indicated in Fig. 49, a buffer is usually needed to prevent the accidental discharge of C due to loading. If the largest peak of $|v_{in}(t)|$ is to be held, a rectifier can precede the peak detector.

A continuous-time MOS peak detector [67] is shown in Fig. 50. Here, the comparator causes the MOSFET to charge C whenever

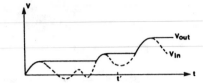

Fig. 48. The waveforms of a peak detector.

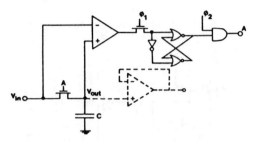

Fig. 49. SC realization of a peak detector.

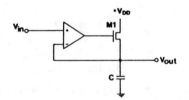

Fig. 50. A MOS continuous-time peak detector.

$v_{in} > v_{out}$. As before, the resetting of C can be performed using a switch, a current source, or an SC resistor.

F. Oscillators

The design of most integrated MOS oscillators to date is based upon principles developed for corresponding active-RC or bipolar emitter-coupled unstable multivibrators. Typical examples can be found in [68]–[70]. The frequency of these continuous-time oscillators can typically be controlled by incorporating a voltage-to-current converter. They have the disadvantage, however, that the frequency is difficult to predict and tends to be temperature sensitive. This is largely due to the difficulty of fabricating a stable and accurate current source without using off-the-chip elements.

A simpler alternative approach is to divide digitally a high-frequency master clock signal, and (if necessary) to suppress the higher harmonics of the resulting signal [60]. This has the advantage of frequency stability, since the master clock is usually crystal controlled. However, such circuits are not easily converted to voltage-controlled operation, and the frequency must be a subharmonic of the clock frequency. Furthermore, if SC filters are used to suppress the higher harmonics, then their output voltages often contain high-frequency components which can cause jitter in the system [70].

It is also possible to replace all resistors in an active-RC oscillator with SC's. The resulting waveform will again contain high-frequency components. The frequency, however, does not need to be a subharmonic of the master clock. It is also well controlled, accurate, and stable, since it depends only on capacitance ratios. An oscillator based on an active-RC phase-shift oscillator prototype is described in [71]. This particular circuit is sensitive to parasitic capacitances; however, the basic approach can be used to realize parasitic-insensitive circuits as well.

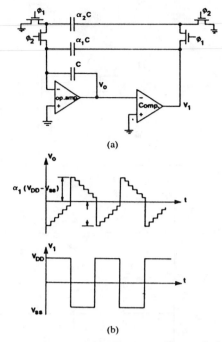

(a)

(b)

Fig. 51. SC oscillator based on the relaxation oscillator principle. (a) Circuit diagram. (b) Waveforms.

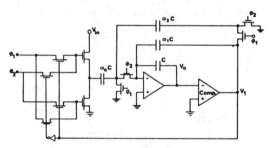

Fig. 52. Voltage-controlled SC oscillator.

An alternative SC oscillator, based the relaxation oscillator principle [72], is shown in Fig. 51. This simple circuit has very good frequency accuracy and, when used as a voltage-controlled oscillator (VCO), good frequency versus voltage linearity as long as the oscillation frequency is much lower than the clock frequency. It can oscillate at frequencies which are not subharmonics of the clock frequency, but will then exhibit some phase jitter which may be as large as one full clock period. The operation of the circuit is as follows. The first op-amp and C act as a charge integrator; the second op-amp as a comparator. Let at some instant the output v_1 of the comparator be V_{SS}, the negative supply voltage. This will be sampled by the inverting capacitor $\alpha_2 C$, and a positive charge fed into C, in each clock period. Hence, v_o, the output of the noninverting integrator stage, decreases in steps of $\alpha_2 V_{SS}$ (Fig. 51(b)). When v_o becomes negative, the comparator output switches to the positive supply voltage V_{DD}. This causes, due to the coupling capacitor $\alpha_1 C$, a negative step of $\alpha_1(V_{DD} - V_{SS})$ in v_o. Thereafter, $\alpha_2 C$ feeds negative charge packets into C, and v_o increases in steps of $\alpha_2 V_{DD}$, until it becomes positive causing the comparator output to switch back to V_{SS}. The cycle is then repeated. For $\alpha_2 \ll \alpha_1$ and $V_{SS} = -V_{DD}$, the oscillation frequency is $f_i \approx (\alpha_2/4\alpha_1)f_c$ where f_c is the clock frequency.

By adding a switched feed-in-capacitor $\alpha_0 C$ (Fig. 52), the circuit can be converted into a VCO. For a positive control

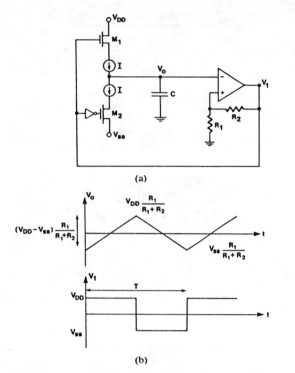

(a)

(b)

Fig. 53. A voltage-controlled oscillator (VCO) whose operation does not require a clock. (a) Circuit diagram. (b) Waveforms.

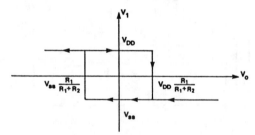

Fig. 54. The dc transfer characteristics of a Schmitt trigger.

voltage ($v_{in} > 0$) the oscillation frequency f_i will increase; for $v_{in} < 0$, it will decrease. Now, for $\alpha_2 \ll \alpha_1$ and $V_{SS} = -V_{DD}$

$$f_i \cong \frac{\alpha_2}{4\alpha_1} f_c + \frac{\alpha_0}{4\alpha_1} \frac{V_{in}}{V_{DD}} f_c \quad (64)$$

results.

A VCO for which f_i is not constrained by a clock is shown in Fig. 53. The comparator with its feedback resistors forms a Schmitt trigger with the dc transfer characteristics shown in Fig. 54. The output v_1 of the comparator controls the current sources I which charge or discharge C. Let $v_1 = V_{DD}$; this implies that $v_o < V_{DD} R_1/(R_1 + R_2)$. Then, M_1 will conduct allowing C to charge. When v_o reaches $V_{DD} R_1/(R_1 + R_2)$, the Schmitt trigger switches, v_1 becomes V_{SS}, M_1 will turn off while M_2 turns on, and C will discharge. The oscillation frequency, for $V_{SS} = -V_{DD}$, will be

$$f_i = \frac{1 + R_2/R_1}{4C} \frac{I}{V_{DD}}. \quad (65)$$

The circuit can be voltage controlled if the current sources are replaced by voltage-to-current converters (Fig. 55). A complete VCO based on this scheme is shown in Fig. 56. Here, transistors M_3–M_7 form the voltage-to-current converter; M_8–M_{11} are the switches which allow or inhibit the current I in M_5 (or M_7) to be

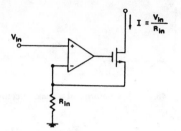

Fig. 55. A voltage-to-current converter.

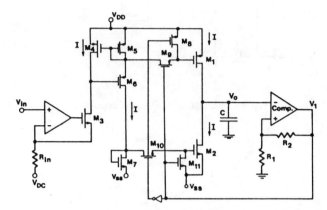

Fig. 56. VCO using the converter of Fig. 55.

mirrored in M_1 (or M_2). For $V_{SS} = -V_{DD}$, the frequency is given by

$$f_i = \frac{1 + R_2/R_1}{4CR_{in}} \frac{V_{in} - V_{DC}}{V_{DD}}. \quad (66)$$

Here V_{DC} is a bias voltage. If the VCO is to oscillate free running (i.e., with $v_{in} = 0$), then V_{DC} must be negative. V_{DC} can be eliminated if a constant-current source is added to bias M_4.

Most oscillators can be made digitally controllable if one of the frequency-determining capacitors is replaced by a PCA. Also, most SC bandpass biquads can be converted into oscillators. This is achieved by designing them to have a large negative pole-Q factor, and adding a nonlinear element which causes the pole-Q to become positive when the amplitude exceeds a threshold value.

V. NONIDEAL EFFECTS IN SWITCHED-CAPACITOR CIRCUITS

Due to the integrated realization of SC circuits, a number of mundane but practically important details must be considered in order to achieve successful fabrication. Some of the potential problems (parasitic capacitances, op-amp offset voltages, element-value inaccuracies, etc.) are mentioned elsewhere in this paper.

Parasitic capacitances to the substrate can usually be rendered ineffective by using the stray-insensitive sections discussed in previous sections. Unless a large (say, 5–10-percent) amplitude response error is acceptable, these techniques must be adhered to. If the errors quoted *are* tolerable, then circuits which are insensitive to bottom-plate (but not top-plate) stray capacitance—such as that shown in Fig. 3—may be used.

Since most stray capacitances couple some circuit nodes and lines to the substrate, special care must be taken to minimize noise injection from the substrate to the signal path. Thus the bottom plates of all capacitors should be connected (or switched) only to op-amp outputs and ground. This eliminates noise injection through the bottom-plate-to-substrate stray capacitances, which are usually the largest parasitic capacitances in an SC circuit.

Whenever SC circuits share the same chip with digital circuits which process large and fast pulses, the physical separation of these two circuit categories should be as complete as possible. It is essential that the digital and analog parts should have separate power supply and ground lines. If CMOS technology is used, then sensitive nodes (such as the op-amp input terminals) can often be shielded by a diffused "well" region connected to the analog ground line or to a low-noise power supply. On some chips, in fact, a large percentage of the silicon area was devoted to the circuitry needed to generate a noise-free substrate bias voltage [47]. In general, for an NMOS chip it is advisable to use separate lines and pins for the substrate bias, the negative bias for the digital circuitry, and the negative supply for the analog circuitry, and to supply low-noise bias voltages to each pin. The metal signal lines can be shielded by including grounded polysilicon layers between the signal line and the substrate, and/or grounded metal lines on both sides of the signal line.

The analog and digital portions of the chip can also be separated by diffusion regions connected to ground or V_{DD}, although the effectiveness of this technique has not been fully tested. It is also very important to establish a good bond between the back surface of the substrate and the package header.

In applications where low-noise performance is crucial (such as in telephone systems), differential circuit design techniques [48] are very effective in minimizing the noise injection from the substrate.

A practical problem which can sometimes cause difficulty in SC circuits is the jitter introduced by the inherent sampled-data character of the signals. In some instances [70] this has made it necessary to use a discrete active-*RC* filter rather than an integrated SC one. This tends to be especially true in systems which adaptively generate internal clock signals using VCO's and are thus asynchronous. The problem can sometimes be solved by careful selection of the clock frequencies, and/or by using on-chip continuous-time *RC* filters. Computer simulation can sometimes reveal probable jitter problems before fabrication.

Clock feedthrough represents one of the most important parasitic effects. If NMOS technology is used, then clock feedthrough can be reduced by connecting a "dummy" switch, with its drain and source tied together and the complement of the clock signal applied to its gate, to those switches which inject clock-feedthrough noise into the signal path [57]. The dimensions of the dummy switch are to be determined such that the injected clock signals cancel. Although the effectiveness of this technique depends on matching the rise and fall times of the clock waveforms (which is easier for slower waveforms), it is typically possible to reduce clock-induced offsets by a factor of about 5 using this method.

For CMOS technology, a similar improvement can be obtained by using transmission gates (parallel-connected n- and p-channel transistors) as switches. The sizes are to be the same for the two transistors, and the clock signals matched complements. The cancellation depends on the dc potential of the switch, which affects the transistors differently.

If the clock-feedthrough reduction obtainable by the described methods is not sufficient, then fully differential [48], [55] or partially differential [56] circuit techniques can be used to achieve further improvement. Finally, in very critical situations, a reset phase can be added to eliminate offsets incurred by clock feedthrough in the preceding stages. This method is similar to that used to minimize the effects of op-amp offset voltages [51]–[53].

At higher frequencies, the response time of the op-amps and the on-resistance of the switches become important [38], [21], [22]. Usually, the speed of the op-amp represents the more severe restriction. From the formulas of [21] and [22], it can be shown that (as a rule-of-thumb) the unit–gain bandwidth of the op-amp should be at least five times larger than the clock frequency. The above rule is valid if no continuous (unswitched) coupling exists among stages. Otherwise, higher order transients exist, and the settling time is extended. Also, the slew rate of the op-amps should be adequate for the largest and fastest signals which the SCF must process.

At lower frequencies the "$1/f$" noise of MOS transistors can be quite large. There are circuit techniques based upon sampling and differencing [48] that can be used to reduce this noise. These techniques are similar to those used to minimize dc offsets. However, they do not reduce (in fact, sometimes even enhance) the wide-band thermal noise of MOS transistors and judgment is required as to their applicability. In any event, prudent choice of transistor sizes is required during the op-amp design to minimize both types of noise. Also, since the out-of-band noise is aliased into the passband by sampling, the bandwidth of the op-amps should not be excessive.

The reader is referred to the companion paper by D. J. Allstot and W. C. Black in this issue for a more in-depth analysis of practical considerations in the design of SC circuits.

VI. Integrated Sampled-Data Analog Systems

In this section, some examples will be described of MOS integrated circuits containing SC stages. Since (in spite of its relatively recent development) there exists a very large number of publications on this technology, only a small sample will be included here, involving recent examples of systems fabricated commercially. This should provide an overview of the state of the art, and a starting point for the compilation of a comprehensive bibliography.

A. Telecommunication Circuits

An important driving force behind the development of analog MOS integrated circuits was the demand generated by the conversion of voice transmission in telephone systems to digital PCM format. This required integrated coder–decoder systems (codecs) which relied heavily on sampled-data analog components. Most of the recently developed codecs included nonlinear DAC's and ADC's, along with SC filters used for antialiasing as well as for 60-Hz rejection, for amplitude equalization, for smoothing, etc., all on a single chip. The chip usually contained digital control logic, voltage amplifiers, and output buffers capable of driving 600- or 900-Ω loads. Some have also included phase-locked loops (PLL's). Representative examples are described in [73]–[76]; a typical system diagram [76] is shown in Fig. 57. The system described in [77] does not include the codec; instead, it incorporated an automatic gain control (AGC) stage and a programmable precision line-balancing network for use in the two-to-four-wire conversion required at the analog-to-digital interface.

Some of the most sophisticated SC integrated circuits have been designed for the detection of dual-tone dialing signals used in telephone systems. The block diagram of such a circuit [78] is shown in Fig. 58. This chip contained more than 40 op-amps which realized SC filters, zero-crossing detectors, and amplitude

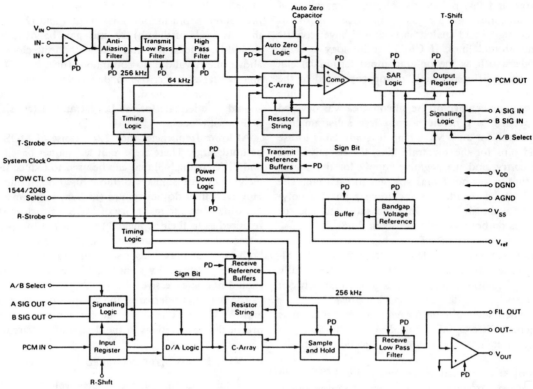

Fig. 57. The block diagram of an integrated coder–decoder (from [76]).

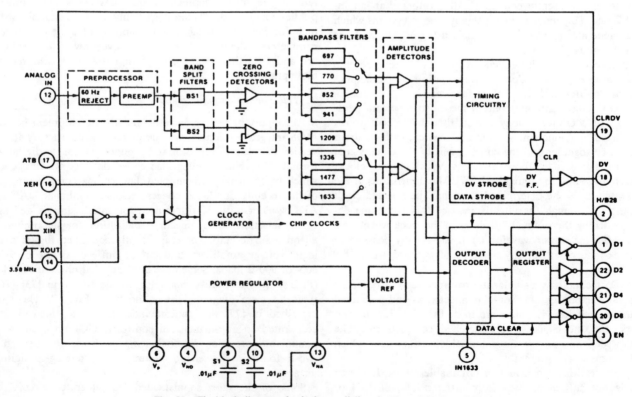

Fig. 58. The block diagram of a dual-tone dialing signal receiver (from [78]).

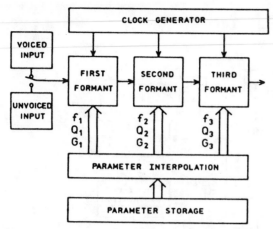

Fig. 59. Block diagram of a typical speech synthesis system.

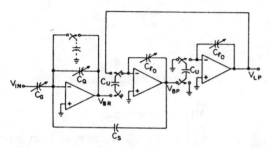

Fig. 60. A programmable SC filter section for the system of Fig. 59.

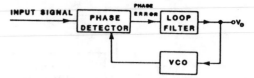

Fig. 61. Block diagram of a phase-locked loop (PLL).

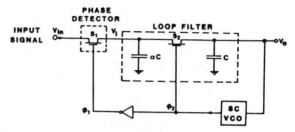

Fig. 62. Schematic diagram of an SC system realizing the PLL of Fig. 61.

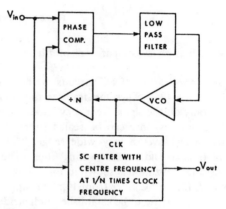

Fig. 63. The block diagram of a tracking filter.

detectors, in addition to a considerable amount of digital circuitry. Another receiver, which is capable of detecting dial pulses as well as dual-tone dial signals, is described in [79]. Besides many of the functions mentioned above, this receiver also contained AGC and all required filters, including the continuous-time antialiasing filters.

Another signal transmission application of SC circuits is in systems used to transmit digital data over analog telephone channels. Such SC modulator/demodulator systems (modems) have been described in [70] and [80]–[82]. These systems have included precision filters, as well as zero-crossing detectors, AGC circuits and, in some cases, [70], [80], even PLL's.

B. Speech Processing Circuits

Speech processing circuits represent some very recent examples of SC circuits. Most of these are used for the realization of speech *synthesis* systems, in which second-order programmable low-pass filters cascaded to simulate the vocal tract [83], [84]. Fig. 59 (from [83]) shows the block diagram of a typical speech synthesis system; Fig. 60 illustrates a possible programmable filter section. An alternative approach is described in [59], where a 9-bit SC multiplying DAC (MDAC) was used as a time-multiplexed multiplier in a lattice realization of the linear predictive coding (LPC) synthesis algorithm in the analog domain.

SC circuits have also been used in speech *recognition*. In the system described in [85], 16 bandpass filters followed by half-wave rectifiers and low-pass filters are integrated. The system performs a spectral analysis of the speech signal. In [83], a discrete prototype of a system is described which calculates the autocorrelation time lags of a speech signal. This calculation is the most time-consuming part of the LPC analysis.

C. Adaptive Networks

Since SC circuits can be digitally programmed, they have many applications in the realization of adaptive networks, including use as adaptive transversal filters, PLL's, and tracking filters.

The key part of an adaptive transversal filter system is the programmable transversal filter. This can be realized using CCD or BBD filters [86]–[88] with sampled-data peripheral circuitry. Such filters require a special double-polysilicon fabrication process.

Alternatively, programmable transversal filtering can also be accomplished by storing the samples of the input signal in analog sample-and-hold stages, and then accessing successively each sample-and-hold by a time-shared MDAC [89]–[91]. Using this approach, a 31-tap programmable transversal filter has been fabricated [91]. It achieves a 50- to 70-dB signal-to-noise ratio, depending on the transfer function programmed.

A PLL [92] consists of a phase detector, a loop filter, and a VCO (Fig. 61). All of these blocks can be realized using SC components, as illustrated schematically in Fig. 62. Hence, many different approaches have been proposed [60], [68], [70], [75], and [93]. However, few of these have in fact been fabricated in integrated SC form. A major reason for this is the jitter caused by the SC components [70]. This could possibly be eliminated by an on-chip very-low-frequency continuous-time low-pass filter; however, to date this has not been tried. It is expected that SC techniques will be used increasingly to realize at least some PLL

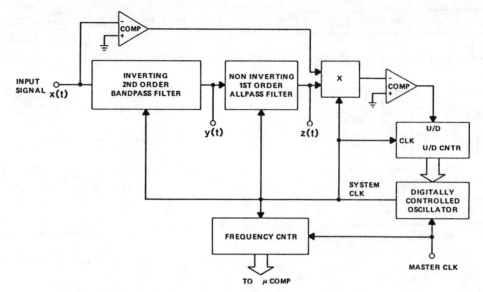

Fig. 64. An alternative tracking filter realization (from [94]).

components, such as phase comparators, prefilters, or even VCO's [70].

A promising application of SC circuits is the realization of tracking filters [60], [94]. Since the transfer function of an SC filter depends only on the signal frequency/clock frequency ratio f/f_c, the frequency response can be scaled simply by changing the clock frequency, over a very wide range. This allows the realization of flexible and stable tracking filters. The block diagram of a possible realization is shown in Fig. 63, where a PLL is used to derive f_c for an SC filter from V_{in}. This approach allows the SC filter characteristic to be established independently of the tracking requirements of the PLL.

An alternative system [94], in which the SC filter is an integral part of the PLL, is shown in a block diagram form in Fig. 64. This approach appears to be promising for the tracking of signals when the signal-to-noise ratio is small.

D. Future Applications

While it is always hazardous to try to predict even the near-term future in integrated electronics, a number of applications seem to be likely candidates for sampled-data analog realization. As SC techniques are extended to the high-frequency range [55], it seems likely that they will be applied in radio and video systems. Some feasibility studies have already been performed [95] on the SC realization of analog delay lines for video signals, and the results are promising. It appears feasible even at the present time to fabricate SC circuits to replace discrete components used in AM and FM communication systems. Indeed, it may be possible to design a single-chip integrated AM stereo radio receiver, complete except for a crystal, antenna, and the speakers.

More applications of sampled-data analog techniques can be expected in speech analysis and data transmission systems. Further development of MOS charge-transfer techniques are likely for A/D and D/A conversion, especially in high-precision applications.

Another likely area for SC circuit applications is music synthesis, which uses various waveform generators and programmable filters that are likely to be easily realizable by sampled-data analog techniques. Finally, sampled-data techniques may also be

applicable in sonar detectors, where tracking filters are often used.

VII. CONCLUSIONS

A few selected circuit design techniques have been described for SC circuits, including both filtering and other signal processing applications. Fully integrated systems applying SC technology were also discussed briefly, and the future potential of the field estimated. It seems likely that, in spite of the rapid advances in the technology of fully digital integrated circuits and the recent appearance of some fully analog ones, SC circuits will continue to represent an important option for the designer of integrated systems for the foreseeable future.

ACKNOWLEDGMENT

The authors are grateful to Dr. S. C. Fan, Dr. M. B. Ghaderi, Dr. T. H. Hsu, and F. J. Wong for critical reading of the manuscript, and for useful discussions.

REFERENCES

[1] G. C. Temes and J. W. La Patra, *Introduction to Circuit Synthesis and Design.* New York: McGraw-Hill, 1977, ch. 7.

[2] R. W. Brodersen, P. R. Gray, and D. A. Hodges, "MOS switched-capacitor filters," *Proc. IEEE*, vol. 67, no. 1, pp. 61–75, Jan. 1979.

[3] D. L. Fried, "Analog sample-data filters," *IEEE J. Solid-State Circuits*, vol. SC-7, pp. 302–304, Aug. 1972.

[4] L. R. Rabiner and B. Gold, *Theory and Application of Digital Signal Processing.* Englewood Cliffs, NJ: Prentice-Hall, 1975.

[5] J. T. Caves, M. A. Copeland, C. F. Rahim, and S. D. Rosenbaum, "Sampled analog filtering using switched capacitors as resistor equivalents," *IEEE J. Solid-State Circuits*, vol. SC-12, pp. 592–600, Dec. 1977.

[6] K. Martin and A. S. Sedra, "Strays-insensitive switched-capacitor filters based on the bilinear z-transform," *Electron. Lett.*, vol. 19, pp. 365–366, June 1979.

[7] R. Gregorian and W. Nicholson, "MOS switched-capacitor filters for a PCM voice codec," *IEEE J. Solid-State Circuits*, vol. SC-14, no. 6, pp. 970–980, Dec. 1979.

[8] B. J. White, G. M. Jacobs, and G. F. Landsburg, "A monolithic dualtone multifrequency receiver," *IEEE J. Solid-State Circuits*, vol. SC-14, no. 6, pp. 991–997, Dec. 1979.

[9] P. E. Fleischer, A. Ganesan, and K. R. Laker, "Parasitic compensated switched-capacitor filters," *IEEE Trans. Circuits Syst.*, vol. CAS-27, pp. 237–244, Apr. 1980.

[10] K. Martin, "Improved circuits for the realization of switched-capacitor

filters," *IEEE Trans. Circuits Syst.*, vol. CAS-27, pp. 237–244, Apr. 1980; also published as BNR Tech. Rep. TRIE 81-78-06, Mar. 1978.

[11] K. Martin and A. S. Sedra, "Exact design of switched-capacitor bandpass filters using coupled-biquad structures," *IEEE Trans. Circuits Syst.*, vol. CAS-27, pp. 469–475, June 1980.

[12] W. Poschenrieder, "Frequenzfilterung durch Netzwerke mit periodisch Schaltern," in *Analyse and Synthese von Netzwerken*, (Proc. NTG-Symp., Stuttgart, Germany, 1966), pp. 220–237.

[13] A. S. Sedra and P. O. Brackett, *Filter Theory and Design: Active and Passive*. Champaign, IL: Matrix Publ., 1978.

[14] M. S. Ghausi and K. R. Laker, *Modern Filter Design*. Englewood Cliffs, NJ: Prentice Hall, 1981.

[15] A. Fettweis, D. Herbst, B. Hoefflinger, J. Pandel, and R. Schweer, "MOS switched capacitor filters using voltage invertor switches," *IEEE Trans. Circuits Syst.*, vol. CAS-27, pp. 527–528, June 1980.

[16] B. Hosticka and G. Moschytz, "Switched-capacitor filters using FDNR like super capacitances," *IEEE Trans. Circuits Syst.*, vol. CAS-27, no. 6, pp. 569–573, June 1980.

[17] J. A. Nossek and G. C. Temes, "Switched-capacitor filter design using bilinear element modeling," *IEEE Trans. Circuits Syst.*, vol. CAS-27, pp. 481–491, June 1980.

[18] H. J. Orchard, "Inductorless filters," *Electron. Lett.*, vol. 2, pp. 224–225, June 1966.

[19] B. J. Hosticka, R. W. Brodersen, and P. R. Gray, "MOS sampled data recursive filters using switched capacitor integrators," *IEEE J. Solid-State Circuits*, vol. SC-12, no. 6, pp. 600–608, Dec. 1977.

[20] H. W. Bode, *Network Analysis and Feedback Amplifier Design*. New York: Van Nostrand, 1945.

[21] G. C. Temes, "Finite amplifier gain and bandwidth effects in switched-capacitor filters," *IEEE J. Solid-State Circuits*, vol. SC-15, no. 3, pp. 358–361, June 1980.

[22] K. Martin and A. S. Sedra, "Effects of the op-amp finite gain and bandwidth on the performance of switched-capacitor filters," *IEEE Trans. Circuits Syst.*, vol. CAS-28, no. 8, pp. 822–829, Aug. 1981.

[23] G. C. Temes, "First-order estimation and precorrection of parasitic loss effects in ladder filters," *IRE Trans. Circuit Theory*, vol. CT-9, pp. 385–400, Dec. 1962.

[24] M. L. Blostein, "Sensitivity analysis of parasitic effects in resistance terminated LC filters," *IEEE Trans. Circuit Theory*, vol. CT-14, pp. 21–25, Mar. 1967.

[25] G. C. Temes and H. J. Orchard, "First-order sensitivity and worst-case analysis of doubly terminated reactance two-ports," *IEEE Trans. Circuits Syst.*, vol. CAS-24, pp. 567–571, Sept. 1977.

[26] L. T. Bruton, "Low-sensitivity digital ladder filters," *IEEE Trans. Circuits Syst.*, vol. CAS-22, no. 3, pp. 168–176, Mar. 1975.

[27] B. Gold and C. M. Rader, *Digital Processing of Signals*. Boston, MA: Lincoln Lab, Mass. Inst. Technol., 1969, p. 95.

[28] G. M. Jacobs, D. J. Allstot, R. W. Brodersen, and P. R. Gray, "Design techniques for MOS switched-capacitor ladder filters," *IEEE Trans. Circuits Syst.*, vol. CAS-25, pp. 1014–1021, Dec. 1978.

[29] F. E. J. Girling and E. F. Good, "Active filters 12: The leap-frog or active ladder synthesis," *Wireless World*, vol. 76, pp. 341–343, July 1970.

[30] P. O. Brackett and A. S. Sedra, "Direct SFG simulation of LC ladder networks with applications to active filter design," *IEEE Trans. Circuits Syst.*, vol. CAS-23, pp. 61–67, Feb. 1976.

[31] K. Martin and A. S. Sedra, "Design of signal-flow graph (SFG) active filters," *IEEE Trans. Circuits Syst.*, vol. CAS-25, no. 4, pp. 185–195, Apr. 1978.

[32] ——, "Designing leap-frog and SFG filters with optimum dynamic range," *Proc. IEEE* (Lett.), vol. 65, pp. 1210–1211, Aug. 1977.

[33] F. Brglez, "Exact nodal analysis of switched capacitor networks with arbitrary switching sequences and general inputs—Part II," in *Proc. 1979 Int. Symp. Circuits and Systems* (Tokyo, Japan), pp. 748–751, July 1979.

[34] H. De Man, J. Rabaey, Arnout, and J. Vandewalle, "Practical implementation of a general computer aided design technique for switched-capacitor filters," *IEEE J. Solid-State Circuits*, vol. SC-15, no. 2, pp. 196–200, Apr. 1980.

[35] G. Müller and G. C. Temes, "A pauper's algorithm for switched-capacitor circuit analysis," *Electron. Lett.*, vol. 17, no. 25, pp. 942–943, Dec. 1981.

[36] K. Martin and A. S. Sedra, "Transfer function deviations due to resistor —SC equivalence assumption in switched-capacitor simulation of LC ladders," *Electron. Lett.* vol. 16, no. 10, pp. 387–389, May 1980.

[37] K. Haug, "Design, analysis and optimization of switched-capacitor filters derived from lumped analog models," *Arch. Elektr. Ubertragung*, vol. 35, pp. 279–287, July/Aug. 1981.

[38] T. C. Choi and R. W. Brodersen, "Considerations for high-frequency switched-capacitor ladder filters," *IEEE Trans. Circuits Syst.*, vol. CAS-27, no. 6, pp. 545–552, June 1980.

[39] M. B. Ghaderi, G. C. Temes, M. S. Lee, and C. Chang, "Bilinear switched-capacitor ladder filters—New results," in *Proc. IEEE Int. Symp. on Circuits and Systems*, pp. 170–174, Apr. 1981.

[40] D. S. Humpherys, *The Analysis Design and Synthesis of Electrical Filters*. Englewood Cliffs, NJ: Prentice-Hall, 1970, pp. 137–147.

[41] M. S. Lee and C. Chang. "Switched-capacitor filters using the LDI and bilinear transformations," *IEEE Trans. Circuits Syst.*, vol. CAS-28, no. 4, pp. 265–270, Apr. 1981.

[42] M. S. Lee, G. C. Temes, C. Chang, and M. G. Ghaderi, "Bilinear switched-capacitor ladder filters," *IEEE Trans. Circuits Syst.*, vol. CAS-28, no. 8, pp. 811–822, Aug. 1981.

[43] M. B. Ghaderi, "New design techniques for switched-capacitor bandpass filters," Ph.D. dissertation, UCLA, 1981.

[44] M. Yoshihiro, A. Nishihara, and T. Yanagisawa, "Low-sensitivity active and digital filters based on the node-voltage simulation of LC ladder structures," presented at the IEEE Int. Symp. on Circuits and Systems, Phoenix, AZ, Apr. 1977.

[45] P. E. Fleischer and K. R. Laker, "A family of active switched-capacitor biquad building blocks," *Bell Syst. Tech. J.*, vol. 58, pp. 2235–2269, Dec. 1979.

[46] M. B. Ghaderi, J. A. Nossek, and G. C. Temes, "Narrow-band switched-capacitor bandpass filters," *IEEE Trans. Circuits Syst.*, vol. CAS-29, no. 8, pp. 557–572, Aug. 1982.

[47] P. R. Gray, D. Senderowicz, H. Ohara, and B. Warren, "A single-chip NMOS dual-channel filter for PCM telephony applications," *IEEE J. Solid-State Circuits*, vol. SC-14, no. 6, pp. 981–991, Dec. 1979.

[48] K. C. Hsieh, P. R. Gray, D. Senderowicz, and D. C. Messerschmidt, "A low-noise chopper-stabilized differential switched-capacitor filtering technique," *IEEE J. Solid-State Circuits*, vol. SC-16, no. 6, pp. 708–715, Dec. 1981.

[49] "S3528 Programmable Low Pass Filter," Advanced Product Description, American Microsystems, Inc., Jan. 1983.

[50] J. McCreary and P. R. Gray, "All-MOS charge redistribution analog-to-digital conversion techniques—Part I," *IEEE J. Solid-State Circuits*, vol. SC-10, pp. 371–379, Dec. 1975.

[51] T. Foxall, R. Whitbread, L. Sellars, A. Aitken, and J. Morris, "A switched-capacitor bandsplit filter using double polysilicon oxide isolated CMOS," in *1980 ISSCC Dig. Tech. Papers*, pp. 90–91, Feb. 1980.

[52] R. Gregorian, "High resolution switched-capacitor D/A converter," *Microelectronics J.*, vol. 12, no. 2, pp. 10–13, Mar./Apr. 1981.

[53] R. W. Brodersen and S. P. Emmons, "Noise in buried channel charge-coupled devices," *IEEE J. Solid-State Circuits*, vol. SC-11, pp. 147–155, Feb. 1976.

[54] R. H. McCharles and D. A. Hodges, "Charge circuits for analog LSI," *IEEE Trans. Circuits Syst.*, vol. CAS-25, pp. 490–497, July 1978.

[55] P. R. Gray, R. W. Brodersen, D. A. Hodges, T. Choi, R. Kaneshiro, and K. Hsieh, "Some practical aspects of switched-capacitor filter design," in *1981 ISCAS Proc.*, pp. 419–422, Apr. 1981.

[56] K. Martin, "New clock feedthrough cancellation technique for analogue MOS switched-capacitor circuits," *Electron. Lett.*, vol. 18, no. 1, pp. 39–40, Jan. 1982.

[57] R. E. Suarez, P. R. Gray, and D. A. Hodges, "All-MOS charge redistribution analog-to-digital conversion techniques—Part II," *IEEE J. Solid-State Circuits*, vol. SC-10, pp. 379–385, Dec. 1975.

[58] D. J. Allstot, R. W. Brodersen, and P. R. Gray, "An electrically programmable analog NMOS second order filter," in *1979 ISSCC Dig. Tech. Papers*, pp. 76–88, Feb. 1979.

[59] R. Gregorian and G. Amir, "An integrated, single-chip, switched-capacitor speech synthesizer," in *1981 ISCAS Proc.*, pp. 733–736, Apr. 1981.

[60] K. Martin and A. S. Sedra, "Switched-capacitor building blocks for adaptive systems," *IEEE Trans. Circuits Syst.*, vol. CAS-28, no. 6, pp. 576–584, June 1981.

[61] J. F. Albarran and D. A. Hodges, "A charge-transfer multiplying digital-to-analog converter," *IEEE J. Solid-State Circuits*, vol. SC-11, pp. 772–779, Dec. 1976.

[62] T. Redfern, J. Connolly, S. Chin, and T. Frederiksen, "A monolithic charge-balancing successive approximation A/D technique," *IEEE J. Solid State Circuits*, vol. SC-14, no. 6, pp. 912–920, Dec. 1979.

[63] B. Fotouhi and D. A. Hodges, "High-resolution A/D conversion in MOS/LSI," *IEEE J. Solid-State Circuits*, vol. SC-14, no. 6, pp. 920–926, Dec. 1979.

[64] A. Dingwall, "Monolithic expandable 6-bit 20 MHz CMOS AID converter," *IEEE J. Solid-State Circuits*, vol. SC-14, no. 6, pp. 926–932, Dec. 1979.

[65] R. McCharles and D. A. Hodges, "Charge circuits for analog LSI," *IEEE Trans. Circuits Syst.*, vol. CAS-25, pp. 490–497, July 1978.

[66] K. Martin, "A high speed, high accuracy pipeline A/D convertor," in *Proc. 15th Asilomar Conf. on Circuits, Systems and Computers*, pp. 489–492, Nov. 1981.

[67] C. Hewes, D. Mayer, R. Hester, W. Eversole, T. Hiri, and R. Pettengill, "A CCD/NMOS channel vocoder," in *Int. Conf. on the Application of CCD's* (San Diego, CA), pp. 3A. 17–3A. 24, Oct. 1978.

[68] H. Khorramabadi, "NMOS phase lock loop," Univ. Calif., Berkeley, Int. Memo UCB/ERL, M77/67, Nov. 1977.

[69] W. Steinhagen and W. Engl, "Design of integrated analog CMOS circuits—A multichannel telemetry transmitter," *IEEE J. Solid-State Circuits*, vol. SC-13, no. 6, pp. 799–805, Dec. 1978.

[70] Y. Haque and V. Saletore, "An MOS phase locked loop for telecommunications applications," in *Proc. 15th Asilomar Conf. on Circuits,*

Systems and Computers, pp. 303–307, Nov., 1981.

[71] E. Vittoz, "Micropower switched-capacitor building blocks for adaptive systems," *IEEE J. Solid-State Circuits*, vol. SC-14, no. 3, pp. 662–624, June 1979.

[72] K. Martin, "A voltage controlled switched-capacitor relaxation oscillator," *IEEE J. Solid-State Circuits*, vol. SC-16, no. 4, pp. 412–414, Aug. 1981.

[73] K. Yamakido, T. Suzuki, H. Shirasu, M. Tanaka, K. Yasunari, J. Sakaguchi, and S. Hogiwara, "A single-chip CMOS filter/codec," *IEEE J. Solid-State Circuits*, vol. SC-16, no. 4, pp. 302–307, Aug. 1981.

[74] D. Marsh, B. Ahuja, T. Misawa, M. Dwarakanath, P. Fleischer, and V. Saari, "A single-chip CMOS PCM codec with filters," *IEEE J. Solid-State Circuits*, vol. SC-16, no. 4, pp. 308–315, Aug. 1981.

[75] A. Iwata, H. Kikuchi, K. Ushcimura, A. Morino, and M. Nakajima, "A single-chip codec with switched-capacitor filters," *IEEE J. Solid-State Circuits*, vol. SC-16, no. 4, pp. 315–321, Aug. 1981.

[76] R. Gregorian, G. Wegner, and W. Nicholson, "An integrated single-chip PCM voice codec with filters," *IEEE J. Solid-State Circuits*, vol. SC-16, no. 4, pp. 322–333, Aug. 1981.

[77] H. El-Sissi, M. Easter, V. Korsky, K. Siemnns, R. Wallace, and W. Siu, "A monolithic NMOS filter and line balance chip," in *1980 ISSCC Dig. Tech. Papers*, pp. 182–183, Feb. 1980.

[78] B. White, G. Jacobs, and G. Landsburg, "A monolithic dual tone multifrequency receiver," *IEEE J. Solid-State Circuits*, vol. SC-14, no. 6, pp. 991–997, Dec., 1979.

[79] P. Fleischer, V. Saari, T. Chien, J. Friend, G. Kraemer, A. Yiannoulos, W. Griffin, and I. Past, "A single-chip dual-tone and dial-pulse signalling receiver," in *1982 ISSCC Dig. Tech. Papers*, pp. 212–213, Feb. 1982.

[80] Y. Kuraishi, T. Makabe, and K. Nakayama, "A single-chip NMOS analog front-end LSI for modems," in *1982 ISSCC Dig. Tech. Papers*, pp. 146–147, Feb. 1982.

[81] L. Lin and H. Tseng, "Monolithic filters for 1200 baud modems," in *1982 ISSCC Dig. Tech. Papers*, pp. 148–149, Feb. 1982.

[82] R. K. MacDonald, M. Goddard, J. L. Schevin, and P. Carbou, "Single-chip 1200-b/s modem melds U.S. and European standards," *Electronics*, pp. 163–167, Jan. 13, 1983.

[83] R. W. Brodersen, P. Hurst, and D. Allstat, "Switched-capacitor applications in speech processing," in *Proc. 1980 ISCAS* (IEEE), pp. 732–737, Apr. 1980.

[84] L. McCready, A. Chowaniec, S. Watson, N. Mehta, and D. Choi, "A monolithic formant-based parametric speech synthesizer," in *1981 ISCAS Proc.*, pp. 986–988, Apr. 1981.

[85] L. Lin, H. Tseng, and D. Cox, "A monolithic audio spectrum analyzer for speech recognition systems," in *ISSCC Dig. Tech. Papers*, pp. 272–273, Feb. 1982.

[86] T. Enomoto, M. Yasumoto, T. Ishihara, and K. Watanabe, "Adaptive equalizer for wideband digital communication networks," in *1982 ISSCC Dig. Tech. Papers*, pp. 150–151, Feb. 1982.

[87] S. Tanaka, H. Tseng, L. Lin, and P. Chen, "An integrated realtime programmable transversal filter," *IEEE J. Solid-State Circuits*, vol. SC-15, no. 6, pp. 978–983, Dec. 1980.

[88] R. Haken, R. Pettengill, and L. Hite, "A general purpose 1024-stage electronically programmable transversal filter," *IEEE J. Solid-State Circuits*, vol. SC-16, no. 6, pp. 984–996, Dec. 1980.

[89] B. Ahaja, M. Copeland, and C. Chan, "A sampled analog MOS LSI adaptive filter," *IEEE J. Solid-State Circuits*, vol. SC-14, no. 1, pp. 148–154, Feb. 1979.

[90] O. Agazzi, D. Hodges, and D. Messerschmitt, "Echo canceller for a 80 kbs baseband modem," in *1982 ISSCC Dig. Tech. Papers*, pp. 144–145, Feb. 1982.

[91] S. Sunter, E. Girczyc, and A. Chowaniec, "A programmable transversal filter for voice-frequency applications," *IEEE J. Solid-State Circuits*, vol. SC-16, no. 4, pp. 367–372, Aug. 1981.

[92] F. M. Gardner, *Phaselock Techniques.* New York: Wiley, 1979.

[93] T. Viswanathan, S. Martaza, V. Syed, J. Berry, and M. Staszel, "Switched-capacitor frequency control loop," *IEEE J. Solid-State Circuits*, vol. SC-17, no. 4, pp. 775–778, Aug. 1982.

[94] K. Martin, "A switched-capacitor realization of a spectral line enhancer," in *1982 ISCAS Proc.*, pp. 229–232, May 1982.

[95] K. Matsui, T. Matsuura, and K. Iwasuki, "2 micron CMOS switched capacitor circuits for analog video LSI," in *1982 ISCAS Proc.* (Rome, Italy), pp. 241–244, May 1982.

A High Performance Low Power CMOS Channel Filter

WILLIAM C. BLACK, JR., MEMBER, IEEE, DAVID J. ALLSTOT, MEMBER, IEEE, AND RAY A. REED, MEMBER, IEEE

Abstract—A new CMOS PCM channel filter is described, which includes transmit and receive filters on a single die. This chip displays an idle-channel noise of typically 0 dBrnC0, a power supply rejection ratio of 40–50 dB at 1 kHz, and a fully operational power dissipation of only 35 mW, making it very cost effective in telecommunication switching systems. The design of this chip, including architectural, switched capacitor filter, and amplifier considerations is described, and typical experimental results are presented.

I. INTRODUCTION

PULSE-CODE-MODULATION (PCM) techniques have become quite common in telephone switching and transmission systems, and are almost universally employed in new systems. In these techniques, the analog subscriber lines are connected with the digital switching network by way of a subscriber-line unit, which performs all necessary interface and conversion functions. Typically, this line unit contains a 2/4 wire interface circuit, three precision filters (line-frequency reject, anti-aliasing, and output smoothing filters), and a Codec circuit, which performs channel encoding and decoding at an 8 kHz rate.

As circuitry to perform these functions must exist for each subscriber line, a significant portion of the total switching system cost rests in each component of the line unit. Years of effort in the area of monolithic A/D and D/A conversion techniques have succeeded in reducing the Codec to a fairly cost-effective integrated form [1]-[4], although attempts to integrate the filter and line-interface functions have, in general, proven less successful. Recently, switched capacitor techniques have offered the potential of very low-cost filters [5]. Many of these techniques have been employed in monolithic channel bank filters, which have often met the required frequency response characteristics [6], [7]. Monolithic realizations to date, however, have suffered from high-noise output, lower power supply rejection, and high-power dissipation, which have severely degraded the cost-effectiveness of these filters in switching systems.

In the device presented here, a number of new circuit techniques are used to provide improved performance over past monolithic implementations, and which, in fact, provides

Manuscript received April 28, 1980; revised August 15, 1980.
W. C. Black, Jr. is with the Electronics Research Laboratory, University of California, Berkeley, CA 94720.
D. J. Ellstot was with the Electronics Research Laboratory, University of California, Berkeley, CA 94720. He is now with MOSTEK, Inc., Carrollton, TX 75006.
R. A. Reed is with National Semiconductor Corporation, Santa Clara, CA 95051.

TABLE I
TYPICAL CHANNEL FILTER REQUIREMENTS FOR AMERICAN AND
EUROPEAN TELEPHONE SYSTEMS

PARAMETER	CONDITION		MAX DEVICE ALLOCATION
GAIN	300Hz-3000Hz		±.125dB
	60Hz	—XMT	—20dB
	4000Hz		—14dB
	>4600Hz	—XMT	—30dB
	>4600Hz	—RCV	—28dB
IDLE CHANNEL NOISE			10dBrnCO
TOTAL HARMONIC DISTORTION	—4dBmO		—50dB
POWER SUPPLY REJECTION	1kHz		MAX
POWER DISSIPATION			MIN

noise and power performance exceeding that of any hybrid-bipolar channel filter reported to date [8].

II. CHANNEL FILTER REQUIREMENTS

The filter circuitry within the subscriber line unit is divided into two separate sections: the transmit filter, which eliminates out of band noise before the incoming signal is encoded (line rejection and anti-aliasing), and the receive filter, which smooths the reconstructed decoder output before it is sent to the subscriber. The receive filter also performs a "$\sin(x)/x$" correction on the decoder output to compensate for the frequency roll-off effect of the decoder's sample and hold circuit.

The exact requirements for these two filters is, to a degree, different for many of the exchanges and transmission systems worldwide. Two sets of standards, however, are most widely used: the AT&T D3 specification and CCITT recommendation G712. A short summary of the filter characteristics which would be necessary to meet both specifications is given in Table I. The values which are listed in this table are generally regarded as the minimum specifications, as many users have more stringent requirements, particularly in the low-frequency rejection and idle-channel noise. Performance parameters which are not listed numerically do not have a specific requirement in either the D3 or CCITT specifications, but may seriously affect the overall economics of a device. In particular, poor power-supply rejection may require improved supply filtering (to the point of a separate *LC* filter or regulator per chip), while a high-power dissipation will greatly inflate power conversion, distribution, and lifetime utility costs and degrade system reliability.

III. FILTER ARCHITECTURE

This filter incorporates all of the subscriber line unit filtering functions on a single CMOS die. A double-polysilicon-gate CMOS process [9] was used to obtain dense, low-power, low-

Reprinted from *IEEE J. Solid-State Circuits*, vol. SC-15, no. 6, pp. 929–938, Dec. 1980.

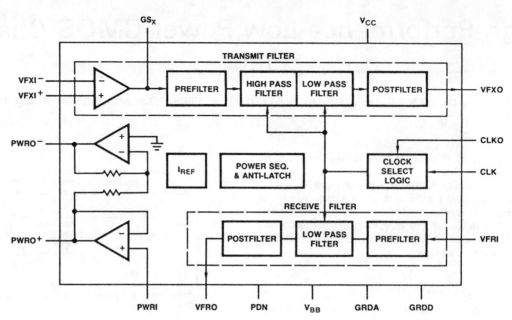

Fig. 1. Block diagram of monolithic channel filter.

noise analog circuitry, yet with relaxed design rules. A block diagram of the filter chip is shown in Fig. 1.

All precision filter sections are composed of bottom-plate switched capacitor ladder filters, which are totally insensitive to switch and substrate parasitic capacitances, and which have been designed for low-power, low-noise, and good supply rejection. Both transmit and receive filters are buffered by low-pass continuous filters which function as input anti-alias and as output smoothing filters, thus making the filters appear to be fully continuous. As opposed to other monolithic implementations of this function, which employ continuous filters on only the transmit side [6], [10] or not at all [11], this chip may be operated in even very noisy board environments and in asynchronous Codec applications with no degradation in performance.

Also included on the die are two balanced power drivers which can directly drive transformer 2/4 wire hybrid circuits, and a programmable clock generator which allows the device to be used with any of the common system clock frequencies: 1.536 MHz, 1.544 MHz, or 2.048 MHz. Under external control, the power drivers and/or the remainder of the chip may be turned off when not in use.

Power levels in this circuit are controlled by a programmable on-chip current reference which is adjusted at wafer probe to provide a low and consistent overall power dissipation, regardless of process variations. Protection from potentially damaging power line and input transients is provided by an on-chip power sequencer, several internal foldback current limiters, and an extensive input-gate clamping network.

IV. SPECIFIC FILTER STRUCTURES

In order to minimize overall component sensitivity, each of the precision filter sections is composed of a ladder network. This type of filter has been shown to display excellent sensitivity properties, and has proven to be easily integrable using switched capacitor techniques [12].

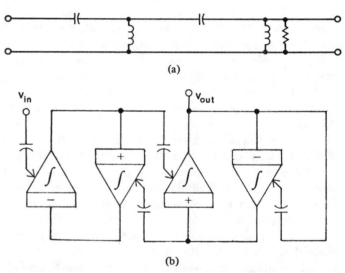

Fig. 2. High-pass ladder filter. (a) Passive RLC prototype. (b) Active ladder implementation using integrators. The capacitively coupled paths into the integrators are tied to inverting integrating nodes.

A. Transmit High-Pass Filter

This filter consists of a fourth-order singly terminated Chebyshev high-pass network which is clocked at 128 kHz. A schematic of the passive RLC circuit and its active ladder equivalent are shown in Fig. 2. This type of circuit was used to provide very good low-frequency rejection, yet with minimal noise introduction and no power-up stability problems. The high-pass clock frequency was ultimately determined by minimum capacitor geometry and power supply rejection considerations (see Section V-B), and was chosen to be as high as possible within a specified total die area. The area of this circuit could not have been made much smaller, even at lower clock frequencies, without degrading supply rejection performance.

The prefilter requirements of the high-pass circuit are considerably reduced from what its clock frequency would appear to indicate. Due to out of phase summation of continuous and sampled voltages in this circuit, the first in-band aliasing terms are actually at 256 kHz. This considerably reduces both the prefiltering circuitry required and the input aliased noise components.

All of the integrators and multiplier-summers were implemented with switched capacitor circuits of the type described in Section V. This enables the frequency response to be determined solely by capacitor ratios and the filter clock frequency.

Whereas in this design, the high-pass filter precedes the transmit low pass, an alternative organization reverses the order [7]. The apparent advantage of that organization is that, by band-limiting the input signal to 3.2 kHz, the following high pass can be clocked at only 8 kHz. This greatly reduces the required capacitor *ratios* for the high pass. However, as described in Section V-B, the power supply rejection is dependent upon the *magnitude* of the integrating capacitance, and hence, there is no savings in die area for a given supply rejection no matter what sampling rate is chosen. The principle disadvantage of this approach is that nearly all of the broad-band noise from the low-pass filter and each high-pass section is aliased into the passband, prior to even being sampled by the codec. The overall in-band dynamic range of the circuit may be further reduced when large, low-frequency components are present along with the voice signal. If these low-frequency components are not eliminated immediately upon entering the filter, internal clipping may occur with even a modest voice-band signal. The dynamic range of the "high-pass last" structure is, therefore, fundamentally lower than with the "high-pass first" configuration, which is employed in this chip and most other similar monolithic filters.

B. Transmit and Receive Low-Pass Filters

Both low-pass filters consist of fifth-order doubly terminated elliptic ladders which are clocked at 512 kHz. The type and order of these filters is very much dictated by the filter rolloff requirements listed in Table I. A diagram of the passive *RLC* prototype and the active equivalent of these circuits are shown in Fig. 3. The required "sin $(x)/x$" correction in the receive filter has been achieved by simply perturbing the filter's integrator coefficients [8] without adding additional hardware or significantly increasing component sensitivity.

The clock frequency of the low-pass filters was chosen to be as high as possible within reasonable op amp settling constraints. The 512 kHz clock used is, in fact, the fastest sampling rate yet reported for a switched capacitor filter. This provides substantially reduced in-band aliasing of broad-band noise (from both internal and external sources), by allowing the fixed frequency pre- and post-filters to perform more efficiently, and by reducing the number of potential aliasing frequencies within a given bandwidth. This considerably improves the effective dynamic range of this circuit, particularly in noisy system environments. This also allows the filter to be used without any degradation in performance in applications which use the transmit and receive sections asynchronously, or which result in the internal filter clock frequency not being an exact multiple of the codec sampling rate. As in the case of the high-pass circuit, each integrator and multiplier-summer is

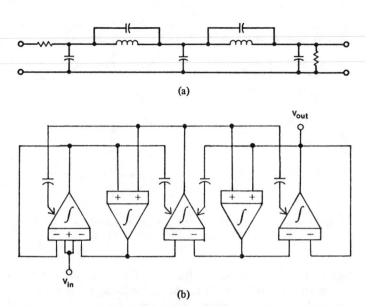

Fig. 3. Low-pass ladder filter. (a) Passive *RLC* prototype. (b) Active ladder implementation.

designed with switched capacitor circuits, eliminating the need for precision resistors of any type.

C. Continuous Filters

Each of the four continuous low-pass filters is composed of a second-order Sallen and Key section which is maximally flat in the voice-band. These filters use implanted-well resistors and MOS capacitors for frequency determining elements, and display a −3 dB frequency of typically 25 kHz. This allows suppression of in-band aliasing terms by nominally 40 dB in the transmit section, and by 53 dB in the receive section. In addition, all clock noise is suppressed by typically 53 dB before being passed on to the Codec or subscriber line, further reducing total channel noise. This latter specification is especially important in 1.544 MHz systems, where the derived on-chip filter clocks are not exact multiples of 8 kHz. In this design for example, the feedthrough at the transmit filter output due to the on-chip 514.67 kHz clock is sampled by the codec to produce an in-band 2.67 kHz difference frequency. As this clock feedthrough is attenuated by 53 dB before being sampled by the codec the 2.67 kHz component is small, so that the idle-channel noise is not significantly degraded. However, if a 257.67 kHz on-chip clock were derived, the in-band difference frequency of 1.67 kHz would only be attenuated by 41 dB before sampling, which would significantly increase the idle-channel noise. In addition to the high sampling rates used internal to this chip, a logic circuit which removes one of every 193 clock pulses [6] or a phase-locked-loop locked to 8 kHz [7] have also been used to avoid this problem.

V. Switched Capacitor Integrators

A. Basic Circuit Design

All of the switched capacitor filters use precision integrators of the type shown in Fig. 4. Of particular note in each of these circuits is that the transfer function from the input to the output is totally insensitive to any of the various parasitic

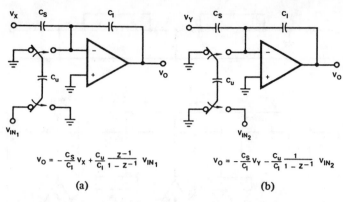

$$v_O = -\frac{C_S}{C_I}v_X + \frac{C_u}{C_I}\frac{z-1}{1-z^{-1}}v_{IN_1}$$

(a)

$$v_O = -\frac{C_S}{C_I}v_Y - \frac{C_u}{C_I}\frac{1}{1-z^{-1}}v_{IN_2}$$

(b)

Fig. 4. Parasitic insensitive switched capacitor integrators. (a) Non-inverting integrator. (b) Inverting integrator.

capacitances.[1] This allows the use of very small switched capacitors (approximately 0.125 pF in the high pass and 0.5 pF in each low pass), and fairly high clock frequencies. In addition, note that each integrator type has a continuous multiplier-summer path available, which may be used to implement transmission zeros [13].

The complete schematic of each precision filter is obtained by simply substituting the switched capacitor integrators shown in Fig. 4 for the continuous integrators shown in Figs. 2 and 3. Integrators with multiple inputs are realized by using a number of switched capacitors in parallel with the same amplifier. The top plates of each capacitor may be tied together, while the bottom plates are sent to switches which are phased according to the sign of the input, as shown in Fig. 4.

B. Power Supply Rejection and Noise

Although switch or substrate parasitic capacitances do not affect the principle transfer functions of the circuits shown in Fig. 4, these parasitics do allow power supply noise to couple into the integrating node. To a first order, the power-supply rejection of these circuits is simply the ratio of the total integrating capacitance to the parasitics on the integrating node, which are coupled to either supply. The effects of these parasitics may be minimized by using minimum geometry switches and by shielding integrating node lines with ground planes wherever possible. These procedures do not eliminate supply coupling, however, requiring that fairly large integrating capacitors be used for good supply rejection. Note that attempting to reduce the magnitude of the integrating capacitance to save area, will directly reduce the supply rejection as well.

In addition to supply rejection, the size of the integrating capacitor also plays a key role in filter noise performance. Because each of the switches in Fig. 4 exhibits a nonzero resistance when turned on, it is a source of thermal noise. This noise is integrated by capacitance C_I, and creates a total effective noise power which is inversely proportional to the integrating capacitance (kT/C noise). Therefore, making the integrating capacitor small will not only lead to poor supply rejection, but will have deleterious effects on filter noise as well. In this design, an average of 15 pF of poly/poly integrat-

[1] The circuit described here has been developed independently by the authors and at least several other workers [14]-[16].

ing capacitance is used in each stage, which is a compromise between supply rejection and noise considerations, and a smaller die size.

VI. AMPLIFIER DESIGNS

Four different amplifier designs are used in this device, each of which has been optimized for a specific circuit application. Two of the designs are transconductance amplifiers, which are used in the switched capacitor integrators and prefilter circuits, and can only drive a capacitive load. The other two circuits are complete operational amplifiers which are used in the post-filters and gain setting circuit (moderate-drive amplifier), and the push-pull power driver (large-drive amplifier). All of the amplifiers use similar input and second gain stages and a common compensation technique, while the two operational amplifiers use similar output stages, but with different sized driver devices. The considerations which were weighed in each design are discussed below.

A. Input Stage

The specifics of the input stage design have been very much dictated by the noise, common-mode range, and supply rejection constraints in the various circuit applications. This latter consideration, power-supply rejection, is fairly subtle in an integrator and is principly responsible for the type and the polarity of the input stage.

The sources of power-supply coupling in integrators are considerably different than those which affect circuits with dc feedback. In most monolithic implementations of this circuit, supply coupling via parasitic capacitance on the integrating node will dominate low-frequency PSRR behavior. This capacitance may be switch or layout related, as previously discussed, or associated with the amplifier used in the integrator. The two amplifier parasitic capacitances which are most important in terms of supply rejection are shown in Fig. 5. It is easy to show that because of these capacitances

$$\frac{\partial V_{out}}{\partial V_{SS}} \simeq \frac{C_{gs}}{C_I}\left[\frac{\partial I}{\partial V_{SS}} \cdot \frac{1}{2g_{m1}} + \frac{\partial V_{T_1}}{\partial V_{SS}}\right] \quad (1)$$

and

$$\frac{\partial V_{out}}{\partial V_{DD}} \simeq -\frac{C_{gd}}{C_I}\left[1 - \frac{\partial I}{\partial V_{DD}} \cdot \frac{1}{2g_{m3}}\right]. \quad (2)$$

The first of these two equations tends to be the more troublesome of the two in that C_{gs} is typically fairly large, and both terms within the brackets are usually of the same sign. This equation further states that attempts to reduce the input stage noise by increasing the device size (and hence, C_{gs}) will proportionately reduce the negative supply rejection. The first of the terms in (1) may be eliminated by employing a supply independent current source, while the second term in (1) may either be reduced by using input devices which are on a very lightly doped substrate, or eliminated by placing the input devices in an isolated well. This latter method was chosen, as it provides a considerable improvement in negative supply rejection over other alternatives in this process. This technique also allows a slight increase in common-mode range, which is useful in three of the four amplifier types.

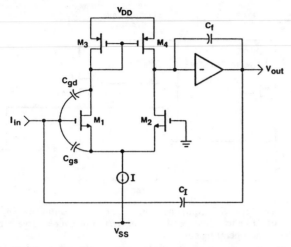

Fig. 5. Simple pole-split amplifier used as an integrator, showing two of the parasitic capacitors which effect low-frequency supply rejection.

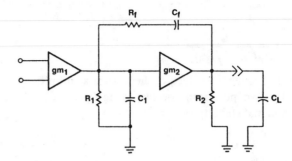

Fig. 6. Pole-split amplifier using RC compensation method.

Since the decision was made to use a supply-independent current source, the transfer function from the positive rail now becomes $-C_{gd}/C_I$. This may result in poor supply rejection for large width input devices, even in a self-aligned process. To eliminate this problem altogether, a cascode circuit was used in the integrator circuits to buffer the drain of the input devices from positive supply variations. Although this cascode configuration reduced the amplifier's common-mode input range, this is of no consequence in a switched capacitor integrator where there is no common-mode requirement. This cascode circuit was not needed in the output buffers and the amplifiers used in the continuous filters, as these circuits displayed lower impedance levels than the simple integrators and/or used dc feedback which greatly reduced the effects of capacitive supply coupling.

B. Supply-Independent Biasing

As mentioned above, some form of supply-independent biasing is desirable in that it allows improved power supply rejection. The supply-independent currents may either be generated separately in each amplifier, or only "mirrored" locally from a common source on the die. This latter method is preferrable in that it not only simplifies the overall amplifier complexity (and reduces the size), but also allows the overall chip power dissipation to be controlled from a single location. By incorporating a fusable link programming circuit into this "master" current source, the overall chip power dissipation may be closely controlled regardless of process variations. This procedure was used in this chip to obtain good power supply rejection and to maintain a small spread in chip power dissipation over large process variations. Note that this procedure is not possible in a depletion-load NMOS process where power dissipation is established by depletion-load currents in each amplifier. As depletion-load currents are intrinsically process sensitive, and not easily adjustable after fabrication, the power dissipation spread in these circuits is typically much larger than with this CMOS chip.

C. Compensation Technique

It is quite important in each of the various amplifier designs that the amplifiers be kept small and dissipate as little power

as possible. This is particularly true in the transconductance amplifiers, which are repeated a large number of times on the chip. These amplifiers have already been reduced in both die size and power by the exclusion of the output stage, and may be further improved by a judicious choice of the compensation technique. It is also important in each amplifier design that the unity-gain bandwidth not greatly exceed what the settling time requirements (as dictated by the clock frequency) indicate is needed. Excess amplifier bandwidth results in more broad-band noise being passed through each stage and, subsequently, being aliased into the passband. In amplifier types which exhibit a poor settling response or which are sensitive to process and/or ambient conditions, a large bandwidth may actually be required in order to guarantee that minimum settling requirements are met. Although noise and power supply rejection requirements force the use of a large input stage transconductance and large capacitive loads, the compensation technique used here still allows the use of small compensation capacitors and small bias currents. Furthermore, these amplifiers are process insensitive, and exhibit a good settling response over widely varying operating conditions.

A common problem in MOS amplifiers, which are compensated in the pole-splitting manner shown in Fig. 5, is the presence of a low-frequency right-half plane zero. This zero is present in bipolar circuits as well, but is typically at a high frequency due to the higher transconductance of bipolar devices. Several methods have been employed in MOS amplifiers to eliminate the effect of this zero. One approach is to drive the compensation capacitor with a voltage follower, which tends to replace the single right-half plane zero with a left-half plane doublet [17], while another is to move the position of the zero by the inclusion of a resistor in series with the compensation capacitor [18]. This latter method tends to be preferable, in that the ultimate die size and power it requires is quite small. This technique is demonstrated in Fig. 6.

In this figure, the effective output impedance of each stage is represented by resistors R_1 and R_2, while the effective capacitive loads on each stage are represented by C_1 and C_L. It may be shown for this circuit (given enough time), that

$$P_1 \simeq \frac{-1}{g_{m2}R_2R_1C_f} \simeq \frac{-g_{m1}}{A_vC_f}$$

$$P_2 \simeq \frac{-g_{m2}C_f}{C_1C_L + C_f(C_1 + C_L)} \simeq \frac{-g_{m2}}{C_L}$$

$$P_3 \simeq \frac{-1}{R_fC_1}$$

and

$$Z_1 = \frac{-1}{R_f C_f - C_f/g_{m2}}. \tag{3}$$

Note that when $R_f = 1/g_{m2}$, the zero is at infinity, leaving a basically two pole response (assuming $C_L \gg C_1$). For this case, minimal stability requires that

$$P_2 \geqslant A_v P_1$$

or

$$C_f \geqslant \frac{g_{m1}}{g_{m2}} C_L. \tag{4}$$

In the integrators used in this chip, noise and supply-rejection considerations have resulted in a fairly large g_{m1} and a large C_L. Thus, a large C_f or a large g_{m2} must normally be used in the amplifier if it is to meet minimal stability requirements. Unfortunately, a large C_f will increase die size and result in a slow settling time, while a large g_{m2} will require increased die size and/or power dissipation. In the compensation technique used in this chip, these unfavorable alternatives are avoided by bringing the zero in from infinity and placing it atop the second pole.[2] This is seen to occur when $R_f = (C_L + C_f)/g_{m2}C_f$, and results in the stability requirement reducing to

$$P_3 \geqslant A_v P_1$$

or eventually

$$C_f \geqslant \sqrt{g_{m1}/g_{m2} \cdot C_1 C_L}. \tag{5}$$

This is a substantially smaller capacitance than that required for the simple pole-split amplifier, allowing a considerable reduction in amplifier size in low-noise, large capacitive load applications. The difficulty with this technique is in maintaining the proper resistance R_f over large process, temperature, and supply variations. If this is not done, the $P_2 Z_1$ doublet will separate, resulting in an increased settling time. A circuit which alleviates this problem is shown in Fig. 7 where a tracking resistance scheme is used. In this circuit, the required series resistance is generated by the drain-to-source impedance of device M_A, whose gate (and hence, resistance) is biased by device M_C, voltage source V_{os2}, and current source I. The input of the second gain stage is represented by V_{in2}, whose quiescent value is equal to V_{os2}, while device M_B and the current source $K \cdot I$ comprise the second gain stage driver and load. Capacitances C_f and C_L are the compensation and load capacitances as before.

Using a simple MOS model, it is easy to show that if

$$\left(\frac{W}{L}\right)_A \simeq \left[\left(\frac{W}{L}\right)_B \cdot \left(\frac{W}{L}\right)_C \cdot K\right]^{1/2} \cdot \frac{C_f}{(C_f + C_L)} \tag{6}$$

then the resistance condition is satisfied, or

$$R_{ds_A} \simeq \frac{(C_f + C_L)}{g_{m2} C_f} \simeq R_f(\text{nom}). \tag{7}$$

[2] Because of the presence of P_3 and other high-frequency poles, the optimal settling condition is actually where Z_1 is at a slightly lower frequency than P_2.

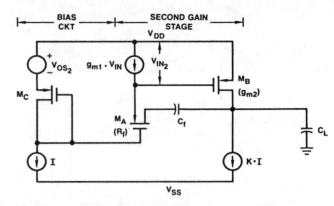

Fig. 7. Tracking RC compensation. Circuit continuously adjusts R (R_{ds} of M_A) to allow good settling response over a large range of process and operating conditions.

Note that (6) is totally independent of process, temperature, and supply variations, and is only a function of relative device sizes, which may be well-matched and easily specified. The use of this technique allows the internal amplifiers to drive up to 45 pF of integrating and load capacitance with only a 9 pF compensation capacitor, yet settle in less than 1 μs and dissipate only about 1 mW.

A schematic of the basic core transconductance amplifier and an associated die photograph are shown in Fig. 8. The device sizes in the second gain stage are different for some of the amplifiers, depending upon the size of the capacitive load. As indicated in the figure, the bias network is distinct from the individual amplifiers and is, in fact, shared over all of the amplifiers within a filter section. The reference current used by each bias generator is supplied by a single supply independent source, as previously discussed.

D. Output Stages

There are two distinct sets of requirements in this filter for an amplifier output stage. One of the applications is for a moderate drive device (10 kΩ and/or 50 pF), while the other is for a very large drive device (300 Ω and/or 500 pF). In each case, the output swings with 10 kΩ loads must exceed ±3.2 V, while the latter must maintain a ±2.5 V swing with 300 Ω loads. Further, each amplifier must display minimum quiescent power and be electronically controllable, entering a tristate power-down mode upon command.

In order to minimize quiescent power, a class B output stage is preferable to a class A design in both of the output amplifier types. This output circuit is usually implemented in conjunction with a class A second-gain stage as shown in Fig. 9(a). This configuration is desirable in that it is efficient, offers low crossover distortion and symmetrical output drive, and is well-behaved under capacitive loading.

A problem with this technique in MOS implementations, however, is that the dynamic range of the circuit is reduced because of body-effect related threshold increases at large output voltages. These threshold increases can easily reduce the output swing to below the previously stated requirements. In this process, the swing towards the positive rail may be improved by placing each of the NMOS devices in a separate p-well, which is shorted to the respective device's source. This

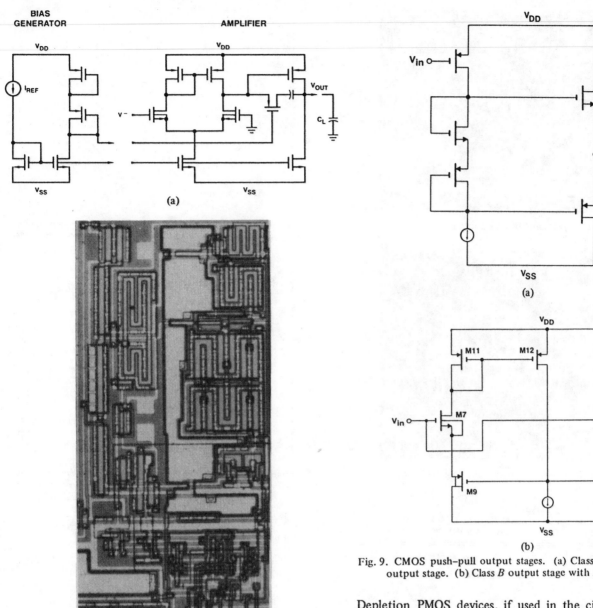

Fig. 8. Basic transconductance amplifier. (a) Schematic, showing amplifier and a bias generator. In switched capacitor integrators, an NMOS cascode is added to the amplifier input stage for increased positive supply rejection. (b) Die photograph showing bias generator (left) and amplifier (right). Bias generator is shared over up to six amplifiers.

Fig. 9. CMOS push–pull output stages. (a) Class *A-B* second gain and output stage. (b) Class *B* output stage with improved swing.

prevents any NMOS threshold increases from occurring, regardless of the output voltage. Maintaining a good swing in the negative direction, however, is more difficult. One approach is to replace the PMOS output source follower with an NMOS pull-down circuit which functionally resembles the PMOS device, but which is not affected by threshold changes at negative output values [17]. This type of circuit requires some form of local feedback from the output node, however, and is difficult to stabilize with large capacitive loads. Another approach is to use depletion PMOS devices, which exhibit a fairly low threshold voltage even with a large body bias.

Depletion PMOS devices, if used in the circuit of Fig. 9(a), however, may allow widely varying currents to flow through the driver devices as the diode-connected PMOS device may not always be in the saturation region. This problem may be alleviated by using an output stage of the type shown in Fig. 9(b). In this circuit, constant quiescent current is maintained, regardless of the output level, by sensing the current through the "dummy" output pair $M7$ and $M9$ by way of the PMOS current mirror $M11$ and $M12$. The current being pulled by $M12$ forces the gates of depletion devices $M9$ and $M10$ to a point which will keep the current through $M7$ and $M9$ constant, regardless of the depletion PMOS threshold voltage. As $M8$ and $M10$ are in parallel with $M7$ and $M9$, the current through these devices is also stabilized, although at a potentially higher level depending upon the device geometries. This type of output stage is used in each of the operational amplifier designs in this chip, where the sizes of the output devices are scaled to meet the different drive requirements. A die photograph of the power driver, which combines the previously discussed transconductance amplifier design with an output stage of this type, is shown in Fig. 10. This circuit may

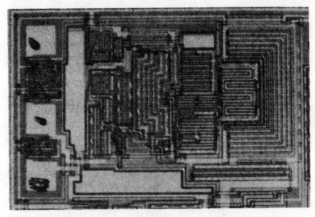

Fig. 10. Power amplifier die photograph.

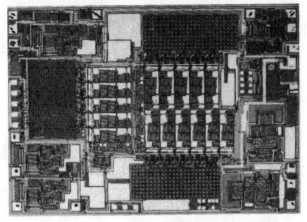

Fig. 11. Die photograph of monolithic channel filter.

TABLE II
MEASURED PERFORMANCE PARAMETERS FOR THE CMOS AMPLIFIER
FAMILY (SUPPLIES ARE ± 4.5 V)

ALL TYPES		CORE AMPLIFIER	
A_V	= 20,000 V/V	OUTPUT SWING	= ±4.0V
$\overline{V_N}$	= 70nV/$\sqrt{\text{Hz}}$ @ 1 KHz	Pd	= 0.8mW
V_{OS}	= 10mV	AREA	= 0.15mm²
CMRR	= 90dB	POWER AMPLIFIER	
PSRR[1]	= 75dB		
CMR[2]	= ±3.5V	OUTPUT SWING	= ±3V (R_L = 300Ω)
1) DC CLOSED LOOP		Pd	= 2.5mW
2) EXCEPT CORE INTEGRATOR		AREA	= .51mm²

be used to directly drive capacitive loads exceeding 1000 pF and/or resistive loads of as little as 300 Ω, yet with a quiescent power of only a few milliwatts. In addition, because the output stage is very efficient, total power dissipation increases less then 3.5 mW per power amplifier when driving a 0 dBm signal into 600 Ω.

VII. EXPERIMENTAL RESULTS

A die photograph of this dual-channel filter chip [19], which includes 21 amplifiers, 14 switched capacitor poles, and 8 continuous active *RC* poles, is shown in Fig. 11. Using a 5-6 μm minimum gate process, the total die size is 129 × 180 mils (3.28 × 4.57 mm). A summary of the amplifier characteristics is presented in Table II. Although the values listed are typical, the worst case degradation of the performance parameters is quite small, principally because of the power-

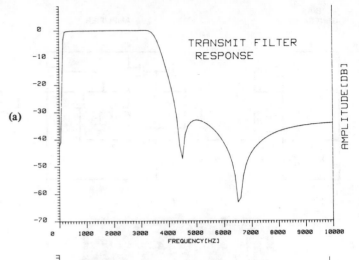

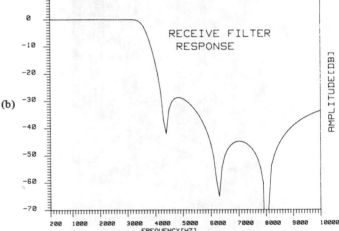

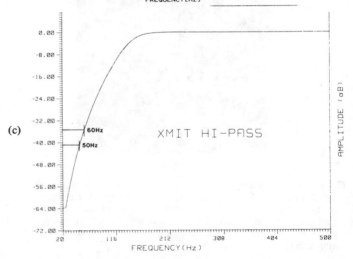

Fig. 12. Measured frequency response of a typical filter. (a) Receive response after sin *x*/*x* correction. (b) Transmit response. (c) Highpass response.

trimming strategy. Fig. 12 shows coarse filter responses of both receive and transmit filters as well as an expanded scale plot of the high-pass response. Note that at 60 Hz, the highpass provides typically 35 dB of rejection, while at 50 Hz, it is 41 dB. Fine resolution plots of the transmit and receive filters are shown in Fig. 13, which both demonstrate passband ripples well within the ±0.125 dB requirement. The idle-channel noise performance of this filter is illustrated in Fig. 14 where (a) shows the typical noise spectral characteristics of a

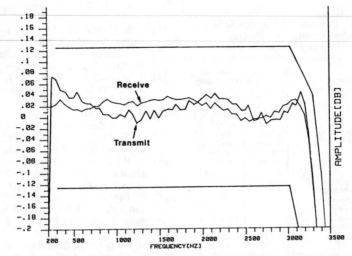

Fig. 13. Measured fine scale transmit and receive responses of a typical filter. Staircase appearance is due to quantization effects in the digital spectrum analyzer.

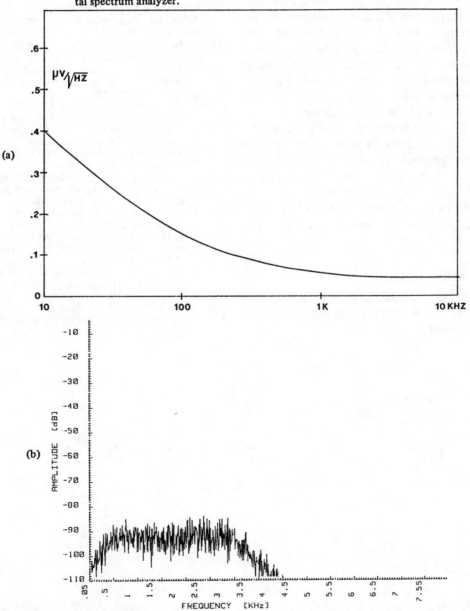

Fig. 14. (a) Measured input-referred noise in unity-gain configuration for transconductance amplifier. (b) Measured *C*-message weighted idle-channel noise at the transmit filter output with a reference of 0.1 V rms and a bandwidth of 10 Hz. Integrated noise in figure is slightly below 0 dBrnC0. Supplies are ±5 V for each figure.

TABLE III
TYPICALLY MEASURED PERFORMANCE PARAMETERS FOR THIS DUAL-CHANNEL CMOS FILTER AND THE ORIGINALLY STATED REQUIREMENTS. ALL SPECIFICATIONS ARE MET OR EXCEEDED

PARAMETER	CONDITION		MAX DEVICE ALLOCATION	TYPICAL PERFORMANCE
GAIN	300Hz-3000Hz		±.125dB	±.06dB
	60Hz	– XMT	– 20dB	– 35dB
	3400Hz		–.7dB	–.5dB
	4000Hz		– 14dB	– 15dB
	>4600Hz	– XMT	– 30dB	– 33dB
	>4600Hz	– RCV	– 28dB	– 29dB
IDLE CHANNEL NOISE			10dBrnCO	0dBrnCO
TOTAL HARMONIC DISTORTION	– 0dBmO		– 50dB	– 55dB
POWER SUPPLY REJECTION	1kHz		MAX	40dB-50dB
POWER DISSIPATION	– WITH POWER AMP ACTIVATED		MIN	
	·0dBm into 600 Ohms			
	Differential			45mW
	Single-ended			39mW
	· Idle Channel			33mW
	– WITH POWER AMP DEACTIVATED			28mW
	– IN POWER DOWN MODE			0.4mW

single transconductance amplifier, and (b) shows a typical C-message weighted noise spectrum of a complete transmit channel, which has an integrated noise of less than 0 dBrnc0. The noise of the receive channel is typically even lower than that of the transmit channel. The noise levels in both the transmit and receive channels remain approximately constant to beyond 85°C. A summary of the overall measured channel filter characteristics is presented in Table III along with the previously stated requirements. The viability of CMOS for designing PCM channel filters is perhaps most vividly demonstrated by the fact that all of the performance parameters presented in this paper were obtained from first revision parts. The precision frequency responses, low idle-channel noise, good power supply rejection, and low-power dissipation make this chip a very cost-effective component in telephone switching and transmission systems.

ACKNOWLEDGMENT

The authors wish to thank S. Patel, J. Wieser, F. Kawamoto, G. Warren, G. Simmons, and R. Thiels for their contributions to the success of this project.

REFERENCES

[1] J. B. Cecil et al., "A two-chip PCM codec for per-channel applications," in ISSCC Dig. of Tech. Papers, Feb. 1978, pp. 176, 177.
[2] G. F. Landsburg and G. Smarandoiu, "A two-chip CMOS codec," in ISSCC Dig. of Tech. Papers, Feb. 1978, pp. 180, 181.
[3] J. M. Huggins, M. E. Hoff, and B. M. Warren, "A single-chip NMOS PCM codec for voice," in ISSCC Dig. of Tech. Papers, Feb. 1978, pp. 178, 179.
[4] K. B. Ohri and M. J. Callahan, Jr., "Integrated PCM codec," IEEE J. Solid-State Circuits, vol. SC-14, pp. 38–46, Feb. 1979.
[5] R. W. Brodersen, P. R. Gray, and D. A. Hodges, "MOS switched capacitor filters," Proc. IEEE, pp. 61–71, Jan. 1979.
[6] P. R. Gray, D. Senderowicz, H. O'Hara, and B. W. Warren, "A single-channel dual-channel filter for PCM telephony applications," in ISSCC Dig. of Tech. Papers, Feb. 1979, pp. 26, 27.
[7] R. Gregorian, Y. A. Haque, R. Mao, R. Blasco, and W. E. Nicholson, Jr., "CMOS switched capacitor filter for a two-chip PCM voice codec," in ISSCC Dig. of Tech. Papers, Feb. 1979, pp. 28, 29.
[8] G. Warren, Nat. Semiconductor Corp., private communication.
[9] G. Simmons, R. Burnley, C. Seaborg, and K. Winter, "P² CMOS microcomputer family attains NMOS performance," Electronics, pp. 111–117, Nov. 1979.
[10] I. A. Young, D. B. Hildebrand, and C. B. Johnson, "A low-power NMOS transmit/receive IC filter for PCM telephony," in ISSCC Dig. of Tech. Papers, Feb. 1980, pp. 184–185.
[11] A. Iwata et al., "PCM CODEC and filter system," in ISSCC Dig. of Tech. Papers, Feb. 1980, pp. 178, 179.
[12] D. J. Allstot, R. W. Brodersen, and P. R. Gray, "MOS switched capacitor ladder filters," IEEE J. Solid-State Circuits, vol. SC-13, pp. 806–814, Dec. 1978.
[13] G. M. Jacobs, D. J. Allstot, R. W. Brodersen, and P. R. Gray, "Design techniques for MOS switched capacitor ladder filters," IEEE Trans. Circuits Syst., pp. 1014–1021, Dec. 1978.
[14] B. White, G. Jacobs, and G. Landsburg, "A monolithic dual-tone multifrequency receiver," IEEE J. Solid-State Circuits, vol. SC-13, pp. 991–997, Dec. 1979.
[15] D. J. Allstot, R. W. Brodersen, and P. R. Gray, "Considerations for the MOS implementations of switched capacitor filters," in Asilomar Dig., Nov. 1978, pp. 684–688.
[16] K. Martin, "Improved circuits for the realization of switched capacitor filters," IEEE Trans. Circuits Syst., Apr. 1980.
[17] Y. P. Tsividis and P. R. Gray, "An internally-compensated NMOS operational amplifier," IEEE J. Solid-State Circuits, vol. SC-13, pp. 748–753, Dec. 1976.
[18] D. Senderowicz, P. R. Gray and D. A. Hodges, "High-performance NMOS operational amplifier," IEEE J. Solid-State Circuits, vol. SC-13, pp. 760–766, Dec. 1978.
[19] W. C. Black, Jr., D. J. Allstot, S. Patel, and J. Weiser, "CMOS PCM channel filter," in ISSCC Dig. of Tech. Papers, Feb. 1980, pp. 84, 85.

Effects of the Op Amp Finite Gain and Bandwidth on the Performance of Switched-Capacitor Filters

KEN MARTIN, MEMBER, IEEE, AND ADEL S. SEDRA, MEMBER, IEEE

Abstract— A pair of complementary strays-insensitive switched-capacitor (SC) integrator circuits are analyzed to determine the errors in their transfer functions due to the finite gain and finite bandwidth of the op amp. The results are used to predict the transfer function deviation of biquadratic filter sections and *LC* ladder simulations. It is shown that while the effect of finite op amp gain is similar to that encountered in active-*RC* filters, SC filters are much more tolerant of the finite op amp bandwidth.[1] However, the relationship between transfer function error and finite op amp bandwidth is an exponential one as contrasted to the linear relationship of active-*RC* filters. Experimental results are presented.

Manuscript received April 16, 1980; revised January 24, 1981. This work was supported in part by the Natural Sciences and Engineering Research Council of Canada under Grant A7394.

K. Martin was with the Department of Electrical Engineering, University of Toronto, Toronto, Ont., Canada. He is now with the Electrical Sciences and Engineering Department, University of California, Los Angeles, CA 90024.

A. S. Sedra is with the Department of Electrical Engineering, University of Toronto, Toronto, Ont., Canada M5S 1A4.

[1]The authors would like to acknowledge Dr. Y. P. Tsividis, who in a private conversation was largely responsible for convincing them of this fact before the present analysis was carried out.

I. INTRODUCTION

THE switched-capacitor (SC) technique enables the design of filters that can be realized in monolithic integrated circuit form using current MOS technology. For this reason considerable effort has been recently directed towards finding suitable circuits and design techniques for SC filters [1]–[8]. Although many of the methods developed for active-*RC* filters can be directly adapted to SC filters, this is not true for the analysis of the effects of the amplifier dynamics on the filter response. In the SC case such an analysis is complicated by the fact that the amplifier dynamics are of continuous-time nature while discrete-time methods have to be used in evaluating the filter transfer function.

This paper considers the effect of the finite op amp gain and bandwidth on the performance of SC filters. As a result of a detailed time-domain analysis, simple formulas

Reprinted from *IEEE Trans. Circuits Syst.*, vol. CAS-28, no. 8, pp. 822–829, Aug. 1981.

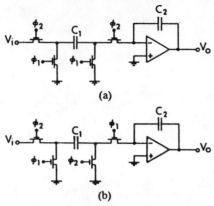

Fig. 1. A pair of strays-insensitive complementary integrators. (a) Inverting. (b) Noninverting.

TABLE I

Errors Due to The Finite dc Gain A_0	For Both Integrators: $\quad m(\omega) = -\frac{1}{A_0}\left(1+\frac{C_1}{C_2}\right)$		$\theta(\omega) = \dfrac{C_1/C_2}{2A_0 \tan\left(\frac{\omega T}{2}\right)}$
	For $(\omega_0 T/2) \ll 1$: $\quad m(\omega_0) \simeq -1/A_0$		$\theta(\omega_0) \simeq 1/A_0$
	Inverting Integrator		**Noninverting Integrator**
Errors Due to The Finite Unity-Gain Bandwidth f_t	$m(\omega) \simeq -e^{-k_1}\left[1-\left(\frac{C_2}{C_1+C_2}\right)\cos\omega T\right]$		$m(\omega) \simeq -e^{-k_1}\left(\frac{C_1}{C_1+C_2}\right)$
	$\theta(\omega) \simeq -e^{-k_1}\left(\frac{C_2}{C_1+C_2}\right)\sin(\omega T)$		$\theta(\omega) \simeq 0$
	For $(\omega_0 T)\ll 1$: $\quad m(\omega_0) \simeq -2\pi\left(\frac{f_s}{f_c}\right)e^{-\pi(f_t/f_c)}$		
	$\theta(\omega_0) \simeq m(\omega_0)$		$\theta(\omega_0) \simeq 0$
$k_1 = \pi\left(\frac{C_2}{C_1+C_2}\right)\left(\frac{f_t}{f_c}\right)$		$f_c = $ clock frequency $= 1/T$	

are presented for the errors in the magnitude and phase of the transfer function of integrator circuits. These error expressions are then used to predict the deviations in the transfer function of biquadratic filter sections and of filters designed as simulations of LC ladder networks. The results should be useful in the design of high-frequency SC filters as well as in predicting the transfer function errors of autiofrequency filters. Although the simplifying assumptions used in the analysis cause the predictions to be only approximate they do allow the designer to obtain quick estimates of the expected errors, reserving the more elaborate analysis method [9] for the final stages in the design process.

II. NONIDEAL RESPONSE OF SC INTEGRATORS

Fig. 1 shows a pair of complementary SC integrators[2] which are completely insensitive to stray capacitances between any node and ground [3]. Unless otherwise specified, it will be assumed throughout this paper that the outputs of these integrator circuits are sampled at the end of clock phase ϕ_2, and that each of the clocks has 50-percent duty cycles. For infinite amplifier gain and bandwidth the inverting integrator of Fig. 1(a) has the ideal transfer function $H_i(\omega)$:

$$H_i(\omega) = \frac{-(C_1/C_2)e^{j(\omega T/2)}}{j2\sin(\omega T/2)} \tag{1}$$

where the clock period $T = 1/f_c$, and f_c is the clock frequency. The ideal transfer function for the noninverting integrator of Fig. 1(b) is

$$H_i(\omega) = \frac{(C_1/C_2)e^{-j(\omega T/2)}}{j2\sin(\omega T/2)}. \tag{2}$$

Both these transfer functions show deviations from the ideal integrator transfer function of the form $(1/j\omega\tau)$. This deviation, however, is not of concern to us here and its effects have been discussed elsewhere [3], [4]. Rather, our objective here is to find the changes in the transfer func-

tions, from the "ideal" given by (1) and (2), due to finite gain and bandwidth. Toward that end we note that the actual transfer function $H_a(\omega)$ of an integrator circuit can be expressed in the form

$$H_a(\omega) = \frac{H_i(\omega)}{[1-m(\omega)]e^{-j\theta(\omega)}} \tag{3}$$

where $m(\omega)$ is the magnitude error and $\theta(\omega)$ is the phase error. Furthermore for small errors; $m(\omega), \theta(\omega) \ll 1$, (3) is approximately equivalent to

$$H_a(\omega) \simeq \frac{H_i(\omega)}{1-m(\omega)-j\theta(\omega)}. \tag{4}$$

As will be shortly seen this form is especially convenient for estimating the gain and phase errors of SC integrators and for evaluating the effects of these errors on the overall filter transfer function.

Assuming that the op amp has a finite dc gain A_0 it can be shown that the actual transfer function of both integrators is given by

$$H_a(\omega) = H_i(\omega) \Bigg/ \left[1+\frac{1}{A_0}\left(1+\frac{C_1}{2C_2}\right)-j\frac{(C_1/C_2)}{2A_0\tan\left(\frac{\omega T}{2}\right)}\right]. \tag{5}$$

Comparing this with (4) one can identify the gain and phase errors due to the finite dc gain A_0. Expressions for these errors $m(\omega)$ and $\theta(\omega)$ are given in Table I. Also given in Table I are approximate expressions for the error terms evaluated at the integrator unity-gain frequency ω_0, which from (1) and (2) is given by

$$\omega_0 = \frac{2}{T}\sin^{-1}\left(\frac{C_1}{2C_2}\right). \tag{6}$$

These approximate expressions are based on the assumption that the signal frequency is much smaller than the clocking frequency; specifically that $(\omega_0 T/2) \ll 1$.

Examination of the error expressions due to finite A_0 indicates that these errors are small and of the same order as those encountered in active-RC filters. The magnitude error of approximately $(-1/A_0)$ is equivalent to having

[2]Northern Telecom Limited has a Canadian Patent for the integrator of Fig. 1(a) in the names of K. Martin and S. Rosenbaum (No. 1088161, Filed April 3, 1978; Issued Nov. 21, 1980).

capacitor ratio inaccuracies of the order of $(1/A_0)$.

To find the effects of the finite op amp bandwidth we carry out a time-domain analysis of both the circuits in Fig. 1 assuming a uniform 6 dB/octave roll-off amplifier gain, that is

$$A(s) \simeq \frac{2\pi f_t}{s}.$$

This analysis is rather tedious and thus is delegated to the Appendix.[3] Assuming that the input is held constant during ϕ_2 and the output is sampled at the end of ϕ_2 the transfer function of the inverting integrator is shown in the Appendix ((A14)) to be

$$H_a(z) = H_i(z) \left\{ 1 - e^{-k_1} + e^{-k_1}\left(\frac{C_2}{C_1+C_2}\right)z^{-1} \right.$$

$$\left. \cdot \left[\frac{1 - e^{-(k_1+k_2)}}{1 - z^{-1}\left(\frac{C_2}{C_1+C_2}\right)e^{-(k_1+k_2)}} \right] \right\} \quad (7)$$

where

$$k_1 = \pi\left(\frac{C_2}{C_1+C_2}\right)\left(\frac{f_t}{f_c}\right) \text{ and } k_2 = \pi\left(\frac{f_t}{f_c}\right).$$

For a clocking frequency f_c no larger than $(f_t/2)$ and for the usual case of $C_1 < C_2$, e^{-k_1} and e^{-k_2} are usually quite small enabling us to make some approximations. Doing this and replacing z by $e^{j\omega T}$, the magnitude and phase errors listed in Table I are obtained. Also given in Table I are approximate expressions for $m(\omega_0)$ and $\theta(\omega_0)$ assuming that $\omega_0 T \ll 1$. Note that although the magnitude of the error terms increases as f_0 approaches $(f_c/2)$, their values will always be smaller than e^{-k_1}. Especially interesting is the exponential relationship between the error terms and the amplifier bandwidth f_t. This should be contrasted with the linear relationship encountered in active-RC filters. The implications of this will be illustrated later on.

The expressions for $m(\omega_0)$ and $\theta(\omega_0)$ in Table I indicate that for given values of f_0 and f_t, the integrator errors can be minimized by using as low a value of f_c as possible (although this would make the prefiltering more difficult).

The transfer function of the noninverting integrator of Fig. 1(b), taking into account the finite f_t of the op amp, can be shown to be approximately given by (see Appendix)

$$H_a(z) \simeq H_i(z)\left[1 - e^{-k_1}\left(\frac{C_1}{C_1+C_2}\right)\right]. \quad (8)$$

Substituting $z = e^{j\omega T}$ results in the magnitude and phase errors given in Table I. Here we note that although the magnitude error is approximately equal to that of the

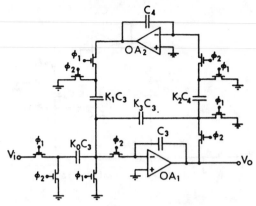

Fig. 2. A second-order bandpass section.

inverting integrator, the phase error is approximately zero. This is a direct consequence of the extra half-clock period available for the integrator to settle before its output is sampled by the inverting integrator input (at the end of ϕ_2, when the two integrators are connected in a loop).

It should be noted that in the cases where more than one input capacitor exist then using the sum of all input capacitances to replace C_1 will increase the accuracy of the formulas for the errors due to finite f_t. Finally it should be mentioned that the analysis presented is not applicable to circuits containing integrators with continuous "feed-ins" (i.e., an unswitched capacitor) unless these feed-ins have input signals which are step functions only.

III. Effect on the Performance of Biquads

Biquad circuits, such as the bandpass circuit in Fig. 2, are formed by connecting an inverting integrator and a noninverting integrator in a feedback loop. When these two integrators have magnitude errors $m_1(\omega)$ and $m_2(\omega)$, then assuming large Q-factors it can be shown that the biquad pole frequency ω_0 undergoes a fractional change given by

$$\frac{\Delta\omega_0}{\omega_0} \simeq \frac{1}{2}[m_1(\omega_0) + m_2(\omega_0)]. \quad (9)$$

Similarly, if the inverting and noninverting integrators have phase errors $\theta_1(\omega)$ and $\theta_2(\omega)$, then the realized pole Q-factor, Q_a, is related to the nominal Q by

$$Q_a \simeq Q/\{1 + Q[\theta_1(\omega_0) + \theta_2(\omega_0)]\}. \quad (10)$$

The change in the value of Q results in a change in the magnitude of the biquad transfer function $|T(j\omega)|$ at resonance given by

$$\frac{|T_a(j\omega_a)|}{|T(j\omega_0)|} \simeq \frac{1}{1 + Q[\theta_1(\omega_0) + \theta_2(\omega_0)]} \quad (11)$$

where T_a denotes the actual transfer function obtained.

The expressions for the integrator errors in Table I can be used in (9)–(11) to predict the frequency and gain errors of the two-integrator-loop biquad. When the op amps have both finite gain and finite bandwidth then simple calculus can be used to show that the total deviation is simply the

[3]An independent derivation of the transfer function of the noninverting integrator has been done by G. C. Temes in the *IEEE J. Solid-State Circuits*, vol. SC-15, pp. 358–361, June 1980.

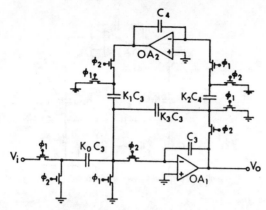

Fig. 3. A second-order bandpass circuit which has the same ideal transfer function as the circuit of Fig. 2 but which is more sensitive to the finite f_t. The analysis in this paper does not apply to this circuit.

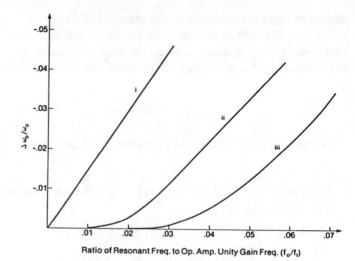

Fig. 4. A comparison of the deviation in pole frequency $\Delta\omega_0/\omega_0$ due to finite f_t for: (i) Tow–Thomas active-RC biquad, (ii) SC biquad with $f_0/f_c = 1/32$, and (iii) SC biquad with $f_0/f_c = 1/12$.

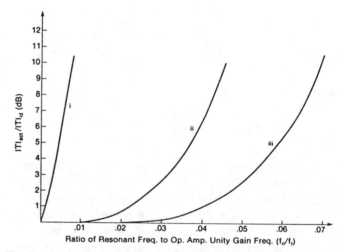

Fig. 5. A comparison of the deviation in resonance gain (due to finite f_t) for second-order bandpass sections with $Q = 25$: (i) Tow–Thomas active-RC biquad, (ii) SC biquad with $f_0/f_c = 1/32$, and (iii) SC biquad with $f_0/f_c = 1/12$.

sum of the deviations caused by the individual nonidealities (assuming small magnitude and phase errors in the integrators). However, care should be exercised in applying these formulas. First, it should be noted that the formulas of Table I have been derived assuming step input signals to the integrator. With reference to Fig. 2 we see that OA1 receives a step input at the beginning of ϕ_2 and OA2 receives a step input at the beginning of ϕ_1 (assuming perfect switches). Thus our formulas can be used for the circuit of Fig. 2 (assuming $K_3 \ll K_1$). There are biquad circuits, however, where this is not the case. An example is the circuit in Fig. 3 which for ideal op amps has a transfer function identical to that of the circuit in Fig. 2. Nevertheless, our analysis cannot be applied to the circuit in Fig. 3. This can be seen by noting that while OA1 still receives a step input at the beginning of ϕ_2, OA2 also receives its input during ϕ_2. If OA1 is not ideal, its output cannot change immediately and the input to OA2 will not be a step but rather an exponential ramp. Thus it should be expected that the circuit of Fig. 3 will have a greater sensitivity to the op amp finite bandwidth than the circuit of Fig. 2.

A second point to be observed in applying the formulas of Table I is that for calculating the finite bandwidth effects all the feed-in capacitances to an integrator should be summed to obtain an effective value for C_1. For example, if we wish to calculate the errors of the inverting integrator in the circuit of Fig. 2, due to the finite f_t, we use the equations in Table I with the effective value of (C_1/C_2) given by

$$\left.\frac{C_1}{C_2}\right|_{\text{effective}} = K_0 + K_1 + K_3.$$

For the noninverting integrator of this circuit we have

$$\left.\frac{C_1}{C_2}\right|_{\text{effective}} = K_2.$$

(Note: The damping capacitor K_3C_3 does not receive a step input which will make the results slightly incorrect, but since this capacitor is usually smaller than the other input capacitors, especially for high-Q filters, this problem has been ignored in order to keep the analysis tractible.)

To gain some insight into the effects of A_0 and f_t on biquad performance we present some numerical data. At the outset we note that if the clocking frequency f_c is at most $(1/5)$ of the op amp's bandwidth f_t the integrator phase and magnitude deviations, caused by the finite f_t, will be less than 0.04 percent for $C_1 < C_2$ (which is almost always the case). Thus for $f_c/f_t < 1/5$, the effects of the finite op amp bandwidth are negligible.

Fig. 4 shows curves of the predicted error ((9)) in pole frequency ($\Delta\omega_0/\omega_0$) versus (f_0/f_t) for the SC biquad of Fig. 2 at two different values of (f_0/f_c). Also shown for comparison is the curve for the Tow–Thomas active-RC circuit [13]. We note that for given values of f_0 and f_t one should use as low a value for f_c as possible. The frequency deviation in the SC circuit is obviously much smaller than that of the corresponding active-RC circuit.

The predicted deviations in resonant-frequency gain ((11)) are depicted in Fig. 5 from which conclusions similar to those mentioned above may be drawn. Finally, Fig. 6 gives the maximum allowable value of Q for a given error

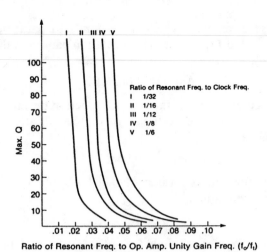

Fig. 6. Maximum allowable biquad Q-factor for less than 10-percent change in gain at resonance (due to finite f_t).

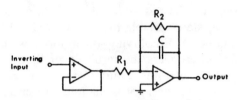

Fig. 7. Circuit used to simulate the finite op amp gain and bandwidth in the experiments.

TABLE II
PREDICTED AND MEASURED TRANSFER FUNCTION DEVIATIONS FOR BANDPASS SC BIQUAD (IDEAL Q-FACTOR $=25$, $f_c=5$ kHz, AND $f_0/f_c=1/12$)

	$A_o=170$, $f_t=\infty$		$A_o=\infty$, $f_t=10$kHz		$A_o=170$, $f_t=10$kHz	
	$\frac{\Delta\omega_o}{\omega_o}$	$\frac{|T_{act}|}{|T_{id}|}$	$\frac{\Delta\omega_o}{\omega_o}$	$\frac{|T_{act}|}{|T_{id}|}$	$\frac{\Delta\omega_o}{\omega_o}$	$\frac{|T_{act}|}{|T_{id}|}$
PREDICTED	−0.007	0.77	−0.007	1.16	−0.014	0.89
MEASURED	−0.008	0.76	−0.009	1.19	−0.015	0.87

in the resonance frequency gain, versus (f_0/f_t). The exponential nature of the effect of finite op amp bandwidth is clearly visible from the curves of Fig. 6. This exponential nature means that in SC filters predistortion for finite bandwidth effects is not practical.

Experimental Verification

A discrete prototype of the second-order SC biquad of Fig. 2 was built using the circuit of Fig. 7 to simulate the nonideal op amp. The nominal transfer function realized had a clock frequency to resonance frequency ratio (f_c/f_0) of 12 and a Q-factor of 25. The gain at resonance was chosen to be unity. Capacitor ratios K_0, K_1, K_2, and K_3 were calculated using formulas given elsewhere [3] to be

$$K_0=0.0210, \quad K_1=0.5204, \quad K_2=0.5204, \quad \text{and} \quad K_3=0.0217.$$

The sampling frequency used was 5 kHz. Originally, the simulated op amps had an infinite dc gain and a unity-gain bandwidth of 400 kHz ($C=66$ pF, $R_1=5.9$ kΩ, and $R_2=\infty$ in Fig. 7). The gain at resonance as well as the frequency of resonance were measured. Next the dc gains of the two op amps were changed to 170 ($R_1=5.9$ kΩ, $R_2=1$ MΩ in Fig. 7) and the resonance gain and resonance frequency were remeasured. For this case the predicted and measured data are given in Table II and show considerable agreement.

Next the simulated op amps were given unity-gain bandwidths of 10 kHz and dc gains of infinity ($R_1=5.9$ kΩ, $R_2=\infty$, and $C=2.7$ nF in Fig. 7), and the gain at resonance along with the frequency of resonance were measured. Again the predicted and measured results shown in Table II agree quite well.

Finally, the op amps were given finite gains ($A_0=170$) and finite bandwidths ($f_t=10$ kHz). The predicted values shown in Table II were calculated by simply adding the contribution due to finite A_0 to that due to finite f_t. From Table II we see that in this case too the theoretical values agree remarkably well with the measured data.

To verify our earlier remarks on the circuit in Fig. 3, this circuit was built using the same capacitor ratios as for the circuit of Fig. 2. While the resonance frequency shift caused by the op amp finite bandwidth was about the same, the Q enhancement (caused by the finite bandwidth) was approximately twice as much.

Finally, it should be mentioned that for an active-RC realization of the same transfer function with 10-kHz op amps, the resonance frequency shift would be -0.04 (5 times larger) while the Q enhancement would be large enough to cause the circuit to be unstable!

IV. EFFECT ON THE PERFORMANCE OF LC LADDER SIMULATIONS

The expressions in Table I can be used in conjunction with formulas from the passive filter literature [10]–[12] to evaluate the deviations in the transfer functions of SC filters designed as simulations of LC ladder prototypes. Specifically, an upper bound on the change in attenuation, $\Delta\alpha_{max}$, due to integrator magnitude errors can be obtained using the formula

$$|\Delta\alpha|_{max} < \frac{8.7|m||\rho|\omega\tau(\omega)}{1-|\rho|^2} \text{ dB} \qquad (12)$$

where $\tau(\omega)$ is the group delay, and ρ is the reflection coefficient. In this formula the magnitude error m corresponds to the uniform tolerance of the reactive elements of the LC ladder. On the other hand, the phase errors of the inverting and noninverting integrators ($\theta_1(\omega)$ and $\theta_2(\omega)$) are equivalent to the parasitic losses in the elements of the LC ladder, and thus give rise to the attenuation deviation

$$\Delta\alpha(\omega)\simeq 4.4\bigg\{[\theta_1(\omega)+\theta_2(\omega)]\omega\tau(\omega)$$
$$+\frac{1}{2}[\theta_1(\omega)-\theta_2(\omega)]\text{ Im}(\rho_1+\rho_2)\bigg\} \text{ dB} \qquad (13)$$

where ρ_1 and ρ_2 are the front-end and back-end reflection coefficients of the LC ladder network, and Im denotes "imaginary part." In the filter passband the second term

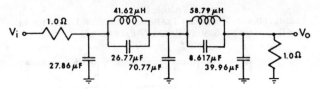

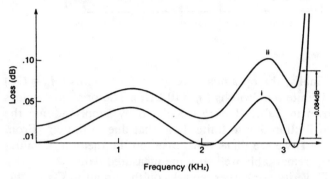

Fig. 8. *LC* ladder prototype for the fifth-order elliptic low-pass filter used in the simulation.

Fig. 9. Attenuation versus frequency obtained from simulation of a switched-capacitor realization of the fifth-order filter whose *LC* prototype is shown in Fig. 8. Clock frequency $f_c = 256$ kHz. The op amp dc gain is (i) 100 000, and (ii) 1000.

on the right-hand side of (13) is usually negligible, leading to the approximate formula

$$\Delta\alpha(\omega) \simeq 4.4[\theta_1(\omega) + \theta_2(\omega)]\omega\tau(\omega) \text{ dB}. \quad (14)$$

To gain some appreciation for the magnitude of errors involved and to verify the theory presented we consider a practical example. Fig. 8 shows *LC* ladder realization of a fifth-order elliptic low-pass filter which meets the following specifications.

Passband: 0 to 3.2 kHz with 0.035-dB ripple.
Stopband: $f \geqslant 4.6$ kHz with 36.3-dB minimum attenuation.

Analysis of this filter results in

$$\rho(3.2 \text{ kHz}) = 0.0896 \quad \tau(3.2 \text{ kHz}) = 355 \text{ } \mu\text{s}.$$

Let us consider a SC realization of this filter using the pair of complementary integrators of Fig. 1 and a clocking frequency of 256 kHz. Assuming that the op amps used have a finite dc gain of 1000, computer simulation[4] results in the modified transfer function shown together with the "ideal" response in Fig. 9. From this plot it can be seen that the maximum deviation in passband response occurs at 3.2 kHz and is approximately equal to 0.084 dB.

Using the formulas of Table I the magnitude and phase errors of the integrators are obtained as

$$m(\omega) \simeq -0.001 \quad \theta(\omega) \simeq 0.001.$$

Substituting for $m(\omega)$ into (12) yields

$$|\Delta\alpha|_{\max} = 0.0056 \text{ dB}.$$

This value is much smaller than the deviation caused by

[4] The simulation was done using the derived formulas for the integrator transfer functions with the op amps assumed to have finite dc gain and infinite bandwidth (i.e., (5)).

the phase error $\theta(\omega)$ which corresponds to parasitic dissipation in the *LC* ladder. This deviation can be evaluated using (14),

$$\Delta\alpha(\omega) = 0.062 \text{ dB}.$$

The two components of the transfer function deviation can be added to obtain an estimate of the overall attenuation deviation as

$$\Delta\alpha(\omega) = 0.068 \text{ dB}$$

which is close to the value of 0.084 dB obtained from computer simulation.

VI. Conclusions

We have presented simple formulas for the deviations in the transfer function of SC integrators caused by the finite gain and bandwidth of op amps. It has been shown that these formulas enable a reasonably accurate prediction of the deviation in the response of a class of SC filters designed using state-variable biquads or *LC* ladder simulation. A number of simplifying assumptions were used in the analysis, nevertheless the resulting predictions should still be useful in obtaining qualitative insights even in the cases where not all the approximations are entirely justified.

Specifically it has been shown that while the effect of finite dc gain is quite similar to that encountered in active-*RC* filters, the effect of finite bandwidth is much smaller in switched-capacitor filters composed of the stray-insensitive integrators of Fig. 1. This enables the design of moderately high frequency high-Q SC filters. However, the exponential nature of the dependence on the op amp finite f_t means that the transfer function deviation increases at a much higher rate than that encountered in active-*RC* filters. For this reason, it might not be practical to predistort for the finite op amp bandwidth.

It has also been shown that to minimize the dependence on the op amp finite bandwidth the clocking frequency should be selected as low as possible. For a given filter passband this implies the need to design using techniques such as the bilinear z-transform.

Appendix
Derivation of the Integrator Transfer Function with Finite Op Amp Bandwidth

A. The Inverting Integrator of Fig. 1(a)

Let the op amp have the transfer function

$$A(j\omega) \simeq -\left(\frac{\omega_t}{j\omega}\right) = \frac{V_0(\omega)}{V_1(\omega)}. \quad (A1)$$

This relationship can be expressed in the time domain as,

$$\frac{dv_0(t)}{dt} = -\omega_t v_1(t). \quad (A2)$$

Thus although $v_1(t)$ can be discontinuous at the switching instants the output signal $v_o(t)$ will be continuous. Examination of the circuit in Fig. 1 reveals that $v_1(t)$ will be discontinuous at the end of ϕ_1, that is at $t = (n-1/2)T$.

During clock phase ϕ_2 we have

$$v_o(t) - v_o\left(n - \frac{1}{2}\right) = \left(\frac{C_1 + C_2}{C_2}\right)\left[v_1(t) - v_1\left(n - \frac{1}{2}\right)^+\right]$$

(A3)

where it has been assumed that $v_i(t)$ remains constant during ϕ_2. Differentiating (A3) gives

$$\frac{dv_o(t)}{dt} = \left(\frac{C_1 + C_2}{C_2}\right)\frac{dv_1(t)}{dt}.$$

(A4)

Substituting (A4) into (A2) and solving results in the value of v_1 at $t = nT$ as

$$v_1(n) = v_1\left(n - \frac{1}{2}\right)^+ e^{-k_1}$$

(A5)

where

$$k_1 = \left(\frac{\omega_t C_2}{C_1 + C_2}\right)\left(\frac{T}{2}\right).$$

(A6)

Substituting the result in (A5) into (A3) gives

$$v_o(n) = v_o\left(n - \frac{1}{2}\right) - \left(\frac{C_1 + C_2}{C_2}\right)(1 - e^{-k_1})v_1\left(n - \frac{1}{2}\right)^+.$$

(A7)

In a similar manner we have during clock phase ϕ_1

$$v_1\left(n - \frac{1}{2}\right)^- = v_1(n-1)e^{-k_2}$$

(A8)

where

$$k_2 = \left(\frac{\omega_t T}{2}\right).$$

(A9)

Also,

$$v_o\left(n - \frac{1}{2}\right) = v_o(n-1) - (1 - e^{-k_2})v_1(n-1).$$

(A10)

At time $t = (n - 1/2)T$ capacitor C_1 is connected to the inverting input terminal of the op amp. The charge on C_1 is instantaneously distributed between C_1 and C_2 according to

$$v_1\left(n - \frac{1}{2}\right)^+ = \left(\frac{C_2}{C_1 + C_2}\right)v_1\left(n - \frac{1}{2}\right)^-$$
$$+ \left(\frac{C_1}{C_1 + C_2}\right)v_I\left(n - \frac{1}{2}\right).$$

(A11)

Using (A8), (A10), and (A11) into (A7) results in

$$v_o(n) = v_o(n-1) - \left(\frac{C_1}{C_2}\right)(1 - e^{-k_1})v_I\left(n - \frac{1}{2}\right)$$
$$- [1 - e^{-(k_1 + k_2)}]v_1(n-1).$$

(A12)

Also, using (A8) and (A11) into (A5) obtains

$$v_1(n) = \left(\frac{C_2}{C_1 + C_2}\right)e^{-(k_1 + k_2)}v_1(n-1)$$
$$+ \left(\frac{C_1}{C_1 + C_2}\right)e^{-k_1}v_I\left(n - \frac{1}{2}\right).$$

(A13)

As the input is assumed constant during ϕ_2 we can substitute

$$v_I\left(n - \frac{1}{2}\right) = v_I(n)$$

in (A12) and (A13). Subsequently we can take the z-transform of both equations and solve for the transfer function $V_o(z)/V_i(z)$,

$$\frac{V_o(z)}{V_i(z)} = \frac{-\left(\frac{C_1}{C_2}\right)}{1 - z^{-1}}\left\{1 - e^{-k_1} + e^{-k_1}\left(\frac{C_2}{C_1 + C_2}\right)z^{-1}\right.$$
$$\left. \cdot \frac{[1 - e^{-(k_1 + k_2)}]}{\left[1 - z^{-1}\left(\frac{C_2}{C_1 + C_2}\right)e^{-(k_1 + k_2)}\right]}\right\}.$$

(A14)

Usually $e^{-(k_1 + k_2)} \ll 1$ and second-order error terms may be ignored to obtain

$$\frac{V_o(z)}{V_i(z)} \simeq \frac{-(C_1/C_2)}{1 - z^{-1}}\left[1 - e^{-k_1} + e^{-k_1}\left(\frac{C_2}{C_1 + C_2}\right)z^{-1}\right].$$

(A15)

Substituting $z = e^{j\omega T}$ results in

$$\frac{V_o(\omega)}{V_i(\omega)} \simeq \frac{-(C_1/C_2)e^{j(\omega T/2)}}{j2\sin(\omega T/2)}\left\{1 + e^{-k_1}\left[\left(\frac{C_2}{C_1 + C_2}\right)\right.\right.$$
$$\left.\left. \cdot \cos\omega T - 1 - j\left(\frac{C_2}{C_1 + C_2}\right)\sin\omega T\right]\right\}.$$

(A16)

B. The Noninverting Integrator of Fig. 1(b)

The transfer function of the noninverting integrator of Fig. 1(b) can be derived using a procedure similar to that in the above. It should be noted, however, that in this case the voltage across C_1 is applied to the op amp inverting input during ϕ_1. Thus the discontinuity in $v_1(t)$ will occur at $t = (n-1)T$. Assuming that the output is sampled at the end of clock phase ϕ_2 the transfer function can be shown to be

$$\frac{V_o(z)}{V_i(z)} = \left[\frac{(C_1/C_2)z^{-1}}{1 - z^{-1}}\right]\left[1 - e^{-k_1}\left(\frac{C_1}{C_1 + C_2}\right)\right.$$
$$- \left(\frac{C_2}{C_1 + C_2}\right)e^{-(k_1 + k_2)}\left[1 + \left(\frac{C_2}{C_1 + C_2}\right)\right.$$
$$\left. \cdot \frac{e^{-(k_1 + k_2)}z^{-1}}{1 - e^{-(k_1 + k_2)}z^{-1}\left(\frac{C_2}{C_1 + C_2}\right)}\right].$$

(A17)

Usually $e^{-k_2} \ll 1$, allowing the approximation

$$\frac{V_o(z)}{V_i(z)} \simeq \left[\frac{(C_1/C_2)z^{-1}}{1 - z^{-1}}\right]\left[1 - e^{-k_1}\left(\frac{C_1}{C_1 + C_2}\right)\right].$$

Substituting $z = e^{j\omega T}$ results in

$$\frac{V_o(\omega)}{V_i(\omega)} \simeq \frac{(C_1/C_2)e^{-j(\omega T/2)}}{2j\sin(\omega T/2)}\left[1 - e^{-k_1}\left(\frac{C_1}{C_1 + C_2}\right)\right].$$

REFERENCES

[1] B. J. Hosticka, R. W. Brodersen, and P. R. Gray, "MOS sampled data recursive filters using switched capacitor integrators," *IEEE J. Solid-State Circuits*, vol. SC-12, pp. 600–608, Dec. 1977.

[2] J. T. Caves, M. A. Copeland, C. F. Rahim and S. D. Rosenbaum, "Sampled analog filtering using switched capacitors as resistor equivalents," *IEEE J. Solid-State Circuits*, vol. SC-12, pp. 592–600, Dec. 1977.

[3] (a) K. Martin, "Switched capacitor filters," Bell Northern Res. Internal Tech. Rep. TR1E81-78-06, Mar. 1978.
(b) ____, "Improved circuits for the realization of switched-capacitor filters," *IEEE Trans. Circuits Syst.*, vol. CAS-27, Apr. 1980.

[4] G. M. Jacobs, D. J. Allstot, R. W. Brodersen, and P. R. Gray, "Design techniques for MOS switched capacitor ladder filters," *IEEE Trans. Circuits Syst.*, vol. CAS-25, pp. 1014–1021, Dec. 1978.

[5] G. C. Temes, H. J. Orchard, and M. Jahanbegloo, "Switched-capacitor filter design using the bilinear z-transform," *IEEE Trans. Circuits Syst.*, vol. CAS-25, pp. 1039–1044, Dec. 1978.

[6] A. Fettweis, "Basic principles of switched-capacitor filters using voltage inverter switches," *AEU*, vol. 33, pp. 13–19, 1979.

[7] K. Martin and A. S. Sedra, "Strays-insensitive switched-capacitor filters based on the bilinear z-transform," *Electron. Lett.*, vol. 15, pp. 365–366, June 1979.

[8] P. Fleischer and K. Laker, "A family of active switched capacitor biquad building blocks," *Bell Syst. Tech. J.*, vol. 58, no. 10, pp. 2235–2269, Oct. 1979.

[9] M. L. Liou and Y. L. Kuo, "Exact analysis of switched capacitor circuits with arbitrary inputs," *IEEE Trans. Circuits Syst.*, vol. CAS-26, pp. 213–223, Apr. 1979.

[10] M. L. Blostein, "Sensitivity analysis of parasitic effects in resistance terminated LC filters," *IEEE Trans. Circuit Theory*, vol. CT-14, pp. 21–25, Mar. 1967.

[11] G. C. Temes and H. J. Orchard, "First-order sensitivity and worst-case analysis of doubly-terminated reactance two-ports," *IEEE Trans. Circuit Theory*, vol. CT-20, pp. 567–571, Sept. 1973.

[12] A. S. Sedra and P. O. Brackett, *Filter Theory and Design: Active and Passive*. Portland, OR: Matrix, 1978, chs. 7 and 8.

[13] L. C. Thomas, "The biquad: Part 1—Some practical design considerations," *IEEE Trans. Circuit Theory*, vol. CT-18, pp. 350–357, May 1971.

High-Frequency CMOS Switched-Capacitor Filters for Communications Application

TAT C. CHOI, RONALD T. KANESHIRO, ROBERT W. BRODERSEN, FELLOW, IEEE, PAUL R. GRAY, FELLOW, IEEE, WILLIAM B. JETT, AND MILTON WILCOX

Abstract —Bandpass filters for communications applications are realized using an 80 MHz differential single-stage CMOS op amp and a fully differential identical-resonator elliptic bandpass ladder filter configuration. Experimental results are given from a CMOS sixth-order 260 kHz elliptic bandpass filter with a Q of 40 clock frequency of 4 MHz, and a power dissipation of 70 mW.

I. INTRODUCTION

IN the past several years, monolithic switched-capacitor filters have been widely applied in the area of commercial communication circuits. These filters can be very accurate since their frequency response depends on capacitor ratios. They are also compatible with current MOS technologies, thus they are useful as interfacing circuits for digital systems [1], [2]. To date, most switched-capacitor filter applications have been limited to the audio range, examples being codec filters for PCM telephony [3]–[6] and filters for speech recognition systems [7]. However, communication systems require filtering above the voiceband. Examples are AM and FM intermediate frequency filtering in radio receivers, TV video processing, channel filters in FDM telephony systems, and filters for data communications. Presently, these filtering functions are realized by passive LC, crystal, or ceramic filters. The extension of switched-capacitor techniques into the several hundred kHz to MHz range would allow the elimination of many of these external components, so that systems, such as communication receivers, could be more fully integrated than is possible now.

This paper describes a design approach for the realization of high-frequency switched-capacitor filters in CMOS technology. While many of the concepts described here could be applied in NMOS technology as well [11], our work focused on CMOS both because of the fundamental advantages in the implementation of high-speed operational amplifiers, and because CMOS is likely to be desirable for other reasons in implementing the highly integrated single-chip subsystems and systems in which these filters would likely be applied.

In Section II, factors limiting the performance of switched-capacitor filters at high frequencies are explored, and design techniques proposed which allow a closer approach to the fundamental limits on achievable performance are examined. In Section III, the design of a fully differential, folded cascode CMOS operational amplifier used in the filters is described. Last, in Section IV, experimental results are given for a 260 kHz high-Q bandpass intended for AM IF filtering applications. In the final section conclusions are reached.

II. DESIGN TECHNIQUES FOR HIGH-FREQUENCY, HIGH-Q BANDPASS FILTERS

A. Introduction

The extension of the range of application of switched-capacitor filters to higher frequencies involves a number of problems, some associated with the high clock rates and resulting severe requirements on operational amplifier settling time, and another set of problems relating to the fact that most of the applications of these filters involve highly selective (high-Q) filter responses. The latter constraint leads to sensitivity problems because of the inherently increased sensitivity of high-Q filters both to the ratios of the capacitors in the filter as well as to the gain and settling behavior of the operational amplifiers. In addition, narrow-band filters require large capacitors ratios, which further aggravate the problems of accuracy and amplifier speed.

Another difficulty arises in the use of elliptic bandpass filters, which are more efficient in realizing the sharp roll-off characteristic often required in high-Q communication filters. The realization of elliptic bandpass filters in active ladder configuration can give rise to unstable dc conditions [2], thus driving the operational amplifier into

Manuscript received June 12, 1983; revised August 17, 1983. This work was supported by the National Science Foundation under Grants ENG-7907055 and DARPA N00039-B1-K-0251, and also by the National Semiconductor Corporation.

T. C. Choi, R. T. Kaneshiro, R. W. Brodersen, and P. R. Gray are with the Department of Electrical Engineering and Computer Sciences and the Electronics Research Laboratory, University of California, Berkeley, CA 94720.

W. B. Jett and M. Wilcox are with National Semiconductor Corporation, Santa Clara, CA 94270.

Reprinted from *IEEE J. Solid-State Circuits*, vol. SC-18, no. 6, pp. 652–664, Dec. 1983.

231

saturation. The successful implementation of high-frequency high-Q filters requires an effective approach which alleviates these problems.

Several authors have explored the use of N-path or pseudo N-path filtering techniques for high-Q filtering [12]–[15]. In the N-path filtering technique, the signal is translated from its initial center frequency to a range centered about zero frequency, low-pass filtered, and then retranslated back up to the original center frequency. In order that all spurious quadrature components introduced by the modulation cancel out, multiple paths are required which are spaced equally in the phase of the modulating signals. The great advantage of this technique is that the center frequency is determined precisely by an externally supplied clock, with no dependence on capacitor ratios, and that the actual filter itself is a low-pass operating at a frequency much lower than the center frequency. The disadvantages are that the modulating signal is at the filter center frequency where carrier feedthrough in the modulators will directly degrade the dynamic range of the filter, that the parallel paths must match to a high degree of accuracy, and that parasitic passbands exist starting at a frequency twice the passband frequency. It appears likely that this type of filter will find wide application in extremely high-Q applications where extreme accuracy of center frequency is paramount, and the more conventional filtering approaches are likely to be used where the required Q is more moderate and dynamic range is paramount. However, because of the wide dynamic range often required in communication filters, we chose a more conventional filtering architecture for this work.

B. Fully Differential Filter Implementation

Since the settling time of the operational amplifier places an upper limit on the allowable clock rate, the operating frequency of a switched-capacitor filter is primarily limited by the transient performance of the operational amplifier. Hence, the selection of an operational amplifier implementation which realizes the required settling time while still achieving sufficient dc gain is perhaps the most important consideration in the implementation of high-frequency filters. Another consideration affecting the choice of operational amplifier configuration is the fact that the effect of charge injection into the signal path from the switch transistors becomes much more important as the clock frequency is increased. This occurs because the switches themselves must be larger so as to minimize the charging time constant, thus increasing the charge injection. This larger clock feedthrough can degrade the effective settling time of the amplifier, degrade the power supply rejection if the clock voltage is dependent on a supply, and give rise to large dc offsets in the filter.

The power supply rejection ratio (PSRR) of the amplifier is particularly important in high-frequency high-Q filters since power supply variations can be coupled to some internal nodes of the operational amplifier and will eventually appear at the output. Because high-Q filters typically take the form of an array of resonators which are weakly coupled to each other, parasitic coupling paths through the power supply can have a strong effect on the response even if they are small in magnitude.

These considerations point strongly to the use of fully differential signal paths and a differential output operational amplifier. The use of differential techniques has been described earlier for voiceband filters [10]. Because it is fully differential, clock noise and power supply variations appear as common mode signals, and therefore will not affect filter response. An additional advantage in the high-frequency case is that the differential-to-single-ended conversion normally required within single-ended operational amplifiers is avoided, thereby improving the bandwidth of the amplifier. The design of such a differential high-speed CMOS amplifier is considered in detail in Section III.

C. Sensitivity in High-Q Bandpass Ladder Filters

As mentioned earlier, most applications in high-frequency communication systems require narrow-band filters ($Q \sim 40$), with a rather tight tolerance in the center frequency accuracy. Compared with voiceband codec filters, these narrow-band filters have pole locations much closer to the $j\omega$ axis. Typical codec filters are 5-pole, 4-zero low-pass filters, with the largest pole Q approximately equal to 3, whereas a sixth-order elliptic bandpass filter with a Q of 40 has individual pole Q's on the order of 100. This means that small variations in the pole locations can cause significant variations in the passband response, and may give rise to a large passband ripple or instability in the worst case. Also, since the filters have narrow bandwidths, a small shift in the center frequency can move the passband outside the frequencies of interest; hence, the filter design must be able to realize a stable center frequency.

To understand more about the sensitivity problem, let us consider a fourth-order all-pole bandpass filter in more detail. A leapfrog realization of the doubly-terminated LC filter in Fig. 1(a) has the signal flowgraph shown in Fig. 1(b). This is typical of leapfrog bandpass structures. The paths denoted by "$1/s\tau_i$" are the integrators, so that a resonator is formed when two such integrators are connected back to back. For this fourth-order filter, there are two resonators coupled together by the coupling paths "a" and "b." Also, "k_1," "k_2" are the termination paths which realize the source and load resistances. Each resonator resonates at a frequency equal to the center frequency of the filter, so that the filter center frequency is only a function of the resonator capacitor ratios and is independent of the coupling capacitor ratios. If we calculate the incremental sensitivity of the transfer function $T(s)$ of this filter, it can be shown that at the passband edge [16]

$$|S_{\tau_i}^{|T|}| \sim Q \qquad \text{where } i = 1, 2, 3, 4 \tag{1}$$

$$|S_x^{|T|}| \sim 1 \qquad \text{where } x = a, b, k_1, k_2. \tag{2}$$

In other words, at the passband edges, the magnitude of the frequency response is about Q times more sensitive to

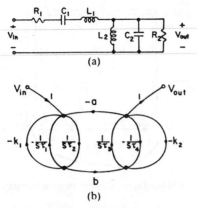

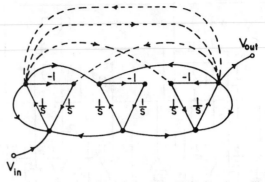

Fig. 1. (a) Fourth-order LC bandpass ladder. (b) Signal flowgraph of fourth-order LC bandpass ladder.

Fig. 2. Signal flowgraph of sixth-order elliptic bandpass filter.

the integrator capacitor ratio than the coupling capacitor ratio.

The significance of this is that the filter should be implemented in such a way that the resonators within the filter have center frequencies which match as closely as possible over all conditions of process variations, temperature variation, supply voltage variation, and so forth. For example, a 0.5 percent variation of all of the resonator center frequencies in the same direction would simply shift the overall response by 0.5 percent without altering its shape. However, a change of frequencies while the others remain constant would grossly distort the passband shape of a filter with an effective Q of, for example, 50.

One approach to achieving a high degree of resonator center frequency matching is to make them identical by making all the individual integrators identical, thus each integrator will have an integrating time constant equal to the reciprocal of the center frequency (in radians). In a switched-capacitor integrator, the integrator time constant τ is defined as

$$\tau = \frac{C_I}{f_s C_S} \tag{3}$$

where C_S is the sampling capacitor, C_I is the integrating capacitor, and f_s is the sample rate. If all the integrators are made identical with a time constant given by the center frequency f_o, then

$$\frac{1}{2\pi f_o} = \frac{C_I}{f_s C_S} \tag{4}$$

or

$$\frac{C_I}{C_S} = \frac{f_s}{2\pi f_o}. \tag{5}$$

Even if the clock rate to center frequency ratio is 20,

$$\frac{C_I}{C_S} \approx 3. \tag{6}$$

With this scheme of identical resonators, all the integrators have the same capacitor ratio, and hence all the

integrator time constants track each other. This is important because f_o is now only a function of one particular capacitor ratio. This ratio is small (6); hence the minimum size capacitor can be on the order of 1 pF without jeopardizing the operational amplifier settling time. With this kind of capacitor size, accuracy of better than half a percent can be easily achieved, and this is more than enough for most center frequency accuracy requirements.

Another obvious advantage of using this scheme is that it greatly simplifies the layout since all integrators are identical. Nevertheless, there is a disadvantage with this method. With identical resonators, the operational amplifier voltage swings are not properly scaled to have equal maxima, resulting in a reduction of dynamic range. In a narrow-band filter with Q on the order of 40, there is a loss of about 6 dB in dynamic range.

D. Component Spreads in High-Q Active Ladder Bandpass Filters

In addition to the sensitivity problem, the fact that high-frequency filters have high Q's also give rise to large spreads in the required capacitor ratios. This stems fundamentally from the fact that the energy stored in the resonator is much larger than the energy transferred into or out of it each cycle. Consider a sixth-order elliptic bandpass filter with a signal flowgraph shown in Fig. 2—the couplings between the resonators have capacitor ratios determined by the Q of the filter as well as the nature of the band edges. Typically, a high Q or a sharp roll off would mean a large ratio. A Q of 40 in this sixth-order elliptic bandpass would require capacitor ratios on the order of 100. In order to maintain adequate ratio accuracy, the minimum size capacitor cannot be too small ($\geqslant 0.1$ pF), and thus the large ratios would require large capacitors which in turn slow down the amplifier response.

There are two kinds of coupling paths between the resonators. A realization of the signal flowgraph in Fig. 2 using switched capacitor integrators is shown in Fig. 3. Here single-ended operational amplifiers are used, but the following analysis is true for fully differential operational amplifiers as well. It is seen that the coupling paths which feed into the outputs of integrators can be realized by feedforward capacitors, whereas those which feed into the

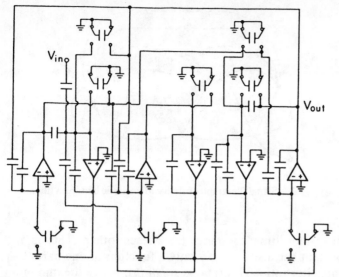

Fig. 3. Six operational amplifier realizations of sixth-order elliptic band-pass filters.

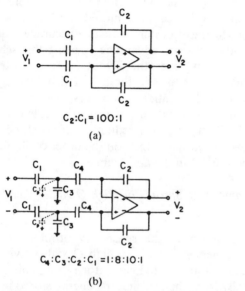

$$C_2 : C_1 = 100:1$$

(a)

$$C_4 : C_3 : C_2 : C_1 = 1:8:10:1$$

(b)

Fig. 4. (a) Direct realization of a gain of 0.01. (b) T-network scheme to realize a gain of 0.01.

inputs of integrators have to be realized via sampling capacitors. For the feedforward capacitors, a T-network scheme can be used to reduce the large coupling ratios required. A straightforward realization of a gain of 0.01 is shown in Fig. 4(a). Here, the gain of the circuit is given by

$$\frac{V_2(s)}{V_1(s)} = -\frac{C_1}{C_2} \qquad (7)$$

and the ratio $C_2 : C_1$ is equal to 100:1. A T-network scheme used to realize a similar gain of 0.01 is shown in Fig. 4(b). Here, the transfer function of the circuit is given by

$$\frac{V_2(s)}{V_1(s)} = -\frac{C_1}{C_1 + C_3 + C_4} \cdot \frac{C_4}{C_2}. \qquad (8)$$

The ratio $C_2 : C_3 : C_1 : C_4$ is equal to 10:8:1:1 for the same

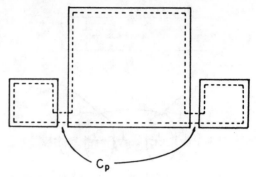

Fig. 5. T-network layout strategy to minimize capacitances.

gain of 0.01. Thus, a maximum capacitor ratio of 100 is reduced to 10 by this method.

The circuit in Fig. 4(b) is sensitive to parasitic capacitors C_p at the center node of the T-network. However, this capacitance can be reduced to a negligible amount by means of proper layout. Fig. 5 shows such a layout of the T-network. Here, the solid lines are the thin-oxide regions, and the dashed lines are the polysilicon of the capacitors (polysilicon to substrate capacitors are assumed). The polysilicon forms the center node of the T-network. Thus, parasitic capacitance C_p consists of only 2 squares of minimum feature size over the field oxide. For the case of Fig. 4(b), if the minimum size capacitors (C_1 and C_4) are equal to 0.1 pF, then C_3 is equal to 0.8 pF. Assuming a 4×4 μm minimum feature over a 0.8 μm field oxide, C_p is approximately equal to 1.4 fF. This C_p adds to C_3, the biggest capacitor of the T-network and 1.4 fF amounts to less than 0.2 percent of 0.8 pF. Hence, the parasitic capacitances are entirely negligible.

This T-network scheme is not applicable towards the coupling paths realized via sampling capacitors because of the additional switching involved and the added clock noise. Thus, it is desirable to convert this type of coupling to feedforward capacitors. This is done by redirecting the signal paths which go into an integrator input to the output of the other integrator of the same resonator. This is shown in Fig. 6(a) and (b). Here, the signal path α which feeds into the input of an integrator in Fig. 6(a) is redirected as shown in Fig. 6(b). However, this integrator output may feed other parts of the circuit as well. Hence, an equal amount of the signal has to be subtracted from those nodes fed by this output. In Fig. 6(b), this is shown by the signal path $\alpha\beta$. Using this method, the signal flowgraph in Fig. 2 can be redrawn in Fig. 7(a). It should be noted that Fig. 7(a) has been simplified by ignoring signal paths which represent a gain factor of 10^{-4} or less of an operational amplifier output. The corresponding switched-capacitor circuit for this flowgraph is shown in Fig. 7(b). Here, fully differential operational amplifiers are used. This circuit has 7 operational amplifiers instead of 6 because the filter output node has been modified and a signal has to be subtracted from it to regain the proper output. An extra operational amplifier is therefore needed to do the subtraction. In this circuit, all the coupling paths are now realized

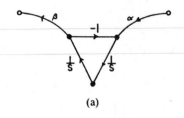

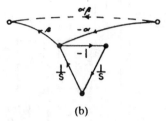

(a)

(b)

Fig. 6. Conversion of coupling paths via sampling capacitors to coupling paths via feedforward capacitors. (a) Original signal flow diagram. (b) Signal flow diagram after conversion.

by feedforward capacitors, and thus T-networks are applicable for reducing the capacitor ratios.

One drawback of the use of continuous coupling paths, such as those represented by T-networks, is the fact that the settling time behavior of the operational amplifiers within the filter is more complex than before, since continuous time paths exist which link the output voltage of one amplifier to the summing node of another during the settling interval. Fortunately, in the case of high-Q filters, the coupling path between operational amplifiers is sufficiently weak that the loop gain around any continuous-path loop is small compared to unity. Under these conditions, it can be shown that the settling time to a given accuracy is lengthened by a factor on the order of two for operational amplifiers whose inputs and outputs are interconnected by continuous time paths.

E. DC Stability in Elliptic Bandpass Active Ladder Filters

Elliptic bandpass filters are very useful in realizing narrow-band filters because of the higher selectivity at the band edges obtained from the transmission zeros. However, implementation of elliptic bandpass filters in the leapfrog configuration can result in unstable dc conditions in the circuit [2]. This problem arises from the presence of inductor loops in the LC ladder (capacitor loops may have similar problems), and is not unique to switched-capacitor filters. To see this, consider the circuits in Fig. 8, where an RC integrator is used to simulate the $I-V$ characteristics of an inductor. Here, active RC integrators are used for demonstration purposes, the same reasoning applied to switched-capacitor integrators. Assuming the operational amplifier is ideal, the transfer function of the integrator is given by

$$\frac{V_{\text{out}}(s)}{V_{\text{in}}(s)} = -\frac{1}{sRC}. \tag{9}$$

The $I-V$ characteristic of an inductor is given by

$$\frac{I(s)}{V(s)} = \frac{1}{sL}. \tag{10}$$

If $I(s)$ is resistively scaled to $V'(s)$, then

$$\frac{V'(s)}{V(s)} = \frac{I(s)R'}{V(s)} = \frac{R'}{sL} \tag{11}$$

where R' is the scaling resistance. Equations (9) and (11) show that, with an ideal operational amplifier, an inductor can be exactly simulated by an active RC integrator, where the RC time constant is related to L by

$$RC = \frac{L}{R'}. \tag{12}$$

The minus sign in (9) is not important since the inductor current can be defined with a different orientation.

If there is an offset voltage V_{os} in the operational amplifier [Fig. 8(a)], then

$$\frac{V_{\text{in}}(s) - V_{os}}{R} = sC(V_{os} - V_{\text{out}}(s)) \tag{13}$$

or

$$V_{\text{out}}(s) = -\frac{V_{\text{in}}(s) - V_{os}}{sRC} + V_{os}. \tag{14}$$

This relationship can be represented by the circuit in Fig. 8(b). If the small offset voltage at the output is ignored, then

$$\frac{V_{\text{out}}(s)}{V_{\text{in}}(s) - V_{os}} \approx -\frac{1}{sRC}. \tag{15}$$

Comparing (15) with (11), this would mean that instead of $V(s)$, $(V(s) - V_{os})$ is being integrated. Hence, in addition to simulating an inductor, there is a small voltage source in series which is equal to the offset voltage of the operational amplifier [Fig. 8(c)]. If such integrators are used to implement an inductor loop as in Fig. 9(a), it will result in the circuit in Fig. 9(b) where there are offset voltage sources in series with the inductors. Since it is very likely that these offsets do not add up to zero, they will cause infinite current in the inductors. In terms of the active equivalent circuit, there will be an operational amplifier driven by a voltage equal to $A(\Sigma_i V_{os})$, where A is the operational amplifier gain. Hence, if the offset voltages do not sum up to zero, then the operational amplifier will be driven into saturation.

The same problem can be analyzed in the context of the active filter. Consider the inductor loop in Fig. 10(a), with inductor voltages and currents defined as shown. This can be part of a doubly-terminated LC ladder. The signal flowgraph representing their $I-V$ characteristic is shown in Fig. 10(b). This is realized by active RC integrators as shown in Fig. 10(c), where V_i's are the integrator inputs, and V_o's are the operational amplifier outputs. Consider the dc condition of the circuit. In order to keep the output

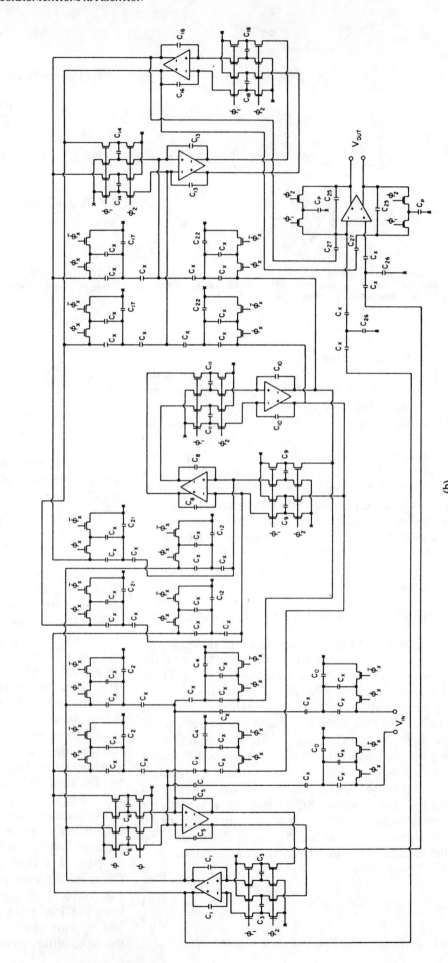

Fig. 7. (a) Signal flowgraph for sixth-order elliptic bandpass filter after converting coupling paths via sampling capacitors to coupling paths via feed-forward capacitors. (b) Seven operational amplifier realizations of sixth-order elliptic bandpass filters.

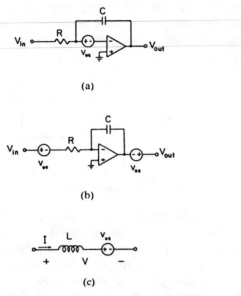

(a)

(b)

(c)

Fig. 8.   (a) An active *RC* integrator with offset voltage in the operational amplifier. (b) Equivalent circuit of Fig. 8(a). (c) Effect of operational amplifier offset voltage in the simulated inductor.

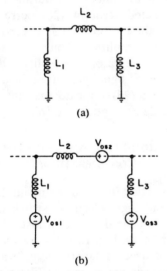

(a)

(b)

Fig. 9.   (a) An inductor loop. (b) An inductor loop simulated by active *RC* integrators with operational amplifier offset voltages.

voltage V_{o1} in active region, its input voltage is given by

$$V_{i1} \approx V_{os1} \qquad (16)$$

where V_{os1} is the offset voltage in the first operational amplifier. Likewise,

$$V_{i3} \approx V_{os3}. \qquad (17)$$

Then

$$V_{o2} = -A_2(V_{os1} + V_{os2} + V_{os3}) \qquad (18)$$

where A_2 is the gain of the second operational amplifier. If the offset voltages of the three operational amplifiers do not add up to zero, then V_{o2} will be driven into saturation.

To solve this problem, the *LC* loops in the passive filter can be broken up by Thevenin equivalent circuits using a technique similar to that by Jacobs *et al.* [8] for low-pass

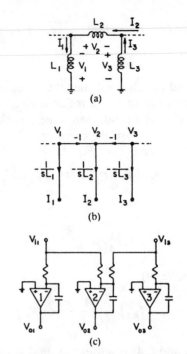

(a)

(b)

(c)

Fig. 10.   (a) An inductor loop. (b) Corresponding signal flowgraph. (c) Realization by active *RC* integrators.

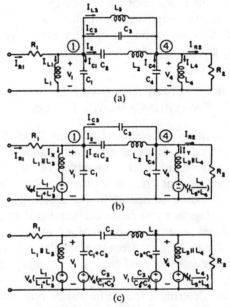

(a)

(b)

(c)

Fig. 11.   (a) Doubly-terminated sixth-order elliptic bandpass *LC* ladder. (b) Inductor L_s represented by voltage-controlled voltage source. (c) Final Thevenin equivalent circuit of Fig. 11(a).

filters. Fig. 11(a) shows a doubly-terminated sixth-order elliptic bandpass filter. By writing the nodal equation at node (1), we have

$$I_{R1} - \frac{V_1}{sL_1} - I_{C1} - I_2 - I_{C3} - \frac{V_1 - V_4}{sL_3} = 0. \qquad (19)$$

Rewriting (19), we have

$$V_1 = (I_{R1} - I_{C1} - I_2 - I_{C3})\frac{sL_1L_3}{L_1+L_3} + V_4\frac{L_1}{L_1+L_3}. \qquad (20)$$

Similarly, writing the nodal equation at node (4) will give

$$V_4 = (I_{C3} + I_2 - I_{C4} - I_{R2}) \frac{sL_3 L_4}{L_3 + L_4} + V_1 \frac{L_4}{L_3 + L_4}.$$
(21)

From (20) and (21), we see that the inductor L_3 can be represented by two voltage-controlled voltage sources, with corresponding change in inductor values, as shown in Fig. 11(b). Likewise, to remove the capacitor C_3, we wrote the nodal equation at node (1) of Fig. 11(b) and have

$$I_{R1} - I_x - I_2 - V_1(sC_1) - (V_1 - V_4)sC_3 = 0 \quad (22)$$

or

$$V_1 = (I_{R1} - I_x - I_2) \frac{1}{s(C_1 + C_3)} + V_4 \frac{C_3}{C_1 + C_3}. \quad (23)$$

Similarly, for node (4)

$$V_4 = (I_2 - I_y - I_{R2}) \frac{1}{s(C_3 + C_4)} + V_1 \frac{C_3}{C_3 + C_4}. \quad (24)$$

Equations (23) and (24) result in the final circuit shown in Fig. 11(c). The shunt inductor L_3 and capacitor C_3 are now replaced by voltage-controlled current sources. The corresponding switched-capacitor circuit for Fig. 11(c) requires only 6 operational amplifiers to implement since the voltage-controlled voltage sources can be easily realized by feedforward paths.

III. THE EFFECTS OF AMPLIFIER NONIDEALITIES ON SWITCHED-CAPACITOR FILTERS

The integrator is the principal building block for active filters realized by the leapfrog or the active ladder synthesis technique. Any nonideal characteristics of the integrator can cause the frequency response of active filters to deviate from their designed response. In particular the finite dc gain of the integrator can cause the passband of the filter to droop. Even a small amount of excess phase in the vicinity of the unity gain frequency of the integrator can result as peaking at the band edge and Q-enhancement for high selectivity filters.

In order to determine the performance requirements of the amplifier that is intended for high-frequency and high-selectivity filters, the relationship between the open-loop gain a_o, and the unity gain bandwidth ω_u of the amplifier and the two poles p_1 and p_2 of the integrator must be established. Here, p_1 is the low-frequency pole caused by the finite amplifier gain, and p_2 is the high-frequency pole attributed to the finite amplifier bandwidth. For continuous-time integrators, there is a straightforward connection between a_o and p_1, and ω_u and p_2 [17]. Unfortunately, such is not the case for switched-capacitor integrators. In order to obtain the exact relationship between the amplifier parameters and the frequency response of the switched-capacitor integrator, the system is first analyzed in the sample-data domain, and the result is then translated to the frequency domain by an appropriate mapping tech-

nique [18]. It can be shown that for a typical LDI switched-capacitor integrator, assuming a single-pole operational amplifier model, the corresponding equivalent integrator poles p_1 and p_2 are given by the following equations:

$$p_1 \approx \frac{1}{T_c} \left[\frac{1}{a_o \dfrac{C_i}{C_s + C_n + C_i}} \right] \quad (25)$$

$$p_2 \approx \frac{1}{T_c} \left[\frac{1}{\exp - mT_c \omega_u} \right] \quad (26)$$

where the terms are defined as follows:
T_c is the period of the clock phases,
a_o is the open-loop dc gain of the amplifier,
C_s, C_i, and C_n are the sampling, integration and the parasitic input capacitance, respectively,
m is the duty cycle of the clocks, and
ω_u is the unity gain frequency of the amplifier.
The bracketed term in (25) is the inverse of the loop gain of the switched-capacitor integrator during the integration phase. The bracketed term in (26) corresponds to the inverse of the settling error that is caused by the finite bandwidth of the amplifier. Thus, the degree of the switched-capacitor integrator nonideality is directly related to the amount of error signal present at the amplifier input. Equations (25) and (26) indicate that the gain and the bandwidth be made as large as possible.

IV. HIGH-SPEED AMPLIFIER DESIGN CONSIDERATIONS

It is clear now that the successful implementation of high-frequency and high-selectivity switched-capacitor filters is dependent on the design of a fast settling amplifier that has sufficient gain. Fig. 12 shows two amplifier topologies that may be used to satisfy the gain and bandwidth requirements. In switched-capacitor filters the amplifiers, which are configured as integrators, drive on-chip capacitive loads; therefore, as long as they are not required to interface with the outside world, they can have a high output impedance. This simplification allows the use of simple two-stage or single-stage amplifier design.

Assume for the sake of simplicity that the active devices in Fig. 12 are characterized by two parameters—transconductance g_m and output resistance r_o—and that they drive ideal current source loads as shown. Under these conditions, it can be shown that both amplifiers have identical low-frequency gain of $(g_m r_o)^2$. Their frequency response, however, is different. The dominant and the nondominant pole locations of the two-stage and the single-stage amplifiers are listed in Table I. If both of these amplifiers are utilized in a closed-loop configuration, such as in a switched-capacitor integrator, it can be shown that the settling behavior is determined by the time constant of the nondominant pole. This is because in a closed-loop configuration with increasing loop gain, the dominant and non-

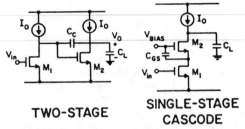

TWO-STAGE

SINGLE-STAGE CASCODE

Fig. 12. Two-stage and single-stage amplifiers.

TABLE I
DOMINANT AND NONDOMINANT POLE LOCATIONS
FOR THE TWO- AND SINGLE-STAGE AMPLIFIERS

	Dominant pole location	Nondominant pole location
Two-stage amplifier	$\dfrac{1}{r_o C_c g_m r_o}$	$\dfrac{g_m}{C_L}$
One-stage amplifier	$\dfrac{1}{r_o C_L g_m r_o}$	$\dfrac{g_m}{C_p}$

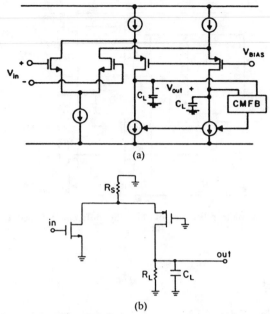

(a)

(b)

Fig. 13. (a) Simplified schematic of the fully differential folded cascode amplifier. (b) Small-signal equivalent-differential mode half-circuit.

dominant poles converge and form a complex pole pair. The settling behavior is determined by the reciprocal of the real part of the complex pole which is approximately equal to $-\frac{1}{2}s_n$, where $-s_n$ is the nondominant pole location in the s-plane. It should be noted that the presence of spurious high-frequency poles tends to push the complex pole pair toward the imaginary axis resulting in a reduced real part σ_p. Even in such a case, the resulting real part of the complex pole is still strongly influenced by the nondominant pole. In light of this information, to achieve a fast settling response, the amplifier must have a very high-frequency nondominant pole.

The ratio of the nondominant pole of the single-stage cascode to the two-stage configuration is given by

$$\frac{s_n\,(\text{one-stage})}{s_n\,(\text{two-stage})} = \frac{C_l}{C_p}. \tag{27}$$

In a switched-capacitor filter environment, C_p, which is equal to the total parasitic capacitance present at the cascode node, is typically one-fifth to one-tenth of the total load capacitance C_l, which the amplifier drives. Thus, the single-stage cascode amplifier exhibits a faster settling behavior than the two-stage amplifier.

To satisfy the gain and speed requirements for the high-frequency and high-selectivity filters, a variant of the single-stage cascode configuration, the folded cascode amplifier is proposed. A simplified schematic is shown in Fig. 13(a). The n-channel devices are the drivers, and the p-channel devices act both as the cascode elements and dc level shifters. The load capacitors C_l function as the compensation capacitor. The small-signal differential-mode half-circuit is depicted in Fig. 13(b). Here, R_s and R_l represent the finite output resistance of the current source loads. In order to maintain a high gain, it is crucial that the output resistance of the current is kept large. The exact

differential-mode voltage gain expression is given in (13):

$$A_v = -(g_{mn}r_{on})(g_{mp}r_{op})\left\{\frac{1}{1+\dfrac{r_{on}}{R_s}+\dfrac{r_{on}}{R_{on}}}\right\}\left\{\frac{1}{1+\dfrac{r_{op}}{R_l}}\right\} \tag{28}$$

where g_{mn} and r_{on} and g_{mp} and r_{op} are the transconductance and the output resistance of the NMOS and PMOS devices, respectively, and R_s and R_l are the finite output resistance of the current source loads, and

$$R_n = \frac{1}{g_{mp}}\left(1+\frac{R_l}{r_{op}}\right)$$

is the effective resistance seen looking into the source terminal of the p-channel cascode device.

If the output resistance of the loads is much larger than that of the active devices, the overall gain approaches the theoretical maximum, which is the combined $g_m r_o$ products of the n- and p-channel devices.

The complete schematic diagram of the fully-differential folded cascode amplifier is shown in Fig. 14. Two PMOS transistors, $MP1$ and $MP1A$, provide the bias current to the amplifier. In order to maintain a reasonably high output resistance, the channel length of these two devices is made longer than that of the cascode elements $MP2$ and $MP2A$. The high impedance current source loads at the output of the amplifier are realized by cascoded current source. $MN1$ through $MN2A$. The resulting output resistance is slightly over 1 MΩ.

The common-mode feedback circuit is an essential part of any fully-differential amplifiers. Without it the common-mode output remains undefined, and the amplifier may drift out of its high gain operating regime. $MN3$ and

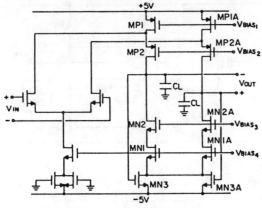

Fig. 14. Complete schematic of the fully differential folded cascode amplifier.

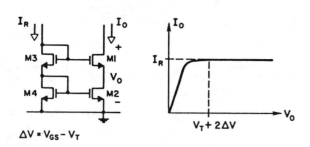

$\Delta V = V_{GS} - V_T$

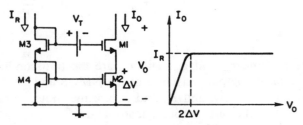

Fig. 15. Concept of the high-swing cascode bias circuitry.

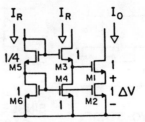

Fig. 16. Practical realization of the high-swing cascode bias.

TABLE II
OPERATIONAL AMPLIFIER PERFORMANCE (± 5 V SUPPLY)

Unity gain frequency (2 pF load)	80 MHz
Setting time to 0.1% of final value (2 pF load 2.5 V steps)	40 ns
Power dissipation	10 mW
Die area	200 mils^2
Open loop gain	1500

Fig. 16. All devices conduct identical biasing current of I_o. They all have the same aspect ratio except for $M5$ whose W/L is $\frac{1}{4}$ that of the others. This causes $M2$ to be biased at the edge of saturation with V_{DS} equal to V_{Dsat}. The voltage across $M1$ and $M2$ can now swing to within 2Δ V from the negative supply rail.

V. EXPERIMENTAL FILTER

An experimental 260 kHz bandpass filter having a selectivity of 40 was designed and fabricated. The specifications of this filter correspond to a typical AM IF filter for a car radio. In order to meet the specifications, a sixth-order elliptic bandpass filter is required. The switched-capacitor implementation is clocked at 4 MHz and through the use of T-networks, the maximum capacitor ratio spread has been reduced from 40 to approximately 8. The block diagram of the switched-capacitor sixth-order elliptic bandpass filter is as shown in Fig. 7. As indicated in Fig. 7(a), a seventh operational amplifier is added here as a summer at the filter output.

To meet the high-frequency high-Q requirements of the filter, a high-speed amplifier was designed and fabricated in CMOS technology. This amplifier employed the fully differential, folded-cascode configuration mentioned in Section IV, resulting in a unity gain bandwidth of 80 MHz and occupying a die area of 200 mils^2. The overall operational amplifier performance is summarized in Table II.

The experimental filter was fabricated using a 4 μm double-poly p-well CMOS technology. The microphotograph of the die is shown in Fig. 17. The frequency response obtained from a representative filter chip is shown in Fig. 18. The traces show the overall frequency response while the bottom traces depict the detailed passband response. The inner traces are for a Q of 40 filter, and the outer traces are for a Q of 20 filter. Both traces were obtained from the same filter chip. The Q has been changed

$MN3A$, which are biased in their triode region, form the common-mode feedback circuit. These two devices sample the common-mode output signal and feed back a correctional common-mode signal into the source terminals of $MN2$ and $MN2A$. The cascode devices, $MN1$ through $MN2A$, amplify this compensating signal to restore the common mode output voltage to its original level.

One of the major drawbacks of the cascode amplifier is its reduced output swing. As shown in Fig. 15 if the cascode devices $M1$ and $M2$ are biased from a diode string $M3$ and $M4$, the voltage across the cascodes can swing only to within $V_T + 2V_{Dsat}$ from the negative rail before $M1$ goes into the triode region. The voltage swing can be improved by inserting a level-shifting dc source whose value is exactly equal to V_T, between the gates of $M1$ and $M3$. The level-shifting voltage source forces $M2$ to be biased at the edge of saturation with a drain-to-source voltage that is equal to V_{Dsat}. Now the voltage across the cascode can swing to $2V_{Dsat}$ from the negative rail before $M1$ is pulled out of its saturation regime of operation. A practical realization of the high swing cascode bias circuit is given in

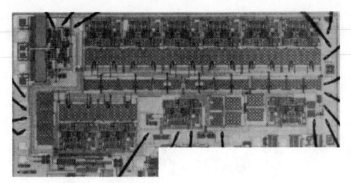

Fig. 17.   Die photo of the experimental filter chip.

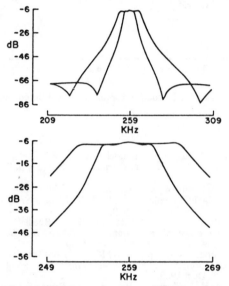

Fig. 18.   Frequency response of the filter chip.

TABLE III
MEASURED RESULTS OF AM IF FILTER AT $V_{DD} = 10$ V

	Specifications	Designed value	Experimental results
Center frequency	260 kHz ± 1%	260 kHz	259 kHz
Ripple content	3 dB max	0.97 dB	1.3 dB
−3 dB-BW	5 kHz min	7.2 kHz	6.5 kHz
Rejection at ± 10 kHz	30 dB min	34.7 dB	38 dB
Stopband rejection	55 dB min	60.4 dB	62 dB
Clock frequency	4 MHz	4 MHz	4 MHz
Gain	−6 dB	−6 dB	−6 dB
Dynamic range			70 dB
PSRR			30 dB
Total in band noise			300 μV_{rms}
Power dissipation			70 mV

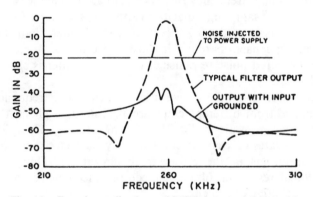

Fig. 19.   Experimentally observed PSRR response of filter chip.

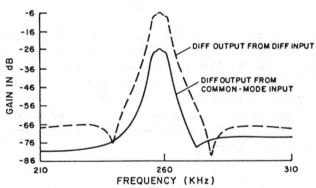

Fig. 20.   Experimental common-mode and differential-mode frequency response.

by switches which alter the amount of attenuation in the termination and the interresonator coupling paths. For both Q's the center frequency remains stable at 260 kHz. The pertinent statistics for this filter are listed in Table III, where we compare the specifications, the designed response, and the experimental results. All the specifications are met and the small discrepancies between the designed And measured results can be explained by small errors in the capacitor ratios.

Table III also shows the dynamic range, average PSRR, total in-band noise, and power dissipation of the experimental filter. The PSRR of the filter was smaller than expected since we would expect a fully differential filter architecture to be insensitive to power supply variations. A plot of the PSRR as a function of frequency is shown in Fig. 19, where we notice that the minimum PSRR in the passband is about 17 dB. This suggested that the circuit was not well balanced, and a close examination of the chip showed nonsymmetrical layouts in certain areas. To verify that this is the case, we measured the common-mode to differential-mode response of the filter by injecting an ac common-mode signal at the input. Indeed, we notice a large common-mode to differential-mode gain as shown in Fig. 20, where we plot the measured frequency response of both the differential-mode signal and the common-mode

signal. From this figure, the common-mode rejection ratio in the passband is only about 20 dB, thus showing an unbalanced circuit which gives rise to the smaller than expected PSRR.

The in-band noise of 300 μV_{rms} is large compared with voiceband low-pass filters, but this is expected of high-Q

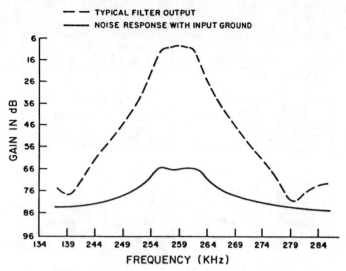

Fig. 21.   Measured noise response of filter chip.

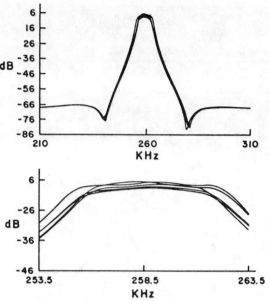

Fig. 22.   Ensemble frequency response of five different filter chips.

bandpass ladders. The fact that high-Q filters have greater noise basically stems from the inherent large gains of the resonators, which is only limited by the terminations, so that the noise contributions are enhanced by the large gains. In high-frequency filters, this situation is made worse because the sampling capacitors cannot be too large. In our case C_S is roughly 0.5 pF, which is smaller than the sampling capacitors found in most codec filters. A plot of the measured noise response of the AM IF filter is shown in Fig. 21.

In addition to the above measurements, the frequency response from different chips are measured to check the variations from die to die. This is shown in Fig. 22, which demonstrates very consistent frequency characteristics with no variations in the center frequency at all. Thus, the identical resonator filter architecture indeed helps to reduce sensitivity problems.

VI. CONCLUSION

A design approach to realize high-frequency and high-selectivity switched-capacitor filters has been presented. An experimental 260 kHz elliptic bandpass filter has been designed and fabricated. From the measured experimental performance of the filter, we can conclude that the switched-capacitor filtering technique is indeed a viable method in implementing high-frequency and high-selectivity monolithic filters.

REFERENCES

[1] R. W. Brodersen, P. R. Gray, and D. A. Hodges, "MOS switched-capacitor filters," *Proc. IEEE*, vol. 67, pp. 61–74, Jan. 1979.
[2] P. R. Gray, R. W. Brodersen, D. A. Hodges, T. C. Choi, R. Kaneshiro, and K. C. Hsieh, "Some practical aspects of switched-capacitor filter design," in *Proc. IEEE Int. Symp. Circuits Syst.*, Apr. 1981, pp. 419–422.
[3] P. R. Gray, D. Senderowicz, H. Ohara, and B. M. Warren, "A single-chip NMOS dual channel filter for PCM telephony applications," *IEEE J. Solid-State Circuits*, vol. SC-14, pp. 981–991, Dec. 1979.
[4] R. Gregorian and W. E. Nicholson, Jr., "CMOS switched-capacitor filters for a PCM voice codec," *IEEE J. Solid-State Circuits*, vol. SC-14, pp. 970–980, Dec. 1979.
[5] W. C. Black, Jr., D. J. Allstot, and R. A. Reed, "A high performance low power CMOS channel filter," *IEEE J. Solid-State Circuits*, vol. SC-15, pp. 929–938, Dec. 1980.
[6] I. A. Young, "A low-power NMOS transmit/receive IC filter for PCM telephony," *IEEE J. Solid-State Circuits*, vol. SC-15, pp. 997–1005, Dec. 1980.
[7] L. T. Lin, H. F. Tseng, D. B. Cox, R. G. Runge, and D. P. Conrad, "A monolithic audio spectrum analyzer for speech recognition systems," in *Dig. Tech. Papers, Int. Solid-State Circuits Conf.*, Feb. 1983, pp. 272–273.
[8] G. M. Jacobs, D. J. Allstot, R. W. Brodersen, and P. R. Gray, "Design techniques for MOS switched-capacitor ladder filters," *IEEE Trans. Circuits Syst.*, vol. CAS-25, pp. 1014–1021, Dec. 1979.
[9] T. C. Choi and R. W. Brodersen, "Considerations for high-frequency switched-capacitor ladder filters," *IEEE Trans. Circuits Syst.*, vol. CAS-27, pp. 545–552, June 1980.
[10] K. C. Hsieh, P. R. Gray, D. Senderowicz, and D. G. Messerschmitt, "A low-noise chopper-stabilized differential switched-capacitor filtering technique," *IEEE J. Solid-State Circuits*, vol. SC-16, pp. 708–715, Dec. 1981.
[11] J. Guinea, "High frequency NMOS switched-capacitor filters," Ph.D. dissertation, University of California, Berkeley, June, 1982.
[12] L. E. Franks and I. W. Sandberg, "An alternative approach to the realization of network transfer functions the N-path filter," *Bell Syst. Tech. J.*, pp. 1321–1350, Sept. 1960.
[13] A. Fettweis and H. Wupper, "A solution to the balancing problem in N-path filters," *IEEE Trans. Circuit Theory*, vol. CT-18, pp. 403–405, May 1971.
[14] M. S. Lee and C. Chang, "Exact synthesis of N-path switched-capacitor filters," in *Proc. IEEE Symp. Circuits Syst.*, Apr. 1981, pp. 166–169.
[15] M. B. Ghaderi, G. C. Temes, and J. A. Nossek, "Switched-capacitor pseudo N-path filters," in *Proc. IEEE Int. Symp. Circuits Syst.*, Apr. 1981, pp. 519–522.
[16] T. C. Choi, "High-frequency CMOS switched-capacitor filters," Ph.D. dissertation, University of California, Berkeley, June 1983.
[17] P. O. Brackett and A. S. Sedra, "Active compensation for high frequency effects in op amp circuits with applications to active *RC* filters," *IEEE Trans. Circuits Syst.*, vol. CAS-23, pp. 68–72, Feb. 1976.
[18] K. Martin and A. S. Sedra, "Effects of the op amp finite gain and bandwidth on the performance of switched-capacitor filters," *IEEE Trans. Circuits Syst.*, pp. 822–830, Aug. 1981.

A Low-Noise Chopper-Stabilized Differential Switched-Capacitor Filtering Technique

KUO-CHIANG HSIEH, PAUL R. GRAY, FELLOW, IEEE, DANIEL SENDEROWICZ, STUDENT MEMBER, IEEE, AND
DAVID G. MESSERSCHMITT, SENIOR MEMBER, IEEE

Abstract—This paper describes the implementation of a wide dynamic range voiceband switched-capacitor filter using a differential chopper-stabilized configuration. The noise behavior of switched-capacitor filters is discussed qualitatively, and the effects of the chopper stabilization on the noise performance is analyzed. Experimental results from a fifth-order low-pass voiceband prototype are presented.

I. INTRODUCTION

SWITCHED-CAPACITOR filtering techniques have been widely applied to voiceband applications requiring dynamic range on the order of 85 dB. Depending on details of design and fabrication, these filters have been limited in dynamic range by operational amplifier noise, thermal noise in the transistor switches, or a combination thereof. In addition, the realization of very high levels of power supply rejection ratio (PSRR) has proven to be a difficult task, with PSRR on the order of 40 dB typical in work reported to date.

This paper describes a switched-capacitor filtering technique which is aimed at improving the dynamic range and power supply rejection of such filters. It differs from conventional approaches in two respects. First, the signal path throughout the filter is fully differential rather than single-ended as in the conventional case. This results in reduced injection of power supply and clock-related signals into the signal path, and also increases the dynamic range since the effective signal swing is doubled. Second, the first stage of the operational amplifier is chopper-stabilized, which reduces one component of the operational amplifier noise, the $1/f$ noise. Assuming that the other sources of noise in the filter are reduced by suitable design of the operational amplifier and choice of capacitor values, this allows the dynamic range of such filters to be extended to beyond 100 dB.

In Section II, the important sources of noise in switched-capacitor filters are discussed from a qualitative point of view. This discussion is included so that the relative importance of $1/f$ noise in the overall noise of a switched-capacitor filter can be better appreciated. In Section III, various circuit alternatives for the reduction of $1/f$ noise are discussed, and the effect of chopper stabilization on the noise spectrum of the switched-capacitor integrator is analyzed. In Sections IV and

Manuscript received May 26, 1981; revised July 15, 1981. This work was supported by the Joint Services Electronics Program under Contract F49620-79-c-0178 and National Science Foundation Grant ENG79-07055.

The authors are with the Department of Electrical Engineering and Computer Sciences and the Electronics Research Laboratory, University of California, Berkeley, CA 94720.

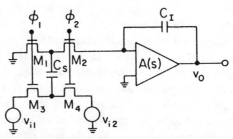

Fig. 1. Typical bottom-plate switched-capacitor integrator.

V, implementation of the differential filter configuration is described, and the design of an experimental fifth-order filter is discussed. In Section VI, experimental results from this circuit are presented.

II. NOISE IN SWITCHED-CAPACITOR INTEGRATORS

The circuit technique described in this paper is aimed at the reduction of the low frequency noise contributed by the operational amplifier. Other important sources of noise are the wide-band thermal noise in the operational amplifiers and the thermal noise in the channels of the transistor switches making up the filter. While the $1/f$ noise is often dominant, its relative importance in the total filter noise is a function of the details of the design of the operational amplifier and the choice of the sampling and integrator capacitor sizes in the filter. The qualitative discussion of noise in switched-capacitor filters below is included so that the relative importance of $1/f$ noise in overall filter noise for a particular design can be better appreciated.

A. Thermal Noise in the MOS Switches

Thermal noise in the MOS transistor switches represents a fundamental limitation on the amount of noise added to the signal as it passes through the integrator for a fixed value of integrating capacitance. These switches correspond to transistors $M1$, $M2$, $M3$, and $M4$ in the example switched-capacitor integrator shown in Fig. 1. It can be shown [5]–[7] that under the assumption that the clock frequency is much larger than the frequency of interest, that the effective baseband equivalent input noise spectral density of the switched-capacitor integrator is exactly the same as that of a continuous time integrator in which a continuous resistor is substituted for the switched-capacitor resistor. Under the same assumption, it can further be shown that the total input referred mean-squared noise voltage between dc and the unity gain frequency of the integrator is simply equal to kT/C_I where C_I

Reprinted from *IEEE J. Solid-State Circuits*, vol. SC-16, no. 6, pp. 708–715, Dec. 1981.

is in this case the value of the integrator capacitor. Thus, the total in-band noise is determined by the integrator capacitor, which is the larger of the two capacitors in Fig. 1 for voice-band switched-capacitor filters with high sampling rates. As a result, this noise source represents a basic limitation on the dynamic range which can be achieved for a given total capacitance.

B. Operational Amplifier Noise

The equivalent input noise spectrum of a typical MOS operational amplifier is shown in Fig. 2. The noise added to the signal path by the operational amplifier can be divided into two components. Flicker noise, or $1/f$ noise, is concentrated at low frequencies and arises from surface states in the channel of the MOS transistors. The magnitude of the low frequency noise component is dependent on the process used, the design of the operational amplifier used, and on the size of the input transistors used in the operational amplifier. Under the assumption that the noise energy associated with the $1/f$ noise lies far below the sampling rate, the effects of this noise component can be analyzed disregarding aliasing effects. Because the $1/f$ noise source is both integrated and translated by the integrator circuit, its effects can be conveniently represented by two noise sources equal in magnitude to the $1/f$ noise of the operational amplifier, one in series with the input of the switched-capacitor integrator and one in series with the output. This is shown in Fig. 3. The objective of the chopper stabilization technique described in this paper is the reduction of this noise source by shifting the noise energy to a higher frequency outside the passband.

The third important noise contribution in the filter is the broad-band thermal noise of the operational amplifier, corresponding to the flat portion of the noise curve shown in Fig. 2. The analysis of the thermal noise of the operational amplifier is complicated by the fact that the bandwidth over which this noise energy lies is much broader than the sampling rate. As a result, the sampling of the next integrator stage causes a portion of the high frequency components of this noise to alias into the passband. The effect of this noise can most easily be visualized by dividing the frequency spectrum into two parts: the range from dc to the unity-gain frequency of the operational amplifier, and the range above the unity-gain frequency of the operational amplifier. For the former, the operational amplifier behaves like a low impedance voltage source when viewed from its own output with an equivalent noise voltage equal to the equivalent input thermal noise voltage of the operational amplifier. Thus, when sampled by the switch and sampling capacitor of the next stage, this energy will be aliased into the passband, and under the assumption that the equivalent input noise is flat in this frequency range and that the time constant of the sampling circuit is long compared to the operational amplifier bandwidth, the effective equivalent input noise density can be found approximately by simply multiplying the broad-band noise energy density by the ratio of the unity-gain frequency of the operational amplifier to the sampling frequency. The effective in-band noise can be represented by inserting a white noise source of this value in series with the output of the integrator, to

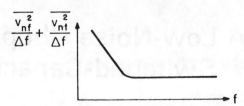

Fig. 2. Typical input equivalent noise voltage spectrum for an MOS operational amplifier.

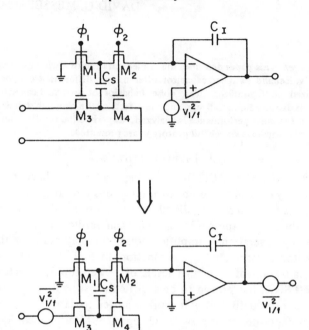

Fig. 3. Representation of operational amplifier $1/f$ noise in the switched-capacitor integrator.

represent the aliasing due to the sampling of the next integrator, and a second source of the same value in series with the input of the integrator. The latter results from the fact that the broad-band noise appears at the summing node of the operational amplifier as well as at the output, and this results in a aliasing of the broad-band noise by the sampling capacitor of the integrator itself.

For frequencies beyond the unity-gain frequency, the amount of noise aliasing that occurs depends on the design of the operational amplifier. In this range, the gain of the operational amplifier is less than unity, and the effect of the feed-band loop around the operational amplifier is negligible. Thus, viewed from its output, the operational amplifier has an output impedance and a noise equivalent resistance which is determined by the circuitry between the compensation point and the output node in the amplifier. If the operational amplifier can be designed so that in this range the output resistance and the noise equivalent resistance at the output are the same, then the increased thermal noise is offset by the fact that the bandwidth over which the noise is sampled by the next stage is reduced by the output resistance, and no additional noise contribution over and above that of the kT/C noise. This situation can be achieved or approximated in one and two stage operational amplifiers which have no output stage and drive the output directly from a high impedance compensation

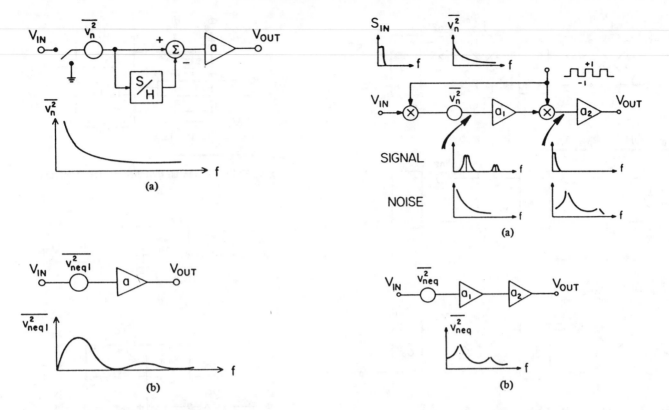

Fig. 4. (a) Concept of correlated double sampling noise reduction. (b) Equivalent input noise of the circuit in (a).

Fig. 5. (a) Concept of chopper stabilization. (b) Equivalent input noise for the circuit in (a).

node. However, in operational amplifiers which utilize some form of output stage, the output resistance is often lower than the equivalent noise resistance in this frequency range, and in this case the operational amplifier thermal noise contribution can be greatly increased because the output noise can be sampled with a large bandwidth.

III. Techniques for the Reduction of $1/f$ Noise

For voiceband applications of switched-capacitor filters, the dominant noise source is often the $1/f$ noise component of the operational amplifier. The $1/f$ noise can be reduced by a number of different methods. One approach is to simply use large input device geometries to reduce the $1/f$ noise associated with these devices. This approach has been widely used in the past, and works particularly well in process technologies which have a low level of surface states at the outset. For processes which have high surface state densities, however, this approach can give uneconomically large input transistor geometries for applications requiring extremely high dynamic range. A second approach is to use buried channel devices so as to remove the channel from the influence of surface states. This approach requires process steps which are not usually included in the standard LSI technologies used to manufacture switched-capacitor filters in high volume. A third approach is to use circuit techniques to translate the noise energy from the baseband to some higher frequency so that it does not contaminate the signal. This approach is the subject of this paper.

One technique for reducing the $1/f$ noise density at low frequencies is the correlated double sampling (CDS) method [8]. This technique is illustrated conceptually in Fig. 4. If the

sample/hold function and the subtractor could be easily incorporated in the operational amplifier without adding additional energy storage elements, then this would be a very desirable approach to $1/f$ noise reduction. However, a practical problem with CDS is that the equivalent input noise is most easily obtained by shorting the input and output nodes of the operational amplifier. This operation requires that the output node of the amplifier slew back and forth between the signal level and the initialized level each clock period. This puts a severe requirement on the operational amplifier settling time.

An alternate approach to $1/f$ noise reduction is chopper stabilization. This technique has been used for many years in the design of precision dc amplifiers. The principle of chopper stabilization is illustrated in Fig. 5. Here, a two-stage amplifier and a voiceband input signal spectrum are shown. Inserted at the input and the output of the first stage are two multipliers which are controlled by a chopping square wave of amplitude +1 and –1.

After the first multiplier, the signal is modulated and translated to the odd harmonic frequencies of the chopping square wave, while the noise is unaffected. After the second multiplier, the signal is demodulated back to the original one, and the noise has been modulated as shown in Fig. 5. This chopping operation results in an equivalent input noise spectrum which is shown in Fig. 5(b), where the $1/f$ noise component has been shifted to the odd harmonic frequencies of the chopping square wave. The $1/f$ noise density at low frequencies is now equal to the "folded-back" noise from those harmonic $1/f$ noise components. Therefore, if the chopper frequency is much higher than the signal bandwidth, the $1/f$

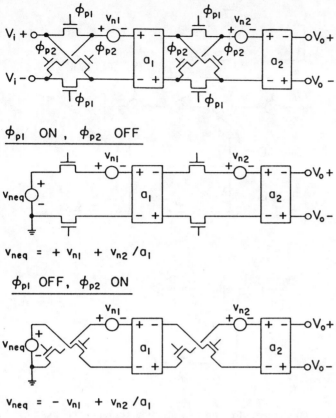

ϕ_{p1} ON, ϕ_{p2} OFF

$$v_{neq} = + v_{n1} + v_{n2}/a_1$$

ϕ_{p1} OFF, ϕ_{p2} ON

$$v_{neq} = - v_{n1} + v_{n2}/a_1$$

Fig. 6. MOS implementation of a chopper-stabilized operational amplifier.

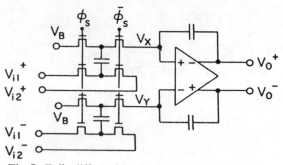

Fig. 7. Fully differential switched-capacitor integrator.

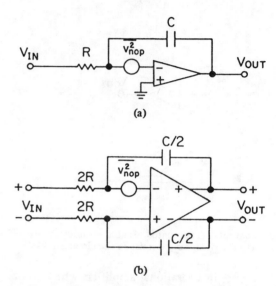

(a)

(b)

Fig. 8. Signal-to-noise comparison between single-ended and differential integrators.

noise in the signal band will be greatly reduced by the use of this technique.

An MOS implementation of the chopper stabilization technique is shown in Fig. 6. The multipliers described before are realized by two cross-coupled switches which are controlled by two nonoverlapping clocks. When ϕ_{p1} is on and ϕ_{p2} is off, the equivalent input noise is equal to the equivalent input noise of the first stage plus that of the second divided by the gain of the first stage. When ϕ_{p1} is off and ϕ_{p2} is on, the equivalent input noise is equal to the negative of this instantaneous value. If the voltage gain of the first stage is high enough, the noise contribution from the second stage can be neglected and the sign of this equivalent input noise changes periodically.

IV. DIFFERENTIAL FILTER IMPLEMENTATION

The dynamic range of a switched-capacitor filter is determined by the ratio of the maximum signal swing giving acceptable distortion to the noise level. Thus, improvements in signal swing result in direct improvement of the dynamic range. In the experimental filter described in this paper, the effective signal swing is doubled relative to a conventional switched-capacitor filter by the use of a differential output integrator. The differential configuration used has the additional advantage that because the signal path is balanced, signals injected due to power supply variations and clock charge injection are greatly reduced.

An example of a bottom-plate fully-differential switched-capacitor integrator is shown in Fig. 7. Two differential input signals are connected to four input nodes of this inte-

grator. Signals v_{i1}^+ and v_{i1}^- form one differential input, and v_{i2}^+ and v_{i2}^- form another differential input. Assuming that the two differential inputs have no common-mode component, the common-mode voltage at the operational amplifier input is set by V_B. This potential can be set with on-chip circuitry to a value which is convenient from the standpoint of the design of the operational amplifier, and can be used to eliminate the necessity of a level shift function within the operational amplifier. In addition to the fact that dynamic range and power supply rejection are improved with this configuration, the fact that both output polarities are available makes the design of elliptic ladder filters much easier.

Two disadvantages to this technique are that the interconnection problem makes circuit layout more complex, and that a differential to single-ended conversion may be necessary in some applications.

The improvement in dynamic range resulting from the use of the differential technique is illustrated in Fig. 8. In both single-ended and differential cases, the transfer functions for the signal from the input to the output are identical. For the operational amplifier noise, the transfer functions from the noise source to the output of the integrator are identical. If the operational amplifier noise is dominant, the result is a 6 dB improvement in S/N ratio, because, for the same power supplies, the effective voltage swing doubles in the differential case. If the thermal noise in the resistors is dominant, the

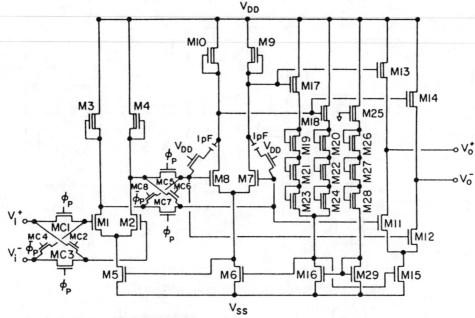

Fig. 9. Schematic diagram of differential chopper-stabilized operational amplifier.

dynamic range is the same for the same amount of total integrating capacitance in both cases.

V. Experimental Differential Chopper-Stabilized Operational Amplifier and Filter

An experimental differential switched-capacitor low-pass filter was fabricated to investigate the use of chopper stabilization for low frequency noise reduction. The operational amplifier will be described first, then the filter configuration used.

The key design problem in the implementation of a filter of the type described above is the realization of an operational amplifier which has differential outputs with a well-defined common-mode voltage, and which incorporates the chopper stabilization without undue complexity. While in principle chopper stabilization could be implemented without the simultaneous use of the differential configuration, the differential configuration is particularly amenable for the implementations of the chopper circuitry using balanced cross-coupled analog switches as shown in Fig. 9. Transistors $MC1$–$MC4$ and $MC5$–$MC8$ form two cross-coupled choppers which are controlled by two nonoverlapping clocks. Transistors $M1$–$M5$, $M6$–$M10$, and $M11$–$M15$ are the input, gain, and output stages, respectively. The operating points of the input stage are biased by the common-mode feedback loop through the gain stage. The common-mode output dc voltage and the operating points of the gain stage are set from common-mode feedback circuit $M16$–$M24$, which is biased by a grounded replica reference string consisting of transistors $M25$–$M29$, and uses depletion transistors $M17$–$M24$ as a level shifter to ensure a well-defined common-mode output dc voltage for large differential output swings. The common-mode dc voltage at the input of the operational amplifier can be obtained through the bias voltage V_B. Pole splitting compensation in the gain stage and a feedforward path to the output stage ($M11$ and

$M12$) are used to improve the stability of the operational amplifier when driving a large capacitance load.

The schematic diagram of a bottom-plate fifth-order differential low-pass ladder switched-capacitor filter is shown in Fig. 10. This configuration is used to implement a differential Chebyshev low-pass filter with cutoff frequency at 3400 Hz. This circuit is fully differential-in and differential-out. Five differential chopper-stabilized operational amplifiers with chopper frequency 128 kHz and master clock frequency of 256 kHZ are used. Thus, the $1/f$ noise is translated by the chopper to exactly the Nyquist rate of the filter clock. As a result, no aliasing effect of the $1/f$ noise back into the passband occurs. All the $1/f$ noise components will appear in the frequency domain, equally spaced between two adjacent integer multiples of the filter clock frequencies. The odd harmonic frequencies of the chopper clock, where the shifted $1/f$ noise located, do not interfere with the master filter clock frequency. This choice of chopping frequency can be shown to be optimum in terms of maximum reduction in baseband $1/f$ noise.

The noise at the output of a switched-capacitor filter can be calculated by knowing the transfer function from each noise source (or equivalent noise source) to the output of the filter. This calculation was carried out for the experimental filter by first measuring the equivalent input noise spectrum of the operational amplifier, and then numerically summing the various contributions weighted by the transfer function from that point in the filter to the output. Aliasing of broad-band noise was taken into account as described in Section II. The resulting predicted noise spectrum is shown in Fig. 11, with and without chopper switches operating. As expected, the dominant noise mechanism without chopper stabilization is the $1/f$ noise component. With chopper operating, the dominant noise in this particular filter is the operational amplifier thermal noise. This results primarily from the choice of cur-

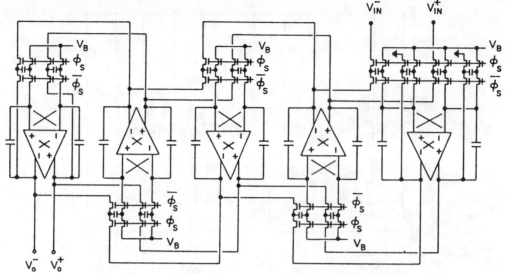

Fig. 10. Differential fifth-order chopper-stabilized filter.

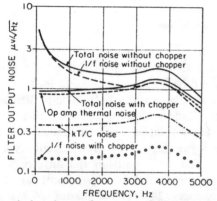

Fig. 11. Theoretical noise performance of the experimental filter, showing each of the three noise contributions with and without chopper stabilization.

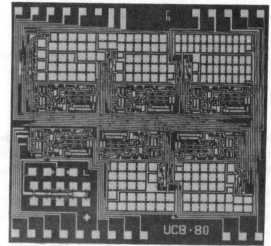

Fig. 12. Die photomicrograph, experimental fifth-order filter.

rent level and W/L ratio in the input transistors of the operational amplifiers, giving a relatively low g_m and a relatively large value of noise equivalent resistance in the frequency range from dc up to the operational amplifier cutoff frequency of 15 MHz. A straightforward modification of this design would give a reduction in this noise source of 8–10 dB, giving the desired result that filter noise be dominated by thermal noise in the switches.

Note that the on resistance of the switches in the chopper in series with the input of the operational amplifier do contribute to the thermal noise of the operational amplifier. This contribution could be significant in a case in which the inherent noise equivalent resistance of the amplifier was very low.

VI. EXPERIMENTAL RESULTS

An experimental prototype differential chopper-stabilized fifth-order Chebyshev filter was designed and fabricated using a metal-gate n-channel MOS process with depletion load. The minimum transistor gate length in the process used was 15 μ. One objective of the experimental filter was to explore the maximum achievable dynamic range, and toward that end a relatively large integrating capacitance averaging 100 pF integrator was chosen. The filter area is about 9600 mils². A die microphotograph is shown in Fig. 12.

The experimental equivalent input noise response of the on-chip test operational amplifier is shown in Fig. 13. The dashed curve is the input referred noise response without chopper stabilization (i.e., the chopper is kept in one of the two possible states). The bottom solid curve is the input referred noise response with chopper frequency at 128 kHz. Notice that in this case the noise at 1 kHz is primarily due to the first folded-back 128 kHz harmonic $1/f$ noise and the thermal noise of the operational amplifier. The total is about 40 dB (100 times in power) less than that without chopper stabilization.

In order to demonstrate further the translation of the $1/f$ noise, the noise was measured with the chopper frequency at one eighth of its nominal value, or 16 kHz, with the result shown as the upper solid curve in Fig. 13. Notice that the noise at 1 kHz is about 12 dB (16 times in power) less than that without chopper stabilization. Also, the $1/f$ noise peak has been shifted to the odd harmonic frequencies of the chopper clock, which are 16 kHz and 48 kHz in Fig. 13. The

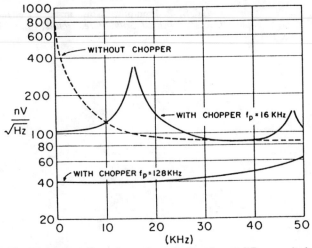

Fig. 13. Experimentally observed operational amplifier equivalent input noise for various chopping frequencies.

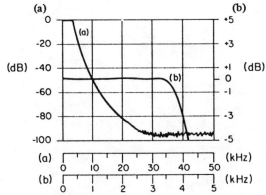

Fig. 14. Experimental filter frequency response. (a) Overall response. (b) Passband response.

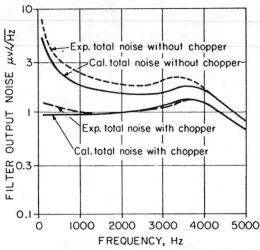

Fig. 15. Experimentally observed filter output noise with and without chopper, compared to theoretically predicted total noise.

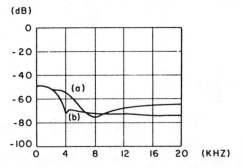

Fig. 16. Experimentally observed PSRR. (a) From positive power supply. (b) From negative power supply.

operational amplifier displayed a gain-bandwidth product of 15 MHz, and power dissipation of 4 mW with + and −7.5 V supplies.

The observed filter frequency response is shown in Fig. 14. The experimental filter output noise response with and without chopper stabilization is shown in Fig. 15. The noise reduction at 1 kHz is about 10 dB. This experimental result agrees closely with the predicted total output spectrum, which is shown on the same graph. The $1/f$ noise is dominant in the filter without chopper operation, and the thermal noise of the operational amplifier is dominant with chopper operation. Despite a 40 dB noise reduction at 1 kHz in the operational amplifier noise when the chopper is operated, only 10 dB of improvement is obtained in the filter noise. This occurs because the aliased thermal noise of the operational amplifier becomes the dominant noise mechanism when the chopper is on.

The experimental positive and negative power supply rejection ratios (PSRR's) for frequency up to 20 kHz are shown in Fig. 16. The minimum PSRR is 50 dB for both supplies in this frequency range. The filter power dissipation is 20 MW. For plus/minus 7.5 V power supplies at 25°C, the experimental filter had a maximum differential output signal swing at 1 kHz and 1 percent total harmonic distortion (THD) equal

to 5 V rms. The filter output noise with C-message weighted was 40 μV rms, giving a dynamic range of 102 dB.

VII. CONCLUSIONS

A differential chopper-stabilized switched-capacitor filtering technique has been described, which extends the dynamic range of the switched-capacitor filters beyond 100 dB. The chopper stabilization greatly reduces the $1/f$ contribution to total filter noise, and is most applicable in applications where very large dynamic range is an objective. The differential configuration results in substantial improvements in power supply rejection ratio. This improvement is of considerable practical significance since power-supply related noise and crosstalk is a major problem in critical applications.

REFERENCES

[1] W. C. Black, D. J. Allstot, S. Patel, and J. Wieser, "A high performance low power CMOS channel filter," *IEEE J. Solid-State Circuits*, vol. SC-15, Dec. 1980.

[2] I. A. Young, "A low-power NMOS transmit/receive IC filter for PCM telephony," *IEEE J. Solid-State Circuits*, vol. SC-15, Dec. 1980.

[3] H. Ohara, P. Gray, W. M. Baxter, C. F. Rahim, and J. L. McCreary, "A precision low power channel filter with on-chip regulation," *IEEE J. Solid-State Circuits*, vol. SC-15, Dec. 1980.

[4] D. J. Allstot, "MOS switched capacitor ladder filters," Ph.D. dissertation, Univ. California, Berkeley, May 1979.

[5] K. C. Hsieh, "Noise limitations in switched-capacitor filters," Ph.D. dissertation, Univ. California, Berkeley, Dec. 1981.

[6] C. Gobet and A. Knob, "Noise analysis of switched capacitor networks," in *Proc. Int. Symp. Circuits Syst.*, Chicago, IL, Apr. 1981.

[7] ——, "Noise generated in switched capacitor networks," *Electron. Lett.*, vol. 16, no. 19, 1980.

[8] R. W. Broderson and S. P. Emmons, "Noise in buried channel charge-coupled devices," in *IEEE J. Solid-State Circuits*, vol. SC-11, Feb. 1976.

A High-Performance Micropower Switched-Capacitor Filter

RINALDO CASTELLO AND PAUL R. GRAY, FELLOW, IEEE

Abstract —MOS technology scaling requires the use of lower supply voltages. Analog circuits operating from a low supply and achieving a sufficiently large dynamic range must be designed if analog/digital interfaces are to be implemented in scaled technologies. This paper describes a high-performance fifth-order low-pass switched-capacitor filter operating from a single 5-V supply. The filter uses a fully differential topology combined with input-to-output class *A B* amplifier design, dynamic biasing, and switched-capacitor common-mode feedback (CMFB). An experimental prototype fabricated in a 5-μm CMOS technology requires only 350 μW of power to meet the PCM channel filter requirements. Typical measured results are: a dynamic range of 92 dB, a supply rejection (PSRR) of 40 dB over the entire Nyquist range, and a total harmonic distortion (THD) of −73 dB for a 2-V rms differential output signal. The chip active area is about 3900 mil^2.

I. INTRODUCTION

THE PERFORMANCE of switched-capacitor filters has steadily improved during the last several years, primarily as a result of improvements in the performance of CMOS operational amplifiers. This improvement has been particularly evident in the PCM channel filter application [1]–[6]. However, the most recent commercial PCM filter implementations still require a power-per-pole of about 1 mW and operate from a ±5 V supply. It has recently been shown [20], [21] that, from a fundamental standpoint, the absolute minimum achievable power dissipation in a voice-band filter with a dynamic range of 90 dB in a 3-μm technology operated from a ±5 V supply is less than 1 μW per pole. A large margin for improvement in power consumption over existing filter designs is therefore possible in principle. The realization of such a reduction, while maintaining high-performance levels, would have important implications in the realization of battery-operated analog/digital interfaces.

A second important consideration in the realization of switched-capacitor filters is the fact that the technological scaling of the mainstream MOS technologies dictates the use of lower power-supply voltages [7]. This fact, and the need for an analog/digital compatible technology, create a strong motivation for developing new analog circuit techniques suitable for low-voltage operation.

Recently several circuit approaches to the implementation of low-power MOS switched-capacitor filters have

Manuscript received October 25, 1985; revised August 1, 1985. This work was supported by NSG Grants ECS-8023872 and ECS-8100012.
The authors are with the Department of Electrical Engineering and Computer Science, University of California, Berkeley, Berkeley, CA 94720.

been described [8]–[14], some of them operating off a low power-supply voltage. These, however, have been intended for applications requiring limited dynamic range and power-supply rejection, and are not suitable for high-performance applications such as PCM telephony. This paper describes a fifth-order CMOS PCM channel filter operated from a single 5-V supply and dissipating about 70-μW per pole which embodies a combination of circuit techniques including input-to-output class *A B* amplifier design, fully differential topology, dynamic biasing, and switched-capacitor common-mode feedback. These techniques provide performance comparable or improved with respect to current 10-V commercial realizations, while dissipating much less power and operating on a 5-V power supply.

The paper is organized as follows. In Section II the fundamental limit to the achievable power dissipation for a low-pass switched-capacitor filter of given dynamic range is computed and compared with the actual value in commercially available devices. In Section III a new class AB operational amplifier, which represents the core of the filter reported in this paper, is described in detail. Section IV discusses the structure of the fifth-order switched-capacitor low-pass filter prototype which has been used to test the level of performance achievable with the new design. Finally, Section V presents some experimental results for both the filter and the op amp which demonstrates that high performance in analog circuits can be preserved when the supply voltage is reduced.

II. LIMITS TO POWER DISSIPATION AND DYNAMIC RANGE

Power consumption reduction is always a major issue in VLSI systems. Present commercial switched-capacitor filters consume far more power than the theoretical minimum required. For this reason, this section examines some fundamental limitations to the achievable minimum power dissipation for a low-pass switched-capacitor filter of given dynamic range. This analysis has been carried out in detail in a previous paper [20], [21], and in this section only the main results are summarized stressing their intuitive interpretation.

In present day switched-capacitor filters, the quiescent dc bias power drawn by the operational amplifiers is normally much larger than the dynamic power drawn from

Reprinted from *IEEE J. Solid-State Circuits*, vol. SC-20, no. 6, pp. 1122–1132, Dec. 1985.

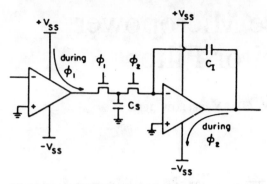

Fig. 1. Minimum power dissipation for a switched-capacitor integrator.

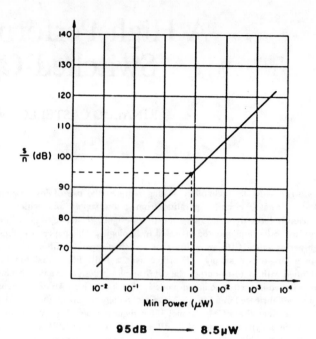

$$95\,dB \longrightarrow 8.5\,\mu W$$

Fig. 2. Minimum power dissipation versus dynamic range for a 3.4-kHz fifth-order low-pass switched-capacitor filter.

the supply in order to charge and discharge the sampling and integrating capacitor within the filter. We assume for this analysis, however, that, through improved circuit techniques, the op amp static power dissipation can eventually be reduced to the point of being negligible with respect to the dynamic power. This condition can be approached, in principle, by appropriate use of class B amplifier architectures. The situation is then analogous to that of a CMOS logic gate. Fig. 1 shows the flow of power from the supplies during one clock period for a switched-capacitor integrator. During phase 1 the positive supply charges the sampling capacitor via the previous switched-capacitor integrator, and during phase 2 the energy stored on the sampling capacitor is discharged by the feedback action of the second op amp through the integrating capacitor to the negative supply. Assuming that a full swing sinusoid of frequency f is present at the output of the integrator, the minimum power dissipation is proportional to the maximum energy that can be stored on the integrating capacitor times the frequency of the signal. Here it is important to emphasize that the power dissipation is proportional to the frequency of the signal and not to the frequency of the clock as in a CMOS gate. As an example this power is on the order of 1–2 μW per integrator for a typical switched-capacitor voice-band filter.

In order to reduce the power consumption, for a given frequency band of interest, the size of the integrating capacitor should then be reduced. The integrating capacitor is, however, constrained to be larger than some minimum value in order to achieve the required dynamic range due to the kT/C noise contribution. This should appear intuitively correct if it is noticed that, since in a switched-capacitor filter signals are represented as energy stored on capacitors, increasing the maximum energy that can be stored is equivalent to increasing the maximum signal energy level that the filter can process and therefore is equivalent to increasing the filter dynamic range. In fact, the dynamic range is related to the ratio of maximum stored energy to thermal energy, kT dictating a minimum capacitor size for a given supply voltage to achieve the required filter dynamic range.

By combining these two results, and applying them to the case of a low-pass filter implemented using switched-capacitor integrators, a plot of achievable signal-to-noise ratio versus minimum power dissipation for a low-pass

filter can be obtained given the value of the filter band edge. In Fig. 2 the case of a fifth-order filter with a 3.4-kHz bandwidth is shown. A dynamic range of 95 dB requires only 8.5 μW of power. Commercially available switched-capacitor PCM filters of similar performance use almost three orders of magnitude more power. Part of the discrepancy between theoretical and actual results is due to the fact that the above analysis neglects the noise contribution associated with the op amp (white and $1/f$ noise). This, however, can only account for a factor of two or so. A very large improvement is therefore feasible if op amps which are more efficient and achieve a signal swing closer to the supply voltage are designed.

III. OPERATIONAL AMPLIFIER ARCHITECTURE

As just illustrated, op amp power consumption tends to dominate in typical filters. Next the design of a low power class AB op amp is described.

The new circuit configuration is based on four main concepts. The first is fully differential architecture which results in better swing and supply rejection. A class AB structure is used for low power dissipation. A single-stage topology with dynamically biased cascode is used for optimum current drive capability and output swing. A switched-capacitor common-mode feedback is used for minimum power. These feature will be discussed in detail in the following sections.

A. Fully Differential Topology

A fully differential switched-capacitor integrator is shown in Fig. 3. The two input voltages V_{in}^+ and V_{in}^- are symmetrical with respect to the common-mode input voltage V_{cmi}, and the two output voltages V_o^+ and V_o^- are symmetrical

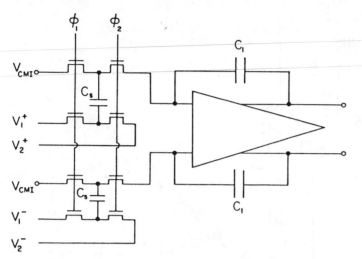

Fig. 3. Fully differential switched-capacitor integrator.

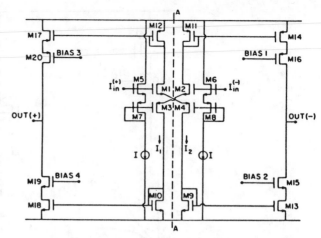

Fig. 4. Simplified schematic of the class AB amplifier.

with respect to the common-mode output voltage V_{cmo}. This structure doubles the output swing, which is of paramount importance for low-voltage applications, and provides a reduction on the sensitivity to supply and clock noise so that good PSRR can be obtained without having to decouple the input summing nodes from the supplies as it is instead required in the single-ended case. This results in a simplified input structure much more suitable for low-voltage design. Furthermore, a fully differential approach allows the designer to independently choose the value of the input and output common-mode voltages (V_{cmi} and V_{cmo}) for optimum performance. Although for maximum swing V_{cmo} should be equal to half of the total supply voltage, the same may not be the case for V_{cmi}. In the present design, in fact, V_{cmi} is higher than V_{cmo} by an n-channel threshold voltage.

The main disadvantage of the fully differential approach is the increased power and area requirement; however, in this design an efficient common-mode feedback (CMFB) circuit limits the power consumption increase to approximately 40 percent, while the total area increase is about 60–70 percent, with respect to a corresponding single-ended realization.

B. Class AB Single-Stage Configuration

Reduction of the quiescent power consumed by the op amp while retaining sufficient speed can be obtained by using a class AB configuration. Fig. 4 shows a simplified schematic of the class AB amplifier used in the filter without CMFB circuit. This circuit is a modified version of a previously proposed structure [18]. The circuit is perfectly symmetric about the axis $A-A$. For zero applied differential input signal, the two matched current sources I uniquely define the circuit quiescent current level. In fact, if for simplicity it is assumed that the four NMOS input devices are identical, and the same is true for the four PMOS devices, then $I_1 = I_2 = I$. Furthermore, since all current mirrors have a gain of 1, the quiescent current in the output branches is also equal to I. It follows, therefore

that the quiescent power consumption in the circuit is precisely controlled by the two matched current sources in the input stage. The dynamic behavior of the circuit is shown in Fig. 5.

In response to a large positive differential input signal, current I_1 goes practically to zero. As a consequence, half of the devices in the circuit become cut off and have not been shown on Fig. 5(a). Current I_2, on the other hand, increases to a peak value which, in principle, is only limited by the value of the input voltage applied. The same current is mirrored to the outputs and can quickly charge and discharge the load capacitance. The comparison between the current–voltage characteristic of a class A and a class AB amplifier can be seen from Fig. 5(b). Notice that, in the example shown, for an input voltage of 300 mV the current in the class AB circuit is already several times larger than the current in the class A circuit. These values are representative of the actual behavior of the circuit reported in this paper.

Although in the above consideration it was assumed that the peak value for current I_2 in the class AB circuit of Fig. 5 is only a function of the applied input voltage, in practice another limiting factor is the total supply voltage. In fact as the current level increases, the sum of the voltage drops across devices $M1$, $M4$, $M9$, and $M14$ in Fig. 5 also increases, until it is equal to the total supply voltage. At this point, some of the devices ($M1$ or $M4$ or both) enter the linear region of operation, and the current level becomes practically constant independently of the value of the input voltage. This problem becomes more and more severe as the supply voltage is reduced, and represents the limiting factor to the maximum achievable peak current for a 5-V total supply voltage. The achievable value for the peak current is also strongly dependent on the value of the input common-mode voltage V_{cmi}. An optimum choice of the value of V_{cmi} is important in order to obtain the best possible performance in the op amp. In fact, increasing V_{cmi} by one n-channel threshold voltage above the middle point between the two supplies, as allowed by the fully differential configuration, gives more than a threefold increase on the achievable peak current level.

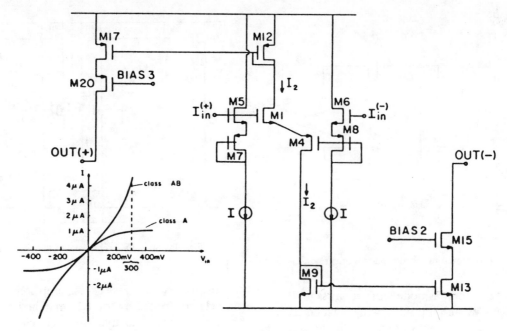

Fig. 5. Active portion of the amplifier for a positive input signal.

By utilizing a class AB configuration, a saving on the quiescent power dissipation for a given speed can therefore be achieved. Furthermore, the low quiescent current level on the output devices improves the voltage swing, and gives a larger dc gain. The class AB structure, however, has also some disadvantages. In particular it tends to be more complicated, and makes the problem of designing the CMFB circuit more difficult.

A single-stage configuration is particularly suitable for class AB operation. Also it has good power-supply rejection at high frequencies (beyond the dominant pole) and gives no high-frequency second-stage noise contribution, an effect which can greatly reduce the dynamic range of a sampled data system due to aliasing effects. Furthermore, in the present design the load capacitance is enough to guarantee stable closed-loop response, so that no extra compensation is required. The main drawback of the single-stage topology, particularly for low-voltage applications, is the reduced output swing due to the cascode devices. However, as shown below, the significance of this problem can be reduced by careful design.

In order to take full advantage of the class AB structure the amplifier must be able to deliver all of the peak input current to the load, without unacceptably compromising the output voltage swing. This requires use of a novel biasing scheme for the cascode devices as explained in the following section.

C. Dynamic Biasing

The necessity of adaptively biasing the cascode current mirrors in the amplifier is illustrated by the circuit of Fig. 6. Fig. 6(a) shows the two output devices of the NMOS cascode current mirror, which is active for a positive input signal. Fig. 6(b) is a table of value for the $V_{GS} - V_T$ of the two devices together with the minimum value of V_{BIAS2}

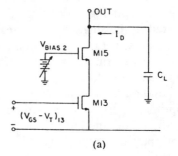

(a)

	Quiescent	Max
$(V_{GS} - V_T)_{13}$	180 mV	1.2 Volts
$(V_{GS} - V_T)_{15}$	100 mV	700 mV
Required V_{BIAS2}	$V_T + 280$ mV	$V_T + 1.9$ Volts

(b)

Fig. 6. Output devices of an NMOS current mirror.

required to keep both devices in saturation for two different conditions of operation. In the first column the current through the two devices is equal to the quiescent value. In the second column the current through the two devices is equal to the peak value that occurs during the transient resulting from applying a full swing signal at the input of the circuit. The numbers on the table are representative of the actual values for the amplifier reported in this paper.

To obtain a large output swing V_{BIAS2} must be as small as possible, and the minimum possible value that guarantees proper operation of the circuit is, from the first column of Fig. 6(b), equal to 280 mV more than the value of the threshold voltage of device $M13$. This is because to obtain the full gain of the circuit it is necessary to insure

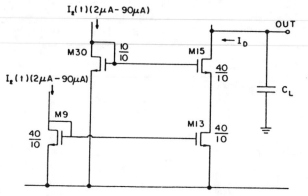

Fig. 7. Complete schematic of NMOS dynamically biased current mirror.

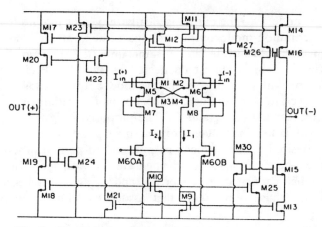

Fig. 8. Detailed schematic of the entire amplifier without CMFB.

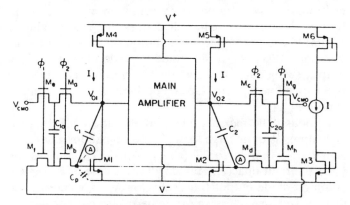

Fig. 9. Simplified dynamic common-mode feedback circuit.

that both $M13$ and $M15$ are in saturation at least in the quiescent state, i.e., when all transients have died away. For this value of V_{BIAS2}, however, as the output current increases $M13$ quickly enters the linear region of operation and the current limits to a value much smaller than the peak current in the input stage.

To guarantee full current driving capability to the load, $M13$ must be kept in the saturation region of operation during the entire transient. This require a value of V_{BIAS2} greater than or equal to the value shown in the second column of Fig. 6(b), i.e., 1.9 V more than the value of the threshold voltage of $M13$. Notice that V_{BIAS2} in the two cases differs by more than 1.6 V. In a fixed bias condition, therefore, there is a basic conflict between maximum current driving and maximum swing and whichever value of the bias voltage V_{BIAS2} is used between the two extreme cases shown on the table will result in a severe degradation of the overall performance of the circuit.

The solution to the problem is to vary V_{BIAS2} during the transient to simultaneously obtain optimum swing and full current drive capability. A simple way to achieve such a variable bias scheme is shown in Fig. 7. The input current $I_2(t)$ during the transient is assumed to varies from 2 μA to a peak value of 90 μA as an example. $M9$, $M13$, and $M15$ make up the NMOS current mirror and for simplicity are shown to have the same size. $M30$ has an aspect ratio 4 times smaller than the other devices. Assuming that the current on $M30$ and $M9$ is the same, a high swing cascode [17] current mirror results where $M13$ is biased at the edge of the linear region with its drain one threshold voltage lower than its gate. This bias condition gives the maximum possible output swing while, at the same time, guarantees a current gain of 1 in the mirror and a very large output resistance at the output node. By forcing $M30$ and $M9$ to carry the same current during the all transient, the full 90-μA current driving capability is achieved together with the same output swing of a fixed bias cascode biased to carry only 2 μA of current. Since the voltage at the gate of the cascode device varies during transients, such a biasing scheme is called "dynamic biasing." Notice that, in actuality, the aspect ratio of $M30$ must be smaller than shown in the above example due to body effects and to guarantee

some margin of safety in the V_{DS} of $M13$ even in the presence of process variations.

Fig. 8 shows the entire amplifier schematic without common-mode feedback and illustrates how the two tracking currents, that are necessary for the proper operation of the dynamic biased current mirrors, can be easily generated. There are two p-type and two n-type dynamically biased cascode current mirrors at the top and bottom of the figure. As an example, consider the bottom right one. The current in $M30$ is forced to track the current in $M9$ by mirroring I_2 from the top via current mirror $M12$, $M27$.

D. Dynamic Common-Mode Feedback

The main drawback associated with the fully differential approach is the need for a CMFB circuit. Besides requiring extra area and power, the CMFB circuit limits the output swing, increases the noise, and slows down the op. amp. These are particularly undesirable effects in a low-voltage low-power system. The design reported here uses a switched-capacitor CMFB circuit similar to one proposed by Senderowicz et al. [3]. A simplified schematic of the circuit architecture used in this design is shown in Fig. 9. Capacitors C1 and C2 provide an ac feedback path from the two outputs of the op amp V_{o1} and V_{o2} to the feedback node A. The common-mode gain from node A to the output is very large and negative, while the differential gain from node A to the output is ideally zero. This implies that the common-mode output voltage is kept at an almost constant value while, at the same time, the op amp differential gain is almost uneffected by the CMFB circuit. The dc value of the common-mode output voltage is, however, not well defined. It depends on the initial voltages across capacitors C_1 and C_2. The purpose of C_{1a} and C_{2a} is to establish the voltage drops across C_1 and C_2 that gives the desired common-mode output, and to periodically restore these voltages to compensate for leakages. In a switched-capacitor integrator C_{1a} and C_{2a} are switched-in on the phase opposite to that associated with the input signal.

This CMFB circuit is particularly suited for low-voltage low-power applications for two main reasons. First, it does not require any extra power consumption, with the exception of the replica circuit that defines the proper value of V_A ($M3$, I, and $M6$) which is, however, shared by all the op amps. Second, it does not degrade the differential output swing since the level shift operation performed by the capacitor C_1 and C_2 is not limited by the voltage supplies.

Although the maximum current that the CMFB circuit can supply is quite large for a positive signal, for a negative signal it is limited to $2I$ (about 2 µA). This makes the CMFB circuit unacceptably slow for the case of a negative common-mode output transient. However, by chosing the device sizes in such a way that the p-type current mirrors are faster than the n-type current mirrors, all common-mode output transients are guaranteed to have a positive polarity and the CMFB circuit is guaranteed to always work at its maximum speed. It turns out that the above choice of device channel lengths is also desirable from noise and speed considerations, as explained in the next section.

E. Noise Considerations

The noise performance of the amplifier is of particular concern since both low voltage supply and low power consumption tend to degrade the dynamic range. For optimum noise performance the input devices should be as large as possible. Furthermore, the input structure should be as simple as possible, and the input referred noise due to all the devices, other than the input ones, should be made as small as possible (ideally negligible).

In this design, the input structure ($M1-M8$) is much more complicated than the classical source-coupled pair (eight devices instead of two). However, the noise power associated with each one of the eight input devices should be divided by 4, when referred back to the input node, since their noise propagates only through one of the two signal paths while the input signal propagates through both. The overall input referred noise produced by $M1-M8$ is, therefore, equivalent to that of one n plus one p device which is comparable to that of a source-coupled pair.

The noise contribution of all the devices, other than the input ones, is next considered. First notice that, as in the

TABLE I
AMPLIFIER DEVICE SIZES

DEVICE	Z(µm)	L(µm)
M 1	180	6
M 2	180	6
M 3	140	6
M 4	140	6
M 5	150	6
M 6	150	6
M 7	200	6
M 8	200	6
M 9	22	10
M10	22	10
M11	29	7
M12	29	7
M13	22	10
M14	29	7
M15	22	6
M16	29	6
M17	29	7
M18	22	10
M19	22	6
M20	29	6
M21	20	9
M22	6	12
M23	28	6
M24	6	14
M25	20	9
M26	6	12
M27	28	6
M30	6	14

TABLE II
AMPLIFIER SPECIFICATIONS

CORE AMPLIFIER SPECIFICATIONS (0-5 Volts Supply) 100µW Quiescent Power Dissipation	
DIFFERENTIAL GAIN	> 10.000 *
UNITY GAIN FREQUENCY	2 MHz *
NOISE	140 nV/√Hz 1KHz 50 nV/√Hz white
OUTPUT SWING	0 5 Volts from Supply *
AREA	300 mils²

* inferred from filter measurement

case of the input devices, they only affect one signal path and therefore their noise power contribution should be divided by 4 when referred back to the input. Notice that this would not be the case if the amplifier was operated single ended. Furthermore the ratio of the noise power contributed by one of these devices over that of an input transistor is inversely proportional to the ratio of their transconductances for the white component and inversely proportional to the ratio of the square of the channel lengths for the $1/f$ component [19].

In order to simultaneously reduce both kinds of noise, the channel length of the current-mirror devices should be made as long as possible. This increases the amplifier voltage gain but reduces the frequency of the first non-dominant pole. However, because of the fact that p-type devices are slower but less noisy than n-type devices [14], it is possible to use devices close to minimum size in the p-type current mirrors (while the devices in the n-type

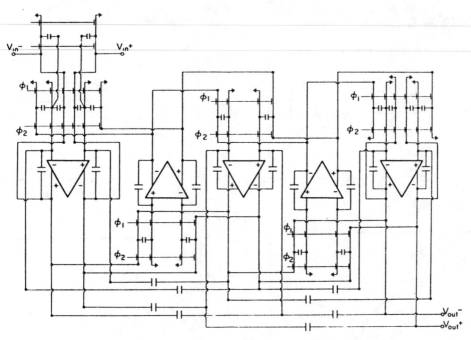

Fig. 10.   Full schematic of the low-pass filter prototype.

current mirrors are longer). This achieves good speed in the circuit without appreciably degrading the overall noise performance. In this design the noise contributed by all devices other than the input ones is only about 15 percent of the total for both $1/f$ and white components while at the same time the frequency of the second pole is kept within 40 percent of the value achievable by using all minimum channel length devices. It is interesting to note that, to achieve this result, the device length is such that the n-type current mirrors are slower than the p-type ones.

As a final point notice that since the cascode devices $M15$, $M16$ give a negligible noise contribution, their channel length can be made very short (consistent with the gain requirement) thereby improving the frequency response.

F. Op Amp Summary

The device sizes for the circuit of Fig. 8 are shown in Table I, and the main amplifier performance for a total supply of 5 V and 100-μA power dissipation is shown in Table II. Some of the entries on the table where not measured directly but inferred from the filter results. Notice the very good output swing particularly for a cascoded output. The relatively large area is due to the fully differential topology. In fact, the dynamic common-mode feedback circuit alone takes more than one-third of the total amplifier area.

Fig. 11.   Chip microphotograph.

IV.   PROTOTYPE FILTER DESCRIPTION

In order to test the performance achievable with the new op amp, a low-pass switched-capacitor filter prototype was built and will be described next.

A full schematic for the realized low-pass switched-capacitor filter is shown in Fig. 10. It is a fifth-order elliptic filter with four transmission zeros that requires a total of five op amps. The filter uses the standard active ladder architecture, for its low sensitivity to parameter variations, [15] and utilizes parasitic-free bottom-plate switched-capacitor integrators [4]. The 6-dB signal loss associated with the ladder structure is compensated by

±2.5V, 25°C, f_clock = 128KHz

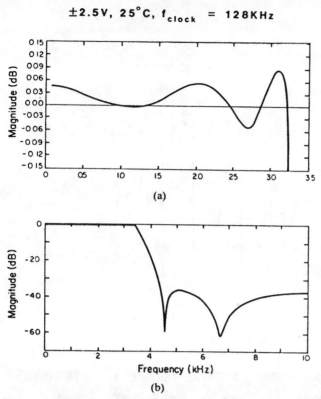

(a)

(b)

Fig. 12. (a) Detailed passband response. (b) Coarse frequency response of the filter.

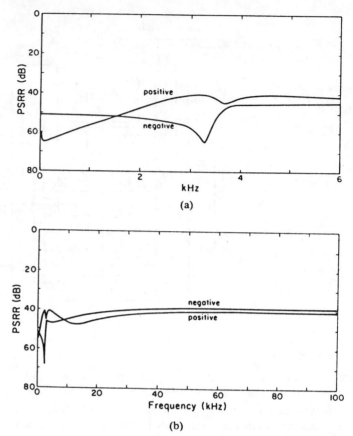

(a)

(b)

Fig. 13. Positive and negative PSRR: (a) in the 0–6 kHz range, and (b) in the 1–100 kHz range.

adding an extra sampling capacitor at the input which gives a gain close to 0 dB in the passband. This causes some peaking (less than 6 dB) at the internal nodes near the bandedge which degrades the filter linearity for large inputs at frequencies close to the band edge. On the other hand, no extra amplification is needed at the output and the overall noise is reduced.

The two-phase clock is externally supplied with two on-chip inverters providing the two-clock complements necessary to drive CMOS transmission gates which are implemented at all output node due to the large swing. The input is supplied differentially to the chip, and the differential to single-ended conversion is done off chip.

In order to be able to provide the output signal off-chip, a new low-voltage low-power buffer amplifier was designed and integrated on the chip prototype. The amplifier was designed to drive a capacitive load of up to 100 pF and/or a resistive load of 10 kΩ or more with a power dissipation from a 5-V supply of only 350 μW, and a settling time to an accuracy of 0.1 percent for a 2.5-V step of less than 3 μS.

A microphotograph of the experimental chip is shown in Fig. 11. A p-well 5-μm CMOS process was used. The process allows for a single level of metal and has a minimum channel length of 5 μm, device oxide thickness of 400 Å, and capacitor oxide thickness of 1000 Å. The chip layout is almost perfectly symmetrical to maximize cancellation of the spurious signals coupled into the system. Some small asymmetries were impossible to avoid (cross-coupled devices), but they were all limited to the metal

layer. The power level in both the filter and the amplifier can be externally controlled.

The central part of the picture is the fifth-order filter which has an active area of 3500 mil^2. In the bottom part of the picture are visible the two amplifiers designed to buffer the output of the filter from outside the chip. Two circuits are necessary because the signal is taken off-chip differentially. The top portion of the chip shows some test structure. Although the test buffer amplifier was functional and showed performance corresponding to the design values [16], due to some layout errors in the interconnections of the buffers at the filter output these circuits had to be bypassed and all the measurements reported were taken using some source followers externally biased as output buffers.

V. EXPERIMENTAL RESULTS

Both detailed and coarse filter response are shown on Fig. 12 for a total power consumption of 350 μW. The supply used is ±2.5 V, the clock frequency is 128 kHz. The total in-band ripple is 0.13 dB. The transmission zeros are at 4.5 and 6.7 kHz and the attenuation in the stopband is more than 35 dB. This agrees well with the simulation results obtained from the DIANA program [22]. The power supply rejection ratio for both supplies is shown in Fig. 13(a) for the frequency interval 0–6 kHz and in Fig. 13(b) for the interval 1–100 kHz. At 1 kHz the PSRR is well

±10% Supply Voltage Variation

DISTORTION

1 KHz input signal

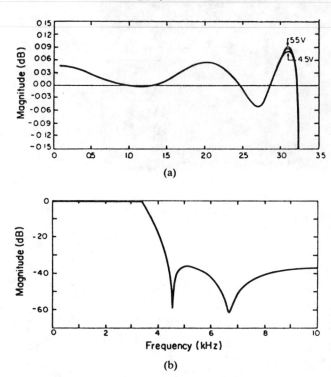

Fig. 14. Changes in the (a) passband response and (b) overall filter response for ±10 percent variation in the supply voltage.

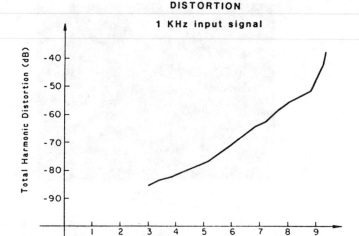

Fig. 15. Total harmonic distortion as a function of the output voltage for a 1-kHz signal.

above 50 dB for both supplies and stays close to 40 dB up to 100 kHz.

The effect of a ±10 percent variation in the total supply voltage is shown in Fig. 14. The only variation that can be detected is in the detail passband plot and it occurs at the bandedge; however, it is only about ±0.01 dB. No appreciable change can be seen on a coarse scale. The position of the zeros is essentially unaffected by the change in supply voltage. The total C-message weighted integrated noise is 70 μV.

The total harmonic distortion for a 2-V rms differential output at 1 kHz is about −73 dB. The good linearity of the filter is further shown in Fig. 15 where the total harmonic distortion (THD) at the output for the nominal supply voltage of 5 V and a 1-kHz input signal is plotted versus the output signal amplitude. The THD stays below −40 dB up to a differential output of approximately 4.6-V peak (3.3 V rms), i.e., 200 mV from both supply rails. The large output swing is primarily due to the use of dynamic biasing for the cascode devices and to the fact that the CMFB circuit behaves linearly even for signals which are larger than the supplies. The linearity of the CMFB circuit also helps produce the low distortion achieved in the filter.

A summary of the achieved filter performance is shown in Table III. The operating conditions are a total supply voltage of 5 V, a clock rate of 128 kHz, and a power dissipation of 350 μW. The total measured C-message weighted noise of 70-μV rms combined with the maximum differential output swing that gives less than 1-percent THD for a 1-kHz signal of 3.3-V rms gives a dynamic range of 93 dB, which is comparable with the value achieved

TABLE III
SUMMARY OF THE FILTER PERFORMANCE

25°C ±2.5 V f clk = 128KHz		
PARAMETER	CONDITION	VALUE
MINIMUM POWER DISSIPATION	—	350μW
P.S.R.R.	1KHz +SUPPLY 1KHz −SUPPLY	56 dB 52 dB
TOTAL HARMONIC DISTORTION	2V rms differential output 1KHz	73 dB
IDLE NOISE	CMESSAGE WEIGHTED	70 μV
OUTPUT SWING DIFFERENTIAL	<1%·THD	3.1(RMS)V
DYNAMIC RANGE	—	93.dB

by typical commercially manufactured filters operated from ±5-V supplies and requiring 10–15 times more power than this device. Another point of interest is the low distortion achieved in the filter −73 dB for a 2-V rms differential output signal. One reason for this is the use of the fully differential topology as demonstrated by the results of Fig. 16, which shows the output spectrum for a 1-kHz pure sinusoidal input that gives a 4.4-V peak-to-peak differential output. This plot is obtained by feeding one of the two filter outputs directly to the spectrum analyzer without passing through the differential to single-ended converter. As can be seen, the only appreciable harmonic is the second one. On this plot the harmonic content of the differential output is depicted. The second harmonic is totally canceled out because of the symmetry of the struc-

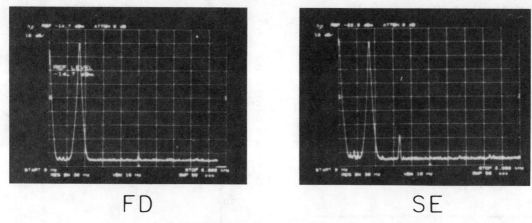

FD   SE

Fig. 16. Comparison between the harmonic distortion for a fully differential and a single-ended output for a differential output voltage of 4.4 p-p V.

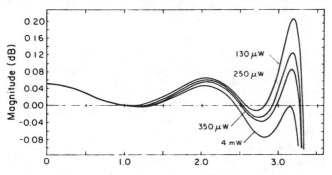

Fig. 17. Variation of the passband response for different power levels.

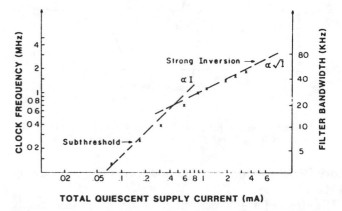

Fig. 18. Maximum clock frequency versus required supply current.

ture. On the other hand, a small amount of third harmonic is now present. The total improvement in THD from right to left is more than 12 dB. The clock feedthrough for the two cases was also compared. For grounded inputs, which gives matched signal paths, the clock feedthrough in the fully differential case is 30 dB less than in the single-ended case.

Fig. 17 shows the change in the shape of the passband for different values of the total power dissipation. For very low power the filter is still functional but the op amp is not capable to settling very accurately within one phase of the clock and peaking occurs. However, channel filter requirement are met over a change in the power level of more than 40 to 1. Here the power level was varied by varying the bias current to the op amps with an external resistor.

The minimum value of the total supply voltage required for proper operation is approximately 3 V. A smaller value could be used if a low threshold process had been used instead of the conventional (not scaled) process featuring approximately ± 0.8-V thresholds.

Finally, Fig. 18 shows the total amount of supply quiescent current that is necessary for the filter to operate properly when the clock rate is increased. This experiment was carried out by chosing a value of the clock rate and then increasing the current level in the filter until the required ripple in the passband was obtained. The last point to the right shows that with a total supply quiescent current of about 3 mA, a 50-kHz filter could be obtained.

At this clock rate (almost 2 MHz) the op amp must be able to accurately settle in about 200 ns. The change on the slope of the curve indicates that the input devices of the op amp move from the subthreshold to strong inversion as the current is increased from 60 μA to 3 mA.

VI. CONCLUSIONS

An experimental switched-capacitor filter has been described which shows that a dynamic range adequate for communications applications can be achieved when a total supply voltage of 5 V is used. Good PSRR up to high frequency and low power dissipation make this approach very suitable for operation as a part of a large digital/analog chip where noise immunity and power consumption reduction are of paramount importance. This circuit was implemented using 5-μm technology and requires 700 mil^2/pole. It is, however, projected that by using a 3-μm technology an area per pole of less than 200 mil^2 can be achieved without compromising the level of performance.

ACKNOWLEDGMENT

The cooperation provided by INTEL Corporation with the fabrication of one of the chip prototypes is gratefully acknowledged. The authors also acknowledge the contribution of M. Wong and W. E. Matthews with the layout and testing of the prototype.

REFERENCES

[1] W. C. Black, Jr., D. J. Allstot, and R. A. Reed, "A high-performance low-power CMOS channel filter," *IEEE J. Solid-State Circuits*, vol. SC-15, pp. 929–938, Dec. 1980.

[2] Y. A. Haque, R. Gregorian, D. Blasco, R. Mao, and W. Nicholson, "A two-chip PCM codec with filters," *IEEE J. Solid-State Circuits*, vol. SC-14, pp. 961–969, Dec. 1979.

[3] D. Senderowicz, S. F. Dreyer, J. H. Huggins, C. F. Rahim, and C. A. Laber, "A family of differential NMOS analog circuits for a PCM codec filter chip," *IEEE J. Solid-State Circuits*, vol. SC-17, pp. 1014–1023, Dec. 1982.

[4] R. Gregorian and W. A. Nicholson Jr., "CMOS switched-capacitor filters for a PCM voice CODEC," *IEEE J. Solid-State Circuits*, vol. SC-14, pp. 970–980, Dec. 1979.

[5] H. Ohara, P. R. Gray, W. M. Baxter, C. F. Rahim, and J. L. McCreary "A precision low-power PCM channel filter with on chip power supply regulation," *IEEE J. Solid-State Circuits*, vol. SC-15, pp. 1005–1013, Dec. 1980.

[6] D. G. Marsh, B. K. Ahuja, M. R. Dwarakanath, P. E. Fleisher, and V. R. Saari, "A single-chip CMOS PCM CODEC with filters," *IEEE J. Solid-State Circuits*, vol. SC-16, pp. 308–315, Aug. 1981.

[7] S. Wong and C. A. T. Salama, "Impact of scaling on MOS analog performance," *IEEE J. Solid-State Circuits*, vol. SC-18, pp. 106–114, Feb. 1983.

[8] B. J. Hosticka, "Dynamic CMOS amplifiers," *IEEE J. Solid-State Circuits*, vol. SC-15, pp. 887–894, Oct. 1980.

[9] M. G. Degrauwe, J. Rijmenants, E. A. Vittoz, and H. J. De Man, "Adaptive biasing CMOS amplifiers," *IEEE J. Solid-State Circuits*, vol. SC-17, pp. 522–528, June 1982.

[10] F. Krummenacher, "Micropower SC biquadratic cell," *IEEE J. Solid-State Circuits*, vol. SC-17, pp. 507–512, June 1982.

[11] M. G. Degrauwe and W. C. Sansen, "A multipurpose micropower SC filter," *IEEE J. Solid-State Circuits*, vol. SC-19, pp. 343–348, June 1984.

[12] H. Pinier, F. Krummenacher, and V. Valencic, "A μP sixth-order SC leapfrog low-pass filter," in *Proc. ESSCIRC '82*, Sept. 1982, pp. 223–225.

[13] B. J. Hosticka, D. Herbst, B. Hoefflinger, U. Kleine, J. Pandel, and R. Schweer, "Real-time programmable SC bandpass filter," *IEEE J. Solid-State Circuits*, vol. SC-17, pp. 499–506, June 1982.

[14] E. Vittoz and F. Krummenacher, "Micropower SC filters in Si-gate CMOS technology," in *Proc. ECCTD '80* (Warsaw, Poland), vol. 1, Sept. 1980, pp. 61–72.

[15] D. J. Allstot, "MOS switched capacitor ladder filter," Ph. D. dissertation, Univ. of California, Berkeley, May 1979.

[16] R. Castello, "Low-voltage low-power S.C. signal processing technique," Ph. D. dissertation, ERL Memo. M84/67, Univ. of California, Berkeley, Aug. 1984.

[17] P. R. Gray and R. G. Meyer, *Analysis and Design of Analog Integrated Circuits*. New York: Wiley, 1977.

[18] W. C. Black, "High speed CMOS A/D conversion technique," Ph. D. Dissertation, Univ. of California, Berkeley, Nov. 1980.

[19] J. C. Bertails, "Low-frequency noise considerations for MOS amplifier design," *IEEE J. Solid-State Circuits*, vol. SC-14, pp. 774–776, Aug. 1979.

[20] R. Castello and P. R. Gray, "Performance limitations in switched-capacitor filters," *IEEE Trans. Circuits Syst.* vol. CAS-32, no. 10, pp. 865–876, Sept. 1985.

[21] P. R. Gray and R. Castello, "Performance limitations in switched-capacitor filters," in *Proc. IEEE Int. Symp. Circuits Syst.* (Kyoto, Japan), June 85.

[22] H. DeMan, J. Rabaey, L. Claesen, and J. Vandewalle, "DIANA-SC: A complete CAD system for switched-capacitor filters," in *ESSCIRC Dig. Tech. Papers*, Sept. 1981, pp. 130–133.

Performance Limitations in Switched-Capacitor Filters

RINALDO CASTELLO AND PAUL R. GRAY, FELLOW, IEEE

Abstract —Switched-capacitor (SC) filters continue to improve in performance mainly through progress in the design of MOS operational amplifiers (op amps). Ultimate limits to achievable filter performance, however, stem from factors more fundamental than op amp nonidealities, factors independent of process and circuit improvements. This paper develops, from certain basic assumptions, ultimate limits on dynamic range, chip area, and power consumption in SC integrators and low-pass filters. For integrators, minimum area and power requirements are shown to vary as the square of desired dynamic range. Some physically realistic approximations lead to expressions relating filter area, power consumption, and dynamic range which involve only fundamental process parameters, supply voltage and filter cut-off frequency. Comparison with actual performance in typical commercially manufactured SC filters suggests that there is still a strong motivation in improving op amp specifications. A typical commercial fifth-order voiceband filter operating from a ±5-V supply with a dynamic range of 95 dB consumes approximately 5 mW and requires an area of approximately 5000 mil² compared with the theoretical minima of 8.5 μW and 11.2 mil², respectively.

I. INTRODUCTION

DESPITE their relatively short history, switched-capacitor (SC) circuits are already fairly mature. Most of their performance characteristics have improved substantially since the first monolithic SC filters using SC integrators were designed and fabricated in 1977 [1], [2]. In particular, the power dissipated per pole has been reduced from about 10 or 20 mW in the first NMOS prototypes to less than 1 mW in the CMOS filters in production today [3]. These figures refer to general purpose systems working from a ±5-V supply and with clock rates of 128 kHz or more. For special purpose applications, on the other hand, much smaller values of power dissipation have been achieved using dynamic techniques which minimize standby power [4]–[7]. Another aspect that has been extensively investigated is the improvement of the dynamic range of SC filters. To this end techniques such as fully differential architectures and noise frequency translation via chopper stabilization have been proposed. This has produced filters with reported dynamic range above 100 dB [8]. To achieve such a result, however, a large increase in the chip area occupied by the filter was necessary. Finally, the total die area occupied by an SC filter has been substantially

reduced. This has allowed the integration on a single chip of many SC filters together with other components [9]–[12].

Almost all of these results have been achieved by improving the performance of the operational amplifiers (op amps) in the filter [3], [13]–[15]. It is likely that, through technological scaling and better circuit design, future CMOS op amps will require steadily decreasing amounts of area and power. Ultimately, however, the reduction on physical size and power dissipation will be limited by the fact that the signals within the filter are represented as energy stored on capacitors. The dynamic range of the filter is related to the ratio of stored energy to thermal energy, kT, dictating a minimum capacitor size for a given filter dynamic range. This, in turn, dictates a minimum power dissipation since this capacitance must be charged and discharged at the signal frequency. Such limitations cannot be overcome by circuit or process improvements; therefore, they determine the ultimately achievable performance attainable in such filters. The objective of this paper is to identify and analyze these ultimate limits on the performance of SC filters, with particular emphasis on power dissipation, silicon die area, and dynamic range.

In Section II, the minimum achievable area and power dissipation of a SC integrator with a given dynamic range and signal frequency are calculated. It is shown that both the minimum power and minimum area vary proportionally with the square of the dynamic range.

In Section III, this analysis is extended to the case of a low-pass SC filter. The results, while intuitively interesting, are a function of the particular filter under consideration and cannot be related to each other in a general way. By introducing additional approximations, which in most practical cases cause only a small error, and normalizing the results to the order of the filter, several simple relationship are obtained. Logarithmic plots showing the dependence of the minimum area and power requirement versus the achievable dynamic range are also provided. Comparison is made between these limits and commercially available devices.

Finally, in Section IV the effect of the op amp non-idealities which were ignored in the derivation of the previous sections are considered. Upper bounds for the absolute minimum power, area, and noise are obtained with reference to a particularly simple but realistic op amp config-

Manuscript received June 18, 1984; revised March 3, 1985. This work was supported by the National Science Foundation under Grant ECS-8023872 and under Grant ECS-8100012.

The authors are with the Department of Electrical Engineering and Computer Science, University of California, Berkeley, CA 94720.

Reprinted from *IEEE Trans. Circuits Syst.*, vol. CAS-32, no. 9, pp. 865–876, Sept. 1985.

262

uration. These results indicate that the op amp limitations should not affect the ultimate filter performance in most practical cases.

II. PERFORMANCE LIMIT FOR THE IDEAL INTEGRATOR

In this section the SC integrator is analyzed to obtain limits for the minimum power consumption and chip area requirement for a given dynamic range and maximum signal frequency. The following calculations refer to the so called differential bottom plate integrator shown in Fig. 1. Such a circuit was chosen for the sake of concreteness, and because it is insensitive to parasitic capacitance and has been used extensively commercially [3], [8], [16]. However the extension of this theory to other SC integrator configurations is straightforward and yields similar results.

The following assumptions will be used throughout the paper:

1) The operational amplifier in the integrator is assumed to be ideal in the sense that it does not contribute any noise to the filter, it does not consume any dc power, and it occupies no chip area. The reason for such drastic assumptions is that there are no fundamental limits, identifiable *a priori*, for the minimum value that can be achieved, via process and/or circuit design improvements, for any of these op amp nonidealities.

The only potential exception to this comes from the op amp white noise. It has, however, been shown [17] that its contribution, when not negligible, can be added to that of the kT/C noise since both can be represented in the same way. In this paper, the op amp white noise is neglected for the sake of simplicity; however, because of the above considerations, the following analysis can be easily extended to include it, if a specific op amp configuration is given. In Section IV the validity of these assumptions will be discussed in more detail.

2) The integrating capacitor is assumed to be much larger than the sampling capacitor, i.e.,

$$\frac{C_i}{C_s} \gg 1 \tag{2.1}$$

where C_s and C_i are the sampling and integrating capacitors, as shown in Fig. 1. Making use of the following basic equation for the SC integrator [19]:

$$\frac{C_s}{C_i} = \frac{2\pi f_{unity}}{f_{clock}} \tag{2.2}$$

where f_{unity} is the unity gain frequency of the integrator and f_{clock} is the clock frequency, condition (2.1) becomes

$$\frac{f_{clock}}{f_{unity}} \gg 2\pi. \tag{2.3}$$

Assumption (2.3) is almost always valid if the integrator is part of a low-pass voiceband SC filter. In such a case, in fact, each integrator has a unity-gain frequency which is comparable in value with the band edge of the filter, while

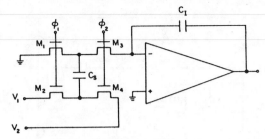

Fig. 1. Bottom plate SC integrator.

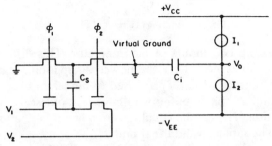

Fig. 2. Circuit used to compute the power drawn from the supplies.

the clock frequency is typically many times larger than the filter band edge to ease anti-aliasing requirements.

On the basis of these assumptions, the absolute minimum integrator area is approximately equal to the area of the integrating capacitor C_i. Assuming a symmetrical power supply equal to $\pm V_s$ volts, a capacitor dielectric with a maximum electric field before break-down equal to E_{max}, and a dielectric constant equal to ϵ_{diel}, the minimum thickness of the capacitor is

$$t_{min} = 2\frac{V_s}{E_{max}}. \tag{2.4}$$

Here the somewhat unrealistic assumption is made that the capacitor oxide thickness can be scaled down so that the peak field approaches E_{max}. Actually, the maximum usable field is a factor of two to four smaller than this because of reliability and yield considerations. The minimum area required to realize a capacitor of value C_i is, therefore,

$$\text{AREA}_{min} = \frac{t_{min}C_i}{\epsilon_{diel}} = \frac{2V_sC_i}{E_{max}\epsilon_{diel}} \tag{2.5}$$

The maximum amount of energy that can be stored in the integrator E_{max} is given by

$$E_{max} = \tfrac{1}{2}(2V_s)^2 C_i = 2V_s^2 C_i. \tag{2.6}$$

Substituting (2.6) into (2.5) gives the minimum area as a function of the maximum stored energy:

$$\text{AREA}_{min} = \frac{E_{max}}{V_s E_{max}\epsilon_{diel}} \tag{2.7}$$

Next, the minimum achievable power consumption is computed. To this end, the integrator of Fig. 1 can be represented as in Fig. 2. Furthermore the left-hand side of the circuit of Fig. 2 can be modified as shown in Fig. 3 with $V_{in} = V_1 - V_2$. The only potential source of error in such a substitution is due to the phase difference existing

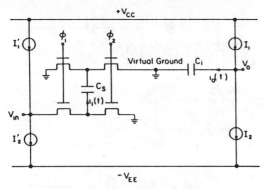

Fig. 3. Equivalent circuit of Fig. 2.

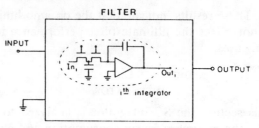

Fig. 4. S.C. integrator imbedded in a filter.

between the two inputs of the integrator. This difference, however, does not effect the power dissipation. The two current sources I_1 and I_2 are used to model an ideal class B op amp. If the amplifier is to have zero quiescent power dissipation, I_1 must be equal to zero when $I_2 \neq 0$ and vice versa. The same is valid for I_1' and I_2'.

The total power dissipation is given by the amount of energy per unit time drawn from the supplies by the two portions of the circuit, i.e.,

1) The amount of energy that C_s draws from one supply and then dumps into the amplifier virtual ground.

2) The amount of energy that C_I draws from the other supply, through the action of the op amp, to be dumped also into the virtual ground. Assuming that the input signal v_i is a pure sinusoid with frequency f and peak amplitude V_i, the energy dissipated during one period of the signal can be computed as in Appendix 1 and is equal to

$$E_{cycle} = \frac{4}{\pi} V_i V_s C_s \frac{f_{clock}}{f} \qquad (2.8)$$

where V_i is the input signal peak value and f is the input signal frequency. The average power dissipation is obtained multiplying the energy per cycle by the frequency of the signal, i.e.,

$$P = \frac{4}{\pi} V_i V_s C_s f_{clock}. \qquad (2.9)$$

The input to output transfer function of the integrator is given by the following relation [19]:

$$V_i = \frac{f V_o}{f_{unity}} = \frac{2\pi f C_i}{f_{clock} C_s} V_o. \qquad (2.10)$$

From (2.10) it can be seen that for an in-band signal, i.e., $f \leqslant f_{unity}$, the maximum amplitude of the input signal that does not cause clipping at the output is a function of the input frequency. For this reason, it is convenient to express the power consumption as a function of the output signal amplitude. This can be done by substituting (2.10) into (2.9) obtaining

$$P = 8 V_s V_o f C_i. \qquad (2.11)$$

For a maximum amplitude sinusoid at the output, i.e., $V_o = V_s$ the power dissipation becomes

$$P = 8 f C_i V_s^2 \qquad (2.12)$$

Using (2.6) into (2.12) gives

$$P = 4 E_{max} f. \qquad (2.13)$$

The minimum power consumption for a full swing sinusoidal output is, therefore, proportional to the maximum energy stored in the integrator times the frequency of the signal.

Lastly, the dynamic range is considered. While the relationship between power consumption, area, and signal frequency can be derived for a stand-alone SC integrator, the dynamic range is a function of the surrounding circuit topology. The SC integrator is used within a filter, and the output noise of the filter depends both on the amount of noise produced by the integrator and the transfer function from the point of injection to the filter output, as shown in Fig. 4. A number of authors have analyzed noise effects in switched capacitor filters [8], [23], [25]. The most important contributions are the $1/f$ noise of the operational amplifier, the broadband thermal noise of the operational amplifier, and the thermal noise in the switches kT/C. Most commercially available filters at the present time are $1/f$ noise limited, but some approach kT/C noise limits.

While operational amplifier noise is of great practical importance, it is not a fundamental limit to filter dynamic range. $1/f$ noise can be reduced to a negligible level by either technological improvements or through circuit techniques such as chopper stabilization. Further it can be shown [8] that if the operational amplifier is designed such that under closed-loop conditions its output resistance and its noise equivalent resistance are equal, then the op amp thermal noise does not contribute any additional noise over and above the kT/C noise discussed below. The matched resistance condition is closely approximated in folded cascode op amps, for example. The analysis below is focused on the kT/C noise contributed by the MOS switches.

The thermal noise contributed by the left-hand side switch is sampled by C_s every clock cycle as the switch turns off. The signal appearing across capacitor C_s is, therefore, a sampled first-order low-pass filtered white noise. It has been shown that for a properly operating SC circuit, i.e. the time constants associated with the switches and the capacitors are much smaller than the clock period, such a discrete random process has a white spectral distribution and a total noise power (variance) equal to kT/C_s [24]. Thus for this noise mechanism, the SC filter can be characterized as a linear discrete time system in which noise samples are added to the signal at each integrator input. Discrete-time linear system theory can, therefore, be used. The noise variance of the output samples, n_i^2, resulting from one such source is [22]

$$n_i^2 = \frac{kT}{C_s} \sum_{m=0}^{\infty} h^2(m) \qquad (2.14)$$

where $h(m)$ is the impulse response from the noise source to the output. Using Parseval's theorem [22] the output sample noise variance can also be expressed in terms of the frequency response from the integrator input to the output

$$n_i^2 = \frac{kT}{C_s} \frac{1}{2\pi} \int_{-\pi}^{\pi} H(e^{j\omega}) H(e^{-j\omega}) \, d\omega \qquad (2.15)$$

where $H(e^{j\omega})$ is the z transform of $h(m)$ evaluated on the unit circle in the z plane. It is convenient to define the effective noise bandwidth, B_o, as

$$B_o = \frac{f_{\text{clock}}}{2\pi} \int_{-\pi}^{\pi} H(e^{j\omega}) H(e^{-j\omega}) \, d\omega. \qquad (2.16)$$

Making use of (2.2), (2.15), and (2.16), can be written as

$$n_i^2 = \frac{1}{2\pi} \frac{kT}{C_i} \frac{B_o}{f_{\text{unity}}} \qquad (2.16b)$$

where C_i is the integrating capacitor and f_{unity} is the integrator unity-gain frequency. The quantity B_o is the effective noise bandwidth from the input of the SC integrator to the integrator output for the particular filter configuration considered. It is equal to the integral of the magnitude squared of the frequency response of the sampled data circuit from the integrator input to the filter output, taken around the unit circle. For low-pass filters where the clock rate is far above the passband, this is equivalent to the frequency-domain integral over the passband of the transfer function from the integrator input to the output for the continuous equivalent circuit. In the following B_o will be called the noise bandwidth to the output.

The noise contributed by the right-hand side switch is also sampled by C_s. However, in this case, the resulting signal cannot rigorously be considered to be a first-order low-pass filtered noise. The reason is that the circuit through which the white noise of the switch is sampled does not have a single pole roll-off since it also contains the op amp. For the case of an op amp with infinite bandwidth, the two switches have equal contribution to the output noise and the total output noise, n_o^2, becomes

$$n_o^2 = \frac{1}{\pi} \frac{kT}{C_i} \frac{B_o}{f_{\text{unity}}}. \qquad (2.17)$$

The quantity n_i^2 calculated in (2.17) is the variance which will be observed in the samples at the output of the filter. The analog output of the filter is produced, in effect, by putting these samples through a first-order sample/hold function, which is inherent in the operation of the final integrator in the filter. Following this operation, the total noise energy is distributed in the frequency band between zero and one-half the sample rate, with the spectrum modified by the appropriate $\sin(x)/x$ function as well as by the details of the response of the filter. In practical

high-order filters, most of this energy is concentrated within the passband. Noise energy will also be present in the output at higher frequencies but these components will be less important both because of the $\sin(x)/x$ response envelope and because of band limiting by subsequent continuous circuitry.

Assuming that the maximum undistorted output signal is approximately equal to the supply voltage V_s, i.e., $\sqrt{2} V_s$ rms, the dynamic range, (DR), of the integrator becomes

$$(\text{DR})^2 = \frac{s^2}{n^2} = \frac{\pi}{2} \frac{V_s^2 C_i}{kT} \frac{f_{\text{unity}}}{B_o} \qquad (2.18)$$

Equation (2.18) can be rewritten as follows by making use of (2.6):

$$(\text{DR})^2 = \frac{\pi}{4} \frac{E_{\text{max}}}{kT} \frac{f_{\text{unity}}}{B_o} \qquad (2.19)$$

Equation (2.19) is particularly useful and suggests that the square of the dynamic range is given by the ratio between the maximum energy stored in the integrator and the thermal energy kT, modified by the ratio between the noise bandwidth to the output and the unit gain bandwidth of the integrator.

As an example, consider the unity gain feedback circuit shown in Fig. 5. This is the simplest configuration in which an SC integrator can be operated. It corresponds to a first-order low-pass filter whose z domain transfer function from C_s to the output is given by

$$H(z) = \frac{\dfrac{C_s}{C_i}}{1 - z^{-1} + \dfrac{C_s}{C_i}}. \qquad (2.20)$$

An approximate plot of $|H(e^{j\omega T})|$ is shown in Fig. 5(b). In this simple case, B_o can be easily computed by making use of the Cauchy residue theorem with the following result:

$$B_o = f_{\text{clock}} \frac{C_s}{C_s + 2C_i}. \qquad (2.21)$$

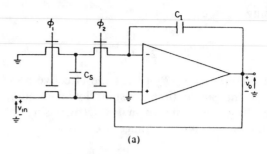

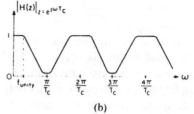

(a)

(b)

Fig. 5. On a pole SC filter.

Assuming $C_s \ll C_i$;

$$B_o = \frac{f_{clock}}{2} \frac{C_s}{C_i} = \pi f_{unity}. \qquad (2.22)$$

For this simple case, B_o is just the effective noise bandwidth of the single time constant low-pass filter whose transfer function is shown in Fig. 5(b). Using the above result the circuit dynamic range becomes

$$(DR)^2 = \frac{V_s^2 C_i}{2kT} = \frac{E_{max}}{2kT} \qquad (2.23)$$

Note that this ratio is simply the maximum energy stored on the integrating capacitor divided by kT. This result has strong implications for the ultimate limit on the ability to scale switched capacitor filters with technological feature size. In effect, silicon dioxide can only store a certain amount of energy per unit volume as dictated by the maximum field strength of the dielectric. For a given oxide thickness and power supply voltage, this dictates a maximum energy storage per unit area, which dictates a minimum area for a given dynamic range. Such a minimum value can be computed by combining (2.19) and (2.7) to give

$$(DR)^2 = \frac{\pi}{4} \frac{V_s \epsilon_{diel} E_{max} AREA}{kT} \frac{f_{unity}}{B_o}. \qquad (2.24)$$

This indicates that the ultimately achievable dynamic range is proportional to the square root of the product of the power supply voltage and the area. The assumption has been made in this analysis that the capacitor oxide thickness is decreased along with the power supply voltage to keep the peak field at E_{max}.

Since the absolute minimum achievable level of power dissipation is proportional to E_{max}, as was shown in (2.13), a relationship similar to (2.24) between dynamic range and power consumption must exist. Mathematically such a relationship can be obtained by combining (2.13) and (2.19) to obtain the following result:

$$(DR)^2 = \frac{\pi}{16} \frac{P}{kTB_o} \frac{f_{unity}}{f}. \qquad (2.25)$$

Thus the dynamic range is proportional to the square root of the minimum power dissipation necessary to charge and discharge the sampling and integrating capacitors from the power supply.

Notice that (2.25) is only valid for $f \leqslant f_{unity}$ since, outside this range, the gain of the integrator is less than 1 and therefore it is not possible to have $V_o = V_s$ for an input signal v_i smaller than the supply voltage. It is easy to see that the absolute maximum for $P(P_{max})$, when both V_i and V_o are not allowed to exceed the supply voltage, corresponds to $f = f_{unity}$. In this case (2.24) becomes

$$(DR)^2 = \frac{\pi}{16} \frac{P_{max}}{kTB_o}. \qquad (2.26)$$

It can be shown that (2.24)–(2.26) are valid for both single ended and fully differential integrators [17].

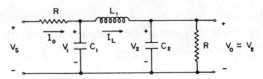

Fig. 6. Passive ladder prototype for a third-order low-pass filter.

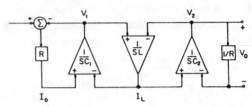

Fig. 7. Active equivalent for the circuit of Fig. 6.

III. APPLICATION TO LOW-PASS LADDER SC FILTERS

In this section, the previous analysis is extended to the case of low-pass ladder filters implemented with SC integrators.

To apply the results of the previous section to the entire filter, it is first shown that there is a one-to-one correspondence between the order of the filter (number of poles) and the number of integrators required to realize it. This is easily done with the help of a simple example.

Fig. 6 shows the passive ladder prototype for a third-order low-pass filter. This circuit can be represented in terms of integrator summers and multipliers as in Fig. 7 [20]. The flow diagram of Fig. 7 shows that each integrator output corresponds to one of the state variables of the filter, i.e., a voltage across a capacitor or a current through an inductor. Therefore the number of integrators will be equal to the number of state variables, which also coincides with the order of the filter.

The above situation can be generalized to an nth-order structure as long as the number of state variables coincides with the number of reactive elements. Even when this is not the case (due to the presence of loops of capacitors or cutsets of inductors), however, it is still possible to modify the passive prototype so that the number of integrators will coincide with the order of the filter by introducing some voltage-controlled voltage sources in the circuit [19]. In a SC implementation, such controlled generators can be realized by simply substituting the basic integrator of Fig. 1 with the integrator-summer of Fig. 8 [19]. All the results of Section II can be extended with no changes to the structure of Fig. 8 if the extra area due to capacitor C_3 is neglected.

From the above considerations and from the results of Section II follows that the minimum amount of area required for a nth-order SC filter is

$$AREA_{tot} = 2 \frac{V_s}{E_{max} \epsilon_{diel}} \sum_{i=1}^{n} C_i = \frac{\sum_{i=1}^{n} E_{max_i}}{V_s E_{max} \epsilon_{diel}} \qquad (3.1)$$

where E_{max_i} is the maximum amount of energy that can be stored in the ith integrator.

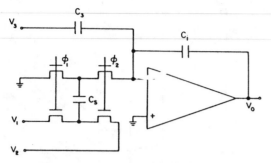

Fig. 8. Bottom plate integrator/summer.

TABLE I
PARAMETER VALUES FOR THE INTEL 2912 PCM LOW-PASS FILTER
($f_{max} = 3.4$ kHz, $1/f_{max} = 294$ μs, $n = 5$)

i^{th} Integrator	f_i (kHz)	$1/f_i$ (μs)
$i = 1$	4.715	212.1
$i = 2$	2.227	449
$i = 3$	5.186	192.8
$i = 4$	4.0415	247.4
$i = 5$	2.965	337.3
$\Sigma / 5$	3.826	287.5

To compute the total power dissipated in the filter for a sinusoidal input of frequency f and peak amplitude equal to the supply voltage V_s, use can be made of (2.11) provided that the gain from the input of the filter to the output of the ith integrators, $G_i(f)$, is known for each integrator. This gives

$$P_{tot} = 8fV_s^2 \sum_{i=1}^{n} G_i(f)C_i = 4f \sum_{i=1}^{n} G_i(f) E_{max_i} \quad (3.2)$$

where f_i^{unity} is the unity gain frequency of the ith integrator. Notice that in (3.2) it is implicitly assumed that $G_i(f) \leqslant 1$ for every i. This condition must be true to be able to process a full swing (supply to supply) input signal without unacceptably large distortion. This assumption will be discussed further later in the paper.

Finally the total output noise contribution can be computed from (2.17), provided that the value of the noise bandwidth from the input of integrator i to the output of the filter, B_i, is known for all the n integrators.

$$n_{tot}^2 = \frac{1}{\pi} kT \sum_{i=1}^{n} \frac{B_i}{f_i^{unity}C_i} . \quad (3.3)$$

This gives for the filter dynamic range

$$(DR_{tot})^2 = \frac{\pi V_s^2}{2kT \sum_{i=1}^{n} \frac{B_i}{f_i^{unity}C_i}} = \frac{\frac{\pi}{4}}{\sum_{i=1}^{n} \frac{kT}{E_{max_i}} \frac{B_i}{f_i^{unity}}} \quad (3.4)$$

where the in-band input-to-output gain of the filter has been assumed to be equal to one and the maximum output swing to be equal to the supply voltage ($\sqrt{2} V_s$ rms). This is usually the case in a SC low-pass filter since the 6 dB in-band loss of the passive prototype can be easily eliminated by making the capacitor that samples the input voltage twice as big as the other sampling capacitor in the first integrator.

An alternate way to express the dynamic range in terms of the sampling capacitors, which will be useful later, is shown in (3.5)

$$(DR_{tot})^2 = \frac{V_s^2 f_{clock}}{4kT \sum_{i=1}^{n} \frac{B_i}{C_{s_i}}} \quad (3.5)$$

where C_{s_i} is the sampling capacitor in the ith integrator.

The above equations involve few approximations and can be used if all of the required parameters are known. The results, however, are a function of the particular filter design adopted and are in a form that does not show any particular relationship between the various performance aspects. More insight on the problem can be gained by introducing some approximations.

First, it is assumed that the sampling capacitors are identical for all the integrators, with the exception of the one that samples the input voltage, which was assumed to be twice as big as the others.

Second, the following approximation is introduced:

$$\sum_{i=1}^{n} \frac{1}{f_i^{unity}} \approx \frac{n}{f_{max}} \quad (3.6)$$

where f_{max} is the band edge of the filter. Physically, this means that the average value of the time constants of all the integrators in the filter coincide with the time constant associated with the band edge of the filter. In a typical low-pass ladder filter, the error introduced by (3.6) rarely exceeds ± 30 percent.

An example of the size of the error introduced by the above approximation can be obtained by substituting into (3.6) the values of the integrator's unity gain frequency and the value of the filter band edge for a commercial filter. Table I, for example, shows the values of the integrator's unity gain frequency together with f_{max} for the INTEL 2912 which is a PCM low-pass filter composed of 5 integrators and realizing a 5 poles 4 zeros transfer function. For these data, the approximation introduces only about 2-percent error.

From (3.1), (2.2), and (3.0) it follows that

$$AREA_{tot} = \frac{2V_s n}{E_{max} \epsilon_{diel}} \frac{f_{clock} C_s}{2\pi f_{max}}$$

$$= \frac{2V_s n}{E_{max} \epsilon_{diel}} C_I = \frac{n E_{max}^I}{E_{max} \epsilon_{diel} V_s} \quad (3.7)$$

where the following two definitions have been introduced:

$$C_I = \frac{f_{clock} C_s}{2\pi f_{max}} \quad (3.8)$$

$$E_{max}^I = 2V_s^2 C_I. \quad (3.9)$$

From (2.2) C_I can be interpreted as the integrating capacitance necessary to obtain an integrator whose unity gain frequency is f_{max}; ϵ_{max}^I is the maximum amount of energy that can be stored in C_I.

In order to obtain a single numerical value for the power dissipated in the filter, (3.2) is evaluated for $f = f_{\max}$; thereby obtaining an upper bound for the minimum power requirement. At that frequency, for a properly designed filter, the gain from the input to each intermediate node can be assumed approximately to be equal to 1, i.e., $G_i(f_{\max}) = 1$ for $i = 1, \cdots, n$ [20].

To understand why this is, in most cases, a good approximation, notice that, to avoid saturation, which will reduce the maximum usable amplitude of the input signal, the gain from the input to each internal node must be less than or equal to one for all frequencies. On the other hand, the value of the gain from all the intermediate nodes to the output should be minimum, to minimize the total output noise contribution. A good compromise between these two requirements is to set the peak value of each intermediate gain to one. Since peaking typically occurs in the proximity of the band edge, the above assumption is usually justified.

Equation (3.6) can be substituted in (3.2) to give

$$P_{\text{tot}}(f_{\max}) = 8nV_s^2 f_{\max} C_I = 4n E_{\max}^I f_{\max}. \quad (3.10)$$

The total output noise is obtained by using (3.3), with the condition that all the sampling capacitors are equal:

$$n_{\text{tot}}^2 = \frac{4kT}{C_s f_{\text{clock}}} \sum_{i=1}^{n} B_i = \frac{2}{\pi} \frac{kT}{C_I f_{\max}} \sum_{i=1}^{n} B_i. \quad (3.11)$$

This implies a dynamic range for the filter (DR_{tot}) of

$$(\text{DR}_{\text{tot}})^2 = \frac{\pi V_s^2 C_I f_{\max}}{2kT \sum\limits_{i=1}^{n} B_i} = \frac{\pi}{4} \frac{E_{\max}^I}{kT} \frac{f_{\max}}{\sum\limits_{i=1}^{n} B_i}. \quad (3.12)$$

Equations (3.7), (3.10), and (3.12) can be normalized to obtain the equivalent area, power, and dynamic range per pole as follows:

$$\text{AREA}_{\text{pole}} = \frac{\text{AREA}_{\text{tot}}}{n} = \frac{E_{\max}^I}{E_{\max} \epsilon_{\text{diel}} V_s} \quad (3.13)$$

$$P_{\text{pole}} = \frac{P_{\text{tot}}}{n} = 4 E_{\max}^I f_{\max} \quad (3.14)$$

$$(\text{DR}_{\text{pole}})^2 = n(\text{DR}_{\text{tot}})^2 = \frac{\pi}{4} \frac{E_{\max}^I}{kT} \frac{f_{\max}}{\frac{1}{n}\sum\limits_{i=1}^{n} B_i}. \quad (3.15)$$

Equation (3.13), (3.14), and (3.15) can be related to each other in the same way as it was done for (2.7), (2.13), and (2.19) to obtain

$$(\text{DR}_{\text{pole}})^2 = \frac{\pi}{16} \frac{P_{\text{pole}}}{kT \frac{1}{n}\sum\limits_{i=1}^{n} B_i} \quad (3.16)$$

$$(\text{DR}_{\text{pole}})^2 = \frac{\pi}{4} \frac{V_s \epsilon_{\text{diel}} E_{\max} \text{AREA}_{\text{pole}}}{kT} \frac{f_{\max}}{\frac{1}{n}\sum\limits_{i=1}^{n} B_i}. \quad (3.17)$$

Comparing (3.16) and (3.17) with (2.24) and (2.25) it can be seen that they have the same physical interpretation with $f_{\max}$ and $(1/n)\sum_{i=1}^{n} B_i$ playing the roles of f_{unity} and B_o, respectively.

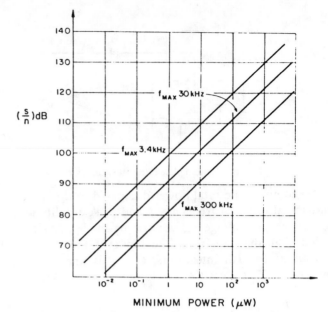

Fig. 9. Maximum dynamic range versus power dissipation for different values of $f_{\max}$.

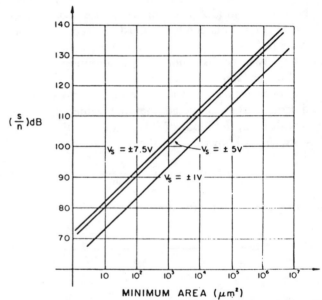

Fig. 10. Maximum dynamic range versus chip area for different values of V_s.

In (3.16) and (3.17), $(1/n)\sum_{i=1}^{n} B_i$ is the only term that depends on the particular circuit architecture used. It turns out, however, that in practical cases, its value is relatively constant. In fact the following approximation can be introduced

$$\frac{1}{n} \sum_{i=1}^{n} B_i = \delta 2 f_{\max} \quad (3.18)$$

where δ is a parameter that depends on the particular filter implementation whose average value can be assumed to be equal to 0.75 with a worst case inaccuracy of about ± 40 percent. The factor of 2 is caused by the fact that B_o is defined over both positive and negative frequencies. For the filter of Table I, for instance, δ is equal to 0.9.

Using (3.18) with $\delta = 0.75$ in (3.16) and (3.17) gives

$$(\text{DR}_{\text{pole}})^2 = \frac{\pi}{24} \frac{P_{\text{pole}}}{kT f_{\max}} \quad (3.19)$$

268

$$(DR_{pole})^2 = \frac{\pi}{6} \frac{V_s \epsilon_{diel} E_{max} AREA_{pole}}{kT} \quad (3.20)$$

From (3.19) and (3.20), the logarithm of P_{pole} and A_{pole} can be plotted versus the achievable dynamic range, DR_{pole}, expressed in decibels with f_{max} or V_s used as a parameter respectively. This is shown in Figs. 9 and 10 in the case that the capacitor dielectric is silicon dioxide with $E_{max} = 5$ 10^6 V/cm. The plots of Figs 9 and 10 can be used for both single ended and fully differential filter configurations. On the basis of the above results, the power consumed and the area occupied by any low-pass SC filter can easily be compared with the theoretical minima.

One fundamental limit which was ignored in this derivation is the uncertainty in capacitor edge definition associated with the photolithographic process. This uncertainty results in a lower limit on the capacitor size that can be used in a given filter application with a given tolerance on the frequency response of the filter. This consideration may or may not require the capacitor sizes to be larger than that dictated by the noise considerations, depending on details of the filter sensitivity, filter frequency response tolerances, and the required dynamic range. For a given set of filter performance requirements, the importance of capacitor matching considerations in selecting capacitor size should decrease with technological scaling since such scaling necessarily involves a steady improvement in photolithographic resolution. This in turn implies a steady improvement in the lithography-limited matching of capacitor structures of a given physical size.

3.1. Comparison of Fundamental Limits with Typical Present-Day Filter Performance

In Table II, the predicted minimum area and power dissipation of a fifth-order elliptic low-pass ladder with 95-dB dynamic range is compared with the approximate parameters for several commercially available devices. The capacitor areas used in typical commercial devices are nearly two orders of magnitude larger than the predicted minimum area, and the power dissipation is also more than two orders of magnitude larger. A recently published class B micropower fifth-order filter [26] achieved a power dissipation somewhat closer to the minimum value, but still more than an order of magnitude higher. These differences are surprisingly large and deserve comment. With regard to the capacitor area, several factors come into play.

a) Actual commercial practice is to use a capacitor dielectric thickness which is such that the maximum field stress is about one-fifth of the maximum value. This is done for reliability and yield reasons.

b) Actual commercial filters display an internal signal swing which is not equal to the supply voltage but is somewhat lower, involving a sacrifice of perhaps 3–6 dB in dynamic range. This, in turn, translates to approximately a factor of two increase in capacitance for a given dynamic range.

c) The capacitor arrays of commercial filters also include summing capacitors in addition to integrating capacitors which increase the total capacitance area by a factor of about 1.5.

TABLE II
COMPARISON OF PERFORMANCE* OF SEVERAL COMMERCIALLY AVAILABLE FIFTH-ORDER PCM FILTERS WITH THEORETICAL MINIMA

FILTER TYPE	GROSS FILTER AREA**	GROSS CAPACITOR AREA ***	POWER DISSIPATION (TYPICAL)
INTEL 2912 (1977) (±5 V)	5600 mil² (3.5 10⁶ μm²)	1440 mil² (9 10⁵ μm²)	40 mW
NATIONAL TP3040 (1978) (±5 V)	3500 mil² (2.2 10⁶ μm²)	1250 mil² (7.8 10⁵ μm²)	4 mW
INTEL 2912A (1979) (±5 V)	4000 mil² (2.5 10⁶ μm²)	1200 mil² (7.5 10⁵ μm²)	4 mW
INTEL 29C51 (1984) (±5 V)	2500 mil² (1.6 10⁶ μm²)	1000 mil² (6.5 10⁵ μm²)	3 mW
CASTELLO et. al. (1985) (±2.5 V)	3400 mil² (2.1 10⁶ μm²)	700 mil² (4.4 10⁵ μm²)	0.35 mW
THEORETICAL MINIMUM (±5 V)		11.6 mil² (7.3 10³ μm²)	0.008 mW
THEORETICAL MINIMUM (±2.5 V)		23.2 mil² (3.6 10³ μm²)	0.008 mW

* All Figures are Approximate

** Gross Filter Area is Area of Polygon Inclosing Filter but not Including Clock Generators or Any Bussing to/from Filter.

d) In order to preserve ratio matching over process variations the capacitor arrays used in actual filters are made up to unit elements. This involves an area sacrifice on the order of a factor of 1.5.

e) Most commercial filters are not kT/C limited in their noise performance. Instead they are designed in a range in which both kT/C and op amp noise are contributing significantly to the noise. The actual noise performance in these filters might typically be 3 dB higher than the kT/C limited performance involving a sacrifice of about 3 dB in dynamic range. This corresponds to an equivalent factor of two increase in the capacitor size for a given dynamic range.

These factors, taken together, account for approximately a factor of 100 in capacitor area, which is about the order of discrepancy between the predicted absolute minimum area and that actually observed in commercial filters.

The comparison above sheds light on the degree to which SC filters can be scaled with technological feature size. First, as can be seen in Table II, most of the area of the commercially available filters is occupied by the operational amplifiers, switches, and interconnections. Assuming the $1/f$ noise problem is solved using circuit or technological means, these circuits can be directly scaled with feature size. With regard to the scaling of the capacitors themselves, it appears that by reducing the op amp noise, producing op amps that swing closer to the supply, and operating the capacitors closer to their maximum field strength, substantial reductions in capacitor area on the order of 5 are possible. These considerations imply that current commercially available filters are perhaps a factor of 10 away from the ultimate minimum possible physical size in the limit of a scaled 5-V technology.

With regard to power dissipation, commercial filters all utilize operational amplifiers with quiescent power dissipation much larger than the dissipation required to simply charge and discharge the integrating capacitors in the filter. The use of class B amplifiers [26] can greatly reduce the power, but still requires the use of significant quiescent current precluding achievement of the minimum possible power dissipation. As a practical matter, there seems to be relatively few applications in which a per-pole switched capacitor filter power dissipation below 0.1 mW is important, if for no other reason than driving an analog signal off the chip into a realistic off-chip impedance requires much more power than that.

IV. EFFECT OF AMPLIFIER NONIDEALITIES

As stated in Section II, all of the above results were based on the assumption of having an op amp with ideal characteristics, i.e. zero power consumption, zero area, zero noise contribution. Such an ideal situation was to be achieved by continuously scaling the feature size, provided that the $1/f$ noise could be eliminated by some technique such as chopper stabilization. In actuality practical constraints will result in other limitations on the level of op amp performance achievable. The ultimate minimum value for the above op amp characteristics is difficult to define. It is however possible, based on a simple model, to obtain upper bounds for the limiting values of the above quantities. This is done in the following. The obtained results show that the op amp fundamental limitations should not substantially affect the ultimate performance of the filter in all practical cases.

4.1. Power Dissipation

In the following section the minimum amount of power requested by the op amp for a given clock frequency is compared with the result of (2.12). The minimum op amp power consumption is obtained under the following assumptions:

1) The limiting factor in the op amp settling time (T_{set}) is given by the linear portion of the step response as opposed to the slewing portion. As a consequence the following equation is valid

$$T_{set} = \delta\tau \qquad (4.1)$$

where δ is a number (typically between 5 and 10) that depends on the accuracy required in the step response, and τ is the time constant of the closed loop step response of the op amp. (A single pole step response is assumed.) If $C_I \gg C_s$ and if no large capacitance is attached at the integrator summing node, it follows that

$$\tau \approx \frac{1}{\omega_u} \qquad (4.2)$$

where ω_u is the unity-gain frequency of the amplifier. The above assumption is quite reasonable since class A/B amplifiers that do not exhibit any slewing behavior and have a power dissipation which is only a few percent higher than their stand by values can be employed.

2) The devices are operated in the subthreshold region. This corresponds to the maximum possible transconductance for a certain current level, I, i.e.,

$$\frac{gm}{I} = \frac{q}{nkT} \qquad (4.3)$$

where n is the subthreshold slope factor whose value is typically between 1 and 2.

3) The time allowed for the op amp to settle is assumed to be $(1/2f_{clock})$, i.e., a 50-percent duty cycle is assumed.

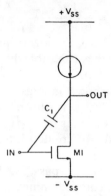

Fig. 11. Simple inverter amplifier.

4) A simple inverter like structure of Fig. 11 is assumed for the op amp. If extra gain is required, cascode devices may be used.

5) The load capacitance of the integrator is assumed to be equal to $2C_s$ due to the sampling capacitor of the next stage plus the effective capacitive load at the output from the feedback circuit which is the series combination of C_I and C_s.

From assumptions 4) and 5) follows that

$$\omega_u = \frac{gm_I}{2C_s} \qquad (4.4)$$

where gm_I is the transconductance of the driver device M1. Using assumption 3) in (4.4) gives

$$\frac{gm_I}{2C_s} \geq 2\delta f_{clock}. \qquad (4.5)$$

The absolute minimum value of gm_I (gm_{min}) is

$$gm_{min} = 4\delta C_s f_{clock}. \qquad (4.6)$$

Using assumption 2), the absolute minimum stand-by current level I_{min} becomes

$$I_{min} = 4n\delta C_s f_{clock}\frac{kT}{q} \qquad (4.7)$$

which gives a minimum power consumption P_{min} of

$$P_{min} = 8n\delta V_s C_s f_{clock}\frac{kT}{q}. \qquad (4.8)$$

Using (2.2) in (4.8) gives

$$P_{min} = 16\pi n\delta V_s C_i f_{unity}\frac{kT}{q}. \qquad (4.9)$$

Comparing the above result with the result of (2.12) in which the signal frequency is assumed to be f_{unity} gives the following result

$$\frac{P}{P_{min}} = \frac{8f_{unity}C_i V_s^2}{16\pi n\delta V_s C_i f_{unity}\frac{kT}{q}} = \frac{V_s}{2\pi\delta n\frac{kT}{q}}. \qquad (4.10)$$

For $n = 1.5$ and $\delta = 7$, (4.10) gives

$$\frac{P}{P_{min}} = \frac{V_s}{21\pi\frac{kT}{q}} = \frac{V_s}{1.7 \text{ V}}. \qquad (4.11)$$

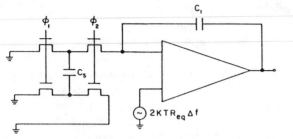

Fig. 12.   Model for the op amp thermal noise.

From (4.11) follows that, for a ± 5-V supply, the simple op amp of Fig. 12 can achieve the required speed of operation while dissipating a power that is only about 35 percent of the power required to charge and discharge the sampling and integrating capacitors. As a consequence, (2.12) gives the absolute minimum power required by an SC integrator with an error that is guaranteed to be less than or equal to about 35 percent. Ideally, at least, such an error should be much less than the above value since, from a fundamental stand point, the absolute minimum power required by the op amp is considerably less than the value given by (4.9) because the unity-gain bandwidth of the simple structure of Fig. 11 does not approach the fundamental frequency limit of MOS transistor M1 which is given be the inherent f_T of the device for the particular bias condition used. Because the parasitic capacitance of M1 is typically much smaller than the load capacitance C_s, it is, at least conceptually, possible to increase the value of the unity-gain bandwidth of an op amp up to a larger fraction of the f_T of the devices used. One possible way to reach such a goal for the simple structure of Fig. 11 is to use positive feedback around M1 in order to obtain a larger transconductance for the same value of the current level and device size.

All of the above considerations suggest that the ultimate limit in the power dissipation of an SC integrator does not come from the op amp. This conclusion is consistent with the assumption of Section II.

4.2. Amplifier Noise and Finite Bandwidth

The only fundamental noise associated with the op amp is the white noise. As was said in Section II, this noise component can be expressed in the same form as the kT/C one. The relative importance of the white noise with respect to the kT/C is considered in this section. The total noise of an SC integrator (both MOS switches and op amp contribution) depends on the relative value of the op amp unity gain bandwidth, ω_u, and the cutoff frequency of the low-pass filter formed by the switch resistance and the sampling capacitor ω_{on} whose value is given by $\omega_{on} = 1/R_{on\,i}C_s$ where $R_{on\,i}$ is the on resistance of the ith MOS switch. With reference to Fig. 12 two extreme cases exist. The first case is the one considered in Section II where an infinite op amp bandwidth has been assumed. This gives equal noise contributions for both the left- and right-hand side switches and a negligible contribution from the op amp assuming a finite total noise energy in the amplifier.

In the other extreme case, the op amp bandwidth is assumed to be much smaller than ω_{on}. By utilizing the results of the noise analysis performed by Gobet and Knob [25], both the noise contributed by the right-hand side switches (n_R^2) and the noise contributed by the op amp (n_{OP}^2) can be expressed as a fraction of the noise contributed by the left-hand side switches (n_L^2), which was calculated in Section II. The final results are as follows:

$$\frac{n_R^2}{n_L^2} \approx \frac{\omega_u}{\omega_{on}} = \frac{gm_I}{2C_s} R_{on2} C_S = \frac{gm_I R_{on2}}{2} \quad (4.12a)$$

$$\frac{n_{OP}^2}{n_L^2} \approx \frac{\dfrac{R_{eq}}{R_{on2}}}{\dfrac{\omega_{on}}{\omega_u}} = \frac{gm_I R_{eq}}{2}. \quad (4.12b)$$

In the derivation of both (4.12a) and (4.12b), (4.4) was used and R_{eq} is the equivalent input noise resistance of the op amp. From (4.12a) and making use of the assumption that $\omega_{on} \gg \omega_u$, it can be concluded that the contribution of the right-hand switches is negligible. On the other hand, assuming that the op amp noise is contributed primarily by the input device M1 and that no high-frequency second stage noise contribution occurs, it follows that $R_{eq} = (2/3)(1/gm_I)$ and (4.12b) gives $(n_{OP}^2/n_L^2) \approx (1/3)$. Between the two extreme cases there is only about 30-percent change in the total output noise contribution; furthermore, if a source coupled pair is assumed at the op amp input, the change is reduced to only about 15 percent. From the above results, it seems reasonable to conclude that in any practical situation (2.17) will be reasonably accurate.

In reality one more potential source of noise degradation exists when the output of the SC integrator is sampled by another circuit of the same kind, which is the case in any SC filter configuration. This is due to the continuous time noise that is transmitted to the output by the amplifier independent of clock phase. Such a wide-band component can be aliased into the baseband by the next stage sampling operation. Fortunately it can be shown that for the case of a single stage transconductance amplifier the above noise contribution combines with the thermal noise of the MOS switches of the following stage in such a way that the variance of the total noise sampled is unchanged.

4.3. Amplifier Area

In this section the minimum amount of area required for the op amp is compared with the result of Section II. The simple structure of Fig. 11 is again assumed. The area of the amplifier is assumed to be approximately equal to the area of M1, i.e., the load device is assumed to be much smaller than M1. The total area of the transistor is assumed to be equal to β times the area of its gate. Typical values for β can be taken to be between 2 and 5. In the following analysis $\beta = 3$ is used to obtain numerical results. Such a value can be achieved in practice by folding the transistor many times in order to reduce source and drain diffusion area. M1 is assumed to be operating in subthreshold. This is done to be consistent with the assumption used in the section dealing with the op amp power and also to ensure a reasonable amount of gain in the amplifier.

The maximum current level in weak inversion for a given aspect ratio is roughly given by [25]

$$I = \mu C_{ox} \frac{Z}{L} \left(\frac{kT}{q} \right)^2. \qquad (4.13)$$

Equation (4.13) defines the minimum value of the aspect ratio Z/L of an MOS transistor for which the device is still operating in weak inversion for any given current level. Minimum aspect ratio corresponds to minimum gate area for a given technology; therefore, the above condition is used in the following calculation. Combining (4.3) with (4.13) an expression for the device transconductance is obtained:

$$gm = \mu C_{ox} \frac{Z}{L} \frac{kT}{qn} \qquad (4.14)$$

Substituting in (4.6) for gm_{min} the expression of (4.14) and making use of (2.2) gives

$$\mu C_{ox} \frac{Z}{L} \frac{kT}{qn} = 4\delta C_s f_{clock} = 8\pi\delta C_I f_{unity}. \qquad (4.15)$$

Multipling both sides of (4.15) by L^2 and solving for the gate area of M1, i.e., $Z \times L$ gives

$$Z \times L = \frac{8\pi\delta qn f_{unity} L^2}{kT\mu} \frac{C_I}{C_{ox}}. \qquad (4.16)$$

Noticing that C_I/C_{ox} is just the area of C_I, follows that

$$\frac{Z \times L}{\text{Area of } C_I} = \frac{8\pi\delta qn f_{unity} L^2}{kT\mu}. \qquad (4.17)$$

The op amp area was assumed to be equal to β times the area of the gate of M1; therefore,

$$\frac{\text{Area of Op Amp}}{\text{Area of } C_I} = \frac{\beta 8\pi\delta qn f_{unity} L^2}{kT\mu}. \qquad (4.18)$$

Using $\mu = 800$ cm^2/V·s, $n = 1.5$, $\delta = 7$, $\beta = 3$ in (4.18) gives

$$\frac{\text{Area of Op Amp}}{\text{Area of } C_I} = 38 L^2 f_{unity}. \qquad (4.19)$$

Assuming a 1-μm minimum channel length technology, it follows from (4.19) that

$$\frac{\text{Area of Op Amp}}{\text{Area of } C_I} = 1, \qquad \text{for } f_{unity} \approx 2.6 \text{ MHz.}$$

These results show that for future scaled technologies, i.e., 1 μm or less minimum channel length, the dominant factor in determining the ultimate limits in the minimum achievable area of an SC integrator is given by the size of the integrating capacitor and not by the op amp area up to filter bandwidths well into the megahertz range. This again is consistent with the assumptions of Section II.

V. CONCLUSIONS

In this paper the fundamental limitations on power dissipation and capacitor size in a SC low-pass filter for a given dynamic range have been analyzed. The results show that both the minimum achievable power dissipation and

the minimum capacitor area vary as the square of the required dynamic range.

By introducing approximations that, in most practical cases, cause only a small error, simple relationships between the dynamic range and the minimum capacitor area and power necessary to achieve in low-pass ladder filters are obtained. These results show that the minimum capacitor area and power consumption for a fifth-order voiceband filter with 95 dB of dynamic range and ± 5-V supply are approximately 7300 μm^2 and 8.5 μW, respectively.

A comparison between the performance achieved by several commercially available SC voiceband PCM filters and the theoretical minima, indicates that for a 5-V technology scaling of technological feature size should allow the reduction of the filter area by a factor of approximately 10 while preserving the dynamic range performance before fundamental limits are encountered. With regard to power dissipation, a voiceband filter using less than 0.1 mW per-pole has already been reported [26]. Although further reductions are certainly feasible, there seems to be relatively few practical applications where this is important.

APPENDIX 1

With reference to Fig. 3 the energy drawn from the supplies during one clock is first computed.

Assuming a positive input signal $v_i(t)$ (the result is dual for a negative one) and calling $i_i(t)$ the current in C_s and $i_o(t)$ the current in C_i, as shown in Fig. 3, then $I_i(t) = I_1'$ and $I_o(t) = -I_2$. The amount of energy drawn from the supplies during one clock period, E_{clock}, is given by

$$E_{clock} = V_{CC} \int_{nT}^{(n+1)T} I_1'(t) \, dt - V_{EE} \int_{nT}^{(n+1)T} I_2(t) \, dt$$

$$= V_{CC} \{ Q_S[(n+1)T] - Q_S[nT] \}$$

$$\quad - V_{EE} \{ Q_i[(n+1)T] - Q_i[nT] \} \qquad (A.1)$$

where $Q_S(nT)(Q_i(nT))$ is the charge on $C_S(C_i)$ at $t = nT$. Assuming that ϕ_2 is on for $nT \leqslant t < (n+\frac{1}{2})T$ that ϕ_1 is on for $(n+\frac{1}{2})T \leqslant t < nT$ then

$$Q_S(nT) = 0$$

$$Q_S[(n+1)T] = C_S V_i[(n+1)T]. \qquad (A.2)$$

From charge conservation at the amplifier summing node follows that

$$Q_i[(n+1)T] - Q_i[nT] = -C_S V_i[nT]. \qquad (A.3)$$

The total energy drown from the two supplies during one clock cycle is, therefore,

$$E_{clock} = V_{CC} C_S v_i[(n+1)T] + V_{EE} C_S v_i[nT]. \qquad (A.4)$$

For a sinusoidal input signal of peak amplitude V_i and frequency f the total amount of energy drawn during a full cycle of the input signal, ϵ_{cycle}, is

$$E_{cycle} = 2 C_S V_i (V_{CC} + V_{EE}) \sum_{n=0}^{M} \sin\left(\frac{\pi}{n} M \right) \qquad (A.5)$$

where $M = f_{clock}/2f$. For simplicity in the following M is assume to be an integer number.

The summation appearing in (A.4) is evaluated below. Noticing that $\sin x = \mathrm{Im}[e^{jx}]$ it follows that

$$\sum_{n=0}^{M} \sin\left(n\frac{\pi}{M}\right) = \mathrm{Im}\left[\sum_{n=0}^{M} e^{jn(\pi/M)}\right]$$

$$= \mathrm{Im}\left[\frac{1 - e^{j(\pi/M)(M+1)}}{1 - e^{j(\pi/M)}}\right] = \mathrm{Im}\left[\frac{1 + e^{j(\pi/M)}}{1 - e^{j(\pi/M)}}\right]$$

$$= \frac{\sin\left(\dfrac{\pi}{M}\right)}{\left(1 - \cos\left(\dfrac{\pi}{M}\right)\right)} = \cot\left(\frac{\pi}{2M}\right). \qquad (A.6)$$

Since $\cot(x) \approx 1/x$ if $x \ll 1$ by making use of (2.2) in the above result it follows that

$$\sum_{n=0}^{M} \sin\left(n\frac{\pi}{M}\right) \approx \frac{f_{\mathrm{clock}}}{\pi f}. \qquad (A.7)$$

Substituting (A.7) in (A.5) gives

$$E_{\mathrm{cycle}} = \frac{2}{\pi} C_S V_i (V_{\mathrm{CC}} + V_{\mathrm{EE}}) \frac{f_{\mathrm{clock}}}{f}. \qquad (A.8)$$

Assuming to use symmetrical supplies, i.e., $V_{\mathrm{CC}} = V_{\mathrm{EE}} = V_s$ the final result is obtained:

$$E_{\mathrm{cycle}} = \frac{4}{\pi} C_S V_i V_s \frac{f_{\mathrm{clock}}}{f}. \qquad (A.9)$$

REFERENCES

[1] B. J. Hosticka, R. W. Brodersen, and P. R. Gray, "MOS sampled data recursive filter using switched capacitor integrator," *IEEE J. Solid-State Circuits*, vol. SC-12, pp. 600–608, Dec. 1977.

[2] J. T. Caves, M. A. Copland, C. F. Rahim, and S. D. Rosenbaum, "Sampled analog filtering using switched capacitors as resistor equivalent," *IEEE J. Solid-State Circuits*, vol. SC-12, pp. 592–599, Dec. 1977.

[3] W. C. Black, Jr., D. J. Allstot, and R. A. Reed, "A high performance low power CMOS channel filter," *IEEE J. Solid-State Circuits*.

[4] F. Krummenacher, "Micropower switched capacitor biquadratic cell," *IEEE J. Solid-State Circuits*, vol. SC-17, pp. 507–512, June 1982.

[5] B. J. Hosticka, D. Herbst, and B. Hoefflinger, U. Kleine, J. Pandel, and R. Schweer, "Real-time programmable SC bandpass filter," *IEEE J. Solid-State Circuits*, vol. SC-17, pp. 499–506, June 1982.

[6] B. J. Hosticka, "Dynamic CMOS Amplifiers," *IEEE J. Solid-State Circuits*, vol. SC-15, pp. 887–894, Oct. 1980.

[7] M. G. Degrauwe, J. Rijmenants, E. A. Vittoz, and H. J. De Man, "Adaptive biasing CMOS amplifiers," *IEEE J. Solid-State Circuits*, vol. SC-17, pp. 522–528, June 1982.

[8] K. C. Hsieh, and P. R. Gray, D. Senderowicz, and D. Messer-schmitt "A low-noise chopper-stabilized differential switched capacitor filtering technique," *IEEE J. Solid-State Circuits*, vol. SC-16 pp. 708–715, Dec. 1981.

[9] L. T. Lin, H. F. Tseung, D. B. Cox, S. S. Viglione, D. P. Conrad, and R. G. Runge, "A monolithic audio spectrum analyzer," *IEEE J. Solid-State Circuits*, vol. SC-18, pp. 40–45, Feb. 1983.

[10] Y. Kuraishi, T. Makabe, and K. Nakayama, "A single-chip analog front-end LSI for modems," *IEEE J. Solid-State Circuits*, vol. SC-17, pp. 1039–1044, Dec. 1982.

[11] Y. A. Haque, R. Gregorian, D. Blasco, R. Mao, and W. Nicholson, "A two-chip PCM codec with filters," *IEEE J. Solid-State Circuits*, vol. SC-14, pp. 961–969, Dec. 1979.

[12] D. Senderowicz, S. F. Dreyer, J. H. Huggins, C. F. Rahim, and C. A. Laber, "A family of differential NMOS analog circuits for a PCM codec filter chip," *IEEE J. Solid-State Circuits*, vol. SC-17, pp. 1014–1023, Dec. 1982.

[13] D. Senderowicz, D. A. Hodges, and P. R. Gray, "A high-performance NMOS operational amplifier," *IEEE J. Solid-State Circuits*, vol. SC-13, pp. 760–768, Dec. 1978.

[14] Y. P. Tsividis and P. R. Gray, "An integrated NMOS operational amplifier with internal compensation," *IEEE J. Solid-State Circuits*, vol. SC-11, pp. 748–753, Dec. 1976.

[15] D. Senderowicz and J. H. Huggins, "A low-noise NMOS operational amplifier," *IEEE J. Solid-State Circuits*, vol. SC-17, pp. 999–1008, Dec. 1982.

[16] R. Gregorian and W. A. Nicholson Jr., "CMOS switched-capacitor filters for a PCM voice CODEC," *IEEE J. Solid-State Circuits*, vol. SC-14, pp. 970–980, Dec. 1979.

[17] K. C. Hsieh, "Noise limitations in switched-capacitor filters," *Ph. D. dissertation*, Univ. of California, Berkeley, CA.

[18] T. C. Choi and R. W. Brodersen, "Considerations for high-frequency switched-capacitor ladder filters," *IEEE Trans. Circuits Syst.*, vol. CAS-27, pp. 545–552, June 1980.

[19] R. W. Brodersen, P. R. Gray, and D. A. Hodges, "MOS switched-capacitor filters," *Proc. IEEE*, pp. 61–71 Jan. 1979.

[20] D. J. Allstot, "MOS switched capacitor ladder filter," *Ph. D. dissertation*, Univ. of California, Berkeley, May 1979.

[21] P. R. Gray, and R. G. Meyer, *Analysis and Design of Analog Integrated Circuits*. New York: Wiley, 1977.

[22] L. R. Rabiner, and B. Gold, "Theory and Application of Digital Signal Processing. *Englewood Cliff*: Prentice-Hall, 1975.

[23] C. A. Gobet and A. Knob, "Noise analysis of switched capacitor networks," *IEEE Trans. Circuits Syst.*, vol. CAS-30, pp. 37–43, Jan. 1983.

[24] C. A. Gobet, "Spectral distribution of a sampled 1st-order lowpass filtered white noise," *Electron. Lett.*, vol. 17, no. 19, pp. 720–721, Sept. 1981.

[25] E. Vittoz and F. Krummenacher, "Micropower SC Filters in Si-Gate CMOS Technology," in *Proc. ECCTD'80*, Warsaw, vol. 1, pp. 61–72, Sept. 1980.

[26] R. Castello, and P. R. Gray, "A 350 μW fifth-order low-pass switched-capacitor filter," in *Dig. Tech.*, New York, Feb. 1985.

Fully Integrated Active *RC* Filters in MOS Technology

MIHAI BANU, STUDENT MEMBER, IEEE, AND YANNIS TSIVIDIS, SENIOR MEMBER, IEEE

Abstract—A fully integrated continuous-time low-pass filter has been fabricated in CMOS technology. The device implements an active *RC* network using integrated capacitors and MOS transistors operated in the nonsaturation region as voltage-controlled resistors. The filter topology is fully balanced for good linearity and for good power supply rejection. The cutoff frequency is voltage adjustable around 3 kHz allowing compensation for process and temperature variations. With ±5 V power supplies a dynamic range of over 94 dB has been achieved.

I. INTRODUCTION

THE basic problem that has hindered the straightforward monolithic realization of active *RC* filters has been the unpredictability of element values due to fabrication process and temperature variations. These variations can cause time constant errors of over 100 percent. Therefore, approaches based on fixed time constant implementation relying on absolute element values are naturally doomed to failure in the context of IC processes available today. However, these processes do allow for accurate realization of element value ratios and hence, of time constant ratios. If these time constants are voltage- or current-controlled, they can be automatically tuned to predetermined values by an on-chip control system; the latter uses an off-chip reference, such as a crystal clock [1], [2] or an external resistor [3]. Control circuits for achieving this automatic tuning have been discussed elsewhere [1], [2], [4], [5] and are not considered here; instead, we will focus on the implementation of the filters themselves.

Several filters have been proposed based on the above principle, and utilizing no switching. As adjustable elements they have used gyrators with variable junction capacitance [2], variable transconductance elements [1], [3], [6], and special operational amplifiers with a variable junction capacitor [7]. A common characteristic of all these approaches is that, like switched-capacitor filters, they are based on nonstandard techniques requiring specialized design. In contrast to this, the filters most extensively studied and best understood, namely those consisting of resistors, capacitors, and operational amplifiers (henceforth referred to as active *RC*), had not yielded to integration so far. In this paper, a method will be presented to integrate such filters on a chip, attaining excellent performance [8].

In active *RC* networks, the time constants can be made adjustable by using voltage-controlled resistors. MOS technology is naturally suited for this approach. Indeed, in this technology high quality capacitors and operational amplifiers are routinely implemented and the MOS transistor, operated in the nonsaturation region, is basically a voltage-controlled resistor. However, there is a potential major drawback in the arbitrary replacement of resistors with transistors. The strong nonlinearity of the MOSFET prohibits the use of large signals, limiting drastically the dynamic range [9]. In the following sections it will be shown how, by proper use of fully balanced networks, the bothersome transistor nonlinearities can be cancelled out, allowing the implementation of filters with wide dynamic range. The resulting filters operate in continuous time utilizing no switching, and thus they have none of the disadvantages associated with analog sampled-data filters. Specifically, there is no need for input antialiasing and output smoothing filters, no Nyquist rate limitations, no potential for operational amplifier high-frequency noise being aliased into the baseband, and no extraneous components at the output due to clock feedthrough. Further, in cases where switched-capacitor or digital filters are desirable, the filters proposed here represent an attractive candidate for implementing input antialiasing and output smoothing filters.

II. TECHNIQUE FOR THE LINEARIZATION OF MOSFET CHARACTERISTICS

A. Transistor Nonlinearities

The drain current of an n-channel MOS transistor in nonsaturation is given by [10]

$$I_D = 2K \left\{ (V_C - V_B - V_{FB} - \phi_B)(V_1 - V_2) \right.$$

$$- \frac{1}{2} \left[(V_1 - V_B)^2 - (V_2 - V_B)^2 \right]$$

$$\left. - \frac{2}{3} \gamma \left[(V_1 - V_B + \phi_B)^{3/2} - (V_2 - V_B + \phi_B)^{3/2} \right] \right\}$$

$$(1a)$$

Manuscript received April 5, 1983; revised July 14, 1983 and August 20, 1983.

The authors are with the Department of Electrical Engineering, Columbia University, New York, NY 10027 and Bell Laboratories, Murray Hill, NJ 07974.

Reprinted from *IEEE J. Solid-State Circuits*, vol. SC-18, no. 6, pp. 644–651, Dec. 1983.

with

$$\gamma = \frac{1}{C'_{ox}} (2qN_A\epsilon_s)^{1/2} \tag{1b}$$

$$K = \frac{1}{2}\mu C'_{ox}\frac{W}{L} \tag{1c}$$

where I_D is the drain current in the triode region, V_C, V_B, V_1, V_2 are the gate, substrate, drain, and source potentials with respect to ground, W and L are the channel width and length, μ is the carrier effective mobility in the channel, V_{FB} is the flat-band voltage, N_A is the substrate doping concentration, C'_{ox} is the gate oxide capacitance per unit area, ϵ_s is the silicon dielectric constant, q is the electron charge and ϕ_B is the approximate surface potential in strong inversion for zero backgate bias (classically, this potential has been taken to be $2\phi_F$ with ϕ_F the Fermi potential, but ϕ_B is actually higher by several kT/q [11]). It is assumed that the source and the drain voltages V_1 and V_2 never become too low to forward bias the drain and the source junctions and never become too high to drive the device into saturation. The ground potential is defined such that V_1 and V_2 vary around zero. In that case the substrate voltage should be negative in order to keep the drain and the source junctions reversed biased (such a definition for ground potential is convenient when two power supplies of opposite values are present).

The 3/2 power terms in (1a) can be expanded in Taylor series with respect to V_1 and V_2. Then I_D can be written in the general form

$$I_D = K\left[a_1(V_1 - V_2) + a_2(V_1^2 - V_2^2) + a_3(V_1^3 - V_2^3) + \cdots\right] \tag{2}$$

where the coefficients a_1 are independent of V_1 and V_2 and are functions of the gate and substrate potentials (V_C and V_B) and all the process and physical parameters involved in the making of the device. The inverse of (Ka_1) is the small-signal resistance R of the transistor; it can be shown that

$$R = \frac{1}{Ka_1} = \left[\mu C'_{ox}\frac{W}{L}(V_C - V_T)\right]^{-1} \tag{3}$$

where V_T is the threshold voltage corresponding to $-V_B$ backgate bias. The value of R may be varied with V_C (henceforth called the control voltage); therefore, for small signals, the MOSFET can be used as a voltage-controlled resistor.

An indication on how the nonlinear higher order terms in (2) affect the transistor characteristics is given by the relative magnitude of the coefficients a_i compared to a_1. Fig. 1 shows the ratios $a_2/a_1, a_3/a_1$, etc., for common process parameters, as computed from (1a). A typical practical situation is illustrated by taking $V_2 = 0$ V, $V_1 = 1$ V, $V_C - V_T = 2$ V, $\mu C'_{ox}(W/L) = 10$ μA/V^2 and $V_B = -5$ V (for usual power supplies of ± 5 V the n-channel transistor substrate is considered connected to the minimum available potential). Then, the first term in the right-hand side of

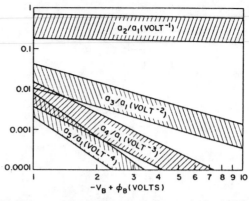

Fig. 1. Coefficients of the nonlinear components in the MOS transistor characteristics normalized to a_1 for different substrate biases and different control (gate) voltages. Each band corresponds to 4 V $< V_C - V_T < 1$ V (bottom of band for 4 V); $\phi_B = 0.7$ V, $\gamma = 1$ V$^{1/2}$.

(2) is 20 μA, the second term is -6 μA, the third term is 3×10^{-2} μA, the forth one is -2×10^{-3} μA, etc. It is seen that the dominant deviation from linearity comes from the second-order term. In the next subsection a simple circuit technique for the cancellation of this term will be presented.

B. Nonlinearity Cancellation

The proposed technique will be illustrated by developing an active *RC* integrator in which MOSFET's are used instead of resistors. The classical implementation is shown in Fig. 2(a). The gain factor of this integrator is given by $(1/RC)$. Consider replacing the resistor with a MOSFET whose small-signal channel resistance is R, as shown in Fig. 2(b). Using the relation (2) with $Ka_1 = 1/R$ we have

$$V_{out} = -\frac{1}{RC}\int_{-\infty}^{t} V_{in}\, dt'$$
$$-\frac{K}{C}\int_{-\infty}^{t}\left(a_2 V_{in}^2 + a_3 V_{in}^3 + \cdots\right)dt'. \tag{4}$$

The first term in (4) represents the ideal response of the integrator [identical with the response of the circuit of Fig. 2(a)] and the second one represents the error due to transistor nonlinearities. For large signals, the error term in the output becomes significant and produces excessive second-order harmonic distortion which limits the dynamic range. Considering the same transistor as in the example of the previous paragraph ($V_C - V_T = 2$ V, $V_B = -5$ V, coefficients a_i given as in Fig. 1) the total harmonic distortion (THD) is approximately 7.5 percent for an input signal of 2 V$_{p-p}$. If THD of less than 1 percent is desired, it is easy to calculate that input signals no larger than 250 mV$_{p-p}$ should be applied. However, if the $a_2 V_{in}^2$ term in (4) were cancelled, the remaining integrator output error due to the higher order terms would be considerably less, resulting in only 0.03 percent THD for the same device and 2 V$_{p-p}$ signal level. A partial cancellation of this type was effectively accomplished in a circuit proposed quite early [12] which implemented a grounded linearized resistor with parasitic dc paths. A fully floating linearized resistor without dc

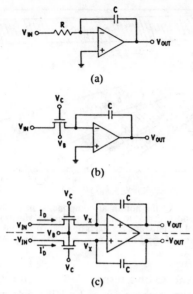

Fig. 2. (a) Classical active *RC* integrator. (b) Small-signal active *RC* integrator with variable gain factor. (c) Large-signal active *RC* integrator with variable gain factor realized as a fully balanced circuit.

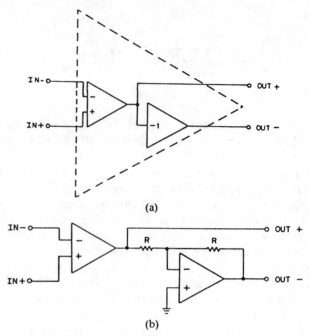

Fig. 3. (a) Definition of balanced output operational amplifier. (b) A possible implementation of the balanced output operational amplifier.

parasitic paths was proposed recently [13]. Both schemes in [12] and [13] use extra circuitry in addition to the MOS transistor to implement the device linearization. It will be shown below that overall linearity can be accomplished without necessarily having to linearize every transistor separately and without the need for any extra circuitry; linearization schemes are considered further elsewhere [14]–[16].

The cancellation of the second-order term in the output of the circuit of Fig. 2(b) can be accomplished simply by using the fully balanced version of the latter as shown in Fig. 2(c). It consists of two identical capacitors, two identical transistors with identical gate and substrate bias, and one operational amplifier whose output voltages are required to be always symmetric with respect to ground (V_{out} and $-V_{\text{out}}$). The two input voltages are also assumed balanced (V_{in} and $-V_{\text{in}}$). One can regard this scheme as the combination of the circuit in Fig. 2(b) with its own mirror image taken with respect to the symmetry axis shown in Fig. 2(c).

The fully balanced integrator will now be analyzed. Assuming infinite op amp gain and zero offset voltage, the two inputs of the operational amplifier are at the same potential V_x (not virtual ground in general). Writing the *KVL* equations for the two outputs we have

$$V_{\text{out}}(t) = -\frac{1}{C} \int_{-\infty}^{t} I_D \, dt' + V_x \tag{5a}$$

$$-V_{\text{out}}(t) = -\frac{1}{C} \int_{-\infty}^{t} I_D' \, dt' + V_x. \tag{5b}$$

The solution for V_{out} is obtained by subtracting (5b) from (5a):

$$V_{\text{out}}(t) = -\frac{1}{2C} \int_{-\infty}^{t} \left(I_D - I_D' \right) dt'. \tag{6}$$

The values of the currents I_D and I_D' are given according to (2):

$$I_D = K \left\{ a_1 [V_{\text{in}} - V_x] + a_2 [V_{\text{in}}^2 - V_x^2] \right. $$
$$\left. + a_3 [V_{\text{in}}^3 - V_x^3] + \cdots \right\} \tag{7a}$$

$$I_D' = K \left\{ a_1 [(-V_{\text{in}}) - V_x] + a_2 [(-V_{\text{in}})^2 - V_x^2] \right. $$
$$\left. + a_3 [(-V_{\text{in}})^3 - V_x^3] + \cdots \right\}. \tag{7b}$$

When (7b) is subtracted from (7a), all the even order terms in V_{in} and all the terms in V_x cancel out:

$$I_D - I_D' = 2K \left[a_1 V_{\text{in}} + a_3 V_{\text{in}}^3 + a_5 V_{\text{in}}^5 + \cdots \right]. \tag{8}$$

Since the terms containing $a_3, a_5, \cdots$ are much smaller than the linear one (see previous subsection), the right-hand side of (8) is practically linear in V_{in}. Using (8) in (6), we obtain

$$V_{\text{out}}(t) \cong -\frac{1}{RC} \int_{-\infty}^{t} V_{\text{in}} \, dt'. \tag{9}$$

This result proves that the fully balanced integrator of Fig. 2(c) has practically the same transfer characteristic as the circuit of Fig. 2(a), even for large signals. It is emphasized that the circuit developed is not a differential scheme since it does not act on the difference of two arbitrary signals, say V_a and V_b; instead, $V_a = V_{\text{in}}$ and $V_b = -V_{\text{in}}$. The circuit thus accepts *one* balanced input (V_{in} and $-V_{\text{in}}$) and produces *one* balanced output (V_{out} and $-V_{\text{out}}$). The requirement of the operational amplifier to balance the output can be met as shown in Fig. 3(a). The circuit contains a high gain stage followed by an inverter. Clearly, no matter how it is connected in an external network, the two outputs will always be balanced. Fig. 3(b) shows a simple practical implementation.

In many filter applications multiple input integrators are

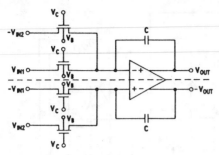

Fig. 4. Fully balanced differential integrator.

needed. Fig. 4 shows a fully balanced differential integrator. This time the circuit operates on two independent large signals: V_{in1} and V_{in2}, performing the integral of their difference. In a completely analogous fashion to our previous analysis, it can be shown that the transistors of each pair cancel their nonlinearities among themselves independently of the others, just as before. The same principle can be used to implement various other linearized filter building blocks of the active *RC* type, such as [16] summers (replacing the capacitors with MOS transistors), differentiators (interchanging the capacitors with the transistors), variable gain amplifiers (replacing the capacitors with resistors), etc. These filter building blocks are input–output compatible and can be connected to each other without any additional interfacing circuitry. Due to the automatic balancing at the output of the operational amplifiers, signal inversion is available at any node pair. The filters thus obtained are naturally balanced from input to output which is a desirable feature for good power supply rejection and increased signal level.

III. NONIDEALITIES

When the ideas introduced in the previous section are applied in practice, the actual circuits will exist in the presence of many nonideal conditions such as device mismatches, errors in the signal balancing, operational amplifier offsets, etc. Analyzing the effects of the latter, it can be shown that the proposed circuits still exhibit good performance [17]; a similar conclusion is suggested by the experimental results to be presented in this paper.

In the previous section only the resistive part of the channel was considered and therefore, strictly speaking, the results derived there are valid only at dc. In fact there are parasitic distributed capacitances from the channel to the gate and to the substrate. These capacitances together with the channel resistance form an intrinsic time constant of the device which will influence the frequency response of the filter in which the MOSFET is used. In order to minimize this effect, the filter should be designed such that the intrinsic transistor time constants are much smaller than the actual time constants needed to be implemented. This strategy usually requires the use of transistors with channel length not larger than several hundred μm [17], or the use of large capacitance values. That is why this filtering technique is more naturally suited for high fre-

quencies where shorter transistors are required.

It was mentioned before that the signal at the balanced operational amplifier inputs [V_x in Fig. 2(c)] is not at virtual ground due to the nonlinearities of the MOS transistors. The question arises whether this introduces any complications in the design of the op amps with respect to the input common mode rejection. It can be calculated [16] that for usual cases, the signal at the amplifier summing nodes is no more than a fifth of the integrator input signal. This indicates that the op amp common mode input signal is relatively small and does not pose any special design problems.

IV. FILTER DESIGN AND CMOS IMPLEMENTATION

Based on the principles presented before, we have designed and fabricated a fifth-order low-pass filter in CMOS technology. The circuit, whose diagram is shown in Fig. 5, is the fully balanced version of a standard active *RC* ladder filter. It is synthesized from multiple-input balanced integrators identical with the one developed before (see Fig. 4). The gates of all transistors used as resistors are connected to the same potential V_C (control voltage). Varying V_C, one can change the frequency response in a manner equivalent to frequency axis scaling.

The process used is a Bell Laboratories standard twin-tub CMOS [18], having the capability to implement capacitors between the gate polysilicon and a second polysilicon layer. The design rules were based on 3.5 μm minimum linewidth.

Since MOSFET's and poly1-to-poly2 capacitors are readily available the only remaining circuit element needed is the balanced-output operational amplifier. The key specification in the design of this element is the proper matching of its outputs; they should be balanced within 1 percent [17] for low filter distortion. For speedy and reliable first-time design, the straightforward implementation shown in Fig. 3(b) was chosen. It uses two regular operational amplifiers and two p-tub resistors matched to each other. One problem with the use of p-tub resistors is their nonlinearity. to reduce this effect, they had to be designed with relatively large dimensions: 40×600 μm for approximately 35 kΩ; in this twin-tub CMOS technology the p-tub resistor nonlinearity comes from its width and length modulation with the signal (due to depletion regions in the junction between the p-tub and the n-tub). The operational amplifiers used are based on a previously designed circuit [19]. With respect to these op amps, two design factors had to be considered; capability to drive resistors and stability when connected as in Fig. 3(b) and in the actual filter. Since the filter MOST resistors are of the order of several MΩ (see below), the effective loads of the operational amplifiers are given by the p-tub resistors used in the inverters [see Fig. 3(b)]. Therefore, the output stage of the op amps was designed to drive low impedances which was reflected in a power dissipation of 2 mW per amplifier. Good output driving capability also increases the amplifier phase margin. This is desirable in order to avoid possible

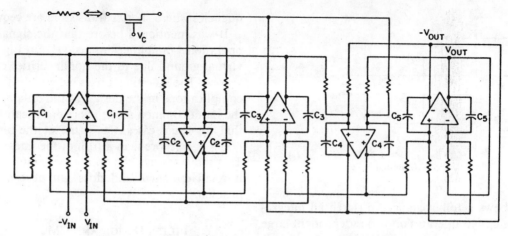

Fig. 5.   Fully balanced active RC ladder filter (fifth-order all-pole).

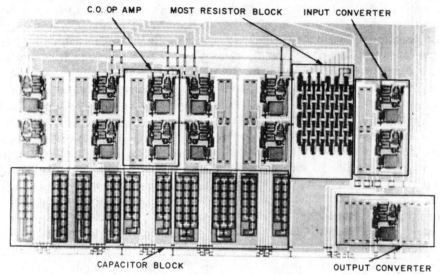

Fig. 6.   Chip microphotograph.

stability problems caused by cascading two operational amplifiers inside a feedback loop. For the same reason, each op amp had to be overcompensated.

The pole locations of the filter were chosen to implement a Chebyshev transfer function with 0.1 dB passband ripple and a cutoff frequency tunable around 3 kHz. These specification require the implementation of large time constants. For practical values of total integrated capacitance, large resistance values must be used (on the order of a couple of MΩ). They were implemented with p-channel MOSFET's having very small shape factors: $W/L = 0.01$ realized with $W = 4$ μm and $L = 400$ μm (p-channel devices were used because they have smaller mobility than the n-channel ones, and thus larger resistances). For such device sizes, the parasitic capacitances to the gate and to the substrate cannot be neglected despite the fact that the actual filter time constants were designed to have much larger values than the transistor intrinsic time constants [16]. This distributed capacitance effect can be simulated simply by modeling each transistor as a series combination of many shorter transistors. Using such simulation it was found that the previous effect produces a peaking of al-

most 1 dB near the band edge. An empirical optimization was performed by perturbing the value of each capacitor separately and observing the effect on the computer simulated frequency response. The peaking could be eliminated very easily by increasing C_2 and C_4 by the same properly chosen amount. However, by doing this, the passband ripple was increased to about 0.2 dB near the band edge. It should be mentioned that this simulation predicted very well the behavior of the actual device and the above compensation worked as expected (see Section V).

Fig. 6 shows a microphotograph of the experimental chip. For precise control over the shape of the frequency response, proper accuracy in the ratios between the filter time constants had to be insured (1 percent accuracy is usually enough to give passband deviations of less than 0.1 dB for filters of this type because they are naturally insensitive to parameter variations). This was accomplished by matching both the resistors and the capacitors. Because there is better control over the latter, the MOST resistors were designed to be identical and the time constant ratios were implemented as capacitor ratios. In order to improve the MOST resistor matching, a layout strategy was used

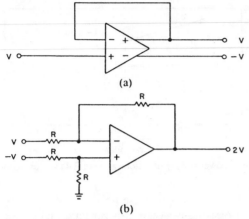

Fig. 7. (a) Unbalanced-to-balanced input converter. (b) Balanced-to-unbalanced output converter.

where each resistor was used split into four identical parts and the parts were interleaved with all the others (see Fig. 6). In this way local process variations influence evenly all resistors. The capacitors were fabricated over a grounded p-tub to reduce the power supply coupling through the substrate.

Included in the chip was also a "single-ended to balanced" input converter and a "balanced to single-ended" output converter. These were implemented with p-tub resistors, according to the diagrams shown in Fig. 7. The active chip area including the converters is 4 mm².

V. EXPERIMENTAL RESULTS

All the data to be presented were measured for the complete system including the balanced filter and the input and output converters. The filter dc gain is 0 dB and the power supplies used were ±5 V.

Figs. 8 and 9 show the frequency response of the filter with stopband and passband details. Input signals of up to 10 MHz were also applied and the filter stopband rejection was found always better than 60 dB; this reflects the continuous time nature of the system. Varying the control voltage V_C, the frequency response could be changed as shown in Fig. 10. Clearly, this effect is similar to a frequency axis scaling. A more exact characterization of the frequency response dependence on V_C is contained in Fig. 11; the experimentally observed cutoff frequency (-3 dB point) is plotted versus the control voltage. It should not be surprising that the curve is a straight line because the dependence of $1/R$ in V_C in (3) is practically of the first degree (neglecting the small variation of mobility with the gate voltage).

The linearity performance of the filter is illustrated in Fig. 12. Here, the measured total harmonic distortion is shown as a function of the signal level for different values of $V_C - V_T$. In generating these plots, the input signal frequency was taken at one-third of the respective filter bandwidth such that the second and the third harmonic distortion components fall into the passband. The points

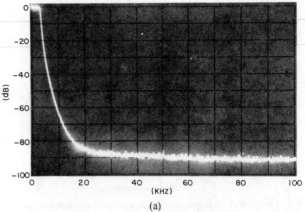

(a)

Fig. 8. Filter frequency response.

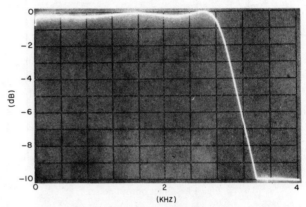

Fig. 9. Passband details of filter frequency response.

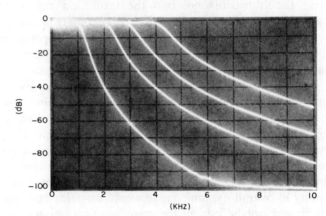

Fig. 10. Frequency response for various control voltages.

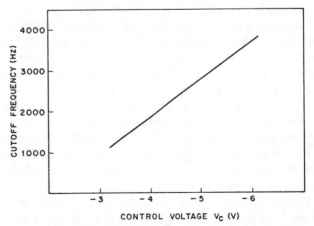

Fig. 11. Cutoff frequency (-3 dB point) versus control voltage.

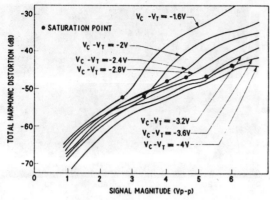

Fig. 12. Total harmonic distortion for various control voltages.

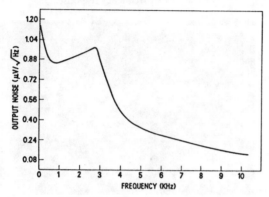

Fig. 13. Filter output noise.

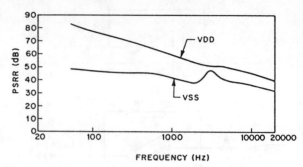

Fig. 14. Power supply rejection ratio (PSRR) versus frequency.

TABLE I
PERFORMANCE PARAMETERS OF THE FILTER INCLUDING
INPUT AND OUTPUT CONVERTERS

Power supply voltages	± 5 V
C-message weighted noise	34 μV rms
Wide band noise (2 Hz–12 MHz)	50 μV rms
Dynamic range[a] (based on -40 dB THD and C-message weighted noise)	94–96 dB
PSRR (1 kHz)	
$+V_{DD}$	-60 dB
$-V_{SS}$	-40 dB
Power dissipation	20 mW

[a]Depending on the value of control voltage.

where the transistors get into the saturation region of operation (shown on the distortion curves) mark a change in the shape of these curves. If a certain application requires the use of large signals with THD smaller than that indicated by Fig. 12 (e.g., -50 dB THD at signals of 6 V_{p-p}), the input signal could be first attenuated by an appropriate factor, then passed through the filter and finally amplified by the same factor. From the curves of Fig. 12 notice that a prefiltering attenuation and post-filtering amplification by only a factor of 2 would accomplish at least 10 dB improvement in THD. Such a strategy would inherently increase the output noise level and a penalty in overall dynamic range would be paid; however, this should not be critical for many applications because the starting dynamic range of the filter itself is very large (see below).

The output noise of the filter was measured to be 34 μV rms for C-message weighting, 42 μV rms for 3 kHz-flat weighting and 50 μV rms for 15 kHz-flat weighting. The wide band noise (2 Hz–12 MHz) remains at 50 μV rms which shows that practically all of it comes at low frequencies. This can be verified from the noise spectrum plotted in Fig. 13. Comparing these experimental results with simulations indicates that, in this implementation, the noise is mostly due to the resistor thermal noise at all but very low frequencies, where the $1/f$ noise of the operational amplifiers becomes predominant.

The power supply rejection performance is shown in Fig. 14 for both positive and negative supplies. It is emphasized that the PSRR is dominated by the performance of the balanced operational amplifiers and not by the other components of the filter. This is indicated by the fact that the curves of Fig. 14 are similar to those of a single balanced output operational amplifier. In addition, the coupling from the negative supply (V_{SS}) can only come from the operational amplifiers because the MOST resistors are p-channel (substrate connected to V_{DD}) and all the capacitors were fabricated over a grounded p-tub.

The power dissipation was 20 mW. This reflects the fact that two regular operational amplifiers were used for each integrator.

A summary of the filter performance is shown in Table I.

VI. CONCLUSIONS

A method has been described for the integration of active RC filters with wide dynamic range in MOS technology. MOSFET's operated in nonsaturation are used as voltage-controlled resistors and are linearized with a simple circuit technique based on fully balanced operation. As an application, a CMOS fifth-order low-pass filter for voiceband was fabricated. The filter works in continuous-time (no switching) and therefore offers natural advantages over sampled-data systems, such as no need for antialiasing or smoothing filtering, no Nyquist rate limitations, no switching noise, and no operational amplifier noise aliasing into the baseband. In addition, the filter design is straightforward, based on well established active RC filter theory. The simplicity and predictability of the design were reflected in successful results from the first trial. The behavior of the actual implementation was in good agreement with computer simulations and analytical results. The filter exhibited low noise and good linearity, resulting in wide dynamic range.

REFERENCES

[1] K. S. Tan and P. R. Gray, "Fully integrated analog filters using bipolar JFET technology," *IEEE J. Solid-State Circuits*, vol. SC-13, pp. 814–821, Dec. 1978.

[2] K. W. Moulding, J. R. Quartly, P. J. Rankin, R. S. Thompson, and G. A. Wilson, "Gyrator video filter IC with automatic tuning," *IEEE J. Solid-State Circuits*, vol. SC-15, pp. 963–968, Dec. 1980.

[3] J. O. Voorman, W. H. A. Bruls, and P. J. Barth, "Integration of analog filters in a bipolar process," *IEEE J. Solid-State Circuits*, vol. SC-17, pp. 713–722, Aug. 1982.

[4] D. Senderowicz, D. A. Hodges, and P. R. Gray, "An NMOS integrated vector-locked loop," in *Proc. Int. Symp. Circuits Syst.*, May 1982, pp. 1164–1167.

[5] Y. Tsividis, "Self-tuned filters," *IEE Electron. Lett.*, vol. 17, no. 12, pp. 406–407, June 1981.

[6] K. Fukahori, "A bipolar voltage-controlled tunable filter," *IEEE J. Solid-State Circuits*, vol. SC-16, pp. 729–737, Dec. 1981.

[7] R. Schaumann and C. F. Chiou, "Design of integrated analog active filters," in *Proc. 1981 Europ. Conf. Circuit Theory Des.*, The Hague, The Netherlands, Aug. 1981, pp. 407–411.

[8] M. Banu and Y. Tsividis, "Fully integrated active *RC* filters in CMOS technology," in *Dig. Tech. Papers, IEEE Int. Solid-State Circuits Conf.*, New York, Feb. 1983, pp. 244–245.

[9] R. L. Geiger, P. E. Allen, and D. T. Ngo, "Switched-resistor filters —A continuous time approach to monolithic MOS filter design," *IEEE Trans. Circuits Syst.*, vol. CAS-29, pp. 306–315, May 1982.

[10] W. M. Penney and L. Lau, Eds., *MOS Integrated Circuits*. New York: Van Nostrand-Reinhold, 1972.

[11] Y. Tsividis, "Problems with precision modeling for analog MOS LSI," in *Proc. Int. Electron Devices Meeting*, San Francisco, CA, 1982.

[12] A. Bilotti, "Operation of a MOS transistor as a variable resistor," *Proc. IEEE*, vol. 54, pp. 1093–1094, Aug. 1966.

[13] M. Banu and Y. Tsividis, "Floating voltage-controlled resistors in CMOS technology," *IEE Electron. Lett.*, vol. 18, no. 15, pp. 678–679, July 1982.

[14] ——, "Fully integrated active *RC* filters," in *Proc. Symp. Circuits Syst.*, May 1983, pp. 602–605.

[15] Y. Tsividis and M. Banu, "Integrated nonswitched active *RC* filters with wide dynamic range," in *Proc. Europ. Conf. Circuit Theory and Design*, Stuttgard, Germany, Sept. 1983.

[16] M. Banu and Y. Tsividis, to be published.

[17] ——, "Detailed analysis of nonidealities in MOS fully integrated active *RC* filters based on balanced networks," to be published.

[18] L. C. Parillo, R. S. Payne, R. E. Davis, G. W. Reutlinger, and R. L Field, "Twin-tub CMOS—A technology for VLSI circuits," *IEEE Int. Electron Device Meeting, Tech. Dig.*, Dec. 1980, pp. 752–755.

[19] V. R. Saari, "Low power, high drive CMOS operational amplifiers," *IEEE J. Solid-State Circuits*, vol. SC-18, pp. 121–127, Feb. 1983.

A Self-Calibrating 15 Bit CMOS A/D Converter

HAE-SEUNG LEE, STUDENT MEMBER, IEEE, DAVID A. HODGES, FELLOW, IEEE, AND PAUL R. GRAY, FELLOW, IEEE

Abstract—A self-calibrating analog-to-digital converter employing binary weighted capacitors and resistor strings is described. Linearity errors are corrected by a simple digital algorithm. A folded cascode CMOS comparator resolves 30 μV in 3 μs. An experimental converter fabricated using a 6 μm gate CMOS process demonstrates 15 bit resolution and linearity at a 12 kHz sampling rate.

I. INTRODUCTION

CONVENTIONAL successive-approximation analog-to-digital conversion techniques require precise component matching to realize high resolution and linearity. Fast, high-resolution A/D converters usually have been realized in the form of hybrid circuits, where two or more different technologies can be combined for optimum performance. A complex thin-film resistor process and laser trimming are often used to provide necessary component matching. Mechanical stress during packaging and long term drift of laser trimmed components often pose serious problems for such converters.

In this paper, a self-calibration technique which enables a monolithic implementation of very high-performance analog-to-digital converters is described. This technique is based on a binary weighted capacitor array DAC [1], [2] and a resistor string DAC [3], [4]. During the calibration cycle, typically performed after each powerup, the ratio errors of the capacitors are measured and stored in a RAM. During the subsequent normal conversion cycles, these data are used to correct for the element matching errors of the capacitor array.

An A/D converter employing this technique can be implemented using standard CMOS or NMOS technology.

Manuscript received April 27, 1984; revised August 6, 1984. This research was supported in part by the National Science Foundation under Grants ECS-8023872 and ECS-8310442, and by the 1981–1982 State of California MICRO program with grants from Fairchild Semiconductor, Racal-Vadic, and GTE-Lenkurt.
H.-S. Lee was with the Department of Electrical Engineering and Computer Sciences and the Electronics Research Laboratory, University of California, Berkeley, CA 94720. He is now with the Department of Electrical Engineering and Computer Science, Massachusetts Institute of Technology, Cambridge, MA 02139.
D. A. Hodges and P. R. Gray are with Department of Electrical Engineering and Computer Sciences and the Electronics Research Laboratory, University of California, Berkeley, CA 94720.

Laser trimming is not needed. A major advantage of this technique is that no special fabrication technology is required, and no special trimming or testing equipment is needed. The calibration can be performed at any time without the use of external components. Any long term variation can be corrected by calibrating at appropriate intervals.[1]

II. SELF-CALIBRATION TECHNIQUE

A block diagram of a self-calibrating A/D converter is shown in Fig. 1. This circuit consists of an N bit capacitor array *main DAC*, an M bit resistor string *sub DAC*, and a resistor string *calibration DAC*, which must have a few more bits of resolution than the sub DAC. Digital control circuits govern capacitor switching during the calibration cycle and store the nonlinearity correction terms in data registers. The ratio errors of the sub DAC, and overall quantization errors, accumulate during digital computation of error voltages. To overcome these errors for a 16 bit converter, 2 bits of additional resolution are needed during the calibration cycle.[2]

Fig. 2(a) shows an N bit weighted capacitor DAC. Suppose that each weighted capacitor C_n has a normalized ratio error of $(1 + \epsilon_n)$ relative to the ideal value due to process variations:

$$C_n = 2^{n-1}C(1 + \epsilon_n), \qquad n = 1A, 1B, \cdots, N. \quad (1)$$

The total linearity error consists of contributions from each capacitor ratio error. These contributions, *error voltages*, can be found to be

$$V_{\epsilon n} = \frac{V_{\text{ref}}}{2^N} 2^{n-1} \epsilon_n \qquad n = 1B, 2, \cdots, N \quad (2)$$

where the subscript n corresponds to the capacitor C_n.

---

[1] We have not observed any long-term drift in MOS capacitor ratios in the course of this work.
[2] In principle, one extra bit is adequate to achieve final linearity within 1 LSB of an ideal straight line, or within 1/2 LSB of an ideal staircase converter response. In practice, two extra bits are necessary to have a margin of safety.

Reprinted from *IEEE J. Solid-State Circuits*, vol. SC-19, no. 6, pp. 813–819, Dec. 1984.

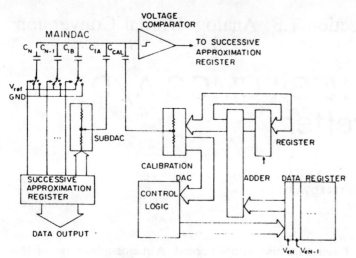

Fig. 1. Block diagram of self-calibrating A/D converter. Digital control logic amounts to a total of about 400 gates and 120 bits of RAM.

Then, the total linearity error V_{error} becomes

$$V_{\text{error}} = \sum_{i=1B}^{N} V_{\epsilon i} D_i. \qquad (3)$$

During the calibration cycle, individual error voltages $V_{\epsilon n}$'s are measured and digitized by the calibration DAC and then stored in the RAM. During the normal conversion cycle, the total error voltage V_{error} is computed by (3) in digital form, and converted to analog voltage by the same calibration DAC. This total error voltage is subtracted from the main DAC through the coupling capacitor C_{cal} to correct the initial linearity error.

The calibration cycle begins my measuring the error voltage due to the MSB capacitor C_N. This is done by sampling the reference voltage V_{ref} on all the capacitors except the MSB capacitor, as shown in Fig. 2(b). Next, charge is redistributed by reversing the switching configuration, as shown in Fig. 2(c). If the MSB capacitor is perfect, it will have exactly half of the total array capacitance. Thus, the top plate voltage is left unchanged by the charge redistribution. A ratio error in the MSB capacitor will cause a small change in the top plate voltage after the charge redistribution. This *residual voltage* V_{xN} is a direct measure of the error voltage corresponding to the error in the ratio of the MSB capacitance to the total array capacitance:

$$V_{xN} = 2V_{\epsilon N}. \qquad (4)$$

Similarly, errors due to smaller capacitors are measured. In each case, a successive approximation search using the sub DAC is employed.

It can be shown easily that the general relation between residual voltages (V_{xn}'s) and the error voltages ($V_{\epsilon n}$'s) is

$$V_{\epsilon n} = \frac{1}{2}\left(V_{xn} - \sum_{i=n+1}^{N} V_{\epsilon i}\right), \qquad n=1B,2,\cdots,N-1.$$
$$(5)$$

This computation is performed in the digital domain by the

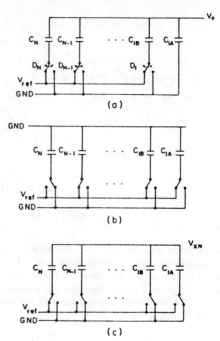

Fig. 2. Binary weighted capacitor array DAC. C_N is the largest capacitor; ideally it has half of the total array capacitance. C_{1A} and C_{1B} are the two smallest capacitors and are equal in value. (b) The self-calibration cycle begins with the MSB capacitor precharged to zero, while all other capacitors are charged in parallel to V_{ref}. (c) The top plate is allowed to float; all bottom switches are reversed. The top plate voltage V_{xN} (ideally zero) is a direct measure of error in the ratio of C_N to total array capacitance. Capacitors C_N through C_2 are evaluated by successive applications of this same process.

logic circuitry.

$$DV_{\epsilon N} = \frac{DV_{xN}}{2} \qquad (6)$$

$$DV_{\epsilon n} = \frac{1}{2}\left(DV_{xn} - \sum_{i=n+1}^{N} DV_{\epsilon i}\right), \qquad n=1B,2,\cdots,N-1$$
$$(7)$$

where $DV_{\epsilon n}$ and DV_{xn} stand for digitized error voltages (*correction terms*) and digitized residual voltages, respectively.

Therefore, by digitizing residual voltages using the sub DAC, correction terms $DV_{\epsilon N}$, $DV_{\epsilon N-1}, \cdots, DV_{\epsilon 1B}$ can be computed subsequently by (6) and (7) using a two's complement adder and a shift register. All these correction terms are stored in digital memory.

During subsequent normal conversion cycles, the calibration logic is disengaged. The converter works the same way as an ordinary successive-approximation converter, except that error-correction voltages are added or subtracted by proper adjustment of the calibration DAC digital input code. When the nth bit is being tested, corresponding correction term $DV_{\epsilon n}$ is added to the correction terms accumulated from the first bit (MSB) through the $(n-1)$th bit. If the bit decision is 1, the added result is stored in the accumulator. Otherwise, $DV_{\epsilon n}$ is dropped, leaving the accumulator with the previous result. The content of the accumulator is converted to an analog voltage by the calibration DAC. This voltage is then subtracted from the main DAC output voltage through the capacitor C_{cal}. The overall operation precisely cancels the nonlinearity due to capacitor mismatches by subtracting V_{error} in (3) from the main DAC. The only extra operation involved in a normal conversion cycle is one two's complement addition. For more detailed mathematical formulation of the self-calibration algorithm, refer to the earlier paper [5].

III. HIGH RESOLUTION COMPARATOR

To realize a fast, high-resolution A/D converter, a high-performance comparator is essential. For example, to achieve 16 bit resolution and at a 50 kHz conversion rate with ±4 V reference voltages, the comparator should resolve 60 μV (1/2 LSB) in about 500 ns. During the calibration cycle, 15 μV resolution is needed to provide 18 bit digital resolution. However, a longer delay is permitted during the calibration cycle. Also, the effect of random noise in the comparator can be reduced by performing the calibration many times and averaging the results.

A two-stage amplifier is chosen for the comparator for easy offset cancellation. The amplifier section of the comparator should provide a gain of more than 100 000 with the minimum delay possible and peak input referred noise less than 60 μV.

When this amplifier is operated in the open-loop configuration, the dominant open-loop poles are determined by the parasitic capacitances at the high impedance nodes. The effect of large capacitive parasitics can be removed from the critical nodes by using cascode stages and source follower stages. A folded cascode amplifier (feasible in CMOS but not in NMOS) provides all the advantages of an ordinary cascode amplifier, while requiring less supply voltage. A schematic diagram of the amplifier employing a folded cascode input stage is shown in Fig. 3. Parasitic capacitances from the large input transistor $M1$ and $M2$ are isolated from the sensitive output nodes by the cascode transistors $M8$ and $M9$. Gate-to-source capacitance and Miller multiplied capacitance of $M6$ are decoupled by the source follower stage consisting of $M11$ and $M12$. A cascode circuit is used for the output stage for the same reason.

The effect of comparator offset voltage is cancelled by closing the feedback switch $M19$. However, charging the large DAC capacitor using only the bias current of the amplifier output stage would be very slow. To speed up the charging, a large switch $M20$ is added. This transistor brings the capacitor voltage to ground potential very quickly. After this switch opens, the offset voltage will be very small (several tens of millivolts at most), reducing the total time needed to settle to within 1/2 LSB of final value. After $M20$ is turned off and the capacitor voltage is at near ground, $M19$ is turned on, sampling the offset voltage on the capacitor. During this time the compensation capacitor C_c is connected to stabilize the amplifier. A 200 pF compensation capacitor would be needed for a 120 pF capacitor array load if the transconductances of the first and the second stages are comparable. A capacitor of this size is impractical for implementation on chip and would also severely limit the comparator slew rate and settling time.

A pole-zero cancellation technique is used to reduce the size of the compensation capacitor as well as to improve

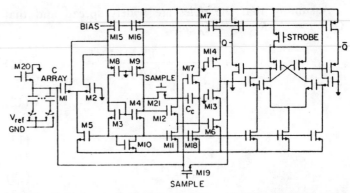

Fig. 3. Folded cascode CMOS comparator with high-speed latch. Input stage includes $M1$–$M4$, $M8$, $M9$, while $M11$, $M12$ form a level shifter. Output stage is made up of $M6$, $M7$. Unnumbered devices form a strobed output latch with sensitivity of about 100 mV.

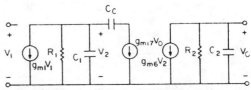

Fig. 4. Equivalent circuit of closed-loop comparator. C_1 is the parasitic capacitance and the drain of $M9$; C_c is the compensation capacitor; C_2 is the load capacitance at the drain of $M13$.

closed-loop performance. A left half-plane zero is introduced by the source follower stage consisting of $M17$ and $M18$. This zero is located so as to cancel the nondominant pole.

Consider the simplified equivalent circuit of Fig. 4. In this equivalent circuit, the common gate stage of $M8$ and $M9$, as well as the source follower stage of $M11$ and $M12$, are omitted since the poles and zeros associated with these stages lie well beyond the range of interest. C_1 is the parasitic capacitance at the output of the first stage (drain of $M9$) and C_2 represents the load capacitance at the drain of $M13$. The node voltage equations for this equivalent circuit are

$$gm_1 V_1 + \left(\frac{V_2}{R_1} + sC_1\right)V_2 + \frac{sC_c}{1 + s\frac{C_c}{gm_{17}}}(V_2 - V_o) = 0 \quad (8)$$

$$gm_6 V_2 + \left(\frac{1}{R_2} + sC_2\right)V_o = 0. \quad (9)$$

Solving for V_o/V_1

$$\frac{V_o}{V_1} = -\frac{a}{1 + bs + cs^2 + ds^3}\left(1 + s\frac{C_c}{gm_{17}}\right) \quad (10)$$

where

$$a = gm_1 gm_6 R_1 R_2$$

$$b = R_1(C_1 + C_c) + R_2 C_2 + gm_6 R_1 R_2 C_c + \frac{C_c}{gm_{17}}$$

$$c = R_1 R_2 C_2 (C_1 + C_2) + \frac{C_c}{gm_{17}}(R_1 C_1 + R_2 C_2)$$

$$d = \frac{R_1 C_1 C_c C_2 R_2}{gm_{17}}.$$

Assuming that the poles are widely spread and that $C_c, C_2 \gg C_1, R_1, R_2 \gg 1/gm_{17}$,

$$p_1 = -\frac{1}{gm_6 R_1 R_2 C_c} \qquad (11)$$

$$p_2 = -\frac{gm_6}{C_2} \qquad (12)$$

$$p_3 = -\frac{gm_{17}}{C_1} \qquad (13)$$

$$z = -\frac{gm_{17}}{C_c}. \qquad (14)$$

To eliminate the nondominant pole p_2, we require that

$$z = p_2 \qquad (15)$$

or

$$\frac{gm_{17}}{gm_6} = \frac{C_c}{C_2}. \qquad (16)$$

To provide a 60° phase margin at the unity gain frequency ω_1, the compensation capacitor should be selected such that

$$\frac{p_3}{\omega_1} = \tan^{-1}60° = \sqrt{3} \qquad (17)$$

or

$$\frac{gm_6 C_c C_c}{gm_1 C_1 C_L} = \sqrt{3}. \qquad (18)$$

The major component of C_1 is the junction capacitance between the compensation capacitor bottom plate and the substrate. Our capacitors were formed between polysilicon and n^+ diffusion into the substrate. Using the measured value for the zero-bias n^+–p junction capacitance $C_{j0} = 0.64 \times 10^{-4}$ pF/μm^2, $\varphi = 0.8$ V and junction reverse bias $V_a = 3$ V.

$$C_j = \frac{C_{j0}}{\sqrt{1 + \frac{V_a}{\varphi}}} = 3 \times 10^{-5} \text{ pF/}\mu\text{m}^2. $$

For a 1000 Å capacitor dielectric oxide, C_{ox} is

$$C_{ox} = 3.45 \times 10^{-4} \text{ pF/}\mu\text{m}^2.$$

Thus,

$$\frac{C_c}{C_1} = \frac{C_{ox}}{C_j} \approx 9.$$

Using this value, (18) becomes

$$\frac{C_c}{C_L} \approx 0.2 \frac{gm_1}{gm_6}.$$

For $gm_1 \approx gm_6$ and $C_L = 120$ pF:

$$C_c = 24 \text{ pF}.$$

Compared with 200 pF required for the simple pole-splitting compensation, this is a great reduction.

The transconductances of n-channel transistors $M8$ and $M17$ should be scaled in proportion to the ratio between the compensation capacitor and the array capacitor. The transconductances of transistors may be scaled by scaling the aspect ratio. The transconductances of $M6$ and $M17$ are

$$gm_6 = \left\{ 2K \left(\frac{Z_6}{L}\right) I_{D6} \right\}^{1/2} \qquad (19)$$

$$gm_{17} = \left\{ 2K \left(\frac{Z_{17}}{L}\right) I_{D17} \right\}^{1/2} \qquad (20)$$

The drain current of $M17$ is

$$I_{D17} = I_{D18} = \frac{K}{2}\left(\frac{Z_{18}}{L}\right)(V_{GS18} - V_T)^2. \qquad (21)$$

Thus,

$$gm_{17} = \left\{ K^2\left(\frac{Z_{18}}{L}\right)\left(\frac{Z_{17}}{L}\right)(V_{GS18} - V_T)^2 \right\}^{1/2}$$

$$= K(V_{GS18} - V_T)\left(\frac{Z_{17}}{L}\right)^{1/2}\left(\frac{Z_{18}}{L}\right)^{1/2}. \qquad (22)$$

Since $V_{GS8} - V_T = V_{GS18} - V_T$ for $V_i = 0$,

$$\frac{gm_{17}}{gm_6} = \frac{\left(\frac{Z_{17}}{L}\right)^{1/2}\left(\frac{Z_{18}}{L}\right)^{1/2}}{\left(\frac{Z_6}{L}\right)}. \qquad (23)$$

Assuming $M17$ and $M18$ are identical,

$$\frac{gm_{17}}{gm_6} = \frac{\left(\frac{Z_{17}}{L}\right)}{\left(\frac{Z_6}{L}\right)} = \frac{Z_{17}}{Z_6}. \qquad (24)$$

This equation shows that the ratio of the transconductances can be scaled by directly scaling the widths of the two transistors. In reality, due to the different body bias, the transconductances of $M6$ and $M17$ cannot match precisely. The result of the mismatch will be a slow settling component in the transient response [6]. The initial amplitude of the slow settling component can be shown to be

$$V_{SS}(0) = V_i \frac{\Delta\omega}{\omega_1} \qquad (25)$$

where $V_{SS}(0)$ is the initial amplitude of the slow settling component, V_i is the input voltage. $\Delta\omega$ is mismatch between the pole and the zero locations, and ω_1 is the unity gain bandwidth of the amplifier. The input voltage V_i is the offset voltage of the comparator plus the charge injection from the big grounding switch, which will not exceed 100 mV. Assuming 20 percent mismatch of transconductances and unity gain frequency $\omega_1/2\pi$ of 4 MHz, the initial amplitude of the slow settling component is less than 2 mV. Approximately 5 time constants of the slow settling component would be required to attain less than 20 μV error. SPICE simulations shows that 1.5μs is needed for the amplifier to settle within 20 μV of the final value. This is acceptable because the loop is closed only once per 20 μs conversion interval.

The closed loop offset sampling does not provide complete offset cancellation due to the charge injection from the small feedback switch $M19$. This residual offset voltage is related to the gate overlap capacitance C_{OL}, the gate area of the switch WL and the voltage levels applied to the gate

$$\Delta V_{OS} = -\frac{C_{OL}}{C_{TOT}}(V_T - V_{GS}(\text{OFF}))$$

$$-\frac{WLC_{ox}}{2C_{TOT}}(V_{GS}(\text{ON}) - V_T). \quad (26)$$

This residual offset voltage for the prototype converter was measured to be about 1.2 mV. This offset voltage is constant as long as the voltage levels are reasonably constant. A 200 mV drift in the threshold voltage V_T or gate-to-source voltages $V_{GS}(\text{ON})$ and $V_{GS}(\text{OFF})$ changes the offset voltage by only 40 μV. The residual offset voltage can be digitized by the calibration DAC and stored in a RAM. This offset data can later be used for a simple digital correction of the residual offset voltage. Experimentally, the offset was reduced to less than 60 μV after digital correction. A detailed description of the digital offset correction technique is presented in a separate paper [7].

The amplifier is followed by a high-speed latch as shown in the schematic diagram. The regeneration is activated by a *strobe* signal, which is applied between the clock edges by an analog delay circuit to avoid digital switching noise. A source follower stage is added between the amplifier and the latch to prevent feedthrough from the latch to the high-impedance output of the amplifier.

IV. EXPERIMENTAL RESULTS

The die photograph of the experimental chip is shown in Fig. 5. This chip contains a 10 bit plus sign capacitor array main DAC, a 5 bit resistor string sub DAC, and a 7 bit resistor string calibration DAC. The successive-approximation and control logic circuits are implemented off chip for simplicity and flexibility.

The capacitors are formed using a heavily doped polysilicon top plate and implanted bottom plate. n-type doping in both places is on the order of $10^{20}/\text{cm}^3$ in order to obtain a voltage coefficient of capacitance smaller than 20 parts per million per volt [8]. The array is made up of 1024 unit capacitors. The capacitance of the unit capacitor is defined by the polysilicon top plate which is 18×18 μm. The MSB capacitor is split into two identical halves and located at the sides of the array to reduce the effect of any possible gradient in etching or oxide thickness.

The resistor strings are formed by source-drain implantation of n-channel transistors. The size of the unit resistor is 20×5 μm. Tree decoders are included in both the sub

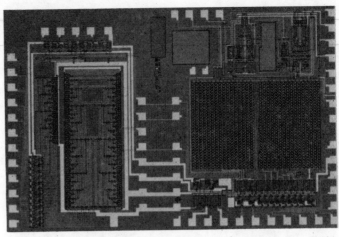

Fig. 5. Photograph of the die, fabricated at UC Berkeley. Overall chip dimensions are 2.8×4.2 mm in a 6 μm silicon-gate CMOS process.

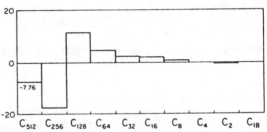

Fig. 6. Typical capacitor ratio error measured by the calibration circuits. Vertical axis is in unit of 1 LSB at the 16 bit level. The large error in C_{256} is due to a mask layout error.

DAC and the calibration DAC. To minimize mismatches due to possible contact resistance variations, contacts are avoided in the unit resistors. The interconnections between the resistors and the tree decoder are made in continuous n+ material [2].

The total die area of the prototype chip excluding bonding pads is 7.5 mm^2. A complete 16 bit converter with all the logic circuits would occupy about 15 mm^2 based on 5 μm design rules.

Fig. 6 shows a typical capacitor ratio error measured using the calibration circuits. To reduce the effect of random noise, each residual voltage is measured 16 times and the average value is used to compute the correction terms. The results are stated in terms of 1 LSB at a 16 bit level. As was shown in the previous section, each correction term is the direct measure of the ratio error. The ratio error shown in Fig. 6 corresponds to initial matching of a 9 bit plus sign. The large ratio errors of the larger capacitors were found to be systematic. We believe this is due to mask design which had varying etch widths for the polysilicon top plate. First-order cancellation of capacitor ratio errors due to etching effects can be obtained by use of design rules which provide equal-width etch channels around the perimeter of all unit capacitors.

A summary of the measured performance of the comparator is shown in Table I. The 3 μs worst-case delay is mainly the intentional timing delay between main DAC

TABLE I
MEASURED PERFORMANCE PARAMETERS OF COMPARATOR

Supply Voltages	±5	V
Resolution	30	μV
Worst Case Delay	3	μs
RMS Input Referred Noise	20	μV
Power Dissipation	10	mW
Closed Loop Settling Time		
(to 20 μV, 120 pF Load)	1.5	μs
Die Area	0.65	mm^2

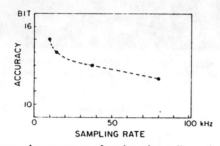

Fig. 7. Measured accuracy as a function of sampling rate for the experimental chip.

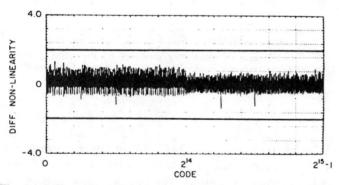

Fig. 8. Differential nonlinearity measured using a code-density test for the experimental converter operating at full speed.

switching time and the strobe time for the comparator output latch. Excessive digital switching noise was being coupled into the critical analog portion of the circuit. A delay of about 2.95 μs was deliberately introduced before strobing the latch to allow switching noise to decay to a sufficiently small level to achieve 15 bit conversion.

The speed-accuracy performance of the complete converter is shown in Fig. 7. 15 bit accuracy was obtained at 12 kHz sampling rate. Due to digital switching noise, the accuracy dropped down to 12 bits at 80 kHz sampling rate. Some of the switching noise may be coupled into the comparator via inductance (~ 10–20 nH) in package traces. If analog and digital circuits are fabricated on one chip, problems due to such inductances and the associated noise should be reduced.

Although the prototype converter was built for 16 bits, 15 bits was the maximum integral linearity achieved. This is the result of unexpectedly poor capacitor matching in the main DAC. To overcome this, the calibration range was increased by doubling the coupling capacitor between the main DAC and the calibration DAC at the cost of the

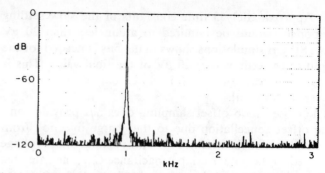

Fig. 9 Measured harmonic distortion of the converter obtained by computation of a 4096 point fast Fourier transform (FFT). Integral nonlinearity larger than about 1 LSB at 15 bits would show up as a noticeable spectral component at the second or third harmonic frequency.

TABLE II
PERFORMANCE CHARACTERISTICS OF CONVERTER MEASURED AT
ROOM TEMPERATURE

Supply Voltages	±5	V
Resolution	15	Bits
Linearity	15	Bits
Offset	< ±1/4	LSB
Conversion Time		
(for ±1/2 LSB Linearity)		
12 Bit	12	μs
15 Bit	80	μs
RMS Noise	40	μV
Power Dissipation		
(excludes logic)	20	mW
Die Area (excludes logic)	7.5	mm^2

calibration resolution. The calibration is now performed at a 17 bit level due to the increase in the coupling capacitor value, and 15 bits is the maximum linearity possible with this level of calibration resolution.

The plot of differential nonlinearity for all 32 768 codes is shown in Fig. 8. The data are obtained by a code density test [9]. In this test, the data were collected while the converter was running at a 12 kHz sampling rate with a 1 kHz sine wave input signal.

Integral nonlinearity was measured statically at the major carries. The maximum error observed was 1.6 LSB at 16 bits. For the interest of audio signal processing, fast Fourier transform was performed on the 4096 digital output codes obtained by sampling a 1 kHz input sine wave. As is shown in Fig. 9, no second or third harmonic distortion is observable.

The performance of the converter is summarized in Table II.

V. CONCLUSIONS

A self-calibrating A/D converter utilizing a capacitor array main DAC and a resistor string sub DAC is described. After the calibration, linearity of the converter was increased to 15 bits from 10 bit initial linearity. In addition to the closed loop offset cancellation, a simple digital offset correction technique is used to reduce the system offset to less than 60 μV. A folded cascode CMOS comparator enables a fast high resolution conversion. A conversion rate of 12 kHz was achieved experimentally. A maximum conversion rate of 80 kHz was obtained at a 12 bit linearity level. The linearity at this speed should be improved by proper isolation of the analog circuits from the noisy digital circuits.

ACKNOWLEDGMENT

We gratefully acknowledge Dr. J. McCreary for helpful suggestions at the outset of this work, J. Doernberg for development of the test system, and Intel Corporation and Reticon Corporation for technical assistance. We appreciate helpful suggestions from reviewers.

REFERENCES

[1] J. L. McCreary and P. R. Gray, "All MOS charge redistribution analog-to-digital conversion techniques—Part I," *IEEE J. Solid-State Circuits*, vol. SC-10, pp. 371–379, Dec. 1975.

[2] A. R. Hamade, "A single chip all-MOS 8-bit A/D converter," *IEEE J. Solid-State Circuits*, vol. SC-13, pp. 785–791, Dec. 1978.

[3] P. R. Gray, J. L. McCreary, and D. A. Hodges, "Weighted capacitor analog/digital converting apparatus and method," U.S. Patent 4129863, Oct. 1977.

[4] D. A. Hodges, P. R. Gray, and J. L. McCreary, "Weighted capacitor analog digital converting apparatus and method," U.S. Patent 4200863, Dec. 1978.

[5] H. S. Lee and D. A. Hodges, "Self-calibration technique for A/D converters," *IEEE Trans. Circuits Syst.*, vol. CAS-30, pp. 188–190, Mar. 1983.

[6] B. Y. Kamath, R. G. Meyer, and P. R. Gray, "Relationship between frequency response and settling time of operational amplifiers," *IEEE J. Solid-State Circuits*, vol. SC-9, pp. 347–352, Dec. 1974.

[7] H. S. Lee and D. A. Hodges, "Accuracy consideration in self-calibrating A/D converters," submitted to *IEEE Trans. Circuits Syst.*

[8] J. L. McCreary, "Matching properties, and voltage and temperature dependence of MOS capacitors," *IEEE J. Solid-State Circuits*, vol. SC-16, pp. 608–616, Dec. 1981.

[9] J. Doernberg, H.-S. Lee, and D. A. Hodges, "Full-speed testing of A/D converters," *IEEE J. Solid-State Circuits*, this issue, pp. 820–827.

A Ratio-Independent Algorithmic Analog-to-Digital Conversion Technique

PING WAI LI, MICHAEL J. CHIN, PAUL R. GRAY, FELLOW, IEEE, AND RINALDO CASTELLO

Abstract — An algorithmic analog-to-digital conversion technique is described which is capable of achieving high-resolution conversion without the use of matched capacitors in an MOS technology. The exact integral multiplication of the signal required by the conversion is realized through an algorithmic circuit method which involves charge summing with an MOS integrator and exchange of capacitors. A first-order cancellation of the charge injection effect from MOS transistor switches is attained with a combination of differential circuit implementation and an optimum timing scheme. An experimental prototype has been fabricated with a standard 5 μm n-well CMOS process. It achieves 12 bit resolution at a sampling rate of 8 kHz. The analog chip area measures 2400 mils².

I. INTRODUCTION

CONVENTIONALLY, most moderate speed analog-to-digital conversion techniques have required the matching of passive or active components to an accuracy comparable to the linearity of the conversion. This requirement for matching has two drawbacks. First, the maximum achievable resolution of an analog-to-digital converter is limited by the process technology. Second, to obtain matching in components, their geometrical size, and therefore the area of the circuit, has to increase. To overcome these limitations, component adjustment techniques such as laser trimming of thin-film resistor networks have been introduced. Although these methods yield components matched to extremely high precision, they have the disadvantages that extra area is needed for placement of the components to be trimmed or a different substrate has to be used for the passive components. Another approach is to develop circuit techniques which eliminate the component matching requirement. One example is the dynamic element matching technique [1], which uses the time-multiplexing principle for attaining exact binary weighing of current sources. The current glitches generated by the multiplexing are low-pass filtered by large capacitors attached to the circuit externally. Another example is the self-calibration method [2], which uses digital methods to calibrate and eliminate the ratio errors in a capacitor array. With this method, extremely high resolution can be obtained at the cost of an extra calibration cycle and a fairly large area. This paper describes an algorithmic circuit technique which allows accurate analog-to-digital conver-

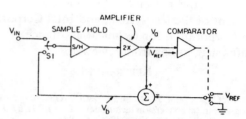

Fig. 1. Block diagram of the algorithmic A/D converter.

sion to be achieved without the use of matched capacitors in an MOS technology. The converter uses the cyclic, also called the recirculating or algorithmic, conversion technique and it includes two operational amplifiers, a comparator, ten capacitors, and a number of minimum size transistor switches which can be accommodated within an area of 2400 mils².

In Section I of this paper, the principle and operation of the algorithmic A/D converter is first reviewed. The limiting factors for the accuracy of an algorithmic A/D converter are also studied using an early implementation as an example. Section II discusses the algorithmic circuit technique which allows exact integral gain to be obtained with an MOS gain block. Section III investigates the errors caused by the charge injection effect. It can be shown that a first-order cancellation of the error is possible. Following this, the circuit implementation of the algorithmic A/D is given in Section IV. The fully differential class A/B operational amplifier, which is vital to the operation of the converter, is described in Section V. Finally, in Section VI some experimental results are presented.

II. ALGORITHMIC A/D CONVERSION PRINCIPLE

The algorithmic analog-to-digital converter, also known as the cyclic or recirculating converter, has been known and utilized in various forms since the 1960's [3]. It was first realized by Hornak [4] in a partially integrated form, using a transformer to achieve the gain of two, in 1975. Subsequently, McCharles [5] achieved full integration of the analog portion of the converter using a metal gate CMOS technology. Recently, Webb *et al.* integrated the converter using bipolar technology for the analog circuit portion and MOS technology for the digital controller portion [6]. A block diagram of the converter is shown in Fig. 1. The algorithmic A/D converter consists of an

Manuscript received May 7, 1984; revised July 23, 1984. This work was supported by Siemens AG.

The authors are with the Department of Electrical Engineering and Computer Sciences and the Electronics Research Laboratory, University of California, Berkeley, CA 94720.

Reprinted from *IEEE J. Solid-State Circuits*, vol. SC-19, no. 6, pp. 828–836, Dec. 1984.

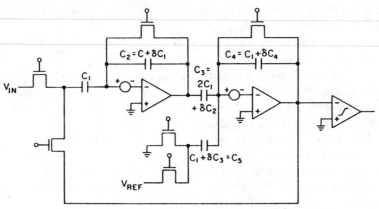

Fig. 2. Example of an implementation of the algorithmic A/D converter in an MOS technology.

analog signal loop which contains
1) a sample-and-hold amplifier.
2) a multiply-by-two amplifier.
3) a comparator, and a
4) reference subtraction circuit.

The operation of the converter consists of first sampling the input signal onto the sample/hold amplifier. This is done by selecting the input signal instead of the loop signal using the select switch $S1$. The input signal is then passed to the multiply-by-two amplifier where it is amplified. To extract the digital information from the input signal, the resultant signal, denoted V_a, is compared to the reference. If it is larger than the reference, the corresponding bit is set to 1 and the reference is then subtracted off from V_a. Otherwise, this bit is set to 0 and the signal V_a is kept unchanged. The resultant signal, denoted V_b, is then transferred, by means of switch $S1$, back into the analog loop for further processing. This process continues until the desired number of bits have been obtained, whereupon a new sampled value of the input signal will be processed. Thus, the digital data come out from the converter in a serial manner, the most significant bit first.

From the previous discussion, it is apparent that the algorithmic A/D converter can be constructed with very little precision hardware. Its implementation in a monolithic technology can therefore be relatively area-sparing. It also possesses inherent sample/hold capability because the sample/hold amplifier is an integral part of the converter. Finally, it possesses floating-point operation capability, i.e., the input signal can be amplified 2^n times before the A/D conversion commences. These properties are very desirable for the design of single-chip complete data acquisition systems.

Loop Nonidealities

Fig. 2 shows an example of a single-ended implementation of the algorithmic A/D converter designed for an MOS technology. Both the sample/hold and the multiply-by-two amplifiers have been realized using *matched* capacitors and MOS operational amplifiers [5]. The reference subtraction circuit is incorporated into the multiply-by-two

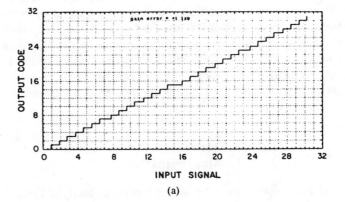

(a)

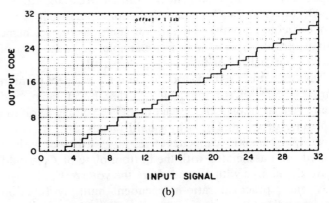

(b)

Fig. 3. Simulated transfer characteristics of a 5 bit algorithmic A/D converter in the presence of (a) loop gain error equal to 1 lsb of conversion and (b) loop offset error equal to 1 lsb of conversion.

amplifier by attaching the capacitor C_5 to the input summing node of the operational amplifier. In this implementation, because the gain values are defined by the ratio of capacitors (assuming the op amps to be ideal), they can be only as accurate as the capacitors are matched, typically 8–9 bits for a standard MOS process. Thus, the signal, in traversing the loop, suffers from a loop gain error which is the sum of the gain errors of the sample/hold and the multiply-by-two amplifiers. Similarly, the input offset voltages of the operational amplifiers, which are on the order of 10 mV, are added to the loop signal during each cycle. This gives rise to another type of error, the loop offset error. Finally, because the MOS transistors inject charges into their surrounding nodes when they are turned off, they give rise to a charge injection effect which contributes to both the loop gain and the loop offset errors.

The effect of these errors on the algorithmic A/D converter can be illustrated by a computer-simulated plot of the transfer characteristics. Fig. 3(a) shows the transfer characteristics of a 5 bit converter with a loop gain error equal to 1 lsb of the conversion, everything else being ideal. It can be seen that a large differential nonlinearity error occurs at the first and second major carry positions. If the loop gain error becomes larger, missing codes start to appear. Fig. 3(b) shows a similar plot of the transfer characteristics for the same converter with loop offset error equal to 1 lsb of the conversion. The same differential nonlinearity error occurs. Also, a missing code has appeared at the first major carry point. In addition to this, an offset appears at the origin which is almost equal to twice the value of the loop offset error. In general, the loop offset error tends to be more severe than the loop gain error in causing undesirable loop characteristics. But in order to achieve near ideal analog-to-digital conversion, both errors should be reduced to a magnitude less than 0.5 lsb of the conversion.

III. CAPACITOR RATIO-INDEPENDENT MULTIPLICATION

If exact gain could be defined without using a ratioed capacitor, then the loop gain error in an MOS algorithmic A/D converter can be eliminated. Conceptually, it is not too difficult to realize an integral multiplication scheme using unratioed capacitors. For example, an exact gain of two can be realized as follows [Fig. 4(a), (b)].

Take two unequal capacitors C_1 and C_2, connect them in parallel to a voltage source V_{signal}, disconnect them, and then stack them on top of each other. The voltage V_0 on the top plate of C_1 in Fig. 4(b) should ideally $2V_{signal}$. However, the parasitic capacitance $C_{parasitics}$ (drawn with a dotted line) associated with the bottom plate of C_1 would share the charge with C_2 and alters the voltage V_0.

A more practical ratio-independent multiply function can be realized by using an integrator. Consider the bottom plate switched capacitor integrator as shown in Fig. 5(a), (b). Assuming the input signal V_{in} to be constant during the integration, the result of integrating this signal n times onto the integrating capacitor C_i would be

$$V_0 = n \frac{C_s}{C_i} V_{in}.$$

To obtain a capacitor ratio-independent multiplication, the position of the integrating capacitor and the sampling capacitor are exchanged. As a result of this capacitor exchange, the charge which was residing in C_i is transferred back onto C_s, and thus results in an output voltage equal to

$$V_0 = nV_{in}.$$

The integrating capacitor functions purely as a inter-

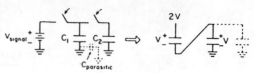

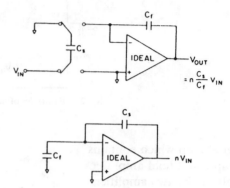

Fig. 4. Conceptual approach to ratio-independent multiply-by-two with capacitors.

Fig. 5. Parasitic insensitive ratio-independent multiply-by-n using a switched capacitor integrator.

mediate storage for the charge sampled onto the sampling capacitor. If the op amp settles within half a clock cycle, then this multiplication requires n clock cycles to complete. This multiplication scheme was proposed independently by Lee [7].

Sources of Error

In the previous discussion of the ratio-independent multiplication, it was tacitly assumed that all components were ideal. Actually, the following effects limit a practical realization of the scheme.
1) Low operational amplifier gain.
2) Finite input offset voltage in the operational amplifier.
3) Charge injection effect from MOS transistor switches.
4) Capacitance voltage dependence.
Depending on the exact circuit used, any or all of these error sources can be significant.

Operational Amplifier Gain

High op amp gain is needed to ensure complete charge transfer between the capacitors. For example, for a 12 bit algorithmic A/D converter, the minimum gain required is about 20 000 or 86 dB. To accommodate this gain requirement, a single-stage CMOS operational amplifier has been designed which achieves a dc gain larger than 40 000 or 92 dB. This design will be covered in a later section.

Operational Amplifier Offset

The equivalent input offset voltage in an operational amplifier can usually be cancelled by storing it on the integrating capacitor during the initialization. It therefore does not pose any serious problem for the multiplication scheme.

Charge Injection Effect

When an MOS transistor switch is turned on, a quantity of charge is stored in its channel. Subsequently, this charge is injected into the surrounding circuit nodes when the transistor is turned off. This phenomenon is commonly known as the charge injection effect. The magnitude of this charge can be expressed in a first-order equation as follows:

$$Q = C_{ox}(V_g - V_t) - C_{ox}V_s$$
$$= Q_1 - Q_2$$

where

Q: charge stored in channel
C_{ox}: gate capacitance of the transistor
V_g: gate voltage
V_t: threshold voltage of MOS transistor
$V_s = V_d$: source or drain voltage
$Q_1 = C_{ox}(V_g - V_t)$
$Q_2 = C_{ox}V_s$.

For simplicity, the back gate bias dependence of the threshold voltage has been neglected and the gate–source and gate–drain overlap capacitances are taken to be zero. Notice that this charge has been split into two components, Q_1 and Q_2. By doing this, the influence of the gate voltage and the source/drain voltage on the channel charge of the transistor switch can be better identified. The first component Q_1, which shall henceforth be called the charge injection offset, is dependent on the gate voltage only, independent of the drain/source voltage, and potentially gives rise to an offset error in a circuit. The second component Q_2, which is called the voltage-dependent charge injection, is dependent on the source/drain voltage and can give rise to a gain error in a circuit if proper measures are not taken to eliminate it.

Effect of Charge Injection

To illustrate the effect of charge injection on a circuit, consider a simple MOS gain block in the transition from the sample to the hold mode. Fig. 6 shows the gain block which consists of an MOS op amp, a sampling capacitor C_1, a feedback capacitor C_2, and three n-channel MOS transistor switches $M1-M3$. During the transition, transistors $M1$ and $M3$ must be turned off and $M2$ turned on. For controlling the transistors, the gates of $M1$ and $M3$ are usually tied to the same control line or the same switching waveform. Now if the input voltage V_{in} is at a potential higher than the ground potential, transistor $M1$ will be turned off first at time t_1, as shown in Fig. 7(a), causing the channel charge to be injected back into the voltage source node A and the op amp summing node B. The state of the circuit between time t_1 and t_0 where $M3$ turns off is illustrated in Fig. 7(b). The partition of the channel charge in $M1$ into source and drain is a complex function of the control voltage waveform and the node

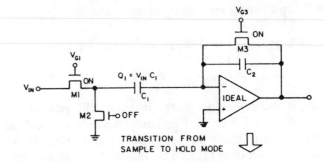
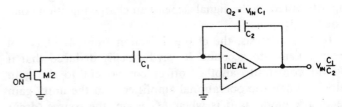

Fig. 6. MOS gain block shown in the transition from the sample to the hold mode.

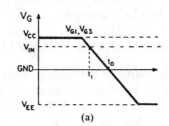

(a)

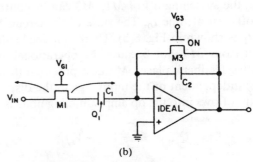

(b)

Fig. 7. (a) Gate control voltage waveforms, $M1$ and $M3$ are controlled using the same voltage source. (b) Equivalent circuit during the transition. Arrow denotes charge injection current.

impedances [8], [9], but in this analysis, it is assumed that the charge is split into two equal halves. Since transistor $M3$ is still conducting when the charge injection takes place, the charge injection will displace from the sampling capacitor an amount of charge equal to

$$Q_{inj} = -\frac{1}{2}C_{ox}(V_g - V_t) + \frac{1}{2}C_{ox}V_{in}$$

so that the resultant sampling charge Q_s becomes

$$Q_s = C_1 V_{in} + \frac{C_{ox}}{2}V_{in} - \frac{C_{ox}}{2}(V_g - V_t)$$
$$= \left(C_1 + \frac{C_{ox}}{2}\right)V_{in} - \frac{C_{ox}}{2}(V_g - V_t).$$

Subsequently, when $M2$ is turned on, Q_s is transferred

onto C_2 and causes an output voltage equal to

$$V_0 = \left(\frac{C_1}{C_2} + \frac{C_{ox}}{2C_2} \right) V_{in} - \frac{C_{ox}}{2C_2} (V_g - V_t)$$

$$= A'V_{in} + V_{offset}.$$

Ideally, if there were no charge injection, the gain of the circuit A would be C_1/C_2. Thus, through the charge injection effect, the gain becomes A' and there is an additional offset term V_{offset}. The alteration of the gain can be attributed to the signal-dependent charge injection component Q_2.

In this analysis, the charge injection from M_3 has not been taken into account. It can be easily deduced that it would contribute another offset component to the gain block. When the operational amplifier is in the unity gain feedback mode as it is when M_3 is on, the source/drain voltage of M_3 is ideally zero, and thus is independent of V_{in}. This means that under all circumstances, the charge injected by M_3 will be a constant.

Cancellation of Signal-Dependent Charge Injection

By proper scheduling of the timing control sequence, the signal-dependent charge injection component and therefore the gain error can be eliminated. The timing diagram for achieving this is depicted in Fig. 8(a). Notice that by delaying the switching off of $M1$, $M3$ can be controlled to switch off first at time t_0. The state of the circuit between t_0 and t_1 is shown in Fig. 8(b). There is indeed still charge injection from $M3$, but because the operational amplifier in negative feedback places $M3$ at a potential (0 for ideal op amp) independent of the input signal, the displacement charge that flows into the op amp summing node is

$$Q_{inj} = - \frac{C_{ox}}{2} (V_g - V_t),$$

and depending on the speed of the operational amplifier and the node impedances of the circuit, this charge may distribute itself between the sampling capacitor C_1 and the feedback capacitor C_2 with varying proportions. But once M_2 is turned on, the feedback action of the operational amplifier forces all charge to reside on C_2. As a result, the output voltage during the hold mode becomes

$$V_0 = \frac{C_1}{C_2} V_{in} - \frac{C_{ox}}{2C_2} (V_g - V_t)$$

and the gain is equal to the ideal gain.

Cancellation of Charge Injection Offset

The use of fully differential circuitry in switched capacitor filters [10] has contributed both to the increase of dynamic range and ease of design. In addition, it leads to a first-order cancellation of the charge injection offset. This can be explained with Fig. 9(a) which shows the differential implementation of the MOS gain block that was described earlier. It consists of two sets of capacitors $C_1 C_3$

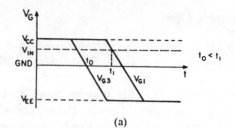

(a)

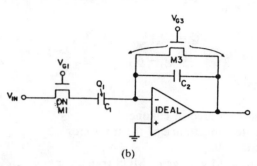

(b)

Fig. 8. (a) Improved switching sequence. (b) Equivalent circuit during switching.

and $C_2 C_4$ which constitute the two signal paths. The MOS transistor switches are duplicated in the two paths. In the transition from the sample to the hold mode as shown in Fig. 9(b), the transistor pairs $M1$ $M3$ and $M2$ $M4$ are turned off first. But instead of connecting the terminals of the capacitors to ground, they are connected together by means of the transistor $M5$. In making use of the proper timing sequence explained in the last section, transistors $M3$ and $M4$ both inject charge into the operational amplifier summing nodes and displace charges in C_1 and C_2. This charge injection is, however, common to both channels of the differential circuit, and it appears, therefore, as a common mode signal to the operational amplifier which is then suppressed by the common mode feedback circuit in the op amp. The output is not affected if transistors $M3$ and $M4$ have the same gate capacitance and overlap capacitances. In actuality, the transistors can be mismatched by as much as 10–20 percent in their capacitances and the cancellation is only approximate. The resultant offset error voltage in this case is

$$V_{offset} = \delta C_{ox} (V_g - V_t)$$

where δC_{ox} is the difference in the gate capacitances.

This represents a significant reduction in the error due to the charge injection from MOS transistor.

Capacitance Voltage Coefficient Cancellation

The capacitor used in an MOS process is usually voltage dependent. This voltage dependence can be expressed as a Taylor series

$$C(V) = C_0 \left(1 + \alpha_0 V + \alpha_1 V^2 + \cdots \right)$$

where C_0 is the zero voltage capacitance and $\alpha_0, \alpha_1, \cdots$ are the capacitance voltage coefficients.

Normally, this capacitance voltage dependence introduces nonlinearities into the circuit which are intolerable

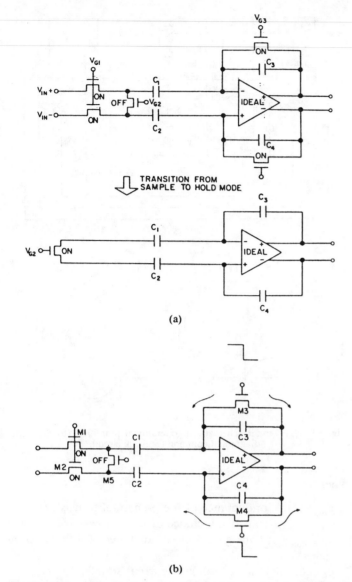

Fig. 9. Differential circuit used to obtained first-order cancellation of charge injection from MOS transistors.

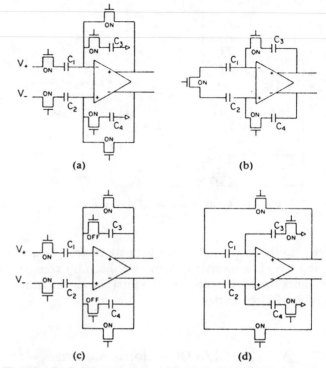

Fig. 10. Ratio-independent multiply-by-two switching sequence. (a) sample V_{in}. (b) Charge transfer. (c) Sample V_{in}. (d) Transfer charge and exchange capacitors.

for high-precision circuit operations. However, the same differential circuit which is used to cancel the charge injection effect also gives a cancellation of odd order capacitance voltage coefficients of the MOS capacitors. This is because an increase of voltage or capacitance in one channel is accompanied by an equal but opposite decrease of voltage or capacitance in the other channel of the circuit. These changes combine to give overall capacitance voltage coefficients equal to the difference of the odd-order capacitance voltage coefficients of the capacitors in the two channels. The even-order capacitance voltage coefficients are summed together. Since $\alpha_1, \alpha_2, \cdots$ are much smaller than α_0, they can be neglected in a first-order analysis and the net result is a voltage dependence of $\delta\alpha_0$.

Ratio-Independent Multiply-by-Two Switching Algorithm

Summarizing the foregoing discussions, the ratio-independent multiply-by-two function can be realized using a differential gain block or integrator with a switching se-

quence as shown in Fig. 10(a)–(d). Assuming that the input differential voltage $V_+ - V_-$ is constant during the whole operation, it is first sampled onto the sampling capacitors C_1 and C_2 in the first clock cycle [Fig. 10(a)]. This signal charge is then transferred onto the integrating capacitors C_3 and C_4 during the second clock cycle by turning on the switch $M3$ [Fig. 10(b)]. Subsequently, C_3 and C_4 have to be separated from the op amp feedback loop to prepare for the second signal sampling. To accomplish this task without introducing signal-dependent charge injection, the capacitors are separated from the op amp input nodes. Another signal sample is then taken [Fig. 10(c)], and during the fourth or the last clock cycle, the first signal charge is redistributed back onto C_1 and C_2 from C_3 and C_4 [Fig. 10(d)]. The operation takes four op amp settling time periods which determine the length of the basic clock cycle.

IV. CIRCUIT IMPLEMENTATION OF ALGORITHMIC A/D CONVERTER

Fig. 11 shows the circuit schematic of the A/D converter. It contains two operational amplifiers, an ac coupled comparator with latch, ten capacitors, and 35 transistors. The differential signal is introduced through the selection switches $M1$ and $M2$ and is sampled onto the sample/hold amplifier. While this voltage value is held, it is multiplied by two by the next amplifying block using the ratio-independent algorithm that has been described. Two complementary reference voltages V_{r+} and V_{r-} are provided. With C_3 and C_4, they allow both the subtraction and addition of reference. Switches $M32-M35$ feed the signal

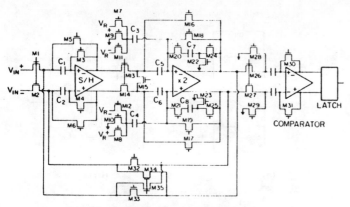

Fig. 11. Circuit schematic of A/D converter.

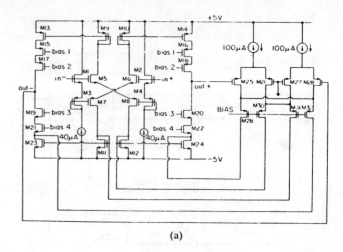

(a)

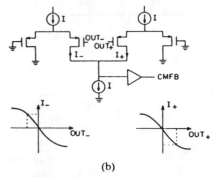

(b)

Fig. 12. (a) Circuit schematic of operational amplifier. (b) Simplified diagram of common feedback circuit.

back into the loop and provide for path reversal as needed by some A/D conversion algorithms. Each bit conversion consumes six clock cycles, and the total conversion time for 12 bit is 72 clock cycles.

V. Class A/B Operational Amplifier

The MOS operational amplifier used in the algorithmic A/D converter must possess high dc gain and fast settling time to an accuracy required by the system. High gain could, of course, be attained by a two gain-stages architecture. But there would certainly be some tradeoff with the maximum bandwidth that can be obtained because of the additional parasitics introduced through the complicated design. A single gain-stage architecture is therefore chosen. The schematic diagram of the operational amplifier is shown in Fig. 12(a). The input consists of four transistors cross coupled at their sources. The input differential is applied directly to the gates of the n-channel transistor $M5$, $M6$ and the opposite p-channel transistor $M7$, $M8$ by means of the source followers formed by $M1-M4$. The input signal is converted into a differential current which is then mirrored to the output through the current mirrors $M9M13$, $M10M14$, $M11M23$, and $M12M24$. The differential transconductance of the input stage $g_{m\,\mathrm{diff}}$, neglecting the influence of the finite conductance of the mirror transistors $M9-M12$, is

$$g_{m\,\mathrm{diff}} = \frac{g_{mn}g'_{mp}}{g_{mn} + g'_{mp}}$$

where g_{mn} is the transconductance of the input n-channel transistor and g'_{mp} is the reduced transconductance of the input p-channel transistor.

Here the transconductance of the p transistor has been reduced by the transfer characteristics of the source follower:

$$g'_{mp} = g_{mp} \frac{g_{mn\,\mathrm{follower}}}{g_{mn\,\mathrm{follower}} + g_{mnb\,\mathrm{follower}} + g_0}$$

where

g_{mp}: transconductance of the p-channel transistor

$g_{mn\,\mathrm{follower}}$: transconductance of the follower transistor

$g_{mnb\,\mathrm{follower}}$: back gate transconductance of follower transistor

g_0: conductance of follower current source.

To increase the output node resistance, triple cascoded transistors are used. The value of this impedance is

$$r_{\mathrm{out}} = (g_{m\,\mathrm{cas}1}r_{o\,\mathrm{cas}1})(g_{m\,\mathrm{cas}2}r_{o\,\mathrm{cas}2})r_{ocs}$$

with

$g_{m\,\mathrm{cas}\,n}$: transconductance of cascode transistor n

$r_{o\,\mathrm{cas}\,n}$: output impedance of cascode transistor n

r_{ocs}: output impedance of current source.

Thus, the overall low-frequency gain of the amplifier is

$$A_0 = g_{m\,\mathrm{diff}}r_0.$$

The operational amplifier operates in a class A/B mode. To obtain high gain, the dc quiescent current of the operational amplifier is kept low. But because the input circuit configuration is not current limited, the slewing at the output is extremely fast. The op amp is compensated by attaching 2 pF capacitors between its output nodes and ground.

A balanced common feedback circuit has been used for this operational amplifier to improve the common mode

Fig. 13.   Die photograph.

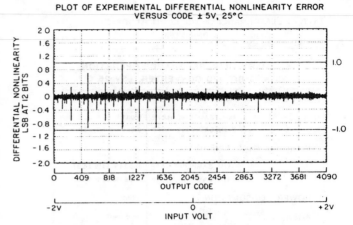

PLOT OF EXPERIMENTAL DIFFERENTIAL NONLINEARITY ERROR
VERSUS CODE ± 5V, 25°C

Fig. 15.   Plot of experimentally obtained differential nonlinearity error versus code with device operating as a 12 bit A/D converter.

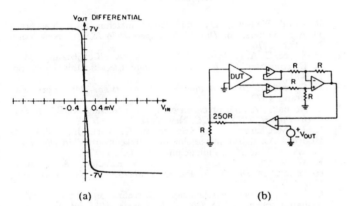

(a)   (b)

Fig. 14.   (a) Operational amplifier transfer characteristics. (b) Test circuit.

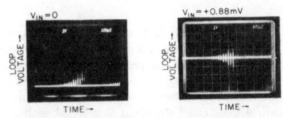

Fig. 16.   Output voltage waveform of the analog loop used as an instrumentation amplifier with gain equal to 2^{12}. (a) Input grounded. (b) Input voltage equal to 0.88 mV.

rejection. The principle of operation can be explained with the simplified schematic in Fig. 12(b). Two identical MOS differential pairs composed of p-channel MOS transistors are connected in the configuration with their drains tied to ground and together to a current source as shown. The common mode feedback signal V_{cmfb} can only be 0 if

$$V_{out+} = -V_{out-},$$

even in the presence of nonlinear transfer characteristics of the MOS differential stage.

VI.   EXPERIMENTAL RESULT

The die photo of the chip is shown in Fig. 13. The signal lines and the power supply lines have been partitioned carefully to avoid noise coupling into the system. The MOS capacitors are poly/diffusion capacitors approximately of 4 pF, and no effort has been made to match them by using process insensitive geometries. The total area of the chip is 2400 mils². The transfer characteristics of the op amp are shown in Fig. 14(a). They are measured using the experimental setup in Fig. 14(b). For a supply voltage of ±5 V, the minimum

gain is 40 000 within an output voltage range of ±3.5 V. The area is about 400 mils².

The dynamic behavior of the converter has been tested using a code density test developed by Doernberg [11]. A plot of the differential nonlinearity error versus code of a typical device operating as an 12 bit A/D converter is shown in Fig. 15. The input signal frequency was 1 kHz and the converter is operated at an 8 kHz sampling rate. The device shows no missing codes at the 12 bit level, although there are occurrences of nonlinearity errors close to 0.9 lsb at the major carry points. The asymmetrical shape of the plot is due to residual loop offset error, which is defined to be the offset voltage accumulated when the signal passes once through the loop. This can be substantiated by oscilloscope photographs of the loop voltage. Fig. 16 show these voltage waveforms with the converter programmed to operate as a cyclic instrumentation amplifier, amplifying the input signal by 2^{12} times in the analog loop. In this operating mode, the input signal is sampled and multiplied by two every time it passes through the analog loop. With the input grounded, the voltage waveform is skewed towards one side. However, when a input voltage of 0.9 mV is applied, the symmetry of the loop voltage waveform is restored. Because the equivalent input offset voltage is twice the loop offset voltage, the residual loop offset is approximately equal to 0.45 mV. This loop offset error is probably still the most important limiting factor in the accuracy of the algorithmic A/D converter.

TABLE I
TYPICAL PERFORMANCE OF THE 12 BIT ALGORITHMIC A/D
CONVERTER

SUMMARY OF PERFORMANCE OPERATION AS A 12 BIT A/D CONVERTER. ±5V 25 C	
DIFFERENTIAL NON - LINEARITY :	
4kHz SAMPLING RATE	0.019% (0.8 LSB)
8kHz SAMPLING RATE	0.022% (0.9 LSB)
INTEGRAL NON - LINEARITY :	
4kHz SAMPLING RATE	0.034% (1.5 LSB)
8kHz SAMPLING RATE	0.081% (3.2 LSB)
POWER CONSUMPTION	17mW
AREA	2400mil²
SUPPLY VOLTAGE RANGE	±4.5V - 6.5V
TECHNOLOGY	5µ POLYSILICON GATE CMOS

The performance of the A/D converter is summarized in Table I. The performance was found to conform well to computer simulations. It is expected that higher resolution can be obtained by using an improved offset cancellation technique and putting the control logic on the same chip. This control logic can be realized as a ROM with a 1 bit address feedback. Because the operating speed of the converter is limited primarily by the settling time of the operational amplifier, a much higher sampling rate can be achieved by increasing the bandwidth of the operational amplifier. If power consumption is not a major concern and the circuit load capacitances are of controllable sizes, a class A single-stage operational amplifier may have a speed advantage. With the settling time below 1 μs, sampling speeds up to 16–20 kHz can be obtained.

VII. CONCLUSION

A technique for A/D conversion has been described which uses a ratio-independent switching algorithm to realize the algorithmic A/D converter without the use of a matched capacitor in an MOS technology. A prototype has been built with a 5 μm CMOS polysilicon gate technology. The converter achieves 12 bit resolution and a maximum sampling rate of 8 kHz. The area of the analog circuitry is 2400 mils^2. It is estimated that the converter, including all the control logic, can be integrated in less than 3000 mils^2 with the same technology. Higher resolution can be obtained with a better offset cancellation technique. This type of converter can be integrated as peripheral function blocks in a mixed analog/digital system where small area is of paramount importance.

ACKNOWLEDGMENT

The authors wish to thank S. Tam and T. Choi for providing technical assistance during the device fabrication and J. Doernberg for setting up the test system. Comments from Prof. D. A. Hodges have been most helpful.

REFERENCES

[1] R. Van de Plaasche, "Dynamic element matching for high accuracy D/A converters," in *Dig. Tech. Papers 1976 ISSCC*, New York, NY, Feb. 1976.

[2] H.-S. Lee, D. A. Hodges, and P. R. Gray, "A self-calibrating CMOS A/D converter," *IEEE J. Solid-State Circuits*, this issue, pp. 813–819.

[3] H. Schmid, *Electronic Analog Digital Conversion.* New York: Van Nostrand, 1970.

[4] T. Hornak, "A high precision component tolerant ADC," in *Dig. Tech. Papers, 1975 ISSCC*, New York, NY, Feb. 1975.

[5] R. H. McCharles, V. A. Saletore, W. C. Black, Jr., and D. A. Hodges, "An algorithmic analog-to-digital converter," in *Dig. Tech. Papers, 1977 ISSCC*, Philadelphia, PA, Feb. 1977.

[6] R. W. Webb, F. R. Cooper, and R. W. Randlett, "A 12b A/D converter," in *Dig. Tech. Papers, 1981 ISSCC*, New York, NY, Feb. 1981.

[7] C. C. Lee, "A new switched-capacitor realization for cyclic analog-to-digital converter," in *Dig. Tech. Papers, 1983 Int. Symp. Circuits and Syst.*, Newport Beach, CA, May 1983.

[8] B. J. Shu, "Switch-induced error voltage on a switched capacitor," M.S. thesis, Univ. California, Berkeley, June 1983.

[9] L. Bienstman and H. J. DeMan, "An eight channel 8 bit microprocessor compatible NMOS D/A converter with programmable scaling." *IEEE J. Solid-State Circuits*, vol. SC-15, pp. 1051–1059, Dec. 1980.

[10] K. C. Hsieh, P. R. Gray, D. Senderowicz, and D. G. Messerschmitt, "A low noise chopper stabilized differential switched-capacitor filtering technique," *IEEE J. Solid-State Circuits*, vol. SC-16, pp. 708–716, Dec. 1981.

[11] J. Doernberg, H.-S. Lee, and D. A. Hodges, "Full-speed testing of A/D converters," *IEEE J. Solid-State Circuits*, this issue, pp. 820–827.

A Per-Channel A/D Converter Having 15-Segment
μ-255 Companding

JAMES C. CANDY, SENIOR MEMBER, IEEE, WILLIAM H. NINKE, MEMBER, IEEE,
AND BRUCE A. WOOLEY, MEMBER, IEEE.

Abstract—This paper describes a companded analog-to-digital (A/D) converter for voiceband signals that is simple and potentially inexpensive. The converter uses only 18 coarsely spaced analog levels. Fine resolution is obtained by oscillating between these levels at an increased speed and averaging the result over a Nyquist interval. The companding used in the converter is effectively the same as that of μ-255 pulse-code modulation (PCM).

In the encoding process a one-bit code is generated at 256 000 samples/s. This 1-bit per sample signal can be transmitted and decoded directly, or a simple digital circuit will produce a 13-bit, 8-kHz linear PCM signal that can be compressed to 8-bit companded PCM format. In this paper the basic operation of the 1-bit coder is described and its performance when connected to a 1-bit decoder is illustrated. Methods for obtaining both linear and compressed PCM are then presented, and the properties of these PCM signals with respect to noise, gain tracking, and harmonic content are described. Relative insensitivity to circuit component variations, absence of analog gates, along with the need to generate only a few analog levels, make the coder especially well suited to integrated circuit realization.

I. INTRODUCTION

PULSE-CODE MODULATION (PCM) codecs used in the telephone network for voiceband signals typically provide signal-to-noise ratios in excess of 30 dB over a dynamic range of 40 dB. Such performance, commensurate with amplitude resolution of 1 in 8000, generally has been achieved economi-cally by sharing a single codec among 24 or more telephone channels. However, having a separate codec for each channel offers a number of advantages. Per-channel conversion between analog and digital signals eliminates the need for multiplexing analog signals, thereby avoiding many noise and crosstalk problems at low-signal levels. It may also provide increased system flexibility, greater overall reliability, lower cost for partially equipped installation, and simplified maintenance for multichannel terminals. Moreover, it may make possible new organizations of communication equipment. The principal objection to per-channel PCM codecs has been expense. The per-channel encoder proposed in this paper and the decoder described in [1] are offered as an answer to this objection. They are particularly well suited for fabrication using conventional integrated circuit processing techniques, and ultimately need cost no more than a few dollars.

Delta modulation is a well-known inexpensive coding system. It has often been considered for telephone use but has found few applications because of the low-channel efficiency. For example, in linear delta modulation bit rates above 1 Mbit/s have been necessary to obtain the required resolution over a sufficiently wide dynamic range [2]. This rate has been reduced to about 50 kbits/s by adapting step sizes to suit the rate of change of signal amplitude [3], [4]. Such adaption has been chosen specifically for voice-type signals and may not be suitable for data. It also is not in accord with the logarithmic amplitude companding commonly used in the digital trunk network. Efficient digital conversion between a delta modulator signal and PCM [15], or any format with sampling at the

Paper approved by the Associate Editor for Data Communication Systems of the IEEE Communications Society for publication after presentation at the 1974 National Telecommunications Conference, San Diego, CA, December 1974. Manuscript received July 7, 1975; revised September 5, 1975.

The authors are with Bell Laboratories, Holmdel, NJ 07733.

Reprinted from *IEEE Trans. Commun.*, vol. COM-24, no. 1, pp. 33–42, Jan. 1976.

299

Nyquist rate, requires a high-grade digital low-pass filter or equivalent. Such circuits are more expensive than the simple accumulator that we are using.

The encoder proposed here is related to delta modulators, but it avoids many of delta modulation's most troublesome disadvantages. As in delta modulation, the input signal is compared with a quantized approximation of itself and the difference is used to change the quantized value. However, digital integration is used to reconstruct the quantized signal, thereby avoiding the problems of drift and step-size imbalance that are characteristic of ordinary delta modulation. Fine resolution is obtained in the coding by causing the quantization to oscillate between coarsely spaced levels and averaging the result over a Nyquist interval, an action that has been named interpolation. The need for very small step sizes is eliminated along with attendant sensitivity to noise and component imperfections. Spacing the quantization levels logarithmically provides the encoder with a wide dynamic range and compatibility with the μ-255 compression commonly used in PCM systems.

The circuits described are unusually tolerant of imperfection in analog components: resolutions of 1 in 10^4 are achieved with a two percent standard deviation in critical resistor ratios. Moreover, the coding is timed by a single regular clock, the frequency of which needs to be held constant only to one part in 10^3 to clearly meet codec requirements for gain stability and distortion. With a clock frequency of 256 kHz, a one-bit code is generated at a 256 ksamples/s and is easily converted first to a 13-bit linear PCM format at 8 ksamples/s and then to the compressed PCM 64 kbit/s format commonly used for toll transmission. The 1-bit signal from this interpolative encoder can be transmitted and decoded directly, and clock frequencies and bit rates down to 40 000 events/s continue to provide reasonably good quality reproduction of speech. These low-bit rates may be attractive in local loop and customer equipment applications.

There now follows a description of the one-bit interpolative encoder and a related decoder, with illustrations of their response to sinusoidal inputs. Techniques for processing signals from the interpolative encoder to obtain both uniform and compressed PCM codes are then explained. Finally the performance of the encoder with respect to gain tracking, noise level, and harmonic distortion is described. The circuit of the encoder is described in the Appendix.

II. A ONE-BIT INTERPOLATIVE CODER

A recent paper [5] described a multibit analog-to-digital (A/D) converter that achieved fine resolution with relatively coarse quantization by means of interpolation. The philosophy was to generate a quantized signal oscillating at many times the Nyquist rate of the input signal, whose short-term average closely approximates the short-term average of the input. Digital representations of the coarsely quantized values occurring during a Nyquist interval were averaged to give a finely quantized representation of the input. A sample-and-hold circuit is unnecessary at the input to these converters.

Fig. 1 shows an adaptation of interpolative quantization to one-bit coding. In the diagram, the analog input x is compared

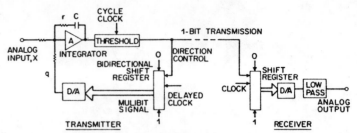

Fig. 1. Outline of an interpolative coder transmitting 1-bit words to a receiver.

with its quantized representation q. The difference, $(x - q)$, when integrated by amplifier A activates a threshold circuit that provides an indication of polarity. When the integrator output is positive the quantized signal q moves up by one level; it moves down when the integrator output is negative. A resistive digital-to-analog (D/A) network driven by a bi-directional shift register provides the quantized approximation to the input. The shift register acts an an accumulator, being fed with ONES at the lower port and ZEROS at the upper; it fills and empties with ONES in much the same way that the integrating capacitor in a delta modulator fills and empties with charge. Digital storage in a shift register has four major advantages over capacitive integration: no leakage is associated with it, each step size is substantially independent of the direction that the signal changes, the digital code representing the accumulated value can be read out directly and, most importantly, the levels of the quantized signal can easily be compressed by appropriate design of the resistive D/A network. An apparent disadvantage is the need for a very large shift register to have sufficiently fine resolution in the quantized signal. This need is avoided by using the interpolating action provided by integrator A, and by cycling the circuit much faster than the Nyquist rate.

The integrator in the forward path of the feedback loop [5], [9] tries to keep the integral of the quantized signal q equal to the integral of the input; consequently the amplitude of q oscillates between levels holding its local average value nearly equal to x. Fig. 2 illustrates the response of a circuit in which q is restricted to odd integers and the input is a constant 2.75. Notice that q jumps between three levels and its average over any 16 cycles equals 2.75. Oscillation between three levels rather than two is necessary because the binary decision requires a step-up or a step-down in amplitude: q must always change, it cannot remain still like the quantization described elsewhere [1], [5].

Signal values corresponding to Fig. 2 are given in Table I. The threshold circuit in the encoder responds to the polarity of the signal formed by an accumulation of the error plus a term proportional to the present error; addition of the present error term is needed to ensure stability, i.e., for oscillations of q to remain bounded within a few levels of the input amplitude. In the encoder of Fig. 1 the present error term is realized by the resistor r which is in series with the integrating capacitor C. The voltage across C represents the accumulated error, and the voltage across r is proportional to the present error. In the example of Table I the time constant rC is chosen to result in the present error and the accumulated error being added in equal proportions. This proportion minimizes the in-

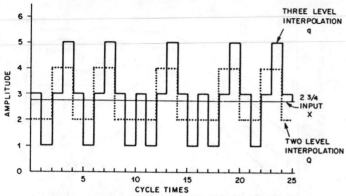

Fig. 2. Response of the code to a constant input $X = 2.75$. q is the feedback signal and Q its average over adjacent cycles.

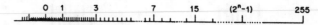

Fig. 3. Quantization levels for a version of 15-segment 255-μ law companding.

TABLE I

n	Input X_n	Error $\epsilon_n = (X_n - q_{n-1})$	$\Sigma\epsilon_n$	Integrator Output $\epsilon_n + \Sigma\epsilon_n$	Code	q_n
0			0			3
1	2.75	-.25	-.25	-.50	0	1
2	2.75	1.75	1.50	3.25	1	3
3	2.75	-.25	1.25	1.00	1	5
4	2.75	-2.25	-1.00	-3.25	0	3
5	2.75	-.25	-1.25	-1.50	0	1
6	2.75	1.75	.50	2.25	1	3
7	2.75	-.25	.27	0	1	5
8	2.75	-2.00	-2.00	-4.25	0	3
9	2.75	-2.25	-2.25	-2.50	0	1
10	2.75	1.75	-.50	1.25	1	3
11	2.75	-.25	-.75	-1.00	0	1
12	2.75	1.75	1.00	2.75	1	3
13	2.75	-.25	.75	.50	1	5
14	2.75	-2.25	-1.50	-3.75	0	3
15	2.75	-.25	-1.75	-2.00	0	1
16	2.75	1.75	0	1.75	1	3

band quantization noise. Similar experience has been reported elsewhere [6], [7], [10].

An average of two consecutive values of q in Fig. 2 provides a new quantized signal Q shown by the dotted curve. It equals two for 5/8 of the time and four for the remaining time, a simple interpolation of the value 2.75 from the levels two and four. Successive pairs of values in the interpolation waveform average together to provide the two-level pattern previously described for other interpolative codecs [1], [5]. As explained later, this approach simplifies the generation of PCM values from the encoder.

III. A CODER WITH SEGMENTED COMPRESSION [11], [12]

The technique of interpolating values spaced uniformly between fixed quantization levels has a natural application to segmented companding of the kind used for transmitting speech. A form of fifteen segment μ-law companding divides the range of signal magnitudes into eight segments, each one

being twice as wide as its lower neighbor. For example, Fig. 3 shows segment boundaries at

$$Q \in \{\pm Q_n\} \qquad (1)$$

where

$$Q_n = (2^n - 1), \qquad 0 \leqslant n \leqslant 8.$$

This type of companding can be achieved with an encoder that has actual quantization levels set at the segment boundaries and has the capability of interpolating 16 levels between adjacent boundaries.

The interpolative codec in Fig. 1 will provide such companding if it is cycled at 32 times the desired PCM word rate and has the appropriate amplitudes generated in the D/A circuit. To provide values of Q at the segment boundaries given by (1), the following set of amplitudes has been chosen for the feedback signal q:

$$q \in \{\pm q_n\} \qquad (2)$$

where

$$q_n = (2^{n+1} - 3)/3, \qquad 0 \leqslant n \leqslant 9.$$

This set meets the requirement that the average of any two adjacent values corresponds to a segment boundary since

$$q_n + q_{n+1} = 2Q_n, \qquad 0 \leqslant n \leqslant 8. \qquad (3)$$

It also has the desirable property that the difference between adjacent pairs of levels differ by a factor of 2,

$$(q_{n+2} - q_{n+1}) = 2(q_{n+1} - q_n), \qquad 0 \leqslant n \leqslant 7. \qquad (4)$$

Requirement (3) is essential; it ensures that, in the steady state with constant input, the average of any 32 consecutive values can interpolate 16 intervals within a segment. Property (4) permits realization of the D/A network with a simple R-$2R$ ladder; this is especially useful for simplifying the integrated circuit realization of the encoder.

Fig. 4 shows the quantized amplitude generated in response to a sinusoidal input by a simulation of the encoder. It is seen to oscillate but remains bounded to within three levels of the input amplitude. Stability of these limit cycles is achieved by including the resistor r in series with the integrating capacitor C, the time constant rC being approximately equal to the cycle period.

An experimental realization of the companded encoder was implemented with commercial integrated circuit parts and discrete components, including a ladder network of ±1 percent tolerance resistors and voltage sources. An outline of this encoder realization is described in the Appendix. Two methods of decoding the digital signal are considered. The simpler method, shown in Fig. 1, communicates the threshold decision to a remote shift register and D/A circuit that reproduces the

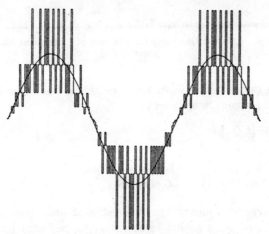

Fig. 4. Response of a simulated coder with 256-kHz cycle rate to a 2-kHz sinewave.

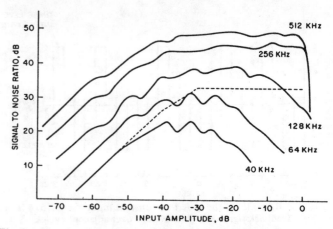

Fig. 5. Signal-to-noise ratio as a function of the amplitude of a 1-kHz input, for specified cycle rates (1 percent circuit tolerances).

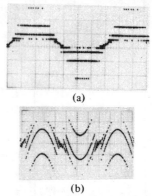

(a)

(b)

Fig. 6. Waveforms of the quantized signal q and the error signal $(x - q)$ for a large input.

quantized signal; this reproduced signal is then low-pass filtered to provide the analog output. In the second method, considered in the following section, a binary code taken from the shift register is accumulated in order to obtain a PCM signal; the PCM signal is then decoded to provide the output.

Fig. 5 is a graph of signal-to-noise ratios that have been measured when transmitting the threshold decision to a remote decoder followed by a low-pass filter whose response approximates the C-message weighting function. Curves are shown for several cycle rates; in each case the value of the integrating capacitor was adjusted to keep the time constant rC approximately equal to the cycle period.

The dashed lines drawn in Fig. 5 show the minimum requirement for $D3$ channel banks; this requirement can be met with cycle rates greater than 128 kHz. The degradation in performance at high amplitudes and low-cycle rates is caused by slope overloading. This effect occurs when the shift register is unable to step through its range fast enough to track the signal, as is illustrated in Fig. 6. Methods for avoiding slope overloading in delta modulators are well known, and related methods have application to this encoder.

It is evident in Fig. 5 that, at the intended operating rate of 256 kHz, noise level requirements are met easily. However, it will be seen that the 12-dB margin is reduced to about 3 dB when the digital signal is converted to the 64-kbit PCM format commonly used in toll transmissions.

Important performance requirements other than signal-to-noise ratio are those for gain tracking and harmonic distortion. These requirements are discussed in Section V, after the methods for obtaining PCM are explained.

IV. PCM SIGNALS

A. Digital Accumulation

The 1-bit signal generated by the threshold decision is potentially useful for local communication; it is easily generated, requires no framing, and can be coded so as to be reasonably tolerant of transmission error [13]. It is, however, unsuitable for communication over long distances, in part because of its inefficiency. Thus, perhaps the most attractive feature of the 1-bit signal is the ease with which it may be converted to PCM, particularly to a 64-kbit format that is compatible with digital toll networks.

Reference [5] describes a method for obtaining PCM from an interpolative encoder; it explains how an average of the quantized amplitude over a Nyquist interval is a good representation of the input. In that work, a binary representation of a quantized signal was accumulated to produce a PCM signal. A similar process can be used with the present encoder.

The first step in the conversion to PCM is the generation of a binary representation of the quantized amplitude q. In the encoder circuit described in the Appendix, the shift register holds a code that determines only the magnitude of q. The polarity is determined by the state of an auxiliary toggle circuit. Table II lists, in its second column, the allowed magnitudes of q, and in the first column lists the corresponding code in the shift register. This code could be converted directly into a binary representation of the amplitudes, but an easier method is available. Recall that the code values are scanned by stepping up or down one at a time, and that the average of any two consecutive amplitudes corresponds to a value of one of the segment boundaries defined in (1). Column 3 in Table II lists these average values, and it can be seen that the binary code corresponding to their magnitude is the code present in the shift register for the larger of the two values that are averaged. The process for generating PCM is therefore a simple one; an outline of the circuit used is given in Fig. 7.

At the top of Fig. 7 is a shift register $SR1$ that may be part

TABLE II

SHIFT DIRECTION MSB - TOWARD - LSB

REGISTER SRI CONTENT n:m CODE (8 BIT)	QUANTIZED SIGNAL LEVELS q_n	AVERAGE OF ADJACENT VALUES	BINARY VALUE OF THE AVERAGE
0 0 0 0 0 0 0 0	1/3	0	0 0 0 0 0 0 0 0 0
0 0 0 0 0 0 0 1	1 2/3	1	0 0 0 0 0 0 0 0 1
0 0 0 0 0 0 1 1	4 1/3	3	0 0 0 0 0 0 0 1 1
0 0 0 0 0 1 1 1	9 2/3	7	0 0 0 0 0 0 1 1 1
0 0 0 0 1 1 1 1	20 1/3	15	0 0 0 0 0 1 1 1 1
0 0 0 1 1 1 1 1	41 2/3	31	0 0 0 0 1 1 1 1 1
0 0 1 1 1 1 1 1	84 1/3	63	0 0 0 1 1 1 1 1 1
0 1 1 1 1 1 1 1	169 2/3	127	0 0 1 1 1 1 1 1 1
1 1 1 1 1 1 1 1	340 1/3	255	0 1 1 1 1 1 1 1 1

TABLE III

SEGMENT BOUNDARY	SIGN-MAGNITUDE BINARY CODE (9 BIT)	2s-COMPLEMENT (13 BIT)
255	1 1 1 1 1 1 1 1 1	0 0 0 0 0 1 1 1 1 1 1 1 1
127	1 0 1 1 1 1 1 1 1	0 0 0 0 0 0 1 1 1 1 1 1 1
63	1 0 0 1 1 1 1 1 1	0 0 0 0 0 0 0 1 1 1 1 1 1
31	1 0 0 0 1 1 1 1 1	0 0 0 0 0 0 0 0 1 1 1 1 1
15	1 0 0 0 0 1 1 1 1	0 0 0 0 0 0 0 0 0 1 1 1 1
7	1 0 0 0 0 0 1 1 1	0 0 0 0 0 0 0 0 0 0 1 1 1
3	1 0 0 0 0 0 0 1 1	0 0 0 0 0 0 0 0 0 0 0 1 1
1	1 0 0 0 0 0 0 0 1	0 0 0 0 0 0 0 0 0 0 0 0 1
0	1 0 0 0 0 0 0 0 0 0 0 0 0 0 0 0 0 0	0 0 0 0 0 0 0 0 0 0 0 0 0
-1	0 0 0 0 0 0 0 0 1	1 1 1 1 1 1 1 1 1 1 1 1 1
-3	0 0 0 0 0 0 0 1 1	1 1 1 1 1 1 1 1 1 1 1 0 1
-7	0 0 0 0 0 0 1 1 1	1 1 1 1 1 1 1 1 1 1 0 0 1
-15	0 0 0 0 0 1 1 1 1	1 1 1 1 1 1 1 1 1 0 0 0 1
-31	0 0 0 0 1 1 1 1 1	1 1 1 1 1 1 1 1 0 0 0 0 1
-63	0 0 0 1 1 1 1 1 1	1 1 1 1 1 1 1 0 0 0 0 0 1
-127	0 0 1 1 1 1 1 1 1	1 1 1 1 1 1 0 0 0 0 0 0 1
-255	0 1 1 1 1 1 1 1 1	1 1 1 1 1 0 0 0 0 0 0 0 1

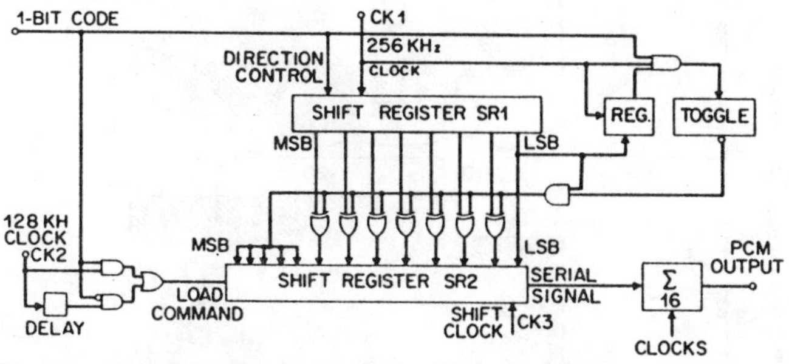

Fig. 7. Circuit for obtaining 13-bit PCM at 8 kHz from the code in the shift register.

of an encoder or, as in this example, a remote decoder. Its content is transferred to a second register $SR2$, at one-half of the cycle rate, by means of clock signal $CK2$, according to the following procedures. If the threshold decision calls for a shift-down in amplitude, then $SR2$ is loaded immediately before shifting occurs. When a shift-up is requested, $SR2$ is loaded after $SR1$ has assumed its new value. The loading clock $CK2$ occurs at one-half the cycle rate so that register $SR2$ receives data that describe the average of consecutive pairs of quantized amplitudes. The logic gates interposed between the registers are controlled by the polarity bit in a way that causes a two's complement, rather than sign magnitude, code to enter $SR2$. This code is listed in the last column of Table III. After each loading, shift register $SR2$ empties its content into a serial accumulator to generate a sum that is periodically dumped to provide the PCM output. A parallel transfer and parallel accumulation might also be used.

B. Properties of the PCM Signal

The rate at which PCM words are generated by this scheme should not be less than twice the bandwidth of the analog input. For the projected applications to telephone messages, the word rate would be 8 kHz, with 32 quantized amplitudes contributing to each word. Each word, being an accumulation of 16 9-bit words, will comprise 13 meaningful bits. Other formats are possible simply by changing the output timing; for example, 12-bit words could be dumped from the accumulator at 16 kHz. This format would accommodate wider input bandwidths and thus simplify the requirements for the input analog filter.

We have called the signal generated by this circuit a sparse uniform PCM because the primary coding is compressed; the full resolution of 13 bits is not always achieved. The signal-to-noise ratio measured for an actual circuit generating 13-bit words at 8 kHz is plotted in Fig. 8 for a 1.02 kHz input sinwave. The noise level is about 4 dB greater than that expected from ideal μ-255 coding and decoding. However, a 3-dB penalty is inherent[1] in interpolating encoders of the type described here and in [5]. Thus, the performance of the circuit is very close to theoretical.

The conversion to PCM is simple and robust. Its output in a linear two's complement format has a number of potentialities for telephone use, such as combining signals for conference calls, linear filtering, echo suppression, and compression to a 64-kbit format.

C. Two Forms of μ-255 Companding

H. Kaneko has described two complementary methods for segmenting companding laws, calling them RLA and DLA [14]. In RLA companding, quantization (reconstruction) levels are assigned to the index points shown in Fig. 2, and decision thresholds are set midway between them. In DLA, decision thresholds are assigned to the index points, and quantization levels are midway between these points.

The technique of interpolation requires RLA companding, but existing digital channel banks use DLA companding. Moreover, the channel banks modify the companding near the origin in order to position quantization levels at zero; however, this affects performance only for input amplitudes that are 70 dB less than peak.

In our experiments, both encoder and decoders used RLA companding, but the digital compression of the PCM code could be either RLA or DLA. No changes in signal-to-noise

[1] Y. C. Ching has demonstrated that this penalty can be avoided by employing a certain kind of double accumulation of the digital codes.

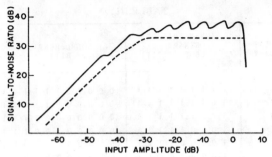

Fig. 8. Signal-to-noise ratio of the 13-bit sparse PCM, for various amplitudes of a 1.02-kHz input.

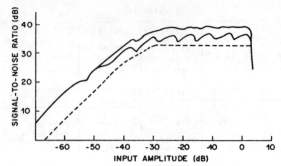

Fig. 9. Signal-to-noise ratio of the compressed PCM: 8-bit words at 8 kHz for 256- and 512-kHz cycle rates.

ratio, gain tracking, or harmonic distortion were detected when switching between RLA and DLA. This is not surprising because we anticipated only a 0.04-dB change in net gain and only a 0.3-dB change of noise level.

It is concluded that interpolating encoders and decoders using RLA companding can be interconnected with devices using DLA companding. Indeed, it is easier to compress the PCM code to DLA format than to RLA. This compression involves the following processes: converting the 13-bit code to sign and magnitude, adding the value 16.5 to the magnitude, and finally locating the most significant bit that is a ONE in the augmented magnitude. The position of the most significant ONE defines the segment, and the following four bits define the interval number.

Fig. 9 shows the signal-to-noise ratios that have been measured on a system composed of 1-bit encoder, code accumulation to 13-bit PCM, compression to 8-bit words and finally decoding with an interpolating decoder [1]. Curves are given for both 256- and 512-kHz cycle rates. The increase in the cycle rate to 512 kHz provides a 3-dB improvement in the PCM signal-to-noise ratio over most of the signal range. At the higher cycling rate noise in the output is largely due to round-off that occurs when the PCM code is compressed to 8-bit words and imperfections in the D/A converter. An auxiliary feedback connection, described in the Appendix, ensures that the compression characteristic of the encoder is centered with respect to the input signal.

For application to digital channel banks it may well be attractive to operate the coder at 512 kHz. This higher cycle rate can be accommodated using standard integrated circuit techniques and requires insignificant increased circuit complexity.

V. PERFORMANCE OF THE CODECS

A. Gain Tracking and Harmonic Content

Interpolative encoders and decoders of the kind described here and in [1] and [5] are very tolerant of imperfections in circuit components. It has been demonstrated that variations in components of the D/A circuit introduce relatively smooth nonlinearity rather than noise into the overall transmission. Thus requirements for accuracy of components are set by the need for adequate gain tracking and lack of harmonic distortion. Figs. 10 and 11 show the gain-tracking error and harmonic content of the output of the entire compressed PCM

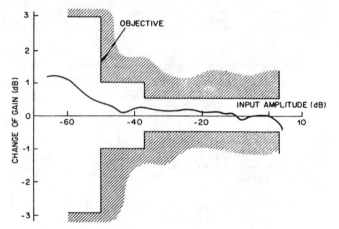

Fig. 10. Gain variations with input amplitude of the entire compressed PCM system.

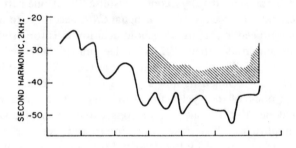
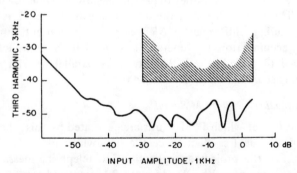

Fig. 11. Harmonic content expressed as a ratio of the input amplitude plotted against input, for the entire PCM system.

codec. These results illustrate that toll network requirements are easily met with the one percent tolerance of components and power supply that were used in the experimental encoder. The effects of changing the value of resistances in the decoding ladder network are discussed in the next section.

B. Sensitivity to Changes in the Ladder Network

Analysis of interpolative coding has shown that toll network requirements can be met with 95 percent certainty when component values in the D/A circuit have standard deviations less than 2 percent. The most critical requirements in this regard are tracking and second harmonic distortion. A demonstration of the one-bit encoder's robustness to imprecision in the D/A network is included in Figs. 12 and 13. These results were obtained by misaligning the D/A circuit in the receiving decoder. Only one resistor in the D/A ladder network was changed at a time; resistors R in legs of the ladder were reduced by 17 percent and resistors $4R$ in rungs of the ladder were reduced by 28 percent. Single resistor variations up to 20 percent appear to be permissible, in that the gain tracking is good to $\frac{1}{2}$ dB and harmonic content is less than 30 dB.

C. Sensitivity to Changes in the Stabilizing Time-Constant rC

In the encoder described here the stabilizing time constant rC was approximately equal to a cycle time, but this value was not critical. Large values of the time constant, $rC > 3T$, make the interpolation process sluggish when responding to amplitudes that barely exceed a segment boundary value. Instability occurs when the time constant is less than one half of the cycle period. Fig. 14 is a graph of signal-to-noise ratio from the one-bit codec for three values of the time constant.

Fig. 15 shows effects of changing rC on the signal-to-noise ratio of the entire PCM codec. The variations of time constant have slightly more significance for the PCM case than those reported in Fig. 14; nonetheless, variations of 2:1 are tolerable. For PCM operation we have found it advantageous to use time constants of the order one and a half times the cycle period because time constants less than a cycle period allow some instability at high frequency, certain components of which contribute to baseband noise when they are sampled at the PCM rate.

VI. SPECTRUM OF THE NOISE

An uncompanded encoder that interpolates values by oscillating between two levels of a coarse quantizer is analyzed in [5]; for frequencies less than half of the cycle rate, the noise in its output is shown to have an rms spectral density given by the first quarter cycle of a sinewave. Measurements of the output of the present encoder for a sinusoidal input indicate that this result is also approximately true for the companded case. Indeed, for frequencies less than a quarter of the cycle rate, the rms noise density is approximately proportional to frequency.

The linear dependence of noise density on frequency at low frequencies provides a convenient means for predicting the signal-to-noise ratio that would be obtained for basebands other than that defined by the C message weighting function. For example, there may be interest in doubling the baseband for communication within a private branch exchange (PBX). Cycling at 256 kHz would then provide a 35-dB signal-to-

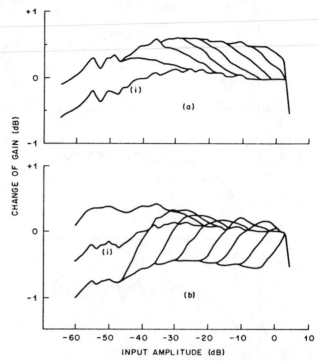

Fig. 12. Variation in gain of the 1-bit codec with changing amplitude of input. Curves (i) are the response when 1 percent components are used. Graph (a) shows the effects of reducing individual resistors in legs of the ladder by 17 percent. Graph (b) shows the effects of reducing individual resistors in rungs of the ladder by 30 percent.

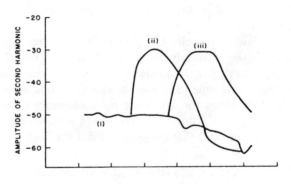

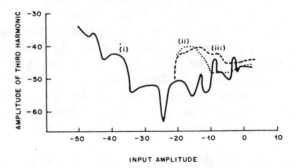

Fig. 13. Amplitudes of second harmonic distortion as a ratio of the input for a 1 kHz input sine wave. Curve (i) is for 1 percent resistors in rungs and 17 percent in legs of the ladder; (ii) and (iii) show effects of reducing rung resistors by 30 percent. Third harmonic distortion is also shown. Curve (i) is for 1 percent resistors in rungs and legs of the ladder; (ii) for reducing a rung resistor by 30 percent and (iii) for reducing a leg resistor by 17 percent.

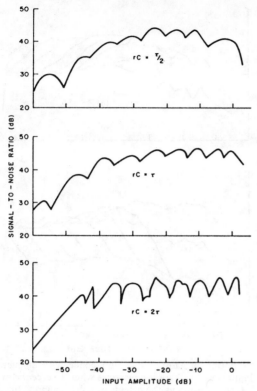

Fig. 14. Effects of changes in the stabilizing time constant rC on signal-to-noise ratio of the 1-bit codec.

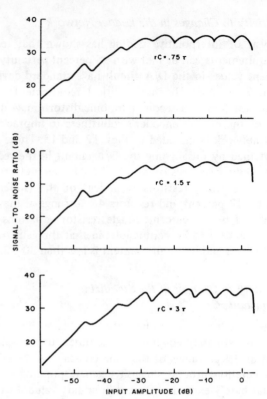

Fig. 15. Effects of changing the stabilizing time constant on the signal-to-noise ratio of the entire PCM system.

noise ratio over a 30-dB amplitude range. This has been confirmed by measurement and it was observed that slope overloading was not a problem for voice signals with the wider frequency band.

At very low frequencies the spectrum of the noise density is approximately inversely proportional to the gain of the integrating filter that accumulates the error signal. Therefore, leakage across the capacitor places a lower bound on the range of frequencies for which the noise spectrum is linear. There is little practical difficulty in achieving leakage time constants in the integrator that are greater than 1 ms.

VII. CONCLUDING REMARKS

We have described an interpolative one-bit encoder that, for telephone signals, is a significant improvement over the one described in [5]. Much of the advantage ensues from introduction of logarithmic companding, which allows use of a single threshold operating at 256 kHz, instead of the 16 thresholds operating at 16 MHz which might be necessary for uniform quantization. This one-bit encoder retains the advantages of being insensitive to variation of component values and requiring only simple accumulation for increasing the word length.

We have shown that an encoder constructed with one percent tolerance components and operating at 256 kHz easily meets requirements for toll transmission; and when the cycling rate is raised to 512 kHz the signal-to-noise ratio of the resulting PCM improves by 3 dB over most of the range of input amplitudes.

No mention has yet been made of the requirements for input filters. In applications where the one-bit code is not

converted to PCM, the requirement for low-pass characteristics are very relaxed because of the high-cycle rate. There may be, however, a need for high-pass filters to eliminate interference from of power lines. Conversion to PCM will require low-pass filters comparable with those presently used for channel banks, but they would be modified, as described in [1], to account for the $\sin x/x$ filtering that is inherent in the digital accumulation process.

The encoder described here and the decoder described in [1] are designed for segmented μ-law companding, but the techniques are equally appropriate for A-law companding. A-law requires that the segment boundaries in (1) be redefined by

$$Q_n[A\text{-law}] = 2^n.$$

The quantization levels (2) should then be set so that

$$q_n[A\text{-law}] = 2^n/3.$$

APPENDIX

Practical Details of a 1-Bit Coder

1) The Circuit: The circuit in Fig. 16 illustrates the important features of the 1-bit coder that was built and tested. At the upper left the analog input signal is fed to one input of a difference amplifier. A quantized representation of the input is applied to the other amplifier input. The difference amplifier and the associated integration filter are designed to function linearly. They are followed by a buffer amplifier that, at its output, clips all but the smallest signal amplitudes. The net voltage gain of these two cascaded amplifiers is about 200 for

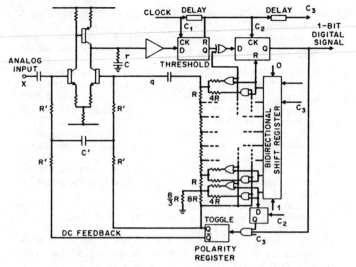

Fig. 16. Outline of the experimental coder that gave the results presented in preceding figures. Resistors R are the legs of the ladder network and resistors $4R$ are the rungs.

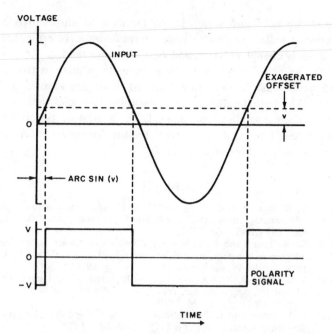

Fig. 17. Influence of dc offset on the duty cycle of the polarity signal.

a small input change of one cycle duration. The output amplitude when compared with the threshold of a D flip-flop, determines the 1-bit digital signal. The digital signal is fed to a holding register via an EXCLUSIVE-OR gate and controls the shifting of the bidirectional shift register.

In this design the bidirectional shift register contains a code that describes only the magnitude of the quantized signal. Whenever the register is empty and a further shift-down is requested, a toggle circuit changes the polarity of the quantized signal. The toggle also then inverts both the output of the shift register and the threshold decision signal controlling that register, thereby preserving negative feedback.

The method used to generate quantized amplitude from the content of the shift register requires no accurate analog gates or carefully matched circuit components. It makes use of the fact that telephone signals have no dc component by capacitively coupling the feedback and using only positive voltages in the ladder network. The effective polarity of the signal is controlled by digital gates. The smallest quantized magnitudes correspond to the state where the OR gates following the register are all on and the AND gates are all off. Increasing positive amplitudes are generated as the row of ONEs enter the shift register and turn on the AND gates in sequence. Effectively, negative amplitudes are generated by inhibiting the AND gates with a ZERO in the toggle and switching off OR gates as ONEs enter the register. When the shift register is empty the $8R$ resistor[2] connected directly to the toggle generates the innermost two levels of the quantized signal $\pm 1/3$. With these means the quantized amplitudes build up as a superposition of binary weighted increments. At no time does a signal increment depend on a difference of two larger quantities.

2) Setting Bias Levels: A liability of companding amplitudes is the need to center inputs with respect to the quantization scale. Bias levels in the coder circuit are established by a separate dc feedback loop that couples the polarity toggle to

[2] Changing the value of this resistor can match the companding to A-law or to that used in D-channel banks.

the input amplifier by way of a smoothing filter. This loop balances the amplifier to a state that makes the digital polarity positive for one half of any long time period. Effects of misalignment are easily calculated for particular cases; an example follows.

A conservative requirement for bias setting is an offset less than $1/32$ of the peak amplitude of any clean input sinewave. This ensures that the offset of the peak amplitude is less than the width of quantization interval. The correction signal, generated in response to such an offset is proportional to the duty ratio of the polarity signal, as illustrated in Fig. 17. If V is the high output level of the polarity toggle, then the average value of the polarity signal is given by

$$\frac{2V}{\pi} \arcsin \left(\frac{1}{32} \right) \approx \frac{V}{50} . \tag{5}$$

This voltage must compensate the offset i_o and v_o of the input amplifier, a requirement that sets the value R' of the resistors in the dc feedback,

$$(2R'i_0 + v_0) < \frac{V}{50} . \tag{6}$$

The capacitance that smooths the feedback must be sufficient to prevent significant distortion of the signal. For example, with sinusoidal input the square wave form generated by the toggle will result in a fixed-amplitude triangular waveform being fed back to the input. The amplitude of this distortion depends on the effective gain between the toggle and the input. In this case the distortion amplitude was restricted to less than one-third percent of the input amplitude. The severest requirement imposed by this restriction occurs for low-frequency inputs having a peak amplitude in the upper half of the second segment (amplitudes of the order

2.5 in the ±255 unit scale). Larger inputs mask the distortion whereas smaller amplitudes have interpolation waveforms that frequently change polarity, and are easily smoothed.

The effective amplitude of the triangular waveform that is fed back must be no more than 10^{-5} of the input signal range. The actual value of the smoothing time constant that meets these specifications depends on a number of circuit parameters, but satisfactory operation usually can be obtained with a time constant of the order 0.1 s.

REFERENCES

[1] G. R. Ritchie, J. C. Candy, and W. H. Ninke, "Interpolative digital-to-analog converters," *IEEE Trans. Commun.*, vol. COM-22, pp. 1797–1806, Nov. 1974.

[2] R. R. Laane and B. T. Murphy, "Delta modulation codec for telephone transmission and switching applications," *Bell Syst. Tech. J.*, vol. 49, pp. 1013–1031, July–Aug. 1970.

[3] I. M. McNair, "Subscriber loop multiplexor—A high pair gain system for upgrading and growth in rural areas," in *Proc. Nat. Electronics Conf.*, vol. 26, 1970.

[4] S. J. Brolin and G. E. Harrington, "The SLC-40 digital carrier subscriber system," in *IEEE INTERCON Dig.*, 1975.

[5] J. C. Candy, "A use of limit cycle oscillations to obtain robust analog-to-digital converters," *IEEE Trans. Commun.*, vol. COM-22, pp. 298–305, Mar. 1974.

[6] C. C. Cutler, "Transmission systems employing quantization," U.S. Patent 2 927 962, Mar. 1960.

[7] F. de Jager, "Deltamodulation, a method of PCM transmission using the 1-unit code," *Phillips Res. Rep.*, no. 7, pp. 442–446, 1952.

[8] H. Inose and Y. Yasuda, "A unity bit coding method by negative feedback," *Proc. IEEE*, vol. 51, pp. 1524–1535, Nov. 1963.

[9] R. C. Brainard and J. C. Candy, "Direct-feedback coders: Design and performance with television signals," *Proc. IEEE*, vol. 57, pp. 776–786, May 1969.

[10] P. T. Nielsen, "On the stability of a double intergration delta modulator," *IEEE Trans. Commun. Technol.* (Concise Papers), vol. COM-19, pp. 364–366, June 1971.

[11] J. C. Candy and B. A. Wooley, "An A/D converter with segmented companding," in *Proc. Nat. Telecommunications Conf.*, Dec. 1974, pp. 388–391.

[12] C. L. Dammann, L. D. McDaniel, and C. L. Maddox, "D2 channel bank: Multiplexing and coding," *Bell Syst. Tech. J.*, vol. 51, pp. 1675–1696, Oct. 1972.

[13] J. C. Candy, "Limiting the propagation of errors in one-bit differential codecs," *Bell Syst. Tech. J.*, vol. 53, pp. 1667–1676, Oct. 1974.

[14] H. Kaneko, "A unified formulation of segment companding laws and synthesis of codecs and digital compandors," *Bell Syst. Tech. J.*, vol. 49, pp. 1555–1588, Sept. 1970.

[15] D. J. Goodman, "The application of delta modulation to analog-to-PCM encoding," *Bell Syst. Tech. J.*, vol. 48, pp. 321–343, Feb. 1969.

A Use of Double Integration in Sigma Delta Modulation

JAMES C. CANDY, FELLOW, IEEE

Abstract—Sigma delta modulation is viewed as a technique that employs integration and feedback to move quantization noise out of baseband. This technique may be iterated by placing feedback loop around feedback loop, but when three or more loops are used the circuit can latch into undesirable overloading modes. In the desired mode, a simple linear theory gives a good description of the modulation even when the quantization has only two levels. A modulator that employs double integration and two-level quantization is easy to implement and is tolerant of parameter variation. At sampling rates of 1 MHz it provides resolution equivalent to 16 bit PCM for voiceband signals. Digital filters that are suitable for converting the modulation to PCM are also described.

I. INTRODUCTION

THIS paper describes the design of a digital modulator that is intended for use in oversampled PCM encoders. These encoders modulate their analog inputs into a simple digital form at high speed; then digital processing transforms the modulation to PCM sampled at the Nyquist rate [1]–[9].

Properties of the preliminary modulation have strong influence on the design of the entire encoder. For example, the resolution of the PCM can be no better than that of the modulation, and the complexity and speed of the digital processor depends on the kind of modulation used and its resolution. The tolerance of the analog circuits employed in the modulator can determine the suitability of the design for integrated circuit implementation and the power consumed by these circuits can be a large part of the power used by the entire encoder. There is, therefore, much incentive to find an efficient modulator, one that provides high resolution (idle channel noise more than 80 dB below peak signal) at moderate sampling rates (less than 1 MHz for 4 kHz telephone signals) yet employs simple robust circuits (tolerances no tighter than ±3 percent).

Early work on oversampled encoders [1], [2] was mostly theoretical and based on delta modulation. Later, practical realization preferred sigma delta modulation but modified it to lower the sampling rate and simplify the digital processing. For a video application, multilevel quantization was used [3] to reduce the modulation rate. For telephone applications, some modulators [4], [5] achieve high resolution by biasing the modulator to an especially quiet state for idle channel operation. One [7] employed triple integration in the sigma delta modulator; another [8] employed digital accumulation and companded quantization levels in the feedback path.

Recent advances in digital integrated circuit technology have greatly reduced the need to have simple digital processing; indeed, now it is feasible [10] to have digital line equalization, echo canceling, digital hybrids, and conferencing on the chip with the codec. These applications, however, place stringent demands on resolution and dynamic range to be provided by the modulator.

The present work explores the advantages of having double integration in a sigma delta modulator. We demonstrate that a particular class of circuits can provide high resolution and be tolerant to imperfection. We explain the reasons for using

Paper approved by the Editor for Signal Processing and Communication Electronics of the IEEE Communications Society for publication without oral presentation. Manuscript received March 28, 1984; revised October 16, 1984.

The author is with AT&T Bell Laboratories, Holmdel, NJ 07733.

multiple integration, and calculate the signal-to-noise ratios. The results are confirmed by simulation and experimental measurements. We show that when more than two integrators are used, the circuit can latch into undesirable modes where its performance is ruined. Finally, we give a design for a digital processor for constructing PCM from this modulation.

II. QUANTIZATION WITH FEEDBACK

Fig. 1(a) shows the circuit of a differential quantizer which is a form of the well-known delta modulator; we will use this circuit to explain our view of feedback modulation. Fig. 1(b) shows a sampled data model of the circuit; it assumes that the A/D and D/A conversion are ideal and that signals are random, so that the quantization may be represented by added noise e and linear gain G (level-spacing/threshold spacing). Accumulation A represents the integration. Mathematical descriptions of related circuits have been presented in several places [1], [3], [4], [12], [14]–[16]. They show that the presence of feedback around the quantizer has three uses, which are summarized below.

Prediction and Preemphasis: The modulated signal M comprises a noise component and a component that is proportional to the rate of change of input amplitude. Modulating a rate of change can be more efficient than modulating the amplitude directly, particularly for video and audio signals whose spectral densities fall with increasing frequency and whose sample values are highly correlated [15], [16]. This improved efficiency can result in decreased sampling rate or a reduction in the number of quantization levels needed for a given resolution.

Control of Overloading: Ordinary PCM quantization overloads by clipping signal amplitudes directly. When this happens to the signal applied to the A/D in Fig. 1, it is the derivative of the input signal that is clipped, resulting in slope overloading of the output signal. Distorting the slope of video [17] and audio signals can be less disturbing than clipping their amplitudes directly. This also can lead to improved efficiency of the modulation.

Noise Shaping: Placing the quantizer in a feedback loop with a filter shapes the spectrum of the modulation noise [3], [13], and at the same time it can decorrelate the noise from the signal. If we assume that the quantization noise e in Fig. 1(b) is white, then the spectral density of the noise in the modulated signal rises with frequency; but after integration the noise is white again at the output. We will see that other circuit configurations [13]–[18] can shape the output noise spectrum to suit particular applications.

When higher order filters are used in place of simple integration [11]–[14], the properties of the modulation are modified. The modulated signal includes components that are proportional to high-order derivatives of the signal. Overloading limits not only the slope but also the rate of change of slope of the signal, and modulation noise rises more steeply with frequency. The restriction on the design of these high-order filters is the need to keep the feedback stable.

All these properties of feedback quantization can influence the design of modulators for oversampled codecs, and in some applications the requirements are in conflict. For example, optimum design of predictors usually calls for leaky integration, but optimum noise shaping calls for long-time constant inte-

Reprinted from *IEEE Trans. Commun.*, vol. COM-33, no. 3, pp. 249–258, Mar. 1985.

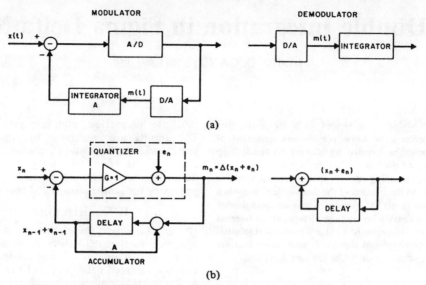

Fig. 1. (a) An example of a differential modulator and demodulator. (b) A sampled data representation of the differential modulation.

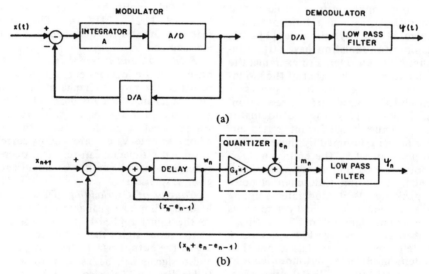

Fig. 2. (a) A sigma delta modulator and demodulator. (b) A sampled data representation of sigma delta modulation.

gration [15]. It is the different emphasis given to these separate properties that accounts for the different filters that have been proposed for feedback quantizers. Our design for the modulator will be based on the requirements of telephone toll-networks.

III. REQUIREMENT OF GENERAL PURPOSE MODULATORS FOR TOLL NETWORK USE

Digital codecs used in the telephone toll-network must accept a wide range of signals, and their design may not rely on properties of restricted classes of signals nor properties of special receivers. We may not rely on there being high correlation between Nyquist samples, nor assume that slope overloading is any more acceptable than clipping amplitudes or that a colored noise is less objectionable than white noise.

We may take advantage of the fact that the signal is band limited, however, by moving quantization noise out of band where it can be removed by appropriate filters. The sigma delta modulator shown in Fig. 2 does this without differentiating the signal: it eliminates the need for integration at the receiver because low-frequency components of the modulated signal represent the input amplitude directly. This structure

has gained favor bcause it is very tolerant of imperfection and mismatch of the two D/A circuits [3]. The structure of Fig. 1 is not so tolerant because its D/A imperfections are multiplied by the large baseband gain of the integrating filter at the receiver.

The next section of this work will be directed at the task of generalizing the filter A used in this modulator for the purpose of moving quantization noise out of the signal band. Applications that want to use signal prediction and special overload characteristics could do so by providing preemphasis and deemphasis filters external to the modulator [15].

IV. SIGMA DELTA MODULATION

The modulator shown in Fig. 2 generates a quantized signal that oscillates between levels, keeping its average equal to the average input. It is easy to show [3] that for active inputs, the spectral density of the noise in the quantized signal is given by

$$N_M(f) = (1 - z^{-1})E(z) = \sigma \sqrt{\frac{2\tau}{3}} \sin(\pi f \tau) \qquad (1)$$

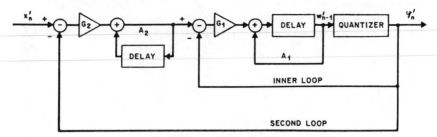

Fig. 3. A quantizer with two feedback loops around it. The gains of all elements are nominally unity.

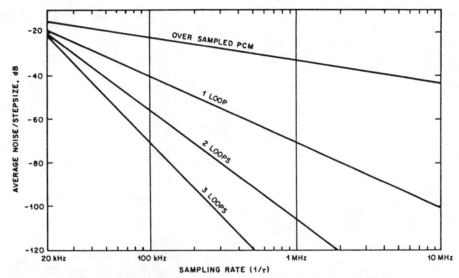

Fig. 4. Quantization noise plotted against sampling frequency for various numbers of feedback loops. Baseband is 3.5 kHz and the noise is referred to the step size: the noise of ordinary PCM is − 10.8 dB on this scale.

where σ is the quantization step size and τ the sampling period. The net noise in baseband $0 \leqslant f < f_0$ is then given approximately by

$$N_{MO} = \frac{\pi e_0}{\sqrt{3}} (2f_0\tau)^{3/2} \qquad (2)$$

provided that $f_0\tau \ll 1$ and $e_0 = \sigma/\sqrt{12}$ is the average noise generated by the quantizer alone. Thus, the resolution can be greatly increased by oversampling and feedback. For example, quantizing a 3.5 kHz signal at 16 MHz with this feedback quantizer reduces the noise e_0 by 96 dB, which is equivalent to a 2^{16} reduction of the step size.

This procedure for increasing the resolution with feedback can be reiterated as illustrated in Fig. 3. Here we have included gains G_n in cascade with each integrator in order to describe circuit imperfections that cause the loop gains to be other than unity. Appendix A shows that the spectral density of the noise in the modulated signal generated by two feedback loops is given by

$$N_M(f) = (1 - z^{-1})^2 E(z) = 2e_0\sqrt{2\tau}(1 - \cos{(2\pi f\tau)}) \qquad (3)$$

when $G_n = 1$. The in-band noise is given approximately by

$$N_{MO} = \frac{\pi^2 e_0}{\sqrt{5}} (2f_0\tau)^{5/2}. \qquad (4)$$

This double feedback increases the resolution by 95 dB for a sampling rate of only 1 MHz and 3.5 kHz baseband. Fig. 4 compares the resolution of PCM to that obtainable with one, two, and three feedback loops. Measurements on circuits with active input signals agree with these calculated values.

The circuits that can be derived by reiterating simple feedback loops of this kind are a very useful class of feedback quantizers with high-order filters [14], [18]. Ritchie points out the penalties that must be paid for using feedback, and they are summarized in the next section.

V. Penalties for Using Feedback

In Fig. 2 the signal applied to the quantizer can be expressed as the input less the noise from the previous cycle: $w_n = x_n - e_{n-1}$, and this noise uses up some of the dynamic range of the quantizer. Overloading may be avoided by adding one extra level to the quantizer because e spans $\pm\sigma/2$. In a similar manner it may be shown that, when the number of feedback loops is $L > 0$, the range of signals applied to the quantizer is increased by $2^{L-1}\sigma$. If this signal exceeds the range of the quantizer, the modulation noise increases, as is illustrated by Fig. 5. For many applications, the increase in noise for large signal values can be tolerated, provided an adequate signal-to-noise ratio is maintained.

Besides requiring additional quantization range, feedback demands increased precision in the gains of the circuits. Gains in the range ±10 percent are usually acceptable for quantizers having a single feedback loop, but more precise gains are needed when additional feedback loops are used. In Fig. 6 we plot the calculated and measured signal-to-noise ratio against values of gains G_n placed in series with each integrator for one, two, and three feedback loops. The measured change of noise with gain is larger and more variable than predicted by calculation. This is because the noise is correlated with signal amplitude in a way that depends on the gains G_n, and this correlation is ignored in the calculations.

The third penalty for using feedback concerns the depend-

311

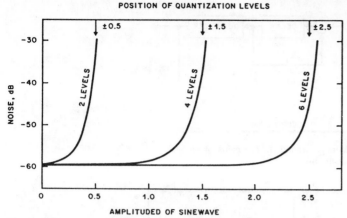

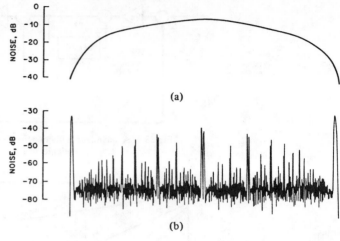

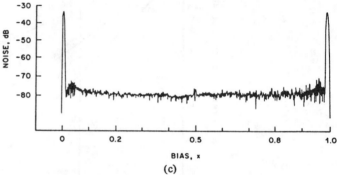

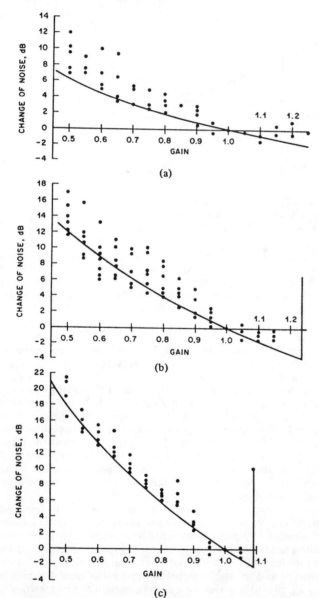

Fig. 5. The noise introduced into sinusoidal signals of various amplitudes by modulation with two feedback loops placed around a quantizer having the stated number of levels. Sampling is at 128 kHz.

Fig. 6. The change in noise with gain placed in cascade with each integrator for multilevel quantization. The reference noise is that measured at unity gain value. The curve gives calculated values that apply for sampling rates that are at least eight times the Nyquist rate. (a), (b), and (c) are for one, two, and three feedback loops, respectively. These circuits become unstable at gains 2, 1.236, and 1.087.

Fig. 7. Graphs of idle channel noise plotted against input bias. (a) Ordinary PCM. (b) Simple sigma delta modulation. (c) Two-level quantization with double feedback. Noise is referred to the step size. The sampling rate is 256 kHz.

ence of modulation noise on signal amplitude. A determination of this dependence for ordinary sigma delta modulation in [19] shows that it can have an important influence on the performance of oversampled codecs. Fig. 7 is a graph of noise plotted against the dc bias, x, applied to the modulator. Fig. 7(a) is the quantization error without feedback; the error is zero for $x = 0$ or 1 which corresponds to the position of two adjacent levels, and elsewhere the error is proportional to the distance to a level. Fig. 7(b) shows the idle channel noise of a quantizer with a single feedback loop. The noise is gathered into a series of narrow peaks, and the large peaks can be an embarrassment in the design of oversampled codecs.

VI. TWO-LEVEL QUANTIZATION

We will demonstrate how the three penalties, described in the previous section, can be avoided by using just two feedback loops and degrading the quantizer to a single threshold circuit that generates a two-level output.

With a single threshold the inconvenience of establishing threshold spacing is removed, and the concept of gain of the quantizer becomes unreal unless other circuit properties provide a calibration for the amplitudes applied to the threshold. We find by experiment that signal levels adjust themselves so that the effective gain of the quantizer compensates for changes in the values of circuit gains G_n. The measured noise shown in Fig. 8 is almost independent of these gains and corresponds to the calculated value for loop gains of unity. Likewise, the change in the attenuation of the signal $Y(\omega)/X(\omega)$ is less than ± 0.05 dB for values of gain $0.5 < G_n < 2$. We find that the penalty of having to establish quantizer gains in one and two feedback loops is substantially eliminated by the use of two-level quantization: signal levels in the circuit automatically adjust themselves to make the effective loop gains unity. With

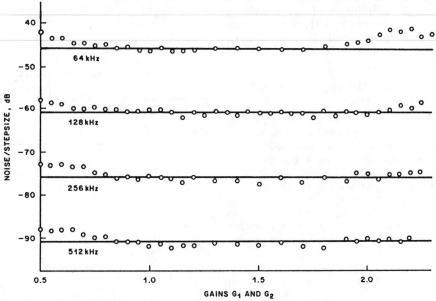

Fig. 8. The noise in double feedback, two-level modulation plotted against gains G_n at various sampling rates. The input was excited with a random signal spanning $\pm 0.1\sigma$. The horizontal lines mark calculated values of noise for unity gain and multilevel quantization with the same step size.

more than two feedback loops, undesirable oscillations spoil the functioning of the circuit when the quantizer has only two levels. We discuss these oscillations in the next section.

The second penalty concerns the loss of quantization range because feedback increases the amplitudes that are applied to the quantizer. With a single threshold there is no limitation on the range of its input amplitude, only the output levels are defined, but we see in Fig. 5 that for two-level quantization the noise increases rapidly with signal amplitude. Fig. 9 presents similar data plotted on logarithmic scales.

The third penalty concerned the correlation of noise with input level. Fig. 7(c) shows a graph of modulation noise plotted against input bias for a modulator utilizing two feedback loops. Comparing it to the graphs in Fig. 7(b), we see that use of two loops substantially decorrelates the noise except at the ends of the range, where the modulation noise peaks in the same fashion that it does in modulators that utilize only one feedback loop. Reference [19] shows that the amplitude of these peaks of noise is given by

$$N_{max} = \sqrt{2} \; (f_0\tau)\sigma \tag{5}$$

which can be large compared to the calculated noise (4). But their width v is narrow:

$$v = f_0\tau\sigma. \tag{6}$$

It appears that the use of two-level quantization and double integration could be the basis of a useful modulator. The next section explains why it is wise to use no more than two integrators.

VII. LIMIT-CYCLES THAT OVERLOAD THE QUANTIZER

The feedback quantizers that we have described cause their outputs to oscillate between levels in a way that keeps their average value equal to the average input. In the desired mode of operation the signals held in the integrators are comparatively small, but when three or more feedback loops are present, other modes can be excited [14]. These modes are characterized by being very noisy and having large-amplitude, low-frequency oscillations in the integrators, which exceed the range

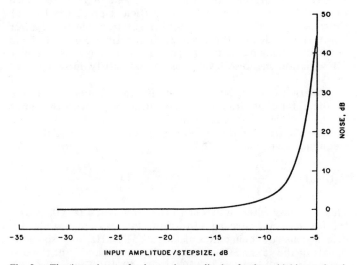

Fig. 9. The dependence of noise on the amplitude of a sinusoidal input signal for two-level modulation with double feedback. The noise is referred to its small signal value. This result applies to sampling rates that are eight or more times the Nyquist rate.

of the quantizer. In particular, when the quantizer has only two levels, the undesirable modes are easily excited and difficult to extinguish. The susceptibility of the circuit to enter undesirable modes has prevented measurement of the resolution of circuits that have triple integration and two-level quantization except for gains G_2, G_3 less than 0.55. Even for gains as low as 0.1, the unwanted modes were self-sustaining after being excited. When two feedback loops are used, however, the contents of the integrators always decay to a small value when excitations are removed, provided that the gains G_n are less than 1.23 for multilevel quantization and less than 2 for two-level quantization.

When modulators function in an undesirable mode, the signals in the integrators are so large, compared to the largest quantization levels, that the inner feedbacks are ineffective. The behavior of the circuit is dominated by the outermost feedback, which, if it contains more than two integrations, is unstable. Stability is regained when the inner feedbacks are made effective by clipping the amplitudes of signals held in the integrators, or by nonlinear feedback [14]. But it is questionable whether the extra resolution obtained in practice can justify the use of these more complicated circuits and the tighter tolerances that they demand.

VIII. IMPLEMENTATION OF THE MODULATORS USING TWO INTEGRATORS

The circuit shown in Fig. 3 is a sampled data model of the modulator; its signals are represented by impulse sample values. This circuit could well be implemented using switched capacitors for accumulation. Implementations in a bipolar technology, however, would prefer to use continuous signals such as those in the circuit of Fig. 10. The analysis of the switched and the continuous circuits in Appendix C shows that their operation is equivalent when the feedback signal $y(t)$ is held constant throughout the sample interval, the time constant RC equals 1.5τ, and the two inputs are related by the expression

$$x_{n\tau}' = \int_{(n-1)\tau}^{n\tau} (2x(t) - x(t - \tau))\, dt. \qquad (7)$$

The analog circuit in Fig. 10 is relatively easy to construct because there are only two main constraints on its design. There is the need to keep signals small in order to conserve power yet have signal levels large enough to swamp noise and imperfection in the threshold circuit, and there is the need to set the time constant RC with sufficient precision. Fig. 11 shows graphs of the modulator's resolution plotted against input amplitude for three values of the time constant. Time constants changing in the range 1.2τ to 1.8τ give less than 1 dB variation in noise level; this should satisfy most applications.

Equation (7) can be used to define the frequency response of the filter that should be placed in cascade with the input of the sampled data circuit in Fig. 3 in order to make its response identical to that of the analog circuit in Fig. 8. That frequency response is given by

$$G(\omega) = \frac{X'(\omega)}{X(\omega)} = (2 - \epsilon^{-j\omega\tau})\, \text{sinc}\,(f\tau). \qquad (8)$$

This low-pass filter, inherent to the circuit of Fig. 10, is useful for reducing aliasing distortion. For example, when 3.5 kHz signals are being modulated at 512 kHz, any spurious signals in the range 508–516 kHz alias into band. But the distortion is small because $|G(508)| < -40$ dB. Table I lists the attenuation of the signals aliased into band for various sampling frequencies; also listed is gain introduced into baseband. This gain could be equalized in the digital processor.

Simulations show that signal amplitudes in the integrators of these two-level feedback modulators can be very large, and real implementations need to limit their size. Clipping their amplitudes speeds the recovery from overload but may increase the noise. Fig. 12 shows that essentially full resolution of two-level quantization is obtained by allowing the integrated signals to swing through at least ±1.0 step sizes. Variation of the modulator's net gain with input amplitude was less than ±0.025 dB.

Fig. 13 shows that leakage in the integration has negligible effect on noise, provided its time constant exceeds $1/2f_0$ seconds. The data in this figure agree with calculations that represent quantization by added noise, and assumes unity gain in the feedback loops.

IX. DESIGN OF THE DIGITAL PROCESSOR

A digital processor will convert the output of the modulator into PCM by smoothing the signal with a digital low-pass filter

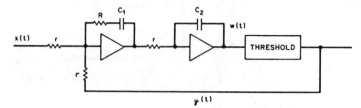

Fig. 10. An analog version of the double integrating modulator.

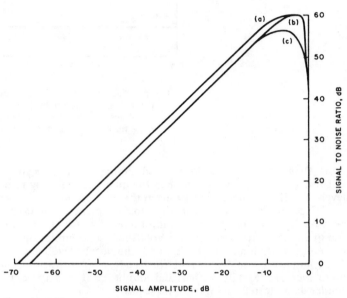

Fig. 11. The signal-to-noise ratio plotted against the amplitude of a sinusoidal input signal for the circuit in Fig. 10. The sampling rate is 256 kHz with 3.5 kHz baseband; the time constant RC_1 is (a) 1.5τ; (b) 2τ; (c) τ.

TABLE I
GAINS OF THE FILTER $(2 - \exp(-j2\pi f\tau))\, \text{sinc}\,(f\tau)$

SAMPLING FREQUENCY	GAIN AT 3.5 kHz	GAIN AT
$1/\tau$ kHz	dB	$(1/\tau - 3.5)$ kHz, dB
64	0.87	-23.9
128	0.24	-30.8
256	0.06	-37.1
512	0.015	-43.2
1024	0.004	-49.3
2048	0.001	-55.3
4096	0.00	-61.4
8192	0.00	-67.4

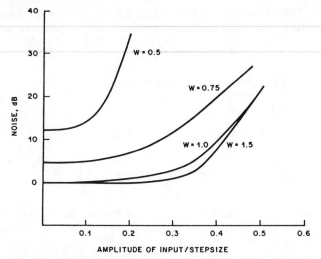

Fig. 12. The effect on noise of clipping the amplitude of signals in the two integrators for various amplitudes of a sinusoidal input. W is the ratio of the clipping level to the step size. Quantization levels are at ± 0.5.

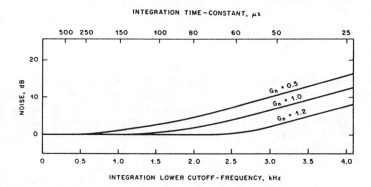

Fig. 13. The dependence of noise on leakage in both integrators for two-level quantization and double feedback. Baseband $f_0 = 3.5$ kHz.

and resampling it at the Nyquist rate. The filter attenuates spurious high-frequency components of the signal and the high-frequency components of the modulation noise, so that resampling does not alias significant noise into baseband.

Reference [9] describes the design of a digital processor that is suitable for use with ordinary sigma delta modulation. It reduces the sampling rate in stages, employing 32 kHz as an intermediate frequency for reaching the 8 kHz Nyquist rate of telephone signals. This technique leads to simple and efficient circuits. Use of double integration in the modulator does not influence the design of the filter with respect to out-of-band signal components, but it does influence the requirements for attenuating modulation noise, because the noise now rises more rapidly with increasing frequency.

The filter used in [9] for lowering the sampling rate to 32 kHz has a triangular-shaped impulse response, and frequency response given by

$$H(\omega) = \left(\frac{\sin fN\tau}{\sin f\tau} \right)^2. \qquad (9)$$

Calculations of noise, in Appendix B, show that such filtering does not provide sufficient attenuation for modulations generated by means of double integration. It is shown that the modulation frequency would need to be raised from 1 to 2.5 MHz in order to make up for use of such inadequate filtering.

A filter that does provide adequate attenuation has frequency response

$$H(\omega) = \left(\frac{\sin fN\tau}{\sin f\tau} \right)^3 \qquad (10)$$

and impulse response

$$h_n = \frac{n(n + 1)}{2} \qquad \text{for } 1 \leqslant n < N$$

$$h_n = \frac{N(N + 1)}{2} + (n - N)(2N - 1 - n)$$
$$\text{for } N \leqslant n < 2N$$

$$h_n = \frac{(3N - n - 1)(3N - n)}{2} \qquad \text{for } 2N \leqslant n < 3N$$

where N is the number of input sample values that occur in one period of the resampling. The duration of the filter impulse response is three resampling periods. Its frequency response has triple zeros at the 32 kHz resampling rate and all harmonics of it. When this filter is used, the noise, aliased into band, results in less than 0.5 dB loss of resolution. This filter can be implemented as an FIR structure that does not require full multipliers because the modulated signal is a 1 bit code. In the final stage of digital processing a sharp cutoff filter is needed. Analysis in Appendix C shows that at least 25 dB of attenuation is needed, which is provided by the filter described in [9]; it comprises two second-order low-pass sections followed by an accumulation and dump.

X. Conclusion

We have demonstrated that a sigma delta modulator with double integration can be designed to provide resolution equivalent to that of 16 bit PCM when modulating 4 kHz signals at 1 MHz, with simple robust circuits. A somewhat higher modulation rate, 2.5 MHz, would permit use of simpler digital processing given by (9) for converting the modulation to PCM.

It is instructive to compare this modulator to the interpolating modulator described in [8]. They provide comparable resolution when the interpolating modulator generates 9 bit words at 256 kHz and the sigma delta modulator generates 1 bit words at 1 MHz. The interpolative modulator has more complex circuits, but the tolerances of the two modulators to imperfection are roughly equivalent. The overriding advantage of the modulator described here is the fact that its quantization is uniform. To obtain comparable resolution from the interpolating modulator without companding its quantization levels would require that it generate 4 bit words at 2 MHz.

There is need to have uniform quantization in order that the encoder can accept the sum of several independent signals without having interaction between them. For example, when digital hybrids and digital conferencing are to be provided, the sum of three or more speech signals may be present in the modulator at one time; then we require that the quantization noise be independent of the signal amplitude. The possible use of triple integration has been rejected because it can latch into noisy modes of operation.

Appendix A
Modulation Noise

Quantization in the modulator shown in Fig. 3 is represented by additive noise e. We assume it is white with spectral power density $2\tau e_0^2$, where e_0 is the noise power in the band of frequencies below the half sampling rate. When the gains G_n are unity, we can describe the modulated signal by the

z-transform expression

$$Y'(z) = z^{-1}X'(z) + (1 - z^{-1})^2 E(z). \tag{11}$$

The noise in this signal has spectral density

$$N_M(f) = 2(1 - \cos(\omega\tau))e_0\sqrt{2\tau}. \tag{12}$$

If N_{M0} is the component of noise in baseband, $0 \leqslant f \leqslant f_0$ and $\omega_0 = 2\pi f_0$, then

$$N_{M0}^2 = \int_0^{f_0} N_M{}^2(f) \, df = \frac{e_0{}^2}{\pi} [6\omega_0\tau$$
$$- 8\sin(\omega_0\tau) + \sin(2\omega_0\tau)].$$

When $f_0\tau$ is small enough that $\sin(2\omega_0\tau)$ can be approximated by the first five terms in its Taylor expansion,

$$N_{M0} = \frac{e_0\pi^2}{\sqrt{5}} (2f_0\tau)^{5/2}. \tag{13}$$

This noise can be made very small by using high modulation rates. Subsequent digital processing, called decimation, lowers the rate without increasing the noise.

APPENDIX B
DESIGN OF THE DECIMATOR

The decimator uses low-pass filters to attenuate out-of-band components of the modulation that will be aliased into band by the resampling. To simplify the design of these filters, there is much advantage in lowering the sampling rate in stages. In the initial stages where the sampling rate is still large compared to the Nyquist rate, it is wise to place zeros of the filters at the new sampling rate and harmonics of it. Suitable sequences of evenly spaced zeros occur in trigonometric functions, and a particularly easy spectral response to implement is $(\sin \omega N\tau / \sin \omega\tau)$. In the time domain it is an averaging of N samples with τ being the period of the input samples and $N\tau$ the period of the output samples.

In [9] a cascade of two such filters having response

$$\frac{1}{N^2}\left(\frac{\sin \omega N\tau}{\sin \omega\tau}\right)^2 = \frac{1}{N^2}\left(\frac{1 - z^{-N}}{1 - z^{-1}}\right)^2 \tag{14}$$

was used to lower sampling rates of sigma delta modulation to 32 kHz. This filter is inadequate for use with modulators employing double integration, as the following analysis will demonstrate. After filtering by (14), the modulation noise in (11) becomes

$$N_D(z) = \frac{e_0\sqrt{2\tau}}{N^2} (1 - z^{-N})^2. \tag{15}$$

Resampling with period $N\tau$ results in

$$N_D(Z) = \frac{e_0\sqrt{2N\tau}}{N^2} (1 - Z^{-1})^2 \tag{16}$$

where $Z^{-1} = z^{-N}$ represents a one-period delay at the new rate. Following arguments similar to (11)–(13), the in-band noise can be approximated by

$$N_{D0} = N^{1/2} \frac{e_0\pi^2}{\sqrt{5}} (2f_0\tau)^{5/2}. \tag{17}$$

Thus, the noise (13) is increased by the root of the decimation ratio N. This loss of resolution could be made up for by increasing the input modulation rate from $1/\tau$ to $N^{1/4}/\tau$.

A better decimating filter is

$$\frac{1}{N^3}\left(\frac{1 - Z^{-N}}{1 - Z^{-1}}\right)^3. \tag{18}$$

It modifies the modulation noise to be

$$N_D(z) = \frac{1}{N^3} (1 - z^{-N})^2 \left(\frac{1 - z^{-N}}{1 - z^{-1}} E(z)\right). \tag{19}$$

The spectral density of the noise following resampling can be obtained by reverting to the time domain. Let $E'(z)$ represent the accumulated noise

$$E'(z) = \left(\frac{1 - z^{-N}}{1 - z^{-1}}\right) E(z) \tag{20}$$

which is equivalent to

$$e'(i\tau) = \sum_{n=0}^{N} e((i - n)\tau). \tag{21}$$

When this is resampled with period $N\tau$, i.e., every Nth sample is retained, there will be no correlation between the samples, if the original samples are uncorrelated. The resampled noise $e'(iN\tau)$ will be white with spectral density $Ne_0\sqrt{2\tau}$. It follows that the noise at the decimator output after resampling can be described as

$$\frac{1}{N^{5/2}} (1 - Z^{-1})^2 e_0\sqrt{2N\tau} \tag{22}$$

and its baseband component for $N\tau f_0 \gg 1$ is approximated by

$$N_{D0} = \frac{e_0\pi^2}{\sqrt{5}} (2f_0\tau)^{5/2}. \tag{23}$$

There is no change from (13). All of the penalty for decimating is represented by the in-band attenuation of the decimating filter which, when equalized, increases the noise.

This decimating filter can be used successfully to lower sampling rates to about four times the Nyquist rate, and its in-band attenuation will be less than 3 dB. Filters with much sharper cutoff characteristics are needed in the final stage of decimation. Their attenuation R must be sufficient to make the modulation noise that is aliased into band small with respect to N_{M0} in (13). That is,

$$\frac{2N\tau}{N^5} e_0{}^2 \int_{f_0}^{\frac{1}{2N\tau}} |R(\omega)|^2 (1 - Z^{-1})^4 \, df$$

$$\ll \frac{e_0{}^2\pi^4}{5} (2f_0\tau)^5 \tag{24}$$

which is satisfied if

$$|R(\omega)| \ll \frac{\pi^2}{\sqrt{30}} (2f_0N\tau)^{5/2}. \tag{25}$$

When $1/N\tau$ is 32 kHz and f_0 is 4 kHz, this requires $R \leqslant -25$ dB. This is easier than the requirement for antialiasing filters in D-channel banks, $R \leqslant -32$ dB. The low-pass filter described in [9] is adequate for use with double integrating modulators.

APPENDIX C
COMPARISON OF A MODULATOR THAT INTEGRATES ANALOG SIGNALS WITH ITS SAMPLED DATA EQUIVALENT

When the modulator in Fig. 3 has input samples $x_n{}'$ and output samples $y_n{}'$ and the signal applied to the quantizer is $w_{n-1}{}'$, a relationship between their values can be expressed as

$$w_N{}' = \sum_{n=0}^{N-1} \sum_{0}^{n} (x_n{}' - y_n{}') - \sum_{n=0}^{N-1} y_n{}' \tag{26}$$

or

$$w_N{}' = \sum_{n=0}^{N-1} (N-n)x_n{}' - \sum_{n=0}^{N-1} (N-n+1)y_n{}' \tag{27}$$

provided all signals are initially zero.

An analog circuit that can have equivalent response is shown in Fig. 10. Here $x(t)$ is a continuous signal and the feedback signal $y(t)$ is held constant throughout each sample interval. To analyze this circuit we make use of the following relationships between integration and summation of sample values.

Lemmas: For $f(t) = 0$ when $t \leqslant 0$ and

$$f_{n\tau} = \frac{1}{\tau} \int_{n\tau}^{(n+1)\tau} f(t)\, dt, \tag{28}$$

$$\frac{1}{\tau} \int_{0}^{n\tau} f(t)\, dt \equiv \sum_{n=0}^{N-1} f_{n\tau} = \sum_{n=0}^{N-1} (N-n)(f_{n\tau} - f_{(n-1)\tau}) \tag{29}$$

and

$$\frac{1}{\tau} \int_{0}^{N\tau} dt \int_{0}^{t} f(t)\, dt = \sum_{n=0}^{N-1} (N\tau - t_n)f_{n\tau} \tag{30}$$

where t_n is the position, in time, of the center of area of the signal waveform during the nth sample interval, i.e.,

$$f_{n\tau}t_n = \frac{1}{\tau} \int_{n\tau}^{(n+1)\tau} tf(t)dt. \tag{31}$$

When we assume that all signals in the circuit are zero for $t \leqslant 0$, we can express the amplitude that is applied to the quantizer in Fig. 3 at the Nth sample time as

$$w(N\tau) = \frac{1}{r^2 C_1 C_2} \int_{0}^{N\tau} \left[\int_{0}^{t} (x(t) - y(t))\, dt \right.$$

$$\left. + RC_1(x(t) - y(t)) \right] dt. \tag{32}$$

Applying results (13)–(15) and noting that because $y(t)$ is held constant, the center of area of its waveform lies mid-

way in the sample time $(n + \frac{1}{2})\tau$, we get

$$\frac{r^2 C_1 C_2}{\tau^2} w(N\tau) = \sum_{n=0}^{N-1} \left(\frac{\tau N - t_n}{\tau} + \frac{RC_1}{\tau} \right) x_{n\tau}$$

$$- \sum_{n=0}^{N-1} \left(N - \left(n + \frac{1}{2} \right) + \frac{RC_1}{\tau} \right) y_{n\tau}. \tag{33}$$

We are primarily interested in modulators that are sampled at very high frequency compared to baseband so that the signal varies little during the sampling interval. To simplify the analysis, we shall now assume that the center of area of the input waveform for each sample lies midway in the interval $t_n = (n + \frac{1}{2})\tau$. Then

$$\frac{r^2 C_1 C_2}{\tau} w(N\tau)$$

$$= \sum_{n=0}^{N-1} (N-n) \left[x_{n\tau} + \left(\frac{RC_1}{\tau} - \frac{1}{2} \right)(x_{n\tau} - x_{(n-1)\tau}) \right]$$

$$- \sum_{n=0}^{N-1} \left(N - n + \left(\frac{RC_1}{\tau} - \frac{1}{2} \right) \right) y_{n\tau}. \tag{34}$$

Comparing this result to (27) we see that they can be equivalent, with the right initial conditions and

$$r^2 C_1 C_2 = \tau^2, \quad RC_1 = 1.5\tau, \quad y_n{}' = y_{n\tau}$$

and

$$x_n{}' = x_{n\tau} + \left(\frac{RC_1}{\tau} - \frac{1}{2} \right)(x_{n\tau} - x_{(n-1)\tau}).$$

This leads to the final result

$$x_n{}' = 2x_{n\tau} - x_{(n-1)\tau} \tag{35}$$

which has spectral equivalent

$$X'(\omega) = (2 - e^{-j\omega\tau}) \operatorname{sinc}(f\tau)X(\omega). \tag{36}$$

REFERENCES

[1] D. J. Goodman, "The application of delta modulation to analog-to-digital PCM encoding," *Bell Syst. Tech. J.,* vol. 48, pp. 321–343, Feb. 1969.

[2] L. D. J. Eggermont, "A single-channel PCM coder with companded DM and bandwidth-restricting filtering," in *Conf. Rec., IEEE Int. Conf. Commun.,* June 1975, vol. III, pp. 40-2–40-6.

[3] J. C. Candy, "A use of limit cycle oscillations to obtain robust analog-to-digital converters," *IEEE Trans. Commun.,* vol. COM-22, pp. 298–305, Mar. 1974.

[4] J. D. Everhard, "A single-channel PCM codec," *IEEE J. Solid-State Circuits,* vol. SC-11, pp. 25–38, Feb. 1979.

[5] T. Misawa, J. E. Iwersen, and J. G. Rush, "A single-chip CODEC with filters, architecture," in *Conf. Rec., IEEE Conf. Commun.,* June 1980, vol. 1, pp. 30.5.1–30.5.6.

[6] J. C. Candy, Y. C. Ching, and D. S. Alexander, "Using triangularly weighted interpolation to get 13-bit PCM from a sigma-delta modulator," *IEEE Trans. Commun.,* vol. COM-24, pp. 1268–1275, Nov. 1976.

[7] L. vanDe Meeberg and D. J. G. Janssen, "PCM codec with on-chip digital filters," in *Conf. Rec., IEEE Int. Conf. Commun.,* June 1980, vol. 2, pp. 30.4.1–30.4.6.

[8] J. C. Candy, W. H. Ninke, and B. A. Wooley, "A per-channel A/D converter having 15-segment μ-255 companding," *IEEE Trans. Commun.,* vol. COM-24, pp. 33–42, Jan. 1976.

[9] J. C. Candy, B. A. Wooley, and O. J. Benjamin, "A voiceband codec with digital filtering," *IEEE Trans. Commun.,* vol. COM-29, pp. 815–830, June 1981.

[10] R. Apfel, H. Ibrahim, and R. Ruebush, "Signal-processing chips enrich telephone line-card architecture," *Electronics,* vol. 55, pp. 113–118, May 5, 1982.

[11] R. Steele, *Delta Modulation Systems.* New York: Wiley, 1975, ch. 3.

[12] F. deJager, "Delta-modulation, a method of PCM transmission using the 1-unit code," *Philips Res. Rep.,* vol. 7, pp. 442–466, 1952.

[13] H. A. Spang and P. M. Schultheiss, "Reduction of quantization noise by use of feedback," *IRE Trans. Commun. Syst.,* vol. CS-10, pp. 373–380, Dec. 1962.

[14] S. K. Tewksbury and R. W. Halloch, "Oversampled, linear predictive and noise-shaping coders of order $N > 1$," *IEEE Trans. Circuits Syst.,* vol. CAS-25, pp. 436–442, July 1978.

[15] R. C. Brainard and J. C. Candy, "Direct-feedback coders: Design and performance with television signals," *Proc. IEEE,* vol. 37, pp. 776–786, May 1969.

[16] B. S. Atal, "Predictive coding of speech at low bit rates," *IEEE Trans. Commun.,* vol. COM-30, pp. 600–614, Apr. 1982.

[17] J. C. Candy and R. H. Bosworth, "Methods for designing differential quantizers based on subjective evaluations of edge business," *Bell Syst. Tech. J.,* vol. 51, pp. 1495–1516, Sept. 1972.

[18] G. R. Ritchie, "Higher order interpolative analog to digital converters," Ph.D. dissertation, Univ. Philadelphia, Pennsylvania, 1977.

[19] J. C. Candy and O. J. Benjamin, "The structure of quantization noise from sigma delta modulation," *IEEE Trans. Commun.,* vol. COM-29, pp. 1316–1323, Sept. 1981.

MOS ADC-Filter Combination That Does Not Require Precision Analog Components*

Max W. Hauser, Paul J. Hurst**, Robert W. Brodersen

University of California

Berkeley, CA

AN OVERSAMPLE-AND-DECIMATE ARCHITECTURE yielding a analog-to-digital interface system with a built-in mask-programmable digital antialias filter in standard 5V digital MOS technology will be reported. Unlike other MOS A/D converter techniques, this approach requires no precision component ratios or precision comparator, nor a double-polysilicon or other special capacitor structure. Prototype devices display approximately 12b converter linearity at an 8kHz output rate with a single 5V power supply and a total die area of 4 8mm^2. The low supply voltage and simplified process requirements are attractive for scaled MOS fabrication technologies. This is intended as a linear (non-companding) analog-interface block for constant-sampling-rate applications such as speech processing.

The system is based on 1b A/D conversion, downsampled 256 times by a 1024-point, finite-impulse-response (FIR) digital lowpass filter. Such a configuration can display a net resolution exceeding the performance of the internal analog components, and since the bulk of the circuitry is digital, it benefits from continued technology scaling. It differs also from *interpolative* A/D circuits[1], in that it requires no internal D/A converter, nor does it produce a coarser quantization with larger inputs.

A delta-sigma-modulator front end, used elsewhere in discrete[2], bipolar[3] and passive grounded-capacitor MOS[4] circuitry, has been realized with a parasitic-insensitive switched-capacitor integrator; Figure 1. The parasitic insensitivity allows

---

*Research supported by DARPA contract N00034-K-0251, and by Fannie and John Hertz Foundation.

**Current address: Silicon Systems, Inc., Nevada City, CA.

[1] Candy, J.C., Ninke, W.H., and Wooley, B.A., "A Per-Channel A/D Converter Having 15-Segment μ-255 Companding", *IEEE Trans. Communications*, Vol. COM-24, No. 1, p. 33-42; Jan. 1976.

[2] Everard, J.D., "A Single-Channel PCM Codec", *IEEE Journal of Solid-State Circuits*, Vol. SC-14, No. 1, p. 25-37; Feb., 1979.

[3] van de Plassche, R.J., "A Sigma-Delta Modulator as an A/D Converter", *IEEE Trans. Circuits and Systems*, Vol. CAS-25, No. 7, p. 510-514; July, 1978.

[4] Misawa, T., Iwersen, J.E., Loporcaro, L. J., and Ruch, J. G., "Single-Chip per Channel Codec with Filters Utilizing Δ-Σ Modulation", *IEEE Journal of Solid-State Circuits*, Vol. SC-16, No. 4, p. 333-341; Aug., 1981.

[5] Candy, J.C., Ching, Y.C., and Alexander, D.S., "Using Triangularly Weighted Interpolation to Get 13-Bit PCM from a Sigma-Delta Modulator", *IEEE Trans. Communications*, Vol. COM-24, No. 11, p. 1268-1275; Nov., 1976.

[6] Ruetz, P.A., Pope, S.P., Solberg, B., and Brodersen, R.W., "Computer Generation of Digital Filter Banks", *ISSCC DIGEST OF TECHNICAL PAPERS*, p. 20-21; Feb., 1984.

good performance from metal/polysilicon capacitors, and the basic circuit requires two capacitors, of similar size, which need not be accurately ratioed.

This front end operates at a sampling rate much higher than the input signal frequencies. It produces a 1b signal that can be expressed as a linear representation of the input (between $+V_{REF}$ and $-V_{REF}$) along with a noise component having a highpass spectrum. A special-purpose digital lowpass filter operates on this 1b signal to remove frequencies above the desired signal bandwidth, and hence most of the quantization noise. Its output is a multibit digital representation at a lower (decimated) sampling rate.

Basic delta-sigma modulator systems generate severe noise components for certain input values; previous designs circumvented this by biasing the input to a low-noise region[4,5] or introducing a squarewave dither signal at a frequency that aliased to dc upon decimation[2]. These methods are unsatisfactory for a general-purpose system that must accommodate dc inputs. Here, a squarewave dither frequency within the decimating filter's stopband randomizes the delta-sigma noise as desired, but does not propagate to the output nor preclude arbitrary dc inputs.

Figure 2 shows the circuit of the switched-capacitor front end in 3μ P-well CMOS technology. Its active area is 0.3mm^2 and its power consumption approximately 2mW. A folded-cascode integrator gain stage drives a single-ended regenerative comparator. In the ϕ_1 clock phase, the integrator output settles and precharges the comparator input. In the ϕ_2 phase, the comparator's positive feedback loop is closed and it latches according to the polarity of the integrator output with respect to a threshold. Regenerative comparators, although fast, are rarely used in analog circuits since their offset voltage is difficult to cancel. In this application, the offset voltage is unimportant, since it does not affect the converter's performance, but merely changes the average voltage at the integrator output. Figure 3 is a die photograph of the front-end circuit.

We have fabricated both CMOS and NMOS front-end designs and concluded that CMOS is far preferable, because of the need for speed and gain in the integrator op amp. Nevertheless, the digital filter is realized as a separate NMOS chip in our prototype circuits, to exploit computer-assisted layout tools[6] now being extended to CMOS.

The architecture of the digital filter (Figure 4) exploits the 256:1 decimation factor and 1b digital input. A 1024-point impulse response, of which a symmetric half is stored in a read-only memory, is distributed to four accumulators. No explicit multiplications are necessary because of the one-bit input signal.

The impulse-response coefficients were developed by computer optimization to satisfy the dual objectives of quantization-noise removal and antialias filtering. The prototypes yield a minimum alias rejection of 46dB, consistent with com-

Reprinted from *1985 IEEE Int. Solid-State Circuits Conf.*, pp. 80–81, 313, Feb. 1985.

puter simulation, and only a simple single-pole prefilter is required before the 2MHz analog input sampling. Because FIR lowpass filters are relatively insensitive to coefficient roundoff, six-bit coefficients are adequate to define the impulse response. The die photograph of Figure 5 shows an area of $4.5mm^2$ in 4μ silicon-gate NMOS technology. We expect a 3μ CMOS version, currently under development, to occupy similar area.

The raw 1b delta-sigma output spectrum from dc to 200kHz appears in Figure 6. Desired signal frequencies extend from dc to 4kHz; the rest is unwanted quantization noise, to be removed by the digital filter. A 500Hz input sinusoid produces a 0dB peak very near the lefthand axis. Smaller peaks on the left are

the fundamental and third harmonic of a 16kHz squarewave dither signal; these too will be removed by the filter. Note the highpass trend in the quantization-noise spectrum.

Figure 7 shows the measured signal-to-noise ration performance. Input is a 500Mz sinusoid; input sampling rate is 2MHz and the output rate is 8kHz. The A/D system maintains a 30dB SNR over a 40dB amplitude range and approaches within 2dB the performance of a trimmed conventional 12b converter tested under the same conditions. Since the oversampled system exhibits its errors as ac noise, rather than dc nonlinearity, an SNR/amplitude plot is a more meaningful measure, for the intended signal-processing applications, than the static linearity specifications associated with standard (dc) A/D converters.

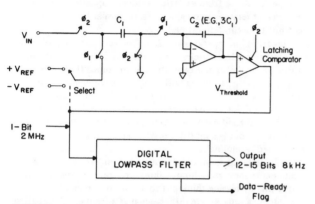

FIGURE 1—System overview showing switched-capacitor input section.

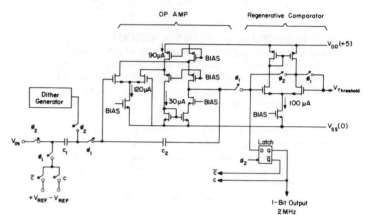

FIGURE 2—Schematic of CMOS switched-capacitor delta-sigma modulator.

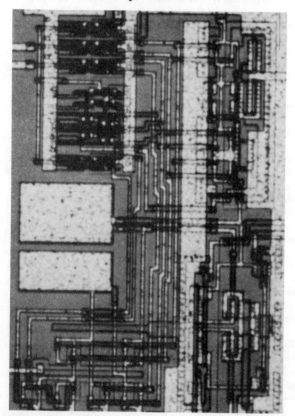

FIGURE 3—Photograph of CMOS switched-capacitor delta-sigma modulator. Dimensions are 0.45mm x 0.65mm.

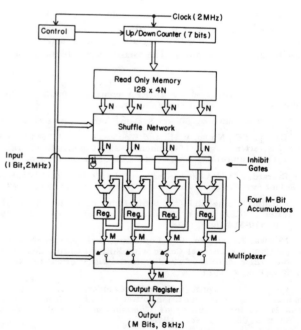

FIGURE 4—Architecture of 1024-point FIR digital filter. Typically N = 6, M = 15.

FIGURE 5—Photograph of 1024-point FIR digital filter.
Dimensions are 1.5mm x 3.0mm.

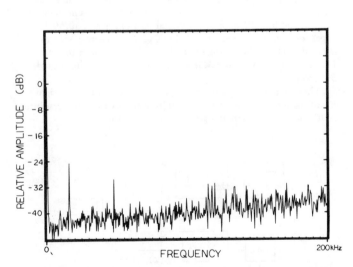

FIGURE 6—Low-frequency detail of 1b modulator output
spectrum.

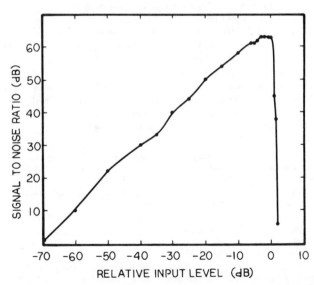

FIGURE 7—SNR performance of modulator-filter combination.
Input is a 500Hz sinusoid.

321

A Voiceband 15b Interpolative Converter Chip Set

Kazuo Yamakido, Sigeo Nishita, Masaru Kokubo, Hirotoshi Shirasu, Ken'ichi Ohwada, Tatsuya Nishihara

Hitachi Central Research Laboratory/Musashi Works

Tokyo, Japan

OVERSAMPLING ANALOG-TO-DIGITAL converters (ADCs) and digital-to-analog converters (DACs) for digital codecs previously developed typically show about 13b resolution, while some make use of nonlinear encoding techniques[1,2,3,4,5]. In practical use, however, linear 15b resolution is necessary in order that unavoidable outband (50/60Hz) input noise and internal digital signal processing, such as filtering and gain control do not degrade system characteristics. Furthermore, A/D and D/A converters should be realized using lower sampling frequencies and a single 5V power supply, since fine pattern VLSI technology is s must for digital codec implementation.

After careful investigation, an interpolative encoding structure with a multilevel quantizer has been adopted: Figure 1. Here, 1024kHz sampling, a 4 level (i.e., $\pm V_{ref}/128$, $\pm V_{ref}/32$) quantizer, and an 8b local DAC were used. The signal and noise transfer function is given by:

$$X(z) = z^{-1}(2 - z^{-1})W(z) + z^{-1}(1 - z^{-1})Q(z)$$

where $X(z)$, $W(z)$ and $Q(z)$ are output signal, input signal and quantization noise, respectively. The combination of noise-shaping with first-order prediction and 4-level quantization can effectively decrease in-band quantization noise while avoiding slope-overload.

The decimation filter converts the sampling frequency to 32kHz, and is composed of cascaded first and second-order Infinite Impulse Response (IIR) filters. To decrease the logic gates, the first one shares a parallel adder with the digital integrator of othe ADC.

The interpolation filter converts the sampling frequency from 32kHz to 512kHz. To restrain round-off noise while using a lower clock pulse frequency, parallel operation, whereby 8b each are allotted for the upper and lower stages, is used, as shown in Figure 2.

The digital interpolative encoder has the same structure as the ADC. A still-lower sampling frequency, 512kHz, was used to relax the severe settling time requirements for the following DAC. Instead, a 9b quantizer and 9b feedback loop filter with a peak limiter, were adopted. This 9b quantizer can be realized using simple inverter gates.

The circuit configuration is shown in Figure 3 for the ADC corresponding to the structure in Figure 1. To save the chip area, the 8b local DAC consists of two selecters, a resistor string with 8 taps and a 4b binary weighted capacitor-array with two additional capacitors. One of these adds the lower 3b voltage generated by the resistor string, and the other subtracts the output voltage of the local DAC from the input signal voltage. The selecters generate the output voltage with both positive and negative polarity using a single voltage reference. A capacitor, C_2, acts as an adder for first-order prediction. Three comparators generate 4-level output through a following logic decoder. The threshold, V_{th}, is set at $1/32\ V_{ref}$. The circuit configuration for the DAC in Figure 2 is similar to that for the local DAC in the ADC.

Chip photographs of samples using 2μ CMOS technology are shown in Figure 4. In the left chip, the ADC is integrated with the first decimation filter in an area $2.0 \times 2.8mm^2$, while in the right chip a DAC and sample-hold circuit are integrated within $2.0 \times 1.9mm^2$. The power supply, reference (V_{ref}) and internal analog ground (VB) voltage are +5V, +1.2V and +2.1V, respectively.

The ADC and DAC performance were verified by computer simulation. Following this, evaluations of sample ADC chips with first decimation filters connected to sample DAC chips were made. Here, the decimation filter was operated with 512kHz clear pulse for the interface into the DAC chip with 512kHz sampling. As shown in Figure 5, the measured signal-to-noise dynamic range was 93dBmOp which corresponds to 15b resolution. Idle channel noise was also -93dBmOp.

To evaluate actual performance when implemented in a system, overall characteristics were measured after placing the chips in a breadboard emulating a portion of a μ-law PCM digital codec. Measured characteristic (Figure 6) shows signal-to-noise ratio for the transmitter with gain settings, from 0 to +7dB, in 3dB steps in the analog portion and 0.25dB steps in the digital portion. Here, 50/60Hz superposition took place up to 70mVOp, which corresponds to -21.7dBmO, on the analog input signal. Power dissipation was 10.7mW for the ADC and 5.5mW for the DAC chip using a single 5V power supply.

These data exceed the severe objective specifications by significant margins, demonstrating the effectiveness of the 15b resolution ADC and DAC for actual digital codec applications.

[1] Wooley, B.A. and Henry, J.L., "An Integrated Per-channel PCM Encoder Based on Interpolation", *IEEE Journal of Solid-State Circuits*, Vol. SC-14, p. 14-20; Feb., 1979.

[2] Kuwahara, H., et. al., "An Interpolative PCM Codec with Multiplexed Digital Filters", *IEEE Journal of Solid-State Circuits*, Vol. SC-15, p. 1014-1021; Dec., 1980.

[3] Candy, J.C. and Wooley, B.A., "A Voiceband Codec with Digital Filtering", *IEEE Trans. Communications*, Vol. COM-29, No. 6, p. 815-830; June, 1981.

[4] Misawa, T., et. al., "A Single-Chip Per-Channel Codec with Filters Utilizing Δ-Σ Modulation", *IEEE Journal of Solid-State Circuits*, Vol. SC-16, No. 4, p. 333-341; Aug., 1981.

[5] Schmid, E. and Reisinger, J., "High Performance Oversampling A/D-Converter for Digital Signal Processing Applications", *ESSCIRC '84 Conference*, p. 201-204; 1984.

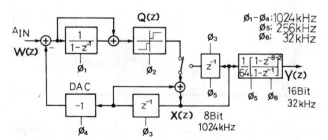

FIGURE 1—Structure for interpolative ADC with 4-level quantizer and decimation filter.

Reprinted from *1986 IEEE Int. Solid-State Circuits Conf.*, pp. 180-181, Feb. 1986.

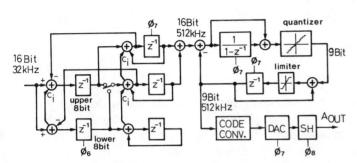

FIGURE 2—Structure for interpolation filter and interpolative DAC with 9b quantizer.

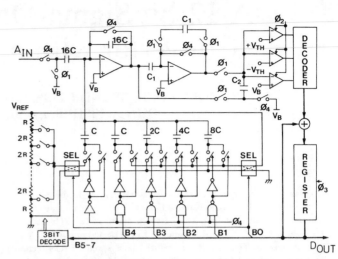

FIGURE 3—Simplified circuit configuration for interpolative ADC.

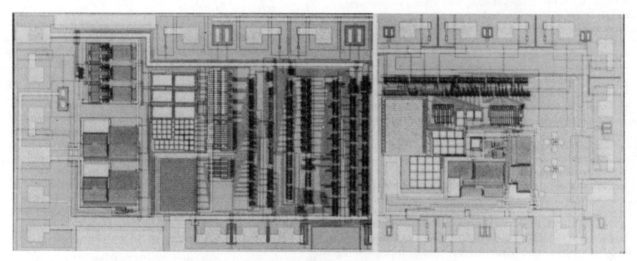

FIGURE 4—Chip photographs of ADC sample and first decimation filter (*left*); and DAC sample with sample-hold (*right*) using 2μ CMOS technology.

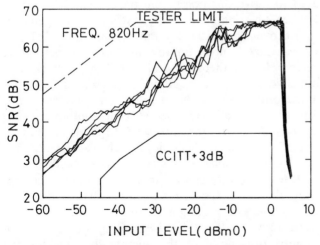

FIGURE 5—Measured SNR of ADC and DAC connected with internal first decimation filter.

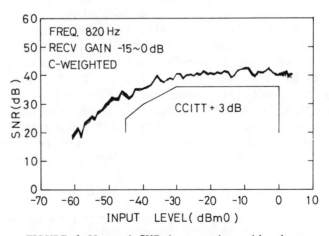

FIGURE 6—Measured SNR for transmitter with gain settings and 50Hz superposition.

A 12-bit Sigma-Delta Analog-to-Digital Converter with a 15-MHz Clock Rate

RUDOLF KOCH, BERND HEISE, FRANZ ECKBAUER, EDUARD ENGELHARDT,
JOHN A. FISHER, MEMBER, IEEE, AND FRANZ PARZEFALL

Abstract —This paper presents a sigma-delta analog-to-digital converter that achieves 12-bit integral and differential linearity and nearly 13-bit resolution without trimming. The baseband width is 120 kHz with a first filter pole at 60 kHz, clock frequency is 15 MHz, and only one 5-V power supply is needed. The circuit was realized in a p-well CMOS technology with 3-μm minimum feature size. Compared with sigma-delta modulators published up to now [1], [2], the input signal frequency and clock rate limit have been increased by one order of magnitude. To achieve this increase, a new integrator concept was developed using bidirectional current sources. The circuit is fully self contained, requiring only a 15-MHz crystal and one blocking capacitor as external elements. This converter was developed as the analog front end of a digital echo cancellation circuit for an integrated services digital network (ISDN).

I. INTRODUCTION

IN an integrated services digital network (ISDN), full-duplex transmission of digital data with a rate of 160 kbit/s over standard two-wire lines is required for the so-called U interface. Either a ping-pong (burst) or a hybrid balancing method is feasible. Line attenuation increases dramatically for higher signal frequencies and is further degraded at longer lengths. It is therefore important to use the transmission method that needs the smallest bandwidth. The hybrid balancing approach requires by a factor of three less bandwidth than the burst method. The bandwidth needed is further reduced by the use of a ternary block code.

In the hybrid balancing method, both parties transmit data simultaneously. The separation of the transmitted from the received signal is achieved with a hybrid. Due to incomplete matching of the hybrid to the line characteristics, a crosstalk between the transmit and receive path occurs. This crosstalk signal can be 35 dB higher in amplitude than the signal received from the far end. Some additional means is therefore needed to remove this crosstalk from the received sum signal. This is done by means of adaptive echo cancellation. In addition to the crosstalk, signal line echos, e.g., from bridge taps, are also cancelled. Our approach employs a digital, adaptive, linear echo canceller (see Fig. 1). Since nonlinearities of the analog blocks (ADC, DAC, transmitter) cannot be compensated

Manuscript received May 9, 1986; revised July 27, 1986. This work was supported in part by the Federal Department of Research and Development, West Germany.
The authors are with Siemens AG, D-8000 Munich 80, Germany.
IEEE Log Number 8610949.

for, the most stringent demand on the analog parts is a linearity of over 70 dB. This demand, together with the large baseband width, makes the development of the analog parts a very challenging task.

II. ANALOG-TO-DIGITAL CONVERTER

A. Functional Principle

A rough specification of the ADC is given in Table I. The extremely high demands on linearity of the ADC, the relatively large baseband width, and the requirement to avoid trimming cannot be fulfilled in a CMOS environment using standard approaches such as simple capacitor or resistor networks. Although one solution might be a self-calibrating ADC, we felt that a sigma-delta modulation converter of second order was a better choice. This choice was proven to meet system specifications and the design parameters were optimized using extensive simulation on the system level.

The functional principles of second-order sigma-delta modulation have been described by Candy [3] and others. A simplified schematic is given in Fig. 2. Two integrators in the forward path, two 1-bit DAC's in the feedback path, and a 1-bit ADC, i.e., a comparator, are used. The dominant noise source of this modulator is the quantization noise of the 1-bit ADC. Oversampling and noise shaping are used to move this noise from the baseband to higher frequencies. Oversampling evenly distributes the quantization noise over the frequency range from dc to half the clock frequency. A noise improvement of 3 dB per octave of oversampling is therefore achieved. The feedback loop contains two integrators, resulting in a high-pass characteristic of second order for the quantization noise. With this filtering, the S/N ratio is improved by an additional 12 dB per octave. With oversampling and filtering, a gain of 15 dB in S/N ratio is theoretically achievable with each octave of oversampling. Our simulations showed that by using ideal elements, a clock rate of 15 MHz resulted in a SNR slightly above 80 dB with a 120-kHz bandwidth. This provides some safety margin in the actual circuit. It is important to notice that the noise shaping is only valid for the quantization noise of the comparator, not for the linearity errors of the DAC's in the feedback path. The performance of the outer DAC is especially crucial for the

Reprinted from *IEEE J. Solid-State Circuits*, vol. SC-21, no. 6, pp. 1003–1010, Dec. 1986.

324

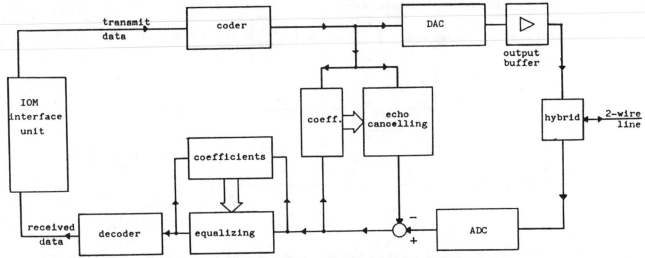

Fig. 1.   Simplified block diagram of a digital echo canceller.

TABLE I
ADC SPECIFICATION

Power supply	5 V
Power consumption	< 20 mW
Signal/(noise + distortion)	> 72 dB
Baseband width	3–120 kHz

performance of the whole modulator. The DAC's linearity must be at least as high as the linearity of the complete circuit reduced by the oversampling ratio. Another problem is the existence of high-frequency noise in the circuit as a natural consequence of the noise shaping principle. Nonlinearities in the circuit give rise to intermodulation products that alias into the baseband.

The output signal of such a sigma-delta modulator is a high-frequency 1-bit signal (in our case the bit rate is 15 MHz). This 1-bit signal contains the analog input signal as its mean value. Digital low-pass filtering is needed to remove the high-frequency noise from the signal. The order of the filter should be higher by one than the order of the noise shaping function. The specified transfer function of that filter is shown in Fig. 3 and can be described by the equation

$$H = (z^0 - 2z^{-64} + 2z^{-192} - z^{-256})/(1 - z^{-1})^3.$$

The filter is realized as a cascade of a 256-stage four-tap FIR part and a third-order IIR part. The complete filter structure is clocked with 15 MHz. At the output, standard multiple bit words with 13 relevant bits are available with this clock rate. Decimation is performed in a separate stage to reduce the data rate to 120 kHz. The low pass and decimation filter were integrated on a separate chip [4]. The performance of a sigma-delta modulator depends heavily on the filter used. All data are only valid for a given filter characteristic.

B. Measured Results

Since the digital 1-bit output signal contains the analog input signal as a mean value, one could apply a low-pass filter between the PDM output and a spectrum analyzer to measure the performance of the modulator. With this approach, however, the output buffer acts as a 1-bit DAC and must have a performance superior to that of the modulator. The design of such a buffer is a rather complex task. Instead, digital samples are taken of the PDM signal. For the measured results given in Figs. 4–7 and in Table II, 4096 samples per measurement were taken with a 15-MHz clock rate and a fast Fourier transform (FFT) was performed to calculate the spectrum. Fig. 4 shows a wideband idle channel output spectrum. No filtering of the signal was performed so the high-frequency noise and the noise shaping behavior can be clearly seen from this measurement. Fig. 5 shows the baseband output spectrum with an 11-kHz sinusoidal input signal of 1-V_{eff} amplitude. A digital low-pass filter, which was integrated on a separate chip, was used for this measurement. The decrease of noise towards higher frequencies is caused by a first pole of the filter at $f = 60$ kHz. The output word of this filter was fed into a test computer and a FFT was performed to calculate the output spectrum.

There is one line of particular interest close to 7.5 MHz in Fig. 4. With zero input and zero offset voltage to the ADC, the output signal would be a periodic stream of ones and zeros corresponding to exactly half the clock frequency, i.e., 7.5 MHz in our case. With a small dc input voltage there would be a few more ones or zeros in the stream that would show up as discrete lines in the spectrum around dc, 7.5 MHz, 15 MHz, and so on. A line in the baseband would reduce the S/N ratio. This effect is a significant problem with sigma-delta modulators of first order where an additional dither signal is often used to smooth out these discrete lines. Our simulations, as well as our measurements, showed that in a modulator of second order, these discrete lines are so far damped and decorrelated [3] that no dither signal is needed. Some measured results are given in Table II. For these measurements a separate filter chip was again used. The output of the filter was fed into a test computer where a FFT was performed and S/THD and S/N were calculated. An 11-kHz sinusoidal input signal was applied to the ADC as described above.

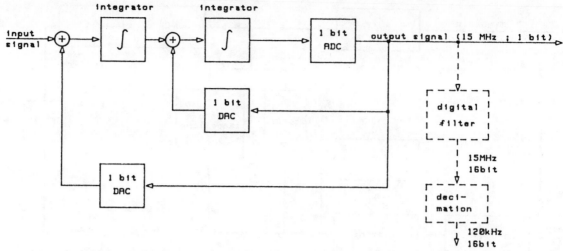

Fig. 2.   Principle circuit schematic of a second-order sigma-delta modulator.

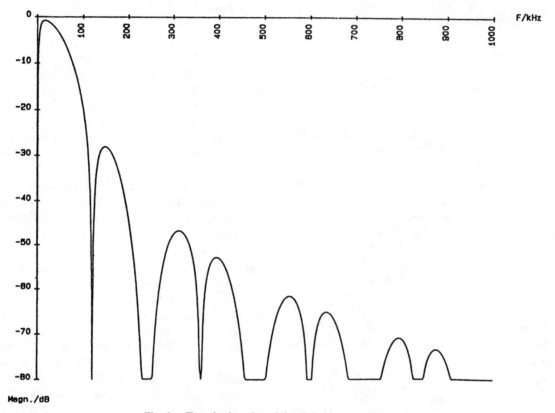

Fig. 3.   Transfer function of the digital low-pass filter.

Figs. 6 and 7 show the S/(N+THD) as function of the input signal amplitude and frequency, respectively. The output signal is evaluated with the low-pass filter and a FFT is performed on the test computer. The measurements for Fig. 6 were taken with an input signal frequency of 11 kHz, as above. With increasing input amplitude, the S/(N+THD) increases until harmonic distortion becomes out of spec. The measurements for Fig. 7 were taken with constant input signal amplitude of 1 V_{eff}. The decrease of the S/(N+THD) above 3 kHz is caused by the first pole of the input stage amplifier. The steeper decrease above 60 kHz results from the digital low-pass filter with its first pole at $f = 60$ kHz. For a 100-kHz input signal of 1-V_{eff}

amplitude we get S/(N+THD) = 55 dB. Once again it is stressed that these data are only valid together with the filter characteristic specified above.

C. Integrator Concept

Two integrators are needed for a sigma-delta modulator of second order (see Fig. 1). Standard realizations of integrators have traditionally used either a resistor–capacitor network or a switched-capacitor configuration. For the modulator described here, the maximum input frequency to these integrators originates from the feedback path and is in the megahertz range. The clock frequency is 15 MHz.

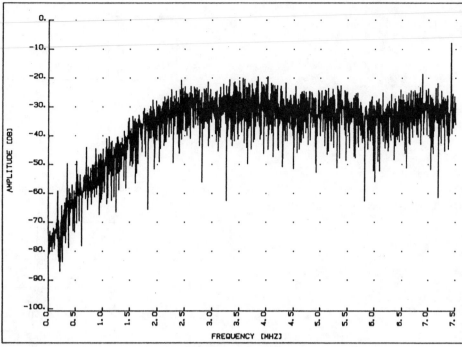

Fig. 4.   Measured idle channel wide-band output spectrum.

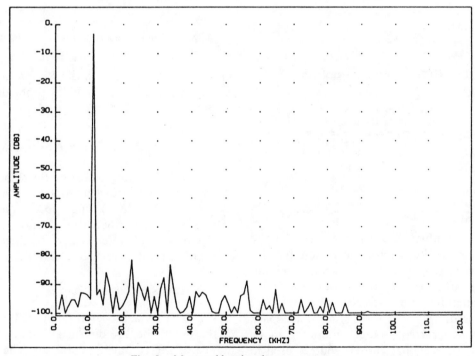

Fig. 5.   Measured baseband output spectrum.

In a switched-capacitor integrator with a clock rate of 15 MHz, op amps with about 100-MHz bandwidth and accordingly high power consumption would be needed to ensure proper settling. On the other hand, no output stage is required for these circuits and fast one-stage amplifiers can be employed. Bandwidth requirements would be greatly reduced, compared to the SC approach, with a standard *RC* integrator. To drive the resistive load, a two-stage amplifier would be required. This type of op amp is much slower than a one-stage amplifier and would exhibit prob-

lems of insufficient slew rate and bandwidth. A new circuit configuration was developed that combines the advantages of both approaches. The principle circuit schematic of the sigma-delta modulator employing this new approach is given in Fig. 8. The integrators are built around bidirectional current sources. With this approach a much smaller bandwidth is needed for the op amps as would be needed for an SC integrator. According to system simulation where linear op amps were assumed, 4 MHz of bandwidth would be sufficient for 80 dB of S/N. Measurements using

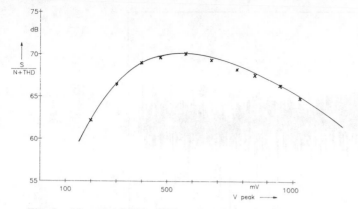

Fig. 6. Measured S/(N + THD) as function of signal amplitude.

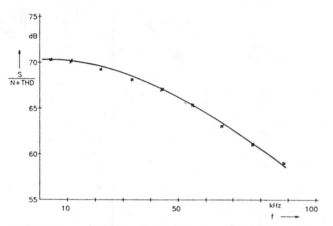

Fig. 7. Measured S/(N + THD) as function of signal frequency.

TABLE II
ADC PERFORMANCE—MEASURED RESULTS

Clock rate	15.36 MHz
Baseband width	3–120 kHz
Eq. input offset voltage	±3 mV
Signal/noise	77 dB
Signal/total harmonic distortion	72 dB
Power consumption	15 mW

variable bandwidth amplifiers showed, however, that 12 MHz is actually needed. We believe the reason for this difference is the presence of frequency-dependent nonlinearities. The op amp has a load of only one transistor gate, hence a one-stage amplifier with high slew rate and large bandwidth can be used. The 1-bit DAC's in the feedback path must be switched current sources in order to be compatible with the bidirectional current sources used in the integrators.

One disadvantage of this approach, compared with SC solutions, is the larger variance of the time constants due to resistor tolerances. The relative variation of the resistors that define the signal current and the feedback current causes different integrator gains for the feedforward and feedback paths. Thus a gain error is introduced into the ADC. However, this gain error is compensated for by a digital AGC in the digital part of the echo canceller. The absolute tolerance of the resistors effects the transfer function of the ADC which, under extreme conditions, can cause stability problems. According to extensive system

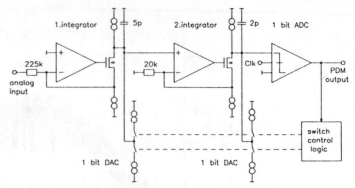

Fig. 8. Sigma-delta modulator realization with current sources.

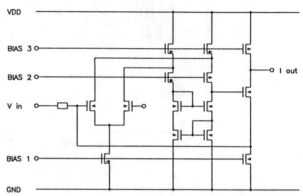

Fig. 9. Circuit schematic of the bidirectional current source.

simulation, a variation in the *RC* product of ±30 percent is permissible without stability problems and without degradation of the S/N ratio.

III. BASIC BUILDING BLOCKS

The basic building blocks of the sigma-delta modulator, the bidirectional current sources, the switched current sources (1-bit DAC's), and the comparator (1-bit ADC) will now be described. In addition to these blocks, a bandgap voltage reference, a biasing network, and an oscillator circuit were also integrated.

A. Bidirectional Current Source

The linearity and S/N ratio of the integrators, especially of the outer one, are of crucial importance for the performance of the complete modulator. The integrators in this circuit receive signals from the feedback path that can be as high as 7.5 MHz. The linearity of the integrators must be high to avoid intermodulation products and to ensure a S/(N + THD) of over 70 dB over the 120-kHz baseband. Power consumption of the op amp must be low because a limit of 20 mW was set for the whole modulator. A good solution to these requirements is the use of a bidirectional current source as the core of the integrator.

Fig. 9 shows a circuit schematic of this current source that is built around a folded cascode op amp. The additional output stage contains two constant current sources of equal value that force a bias current through the output

TABLE III
OP-AMP PERFORMANCE—MEASURED RESULTS

Open-loop gain at dc	80 dB
Unity-gain bandwidth	20 MHz
Phase margin	63°
Slew rate	20 V/μs
Power supply rejection at 1 kHz	55 dB
Common-mode rejection at 1 kHz	80 dB
Random offset voltage	$\pm$1.5 mV
Thermal input noise	25 nV/sqrt (Hz)

TABLE IV
BIDIRECTIONAL CURRENT SOURCE—MEASURED RESULTS

Loop gain at dc	90 dB
Unity-gain bandwidth	18 MHz
Random offset voltage	$\pm$3 mV
Maximum signal current	4.5 μA resp. 7.5 μA
Power consumption	1.2 mW

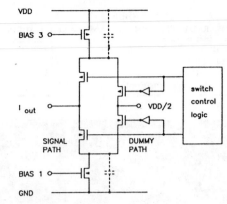

Fig. 10. Circuit schematic of the switched current source.

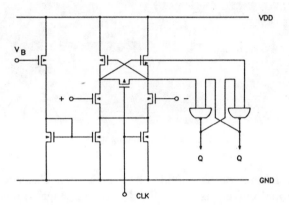

Fig. 11. Circuit schematic of the comparator.

transistor. The input voltage is converted to an input current of positive or negative sign through resistor R. This input current flows as additional current through the output transistor. The current is fed in at the source node and flows out of the output stage at the drain node. In this way, the current through the output transistor is modulated as function of the input voltage. The bias current of the output stage has to be sufficiently higher than the maximum signal current to ensure a low harmonic distortion. With a fast high-gain op amp, a current ratio of about two is sufficient.

The output impedance of the current sources forms a leakage path parallel to the integration capacitor. A too low value of leakage resistance reduces the low-frequency gain of the ADC and, consequently, the noise floor at low frequencies is increased. The minimum value permissible for the leakage resistance is 5 MΩ according to our simulation.

The equivalent offset voltage of the ADC given in Table I is caused in equal parts by the offset voltage of the core amplifier and by mismatch of the upper and lower current sources in the outer integrator. Measured results for the folded cascode core amplifier and the bidirectional current source are given in Tables III and IV.

B. Switched Current Source

Since bidirectional current sources are used in the feedforward path, switched current sources are required in the feedback path to realize the 1-bit DAC's. The linearity and S/N ratio of the outer DAC must be superior to the required performance of the complete modulator. The requirements of the DAC, however, are relaxed by the oversampling effect [3]. In our application, a S/(N + THD) of about 55 dB was needed.

One problem inherent to switched current sources is the charging and discharging of parasitic capacitances. The voltage change on the parasitic capacitor prevents fast switching of the current and would cause a loss of charge on the integration capacitor. To achieve fast switching, no loss of charge, and a high internal resistance of the current

sources, a differential configuration was chosen as shown in Fig. 10. The current is never switched off but shifted between the two branches. The switching transistors are operated in the saturation region to guarantee a high internal resistance of the current source, very much like the differential input stage of an op amp. With this configuration, the parasitic capacitance at the common source node of the switching transistors is kept at a constant voltage and thus causes no loss of charge.

No relevant measured results could be obtained for this switched current source since all attempts at probing were unsuccessful because of loading effects. From the performance of the whole modulator, however, one can conclude that the S/(N + THD) of the current sources is better than the 55 dB mentioned above.

C. Comparator

The third important building block is the 1-bit ADC or comparator. Since it is embedded in two high-gain feedback loops with the switched current sources (see Figs. 2 and 6), its requirements on resolution and offset voltage are moderate (in the range of several tens of millivolts). With a 15-MHz clock, however, the response time is of utmost importance. A latch-type comparator is a good choice for this application. As shown in the circuit schematic in Fig. 11, the comparator consists of a differential stage with a latch as load. In the reset mode, the nodes of the latch are shorted to set it to the astable high-gain

TABLE V
COMPARATOR PERFORMANCE — MEASURED RESULTS

Resolution	2 mV
Response time	15 ns
Random offset	± 3 mV
Averaged power consumption	1.5 mW

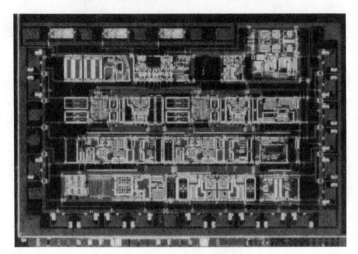

Fig. 12. Chip micrograph.

mode. In the compare mode, these nodes are released and additional current is fed to the differential stage to enable fast switching of the latch. Measured results are given in Table V.

D. Chip Architecture

A chip micrograph is shown in Fig. 12. The chip area of this experimental device is about 3 mm^2. It was realized in a 3-μm p-well CMOS technology. A second layer of poly-silicon was used for poly1–poly2 capacitors and for high ohmic resistors. A channel length of 6 μm was frequently used to avoid short-channel effects.

A structured layout was used in which the circuit is organized into several rows. The bottom row contains the static analog parts, i.e., the bandgap voltage reference, two buffer amplifiers for the reference voltages, and a bias current network. The second row from the bottom contains the midfrequency analog parts, i.e., the bidirectional current sources and a biasing network. The next row up contains the high-frequency switched current sources and the associated bias network. The top row consists of an oscillator and small logic block which run at 15 MHz and two comparators. Only one of the comparators was used in the circuit. The other one was for backup purposes.

IV. SUMMARY

A sigma-delta modulator realized in a 3-μm CMOS technology has been presented in which 12-bit linearity is achieved without trimming. Baseband width and clock rate are one order of magnitude higher than in modulators described up to now. This high-frequency range was obtained by implementation of a new integrator concept, in which requirements of the op amps were significantly reduced compared with *RC* or *SC* integrator approaches.

ACKNOWLEDGMENT

The authors wish to thank Dr. E. Schmid for his suggestions to the basic conept of the modulator and for many stimulating discussions.

REFERENCES

[1] H. L. Fiedler and B. Hoefflinger, "A CMOS pulse density modulator for high resolution A/D converters," *IEEE J. Solid-State Circuits*, vol. SC-19, pp. 995–996, Dec. 1984.
[2] M. W. Hauser, P. J. Hurst, and R. W. Brodersen "MOS ADC-filter combination that does not require precision analog components," in *ISSCC Dig. Tech. Pap.*, Feb. 1985, pp. 80–81.
[3] J. C. Candy, "A use of double integration in sigma delta modulation," *IEEE Trans. Commun.*, vol. COM-33, pp. 249–258, Mar. 1985.
[4] A. Huber *et al.*, "FIR lowpass filter for signal decimation with 15MHz clock frequency," in *Proc. ICASSP 86*, Apr. 1986.

An 8-bit High-Speed CMOS A/D Converter

TOSHIO KUMAMOTO, MASAO NAKAYA, HIROKI HONDA, SOTOJU ASAI, YOICHI AKASAKA,
AND YASUTAKA HORIBA

Abstract — An 8-bit high-speed A/D converter has been developed in a 1.5-μm bulk CMOS double-polysilicon process technology. The design, process technology, and performance of this A/D converter will be described. In order to achieve a high-speed low-power A/D converter, a fine pattern process technology and a new capacitor structure have been introduced and transistor sizes of a chopper-type comparator have been optimized. High-speed (30 MS/s) and low power consumption (60 mW) have been obtained. Computerized evaluations such as the histogram test and the fast Fourier transform (FFT) test have been applied to obtain dynamic performance. The linearity error in dynamic operation is less than ± 1 LSB. Signal-to-peak-noise ratio is 40 dB at a sampling rate of 14.32 MS/s and an input frequency of 1.42 MHz.

I. INTRODUCTION

RECENTLY, digital signal processing is being applied to video systems, such as TV's and VCR's. High-speed A/D converters with low-power dissipation and low cost have consequently been desirable. Up to this time, high-speed 8-bit A/D converters have been developed with both bipolar [1], [2] and CMOS [3], [4] technologies. The power consumption of these A/D converters is more than 100 mW and their die sizes are too large to be integrated with digital circuitry.

This paper describes a high-speed, low-power, 8-bit flash A/D converter fabricated in a 1.5-μm bulk CMOS double-polysilicon process. It operates at a sampling rate of 20 MS/s with 60 mW ($V_{dd} = 3.45$ V), or at a sampling rate of 30 MS/s with 180 mW ($V_{dd} = 4.65$ V).

It has been difficult to attain a video-speed 8-bit A/D converter using bulk CMOS technology because of its parasitic capacitances. This problem has been solved by using a new capacitor structure for an input network, fine pattern technologies, and improvement in the comparator design.

In Sections II and III of this paper, the circuit configuration of this A/D converter and design of its comparator circuit are described. Layout topology of comparators and ladder resistors is described in Section IV. Process technology and a new capacitor structure are described in Section V. In Section VI experimental results and dynamic performance are described.

Manuscript received July 1, 1986; revised August 11, 1986.
The authors are with the LSI Research and Development Laboratory, Mitsubishi Electric Corporation, 4-1 Mizuhara, Itami, Hyogo 664, Japan.
IEEE Log Number 8610948.

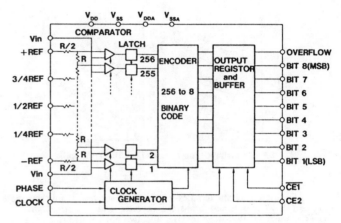

Fig. 1. Block diagram of the A/D converter.

II. CIRCUIT CONFIGURATION

A block diagram of the flash A/D converter is shown in Fig. 1. The A/D converter is composed of 257 ladder resistors, 256 chopper-type comparators, tap detection logic, a 256-to-8 encoder, a clock generator, and nine output buffers (8 bit + overflow). Simultaneously, 256 comparators compare an input voltage with each ladder tap voltage which is made by dividing the reference voltage by the resistor ladder. The minimum ladder tap voltage which exceeds the input voltage is identified by the transition detection logic. The encoder circuit converts the identified ladder tap voltage to a binary code. In this encoder circuit, the gate voltages of load transistors are controlled by a clock signal, which reduces the power dissipation. The nine outputs (including one extra output for overflow) of this encoder circuit are followed by nine output registers and buffers.

Basically, this A/D converter needs two complementary clocks. However, in order to attain the dynamic operation required from each circuit, eight clock signals are made from an external clock by using inverters, NAND, and NOR gates.

III. COMPARATOR CIRCUIT

A. Chopper-Type Comparator

Fig. 2 shows the chopper-type comparator circuitry that uses a two-stage transmission-gate-connected inverter [4].

Reprinted from *IEEE J. Solid-State Circuits*, vol. SC-21, no. 6, pp. 976–982, Dec. 1986.

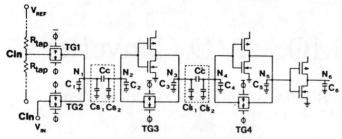

Fig. 2. Chopper-type comparator circuit.

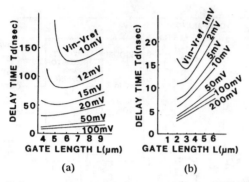

Fig. 3. Simulated characteristics of delay time versus inverter gate length. (a) *TG*-connected one-stage inverter and a normal inverter. (b) *TG*-connected two-stage inverter and two normal inverters.

This type of comparator operates basically with complementary clock signals ϕ and $\bar{\phi}$. During the first period ϕ, the reference ladder tap voltage is applied to node $N1$ and the inverters are autozeroed simultaneously to their equilibrium points because transmission gates $TG1$, $TG3$, and $TG4$ are closed and $TG2$ is open. During the second period $\bar{\phi}$, the input voltage is applied to node $N1$. The difference between the input voltage and the ladder tap voltage is transmitted through the coupling capacitor C_c and is amplified by the inverter stages.

In general, a long gate length of the transistors in the inverting amplifier is required to obtain a high gain, because an inverter with a short gate length has a low voltage gain caused by the channel-length-modulation effect. On the other hand, it is necessary for the gate length to be short in order to obtain high-speed operation. Therefore, it is difficult to attain 8-bit resolution for video frequencies if one uses only a one-stage *TG*-connected inverter. The circuit simulation results for the circuitry are shown in Fig. 3. Fig. 3(a) shows a characteristic of delay time versus inverter gate length in the case where only a one-stage *TG*-connected inverter and a normal inverter are used. Fig. 3(b) has been obtained in the case of a two-stage *TG*-connected inverter and two normal inverters. From this figure, it is clear that a two-stage *TG*-connected inverter is necessary in order to obtain a 10-mV resolution at video speeds.

In the two-stage configuration, the gate length of the transistors in the first *TG*-connected inverter are shortened to 3 μm in order to get high-speed operation. The gain degradation caused by the channel-length modulation of

the first-stage transistors has been avoided by adding two successive inverters whose gate lengths are 5.5 μm. Increase in the current, which is caused by shortening the gate length of the first-stage transistor, is suppressed by decreasing the gate width.

In actual operation, switching noise is likely to be caused by the undesired clock feedthrough in the transmission gates and also interferes with the fast and precise operation of the comparator. To reduce this noise, the gate widths of the p- and n-channel transistors of $TG3$ and $TG4$ have been designed to have the same gate–drain and gate–source capacitances.

B. Successive Comparators with Ladder Resistors

To get high-speed performance in the flash-type A/D converter which uses the chopper-type comparator, it is necessary to reduce the autozeroing and sampling periods. These are the periods that were previously mentioned as the first period ϕ and the second period $\bar{\phi}$, respectively. During the autozeroing period, the resistor tap voltage should be recovered from the transient state, which is caused by the charging or discharging of the coupling capacitor. The recovery time depends on the ladder resistance R_{tap} between the adjacent taps and input capacitance C_{in} of the comparator. The change of reference tap voltage in the distributed-constant-circuit model is given by [5]

$$v(x,t) = V_{ref}x - V_{in} - V_{ref}\frac{2}{\pi}\sum_{n=0}^{\infty}\frac{1}{n(-1)^{n-1}}$$
$$\cdot \exp\left(-\frac{n^2\pi^2}{RC(b-a)^2}t\right)$$
$$\cdot \left\{(a \cdot V_{ref} - V_{in}) \cdot \sin\frac{n\pi(b-x)}{b-a}\right.$$
$$\left. - (b \cdot V_{ref} - V_{in}) \cdot \sin\frac{n\pi(a-x)}{b-a}\right\}$$

where

$$R = 2^N \cdot R_{tap}$$
$$C = 2^N \cdot C_{in}$$

N is a bit number, V_{in} is an input voltage, V_{ref} is the reference voltage, and $v(x, t)$ is a tap voltage as a function of normalized position x and time t. As for voltage boundary conditions, it is assumed that a voltage at the normalized position "a" is $a \cdot V_{ref}$ and a voltage at the normalized position "b" is $b \cdot V_{ref}$. Fig. 4 shows the worst-case ladder recovery error as a function of the recovery time. It is clear that it is necessary for the product of R_{tap} and C_{in} to be less than 3×10^{-13} FΩ to get an 8-bit resolution with a 30-MS/s operation. An equivalent circuit for the input network can be referred to in Fig. 2. To

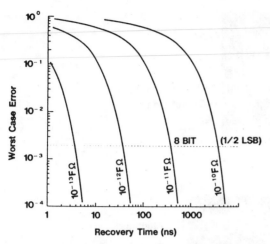

Fig. 4. Numerical transient analysis of the worst-case ladder recovery error.

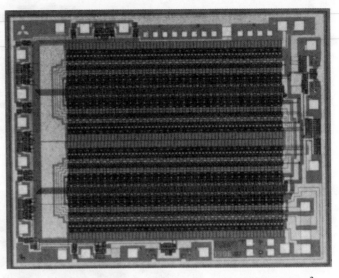

Fig. 6. Photomicrograph of the A/D converter (3.08×2.56 mm^2).

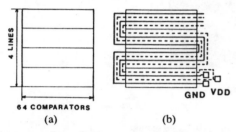

Fig. 5. Layout topology of comparator and ladder resistors. (a) Pattern folding of comparator series. (b) Continuous service layout of analog ground and analog supply voltage lines. The dashed line represents the analog supply voltage line, and the solid line is the analog ground line.

reduce the product of R_{tap} and C_{in}, it is necessary to reduce the parasitic capacitances C_{s1} and C_{s2}, which form a considerable part of C_{in}. On the other hand, to transfer the input voltage without losses, it is necessary to increase the coupling capacitance C_c.

IV. LAYOUT TOPOLOGY OF COMPARATOR AND LADDER RESISTORS

An 8-bit flash A/D converter is composed of 256 comparators. It is necessary to use a folded arrangement of comparator rows because a straight arrangement of the comparators makes the chip too long to be mounted in a conventional package. In this case, the rows were stacked by folding them three times; that is, the series of 256 comparators is arranged in four stacked lines, where each line has 64 comparators, as is shown in Fig. 5(a).

Furthermore, the material of the ladder resistor is doped polysilicon and the ladder resistors at the folding points are connected with small resistance error by increasing the widths of the metal leads that connect the resistors at these points.

Analog ground lines and analog supply voltage lines for the comparators are continuously laid out, as is shown in Fig. 5(b). They have continuous voltage distribution and therefore minimize linearity errors due to ground and supply line voltage drops.

A photomicrograph of the A/D converter is shown in Fig. 6. The chip size is 3.08×2.56 mm^2.

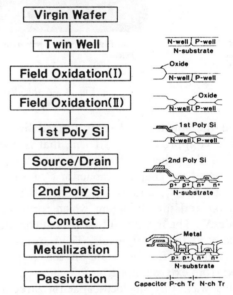

Fig. 7. Process sequence of the A/D converter.

V. PROCESS TECHNOLOGY AND NEW CAPACITOR STRUCTURE

The A/D converter is realized in a 1.5-μm CMOS double-polysilicon and single-aluminum technology. The ladder resistors are made of doped polysilicon whose sheet resistance is 25 Ω/sq and the ladder resistance is approximately 2 Ω. Fig. 7 shows the process sequence of the A/D converter. Field oxide is formed by two oxide layers: thick and thin oxide layers whose thicknesses are approximately 1.1 and 0.6 μm, respectively. The thick oxide and the thin oxide are formed by high-pressure oxidation and by atmospheric oxidation, respectively.

An equivalent circuit for the input network is shown in Fig. 2. To decrease the parasitic capacitances C_{s1} and C_{s2}, and to increase the coupling capacitance C_c, a new capacitor structure is introduced. Fig. 8 shows the structure that gives a small C_{s1} and C_{s2} and large C_c. To realize small C_{s1} and C_{s2}, the thickness of oxide between the first polysili-

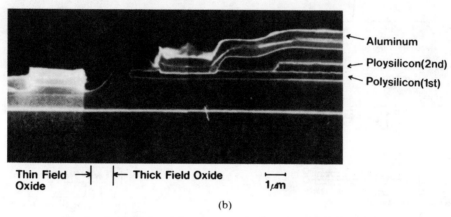

Fig. 8. A shielded-capacitor structure. (a) Top view. (b) Cross section.

con and the substrate which is formed by high-pressure oxidation is approximately 1.1 μm, that is, twice as thick as that of the other field oxide. In addition, the second polysilicon of the coupling capacitor C_c forms the output-side electrode N_2, and is sandwiched between the first polysilicon and the aluminum which form the input-side electrode N_1. The metal and first polysilicon are connected electrically at their peripherals. The output-side electrode which is in the floating state in the $\bar{\phi}$ period is thus shielded from external noise.

In this way, this shielded-capacitor structure has two advantages: 1) decreasing the parasitic capacitance, and 2) shielding from noise at the output-electrode.

VI. EXPERIMENTAL RESULTS

A. Conventional Evaluation

Figs. 9 and 10 show practical test results of the 8-bit A/D converter chip. In both cases, input signals have been A/D-converted by this 8-bit chip under test and then D/A-"reconstructed" by using an appropriate high-speed D/A converter.

Fig. 9 shows the "reconstructed" linear ramp waveform with a sampling rate of 20 MS/s. Fig. 10(a) shows a 1-MHz sine wave reconstructed at a sampling rate of 20 MS/s and a supply voltage V_{dd} of 3.5 V. Fig. 10(b) and (c) shows 1- and 4-MHz sine waves, respectively, reconstructed at a sampling rate of 30 MS/s and a supply voltage V_{dd} of 5.0 V. In these figures, the effect of clock

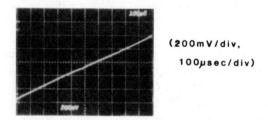

(200mV/div,
100μsec/div)

Fig. 9. D/A-reconstructed waveform of linear ramp at a sampling rate of 20 MS/s and an input frequency of 1 kHz.

noise superposed on the input signal has been observed. This is due to the fact that the input impedance of the chopper-type comparator alternately takes two values according to the two operating periods synchronized with the clock. The noise can be reduced by applying an input signal via an operational amplifier with low output impedance. Fig. 11 shows the power dissipation as a function of the sampling clock frequency and supply voltage. The maximum sampling rate is 30 MS/s at $V_{dd} = 4.65$ V, where the power dissipation is 180 mW. The power consumption is reduced to 60 mW in the condition of $V_{dd} = 3.45$ V and 20-MS/s sampling.

B. Evaluation in a Digital Signal Processing Method

The dynamic performance of the A/D converter has been measured by using the conventional method of converting an analog signal and reconstructing it with a D/A converter. However, this introduces an additional D/A converter error. To prevent this error, a digital signal processing method is introduced.

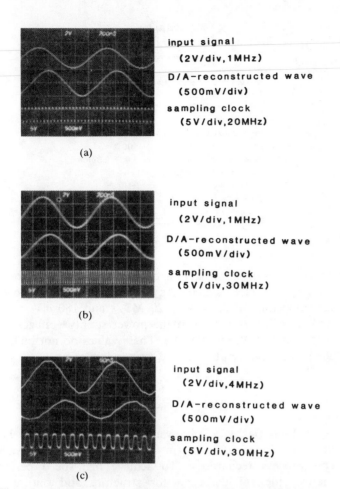

input signal

(2V/div,1MHz)

D/A-reconstructed wave

(500mV/div)

sampling clock

(5V/div,20MHz)

(a)

input signal

(2V/div,1MHz)

D/A-reconstructed wave

(500mV/div)

sampling clock

(5V/div,30MHz)

(b)

input signal

(2V/div,4MHz)

D/A-reconstructed wave

(500mV/div)

sampling clock

(5V/div,30MHz)

(c)

Fig. 10. D/A-reconstructed sine wave. (a) 1-MHz input, 20-MS/s sampling rate, $V_{dd} = 3.5$ V. (b) 1-MHz input, 30-MS/s sampling rate, $V_{dd} = 5.0$ V. (c) 4-MHz input, 30-MS/s sampling rate, $V_{dd} = 5.0$ V.

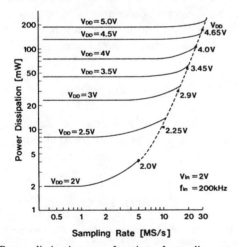

Fig. 11. Power dissipation as a function of sampling rate and power supply voltage.

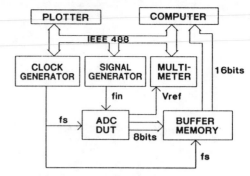

Fig. 12. Measurement setup for a computerized A/D converter test.

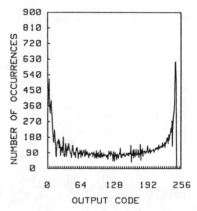

Fig. 13. Histogram plots of the A/D converter at a sampling rate of 14.32 MS/s and an input frequency of 1.43 MHz.

The measurement setup for evaluating the dynamic performance is shown in Fig. 12. In this configuration, the A/D converter output data are stored in a buffer memory, and are analyzed by a computer. Two dynamic tests, the histogram test and the fast Fourier transform (FFT) test, are used for the characterization of the A/D converters.

1. Histogram Test: A sine-wave-based histogram test [6] provides both a localized error description and a global description of the A/D converter. Thirty-thousand samples are enough for rejecting significant degradation in the repeatability of the measurement. At the end of the sampling, the histogram is plotted by taking output codes along the X axis and the frequency of occurrence along the Y axis. Fig. 13 shows an example of the histogram plot. The input frequency is 1.43 MHz at a sampling rate of 14.32 MS/s. The absence of large spikes indicates small differential nonlinearity errors. There are no missing codes since there is no plot which falls on the X axis. The differential nonlinearity at each code transition is also obtained from Fig. 13 using the computer analysis based on the probability density function of the sine wave. Fig. 14 shows the differential nonlinearity error plots from the data in Fig. 13.

2. FFT Test: The FFT is used to characterize performance in the frequency domain. From the frequency-domain representation of the data, the linearity of dynamic transfer function can be measured.

A full-scale sine wave of specified frequency is applied to the device under test. Next, a 1024-point record sampled with a specified sampling rate is analyzed with the FFT calculation [6] by the computer. By summing up the magnitudes of fundamental spectral lines as well as the magnitudes of remaining spectral lines, it is also possible to compute $20 \cdot \log(S/N)$, which equals the S/N ratio expression in decibels. Furthermore, effective bits can be calculated by this S/N ratio [7].

Any input frequency is acceptable in the FFT test when a window function is used. However, this introduces some

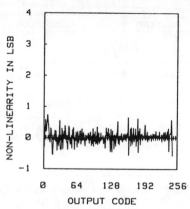

Fig. 14. Differential nonlinearity error plot at a sampling rate of 14.32 MS/s and an input frequency of 1.43 MHz.

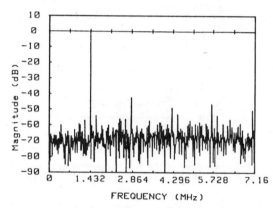

Fig. 15. An example of the FFT test results.

TABLE I
CHARACTERISTICS OF THE A/D CONVERTER

Resolution	8 bits
Maximum Conversion Rate	30MS/s
Power Consumption	60mW at 20MS/s
	180mW at 30MS/s
Accuracy (DC)	±1/2LSB
Input Capacitance	28pF
Ladder Impedance	500Ω
Chip Size	3.08×2.56mm²
Number of Elements	
Transistor	9410
Resistor	257
Capacitor	512

errors because FFT analysis is performed by using the digital data that is changed by the window function. Hence, it is difficult to separate the fundamental from other spectral lines in obtaining the S/N ratio. In order to eliminate the "leakage" of the spectrum without using the window function, input frequency f_{in}, sampling rate f_{sc}, a cycle number M, and data number N are chosen to satisfy the following equation:

$$\frac{1}{f_{sc}} \cdot N = \frac{1}{f_{in}} \cdot M$$

where N is 1024 in this case and M is an appropriate integer number.

Fig. 15 shows an example of FFT test results. This indicates the signal-to-peak-noise ratio of 40 dB (the signal-to-noise ratio is 38 dB and the harmonic-distortion ratio which is calculated up to the fifth harmonic is -39 dB) at a sampling rate of 14.32 MS/s, an input frequency of 1.42 MHz ($M = 101$), and an effective bit (effective linearity) of 6.0 bits.

C. Summary of Characteristics

Typical chip parameters and characteristics of this A/D converter have been summarized in Table I. An input

capacitance of 28 pF is an average value at the 50-percent duty cycle of clock signals. As previously mentioned, power consumptions of 60 mW at 20 MS/s and 180 mW at 30 MS/s are measured when the power supply voltages are 3.45 and 4.65 V, respectively. These values do not include the power consumption of the resistor ladder.

VII. CONCLUSION

An 8-bit flash A/D converter for video frequency has been developed in a 1.5-μm bulk CMOS double-polysilicon process technology. To achieve high-speed performance, the shielded-capacitor structure and optimized two-stage *TG*-connected inverter in the chopper-type comparator are used. Computerized evaluations such as histogram and FFT tests have been applied to obtain dynamic performance. A power consumption of 60 mW at 20 MS/s and a chip size of 3.08 mm × 2.56 mm have been achieved sufficiently for consumer applications.

ACKNOWLEDGMENT

The authors wish to acknowledge Dr. K. Shibayama, Dr. H. Nakata, and Dr. T. Nakano for their support in this project.

REFERENCES

[1] J. G. Peterson, "A monolithic video A/D converter," *IEEE J. Solid-State Circuits*, vol. SC-14, pp. 932–937, Dec. 1979.
[2] M. Hotta, K. Maio, N. Yokozawa, T. Watanabe, and S. Ueda, "A 120-mW, 8bit video-frequency A/D converter with shallow-groove-isolated bipolar VLSI technology," in *Proc. 1984 Symp. VLSI Technol.*, Sept. 1984, pp. 58–59.
[3] T. Tsukada, Y. Nakatani, E. Imaizumi, Y. Toba, and S. Ueda, "CMOS 8b 25MHz flash ADC," in *ISSCC Dig. Tech. Pap.*, WAM 2.7, Feb. 1985, pp. 34–35.
[4] A. G. F. Dingwall and V. Zazzu, "High-speed CMOS A/D and D/A conversion," in *ISCAS Dig. Tech. Pap.*, 1984, pp. 420–424.
[5] M. Nakaya, T. Kumamoto, T. Miki, and Y. Horiba, "Analysis of reference-tap-voltage fluctuation in flash A/D converter," *Trans. ECE Japan*, vol. J69-C, no. 3, pp. 237–244, Mar. 1986.
[6] B. E. Peetz, A. S. Muto, and J. M. Neil, "Measuring waveform recoder performance," *Hewlett-Packard J.*, pp. 21–29, Nov. 1982.
[7] K. Uchida, "Testing the dynamic performance of high-speed A/D converters," in *Dig. IEEE Test Conf.*, Nov. 1982, pp. 435–400.

An 8-MHz CMOS Subranging 8-Bit A/D Converter

ANDREW G. F. DINGWALL, MEMBER, IEEE, AND VICTOR ZAZZU, MEMBER, IEEE

Abstract —An 8-bit subranging converter (ADC) has been realized in a 3-μm silicon gate, double-polysilicon capacitor CMOS process. The ADC uses 31 comparators and is capable of conversion rates to 8 MHz at $V_{DD} = 5$ V. Die size is 3.2×2.2 mm^2.

I. INTRODUCTION

LOW-COST monolithic megahertz-rate A/D converters are critical building blocks for video and growing numbers of digital signal processing applications. Megahertz-rate conversions correspond well with signal processing logic throughputs of CMOS and TTL IC's logic families. At present, single-clock-cycle "flash" ADC's are almost exclusively used in MOS systems at megahertz rates, because the design is straightforward and this approach offers maximum speed. Although the "flash" ADC one-clock-cycle IC's [1], (requiring 2^N comparators for an N-bit comparison) are fast, the large numbers of active tracking comparators result in large die sizes, significant input loading, and power dissipation, and demand carefully designed external circuitry to achieve accuracy

A two-clock-cycle "subranging" approach provides a powerful, lower cost alternative to the single-clock-cycle "flash" ADC when maximum speeds are not necessary. Indeed, highest accuracy high-speed hybrid modules utilize the subranging approach since system problems with "flash" ADC's become unmanageable with more than 9 bits due to the large numbers (≥ 1024) of comparators required. At the 8-bit level, a CMOS subranging design requires eightfold fewer ($2^{1+(N/2)} - 1$) comparators. Direct comparisons of this subranging design with similar flash architectures [2] demonstrate significantly less loading problems, and show better accuracy over its rated speed range. This paper discusses a 3-μm design-rule bulk-silicon CMOS 8-bit subranging ADC design which operates to 8 MHz and with accuracy capability to at least the 9-bit level.

II. ADVANTAGES AND DISADVANTAGES OF THE SUBRANGING DESIGN APPROACH

Table I illustrates trade-offs at the 8-bit level for flash and subranging ADC architectures. Potential performance and cost advantages for the subranging approach are en-

Manuscript received April 12, 1985; revised July 24, 1985.
A. G. F. Dingwall is with RCA Laboratories, Princeton, NJ 08544.
V. Zazzu is with RCA Solid State Division, Somerville, NJ 08876.

TABLE I
TRADE-OFFS IN SUBRANGING AND FLASH 8-BIT FLASH A/D ARCHITECTURES

	Flash (1)	Subranging
Total Comparators	256	31
Clock Cycles/Conversion	1	2
Relative Speed	1	0.5
Rel. Input Loading	1	0.12
Rel. Power Dissipation	1	0.2
Rel. Die Size	1	0.4
Typ. Diff Linearity Error	.4 LSB	.3 LSB
Typ. Integral Linearity Error	.7 LSB	.5 LSB

tirely due to reduced number of comparators and consequent smaller chip size. At the 8-bit level, the necessary 256 comparators in flash designs load the input signal as well as the power and reference supplies. Such loading impacts integral linearity, power dissipation, and input signal drive specifications. The subranging approach can offer a typical fivefold improvement in design margins for these such factors; improved margins translate into improved accuracy.

The subranging approach reduces comparator count from 256 to 31. This produces a relatively small die size with input loading and power dissipation greatly reduced. Such reduced loadings make this part more user friendly and of better accuracy than flashes for 1–8 megasamples per second operation. The major disadvantage of the subranging approach is reduction in throughput rate (one-half that of flash); two clock cycles are required per measurement cycle, since the fine measurement cannot start until the coarse measurement is complete. A second potential disadvantage is poorer differential linearity due to piecing the MSB and LSB resistor ladder networks at 16 common tap points. Experimentally, this has not been a significant factor at the 8-bit level when using a design with the intermeshed ladder architecture, discussed subsequently.

III. CONVENTIONAL SUBRANGING ADC ARCHITECTURES

Fig. 1 illustrates the classical 8-bit subranging (two 4-bit steps) A/D converter architecture generally applied in bipolar designs [3]. This partitioning results in fewer comparators [31] than a 5-bit MSB, 3-bit LSB design which

Reprinted from *IEEE J. Solid-State Circuits*, vol. SC-20, no. 6, pp. 1138–1143, Dec. 1985.

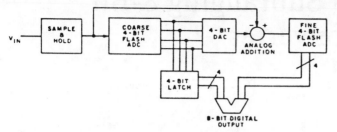

Fig. 1. Conventional 8-bit subranging ADC.

would require 39 comparators. Two clock cycles are required for a complete conversion. In the first cycle, the input signal is sampled and held and the MSB's determination is made. The four MSB's are then applied to a 4-bit DAC having at least 8-bit accuracy. The analog output of the DAC is analog-subtracted from the "held" input signal and finally the (amplified) resultant error voltage is applied to a second ADC which generates the four LSB's of the 8-bit code. While effective, this classical approach is comparatively difficult to implement in CMOS because of the difficulty in performing the analog subtraction of the input signal at megahertz throughput rates.

An alternative subranging ADC architecture which avoids analog subtraction or DAC's was described by Sekino et al. [4]. A 256-tap resistor ladder with 256 differential stages, similar to the inputs of a fully parallel flash architecture, is accessed in two steps. Such multiple input stages are disadvantageous in CMOS implementations since input loading would not be significantly reduced. Reduced speed in this two-clock-cycle architecture would result without the full potential for smaller chip size and reduced input and power supply loading.

IV. THE INTERMESHED LADDER CMOS SUBRANGING ARCHITECTURE

The modified intermeshed ladder architecture, shown in Fig. 2, was developed to maximize speed and to simplify the subranging implementation in CMOS technology. Only 31 input stages and comparators are required as in a classic design. The LSB's determination differs from the first classic approach of Fig. 1, but resembles the Sekino approach in that the need for analog circuits has been eliminated; only high-speed zero-offset switching and an ability to store charge on the inverter gates of the auto-zeroed digital comparators are required. Both functions can be implemented routinely in CMOS operating to tens of megahertz. Circuit simulation indicated that parasitics associated with interconnecting, decoding, and switching LSB tapes in a conventional linear 256-tap CMOS resistor ladder would be large and would severely limit maximum speed. An intermeshed ladder-design layout employs separate polysilicon and diffused resistors for the coarse and fine sections. This combination offers improved switching speed and lower parasitics through use of a compact low capacitance "merged MOS/diffused resistor" fine ladder array.

The central feature of the 8-bit CMOS subranging design, shown in Figs. 2 and 3, is the intermeshed "coarse-fine" resistor network which provides the MSB and LSB reference levels against which the input is tested. As in the conventional architecture, the coarse MSB's ladder has 16 (one for overflow) low resistance sections. A higher resistance, "fine" ladder array is intermeshed (paralleled across) each coarse resistor section; each of the 16 fine ladder sections is tapped at 15 nodes. Several advantages accrue with this architecture. Separate but intermeshed ladders simplify layout, minimize parasitics, allow optimum impedance levels to both sections, and assure monotonicity since the two ladders are common at all MSB/LSB splicing nodes. Further, the 1-of-15 decoding signal required to activate the selection of the LSB's section is directly available at the output logic of the coarse comparator string [2]; thus neither additional cycle time nor circuitry is required for an intermediate recoding to binary to access the LSB ladder of the DAC as is the case in the classical subranging architecture of Fig. 1. Finally, the architecture has shown exceptionally good integral and differential linearity at the splice points because the primary ladder resistances are not significantly loaded due to the small number (≤ 16) of comparators active at any time. Such factors reduce major error sources existing in high comparator-count flash ADC's [2].

V. INTERMESHED LADDER PERFORMANCE

The photomicrograph in Fig. 4 shows details of the intermeshed resistor network. The low-impedance (500-Ω) MSB network is formed of 16 polysilicon resistor sections. The LSB resistor array in parallel with every polysilicon section provides a 16-fold finer resolution for LSB determination. Fine networks are formed of complementary p^+ and n^+ diffusions in a structure merging narrow resistor sections between wider MOS switch source diffusions. The burden of reference accuracy rests with the polysilicon coarse resistor network which must be accurate to an 8-bit level; any deviation from this will affect the integral linearity. The fine resistors however, are only required to match each other to a 4-bit level. The use of complementary MOS switches provides significant cancellation of spuriously injected clock noise. The impedance of each LSB fine ladder section, by design, is more than 32-fold higher than its parallel MSB resistor to ensure less than one-half LSB of interactions during the A/D conversion. Shared output drain diffusions minimize parasitics.

Settling time requirements of the intermeshed resistor array are an important factor in the timing budget near maximum conversion rates. Fig. 5 shows actual ladder settling response for the LSB reference node, as measured with a low-capacitance microprobe, for a full-scale input transition. The measured response roughly approximates a single-pole response with 7-ns time constant and an initial 2-ns turn-on transition. To avoid aliasing, practical A/D systems bandwidth-limit input samples to appreciably less

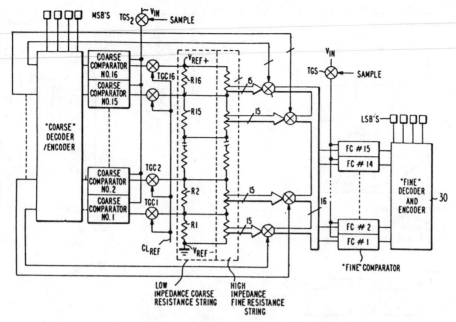

Fig. 2.   Block diagram of CMOS subranging ADC.

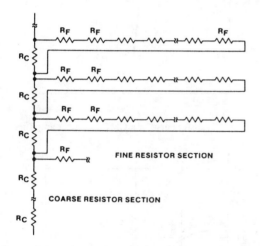

Fig. 3.   Expanded version of intermeshed resistor reference network.

Fig. 4.   Expanded photomicrograph of intermeshed reference network.

than full scale. Approximately, five time constants normally provide suitable settling at the 8-bit accuracy level when inputs are bandwidth limited. At the maximum-rated 8-MHz sample rate, a five-time-constant settling condition of 37 ns represents 30 percent of available 125-ns cycle time.

VI.   TIMING

System timing for the developmental ADC is shown in Fig. 6. The subranging architecture requires four distinct functions:

1) sample unknown;
2) four MSB's determination;
3) four LSB's determination; and
4) output binary data.

The conversion cycle begins as the coarse and fine com-

parators simultaneously sample and hold the unknown input signal on the level shifting capacitor. During this sample interval the comparators are autozeroed. At the completion of the sample cycle, the coarse comparators perform their comparison against 16 reference voltages spaced between V_{ref+} and V_{ref-}. The coarse comparators produce a thermometer bar code, which is recoded by slower binary decoding logic to produce the four output MSB's; the direct 1-of-16 select signal initiates the determination of the four LSB's. Upon selection of the proper fine resistor network, the 15 fine-voltage taps are switched into the fine comparators. Finally, the four LSB and stored MSB data are combined and the resultant data are switched to the three-state outputs.

On-chip timing circuitry automatically optimizes internal clock signals independent of the frequency and duty cycle of the external sample clock signals. This feature assures near optimum performance for users having symmetrical, asymmetrical, or even low-amplitude sine-wave external

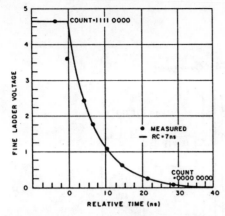

Fig. 5. Measured settling of interdigitated fine ladder network for full-scale 5-V input step versus 7-ns time-constant exponential decay.

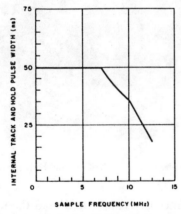

Fig. 6. System timing for CMOS subranging ADC.

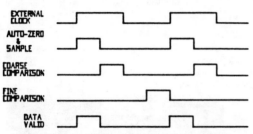

Fig. 7. Reduction of internal track-and-hold/autozero pulsewidth at higher sampling rates.

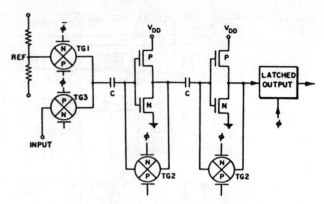

Fig. 8. High-speed CMOS comparator.

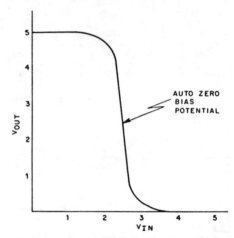

Fig. 9. Transfer curve of a CMOS inverter amplifier.

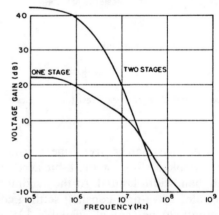

Fig. 10. Computed gain versus frequency bode plot for 3-μm gate-length CMOS autozeroed inverter stages.

clocks. No internal clock signal remains constant in width over all frequencies since its pulsewidth shrinks in a controlled manner with increasing frequency. For low-frequency sample rates, very conservative timings are employed to ensure accuracy. Near maximum frequency rates, internal clock waveforms are internally balanced to minimize overall accuracy loss. This strategy extends useful frequency response without compromising low-frequency performance, and assures an optimum, gradual roll-off in accuracy at highest frequencies of operation.

Fig. 7 illustrates design timing for the input/autozero clocking. At low frequencies, a conservative 50-ns timing sample/autozero timing pulse is employed. As the rated maximum-sample frequency of 8 MHz is approached, the conservative 50-ns timing is progressively reduced to allow

more time for fine ladder settling and comparator settling. Above 10 MHz, the rate of reduction of the sampling is increased more rapidly in a trade-off which ensures best possible operation to 12 MHz.

VII. COMPARATOR

Circuit performance is ultimately determined by simplicity and speed of the autozeroed, sequentially sampled comparator circuit shown in Fig. 8. Identical comparators

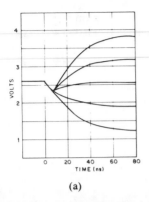

(a)

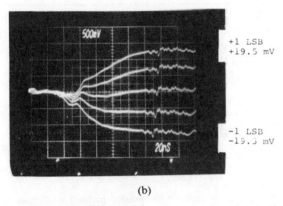

(b)

Fig. 11. Comparison of simulated and measured voltage output of a second-stage autozeroed comparator after application of reference voltage.

are used throughout in both the coarse and fine comparator sections to ensure tracking at the splice points. During the sample period, the unknown input voltage is sampled via $TG3$ and level shift capacitor C and held on the parasitic gate capacitance of the inverter. Simultaneously, the comparator inverters are autozeroed to their toggle point by closing $TG2$.

This establishes the toggle point, which is the point of highest speed and small-signal inverter gain. The transfer curve of a typical size CMOS inverter is shown in Fig. 9.

Standard small-signal analysis of this CMOS inverter, biased at its toggle point, yields expressions for voltage gain A_0 and output impedance in saturation R_0

$$A = V_0/V_1 = (gm_p + gm_n)R_0 \qquad (1)$$

$$R_0 = R_{m0}\|R_{p0} = (rdsp)(rdsn)/(rdsp + rdsn). \qquad (2)$$

For low-frequency operation, typical 3-μm CMOS parameters are $(gm_p + gm_n) = 60$ umho and $R_{n0}\|R_{p0} = 0.2$ MΩ, yielding a predicted small-signal low-frequency voltage gain near 12.

Simulated autozeroed bulk-silicon CMOS inverter gain versus frequency for one and two stages is given by the Bode plot in Fig. 10. Calculated low-frequency small-signal gain of one state is 22 dB (12.5) and two-stage gain is 120 (42 dB) with a unity-gain frequency of 33 MHz. Two gain stages were selected in this design since higher net gain results with input frequencies suitable for this device. The

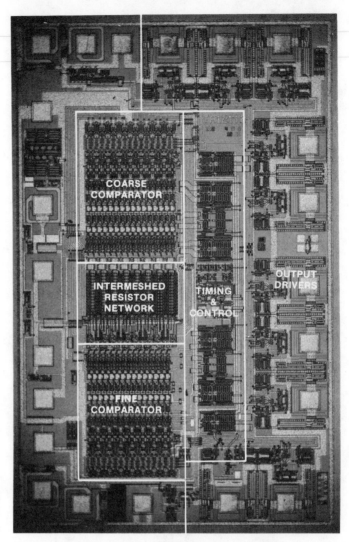

Fig. 12. Chip photomicrograph of TA12172.

resultant experimentally determined ADC 3-dB bandwidth is 30 MHz.

Fig. 11(a) and (b) shows simulated and measured second-stage comparator response for a typical 10-mV (1/2 LSB) input signal steps. Both measurements and simulation suggest that after the reference voltage is applied, 30 ns is sufficient time to produce reliable detection against the previously sampled input to typically 2 mV. At a V_{ref} of 2.5 V the accuracy requirement of the comparator is 5 mV for $\pm 1/2$ LSB operation and this requirement is satisfactorily achieved.

VIII. EXPERIMENTAL

Fig. 12 is a photo of the developmental 8-bit subranging ADC chip. Locations of the MSB and LSB comparators, intermeshed resistor ladders, decoders, and output blocks are delineated. Maximum ratings are 8-MHz sampling rate, although selected devices function to 12 MHz at reduced specifications.

Operation of the chip is illustrated in Fig. 13. Fig. 13(a) compares an input ramp signal with the A/D-digitized and D/A-restored replica at a 5-MHz sample rate, showing

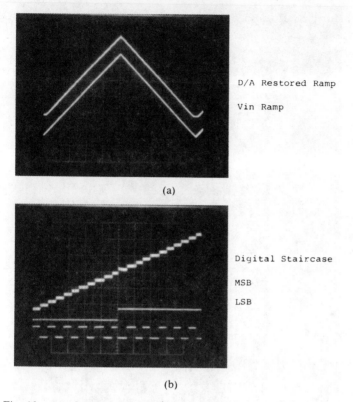

D/A Restored Ramp

Vin Ramp

(a)

Digital Staircase

MSB

LSB

(b)

Fig. 13.   (a)   Input ramp versus D/A restored. (b)   Enlarged section of digital staircase at the midpoint.

TABLE II
TYPICAL MEASURED PERFORMANCE VALUES AT $fs = 5$ MHz and $V_{DD} = 5$ V

Power Dissipation	20mW
Input Current	<.100µA
Differential Linearity	± 0.3 LSB
Integral Linearity	± 0.5 LSB
Input Dynamic Range	0 to V ref
External Reference	3 V to V_{DD}
Output Drive	2 LSSTL Loads
Input Bandwidth (3dB)	30 MHz
Cin	5pf

satisfactory differential linearity and smooth piecings of the separate MSB and LSB measurements. Fig. 13(b) shows an expanded view of differential linearity at a splicing point. Typical chip parameters are given in Table II.

REFERENCES

[1] A. G. F. Dingwall and V. Zazzu, "CMOS on sapphire sparkles for video speed encoder," *Electronics*, vol. 55, no. 9, pp. 137–140, May 5, 1982.
[2] A. G. F. Dingwall, "Monolithic expandable 6 bit 20 MHz CMOS/SOS A/D converter," *IEEE J. Solid-State Circuits*, vol. SC-14, Dec. 1979.
[3] R. J. Van De Plassche and R. E. Van De Grift, "A high speed 7 bit A/D converter," *IEEE J. Solid-State Circuits*, vol. SC-14, pp. 938–943, Dec. 1979.
[4] T. Sekino, "A monolithic 8b two-step paralled ADC without DAC and subtractor circuits," in *ISCC Dig. of Tech. Pap.*, Feb. 1981, pp. 46–47.

A Pipelined 5-Msample/s 9-bit
Analog-to-Digital Converter

STEPHEN H. LEWIS AND PAUL R. GRAY, FELLOW, IEEE

Abstract —A pipelined, 5-Msample/s, 9-bit analog-to-digital (A/D) converter with digital correction has been designed and fabricated in 3-μm CMOS technology. It requires 8500 mil^2, consumes 180 mW, and has an input capacitance of 3 pF. A fully differential architecture is used; only a two-phase nonoverlapping clock is required, and an on-chip sample-and-hold (S/H) amplifier is included.

I. INTRODUCTION

TRADITIONAL designs of high-speed CMOS analog-to-digital (A/D) converters have used parallel (flash) architectures [1]–[13]. While flash architectures usually yield the highest throughput rate, they tend to require large silicon areas because of the many comparators required. An important objective is the realization of high-speed A/D converters in much less area than that required by flash converters so that the A/D interface function can be integrated on the same chip with associated complex, high-speed, image-processing functions. Multistage conversion architectures reduce the required area by reducing the total number of comparators [14]–[19]. Using a pipelined mode of operation in these architectures allows the stages to operate concurrently and makes the maximum throughput rate almost independent of the number of stages. Also, digital correction techniques significantly reduce the sensitivity of the architecture to certain component nonidealities. Pipelined configurations have been previously applied in high-performance board-level converters, but they have not been applied to monolithic CMOS A/D converters because of the difficulty of realizing high-speed interstage sample-and-hold (S/H) gain functions in CMOS technologies. In this paper, an experimental four-stage pipelined A/D converter with digital correction that has 9-bit resolution and 5-Msample/s conversion rate in a 3-μm CMOS technology is described. The experimental converter uses high-speed differential switched-capacitor circuitry to carry out the interstage gain functions.

This paper is divided into four additional parts. In Section II, pipelined A/D architectures are described con-

ceptually, and their advantages over flash and two-step subranging architectures [17] are explained. In Section III, the error sources present in pipelined A/D converters are identified, and the way in which digital correction eliminates the effects of some of these errors is shown. In Section IV, the circuits in an experimental prototype are described. Finally, experimental results from the prototype converters are given in Section V.

II. CONCEPTUAL DESCRIPTION

A block diagram of a general pipelined A/D converter with k stages is shown in Fig. 1. Each stage contains an S/H circuit, a low-resolution A/D subconverter, a low-resolution digital-to-analog (D/A) converter, and a differencing fixed-gain amplifier. In operation, each stage initially samples and holds the output from the previous stage. Each stage then does a low-resolution A/D conversion on the held input, and the code just produced is converted back into an analog signal by a D/A converter. Finally, the D/A converter output is subtracted from the held input, producing a residue that is amplified and sent to the next stage.

The primary potential advantages of the pipelined architecture are high throughput rate and low hardware cost. The high throughput rate of the pipelined architecture stems from concurrent operation of the stages. At any time, the first stage operates on the most recent sample, while the next stage operates on the residue from the previous sample, and so forth. If the A/D subconversions are done with flash converters, a pipelined architecture only needs two clock phases per conversion. Flash architectures also require two clock phases per conversion, one each for sampling and A/D conversion, and use pipelining to do the digital decoding operation. The throughput rate of flash converters is maximized because their pipelined information is entirely digital and can be transferred to 1-bit accuracy in less time than it takes to generate and transfer the analog residue in a pipelined multistage architecture. The area and consequent manufacturing cost of pipelined converters is small compared to those of flash converters, however, because pipelined converters require fewer comparators than flash converters. For example, the 9-bit prototype pipelined converter described in Section IV

Manuscript received May 8, 1987; revised July 24, 1987. This work was supported by DARPA under Contract N00039-C-0107 and by the National Science Foundation under Contract DCI-8603430.

The authors are with the Electronics Research Laboratory, University of California, Berkeley, CA 94720.

IEEE Log Number 8716973.

Reprinted from *IEEE J. Solid-State Circuits*, vol. SC-22, no. 6, pp. 954–961, Dec. 1987.

343

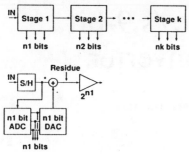

Fig. 1. Block diagram of a general pipelined A/D converter.

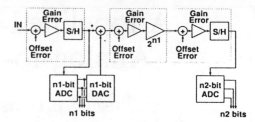

Fig. 2. Block diagram of a two-stage pipelined A/D converter with offset and gain errors.

uses 28 comparators and requires a core area of 8500 mil^2 in a 3-μm CMOS technology. A 9-bit flash converter would use 512 comparators and would be more than ten times larger than the pipelined prototype in the same technology. Not only is the area small for pipelined converters, but also it is linearly related to the resolution because if the necessary accuracy can be achieved through calibration or trimming, the resolution can be increased by adding stages to the end of the pipeline without increasing the number of clock phases required per conversion. In contrast, flash and subranging architectures need exponential, rather than linear, increases in area to increase their resolution and also require trimming or calibration for greater than 8- or 9-bit linearity.

Other advantages of the pipelined architecture stem from the use of S/H amplifiers to isolate the stages. First, because an S/H amplifier can also be used on the input of the A/D converter, pipelined architectures can accurately sample high-frequency input signals. Second, the interstage gains from these amplifiers diminish the effects of non-idealities in all stages after the first stage on the linearity of the entire conversion; furthermore, this allows the converter to use a digital correction technique in which non-linearity in the A/D subconversions has little effect on the overall linearity. This subject is presented in Section III.

The main disadvantage of pipelined A/D converters is that they require the use of operational amplifiers (op amps) to realize parasitic-insensitive S/H amplifiers. Although the S/H amplifiers improve many aspects of the converter performance, the op amps within the S/H amplifiers limit the speed of the pipelined converters. In contrast, op amps are not required in subranging architectures. Because high-speed op amps are difficult to realize, a common goal in the design of subranging A/D converters is to avoid using op amps. If op amps are not used, however, it is impossible to realize parasitic-insensitive S/H amplifiers. The consequent high-frequency input sampling is poor, stage operation is sequential, and tolerance to error sources in stages after the first is unimproved from that of the first stage. Also, flash converters usually do not use an input S/H amplifier because of the difficulty in realizing an op amp in CMOS technologies that is fast enough to drive the inherently large input load. Therefore, flash converters often suffer reduced performance at high input signal frequencies.

III. ERROR SOURCES

The primary error sources present in a pipelined A/D converter are offset errors in the S/H circuits and amplifiers, gain errors in the S/H circuits and amplifiers, A/D subconverter nonlinearity, D/A subconverter nonlinearity, and op-amp settling-time errors. With digital correction, as shown below, the effects of offset, gain, and A/D subconverter nonlinearity are reduced or eliminated; therefore, the D/A converter nonlinearity and op-amp settling-time errors limit the performance of pipelined A/D converters. To begin the error analysis, the effects of offset and gain errors are considered next.

A block diagram of a two-stage pipelined A/D converter with offset and gain errors in each of the S/H circuits and the interstage amplifier is shown as a representative example in Fig. 2. The nonideal S/H circuits and interstage amplifier are replaced by ideal elements in series with gain and offset errors, and each of these replacements is surrounded by a dotted line. The gain error in the first-stage S/H circuit changes the conversion range of the A/D converter and does not affect linearity. The gain errors in the interstage amplifier and second-stage S/H circuit can be combined into one equivalent error that does affect linearity. However, because the interstage gain only has to be accurate enough to preserve the linearity of the stages after the first stage, the effect of this gain error on linearity is small. For example, if both stages in Fig. 2 have 4-bit resolution, and if the only error is in the gain of the interstage amplifier, the interstage amplifier gain should be equal to 16 and must be accurate to within ± 3 percent.

The offset error in the first-stage S/H circuit causes an input-referred offset but does not affect linearity. The offset errors in the interstage amplifier and second-stage S/H circuit can be combined into one equivalent offset that does not affect linearity if digital correction is used. Because addition is commutative, the equivalent offset can be pushed to the left of the first-stage subtractor. To move the equivalent offset to the input branch, where is causes an input-referred offset, an equal but opposite offset must be inserted in the first-stage A/D subconverter branch. As shown below, the effect of the offset in the first-stage A/D subconverter is eliminated by the digital correction.

Next, the effect of nonlinearity in the first-stage A/D subconversion is considered. A block diagram of one stage in a pipelined A/D converter is shown in Fig. 3(a). A 2-bit stage is used as a representative example. Nonlinearity in

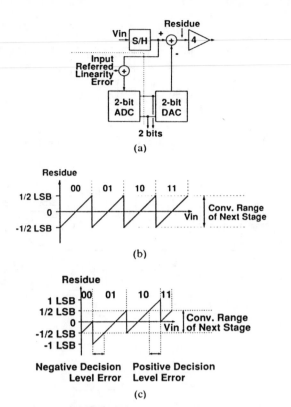

(a)

(b)

Negative Decision   Positive Decision
Level Error   Level Error

(c)

Fig. 3. (a) Block diagram of one 2-bit stage in a pipelined A/D converter. (b) Ideal residue versus input. (c) Residue versus input with A/D subconverter nonlinearity.

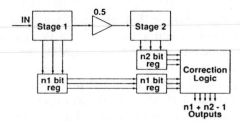

Fig. 4. Block diagram of a two-stage pipelined A/D converter with digital correction.

the A/D subconverter is modeled as an input-referred linearity error. The effect of this nonlinearity is studied by examining plots of the residue versus the input. Two such plots are shown in Fig. 3(b) and (c).

In Fig. 3(b), both the A/D subconverter and the D/A converter are assumed to be ideal. The plot has a sawtooth shape because when the input is between the decision levels determined by the A/D subconverter, the A/D subconverter and D/A converter outputs are constant; therefore, the residue rises with the input. When the input crosses a decision level, the A/D subconverter and D/A converter outputs increase by one least significant bit (LSB) at a 2-bit level, so the residue decreases by 1 LSB. Here, the residue is always between ±1/2 LSB and consists only of the part of the input that is not quantized by the first stage. With the interstage gain equal to 4, the maximum residue is amplified into a full-scale input to the next stage; therefore, the conversion range of the next stage is equal to the maximum residue out of the first stage.

A similar curve is shown in Fig. 3(c) for a case when the A/D subconverter has some nonlinearity, but the D/A converter is still ideal. In this example, two of the A/D subconverter decision levels are shifted, one by −1/2 LSB and the other by +1/2 LSB. When the input crosses a shifted decision level, the residue decreases by 1 LSB. If the decision levels are shifted by less than 1/2 LSB, the residue is always between ±1 LSB. Here, the residue consists of both the unquantized part of the input and the error caused by the A/D subconverter nonlinearity. Be-

cause the D/A converter is assumed to be ideal, these increased residues are accurate for the codes to which they correspond; therefore, at this point, no information is lost. If the interstage gain is still 4, however, information is lost when the larger residues saturate the next stage and produce missing codes in the conversion. Therefore, if the conversion range of the second stage is increased to handle the larger residues, they can be encoded and the errors corrected. This process is called digital correction [20], [21] and is described next.

A block diagram of a two-stage pipelined A/D converter with digital correction is shown in Fig. 4. The new elements in this diagram are the pipelined latches, the digital correction logic circuit, and the amplifier with a gain of 0.5. The amplifier with a gain of 0.5 is conceptual only and is drawn to show that the interstage gain is reduced by a factor of 2 so that nonlinearity error in an amount between ±1/2 LSB at a $n1$-bit level in the first-stage A/D subconversion does not produce residues that saturate the second stage. If the first stage is perfectly linear, only half the conversion range of the second stage is used. Therefore, 1 bit from the second stage is saved to digitally correct the outputs from the first stage; the other $n2-1$ bits from the second stage are added to the overall resolution. After the pipelined latches align the outputs in time so that they correspond to one input, the digital correction block detects overrange in the outputs of the second stage and changes the output of the first stage by 1 LSB at a $n1$-bit level if overrange occurs. Digital correction improves linearity by allowing the converter to postpone decisions on inputs that are near the first-stage A/D subconverter decision levels until the residues from these inputs are amplified to the point where similar nonlinearity in later-stage A/D subconverters is insignificant.

To do the digital correction, a correction logic circuit is required. Also, if flash converters are used in the stages, all stages after the first require twice as many comparators as without digital correction. The logic is simple, however, and none the comparators needs to be offset canceled.

It is shown above that with digital correction, nonlinearity in the A/D subconverters can be corrected if the D/A converter is ideal. Therefore, the D/A converter in the first stage determines the linearity of the entire A/D converter. Such D/A converters can be realized with resistor strings for linearities in the 8–9-bit range. For integral linearity greater than 9 bits, the design of such a D/A converter is not trivial and either requires calibration or trimming. Also, fast settling op amps are required to do analog subtraction and amplification at the sampling rate of the A/D converter. The 3-μm CMOS prototype described in Section IV is able to do these functions at 5 Msample/s. The maximum speed of such processing increases in scaled technologies, and video conversion rates should be achievable in 1.5–2-μm CMOS technologies.

IV. PROTOTYPE

Several important design considerations for the prototype converter are now presented. To minimize design time, assume that all stages are identical. Fast op amps and flash subconverters are used to operate at as high a speed as possible. The most basic architectural decision is to choose the resolution per stage; for efficient use of the conversion range of each stage, this choice determines the corresponding value of interstage gain. To attain maximum throughput rate, the resolution per stage should be small so that the interstage gain is small and the corresponding closed-loop bandwidth of the gain block is large. Conversely, large resolution and corresponding gain per stage are desirable to achieve high linearity because the contributions of nonidealities in all stages after the first are reduced by the combined interstage gain preceding the nonideality. Thus the speed and linearity requirements conflict in determining the optimum resolution per stage. It also can be shown under certain simplifying assumptions that to minimize the amount of required hardware, the optimum resolution per stage is about 3 or 4 bits per stage, which is about midway between the high and low end. This compromise in the resolution per stage keeps both the number of op amps and the number of comparators small. Finally, because the goal of this project was to realize an A/D converter small enough that it could be incorporated within a primarily digital chip, the A/D converter must be able to operate in the presence of large power supply noise caused by the digital circuits. To reduce the sensitivity of the converter to this noise, all analog signal paths in the prototype are fully differential.

To meet these requirements, the prototype is divided into four stages with 3 bits produced per stage. A block diagram of one stage is shown in Fig. 5. The A/D subconversions are done with flash converters, so each stage needs seven comparators. The S/H amplifier block replaces both the S/H circuit and interstage amplifier shown in earlier figures. Because the interstage gain is 4 instead of 8, half the range and one bit from each of the last three stages are saved to digitally correct the outputs of the previous stages. Thus, instead of obtaining 3 bits of resolution from each of these stages, only 2 bits of resolution are obtained from each. The digital correction is done off the chip. In total, 9 bits of resolution are produced, using 28 comparator and four op amps.

The S/H amplifier block is expanded in Fig. 6(a). Fig. 6(b) shows that the clock is divided into two nonoverlapping phases. On clock phase ϕ_1, the input is sampled onto

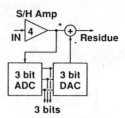

Fig. 5.   Block diagram of one stage in the prototype.

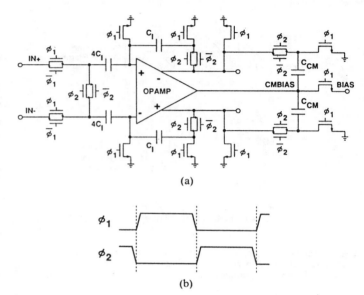

(a)

(b)

Fig. 6.   (a) Schematic of S/H amplifier. (b) Timing diagram of a two-phase nonoverlapping clock.

the $4C_I$ capacitors, and the integrating C_I and common-mode feedback C_{CM} capacitors are reset. On ϕ_2, the left sides of the sampling capacitors are connected together so the difference between the two sampled inputs is amplified by the ratio of the sampling to integrating capacitors. To the extent that the op amp in a closed-loop configuration drives its differential input to zero, the gain is insensitive to parasitic capacitances on either the top or bottom plates of any of these capacitors. Meanwhile, the common-mode feedback (CMFB) capacitors are connected to the outputs of the op amp to start the CMFB circuit. Switched-capacitor CMFB is useful in pipelined A/D converters because pipelined converters inherently allow a clock phase needed to reset the capacitor bias.

As a result of the use of digital correction, the offsets of all the op amps are simply referred to the input of the A/D converter, each in an amount diminished by the combined interstage amplifier gain preceding the offset. Therefore, the op amps do not have to be offset canceled and do not have to be placed in a unity-gain feedback configuration. Since the op amps do not have to be unity-gain stable, their speed can be optimized for a closed loop gain of 4. The op amp, shown in Fig. 7, uses a fully differential, class A/B configuration with dynamic bias. The class A/B structure gives both high slew rate and high gain after slewing. According to simulation, the amplifier dissipates 20 mW and settles in 50 ns to an accuracy of 0.1 percent with a 5-V differential step into a 4-pF load.

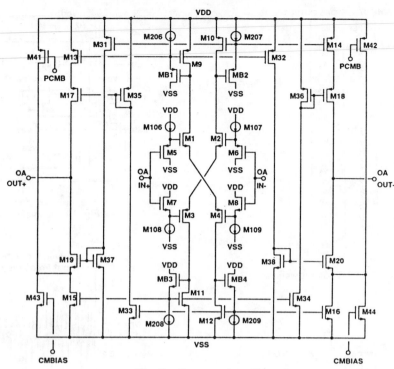

Fig. 7. Op-amp schematic.

The op amp is similar to one reported by Castello and Gray [22], and its operation is now described. Transistors $M_1 - M_4$ form the input stage and generate the class A/B action. Source followers $M_5 - M_8$ are used to bias the input stage so that it conducts some current even for zero differential input. For an increase in the voltage on the positive input and a corresponding decrease on the negative input, the gate-to-source voltages of both M_1 and M_4 increase while those of M_2 and M_3 decrease; therefore, the current in M_1 and M_4 increases and that in M_2 and M_3 decreases from their standby values. Transistors M_9 and M_{13}, M_{10} and M_{14}, M_{11} and M_{15}, and M_{12} and M_{16} form current mirrors that reflect and amplify current from the input branches to the output branches. Cascode transistors $M_{17} - M_{20}$ increase the gain of the op amp by increasing the output resistance of the output nodes to ground. A high-swing dynamic bias circuit composed of transistors $M_{31} - M_{38}$ adjusts the gate bias on the cascode transistors so that the output branches can conduct large currents during slewing and have high swings during settling. Transistors $M_{41} - M_{44}$ together with the C_{CM} capacitors and associated switches in Fig. 6(a), form the CMFB circuit. Because the gates of M_{41} and M_{42} are tied to a constant bias voltage, these transistors are constant-current sources. The gates of M_{43} and M_{44} are connected to the CMBIAS terminal shown in Fig. 6(a). This point is alternatively switched from a bias voltage on ϕ_1 to a capacitively coupled version of the output on ϕ_2. During ϕ_2, the CMBIAS point rises and falls with changes in the common-mode output voltage. This change adjusts the current drawn through M_{43} and M_{44} so that the common-mode output voltage is held constant near 0 V. Note that if the two halves of the differential circuit match perfectly,

changes in the differential output voltage do not change the CMBIAS point.

Because the speed of this op amp is limited by the speed of its current mirrors, wide-band current mirrors are used to increase the speed. To this end, transistors $M_9 - M_{12}$ are not simply diode connected, but instead are buffered by source followers $MB_1 - MB_4$. Because of this change, the currents needed to supply the parasitic capacitance between the gates and sources of the current mirrors at high frequencies come from the power supplies instead of from the input branch. The drawback to this approach is that the drain-to-source voltages of transistors $M_9 - M_{12}$ are increased by the gate-to-source voltages of transistors $MB_1 - MB_4$, respectively. Therefore, input stage transistors $M_1 - M_4$ operate with less drain-to-source voltage than if $M_9 - M_{12}$ were diode connected. As a result, $M_1 - M_4$ enter the triode region for smaller differential inputs than with diode-connected loads, and the amount of current that the input stage can produce while slewing is limited. Because a high-swing dynamic bias circuit is used, this is not a problem for ± 5-V operation; however, for $+5$-V operation, these wide-band current mirrors probably would limit the slew rate of the op amp.

A block diagram of an A/D, D/A subsection is shown in Fig. 8. To save area, one resistor string is shared for both the A/D and D/A functions. The resistor string divides the reference into equal segments and provides the boundaries between these segments as thresholds for a bank of comparators. The comparators are clocked at the end of ϕ_2. On ϕ_1, eight D/A converter outputs are enabled and one is selected based on control signals generated from the comparator outputs ($y_1, \cdots, y_8$). Although Fig. 8 shows a single-ended representation of both the

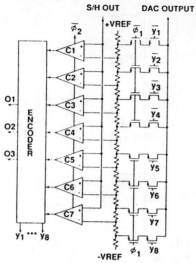

Fig. 8. Block diagram of A/D, D/A subsection.

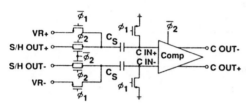

Fig. 9. Connection of comparator with A/D, D/A subsection.

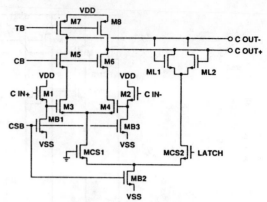

Fig. 10. Comparator schematic.

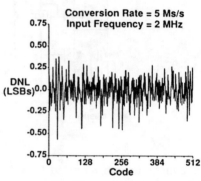

Fig. 11. DNL versus code.

A/D subconverter and D/A converter functions, on the prototype, both functions are fully differential. Therefore, instead of just one D/A converter output, equal and opposite D/A converter outputs are used. Also, each comparator compares a differential input to a differential reference instead of a single-ended input to a single-ended reference.

The connection of a comparator within an A/D, D/A subsection is shown in Fig. 9. The points labeled $VR+$ and $VR-$ are connected to taps on the resistor string that depend on which comparator in the bank is under consideration. For example, for the top comparator, $VR+$ is connected to the most positive A/D subconverter tap, and $VR-$ is connected to the most negative A/D subconverter tap. On clock ϕ_1, the comparator inputs are grounded, and the capacitors sample the differential reference. On ϕ_2, the left sides of the capacitors are connected to the differential input. Ignoring parasitic capacitance, the input to the comparator is then the difference between the differential input and the differential reference. The parasitic capacitances on the inputs to the comparator attenuate the input slightly, but the decision is not affected if the comparator has enough gain. As mentioned in Section III, because of digital correction, no offset cancellation on the comparator is required. Therefore, the comparator is never placed in a feedback loop and does not have to be stable in a closed-loop configuration.

The comparator, shown in Fig. 10, uses a conventional latched-differential-amplifier configuration. Transistors M_1 and M_2 are source followers. Transistors $M_3 - M_8$ form a

differential amplifier, and ML_1 and ML_2 form a latch. Transistors MCS_1 and MCS_2 form a current switch that allows the bias current from MB_2 to flow through either the differential amplifier or the latch. With the latch signal low, the inputs are amplified. Because M_7 and M_8 are biased in the triode region, the gain of the amplifier is only about 20 dB. When the latch signal is raised, the bias current is switched from the amplifier to the latch. During the transition, the parasitic capacitances on the inputs to the latch hold the amplified input. Finally, the latch switches, and the comparison is completed.

V. EXPERIMENTAL RESULTS

As mentioned in Section IV, the digital correction is done off chip. This allows tests to be run to evaluate the need for the correction. Unless stated otherwise, all results are obtained using the full correction; that is, digital correction is applied to the first three stages. The prototype has been tested primarily in two ways [23], [24]: first with a code density test, and second with a signal-to-noise ratio (SNR) test. Both tests have used high- and low-frequency input signals. Results of the code density test are shown in Figs. 11 and 12.

In Fig. 11, differential nonlinearity (DNL) is plotted on the y axis versus code on the x axis for all 512 codes. The conversion rate is 5 Msample/s, and the input frequency is 2 MHz. Because the DNL never goes down to -1 LSB,

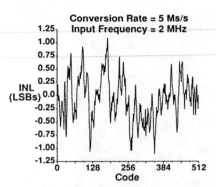

Fig. 12.   INL versus code.

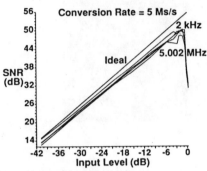

Fig. 13.   SNR versus input level.

TABLE I
DATA SUMMARY OVER INPUT FREQUENCY VARIATION
9-bit Resolution; 5-Msample/s Conversion
Rate; ±5-V Power Supplies

Input Frequency	2 kHz	2 MHz	5.002 MHz
Peak DNL (LSB)	0.5	0.6	0.5
Peak INL (LSB)	1.0	1.1	1.2
Peak SNR (dB)	50	50	49

there are no missing codes. The maximum DNL is less than 0.6 LSB.

In Fig. 12, integral nonlinearity (INL) is plotted on the y axis versus code on the x axis. Again, the conversion rate is 5 Msample/s and the input frequency is 2 MHz. The maximum INL is 1.1 LSB. The nonideality in the curve is caused by both nonlinearity in the first-stage D/A converter and incomplete settling of the first-stage op-amp output.

Under the same conditions as in Figs. 11 and 12 but with the digital correction completely disabled, the maximum DNL and INL are about 10 LSB at a 9-bit level, owing to comparator offsets. If the correction is applied only on the first stage, the maximum DNL and INL drop to about 3 LSB. When digital correction is applied on the first two stages, the maximum DNL is about 0.9 LSB and the maximum INL is about 1.5 LSB; therefore, there are no missing codes in this case. Also, the uncorrected histogram data from the code density test show that there are no codes for which any residue is greater than the reference level for comparator C_1 or less than the reference level for comparator C_7 as labeled in Fig. 8. This means that the maximum absolute value of nonlinearity in an A/D subconversion is less than or equal to 1/4 LSB at a 3-bit level, and the full digital correction range (±1/2 LSB) is not used. Therefore, comparators C_1 and C_7 are not needed in the last three stages.

SNR measurements were made by taking fast Fourier transforms on 1024 samples from the A/D converter at the downsampled rate of 20 kHz while the converter was running at 5 Msample/s. In Fig. 13, SNR is plotted on the y axis versus input level on the x axis for five input

frequencies: 2 kHz, 22 kHz, 202 kHz, 2.002 MHz, and 5.002 MHz. The curve for 5.002 MHz represents a beat frequency test on the converter when compared to the curve for 2 kHz because the converter is running at the difference between these two frequencies or 5 Msample/s. An ideal 9-bit curve is also shown. The peak SNR is around 50 dB instead of 56 dB, as would be expected with a 9-bit converter; this difference is accounted for by distortion generated from the INL for large input signals. When the input signal is reduced in amplitude, the distortion is reduced and the real curves approach the ideal 9-bit curve. Note that there is little difference in the curves for different input frequencies, showing that the first-stage S/H amplifier is able to accurately sample high-frequency input signals.

The results of the code density and SNR tests for variations in the input frequency are summarized in Table I. Peak DNL, INL, and SNR are shown for three input frequencies, and the performance is almost constant. This is important because it shows that the first-stage S/H amplifier is able to accurately sample high-frequency input signals.

A photograph of the core of a prototype chip is shown in Fig. 14. The core is about 50 mil high by 150 mil wide. The stages follow one after another and are identical except that the fourth stage does not have a D/A converter or a subtractor and the two-phase nonoverlapping clock alternates from stage to stage. A test op amp and a test comparator are at the end. The prototype was made by MOSIS in a 3-μm, double-polysilicon, p-well, CMOS process.

VI.   SUMMARY

This paper reports on a prototype pipelined A/D converter with typical characteristics summarized in Table II. In summary, the prototype demonstrates that pipelined architectures and digital correction techniques are of potential interest for high-speed CMOS A/D conversion applications.

ACKNOWLEDGMENT

The authors gratefully acknowledge the help of R. Kavaler and J. Doernberg with the testing of the prototype converters.

Fig. 14.   Photograph of the core of the prototype.

TABLE II
TYPICAL PERFORMANCE: 25°C

Technology	3-u CMOS
Resolution	9 bits
Conversion Rate	5 Ms/s
Area*	8500 mils^2
Power Supplies	±5 V
Power Dissipation	180 mW
Input Capacitance	3 pF
Input Offset	< 1 LSB
CM Input Range	±5 V
DC PSRR	50 dB

*Does not include clock generator, bias generator, reference generator, digital error correction logic, and pads.

REFERENCES

[1] A. G. F. Dingwall, "Monolithic expandable 6 bit 20 MHz CMOS/SOS A/D converter," *IEEE J. Solid-State Circuits*, vol. SC-14, pp. 926–932, Dec. 1979.

[2] J. G. Peterson, "A monolithic video A/D converter," *IEEE J. Solid-State Circuits*, vol. SC-14, pp. 932–937, Dec. 1979.

[3] T. Takemoto *et al.*, "A fully parallel 10-bit A/D converter with video speed," *IEEE J. Solid-State Circuits*, vol. SC-17, pp. 1133–1138, Dec. 1982.

[4] Y. Fujita *et al.*, "A bulk CMOS 20MS/s 7b flash ADC," in *ISSCC Dig. Tech. Papers* (San Francisco, CA), Feb. 1984, pp. 56–57.

[5] A. G. F. Dingwall and V. Zazzu, "High speed CMOS A/D and D/A conversion," in *Dig. Tech. Papers, 1984 IEEE Int. Symp. Circuits Syst.*, May 1984, pp. 420–424.

[6] M. Inoue *et al.*, "A monolithic 8-bit A/D converter with 120 MHz conversion rate," *IEEE J. Solid-State Circuits*, vol. SC-19, pp. 837–841, Dec. 1984.

[7] T. Tsukada *et al.*, "CMOS 8b 25 MHz flash ADC," in *ISSCC Dig. Tech. Papers* (New York, NY), Feb. 1985, pp. 34–35.

[8] A. K. Joy *et al.*, "An inherently monotonic 7-bit CMOS ADC for video applications," *IEEE J. Solid-State Circuits*, vol. SC-21, pp. 436–440, June 1986.

[9] T. Kumamoto *et al.*, "An 8-bit high-speed CMOS A/D converter," *IEEE J. Solid-State Circuits*, vol. SC-21, pp. 976–982, Dec. 1986.

[10] B. Peetz, B. D. Hamilton, and J. Kang, "An 8-bit 250 megasample per second analog-to-digital converter: Operation without a sample and hold," *IEEE. J. Solid-State Circuits*, vol. SC-21, pp. 997–1002, Dec. 1986.

[11] Y. Yoshii *et al.*, "An 8b 350MHz flash ADC," in *ISSCC Dig. Tech. Papers* (New York, NY), Feb. 1987, pp. 96–97.

[12] Y. Akazawa *et al.* "A 400MSPS 8b flash AD conversion LSI," in *ISSCC Dig. Tech. Papers* (New York, NY), Feb. 1987, pp. 98–99.

[13] J. Corcoran *et al.*, "A 1GHz 6b ADC system," in *ISSCC Dig. Tech. Papers* (New York, NY) Feb. 1987, pp. 102–103.

[14] R. J. van de Plassche, "A high-speed 7 bit A/D converter," *IEEE J. Solid-State Circuits*, vol. SC-14, pp. 938–943, Dec. 1979.

[15] R. A. Blauschild, "An 8b 50ns monolithic A/D converter with internal S/H," in *ISSCC Dig. Tech. Papers* (New York, NY) Feb. 1983, pp. 178–179.

[16] R. E. J. van de Grift and R. J. van de Plassche, "A monolithic 8-bit video A/D converter," *IEEE J. Solid-State Circuits*, vol. SC-19, pp. 374–378, June 1984.

[17] A. G. F. Dingwall and V. Zazzu, "An 8-MHz CMOS subranging 8-bit A/D converter," *IEEE J. Solid-State Circuits*, vol. SC-20, pp. 1138–1143, Dec. 1985.

[18] R. E. J. van de Grift and M. van der Veen, "An 8b 50 MHz ADC with folding and interpolation techniques," in *ISSCC Dig. Tech. Papers* (New York, NY), Feb. 1987, pp. 94–95.

[19] S. H. Lewis and P. R. Gray, "A pipelined 5MHz 9b ADC," in *ISSCC Dig. Tech. Papers* (New York, NY), Feb. 1987, pp. 210–211.

[20] O. A. Horna, "A 150 Mbps A/D and D/A conversion system," *Comsat Tech. Rev.*, vol. 2, no. 1, pp. 52–57, 1972.

[21] S. Taylor, "High speed analog-to-digital conversion in integrated circuits," Ph.D. dissertation, Univ. of Calif., Berkeley, pp. 30–42, 1978.

[22] R. Castello and P. R. Gray, "A high-performance micropower switched-capacitor filter," *IEEE J. Solid-State Circuits*, vol. SC-20, pp. 1122–1132, Dec. 1985.

[23] W. A. Kester, "Characterizing and testing A/D and D/A converters for color video applications," *IEEE Trans. Circuits Syst.*, vol. CAS-28, pp. 539–550, July 1978.

[24] J. Doernberg, H. Lee, and D. A. Hodges, "Full-speed testing of A/D converters," *IEEE J. Solid-State Circuits*, vol. SC-19, pp. 820–827, Dec. 1984.

A Monolithic 12b+Sign Successive Approximation A/D Converter

J. John Connolly, Thomas O. Redfern, Sing W. Chin and Thomas M. Frederiksen

National Semiconductor Corp.

Santa Clara, CA

SUCCESSIVE APPROXIMATION A/Ds, with resolutions of 12b and higher, are limited by the performance of the comparator. The comparator problems of accuracy, response time, stability and noise are significantly improved using sampled data-techniques. New A/D systems are made possible by a multiple input comparator which allows an arbitrary number of differential voltage inputs[1]. The successive approximation logic forces the sum of the output voltages from the multiple DACs to equal the analog input voltage to within ± 1/2 LSB. The comparator, Figure 1, makes use of capacitors to scale the multiple differential voltages and to convert them to input charges. Equilibrium is detected when the net change in charge at the summation node is zero.

The sign of the differential analog inputs is determined by the order in which the inputs are sampled relative to the auto-zero time of the comparator. By controlling the order of sampling, negative inputs can be made to appear positive with reference to the DACs. Thus the DACs only require a single positive voltage reference to convert bipolar inputs. This provides an additional bit of resolution and eliminates the inaccuracies inherent in techniques requiring reference polarity inversion. This sign magnitude format is internally converted to 2's complement by subtracting 1LSB and taking the one's complement of the result. The 1LSB subtraction is accomplished by changing the sign of the +1/2LSB DAC offset for negative inputs. This analog subtraction replaces a 13b digital adder and is again accomplished with simple timing changes.

The A/D, Figure 2, uses two 6b DACs. Each 6b DAC consists of two three-bit DACs, a coarse and a fine. The coarse DAC divides the voltage present at its reference input into eight equal voltages. The fine DAC further subdivides the bottom voltage ($V_{REF}/8$) into eight equal sub-voltages. The differential voltage action of the comparator allows the fine DAC to be fitted between any adjacent voltage taps of the coarse DAC. This allows the reference input to be divided into 64 equal voltages which provides the desired 6b DAC. The voltage which is applied to the LS DAC is reduced by 1/8 and the input capacitor to the comparator is also reduced by 1/8 to provide the 64:1 reduction needed for the LS 6-bit group.

Linearity trimming is required to achieve the highest possible yield. Since the untrimmed accuracy of the DACs is at least 9b, only the three MSBs need be trimmed. A correction voltage is supplied to the comparator for each of the eight voltages corresponding to the eight states of the three MSBs. This correction voltage is derived from the LS DAC by a second set of decoding switches and has a range of ± 16 LSBs with a resolution of 1/4LSB. A 7b word from the MS LPROM (Laser-Programmable ROM), Figure 3, determines the sign and magnitude of the cor-

rection voltage. The three MSBs of the SAR latch (not the sign bit) provide address information for this LPROM. One of eight possible trim corrections is set as the successive approximation logic searches the three MSBs.

In the initial calibration, 1LSB of the MS DAC (64 LSBs of the A/D) is used as the basic reference. The LS LPROM is used to adjust the $V_{REF}/8$ input to the LS DAC such that the full scale contribution of this DAC is just equal to 1LSB of the MS DAC.

LPROM trimming has many advantages. The LPROM is not physically associated with the precision resistor networks and will not upset the accuracy of this critical area. Adjusting the resistor network directly with partial laser cuts is efficient, but unfortunately causes changes in the resistor value with time and temperature cycling, due to the development of microcracks which propagate from the damaged area. A complete resistor cut (to an open condition) is more stable and reliable and therefore was used. Electrically forcing corrections greatly reduces test time: no laser cutting is required until the last step in the test sequence. Finally, the repositioning time of the laser is minimized because the programming areas of the LPROMs are X-Y arrays. The details of the LPROM are shown in Figure 3.

This A/D converter has processor I/O capabilities approaching, and in some cases exceeding, board level data acquisition systems. These include direct interfaces to most popular microprocessors* and their derivatives, minicomputers and DMA controllers. Output data format control is also provided. In addition, output signals to control the timing of an external analog multiplexer and a sample and hold circuit are available.

The total operating power is 25mW which includes power drain from V_{REF} (5V), Analog V_{CC} (5V), Digital V_{CC} (5V) and the V_{EE} (−5V) supply. The V_{EE} supply is only required if either of the analog inputs goes below ground.

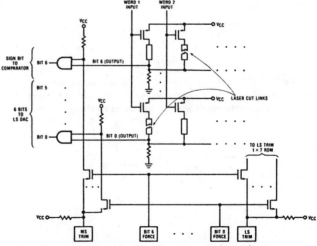

FIGURE 3—Details of LPROM showing SiCr programming links and the forcing inputs used at wafer sort to determine linearity corrections before the links are cut.

---

*8080, Z-80, 6800.

[1] Redfern, T.P., Connolly, J.J., Chin, S.W. and Frederiksen, T.M., "A Monolithic Charge-Balancing Successive-Approximation A/D Technique", *ISSCC DIGEST OF TECHNICAL PAPERS*, p. 176-177; Feb., 1979.

Reprinted from *1980 IEEE Int. Solid-State Circuits Conf.*, vol. 23, pp. 12–13, Feb. 1980.

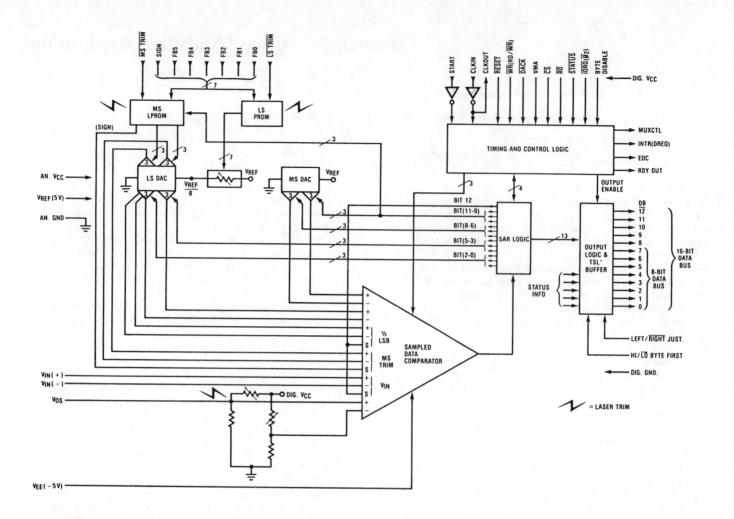

FIGURE 2—Functional diagram of monolithic 13b A/D
converter showing interconnection of multiple input sampled
data comparator to the two six-bit Distributed-DACs
(D DACs).

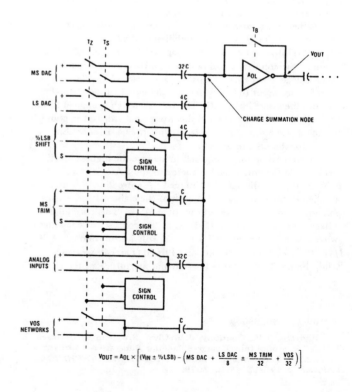

[Right]

FIGURE 1—Multiple differential — voltage input sampled
data comparator using capacitors to both scale the input
voltage differences and convert them to charges.

$$V_{OUT} = A_{OL} \times \left[(V_{IN} \pm \tfrac{1}{2}LSB) - \left(MS\ DAC + \frac{LS\ DAC}{8} \pm \frac{MS\ TRIM}{32} + \frac{VOS}{32} \right) \right]$$

An Error-Correcting 14b/20μs CMOS A/D Converter

Ziya G. Boyacigiller, Basil Weir and Peter D. Bradshaw

Intersil, Inc.

Cupertino, CA

PREVIOUSLY DESIGNED INTEGRATED MOS A/D converters have been limited to low speed and resolution due to slow capacitive settling times and poor initial component accuracies. This paper will cover a 4.1x4.2mm IC affording 14 bits of accuracy in 20μs, with direct bus and logic interface, few external components and the wide temperature range advantages of CMOS A/Ds.

The D/A converter and the comparator are the key elements in high-resolution, high-speed A/D converters. The accuracy of the D/A, the resolution of the comparator, the settling time at the summing node and noise affect the accuracy of the A/D. The speed of the comparator, the D/A and summing junction settling times affect the A/D conversion speed. The techniques used in this converter result in the absence of missing codes and *digitally-calibrated* total accuracy without any analog trimming of components.

The A/D uses a thin-film 17b low accuracy DAC that resembles an R-2R ladder, but uses a radix of about 1.85 instead of 2. The modified radix results in redundant D/A codes which cover the analog output range without missing codes. Each bit value of the DAC is measured using the internal comparator and an accurately calibrated input voltage and is entered as a 17b word into an on-chip EPROM. This technique allows the A/D to use untrimmed, low accuracy DAC components. The gain error is inherently adjusted to less than an LSB during linearity correction. Since parts are programmed after packaging, calibration is unaffected by assembly or burn-in drifts.

An error-correcting algorithm uses the redundant D/A and a successive approximation register. At the $(n)^{th}$ step of the conversion, two trial bits are switched in together, — the $(n)^{th}$ and the $(n+4)^{th}$. If the comparator decides to *keep*, then the $(n)^{th}$ bit stays in and the $(n+4)^{th}$ is removed. If the decision is *drop*, then both bits are removed. An erroneous decision to *keep* is corrected by the removal of the $(n+4)^{th}$ bit which comprises about 8% of the pair. On the other hand a comparator error to *drop* the $(n)^{th}$ bit is corrected later since the succeeding bits in a 1.85 radix DAC add up to about 118% of the $(n)^{th}$ bit. With this algorithm the comparator input needs to settle to only about 3 time-constants compared to more than 12 time-constants normally required for 14-bit accuracy; Figure 2.

A comparator decision to *keep* causes the corresponding bit value to be accessed from the EPROM and added into an accumulator, while a *drop* decision leaves the accumulator contents unchanged. At the end of the conversion the accumulator contains the sum of the values of the bits that were *kept*, which represents the input voltage. 17b of internal resolution are needed to compensate for accumulated errors that can result from the successive additions of the bit values. The result is rounded-off to 14b at the end of the conversion.

Figure 3 shows the basic analog circuit. In the auto-zero mode the feedback switch, S6, connects the comparator's negative input to its output. The positive input is grounded through switch S3. This configuration charges the C_{AZ} capacitor to the offset voltage of the comparator. During the first half of each conversion step the D/A is switched to the next trial code and the summing node is grounded through switches S1 and S3, to reduce the settling time-constant. The comparators are switched into a pseudo-auto-zero configuration, where node voltages are forced close to their quiescent values through switches S4 and S5. During the second half of the conversion, the summing node is disconnected from ground to increase the comparator overdrive voltage. The comparators are released out of the pseudo-auto-zero into the compare mode, providing a high voltage-gain at high speeds to the summing node. The DAC switches, cross-unders and metal lines are compensated for low temperature-drifts. Force and sense lines are provided for the analog-input, reference-input and the analog-ground to make the A/D insensitive to parasitic resistances.

The system, Figure 4, consists of a 17b DAC, an autozeroed comparator, AZC, a 17x17b EPROM, a 17b carry-save full adder, a 15b output latch (14b plus over-range), three-state outputs and control logic. Digital logic is pipe-lined and carried in parallel with the analog comparisons.

The I/O capabilities are designed for direct interface to most popular microprocessors, their derivatives, minicomputers and DMA controllers. The internal output latch stores the previous result during conversion and is treated as a memory, facilitating high-speed processor interface applications and freeing the computer from A/D conversion support tasks.

This A/D converter uses +5/—5V supplies and typically consumes 50mW power.

Acknowledgment

The authors would like to thank L. Evans for his invaluable ideas on the comparator design.

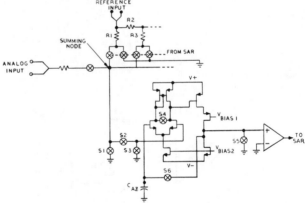

FIGURE 3—Simplified analog circuit diagram.

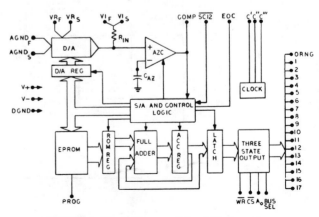

FIGURE 4—Block diagram of the A/D converter.

Reprinted from *1981 IEEE Int. Solid-State Circuits Conf.*, vol. 24, pp. 62–63, Feb. 1981.

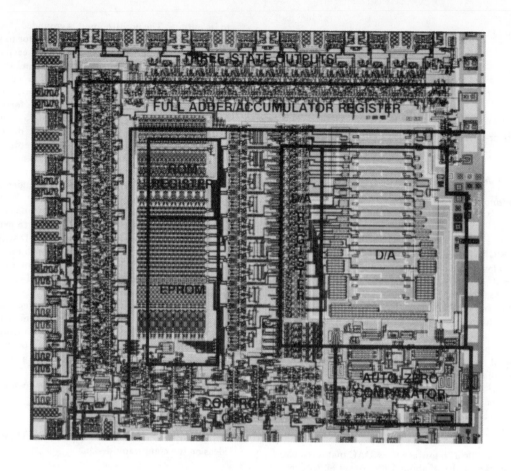

FIGURE 1—Chip photograph.

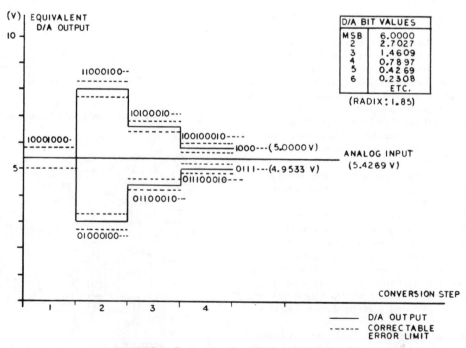

FIGURE 2—Recovering from *keep* and *drop* errors.

Autocalibration cements 16-bit performance

John Croteau, Don Kerth, and Dave Welland
Crystal Semiconductor Corp., 2024 Saint Elmo Rd., Austin, TX 78744; (512) 445-7222.

Not often do solutions to a sticky design problem come wrapped in one neat package. Take analog-to-digital converters, for instance. Engineers designing them into multichannel, fast data-acquisition systems know full well the thorny task of not only achieving accuracy and speed, but also guaranteeing performance over time and during environmental changes. Getting and keeping the needed results usually amounts to an uphill battle. The designer must select the right converter and match it to a compatible sample-and-hold amplifier. Then he must devise a clever calibration scheme to protect speed and accuracy from offset, gain, and linearity errors "thrown in" by the converter and sampling amplifier.

Using microcontrolled autocalibration, a fleet of fast sampling a-d converters corrects gain, offset, and linearity errors. The ICs outperform hybrids.

While manual calibration is always possible, accuracy over time and tempeature can be ensured only with a built-in, system-calibration design. Such a scheme usually requires either dedicated d-a converters assigned to trimming gain and offset errors, or a multiplexer to feed precision reference signals through the system at the right moments. This setup, however, comes at the cost of custom calibration software and pc board space taken up by the extra circuitry.

The end to these problems is now at hand—and in one package. A new breed of monolithic sampling a-d converters are the fastest 12- and 16-bit performers on the scene. Moreover, the chips for the first time incorporate microcontrolled automatic calibration that not only corrects for gain and offset errors, but also for linearity errors—the bane of most converters. Since autocalibration is independent of time or temperature, the chips maintain $\pm 1/2$-LSB accuracy for those three parameters with no missing codes.

Because these smart analog CMOS chips come with an intrinsic sample-and-hold circuit at the input, they act as complete front ends for digital signal processing. What's more, since they can operate with a host, the converters carry a traditional microprocessor interface and three-state output buffers. They can also run as stand-alone devices.

The flagship of the fleet, the CSC5016, leads a squadron of three successive-approximation converters with various resolution and throughput. The chip boasts a 16-bit relative accuracy and no missing codes over rated temperature at sampling rates to 50 kHz (16 μs conversion time). Its siblings, the CSC5012 and CSC-5014, deliver 12- and 14-bit resolution and accuracy, and sample at 63 and 50 kHz. A fourth chip, the CSC5212, calls on a two-step, or half-flash, technique to sample at a lightning fast 1 MHz, with 12-bit accuracy.

The four converters tap the strength of standard 3-μm CMOS processes to shatter the performance limits of even hybrid technology. The basic on-chip

calibration technique—weighted capacitor circuits—hurdles the 10-bit accuracy barrier of uncalibrated capacitor-based devices. In operation, the on-chip microcontroller manipulates on-chip capacitive d-a converters to trim the offset, gain (full-scale), and linearity errors. Control of calibration, however, is up to the user who at any time can order a calibration. Or the chip can be placed in a mode in which calibration occurs continuously in the background.

The converters achieve their accuracies by exploiting the "analog" qualities of CMOS, namely its precise, stable capacitors and fast linear switches. Comparators are auto-zeroed to cancel offsets caused by poor threshold matching or other inherent CMOS limitations. And because the autozeroing operates at well above the $1/f$ corner frequency of the CMOS transistors, the transistor $1/f$ noise "looks" like an offset voltage and is also canceled.

Even without all this, the family's stingy power consumption alone should win supporters. The 16-bit chip needs just 150 mW, and the speedy 12-bit model only 750 mW. This puny power demand represents about a factor of 4 drop from what hybrids call for—and the mandatory sample-and-hold amplifier for the hybrids can double

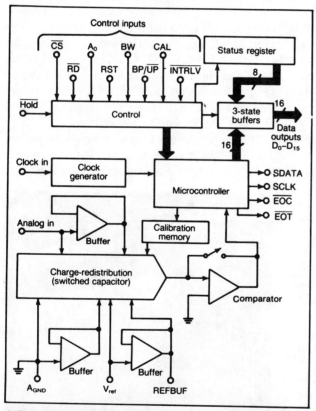

1. Three successive approximation a-d converters ensure no missing codes at 12-, 14-, and 16-bit resolution and accuracy, while running at throughput rates of 63, 50, and 50 kHz, respectively. The secret is a proprietary autocalibration design with a charge redistribution architecture. The CMOS chips also perform the sample-and-hold function.

power needs again. Also, while the CMOS converters run off ± 5 V, the hybrids need ± 15 V.

As an added bonus, the autocalibration maintains specified accuracy forever. For example, the capacitor ratios in all the calibration schemes are arranged so that capacitors can shift in value and the converter will still calibrate accurately. Finally, the bottom line: The installed cost of these IC performers runs about 10% that of hybrids, including design test, qualification, and the component cost of the sample-and-hold and calibration circuitry.

The 14- and 16-bit converters in the CSC501x series get their high throughput by coupling the standard successive-approximation (SAR) algorithm with a charge-redistribution (switched-capacitor) architecture (Fig. 1). This inherent sampling technique eliminates the need for a separate sample-and-hold amplifier either on or off the chip. The user simply feeds the sampling command to the converter's Hold pin. Complex timing signals, such as Sample, Hold, and Convert, are not needed. The circuit freezes the input signal, kicks off a conversion, and then automatically returns to the job of sampling.

The inherent analog memory resulting from the capacitive array structure of these converters plays a key role in allowing the converters to calibrate themselves. To do so demands switching in and out different arrangements of capacitors under the microcontroller's jurisdiction. For each calibration experiment, portions of the array of capacitors forming the d-a converter is charged to V_{ref} in one configuration. The charge is trapped, the array modified, and the voltage resulting from the new configuration at the comparator input indicates which configuration remains. The process continues until the rated linearity is achieved. Conversion and calibration cycles share the same comparator, reference voltages, and control logic.

DYNAMIC ADJUSTMENT

In other words, these three converters attain $\pm 1/2$-LSB relative accuracy and no missing codes by dynamically adjusting the bit weights in their internal d-a converters. Each bit capacitor actually consists of a group, or sub-array, of smaller capacitors that can be switched in parallel with each other as required (Fig. 2). During calibration the chip's microcontroller "knows" that each bit should equal the sum of all less-significant bits, plus 1 LSB. For example, the ideal value of the capacitor for bit 4 is given by $16C = 8C + 4C + 2C + 1C + 1C$. Thus to calibrate bit 4, (B_4), the microcontroller manipulates that sub-array of capacitors. The microcontroller must determine the exact configuration such that B_4 just equals the sum of the capacitance forming B_0 through B_3, plus one dummy LSB, $B0_1$.

Clearly, since each bit's accuracy relies on that of all the lower bits, calibration starts with the LSB and proceeds toward the MSB. Since all bits are switched in and out independently and are free of superposition errors, the linearity of the converters is limited only by the trimming resolution ($1/4$ LSB). Furthermore, the converters maintain better linearity over rated temperature changes than resistor-based designs, even without calibration, because the capacitors inherently track each other better.

Offset adjustments to within $\pm 1/2$ LSB are made with similar subarrays. This calibration is made relative to analog ground rather than V_{ref}, so midscale accuracy is adjusted in the bipolar mode to eliminate bipolar offset. Full-scale accuracy, on the other hand, is inherent in the charge-redistribution design and remains within $\pm 1/2$ LSB.

When incorporated in a system, self-calibration can be initiated in several ways. For example, on power-up logic must be reset and the device initially calibrated. At any later time, the user recalibrates by strobing Chip Select ($\overline{CS}$) low while the Calibrate input (CAL) latches high. The converter keeps calibrating itself until CAL again latches low. Each time CAL latches high, the calibration picks up where it left off during the previous cycle, permitting piecemeal calibration over many cycles.

The converters also run in an interleaved mode, resulting in continuous calibration that is transparent to the user. In this mode, small portions of the calibration occur between each conversion. Conversion time increases by a few percentage points while throughput drops about 20%. Latching, or hardwiring the $\overline{INTRLV}$ pin low, puts the device into that mode. By taking advantage of its microcontroller, the CSC5016 can achieve an accuracy below its own noise floor. Given the converter's 4.5-V input voltage range and noise specification of 40 μV rms, apparently the signal-to-noise ratio should be −92 dB. However, because the microcontroller implements multiple "experiments" per calibration step, followed by a statistical analysis of the results before it makes a decision, the effective calibration noise floor drops to −117 dB. Thus, the microcontroller calibrates beyond the 16-bit level, yielding $\pm 0.001\%$ accuracy and no-missing-codes at 16 bits.

To reach the 12-bit, 1-MHz level, the CSC5212 moves out on its own tack, combining a recycling two-step flash circuit with an input sampling amplifier (See ELECTRONIC DESIGN report, p. 90*). The internal 6-bit flash converter calls on a unique reference generator (not a divider) to create the 64 voltage levels that feed the chip's 64 comparators, all of which work on the sampling amplifier's output signal. After the first pass of that signal through the comparators, the circuitry latches the resulting 6 MSBs and sends the data to a d-a converter-like circuit. The d-a's output—a replica of the analog input, less the quantization error of the flash converter—is summed with the sampling amp's output, amplified 64 times, and applied to the flash converter to derive the

* Page number refers to *Electronic Design*, September 4, 1986.

6 LSBs.

The CSC5212S microcontroller runs autocalibration. That includes zeroing the comparators and the sampling amplifier, adjusting the reference levels, and interrupting the host between calibration cycles if the reference levels change by more than ± 1.25 LSB or if the input exceeds the reference by more than 64 LSBs.

The controller also sports status and control registers with I/O functions similar to those of its siblings. In addition, its data output bus is bidirectional, offering an alternate path to the control register over which the host can initiate calibration or conversion. Finally, output data is available in one or two bytes.

At first glance, the digital I/O of the SAR converters may look conventional. Actually, the family's native intelligence makes it more versatile and more useful than other converters. While converting, the family keeps a running tab of its own status in an 8-bit register. For example, as a conversion cycle begins, status bit S6 latches high. When the cycle is complete, S_6 is cleared and S_7 goes high indicating the converter is tracking (sampling) the input. The user or host retrieves the signal being sampled by strobing $\overline{CS}$ and $\overline{RD}$ low with the Read Address (A_0) low. The 8 status bits, S_0–S_7, now appear on data output pins D_0–D_7, where they may be read by the host.

Memory mapping of the converter occurs when the host applies the LSB from the address bus to the chip's A_0 input, and the decoded address to $\overline{CS}$ (Fig. 3). During a read cycle, the converter polls its status register, retrieving the status word for analysis. The host's software routine ingests this information (the status of each bit) and may decide to recalibrate, convert, or retrieve output data words. However, it is unnecessary to use the status register; traditional End of Convert ($\overline{EOC}$) outputs are avail-

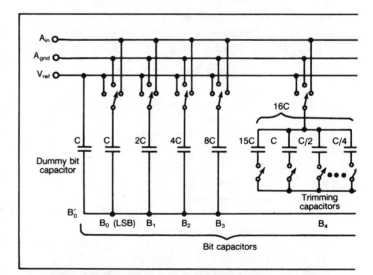

2. These switched-capacitor converters are not only fast but also smart. Autocalibration maintains their accuracy over time and temperature. Linearity is trimmed by adjusting the capacitance for each bit to $1/4$-LSB precision.

able for generating interrupts or telling DMA controllers to dump data directly into memory.

All three converters (CSC5016, CSC5014, and CSC5012) link identically with processor or computer buses and carry identical pinouts. Accordingly, designers can easily upgrade a system simply by dropping a higher accuracy converter into the same socket. Moreover, the converters easily adapt to 8- or 16-bit data buses by hard-wiring of their Bus Width (BW) inputs. With BW hard-wired high, the converter puts its complete data word onto the data pins during each read cycle $\overline{CS}$ and $\overline{RD}$ strobed low, while A_0 is high.

The output data on the pins of the 12- and 14-bit units are left-justified on the 16 data pins. That is, the CSC5012 presents its output on pins D_{15}–D_4 with the MSB on pin D_{15}. Trailing zeros appear on pins D_3–D_0. Since all three converters present their 8-bit status words on pins D_0–D_7, the 16-bit data bus connects directly to D_0–D_{15}. On the other hand, if BW is hardwired (low) to put data on an 8-bit bus, the 8 MSBs appear on D_0—D_7 during the first read cycle following a conversion, and the remaining bits appear on those pins on the next read cycle. The 8-bit data bus is therefore connected to D_7–D_0.

To use the CSC family as stand-alone devices, sans host, just hardwire $\overline{CS}$ and $\overline{RD}$ low, and BW and A_0 high. This read-only configuration puts new data on the output pins at the end of each conversion. Conversions initiate by bringing the Hold-pin low; calibration starts by bringing the CAL pin high. Finally, regardless of the machine that runs the show—a Cray, a PC, a 6900 microcomputer, for instance—the data is available in serial form at the Serial Data pin (SDATA). The output data can be latched on the rising edge of the Serial Clock (SCLK) output.

Given the cost and complexity of high-end converters, few designers in the past would have considered wasting two converters on a single signal. Usually therefore, compromises were made when input signals increased in dynamic range or required greater accuracy in the lower

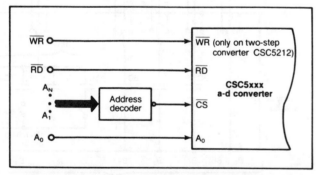

3. Like most modern converters, these speedy CMOS chips interface easily with a host processor. To achieve memory mapping of the converters' status registers and output data, a base address is decoded and fed to the $\overline{CS}$ input. In addition, the least significant address bit is applied to the A_0 input.

Price and availability
The CSC5012 a-d converter is available in 12- and 24-µs (63 and 34 kHz) versions. Prices start at $34 in quantities of 100; sample quantities are available from stock. Evaluation boards for the CSC5012 can be had for $125. Engineering samples of the CSC5014 will be available in November with the CSC5016 available in January. A 400-kHz version of the CSC5212 will have engineering samples in November, followed by a 1-MHz version in February.

portion of their frequency spectrum so that one converter could not do the job by itself. However, these chips make it possible to join 12- and 16-bit converters (Fig. 4). The complete circuit operates off ± 5 V and needs less than 900 mW of power. The 12-bit CSC5212 runs off a 15-MHz system clock, which is divided down to 3.75 MHz for the 16-bit CSC5016; similarly, the 1-MHz sampling clock feeding the 12-bit chip is divided down to 31.25 kHz for the 16-bit device. With a 3.75-MHz master clock and a sample rate of 31.75 kHz, the CSC5016 has enough headroom to operate in the interleave mode, completing a full calibration every two seconds.

To avoid aliasing, the input to the CSC5212 should have no frequency components above 500 kHz (half the sampling rate). Similarly, the CSC5016 demands a 15-kHz antialiasing filter ahead of it. To control harmonic and intermodulation distortion, the sampling clock must have a low jitter, and the analog input should be driven from a precision buffer.

For this task, both converters sample continuously and the DMA controller directs their outputs. Status (A_0) pins of the 12- and 16-bit devices are hardwired high and low, respectively, denying the host their status information. Instead, the End-of Convert (EOC) outputs synchronize the DMA controller to the conversion process. When the host asks the controller for a block of samples from either converter, the controller unmasks the appropriate interrupt pin (INT 0 or INT 1) and takes over the address and the control buses. Upon receiving an interrupt, the DMA chip enables the converters' output buffers, drives their address bus, and latches the data into the designated memory location. The IC then increments the address for the next sample (unless it has already handled the request) and relinquishes control of the bus.

These chips are relatively easy to drive. Their analog input (A_{in}), Analog Ground (A_{GND}), and reference inputs all look like infinite impedances at dc; that is, no dc signal current flows through them. However, transients arise while capacitors are switching to and fro. In particular, displacement currents occur within the CSC5016 as the bit capacitors switch between reference and ground during bit decisions, and at the time sampling starts, when all the capacitors switch to the analog input. To handle this problem and simplify driving, all three inputs (the analog ground appears as a signal source) are buffered internally. Thus the power bus, rather than the signal source, delivers these currents (Fig. 1 again). The buffered reference voltage is brought to the Reference Buffer (REFBUF) pin. There it enlists the aid of an external $0.1\text{-}\mu\text{F}$ capacitor to handle reference voltages up to the positive analog supply voltage, without sacrificing the settling time at each bit decision.

However, since the offset voltages of these buffers would introduce errors, the circuit only rough charges the capacitor array, switching to the unbuffered voltage for fine charging before making a decision. The residual fine-charge displacement currents, although much less than that of the rough-charge currents, require a dynamic impedance of less than 1 Ω at one fourth the clock fre-

quency. The analog input, on the other hand, demands only that the output impedance of the antialiasing filter remain below 400 Ω at 1 MHz, a criterion easily met by low-cost op amps. Acquisition time is a direct function of the impedance of the signal source, and is specified at 4 μs if the dc source resistance is under 200 Ω.

Driving the 12-bit converter is even easier since no switching occurs at the input to the internal sample-and-hold amplifier. The signal source need only drive the chip's mere 5 pF of input capacitance. But since the plus and minus reference inputs are sampled at 1/20th of the clock frequency (750 kHz), they must be supplied from an impedance of under 5 Ω at that frequency. This is easily accomplished by hanging a $0.1\text{-}\mu\text{F}$ capacitor on the $+V_{ref}$ pin; however, the reference buffer (Fig. 4 again) driving both chips does the job nicely.□

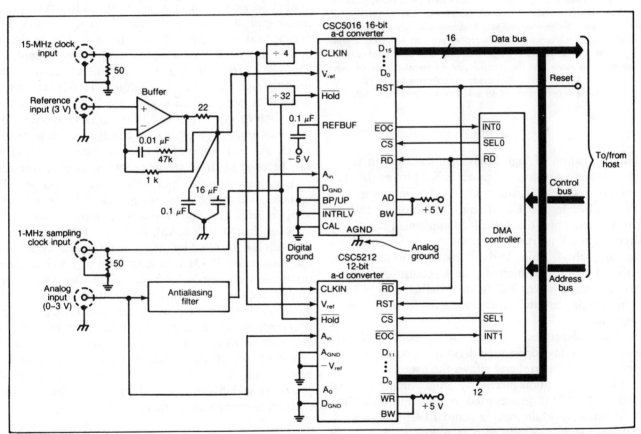

4. By combining the 12- and 16-bit converters in one system, you can analyze signals with frequency components to 500 kHz, over a dynamic range of 72 dB. You can further scrutinize the frequency components of that signal below 15 kHz over a 96-dB dynamic range. Both converters continuously sample the signal while the DMA controller selectively passes the data on to the host.

A CMOS Programmable Self-Calibrating 13-bit Eight-Channel Data Acquisition Peripheral

HARLAN OHARA, HUNG X. NGO, MICHAEL J. ARMSTRONG, MEMBER, IEEE, CHOWDHURY F. RAHIM, MEMBER, IEEE, AND PAUL R. GRAY, FELLOW, IEEE

Abstract—A 13-bit self-calibrating algorithmic A/D converter with sample and hold and eight-channel multiplexer has been implemented in a standard 3-μm CMOS technology. Digital circuitry including a sequencer, instruction RAM, and microprocessor interface have been incorporated to implement a complete data acquisition peripheral. The conversion algorithm, calibration technique, and gain trimming circuitry are described. A bandgap voltage reference with buffered output and additional output voltage proportional to absolute temperature has also been implemented. Experimental data show that the integral and differential linearity meet 1/2 LSB at the 13-bit level; the dynamic performance is such that harmonic distortion of better than 84 dB below the fundamental is achieved.

I. INTRODUCTION

THE realization of high levels of integration in precision data acquisition components has proven difficult for several reasons. Traditional A/D conversion approaches utilize laser-trimmed thin-film resistors and bipolar technology, but it is difficult to implement the digital circuitry required to realize a complete data acquisition system on a single chip. Perhaps more importantly, the wide diversity of applications for data acquisition components and systems makes it difficult to define a single fixed function component that addresses a wide spectrum of applications.

This paper describes a programmable peripheral data acquisition chip developed to address these concerns. The architecture of the peripheral allows the user to tailor the device to his application merely by programming internal registers. Self-calibration was used to allow the realization of 13-bit linearity while using a standard CMOS process. A photomicrograph of the chip is shown in Fig. 1. In Section II, the overall function of the chip is described. In Section III, the self-calibrating A/D conversion technique used is described. In Section IV, the digital architecture is described. In Section V, experimental results are presented.

Manuscript received May 20, 1987; revised July 30, 1987.
H. Ohara, H. X. Ngo, M. J. Armstrong, and C. F. Rahim are with Micro Linear Corporation, San Jose, CA 95131.
P. R. Gray is with the Department of Electrical Engineering and Computer Sciences and the Electronics Research Laboratory, University of California, Berkeley, CA 94720.
IEEE Log Number 8716975.

II. FUNCTIONAL DESCRIPTION

A block diagram of the chip is shown in Fig. 2. The analog section consists of an input multiplexer, programmable-gain amplifier and sample/hold, 13-bit A/D converter, and bandgap voltage reference. The analog input comes in through a four-channel differential multiplexer. The same circuit with a mask option can also be configured for eight single-ended channels. The signal then passes through a programmable-gain amplifier and sample and hold before entering the 13-bit A/D converter. An additional analog block is the bandgap voltage reference, which has two outputs: a buffered reference output which can be used to drive strain gauges and other transducers, and a temperature output which gives access to a PTAT voltage so that the user can get an indication of the die temperature. The digital section mainly consists of a sequencer, instruction RAM, and data RAM. Operation of the chip is controlled by the sequencer which cycles through the instruction RAM locations, each of which contains an instruction requesting an appropriate conversion from the analog section. The conversion results are stored in the data RAM where they can be read out through the microprocessor interface. There is a limit alarm register set and comparator which can generate an interrupt when a programmed set of conditions is met, such as above limit, below limit, in band, and out of band. A timer is also provided which can be used to control the sample rate without microprocessor intervention.

III. SELF-CALIBRATING ALGORITHMIC A/D

A conceptual block diagram of an algorithmic A/D converter is shown in Fig. 3, consisting of a sample and hold, a precision multiply-by-two amplifier, and a comparator [1]–[3]. The A/D algorithm is conceptually very similar to long division performed on binary numbers, where the steps involved are magnitude comparison, conditional subtraction, and shifting the remainder to determine the next bit. A similar series of steps is used to perform A/D conversion.

Reprinted from *IEEE J. Solid-State Circuits*, vol. SC-22, no. 6, pp. 930–938, Dec. 1987.

Fig. 1. Photomicrograph of the chip.

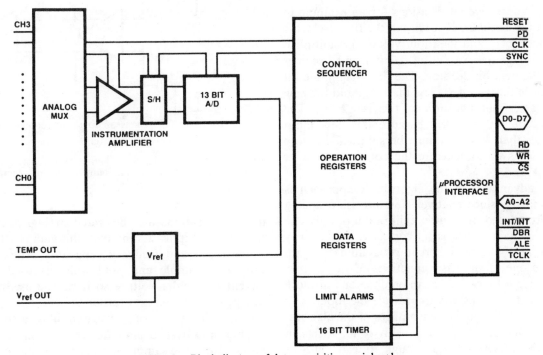

Fig. 2. Block diagram of data acquisition peripheral.

Assuming a positive input voltage for simplicity, the process starts with an initial cycle to determine the sign bit. During this first conversion cycle, the input goes through the sample and hold, is multiplied by two, and the sign bit decision is made. This doubled input voltage is then recirculated and compared against V_{ref} to make the first magnitude bit decision at a decision level of $V_{ref}/2$. If that magnitude bit is a ONE, V_{ref} is subtracted from the input voltage to generate the remainder; otherwise the input voltage is used as the remainder. The remainder voltage is then multiplied by two to be used as the input voltage for the next bit decision. If the magnitude bit is a ZERO, the output voltage itself is used as the next input voltage. This cycle is repeated for each successive bit. As

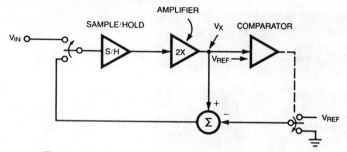

Fig. 3. Block diagram of an algorithmic A/D converter.

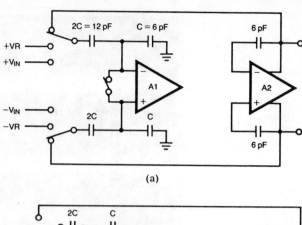

(a)

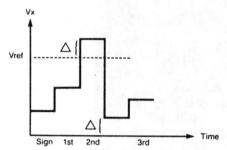

Fig. 4. Remainder voltage as a function of time.

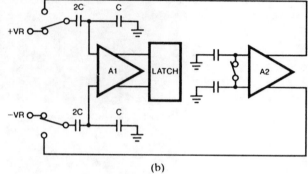

(b)

each bit is determined, it is stored in the digital section, where it will be transferred to the data RAM after converting the output to a two's complement result.

The remainder voltage V_x is shown in Fig. 4 as it changes during successive bit decisions for an example in which the input is about 0.3 V_{ref}. During the first cycle, the sign bit is determined, and the input voltage is multiplied by two; this voltage has a value of 0.6 V_{ref} and becomes the input for the next bit decision. The first magnitude bit is a ZERO because this voltage is less than V_{ref}, and the new input voltage after multiplication by two is 1.2 V_{ref}. The second magnitude bit is a ONE because this new input voltage is larger than V_{ref}; V_{ref} must be subtracted before multiplication by two to generate the correct remainder voltage of 0.4 V_{ref}.

One of the advantages of the algorithmic approach is hardware simplicity, since relatively few precision analog components are needed. It requires calibration of only one parameter; if the gain of two is set correctly, the linearity is assured. It also has an inherent programmable gain function; since one of the basic functions of the loop is multiplying a signal by two, this can be done one or more times before a conversion starts to prescale the input.

There are several potential disadvantages of the algorithmic technique. One problem is a sensitivity to offsets in the loop. If the offset is large enough, it can cause a linearity error because every offset voltage is multiplied during each cycle around the loop. There is also a speed limitation because of the op-amp settling time. To meet the performance goals, the critical design issues were loop offset cancellation, loop gain calibration, and optimization of the op-amp settling time.

For each bit decision, there are three analog operations which must take place. Because of this, it is important to reduce the op-amp settling time to a minimum. These three

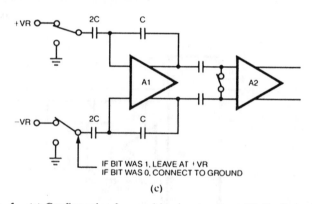

Fig. 5. (a) Configuration for acquiring input voltage. (b) Configuration for making bit decision. (c) Configuration for generating remainder voltage.

analog operations are illustrated in Fig. 5. Amplifier A_1 does the multiplication by two; this gain is set by the ratio of the input and feedback capacitors. Amplifier A_2 is a unity-gain sample and hold which is used to hold the current remainder voltage so it can be used as the input voltage for the next bit decision.

In the first phase of conversion, illustrated in Fig. 5(a), the multiply-by-two amplifier is acquiring the remainder voltage from the sample and hold. For the very first cycle, it will be acquiring the input voltage instead.

In the second phase, shown in Fig. 5(b), the amplifier A_1 is being used open loop as the front end of a comparator to generate the bit decision. V_{ref} is subtracted from the input voltage using a switched-capacitor technique and amplifier A_1 determines the sign of the result. The latch senses the output of the amplifier, and passes the decision result to the digital section. The unity-gain sample and hold is not active in this mode.

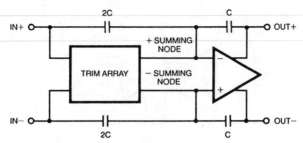

Fig. 6. Gain of two circuit with trim array.

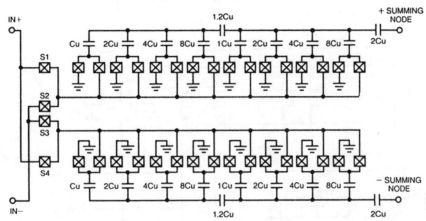

Fig. 7. Gain trim capacitor array schematic.

In the third phase, illustrated in Fig. 5(c), the remainder voltage is being generated so it can be used as the input voltage for the next bit decision. If the magnitude bit was a ONE, V_{ref} would be subtracted from the remainder voltage before the multiplication by two takes place. The remainder voltage is being stored on the sample-and-hold capacitors so it can be used as the input voltage for the next bit.

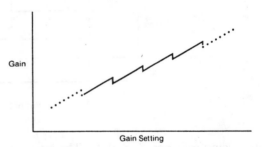

Fig. 8. Typical transfer function of gain trim array.

A. Linearity Calibration Technique

To meet the linearity specification, it is important that the gain of two be set properly. To adjust the gain of two, an 8-bit plus sign trim array was implemented to modify the gain of the multiply-by-two circuit. The calibration algorithm is performed by using a basic property of A/D converters: a conversion with V_{in} set to V_{ref} produces plus full scale. If the gain of two is too small, the remainder voltages will be attenuated as they go around the loop, and the conversion result will be less than full scale. If the gain is too large, the remainder voltages will increase as they go around the loop, and the converter will over-range. The calibration controller performs successive approximation on the nine gain trim bits using this criterion to determine the correct value for each bit. Two techniques are used to increase the accuracy of the calibration procedure: the calibration conversions are done using 16-bit conversions to ensure a negligible error at the 13-bit level, and instead of doing a single conversion to determine each control bit, the results of seven conversions are averaged.

Fig. 6 shows how the trim array is connected to the multiply-by-two amplifier, and Fig. 7 shows the circuit implementation of the trim array itself. To minimize area, a split-array DAC consisting of a main array and a sub-array connected through a coupling capacitor is used. One potential problem of this type of circuit is that the transfer function of the array could be nonmonotonic. Fortunately, monotonicity is not a requirement, but there must be no gaps in the transfer curve. The value of the coupling capacitor has been slightly increased to ensure that the segment endpoints of the transfer curve overlap, as shown in Fig. 8. Switches S_1–S_4 are used to control whether the trim array increases or decreases the gain of two.

B. Offset-Nulled Operational Amplifier

As stated earlier, offsets in the loop cause integral and differential nonlinearity problems. Fig. 9 shows the auto-zero implementation used to reduce the offsets to a negligible level. Each amplifier has an auxiliary input stage which is used to store an offset correction voltage. At the start of each conversion, an auto-zero cycle takes place where the main input stage is grounded, and the auxiliary input stage

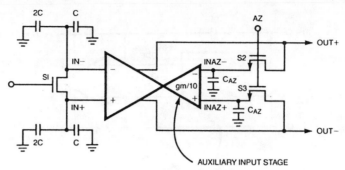

Fig. 9. Implementation of auto-zero function.

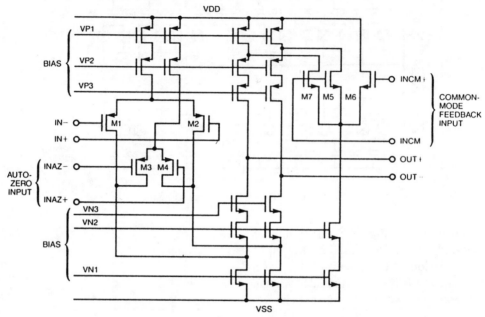

Fig. 10. Simplified schematic of operational amplifier.

is activated using switches S_2 and S_3. The auxiliary input stage forces the output of the amplifier to zero, storing a representation of the offset voltage on the CAZ capacitors. To increase the accuracy, switch S_1 is opened while the auxiliary input stage is active so that any offset caused by charge injection from switch S_1 will also be cancelled. Finally, the auto-zero switches S_2 and S_3 are turned off, and the amplifier is ready for use. The gain of the auxiliary input stage is one-tenth the gain of the main input stage, such that any error due to charge injection from the auto-zero switches or offset in the auxiliary input stage will be reduced by a factor of 10 when referred to the input [4].

A simplified schematic of the operational amplifier is shown in Fig. 10. Since amplifier settling time is very important for this type of converter, a folded-cascode operational transconductance amplifier was used for a well-controlled transient response. Transistors M_1 and M_2 are the main input devices, and M_3 and M_4 form the auxiliary input stage. The auxiliary input devices have one-tenth the transconductance of the main devices. Transistors $M_5 - M_7$ form the common-mode input stage which can be used to set the output common mode of the amplifier to any convenient voltage. Analog ground was

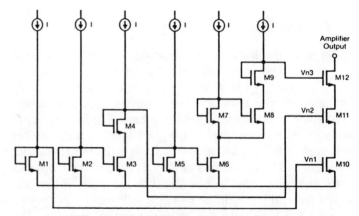

Fig. 11. Simplified schematic of bias circuit.

used for optimum power supply rejection. The output stage uses three output transistors in a cascode configuration to get adequate dc gain. The transistors are biased for maximum swing using a variation of the high-swing cascode biasing previously reported [5]. Fig. 11 is a simplified schematic of the bias circuit. Transistor M_3 is operating in the linear region because it carries the same current as device M_2, but has a larger aspect ratio. The V_{ds} of M_3 acts

TABLE I
TYPICAL AMPLIFIER PERFORMANCE PARAMETERS
(± 5 V; 25°C)

- Open loop gain: 125 dB
- Output voltage swing: within 1.5 V of rails
- Total settling time to 0.01%, 6 volt step: 300 ns
- Power dissipation: 30 mW
- Area: 1700 square mils

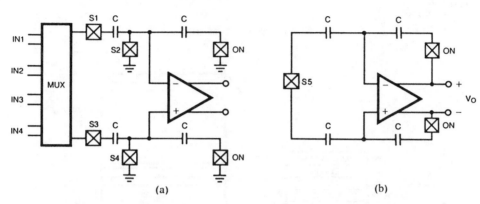

Fig. 12. (a) Input multiplexer during sample mode ($c = 9$ pF). (b) Input multiplexer during hold mode.

as a relatively low-impedance voltage source with a value approximately proportional to the $V_{gs} - V_t$ of transistor M_2; this voltage sets the V_{ds} of the bottom output transistor M_{10}. The V_{n3} generator is similar to the V_{n2} generator, but an additional linear region device M_6 is added to provide the additional voltage required to set the V_{ds} of the second cascode device M_{11}. Using a similar technique, an additional voltage was added to V_{n2} and V_{n3} to ensure adequate V_{ds} even at low current levels. Complementary circuits are used to produce the bias voltages for the p-channel devices. Table I shows typical amplifier performance parameters.

C. Input Sample/Hold

An additional analog block is the input sample and hold, which works with the multiplexer to bring in the input signal. In Fig. 12(a), switches $S_1 - S_4$ are ON; the input voltage goes through the multiplexer and is stored on the two input capacitors. These capacitors are relatively small (9 pF) to allow the device to connect directly to signal sources with high source impedances. When the circuit goes into the hold mode (Fig. 12(b)), switches $S_1 - S_4$ are turned OFF, and switch S_5 turns ON. This forces any difference in charge on the capacitors to generate a current which will then produce an output voltage. Since only the difference in charge stored on the input capacitors causes current flow, this circuit has a good common-mode rejection ratio, and also allows the input voltage range to include the positive and negative power supplies [6]. One potential problem is that errors in the ratio of the input capacitors to the feedback capacitors will cause a gain error. To compensate for this, both the input voltage and the reference voltage pass through this same sample and

hold, and since an A/D conversion is determining the ratio of V_{in} to V_{ref}, multiplying both voltages by a constant factor has no effect on the conversion result.

IV. DIGITAL ARCHITECTURE AND MICROPROCESSOR INTERFACE

The most important element of the digital section is the sequencer, which controls the A/D converter according to the contents of the instruction RAM. The instruction RAM is loaded with one to eight sets of conversion characteristics at system initialization. The run bit is then set to initiate operation. This causes the control sequencer to begin cycling through the instruction RAM, initiating appropriate conversions. Each instruction RAM location has an associated data RAM location where the data from its A/D conversion are stored. The data RAM is double buffered such that the data from the previous sequence of conversions can be read out by the microprocessor or DMA controller while the current sequence is in progress. The attributes the user can specify in the instruction RAM are shown in Fig. 13 as follows.

1) The last bit indicates that this operation is the last operation in the sequence.

2) The alarm-enable bit controls whether the limit alarms are active for this operation.

3) The 3-bit mode select field determines how the conversion starts and stops. For example, whether it runs at full speed, waits for an external sync signal, waits for a timer countdown, etc.

4) The 3-bit channel field specifies which of the channels will be used as the input for the A/D conversion.

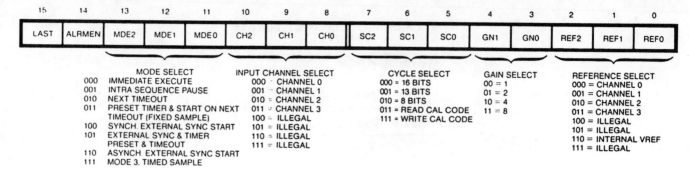

15	14	13	12	11	10	9	8	7	6	5	4	3	2	1	0
LAST	ALRMEN	MDE2	MDE1	MDE0	CH2	CH1	CH0	SC2	SC1	SC0	GN1	GN0	REF2	REF1	REF0

	MODE SELECT	INPUT CHANNEL SELECT	CYCLE SELECT	GAIN SELECT	REFERENCE SELECT
000	IMMEDIATE EXECUTE	000 · CHANNEL 0	000 = 16 BITS	00 = 1	000 = CHANNEL 0
001	INTRA SEQUENCE PAUSE	001 ·· CHANNEL 1	001 = 13 BITS	01 = 2	001 = CHANNEL 1
010	NEXT TIMEOUT	010 = CHANNEL 2	010 = 8 BITS	10 = 4	010 = CHANNEL 2
011	PRESET TIMER & START ON NEXT	011 = CHANNEL 3	011 = READ CAL CODE	11 = 8	011 = CHANNEL 3
	TIMEOUT (FIXED SAMPLE)	100 = ILLEGAL	111 = WRITE CAL CODE		100 = ILLEGAL
100	SYNCH. EXTERNAL SYNC START	101 = ILLEGAL			101 = ILLEGAL
101	EXTERNAL SYNC & TIMER	110 = ILLEGAL			110 = INTERNAL VREF
	PRESET & TIMEOUT	111 = ILLEGAL			111 = ILLEGAL
110	ASYNCH. EXTERNAL SYNC START				
111	MODE 3. TIMED SAMPLE				

Fig. 13. Instruction RAM attributes.

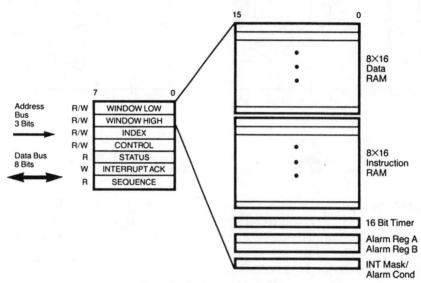

Fig. 14. Register configuration.

5) The cycle select field specifies whether the converter is to be short-cycled. Since each bit is one cycle around the loop, faster conversions can be done by doing fewer cycles. This field is also used to access the calibration register.

6) The 2-bit gain field specifies what prescaling should occur before the conversion begins. This allows gains of 1, 2, 4, and 8.

7) The 3-bit reference field allows the user to select the reference for the A/D conversion. The reference can be any of the input channels or the internal reference.

There are also self-test features included which allow the user to measure the system offset, the reference, and the negative of the reference without requiring any external circuitry.

The register configuration involves two types of registers (Fig. 14). There are seven primary registers which are directly addressed from the CPU. These start and stop the sequencer, indicate the chip status, and allow acknowledging of interrupts. Three of these registers are used to address and access the secondary registers. The secondary registers consist of eight 16-bit data RAM locations, eight 16-bit instruction RAM locations, a 16-bit timer register, two 16-bit alarm threshold registers, and a register which allows setting the interrupt mask and the alarm condition; i.e., in band, out of band, above threshold, and below threshold.

V. Experimental Results

The most significant performance parameter is integral linearity. Integral linearity is an excellent indication of overall converter accuracy; any problem in gain calibration, offset cancellation, or amplifier performance will affect this parameter. Fig. 15 shows a typical integral nonlinearity curve at the 13-bit level after a calibration has been performed. An important issue is how variations in supply voltage and temperature affect the linearity after calibration has been performed. The calibration is intended to only correct for errors in the ratio of the capacitors, and if there is enough margin in the amplifier performance there should be very little variation in linearity as conditions change. Figure 16 shows how the integral nonlinearity varies with a fixed calibration code over variations in power supply and temperature. Another important parameter is harmonic distortion, since it gives an indication of the dynamic performance of the converter, especially the sample and hold circuit. To evaluate the dynamic performance, a 2.5-V peak 2-kHz sine wave was digitized by the converter, which was running 13-bit conversions with a device clock of 8 MHz, resulting in a sampling period of 27.5 μs. After windowing, the output spectrum was generated by performing a Fourier transform (Fig. 17). The resulting spectrum indicates that all

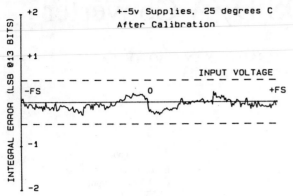

Fig. 15.   Plot of integral nonlinearity.

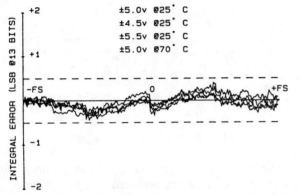

Fig. 16.   Plot of integral nonlinearity as a function of supplies and temperature.

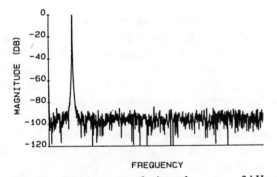

Fig. 17.   Output spectrum for input frequency = 2 kHz.

harmonic components are more than 84 dB below the fundamental. A beat frequency test was also performed with an input frequency of 2 kHz above the sample rate with very little degradation in performance. Additional performance parameters are given in Table II.

TABLE II
TYPICAL DEVICE PERFORMANCE PARAMETERS
(± 5 V; 25°C)

A/D Converter

Integral nonlinearity after cal	0.4 lsb @ 13 bits
Input offset voltage	0.3 mV
Conversion time (includes gain, S/H)	
8 bit, 1× gain	18 μs
13 bit, 1× gain	25 μs
8 bit, 8× gain	21 μs
13 bit, 8× gain	28 μs
Max VREF value	3.2 V

Input sample/hold and programmable amplifier function

Input offset voltage	0.1 mV
CMRR(DC)	>100 dB
PSRR(DC)	>76 dB

Voltage reference

Nominal output value	2.500 V
Drift, 0-70 C	$\pm$18 ppm
PSRR	85 dB
Max output current	3 mA
Rout	0.15 ohm
Thermometer output sensitivity	5 mV/°C

Overall chip

Power dissipation, active	250 mW
Power dissipation, power down	1 mW
System gain accuracy (external Vref)	0.04%
Die size	255×272 mil
Technology	3 micron, 2 poly CMOS
Number of transistors	25,000

ACKNOWLEDGMENT

The authors would like to acknowledge D. Schwan, M. Bisgood, B. Norling, and J. Gianelli for their layout design, A. Grossman for her assistance in logic design and verification, and E. J. Sin and K. Rexroad for their testing assistance.

REFERENCES

[1] R. McCharles and D. A. Hodges, "Charge circuits for analog LSI," *IEEE Trans. Circuits Syst.*, vol. CAS-25, pp. 490–497, July 1978.
[2] P. W. Li, M. Chin, P. R. Gray, and R. Castello, "A ratio-independent algorithmic analog–digital conversion technique," *IEEE J. Solid-State Circuits*, vol. SC-19, pp. 828–836, Dec. 1984.
[3] R. W. Webb, F. R. Cooper, and R. W. Randlett, "A 12b A/D converter," in *ISSCC Dig. Tech. Papers*, Feb. 1980, pp. 54–55.
[4] M. Degrauwe, E. Vittoz, and I. Verbaouwhede, "A micropower CMOS-instrumentation amplifier," *IEEE J. Solid-State Circuits*, vol. SC-20, pp. 805–807, June 1985.
[5] C. A. Laber, C. F. Rahim, S. F. Dreyer, G. T. Uehara, P. T. Kwok, and P. R. Gray, "Design considerations for a high-performance 3-μm CMOS analog standard-cell library," *IEEE J. Solid-State Circuits*, vol. SC-22, pp. 181–189, Apr. 1987.
[6] R. C. Yen and P. R. Gray, "An MOS switched capacitor instrumentation amplifier," *IEEE J. Solid-State Circuits*, vol. SC-17, pp. 1008–1013, Dec. 1982.

An 80-MHz 8-bit CMOS D/A Converter

TAKAHIRO MIKI, YASUYUKI NAKAMURA, MASAO NAKAYA, SOTOJU ASAI,
YOICHI AKASAKA, AND YASUTAKA HORIBA

Abstract —A high-speed 8-bit D/A converter has been fabricated in a 2-μm CMOS technology. In order to achieve high accuracy, a current-cell matrix configuration and a switching sequence named "symmetrical switching" have been introduced. The mismatch problem of small-size transistors has been relaxed by this matrix configuration. The linearity error caused by an undesirable current distribution of the current sources has been reduced by the symmetrical switching. This switching sequence has been developed on the basis of an analysis of the influence of ground-line resistance. In order to realize high-speed operation, a high-speed decoding circuit and a fast-settling current source have been developed. High-speed low-power decoding has been realized by a decoding circuit with one stage of latches. Fast settling time has been attained by designing a current source considering the relationship between the settling time and output impedance. The experimental results have shown that the maximum conversion rate is 80 MHz, a typical dc integral linearity error is 0.38 LSB, a typical dc differential linearity error is 0.22 LSB, and the maximum power consumption is 145 mW. The chip size is 1.85×2.05 mm^2.

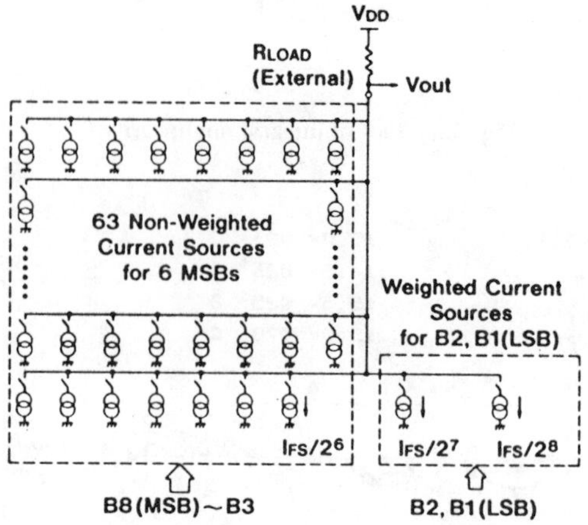

I. INTRODUCTION

ONE OF THE main applications of high-speed D/A converters is in display systems which include consumer video systems such as digital TV and high-definition TV. In these systems, the CMOS D/A converter has advantages of low power, low cost, and I/O compatibility with both TTL and external CMOS circuitry. Another important advantage of a CMOS D/A converter is its capability of being integrated with memories and digital processing IC's for video applications [1]. High-resolution display systems such as high-definition TV, however, need very high-speed converters which are clocked at more than 65 MHz and have resolution of more than 8 bits. The CMOS D/A converters produced up to now [2], [4] have had a low conversion rate, or, if the speed is satisfactory, a high power consumption.

This paper will describe an 80-MHz, 145-mW, 8-bit D/A converter fabricated in a 2-μm CMOS process [3]. A current-cell matrix has been introduced as a basic architecture to achieve high differential linearity and monotonicity. A decoding circuit with a minimum gate delay has been developed to achieve the two-dimensional decoding with neither degradation of speed nor increase of power consumption. Current sources and current switches have been optimally designed by considering the relationship between the settling time and output compliance character-

Manuscript received April 26, 1986; revised June 26, 1986.
The authors are with the LSI Research and Development Laboratory, Mitsubishi Electric Corporation, 4-1 Mizuhara, Itami, Hyogo 664, Japan.
IEEE Log Number 8610619.

Fig. 1. Basic architecture of the DAC.

istics. The linearity error caused by a voltage drop along the analog ground line has been reduced by "symmetrical switching."

II. BASIC ARCHITECTURE

Small transistors which have less than 5-μm gate length are required to realize the fast settling of the current sources (the details will be discussed in Section IV). However, as the device size decreases, the mismatch among transistors increases. This mismatch degrades differential linearity and monotonicity which are necessary in video applications. In order to improve these characteristics, the current for the six MSB's has been generated by 63 non-weighted current sources arranged in a matrix. Fig. 1 shows the basic architecture of the D/A converter. For example, if the digital inputs for the six MSB's correspond to the decimal number of 30, 30 current sources in the matrix are turned on and these outputs are summed up with the outputs of weighted current sources. The output is obtained in the form of the voltage drop across the external load resistor. Since the output of the largest current sources is only 4 LSB, even a 12.5-percent relative mismatch is allowable for retaining a differential linearity error of 0.5 LSB.

III. MATRIX DECODING

To decode the matrix, a region like the shadowed portion in Fig. 2 should be selected. Since this portion is neither a point nor a rectangular, a simple combination of

Reprinted from *IEEE J. Solid-State Circuits*, vol. SC-21, no. 6, pp. 983–988, Dec. 1986.

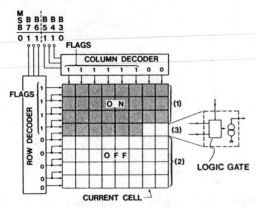

Fig. 2. Two-step decoding.

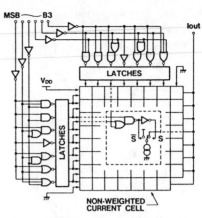

Fig. 3. High-speed decoding circuit.

NOR gates and/or NAND gates cannot attain the decoding. Decoding circuits that realize this complex decoding at speeds less than 20 MHz have already been developed [5], [6]. One solution for the 80-MHz decoding may be to introduce several stages of pipelined latches into these decoding circuits. The inserted latches, however, increase the power consumption. In order to obtain a high-speed decoding circuit with low power consumption, the decoding should be achieved with a minimum number of pipeline stages. In order to reduce the pipelining stages required for 80-MHz decoding, the following decoding circuit of minimum delay has been developed.

Fig. 2 shows the decoding logic achieved by using the minimum number of logic stages. The matrix consists of three types of rows. They are: 1) rows in which all of the current cells are turned on; 2) rows in which all of the current cells are turned off; and 3) a certain row in which current cells are turned on depending upon the column decoder signal. In consideration of these three types of rows, a decoding logic which is carried out in two steps has been developed. The details of the decoding are as follows. In the first step, digital inputs are decoded in the row decoder and column decoder. The number of flags in the columns corresponds to the input value of the column decoder. The number of flags in the rows corresponds to the input value of the row decoder plus one. In the next step, each logic gate in the current cell identifies the row type described above by comparing one row signal with the one next to it. If both of the row signals are at a high level, then the current source is turned on regardless of column signal. If the two row signals are different, then the current source is turned on depending upon the column signal.

Fig. 3 shows the actual decoding circuit. The above-described decoding logic has been achieved by using peripheral decoders and OR–NAND gates in the current cells, i.e., by two logic stages. The inverters at the input node and the inverter inside the current cell are used for buffering and generation of complementary signals. One stage of latches has been inserted between the peripheral decoders and the matrix to suppress the glitch and to enhance the decoding speed. However, since the decoding and the other required functions have been realized by the above-described four gate stages fabricated with 2-μm design

rules, no more stages of latches have been necessary for 80-MHz decoding. This decoding circuit consumes approximately 70 mW and contributes to attaining the high-speed low-power characteristics of this D/A converter.

IV. DESIGN OF CURRENT SOURCE

Fig. 4 shows a circuit diagram of the LSB current source. In the MSB current cell of the matrix and in the current source for $B2$, each of Q_1, Q_2, and Q_3 is formed by four paralleled transistors and two paralleled transistors, respectively. The primary bias voltage V_{G1} is directly applied to the current source transistor Q_1 and determines the full-scale current. The secondary bias voltage V_{G2} is applied through transmission gates. The current source transistor Q_1 and the current switch transistors Q_2 or Q_3 are operated in saturation. The discharge transmission gates TG_3 and TG_4 are formed by single NMOS transistors to realize low-resistance switches with low stray capacitance.

The current source transistor Q_1 has been designed to have a low transconductance in order to enhance the immunity against voltage fluctuation along the ground line (discussed in Section V), and also, against noise at the node of the primary bias voltage V_{G1}. For the same purpose, the cascode-connected configuration for the current source Q_1 as is shown in Fig. 5(b) has not been used. In the cascode configuration, Q_4 and Q_5 are operated in saturation, whereas Q_1 and Q_2 are operated in saturation in the single transistor configuration shown in Fig. 5(a). Since the voltage V_x across Q_4 and Q_5 is lower than the voltage V_{out} across Q_1 and Q_2, Q_4 and Q_5 should be designed to have relatively high transconductance. This high transconductance can cause a low immunity against the voltage fluctuation at the ground line and the bias node.

In order to realize a fast settling time, the stray capacitance at the output node should be small. The stray capacitance at the common source node should also be small to minimize the recovery time of the voltage at this node during the switching transition. This can be achieved by using small-sized transistors for the current source and the current switch. However, short-channel devices de-

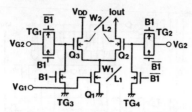

Fig. 4. Circuit diagram of the LSB current source.

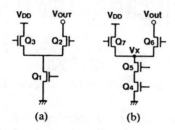

(a)   (b)

Fig. 5. Configuration of current source. (a) Single-transistor configuration. (b) Cascode configuration.

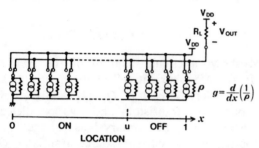

Fig. 6. Model for analyzing influence of output impedance of current sources.

crease the output impedance, and consequently, the linearity is degraded.

Fig. 6 shows a model for analyzing the influence of the output impedance. A one-dimensional current source array is assumed here. The current sources are uniformly arranged from $x = 0$ to $x = 1$. Assuming i is the output current per unit length of the current source array and a constant g is the output conductance of a unit length of the current source array, then the relationship between i and output voltage V_{out} is as follows:

$$i = i_0 - g V_{out} \qquad (1)$$

where i_0 is an output current per unit length of the current source array when $V_{out} = 0$. Assuming u is a ratio of input value to the full scale, the total output current when the input value corresponds to u is given by integrating i from 0 to u. Since i is not a function of x, the output voltage V_{out} is given as

$$V_{out}(u) = R_L \int_0^u i\, dx = R_L i u \qquad (2)$$

where R_L is a load resistance. V_{out} is obtained from (1) and (2) as

$$V_{out}(u) = R_L i_0 \frac{u}{1 + g R_L u}. \qquad (3)$$

Assuming the ideal output $V'_{out}(u)$ is given by a straight

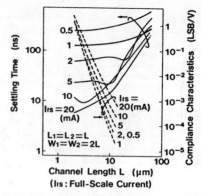

Fig. 7. Simulation result for settling time and compliance characteristics as a function of channel length.

line between zero and full-scale output, that is,

$$V'_{out}(u) = V_{out}(1) u \qquad (4)$$

the deviation of the actual output from the ideal output normalized by the full-scale output is obtained as follows:

$$E(u) = \frac{V_{out}(u) - V'_{out}(u)}{V_{out}(1)} = g R_L \frac{u(1-u)}{1 + g R_L u}. \qquad (5)$$

The maximum of $E(u)$ gives an integral linearity error. Since all current sources are arranged along the unit length, g is equal to the total output conductance when all current sources are connected in parallel. Assuming C_p is a compliance of the full-scale current defined as a normalized fluctuation of the full-scale current caused by a unit voltage change of output node, the relationship between g and C_p is given by

$$C_p \triangleq \frac{(\Delta I_{fs}/I_{fs})}{\Delta V_{out}} = \frac{g}{I_{fs}} \qquad (6)$$

where I_{fs} is a full-scale current. Consequently, the integral linearity error E is approximated as

$$E = \frac{1}{4} g R_L = \frac{1}{4} C_p V_{fs} \qquad (7)$$

where V_{fs} is a full-scale voltage swing. For example, when the full-scale voltage swing is 1 V, an output compliance of 0.4 LSB/V is required in order to suppress the integral linearity error within 0.1 LSB.

The relation of the settling time and the output compliance as a function of channel length has been simulated. The results are shown in Fig. 7. It is assumed that the current source transistor and current switch transistors have the same size and that the channel width is twice the channel length. A load resistance which produces a full-scale swing of 1 V for each full-scale current, and a load capacitance of 20 pF are also assumed. A gate length of less than 5 μm and a full-scale current of more than 10 mA are necessary to achieve a settling time of 12.5 ns. On the other hand, the minimum gate length of 2 μm has a compliance of 1 LSB/V which causes an integral linearity error of 0.25 LSB. The gate length of 2.5 μm has been chosen for L_1 and L_2 to have a margin in both settling time and linearity. Owing to this optimization and the small stray capacitances in the 2-μm design-rule geometries, the fast settling time has been achieved without a serious degradation in the linearity.

V. SYMMETRICAL SWITCHING

As discussed in the previous section, an output current of 10–20 mA is necessary to achieve fast settling time. However, this large current also causes a linearity error. The voltage drop along the ground line caused by the current changes the bias voltage of the current sources, producing a tapered error distribution in the output value of each current source. In a conventional switching sequence of the current source, the tapered distribution results in a significant linearity error. In order to avoid this problem, a switching sequence named "symmetrical switching" has been introduced.

The top of Fig. 8 shows a model for the analysis of this problem. Although a matrix configuration is used in the D/A converter, the current sources are assumed to be arranged in a one-dimensional way for simplicity. The influence of the output impedance discussed in the previous section has been neglected. Nonweighted current sources are uniformly arranged along a ground line from $x = 0$ to $x = 1$. Assuming that $j(x)$ is the output current per unit length of current source array, this current is related to the ground line voltage $v(x)$ as

$$j(x) = j_0 - G_m v(x) \tag{8}$$

where j_0 is the output current per unit length of the current source array when $v = 0$ and the constant G_m is a transconductance between the bias voltage and $j(x)$. Since all current sources are arranged along the unit length, j_0 and G_m are equal to the full-scale current in the case that no voltage drop occurs along the ground line and equal to the transconductance between the full-scale current and the bias voltage, respectively. The voltage drop along the ground line between x_0 and $x_0 - \Delta x$ is given by Ohm's law, that is [7]

$$v(x_0) - v(x_0 - \Delta x) = \Delta x R \int_{x_0}^{1} j(x)\, dx \tag{9}$$

where R is the total resistance of the ground line. From (8) and (9), the following differential equation and boundary conditions are derived:

$$v(x) - \frac{1}{G_m R} \frac{d^2}{dx^2} v(x) = \frac{1}{G_m} j_0 \tag{10}$$

$$v(0) = 0 \tag{11}$$

$$\left. \frac{dv}{dx} \right|_{x=1} = 0. \tag{12}$$

The boundary condition (12) is confirmed by substituting $x_0 = 1$ into (9). The current distribution normalized by an average current is calculated by substituting the solution of

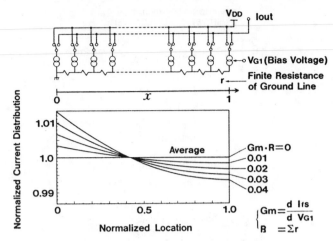

Fig. 8. Current distribution of current sources.

this differential equation into (8), that is

$$j_{nom}(x) = \sqrt{G_m R}\; \frac{\cosh\left\{ \sqrt{G_m R}\,(1-x) \right\}}{\sinh \sqrt{G_m R}}. \tag{13}$$

This normalized current distribution is shown in the bottom of Fig. 8. The outputs of the current sources have a tapered distribution. This tapered distribution produces a significant linearity error in a conventional switching sequence.

Fig. 9 shows an example of a conventional switching sequence and the proposed new switching sequence. If the current sources are connected to the output node sequentially from the left to the right as input value increases, each deviation from the average is accumulated and produces a significant integral linearity error. This error is obtained as the maximum value of the integration of the deviation, that is [7]

$$E_{sq} = \frac{G_m R}{9\sqrt{3}}. \tag{14}$$

For example, assuming that the bias voltage is 2.5 V, the threshold voltage is 0.7 V, the full-scale current is 13.3 mA, and the resistance of the ground line is 2 Ω, the error amounts to 0.2 percent.

In order to avoid this problem, the symmetrical switching shown in Fig. 9 has been introduced. In the symmetrical switching, the current sources are connected symmetrically about the center as the input value increases. The integral linearity error has been reduced by this switching sequence because a deviation from the average of the output current is immediately canceled in the next step. The error caused by this switching sequence is approximated as [7]

$$E_{sy} = \frac{G_m R}{36\sqrt{3}} \tag{15}$$

which is 25 percent of the error caused by the sequential switching. Assuming the same condition as in the sequential switching, the integral linearity error is reduced to 0.05

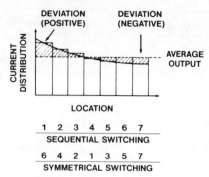

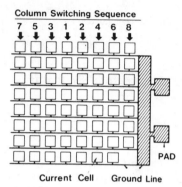

Fig. 9. Symmetrical switching.

Fig. 10. Switching sequence of matrix.

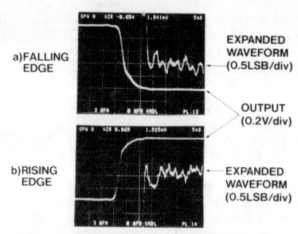

Fig. 11. Settling waveform of DAC.

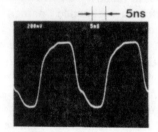

Fig. 12. Full-scale swing at 80-MHz data update rate.

percent by using this switching sequence. On the other hand, the differential linearity error is increased by this sequence. This error is given by the difference between the output at $x = 0$ and at $x = 1$. It is 1.5 percent of the output of the unit current sources in a typical condition. However, this value corresponds to a differential linearity error of only 0.06 LSB in this D/A converter in which current sources of 4 LSB are used in the matrix.

Fig. 10 shows the switching order used in the current-cell matrix. A comb-shaped ground line has been used. The resistances of the backbone and the tooth of the comb are 0.4 and 5 Ω, respectively. Two bonding pads have been allocated to reduce the resistance of the backbone of the comb. The symmetrical switching is introduced into the column switching order by simply rearranging the logic gates in the column decoder shown in Fig. 3. This switching sequence is not used for the row switching order, because the resistance of the backbone is small enough to minimize the voltage drop along itself. However, if the backbone is narrow, it is more effective to introduce this technique into the row switching order also.

VI. PROCESS

This D/A converter has been implemented in a 2-μm double-polysilicon CMOS technology. The first polysilicon forms gate electrodes and interconnection. The second polysilicon has been used only for interconnection. A double-aluminum process would seem to be more flexible in terms of the layout of some interconnections, such as analog ground lines, current output lines, and power supply

lines, in which it is necessary to have small resistances. However, the double-polysilicon process has been adopted because of its advantage of feasible integration of the other analog circuits such as a flash A/D converter which contains capacitors and resistors.

VII. EXPERIMENTAL RESULTS

Fig. 11 shows the full-scale transition of the output, where the full-scale current is 13.3 mA and the external load is 75 Ω. The expanded waveform shows that the settling time to 0.5 LSB is approximately 12.5 ns at both rising and falling edges. The 10–90-percent rise/fall time is less than 5.5 ns. The amplitude of the full-scale swing is constant up to 80 MHz. Fig. 12 shows the output waveform of the full-scale swing at the data-update rate of 80 MHz. This waveform also shows that the decoder is successfully operated at the same frequency because all gates in the decoding circuit alternate their outputs in every swing. The maximum speed of the decoding circuit is approximately 95 MHz. This speed has been measured by observing a sudden change of the waveform and the amplitude. Fig. 13 shows a result of linearity measurement at dc. In this example, an integral linearity error of 0.19 LSB and a differential linearity error of 0.17 LSB have been obtained without trimming. It has been observed that the glitch consists of a single pulse. Its height is 40 mV and its duration is 4.5 ns. The glitch energy is 100 ps·V. This glitch is caused by an internal skew between the row decoder outputs and column decoder outputs. A chip photomicrograph is shown in Fig. 14. Current cells have

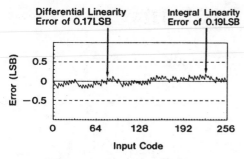

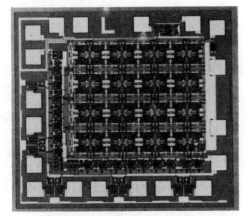

Fig. 13.   Linearity of DAC.

Fig. 14.   Photomicrograph of DAC.

TABLE I
CHARACTERISTICS OF THE DAC

Resolution	8 bit
Settling Time	12.5 ns
Rise/Fall Time	5.5 ns
Integral Linearity Error (DC)	0.38 LSB (Typ.)
Differential Linearity Error (DC)	0.22 LSB (Typ.)
Glitch Energy	100 ps·V
Power Consumption	145 mW
Chip Size	1.85 mm x 2.05 mm

been arranged in an 8×8 matrix. The chip size is 2.05×1.85 mm^2 and the active area is 1.6×1.3 mm^2. The power consumption is 145 mW at a power supply of 5 V. The characteristics of this D/A converter are summarized in Table I.

VIII.   CONCLUSION

An 80-MHz, 145-mW, 8-bit D/A converter has been fabricated in a 2-μm double-polysilicon process. High accuracy has been achieved by adopting a current-cell matrix configuration and introducing a switching sequence named "symmetrical switching." High-speed and low-power decoding is achieved by a decoding circuit with minimum gate delay. Fast-settling and high-impedance current sources have been realized by an optimization of the current source geometry based on an analysis of the relationship between a linearity error and output impedance of the current source.

ACKNOWLEDGMENT

The authors wish to thank Dr. H. Oka, Dr. K. Shibayama, Dr. H. Nakata, and Dr. T. Nakano for their encouragement, and Y. Honda for his help in the implementation of the device.

REFERENCES

[1] M. Yoshimoto, S. Nakagawa, K. Murakami, S. Asai, Y. Akasaka, Y. Nakajima, and Y. Horiba, "A digital processor for decoding of composite TV signals using adaptive filtering," in *ISSCC Dig. Tech. Papers*, Feb. 1986, pp. 152–153.
[2] P. H. Saul, D. W. Howard, and C. J. Greenwood, "An 8b CMOS video DAC," in *ISSCC Dig. Tech. Papers*, Feb. 1985, pp. 32–33.
[3] T. Miki, Y. Nakamura, M. Nakaya, S. Asai, Y. Akasaka, and Y. Horiba, "An 80MHz 8b CMOS D/A converter," in *ISSCC Dig. Tech. Papers*, Feb. 1986, pp. 132–133.
[4] *Electron. Des.*, vol. 33, no. 14, p. 177; also *Electron. Des.*, vol. 33, no. 15, pp. 37–38, June 1985.
[5] K. Hareyama, M. Murayama, and K. Yamaguchi, in *Nat. Conv. Rec. IECE Japan*, Mar. 1982, pp. 2–170.
[6] V. W-K. Shen and D. A. Hodges, "A 60 ns glitch-free NMOS DAC," in *ISSCC Dig. Tech. Papers*, Feb. 1983, pp. 188–189.
[7] T. Miki, Y. Nakamura, M. Nakaya, and Y. Horiba, "Influence of non-zero resistance of analog ground line in D/A converter," *Trans. IECE Japan*, vol. E69, no. 4, pp. 258–260, Apr. 1986.

A 1-GHz 6-bit ADC System

KEN POULTON, MEMBER, IEEE, JOHN J. CORCORAN, MEMBER, IEEE,
AND THOMAS HORNAK, FELLOW, IEEE

Abstract — A two-rank GaAs sample-and-hold (S/H) chip and four
250-MHz silicon digitizers form a 1-GHz 6-bit ADC system. The two-rank
S/H architecture avoids dynamic errors inherent to interleaved ADC's;
accuracy exceeds 5.2 effective bits up to 1-GHz input frequency. Special
attention is paid to avoiding GaAs slow transient errors.

I. INTRODUCTION

THE NEED for high-speed digital waveform acquisition is increasing rapidly. At the same time, high-speed
IC technology is making faster ADC's possible. In this
paper, we describe a 6-bit 1-gigasample/s (1 Gs/s) ADC
which utilizes both bipolar silicon and GaAs MESFET
technologies.

The intended use of this ADC in a digitizing oscilloscope sets a number of requirements: for waveform display, at least 6-bit resolution is required; for transient
capture and good reconstruction of signals with 250-MHz
bandwidth, a sample rate of 1 Gs/s is needed; for capture
of repetitive signals, a bandwidth of 1 GHz is desired.

II. SYSTEM ARCHITECTURE

The hardest of these requirements is achieving a sample
rate of 1 Gs/s. The obvious approach is to build a 1-Gs/s
flash digitizer. This approach was explored, and it was
found that a digitizer built in the available 5-GHz f_T
bipolar technology could not work at that sample rate.
Building the digitizer in GaAs D-MESFET technology was
also not practical, due to both yield and slow transient
problems.

Another approach is to use several (N) digitizers in
parallel, each running at $1/N$ Gs/s in a time-interleaved
fashion. We found that we could build a bipolar digitizer
that would operate at up to 400 Ms/s. We chose to
operate it at 250 Ms/s to allow greater operating margins
and to decrease the demands on the high-speed memories
that follow the digitizers.

The problem with the interleaved architecture is that the
digitizer accuracy degrades badly at input frequencies beyond 100 MHz, so a sample-and-hold (S/H) circuit for
each digitizer is necessary, as shown in Fig. 1. GaAs
MESFET technology was chosen for the S/H function. It

Manuscript received May 12, 1987; revised August 25, 1987.
The authors are with Hewlett-Packard Laboratories, Palo Alto, CA
94304.
IEEE Log Number 8717213.

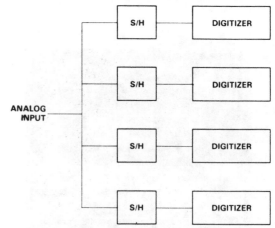

Fig. 1. An interleaved ADC system.

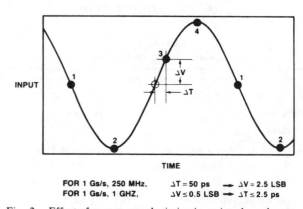

FOR 1 Gs/s, 250 MHz,   ΔT = 50 ps → ΔV = 2.5 LSB
FOR 1 Gs/s, 1 GHZ,   ΔV ≤ 0.5 LSB → ΔT ≤ 2.5 ps

Fig. 2. Effect of uneven sample timing in an interleaved system.

is quite well suited to building very fast S/H circuits,
particularly of the diode-bridge topology. Its advantages
include a very fast Schottky diode, high f_T transistors, and
low parasitic capacitances due to the semi-insulating substrate.

However, using four separate S/H circuits is likely to
introduce timing errors in the sampling of the input. This
problem is shown in Fig. 2: when sampling a high-speed
signal (in this example, a sine wave), any deviation of the
sample timing from precise 1-ns sampling intervals will
appear as an error in the recorded voltage. A typical
deviation of only 50 ps will produce an error of 2.5 least
significant bits (LSB's, 1/64 of the input range for this
6-bit system) when sampling a 250-MHz sine wave. In
order to reach the 1-GHz bandwidth goal with full 6-bit

Reprinted from *IEEE J. Solid-State Circuits*, vol. SC-22, no. 6, pp. 962–970, Dec. 1987.

376

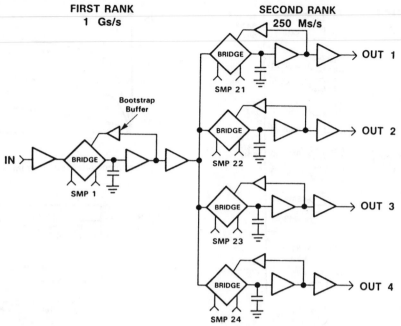

Fig. 3. S/H chip block diagram. All buffers are unity-gain source followers.

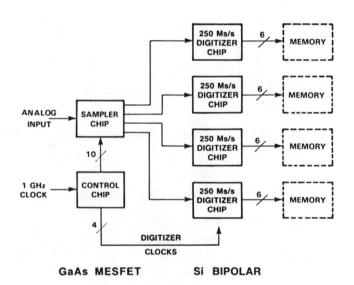

Fig. 4. ADC system block diagram.

accuracy, the sample timing must be regular to within 2.5 ps.

The solution chosen was to use a two-rank S/H architecture. The block diagram of its implementation in GaAs is shown in Fig. 3. (The triangles show buffer stages; the diamonds represent sampling gates [1]. These are sometimes called *track*-and-hold circuits rather than *sample*-and-hold circuits; we shall ignore that distinction of terminology and use the terms interchangeably. in this paper.) The first-rank S/H operates at 1 Gs/s (500 ps sample and 500 ps hold), and presents a sampled-and-held waveform to the four second-rank S/H's. The second-rank samplers operate at 250 Ms/s (1 ns sample and 3 ns hold); they go from track to hold during the hold time of the first-rank sampler, so their input voltage is not moving when they capture their sample. Therefore, the timing

regularity requirement on the second-rank S/H clocks is relaxed from 2.5 ps to about 50 ps. This accuracy can be easily achieved with clocks generated with GaAs logic circuits.

Fig. 4 shows how the system was partitioned into IC's. The five S/H circuits were placed on one GaAs chip, and the clock generation circuits were placed on a second GaAs chip. Each bipolar digitizer is a single chip. In our prototype system, each memory block is an ECL board capable of accepting 1024 data words at up to 400 Mword/s. An NMOS memory chip was later developed to serve this function.

III. GaAs Clock Control Chip

The two GaAs IC's are fabricated in a depletion-only MESFET process [2]. The gate length is 1 μm, f_T is about 12 GHz, and the pinchoff voltage is approximately -2V. This process incorporates a high-density silicon nitride MIM capacitor.

The control chip (Fig. 5) provides amplification and frequency division of the input 1-GHz clock signal, and does the gating necessary to provide all the system clock signals. The input clock, 1 V_{pp} at 1 GHz or below, is amplified by the clock amplifier to 3-V 200-ps rise-time GaAs levels. Frequency division is performed by the two flip-flops connected as a divide-by-four Johnson counter, and the ten gates at the right side of the figure provide the five complementary clocks for the five samplers. All logic functions are implemented in buffered FET logic (BFL) [3].

The timing of the clock signals is shown in Fig. 6. At the top, the 1-GHz clock for the first rank S/H is shown. The next group of traces shows the four 25-percent duty-cycle

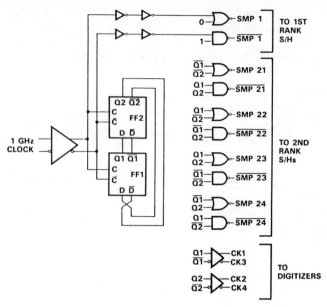

Fig. 5. Control chip block diagram.

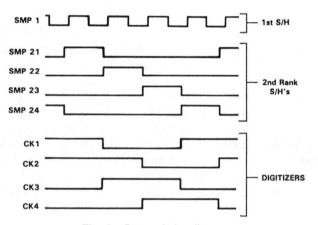

Fig. 6. System timing diagram.

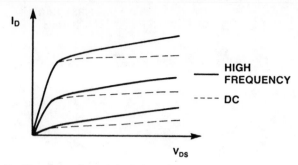

Fig. 7. Frequency-dependent drain conductance of a GaAs MESFET.

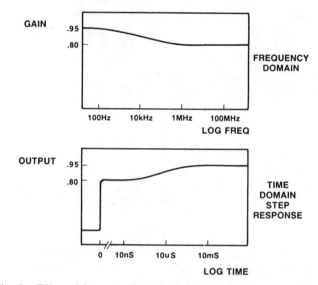

Fig. 8. Effect of frequency-dependent drain conductance on source-follower buffer.

clocks used in the second-rank S/H circuits. At the bottom are the four 50-percent duty-cycle clocks which drive the four digitizers. The progress of one analog sample through the system is highlighted in gray: the first-rank sampler tracks the input signal for half a clock period (500 ps when operating at 1 Gs/s) and then holds the acquired value for half a clock period. For most of that clock period, the second-rank sampler is in the track mode. It is allowed to settle during the first rank's hold time and goes into hold mode just before the first-rank sampler goes into track again. It holds the sample value while the digitizer is propagating that value through its comparators. Finally, on the rising edge of $CK1$, the digitizer comparators are latched and the conversion of this sample is completed.

There is one critical timing requirement met by the control chip: the second-rank samplers must go into the hold mode before the first-rank sampler returns to the track mode. This is accomplished by the introduction of two delay gates (inverters) in the clock path leading to the first-rank sampler.

IV. GaAs Sample-and-Hold Chip

The S/H chip block diagram is shown in Fig. 3. It uses two main subcircuits: a diode-bridge sampling gate (shown as a diamond shape) and a source-follower buffer (represented by the triangles). Each sampler is followed by two source-follower stages to buffer the small hold capacitors (300 fF) from the relatively large capacitances on the outputs.

The major problem encountered in the design of this chip was not in reaching the speed goals, but rather in avoiding low-frequency transients caused by the GaAs FET's [4]. The drain characteristics of a GaAs MESFET are shown in Fig. 7. At high frequency (or in a pulsed measurement), the MESFET shows the characteristic marked by the heavy line. At low frequency, it shows a very different characteristic, shown with the dashed line. In a simple source-follower circuit, this frequency-dependent drain conductance manifests itself as a frequency-dependent gain as seen in Fig. 8: from dc to around 100 Hz, the gain is about 0.95, but at high frequency (above 10 MHz) the gain is only 0.80. The transition between these two regions extends over two to four decades of frequency and tends to shift with temperature, i.e., it does not behave like an RC filter. In the time domain, if the same circuit is

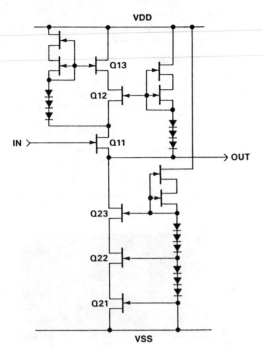

Fig. 9.   Schematic of source-follower buffer.

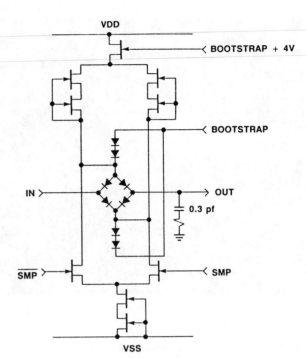

Fig. 10.   Schematic of diode bridge and associated drive circuits.

driven with a fast step of amplitude 1.0, there is a very fast (< 200 ps) rise to 0.80, and, much later, over a period of milliseconds, a rise to the final value of 0.95. This 15-percent gain error is clearly not compatible with a 6-bit ADC system, especially since five source followers are cascaded in a single path through the S/H chip.

Our solution to this problem is a doubly cascoded source follower, as shown in Fig. 9. The primary current source FET, Q_{21}, is cascoded twice by FET's Q_{22} and Q_{23}. The cascoding serves to keep V_{DS} of Q_{21} constant, reducing the effective drain conductance of the current source to the point where the current is essentially no longer a function of the input voltage, and therefore, no longer a function of the frequency of the input signal. The diode level-shift stack at the right maintains a 2-V drain-to-source bias on each FET; it uses a relatively small current.

The follower part of the circuit uses the same elements. Q_{11} is the source-follower FET. To keep its V_{DS} constant, Q_{11}'s drain is bootstrapped (made to track the input voltage shifted by +2 V) by Q_{12} which is in turn bootstrapped by Q_{13}. The bias for the gates of Q_{12} and Q_{13} must follow the signal voltage, so the diode stack which biases Q_{12} is tied to the buffer output. The diode stack which biases Q_{13} is tied to the source of Q_{12} rather than the source of Q_{11}. This improves the high-frequency performance of the source-follower circuit by reducing parasitic capacitances and conductances between the output and V_{DD}.

This source-follower circuit is used for all the buffers on the S/H chip, with only variations in device widths for scaling. It has a slow transient gain error of only 0.5 percent, which is a 30 times improvement over an uncascoded source follower built in this technology.

The Schottky-diode bridge circuit is shown in Fig. 10. (All diodes are MESFET gate-to-drain/source junctions).

The FET differential pair (driven by the *SMP* signals) switches the current from the current source at the bottom to drive the circuit into the sample or hold mode: when *SMP* is high, a current I passes through the diamond-shaped diode bridge, forcing the voltage at OUT to be the same as the voltage at IN. When *SMP* is low, the current I passes through the diodes connected to the BOOTSTRAP node, causing the diode bridge diodes to be reverse biased and breaking the connection between IN and OUT.

Two characteristics of the diodes are critical to the performance of a diode bridge circuit: R_{on} and C_{off}. Since they can be traded off against each other by changing the diode size, their product is a figure of merit for a process' diodes; this GaAs process has an $R_{on}C_{off}$ product of about 1 ps. The ON resistance of the switch formed by the diode bridge will be R_{on}, so the bandwidth will be set by the product $R_{on}C_{hold}$ (neglecting parasitics and the source impedance at IN). The isolation of the switch is determined by the series–shunt–series combination formed by the C_{off} of two reverse-biased bridge diodes in series, with the R_{on} of the diodes connected to BOOTSTRAP forming the shunt element in the middle. The isolation is thus proportional to $C_{hold}/(2\pi f C_{off}^2 R_{on})$.

One limitation of a diode bridge is the "pedestal": a jump in the voltage on the hold capacitor that occurs when the bridge goes from sample to hold. A major cause of the pedestal is the capacitance of the two bridge diodes connected to the hold capacitor: as the bridge is switched off, these two diodes go from a ON bias of 0.8 V to an OFF bias of about −1.6 V. If the node labeled BOOTSTRAP is grounded, these OFF-state biases are dependent on the voltage on the hold capacitor, causing a gain loss of $2C_{off}/(C_{hold} + C_{off})$ (about 9 percent in this design). Furthermore, since the diode capacitances are highly nonlin-

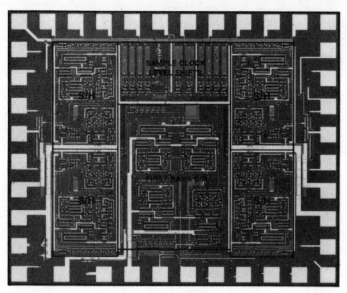

Fig. 11.   Photograph of S/H chip.

TABLE I
S/H CHIP PERFORMANCE

Acquisition Time	400 ps
Hold Pedestal	8 mV (0.4 LSB)
Droop	10 mV/us (0.5 LSB/us)
Aperture Jitter	2.5 ps pp
Gain Flatness dc to 250 MHz	0.4 dB

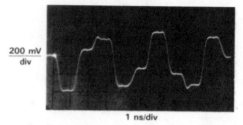

Fig. 12.   Output of first-rank S/H circuit when sampling a 300-MHz sine wave at 1 Gs/s.

ear, significant harmonic distortion of the signal voltage would also occur.

To avoid this, the BOOTSTRAP node is made to follow the voltage on the hold capacitor by the "bootstrap buffer" shown on Fig. 3. The bootstrapping ensures that the voltage changes on the two diodes connected to the hold capacitor are equal and opposite during the fast sample-to-hold transition. Thus the pedestal is cancelled, and the gain and linearity of the S/H is increased.

The impedance seen at the input to the bridge includes the impedance of the load devices of the bridge driver circuit (shown above the BOOTSTRAP node in Fig. 10). Since this load impedance is frequency dependent, the loading on the buffer stage which drives the analog input of the bridge circuit will tend to cause a frequency-dependent gain error. To avoid that, the load devices are cascoded to decrease the loading on the buffer. In addition, the BOOTSTRAP + 4 V input is provided by the bootstrap buffer; it is simply the BOOTSTRAP voltage level shifted up by 4 V to drive a second cascode for the loads. This helps to reduce the frequency-dependent loading on the previous buffer stage even further.

One aspect of the design that is critical for achieving fast acquisition of the input signal is control of resonances due to parasitic inductances. The hold capacitor is shown connected to ground (through a resistor) in Fig. 10, but incurs about 0.5-nH parasitic inductance (on the chip and in a bond wire) on the way. The resistor (30 Ω) was added in series with the ground lead to damp out this resonance. Similar damping and attention to resonances was used in the power supplies.

An S/H chip die photo is shown in Fig. 11. The first-rank bridge and buffers are in the center of the chip; the second-rank bridges and output buffers form the symmetrical blocks in the four corners. Multiple separate ground and power supply pads and buses are used to achieve low inductances in these nodes and to avoid crosstalk between the S/H circuits.

Table I shows some performance measurements of the sampler chip by itself. The acquisition time (time to settle to within 1/2 LSB of steady state on a full-speed input signal) is 400 ps. This is safely less than the 500 ps required for 1-Gs/s operation. The hold pedestal and droop are small, and their effects are reduced even further by an offset calibration procedure. The gain flatness through the whole chip (five source followers and two diode bridges) is 0.4 dB from dc to 250 MHz. Off-chip correction circuits can be used to further improve the gain flatness.

Aperture jitter at 1 Gs/s can be measured by comparing the noise recorded by one of the ADC digitizers with the analog input grounded (through 50 Ω) to the noise recorded when the input signal is a full-scale 1-GHz sine wave sampled near its zero crossing. When the input is a 1-GHz sine wave, an ADC with zero aperture jitter will record only its voltage noise, but an ADC with aperture jitter will record an additional apparent voltage noise, equal to the aperture jitter multiplied by the slew rate (as described in Fig. 2). The noise in our ADC, however, is less than 1 LSB for both of these tests so the noise will appear only when the input voltage is near a digitizer code transition. To attain greater measurement precision, the input frequency is offset from 1 GHz by 10 kHz, producing a slow staircase waveform with short noisy regions at each code transition where two codes appear for a while. The amount of noise is estimated from the fraction of time that the noise shows up on the staircase. Aperture jitter in the first rank S/H is thus measured to be only 2.5 ps peak to peak.

Performance of the first-rank S/H circuit, sampling a 300-MHz sine wave at 1 Gs/s, is shown in Fig. 12.

TABLE III
ADC SYSTEM CHARACTERISTICS

Chip	Sampler	Control	Digitizer
Technology	GaAs D-MESFET 1 um gate	GaAs D-MESFET 1 um gate	Si bipolar 5 GHz
Chip Size	1.4 x 1.7 mm	1.1 x 1.7 mm	2.2 x 3.5 mm
Device Count	460	390	850
Number of Pads	40	35	52
Power	2.9 W	2.5 W	2.5 W

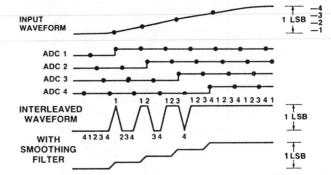

Fig. 13. Increasing the resolution of an interleaved ADC by introduction of offsets and digital filtering.

V. DIGITIZER CHIP

The digitizer, a complete ADC in itself, is implemented in a 5-GHz f_T silicon bipolar process with oxide isolation. It uses a modified flash architecture (sometimes called analog folding) which reduces the transistor count and power when compared to a traditional flash converter design. An on-chip emitter-follower buffer reduces the capacitance seen at the analog input by an order of magnitude from the unbuffered value of 25 pF. This makes it practical to drive the digitizer from the 75-Ω source impedance presented by the S/H chip output buffers and settle within the required 3 ns. The digitizer chip is described in more detail in [5].

VI. CHIP SUMMARY

Various statistics about the three chips are shown in Table II. The GaAs chips, while simple by silicon standards, are complex for analog GaAs. Their power dissipation is relatively high, due to the use of power-hungry depletion-only logic in the control chip and the need to drive relatively low off-chip impedances (50- and 75-Ω lines).

VII. INCREASING THE RESOLUTION

The major disadvantage of an interleaved architecture is the need to match the digitizers well, both in time and in voltage. With real digitizers, there will be regions in the input voltage range near each code transition where the four digitizers will put out differing digital codes, even for a dc input voltage. When the interleaved data are displayed, "LSB noise" patterns are seen when the input is near a transition. The noise is 1 LSB high with a period of four samples.

This problem with interleaving can be turned into an advantage, however. In our ADC, the four digitizers are offset in voltage from each other by successive 1/4-LSB steps. In this way, the four digitizers are interleaved in voltage, as well as in time. In order to take advantage of this fourfold increase in resolution, as well as to eliminate the "LSB noise" which now occurs for any slowly changing input voltage, a nonlinear digital smoothing filter is

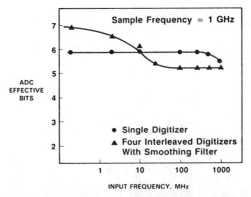

Fig. 14. ADC effective bits of resolution. Single-ADC data represent every fourth point from an interleaved 1-GHz record.

used. It replaces each sample value with the running average of four adjacent samples. To avoid the fourfold bandwidth reduction that a such a filter normally causes, this filter is disabled when the differences between successive samples are larger than one LSB. The result is seen in Fig. 13. The effective resolution for slowly changing input signals is, ideally, increased to 8 bits.

Note that the increase in resolution is due to the 1/4-LSB offsets between the digitizers. Unlike standard oversampling and averaging techniques, this technique is deterministic: if the input voltage moves less than 1/256 of the input range in four sample periods, the resolution is a true 8 bits. Although it provides 8-bit resolution on a sine wave only if the input frequency is below about 1 MHz, it is quite useful in the flat portions of pulse waveforms.

VIII. ADC SYSTEM PERFORMANCE

The effective bits of resolution [6] achieved by this ADC are plotted in Fig. 14. (This and all succeeding data are derived from analysis of the measured digital data from the digitizers.) One trace shows the effective bits achieved using the data from only one of the four digitizers. Effectively sampling at 250 Ms/s, it demonstrates 5.9 effective bits to 250-MHz input frequency, and 5.5 effective bits at 1 GHz. The drop in effective bits at 1 GHz is primarily due to harmonic distortion in the S/H. The other trace shows effective bits for the whole ADC, operating at 1 Gs/s: 5.2 effective bits is realized up to 1-GHz input

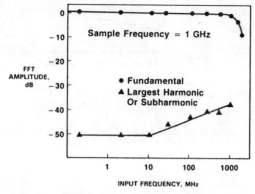

Fig. 15.   ADC frequency response and distortion.

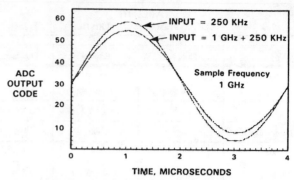

Fig. 17.   Reconstructed sine waves with resolution enhanced by interleaving.

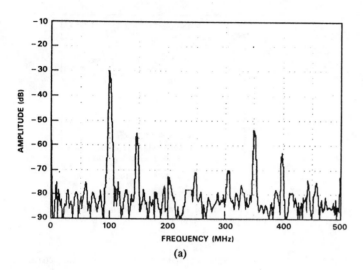

(a)

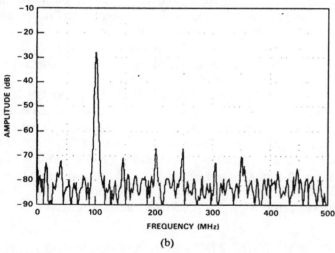

(b)

Fig. 16.   (a) Spurious sampling products observed when first-rank S/H
is forced to remain in track mode. (b) Spurious sampling products
observed when first-rank S/H operates normally.

TABLE III
ADC SYSTEM CHARACTERISTICS

Sample Rate	1 Gs/s (and lower)
Bandwidth (3 dB)	1.7 GHz
Input Range	± 0.64 V
Random Noise	< 5 mV pp < 0.25 LSB
Effective Bits 1 Gs/s, 1 GHz input 1 Gs/s, 200 kHz input	5.2 bits 6.9 bits
Power	16 W

frequency. The decrease in accuracy from the single digitizer case is due to mismatch between the four digitizers,
both in static linearity characteristics, and in the calibration of their voltage references. At lower frequencies, use
of the voltage interleaving and the "LSB noise" smoothing
filter increases performance to about 7 effective bits. This
is one bit less than the ideal 8 bits, again due to nonlinearities and mismatches in the digitizers.

Fig. 15 shows the frequency response of the ADC system. The input signal bandwidth is 1.7 GHz, and is limited
by the S/H. Harmonic distortion is beginning to rise near
the high end of the frequency range, but is still only
−40 dBc at 500 MHz, corresponding to ±0.3 LSB errors.

The utility of the two-rank S/H architecture is demonstrated by its suppression of sampling products related to
one quarter the sampling rate. Fig. 16(a) shows an FFT of
a 1102-MHz sine wave sampled at 1 Gs/s with the first-
rank sampler remaining in sample mode (and therefore
acting only as a buffer). The spurious sampling products of
the input signal around 250 MHz (148 and 352 MHz) have
an amplitude of −25 dBc. Fig. 16(b) shows the FFT with
the first-rank sampler running normally. All spurious components are now below −39 dBc.

Fig. 17 shows the result of a beat frequency test. The
larger sine wave is the result of sampling a 250-kHz sine
wave at 1 Gs/s and displaying the digital data. The
smaller sine wave is the result of sampling a 1.000250-GHz
sine wave and displaying the beat frequency of 250 kHz.
The acquired waveform is free of noise and shows no
evidence of breakup, even at 1-GHz input.

Table III summarizes important characteristics of the
ADC system. "Random noise" means noise exclusive of
ADC quantization noise. This is measured similarly to the
aperture jitter measurement: a low slew rate signal (10
kHz) is applied and the fraction of the acquisition trace
which looks noisy (shows two codes alternating) is measured.

Fig. 18. Photograph of package.

IX. PACKAGE

The six chips that make up one ADC are mounted in the package shown in Fig. 18. In the center, the two GaAs chips are mounted side by side on a one inch square thin-film hybrid substrate. In the corners are the four additional hybrids for the four digitizer chips. The sampler outputs and the digitizer clocks travel from the GaAs IC's to the digitizers on microstrip transmission lines with wire bonds jumping the gaps between the hybrid substrates. The sampler outputs are on 75-Ω lines terminated only by the output impedance of the output buffers. The digitizer clocks are on complementary 75-Ω lines terminated at the digitizer with on-chip resistors. Signal traces come to pads at the edges of the hybrids where they mate through a conductive elastomer to matching pads on the bottom of a PC board. These signal traces are of various impedance levels: 50-Ω lines are used for the high-speed analog and clock inputs and for the digital ECL output data lines, with thin-film termination resistors on the hybrid. Most other signals are slower and thus may be high-impedance lines; power supplies have bypass capacitors on the PC board, the hybrids and, in some cases, on the chips. The whole package is 2.8×3.8 in; the backside is finned to help dissipate the power.

X. SUMMARY

A 1-Gs/s 6-bit ADC system has been demonstrated. It uses GaAs MESFET IC's to provide accurate sampling at up to 1 Gs/s, and four 6-bit silicon bipolar digitizers which are interleaved in voltage as well as in time. A two-rank sample-and-hold architecture is utilized to avoid uneven sample timing due to the interleaved architecture. Frequency-dependent drain characteristics of the GaAs MESFET are circumvented by circuit design techniques. A nonlinear filtering technique is used to increase the effective resolution beyond the 6 bits of the digitizers. Finally, 5.2 effective bits is achieved at up to 1-GHz input frequency, and 6.9 effective bits at low frequencies.

ACKNOWLEDGMENT

The authors wish to acknowledge the contributions of K. Knudsen and R. Kagarlitsky, the Hewlett-Packard Microwave Technology Division, and the Hewlett-Packard Santa Clara Technology Center.

REFERENCES

[1] J. R. Gray and S. C. Kitsopoulos, "A precision sample and hold circuit with subnanosecond switching," *IEEE Trans. Circuit Theory*, vol. CT-11, pp. 389–396, Sept. 1964.
[2] R. L. Van Tuyl *et al.*, "A manufacturing process for analog and digital GaAs integrated circuits," *IEEE Trans. Microwave Theory Tech.*, vol. MTT-30, no. 7, pp. 935–942, July, 1982.
[3] R. L. Van Tuyl *et al.*, "GaAs MESFET logic with 4-GHz clock rate," *IEEE J. Solid-State Circuits*, vol. SC-12, no. 5, pp. 485–496, Oct. 1977.
[4] M. Rocchi, "Status of the surface and bulk parasitic effects limiting the performance of GaAs IC's," *Physica*, vol. 129B, pp. 119–138, 1985.
[5] J. Corcoran and K. Knudsen, "A 400 MHz 6b ADC," in *ISSCC Dig. Tech. Papers*, Feb. 1984, pp. 294–295.
[6] B. Peetz, "Dynamic testing of waveform recorders," *IEEE Trans. Instrum. Meas.*, vol. IM-32, no. 1, pp. 12–17, Mar. 1983.

A Single-Chip CMOS PCM Codec with Filters

DOUGLAS G. MARSH, BHUPENDRA K. AHUJA, TOSHIO MISAWA, SENIOR MEMBER, IEEE,
MIRMIRA R. DWARAKANATH, MEMBER, IEEE, PAUL E. FLEISCHER, SENIOR MEMBER, IEEE,
AND VEIKKO R. SAARI, MEMBER, IEEE

Abstract—A complete PCM codec using charge redistribution and switched capacitor techniques will be described. The device is implemented in a two-level polysilicon CMOS technology using 23.4 mm^2 of active area. It features all the required transmission filters needed for telephony, two on-chip voltage references, TTL compatible digital interfaces, and low-power dissipation. The architecture of the chip allows asynchronous operation, a variable PCM data rate from 100 kbits/s to 4.096 Mbits/s, μ/A law operation via pin selection, and gain selection at either of two levels in each direction.

I. INTRODUCTION

EXISTING telephone systems which use pulse-code modulation for the digital transmission of voiceband signals use a high-speed codec (coder-decoder) multiplexed over many channels. Now, as a result of the rapid and continuing reduction in the cost of complex digital LSI, the move toward digitization has reached the customer premise and switching equipment. It has been recognized for a number of years that using codecs on a per channel basis would result in system design and maintenance advantages in all the above applications. However, the low cost and power dissipation of existing shared codecs [1] (on a per channel basis) and of active thin-film RC filters [2], [3] has presented a strong barrier to per channel codecs. In fact, from an economic standpoint, the separate replacement of either the existing filters with monolithic ones or the complex shared codec and PAM architecture with many per channel codecs cannot be justified. However, replacing both these functions with a single chip does achieve the system advantage at an attractive first cost.

This paper describes a monolithic chip which uses charge redistribution [4], [5] and switched capacitor [6]-[8] techniques to perform the codec and filtering functions. The device, which is implemented in a two-level polysilicon, 5 μm line CMOS technology, has two on-chip trimmed voltage references, TTL compatible digital interfaces, and low-power dissipation. The architecture of the chip allows asynchronous operation, a variable PCM data rate from 100 kbits/s to 4.096 Mbits/s, μ or A law operation via pin selection, and gain selection at either of two levels in each direction.

Manuscript received January 28, 1981; revised April 2, 1981.
D. G. Marsh, P. E. Fleischer, and V. R. Saari are with Bell Laboratories, Holmdel, NJ 07733.
B. K. Ahuja was with Bell Laboratories, Murray Hill, NJ 07974. He is now with the Intel Corporation, Santa Clara, CA 95051.
T. Misawa and M. R. Dwarakanath are with Bell Laboratories, Murray Hill, NJ 07974.

II. CHIP DESIGN

Fig. 1 shows a block diagram of the chip [9]. The transmit side consists of an active RC antialiasing filter followed by a fifth-order elliptic filter, a second-order high-pass notch, a sample-and-hold stage, and the encoder. There is also an auto-zero loop. In the receive direction, the D/A output is sampled-and-held and followed by another fifth-order low-pass filter. Separate bias and reference networks are supplied for each direction of transmission.

A. Filters

Fig. 2 shows the complete transmit low-pass filter. The input active RC filter is a conventional Sallen–Key structure and is needed to reject energy at and above the sampling frequency of the low-pass switched capacitor filter. The 3 dB frequency of the RC filter is nominally at 18 kHz. The resistors are made from 2.5 kΩ per square p-tub material and occupy only about one-fourth the area of the capacitors. Nevertheless, the distributed parasitic capacitance of the resistors is significant, so the capacitor ratio must be set slightly larger than 2:1 to predistort the filter and obtain a maximally flat shape. Since the absolute value of the RC time constant, which may have a 4:1 variation, sets the characteristic of this stage, allowance for that variation must be made. This produces a worst case ripple of only ±0.04 dB at 3 kHz while requiring a 256 kHz sampling rate in switched capacitor filter to ensure the required 32 dB of out-of-band rejection.

The 256 kHz sampling rate for the low-pass filter (LPF) is obtained by sampling on both phases of the clock with two identical valued capacitors C_B and C_U. At the output of the first stage, the signals from the two phases are summed, thereby introducing a transmission zero at 128 kHz. Only a 128 kHz sampling rate is needed thereafter, thus reducing the total capacitance of the filter.

Two additional components C_{B1} and C_{U1} are shown at the input to the LPF. When these are placed in parallel with C_B and C_U, the gain of the input stage is increased.

The filter uses parasitic free toggles throughout [10], and has been designed to equalize its overload characteristics. That is, the gain of each stage is set so that the peak transmission from the input to each output is the same. Since the output stage of the op amp [11] is a high-gain stage with the sources of the devices tied to the power supplies and since the switches on the op amp outputs are dual channel, the op amps can drive nearly to the power rails without clipping. However, the input

Reprinted from *IEEE J. Solid-State Circuits*, vol. SC-16, no. 4, pp. 308–315, Aug. 1981.

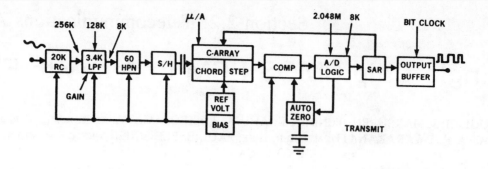

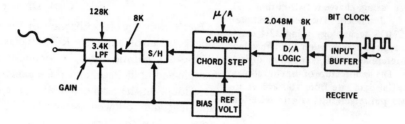

Fig. 1. Block diagram of PCM codec with filters.

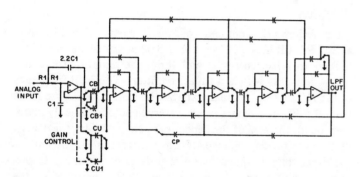

Fig. 2. Transmit 3.4 kHz low-pass filter showing input active *RC* filter and gain control.

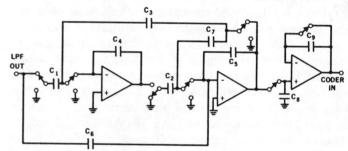

Fig. 3. Transmit 60 Hz high-pass notch followed by 8 kHz sample and hold.

antialiasing filter op amp operates in the voltage follower mode, and is limited by the common mode of the amplifier. For 0 dB gain, the result is that any 60 Hz tone which might overload the LPF will have already overloaded the antialiasing filter. Thus, we can place the high-pass notch at the output of the low-pass filter.

The high-pass notch (HPN), shown in Fig. 3, is a two-amplifier biquad section [10]. Above its cutoff, the signal has a direct feed to the output. Thus, from a noise point of view, the 8 kHz foldovers are in narrow 100 Hz bands, and the total noise of the stage is just that of an unsampled noninverting gain $(1 + C_6/C_5) \approx 2$ plus these narrow foldovers. The output of the HPN drives a rather large capacitive array ($\sim$120 pF) in the A/D. Settling time effects in charging this array could disturb the filter: if the output of the HPN does not fully settle during a sample period, a transmission degradation will result. Therefore, the output of the HPN is buffered from the encoder with a fully balanced sample-and-hold amplifier. This does not significantly degrade system noise performance: whether or not we use the sample-and-hold stage, 8 kHz sampling—with all its associated noise foldovers—occurs at the output of the HPN.

Power supply feedthrough is a significant concern in codec

design. In switched capacitor filters, there are a number of sources of feedthrough, but the dominant ones are due to routing and switches which interface op amp summing nodes. The obvious solutions are to use minimum size switches and as large a minimum capacitor as practical (80 μm × 80 μm or 1.6 pF). In addition, this design uses n-channel switches on the summing nodes, biases the p-tub of the switches from a regulated supply, and uses a grounded p-tub under all the interconnect (Fig. 4). During power down, the PD signal is high and the p-tub bias is switched to the negative power supply in order to avoid substrate injection. Finally, only n-channel input op amps are used in the filter, and the common p-tub of the input devices is tied to their sources. If this were not done, then negative supply variations change the device threshold voltage and therefore the gate-to-source voltage due to backgate effects. Since the positive op amp input gate is grounded, the result is that the sources of the input pair would move with the power supply, and this movement would couple across the gate-to-source capacitance of the negative input device into the filter.

On the receive side, only a low-pass filter is used. The design is similar to the transmit filter except that C_U, C_{U1}, and the switch associated with C_P are omitted. Furthermore, the com-

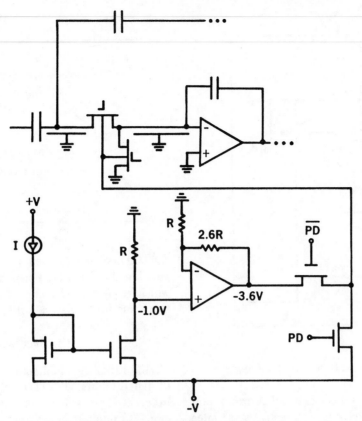

Fig. 4. Circuitry for improvement in power supply rejection ratio.

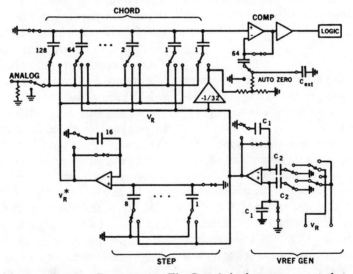

Fig. 5. μ-law A-to-D converter. The D-to-A is the same except that
the comparator is replaced by a sample and hold amplifier, there is no
autozero loop, and the step array is modified.

ponent values are all changed because this stage must have an
x/sinx passband shape. The transfer function of the stage has
actually been modified from the x/sinx characteristic since 16
samples of the D/A output are used, but one of these samples
is taken during the reset interval in the D/A and is therefore
zero valued.

B. Codec

The A/D converter is shown in Fig. 5. Similar designs have
been described in detail previously [5]. Capacitor arrays are
used for both the chords and steps with a buffer amplifier
between them. A differential amplifier is used to generate
either a positive or a negative reference voltage. Local autozeroing is used around both op amps and in the comparator,
and an overall autozero loop—requiring an external capacitor—
which operates independently of the signal is employed. The
comparator consists of a differential stage, a cascode, and a
latch. The chord search algorithm is binary.

The D/A is similar except that a sample-and-hold amplifier
replaces the comparator, there is a minor change in the step

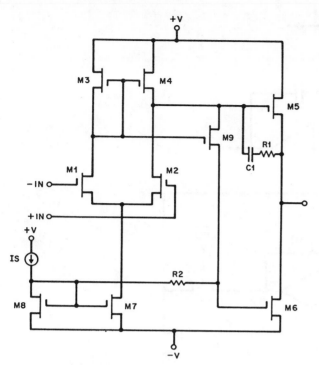

Fig. 6. Simplified schematic of the basic operational amplifier.

generator, and there is no need for the external autozero loop. The D/A output is not 100 percent held; there is a short interval during which the arrays are discharged and the amplifier autozeros are operated. Since the filter takes one sample during that interval, the filter, as mentioned earlier, is predistorted from the $x/\sin x$ shape.

C. Operational Amplifiers

A key to obtaining good overall performance at low-power dissipation lies in the design of the operational amplifiers. Although a detailed description will be presented elsewhere [11], a brief discussion of the design follows.

A simplified schematic of the basic operational amplifier is given in Fig. 6. The design is a two gain-stage amplifier ($M1 - M6$) biased from a current-related voltage source ($IS, M8$). Only one such source is used in each direction of transmission to save power. The input stage uses large gate areas (3400 μm^2 n-channel devices and 800 μm^2 p-channel loads) which keep $1/f$ noise down to about 50 nV/$\sqrt{\text{Hz}}$ at 1 kHz. The transconductance of the input pair and the compensation capacitor C_1 are set to give a 2 MHz unity gain bandwidth and 35 nV/$\sqrt{\text{Hz}}$ of broad-band noise while obtaining a dc gain of 80 dB. Since nearly all amplifiers drive only capacitive loads, the $M6$ current can be made low for those applications. If nothing else were done, this would result only in reduced negative-going output slewing. To ensure that this reduction does not occur, $M9$ and $R2$ are added. Now, as the output slews negatively, $M1$ will turn on hard and $M2$ turns off, resulting in sufficient V_{GS} across $M9$ to turn it on. This pulls the gate of $M6$ up and substantially increases the output pull down current. Once slewing stops, $M9$ goes off, the signal does not drive $M6$, and its current returns to a low level. Resistor $R2$ isolates changes at the gate of $M6$ from the input pair current source $M7$.

When these amplifiers are used in the filter, the p-tubs of $M1$ and $M2$ are tied to their sources for PSRR improvement as mentioned previously. When the amplifiers are used in locations which require good common mode range, the p-tubs are tied to $-V$ and fully balanced differential circuits are used.

III. SPECIAL FEATURES

Table I lists the architectural features of the chip. There are 18 digital inputs, all TTL compatible. Three different buffers are used—eight high speed, four medium speed, and six low speed—to minimize the power dissipation. The level shifting is performed immediately upon entering the chip, rather than at the transmission gates, for two reasons. First, fewer level shifters are needed, and second, logic transients are kept off the ground to minimize idle circuit noise. Twenty percent of the chip power dissipation occurs in these buffers.

Asynchronous operation is achieved by building two completely separate directions of transmission. Sharing the arrays would have added handshaking circuitry and would have increased absolute delay. The addressing structure involves using two signals plus SYNC for each direction, accounting for six of the inputs. Only the SYNC edge needs to be high speed, so it may be shared by several channels.

The data rate may be the master clock (2.048 Mbits/s) or it may be any other rate from 100 kbits/s to 4.096 Mbits/s. The PCM input, the master clocks, the data clocks, and a selection lead account for six inputs. However, the default state of the data rate is that it equals the master clock. Therefore, one packaging option brings out only the master clocks, thus saving pins.

The μ-law is the default state of the device. However, by tying a control pin low, the last plate in the chord array is paralleled with the second to last plate in both the A/D and D/A, and the sign of the odd bits (except the sign bit) is reversed in I/O registers. This results in the implementation of the A-law.

The codec operates in an 8-bit PCM mode, with the intent that any signaling information can be inserted or detected in the common equipment. However, when receiving data, the D/A must still be capable of changing gears to a 7-bit mode when a signaling frame signal is received [12]. The two modes are achieved by allowing half of the last capacitor in the D/A step array to switch from ground to the reference voltage for 8 bits or holding it at ground while forcing the LSB bit high for 7 bits.

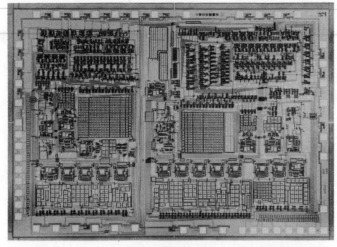

Fig. 7. Photomicrograph of the chip. The active area is 23.4 mm² and the area to the outer edge of the pads and interconnect is 29.3 mm².

TABLE II
POWER DISSIPATION

Input TTL to MOS Level Shifters	12 mW	} Logic
Digital Logic	5 mW	
A/D and D/A	10 mW	} Codec
Bias and References	13 mW*	
Filters	20 mW	
Total	60 mW	

* 5 mW in power off mode.

TABLE III
FILTER PERFORMANCE

	Transmit (G=6.02 dB)	Receive (G=0 dB)
Passband Ripple (300-3000 Hz)	$\leq \pm 0.08$ dB	$\leq \pm 0.04$ dB
Stop Band Rejection ($\geq$4600 Hz)	$\geq$34 dB	>30 dB
Total Harmonic Distortion		
0 dBm0	-63 dB	-60 dB
+3 dBm0	-61 dB	-66 dB
Rejection at		
50 Hz	>25 dB	
60 Hz	>35 dB	
Delay Distortion (1-2.6 kHz)	55 μs	93 μs
Power Supply Rejection		
at 1 kHz		
+5V	44 dB	50 dB
-5V	44 dB	44 dB
at 100 kHz noise in, C-msg out		
+5V	36 dB	41 dB
-5V	39 dB	42 dB
Idle Channel Noise	+5 dBrnC0	+2 dBrnC0

Gain adjustments of 6.02 dB are available in the first stages of the transmit and receive filters. These can be enabled independently.

Two inputs are used in a four state disable. The states are: enable the entire device, the encoder only, the encoder logic only, or power off the device. In the power-off state, however, the references are not turned off, thus improving their reliability. The two references are compensated Zeners, with fusable aluminum links across resistors for trimming.

A photomicrograph of the chip is shown in Fig. 7.

TABLE IV
CODEC PLUS FILTER PERFORMANCE

Power Supply		±5V, 60 mW		
		5 mW (power off)		
Gain Tracking at 1020 Hz,	+3 dBm0			
	to	±0.3 dB		
	-40 dBm0			
	-40 dBm0			
	to	±0.7 dB		
	-55 dBm0			
Signal/Distortion: (C-msg Weighted)				
at 1020 Hz, +3 dBm0 to -30 dBm0		>36 dB		
-40 dBm0		>30 dB		
-45 dBm0		>25 dB		
Idle Channel Noise		10-16 dBrnC0		
Power Supply Rejection				
at 1 kHz		Transmit	Receive	End-to-End
+5V		42 dB	42 dB	38 dB
-5V		41 dB	44 dB	42 dB
at 100 kHz noise in, C-msg out				
+5V		28 dB	45 dB	28 dB
-5V		32 dB	37 dB	31 dB

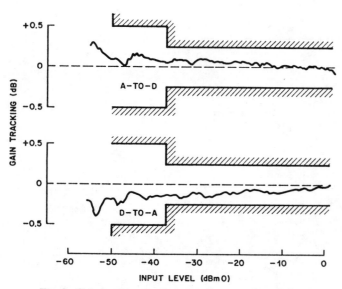

Fig. 8. Gain tracking for each direction of transmission.

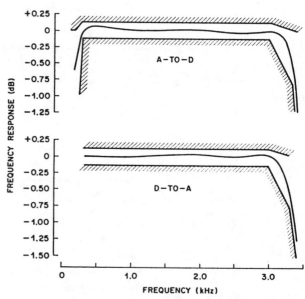

Fig. 9. Frequency response for each direction of transmission.

IV. PERFORMANCE

The device operates between ±5 V supplies at a nominal dissipation of 60 mW with a worst case of 120 mW at −25°C. Table II shows that the 60 mW is spread nearly equally among the three major sections of the device. Table III shows the filter performance. The transmit ripple exceeds the receive ripple for two reasons: at the low band edge, the transmit filter includes variations due to the HPN; at the high band edge, it includes variations due to the active *RC* filter. Harmonic distortion is dominated by the second harmonic. Table IV shows the major characteristics of the complete device. Gain tracking,

signal-to-distortion, and idle channel noise (ICN) are all well within *D*-channel bank requirements [12].

Gain tracking for both analog-to-digital and digital-to-analog are shown in Fig. 8. The measurement was made with a Hewlett–Packard primary multiplex analyzer. Note the small, consistent slopes which cancel each other when the measurement is made back-to-back. This indicates that both chord arrays have an identical small deviation from binary scaling. Fig. 9 shows the passband characteristics for A/D and D/A. The *C*-message weighted *S/N* versus input signal level is shown in Fig. 10. The D/A result is very close to what is expected from

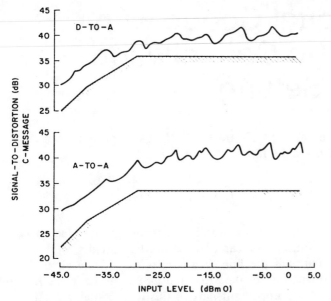

Fig. 10. Signal-to-distortion for D-to-A and end-to-end.

used for the codec and switched capacitors are used for the filters. The architectural features which increase the utility of the device in system applications have been discussed. Performance is sufficiently within the D-channel bank requirements that appropriate portions of the total error budget may be allocated to other parts of the system.

ACKNOWLEDGMENT

The authors are pleased to acknowledge the contributions of T. E. Seidel for process development, D. L. Fraser for consultation on switched capacitor filters and for supplying initial test chips, and J. R. Barner for layout assistance.

the ideal μ-225 law quantization. The end-to-end characteristic degrades by 1-2 dB at -45 dBm0, indicating a slightly nonideal effect in the A/D converter. ICN is dominated by the transmit side.

Power supply rejection in Tables III and IV is given for both a 1 kHz tone and for white noise, band limited to 100 kHz. In the latter case, the measurement is the rejection obtained between the total noise power input compared to the C-message weighted noise output. Thus, both frequency shaping and the aliasing effects are taken into account. The dominant effect is at the input to the A/D converter because of the 8 kHz sampling of the spectrum which appears at the transmit filter output.

V. CONCLUSION

A complete single-chip PCM codec with filters has been described which is fabricated using a two-level polysilicon CMOS technology with 5 μm lines. Charge redistribution is

REFERENCES

[1] D. A. Spires, "The D4 channel bank codec," *IEEE Trans. Circuits Syst.*, vol. CAS-25, pp. 468–475, July 1978.
[2] R. A. Friedenson, R. W. Daniels, R. J. Dow, and P. H. McDonald, "RC active filters for the D3 channel bank," *Bell Syst. Tech. J.*, vol. 54, pp. 507–530, Mar. 1975.
[3] J. J. Friend and W. Worobey, "STAR: A universal active filter," *Bell Lab. Rec.*, pp. 232–236, Sept. 1979.
[4] J. L. McCreary and P. R. Gray, "All MOS charge-redistribution A-D conversion techniques, Part I," *IEEE J. Solid-State Circuits.* vol. SC-10, pp. 371–379, Dec. 1975.
[5] M. R. Dwarakanath and D. G. Marsh, "A two-chip CMOS codec," in *ICC Dig. Tech. Papers*, pp. 11.3.1–4, June 1980.
[6] D. L. Fried, "Analog sampled data filters," *IEEE J. Solid-State Circuits*, vol. SC-7, pp. 302–303, Aug. 1972.
[7] J. T. Caves *et al.*, "Sampled analog filtering using switched capacitors as resistor elements," *IEEE J. Solid-State Circuits*, vol. Sc-12, pp. 592–599, Dec. 1977.
[8] D. J. Allstot, R. W. Broderson, and P. R. Gray, "Fully integrated high-order NMOS sampled data ladder filters," *ISSCC Dig.*, pp. 82, 83, Feb. 1978.
[9] B. K. Ahuja, M. R. Dwarakanath, T. E. Seidel, and D. G. Marsh, "A single chip CMOS PCM CODEC with filters," *ISSCC Dig.*, pp. 226, 227, Feb. 1981.
[10] P. E. Fleischer and K. R. Laker, "A family of active switched capacitor biquad building blocks," *Bell Syst. Tech. J.*, vol. 58, pp. 2235–2270, Dec. 1979.
[11] V. R. Saari, "Fast settling, low power CMOS operational amplifier," to be published.
[12] "The D3 channel bank compatibility specification," AT&T, issue 2, Oct. 14, 1974.

A 3-μm CMOS Digital Codec with Programmable Echo Cancellation and Gain Setting

PAUL DEFRAEYE, MEMBER, IEEE, DIRK RABAEY, MEMBER, IEEE, WIM ROGGEMAN, MEMBER, IEEE, JOHAN YDE, AND LAJOS KISS

Abstract—A 3-μm CMOS digital signal processor (DSP) performs speech signal shaping, programmable echo cancellation and gain setting functions for telephone applications. A/D and D/A conversions are performed by making use of $\Sigma\Delta$ modulators, decimator, and interpolator filter blocks. In addition, the DSP acts as a control interface between the subscriber line interface circuit (SLIC) and the line card controller, the debouncing of eight line status bits included.

I. INTRODUCTION

TEMPERATURE-, tolerance- and ageing independency of filter characteristics are the main advantages of digital signal processing over analog techniques. The most recent developments in this field try to move the barrier between the analog and digital worlds as far as possible in order to keep the analog circuitry very simple and noncritical. Moreover, these digital codecs make it possible to integrate programmable functions. Especially for modern telephone line circuit applications, digital filters, echo cancelers and gain settings have to be adaptable to customer's specifications. In addition, control functions can be added to the digital signal processor (DSP) chip in order to reduce the component count of the system. By so doing, it becomes feasible to build a PCM line circuit with only three custom LSI's per line and to put eight line circuits on a single line card.

The new PCM line circuit is shown in Fig. 1. The circuit is connected to the line by 350-V solid-state switches that have an on-resistance of only 10 Ω. A single-chip subscriber line interface circuit (SLIC) combines several functions such as line feeding, AC impedance synthesis, two to four wire conversion, line status sensing, and relays driving. The analog speech signals are converted by the DSP chip which performs A/D, D/A, echo cancellation and gain setting functions. A 13-bit linear code is transmitted and received via two multiplexed lines to and from a single transcoder chip per line card. After conversion between linear code and A- or μ-law 8-bit PCM, the transcoder chip interfaces the logarithmic PCM signals to the line card

Manuscript received October 29, 1984; revised December 27, 1984.
The authors are with Bell Telephone Manufacturing Company, Central Research Laboratory, Francis Wellesplein 1, B-2018 Antwerpen, Belgium.

controller. This controller performs the interface between the central office and 16 possible line circuits.

Through the line card controller, the line circuit status information for the SLIC and DSP is stored in the DSP bit map. An important part of this data contains coefficients for echo cancellation and gain setting. The DSP transmits the SLIC status information and receives eight line status bits such as ring trip and switch hook. These bits are debounced in a multiplexed mask programmable hit timer and sent to the card controller.

The 12/16-kHz metering bursts are generated for all eight lines on the transcoder chip. This signal is mixed with the speech in the SLIC at the request of the DSP status information.

In Section I, the DSP architecture is described and special emphasis is placed on the duality between the DSP transmit and receive paths. In Section II the $\Sigma\Delta$ modulator as a basic building block for A/D and D/A conversion is discussed. The digital speech path and the control part are described in Sections III and IV, respectively. Section V gives an overview of the implementation aspects and test results.

II. DSP ARCHITECTURE

Fig. 2 depicts the basis building blocks of the signal processing part of the DSP.

The heart of the analog section is a $\Sigma\Delta$ modulator which oversamples the speech signal at 1 MHz with 1-bit resolution. This high sampling rate only requires a simple anti-alias filter with a 50-kHz bandwidth. The differential input section of this filter acts as the first echo cancellation stage in order to avoid overload of the $\Sigma\Delta$ modulator with echo return signal.

The $\Sigma\Delta$ modulator generates a lot of quantization noise but it is shaped in such a way that it is out of the speech band. The subsequent decimation filters reject these high frequency components and reduce the sampling frequency from 1 MHz (1 bit) to 32 kHz (15 bit). Two digital biquad sections perform CCITT low-pass filtering at 3400 Hz. Further decimation to 8 kHz occurs prior to high-pass filtering at 300 Hz using a fifth-order filter. The gain

Reprinted from *IEEE J. Solid-State Circuits*, vol. SC-20, no. 3, pp. 679–687, June 1985.

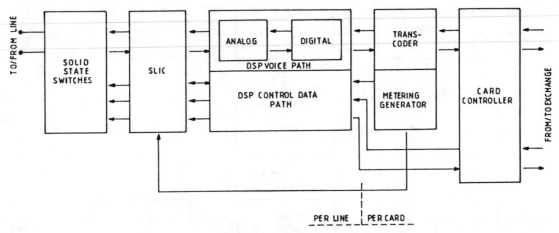

Fig. 1.   Block diagram of the PCM line circuit.

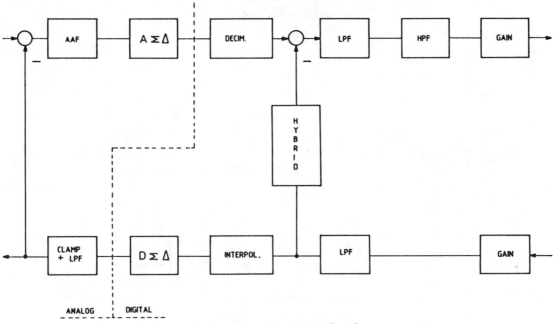

Fig. 2.   The DSP speech path.

setting sections of both the transmit and receive sides are software programmable multipliers. The level adaptations compensate for the tolerances of the analog front end of the line circuit and adjust the gains depending on customer's specifications.

The transmit path is exactly the opposite of the receive section. After gain setting, interpolation to 32 kHz is done by insertion of 0's and low-pass filtering in two biquads and two first-order filter sections. D/A conversion is performed in three steps. Firstly, the sample rate is increased to 1 MHz in the interpolation filter consisting of a triangular and a rectangular finite impulse response filter. Next, the 13-bit samples are reduced in wordlength to 1 bit in a digital $\Sigma\Delta$ modulator performing the same function as the analog $\Sigma\Delta$ modulator used for the A/D conversion. Finally the pulse density modulated (PDM) output drives a switchable clamp to either voltage reference prior to low-pass filtering with a 30-kHz bandwidth.

A software programmable digital hybrid can be tuned to cancel the excess echo return signal that passes through the analog hybrid. This digital echo canceler consists of a 4-tap FIR filter in parallel with a first-order IIR section.

III. THE ANALOG FRONT END

Fig. 3 shows the block diagram of the analog front end of the DSP. The heart of the analog front end is a second-order $\Sigma\Delta$ modulator.

A. The $\Sigma\Delta$ Modulator

For the design of the A/D and D/A conversion systems, our first aim was to reduce the amount of analog circuitry. Secondly, for the parts where an analog implementation is needed, we wanted a robust and parameter insensitive design. Finally we incorporated facilities to correct analog gain variation in the digital part.

As a result of our investigations, a 1-MHz (1-bit) $\Sigma\Delta$ modulator seemed attractive because of its interesting implementation aspects and the simplicity of the anti-alias and the post filters [1]. The block diagram of Fig. 4 depicts a second-order $\Sigma\Delta$ modulator used in the DSP chip. The analog $\Sigma\Delta$ modulator (A/D conversion) is built around two switched capacitor integrators and a comparator. The digital $\Sigma\Delta$ modulator (D/A conversion) has two accumu-

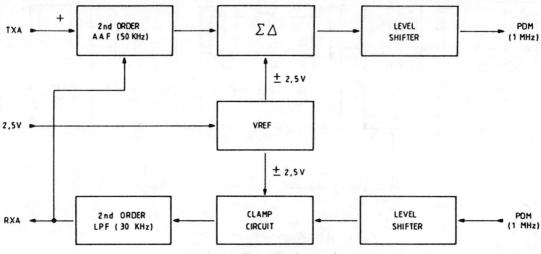

Fig. 3. The analog front end.

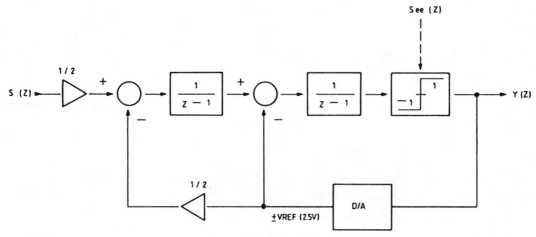

Fig. 4. The $\Sigma\Delta$ modulator.

lators and a sign detector. The feedback loops are easily implemented by bit manipulation techniques because only plus or minus the full dynamic range (2^{13}) has to be subtracted.

The quantization noise of this 1-bit system is extremely large but it is shaped in such a way that most of it is out of band. This is illustrated in Fig. 5, where the noise at the output of a first- and a second-order system are compared. If white noise at the quantizer output is assumed, the second-order system delivers more noise, but it is more out of band. These curves have been verified by breadboard measurements, followed by evaluation of the DSP chip. The theoretical noise figure of 86 dbmp [1] is very near to the 82 dbmp measured in practice.

B. The Anti-Alias Filter

Because the sampling frequency of the $\Sigma\Delta$ modulator is so high, the anti-alias filter is no longer a critical part in the A/D conversion system. We choose a second-order lowpass filter with cutoff frequency at 50 kHz and a tolerance of up to 50 percent gives no problem.

The differential input stage of this filter acts as a first echo cancellation section in order to avoid overload of the

$\Sigma\Delta$ modulator with echo return signal. If the central office- and test balance network impedances are equal, the cancellation effect of this analog hybrid is at least 35 dB. So, in practice, the digital hybrid has only to cancel the remaining return signal left due to mismatch between the echo return and the analog hybrid paths. For all practical combinations of central office- and test balance impedances, a Trans Hybrid Loss (THL) of 20 dB for frequencies between 300 and 600 Hz and 30 dB between 600 and 3400 Hz is guaranteed due the programmability of the digital hybrid section.

C. The Post Filter

The 1-MHz PDM signal, generated by the digital $\Sigma\Delta$ modulator contains a lot of high-frequency noise. This noise has a quadratic characteristic in the frequency domain. Therefore, a second-order post filter is needed to keep the noise within acceptable limits. A 30-kHz bandwidth was shown to be a good nominal cutoff frequency.

Although the cutoff frequency is not critical, a few aspects of this analog design require special care. Firstly, the pulsewidth of a "zero" and a "one" at the input of the

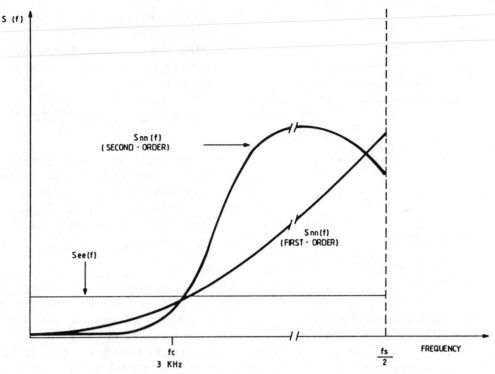

Fig. 5.   The $\Sigma\Delta$ modulator noise-shaping performance.

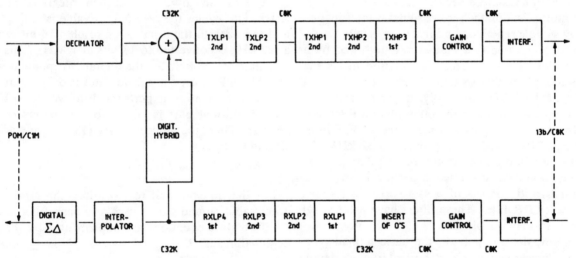

Fig. 6.   The DSP digital signal processing part.

filter have to be exactly equal. Otherwise, offset and signal distortion will occur. For the same reasons, the edges have to be as steep and identical as possible. This is realized by reclocking and buffering the PDM signal in a clamp circuit prior to low-pass filtering.

A second remark concerns the filter itself. Because the noise is high frequency, the slew rate of the opamp is a limiting factor. In order to avoid nonlinear effects, the filter is split into a passive section and a first-order active filter.

IV. THE DIGITAL SPEECH PATH

Fig. 6 gives an overview of the digital speech path. In the transmit part, the 1-bit PDM signal is reduced in sample rate from 1 MHz to 32 kHz with 15-bit resolution. This is

performed in three steps in a bit parallel architecture. The subsequent digital filters have a bit serial organization. The transmit low-pass filter has a tripple function. Firstly, it compensates the attenuation versus frequency of the analog front end and the decimator. Secondly, this block performs CCITT low-pass filtering at 3400 Hz and finally a last decimation step to 8 kHz is realized. All these functions are combined in one multiplexed double biquad filter structure. The same mask programmable filter organization is used as a transmit-high-pass and a receive-low-pass filter. It is discussed in this section in more detail.

The transmit-high-pass filter is a fifth-order filter making use of a biquad pair and a first-order section of the form $(1-z^{-1})$. This system creates a transmission zero at dc and at 50 Hz the attenuation is better than 20 dB, as requested by CCITT. Because the sampling frequency for this filter is only 8 kHz, the same shift-accummulator

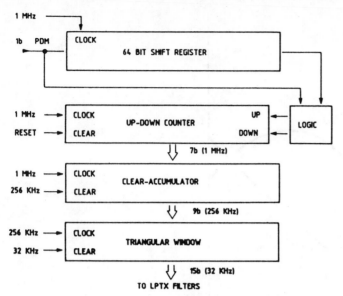

Fig. 7. The decimator.

structure is used for both transmit and receive gain settings. These signal level adaptations are software programmable for two reasons. Firstly, it is possible to compensate the gain variation of the analog front end of each line circuit separately. Secondly, it creates the ability to adapt the signal levels to customer's needs.

After this multiplication the receive signal is interpolated to 32 kHz, by sample repetition and insertion of two zero's. These 32-kHz samples enter a biquad pair followed by a first order section $(1 + z^{-1})$. This combination meets the three requirements for this section. Firstly, it realizes an interpolation of 8–32 kHz. Secondly, it compensates the gain versus frequency decay of the interpolation filter and the analog front end and, finally, it meets the CCITT filter characteristic at 3400 Hz. Interpolation from 32 kHz to 1 MHz is performed in the same 15-bit parallel data path as the digital $\Sigma\Delta$ modulator discussed in Section III-A.

The digital hybrid consists of a 4-tap FIR in parallel with a first-order IIR filter. These filters have a bit serial architecture at 16-kHz sample rate.

The decimator, the interpolator, the basic biquad pair, and the digital hybrid will now be described in more detail.

A. The Decimator

The out of band quantization noise in the 1-bit PDM signal is suppressed in three steps (Fig. 7). Firstly, a 64-tap rectangular FIR filter synthesizes transmission zeros at multiples of 16 kHz. The transfer function $H_1(z)$ can be realized easily by an up/down counter controlled by the values of x_n and x_{n-64}

$$H_1(z) = \sum_{i=0}^{63} z^{-i} = \frac{1 - z^{-64}}{1 - z^{-1}}.$$

Simulation has shown that for this counter a dynamic range of 7 bit needs to be provided.

A second rectangular window synthesizes zeros at multiples of 256 kHz

$$H_2(z) = \sum_{i=0}^{3} z^{-i}.$$

Because decimation from 1 MHz to 256 kHz is done immediately after this filter step, an accumulator clocked at 1 MHz and reset at 256 kHz is used for this section.

Finally, a 16-tap triangular window [2] creates zeros at multiples of 16 kHz

$$H_3(z) = \sum_{i=0}^{7} (z^{-i})^2.$$

This filter is sharp enough to admit further decimation to 32 kHz. When we describe the interpolator, an example of such a triangular window realization will be discussed in detail. At the output of the decimator a 15-bit code is provided to the transmit low-pass filter.

B. The Interpolator

The receive filter output is interpolated from 32 kHz to 1 MHz in two steps. Firstly, interpolation to 256 kHz is performed with a 16-tap triangular window FIR filter. Then, a 4-tap rectangular window is realized by oversampling the interpolator output by a factor of four.

The triangular window is built around a 16-bit accumulator clocked at 1 MHz (Fig. 8). At 32 kHz, the difference between input and output is latched in a hold register. This value is divided by 2^6 and is then integrated at 256 kHz. After eight cycles the new output value is multiplied by 2^3 and fed back to the input. As a result a 16-bit input signal at 32 kHz is converted to 13 bits at 1 MHz. This output is available at 256 kHz for the digital $\Sigma\Delta$ modulator running at 1 MHz. Because the accumulator is only 16 bit wide, the 19-bit accumulation is performed in two steps. Firstly the 6-bit fractional part is added to the 16 bits of the hold register. Then the integer parts (13 bit) are added in a second cycle.

C. The Basic Filter Structure

For the transmit and receive filters the same mask-programmable biquad pair is used as a basic filter architecture [3]. Two state equations are implemented by the structure depicted in Fig. 9

$$y_n = a_0 x_n + a_1 x_{n-1} + a_2 x_{n-2} - b_1 y_{n-1} - b_2 y_{n-2}$$
$$z_n = a_3 y_n + a_4 y_{n-1} + a_5 y_{n-2} - b_3 z_{n-1} - b_4 z_{n-2}.$$

The signal data are stored in shift registers. Via a multiplexer, five data words are applied bit serially as ROM address. The ROM contains all combinations of the filter coefficients and these partial products are accumulated in a shift accumulator. The ROM words are 13 bit wide: three integer bits and nine fractional bits. The speech samples are truncated to 21 integer bits providing enough headroom to exclude eventual overflows. The calculation of one filter cycle takes 32 2-MHz clock periods. So, for the low-pass filters two biquads at 32 kHz can be multiplexed. Because the high-pass filters are running at 8 kHz, six cycles are available for other applications. Two of them are used for the gain setting multiplications. These coefficients are loaded in latches under software control.

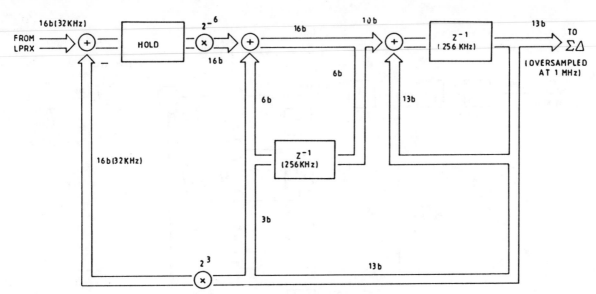

Fig. 8.   The interpolator.

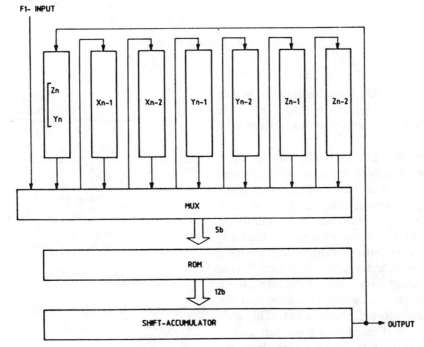

Fig. 9.   The basic filter structure.

D. The Digital Hybrid

The digital hybrid consists of a 4-tap FIR filter in parallel with a first-order IIR filter (Fig. 10). Due to ac-couplings and the synthesis of complex central office impedances in the analog front end, the IIR filter is needed to match the large echo return signal delay at low frequencies (< 300 Hz). At higher frequencies (300-3400 Hz) flat signal delay can be assumed and the FIR filter only synthesizes the smooth amplitude response of the echo return path including interpolator, digital $\Sigma\Delta$ modulator, analog front end, and decimator.

Because the input of the digital hybrid is decimated from 32 to 16 kHz, each filter section can be constructed around one serial Booth multiplier performing one calculation in 16 μs [4]. The FIR filter coefficients are signed, 2-bit integer, 4-bit fractional. The IIR filter coefficients are 6-bit fractional and unsigned because only low-pass filter characteristics have to be generated. Finally the results of the two hybrid sections are added and interpolated again to 32 kHz. This signal is subtracted from the echo return signal at the decimator output.

V.   THE CONTROL PART OF THE DSP

An important feature of a dedicated digital codec is that the control circuitry can also be integrated together with the signal-processing part. In many cases a lot of external

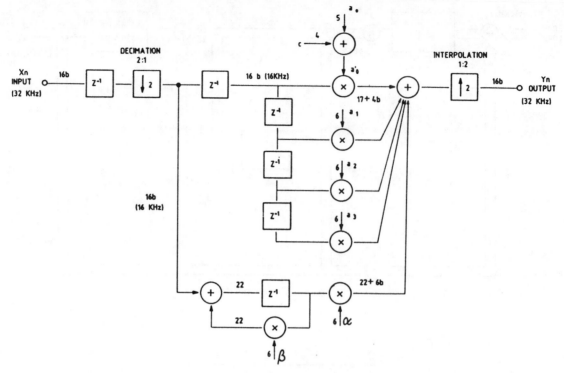

Fig. 10. The digital hybrid.

components can be saved in this way. In the case of our new line circuit, it was advantageous to integrate all the control logic that is needed per subscriber line on the DSP chip.

From the card controller the DSP receives in every 8-kHz frame a bit map of eight bytes over a multiplexed 4-MHz link. A part of this data contains status information for the DSP, e.g., power-up bit, tri-state bit, and speech channel assignment. The bit map also includes an address and data field for an auxiliary bit map containing hybrid coefficients and gain setting multiplicands. With a special protocol these seven bytes are transmitted during several frames at the initialization phase of the line circuit.

Two bytes of the bit map information are sent to the SLIC over a bidirectional 256-kHz current link. This data contains SLIC status information such as power-up, metering request, and relays drive state. Via the same link, the SLIC retransmits 16 bits containing 4 bits of line status information: switch hook, ring trip, two-party line, and overcurrent detect. The DSP also senses 4 bits of line circuit status information on the line card. This line status byte is debounced in an 8-bit hit timer and sent to the card controller over a 1-MHz multiplexed link.

The debouncer is in fact a multiplexed counter which is enabled on the status change of the corresponding bit. When the new status is held longer than the debounce time, the state of the output of the debouncer is changed. In order to have a flexible structure, the hit times of each bit are stored in ROM. Counting is performed by fetching the previous value out of a RAM and incrementing it bit-serially using only one half adder cell. This ROM/RAM combination yields a linearly expandable layout in relation

TABLE I
TECHNOLOGY CHARACTERISTICS

- 10 V N-WELL CMOS

- Full 3 μm :
  . Gate Length (layout) : 3 μm
  . Gate Length (effective) : 2.3 μm
  . Poly Pitch : 6 μm
  . Metal Pitch : 7 μm
  . Active Area Pitch : 7 μm

- Gate Oxide Thickness : 425 A°

- Double Poly
  . Interpoly Thickness : 800 A°
  . Sheet Resistance : 20 Ω/□

- Chip Area : 33 mm²

to the number of bits. For the 8-bit debouncer an area of only 1 mm² was needed in a 3-μm CMOS technology.

VI. IMPLEMENTATION ASPECTS AND TEST RESULTS

For the DSP, a high-performance 3-μm CMOS technology was used. The most important characteristics of this technology are presented in Table I. A second polysilicon layer is used only for capacitors in order to allow optimal filter performance.

A high packing density is obtained by using a dynamic CMOS design approach [5]. As a typical example of this type of logic a dynamic full adder cell is depicted in Fig. 11. This cell can be used in an accumulator alternated with its CMOS complement where NMOS and PMOS device functions are interchanged. This technique guarantees a race-free logic with a small number of devices per cell and only one inverting buffer between the cells.

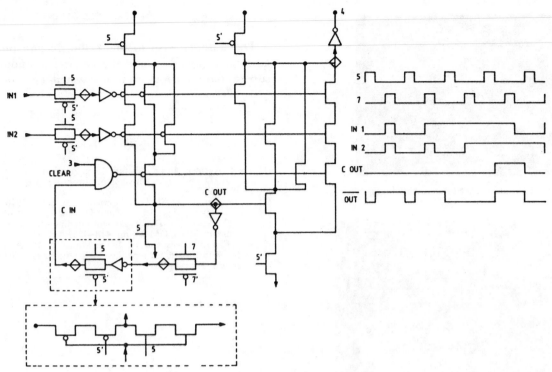

Fig. 11. A dynamic CMOS full adder cell.

TABLE II
TEST RESULTS

SYMBOL	PARAMETER		VALUE	UNIT
I_{VDD}	CURRENT DRAWN AT +5V		3	mA
	STAND BY		1	mA
I_{VSS}	CURRENT DRAWN AT -5V		10	mA
	STAND BY		3	mA
I_{CNTX}	IDLE NOISE TX PATH		-74	dbmp
I_{CNRX}	IDLE NOISE RX PATH		-82	dbmp
HD_{TX}	IN BAND HARM. DIST. TX		-52	dB
HD_{RX}	IN BAND HARM. DIST. RX		-52	dB
GT_{TX}	GAIN TRACKING :	+ 3 TO -40 dbmo	$\pm$ 0.15	dB
		-40 TO -50 dbmo	$\pm$ 0.3	dB
		-50 TO -55 dbmo	$\pm$ 1.0	dB
GT_{RX}	GAIN TRACKING :	+ 3 TO -40 dbmo	$\pm$ 0.1	dB
		-40 TO -50 dbmo	$\pm$ 0.2	dB
		-50 TO -55 dbmo	$\pm$ 0.5	dB
GA_{TX}	GAIN ADJUSTMENT TX,	RANGE	-7 TO 6	dB
		STEP	.26 $\pm$ .1	dB
GA_{RX}	GAIN ADJUSTMENT RX,	RANGE	-13.4 TO 4.5	dB
		STEP	.25 $\pm$ .1	dB
AERL	AVERAGE ECHO RETURN LOSS		>35	dB
	(300 - 3400 Hz)			

Using these techniques, the complete chip contains 23 000 devices in an area of only 33 mm². A 10-V analog part of 7.5 mm² is included in this area. Power consumption has been kept below 75 mW. Table II summarizes some test results of this high density circuit and Fig. 12 shows a scope picture of the receive filter transfer characteristic. These results, combined with the performance of our single chip SLIC, show that CCITT specifications are met and enough flexibility is obtained to satisfy customer's needs.

VII. CONCLUSIONS

A high-performance DSP chip has been described. The 33-mm² 3-μm CMOS device combines the functions of digital codec, digital echo canceller, and signal level adaptor. In addition, the DSP acts as a control interface between the SLIC and the line card controller. Fig. 13 shows a microphotograph of this 23K transistor device. Power consumption has been kept below 75 mW.

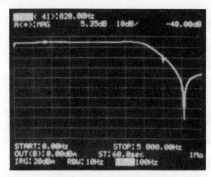

Fig. 12.   The receive filter amplitude response.

Fig. 13.   A DSP microphotograph.

ACKNOWLEDGMENT

The authors wish to thank the Belgian public network administration RTT and the IWONL foundation for their substantial contribution in the system development.

REFERENCES

[1] B. P. Agrawal and K. Shenoi, "Specification—based design of $\Sigma\Delta$M for A/D and D/A conversion," in *Proc. ICA SSP 82*, vol. 3, pp. 1980–1983.
[2] Candy *et al.*, *IEEE Trans. Commun.*, p. 126E, 1976.
[3] A. Antoniou, Digital Filters and Design.   New York: McGraw Hill.
[4] R. F. Lyon, "Two's complement pipeline multipliers," *IEEE Trans. Commun.*, vol. COM-24, pp. 418–425, Apr. 1976.
[5] N. F. Goncalves and H. J. De Man, "Nora: A racefree dynamic CMOS technique for pipelined logic structures," *IEEE J. Solid-State Circuits*, June 1983.

A 160-kb/s Digital Subscriber Loop Transceiver with Memory Compensation Echo Canceller

ROGER P. COLBECK, MEMBER, IEEE, AND PETER B. GILLINGHAM, MEMBER, IEEE

Abstract —A full-duplex transceiver chip incorporating an adaptive echo cancelling modem and a 2.048-Mb/s serial interface is described. The device provides a full-duplex communication link at 160 or 80 kb/s on up to 4 or 5 km, respectively, of 0.5-mm twisted-pair cable. Full integration is achieved through the use of RAM-based sign-algorithm echo–cancellation, biphase line code, a fixed switched-capacitor equalizer and a digital phase locked loop. The paper emphasizes system design considerations and a chip architecture minimizing power dissipation, silicon area and off-chip components. A double poly 3-μm CMOS technology is used to implement the 5-V 22-pin device which dissipates less that 50 mW and occupies 27.7 mm^2.

I. INTRODUCTION

TO SATISFY the requirements for an Integrated Services Digital Network (ISDN), a cost-effective solution to high speed data communication over existing twisted-pair cable is necessary. Two competing techniques, the time compression multiplexing (TCM) ping-pong system and the full-duplex echo cancelling scheme, have emerged to meet these requirements. Although in theory the latter can achieve longer line lengths, it has been argued that echo cancellation demands greater circuit complexity and higher precision signal processing. Until recently the development of integrated digital echo cancelling hybrids has been restricted to test chips [1], while several fully integrated TCM chips have already been announced [2], [3]. A RAM based echo canceller design [4], [5], which combines an easily implemented algorithm with a simple analog front end, led to the development of a single chip transceiver [6].

The major requirements for a digital transceiver chip are high performance over a variety of line lengths, gauge changes and bridge taps, low-power dissipation to enable remote power feeding, and small die area to minimize cost. This paper describes the design of a CMOS chip which satisfies these conditions, providing a high speed communication link over twisted pair in both the PABX environment and in the medium range loop plant. Section II introduces the transmit side of the chip including the serial telecommunication bus interface, and discusses the choice

of line-code. Section III describes the receive path filters and clock recovery circuitry. We discuss the sign algorithm RAM-based echo cancellation technique as it applies to twisted pair transmission and show some simulation results in Section IV. Finally, in Section V, the overall performance of the device on a variety of lines is summarized.

II. TRANSMIT PATH

Fig. 1 shows a block diagram of the transceiver. A 2.048-Mb/s serial PCM highway supplies the transmit data to the chip. It contains 32 8-bit channels cycling at 8 kHz, only 3 or 4 of which are utilized by the transceiver. If a line rate of 160 kb/s is selected, two 64 kb/s B-channels, one 16 kb/s D-channel, and one 64 kb/s C-channel are extracted from the PCM highway. When the 80 kb/s line rate is enabled only 3 channels of information on the serial PCM highway are used, a single 64 kb/s B-channel, an 8 kb/s D-channel and the C-channel. The B-channels are typically assigned to user voice or data, while the D-channel is normally reserved for signalling information. The C-channel is not transmitted over the line interface, but used internally to configure the chip for various modes of operation. These include selection of 80 or 160 kb/s line rates, accelerated echo canceller convergence, and a suite of test functions spanning from system oriented analog and digital loopback to modes allowing every functional block of the circuit to be tested individually. A separate PCM highway carries the received data stream. The B and D channels are passed through transparently from the line interface while the C-channel holds the internal status of the chip, indicating SYNC and the quality of the received data. For applications not requiring the 2.048-MHz interface the chip can be defaulted into baseband modem operation via external mode select pins. The mode select pins also program the part for SET or CO operation.

The serial transmit data stream is then scrambled before insertion of a SYNC bit through a pseudorandom sequence generator having the polynomial $1 + x^{-3} + x^{-5}$. This ensures that data will not emulate the SYNC bit, causing the receiver to lock on out of phase to the true SYNC. Data is scrambled again with polynomial $1 + x^{-6} + x^{-7}$ in the

Manuscript received April 23, 1985; revised September 30, 1985.
The authors are with Mitel Corporation, Kanata, Ont., Canada K2K 1X3.
IEEE Log Number 8406484.

Reprinted from *IEEE J. Solid-State Circuits*, vol. SC-21, no. 1, pp. 65–72, Feb. 1986.

401

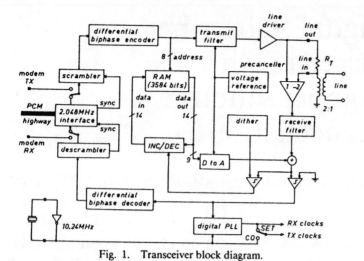

Fig. 1. Transceiver block diagram.

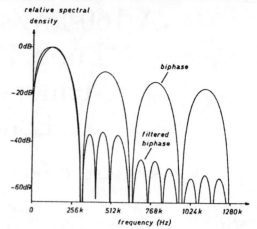

Fig. 2. Simulated TX filter response for 160-kb/s data rate.

SET device or $1 + x^{-4} + x^{-9}$ for the CO. Scrambling is necessary for predictable behavior of the echo canceller. Different scrambler polynomials in SET and CO must be used to reduce correlation between the near and far-end signals which could result in the echo canceller converging to far-end data.

Next, the data is differentially biphase encoded. Differential encoding of the baseband serial data makes the receiver polarity independent. The tip and ring leads can be reversed without effect on data recovery. The biphase line-code was chosen for a number of reasons. First, it has the desirable properties of no dc component, necessary for line powered terminals, and a transition within each baud which allows easy timing recovery. Second, the primary spectral lobe of 80 or 160 kb/s biphase falls within the usable bandwidth of twisted pair cable so that a simple fixed equalizer can be used. Finally, biphase has a short impulse response which minimizes the size of the echo canceller RAM.

The differentially encoded biphase is conditioned through a 15-stage switched-capacitor FIR filter to remove unnecessary spectral content in the transmitted output. The filter is clocked at sixteen times the baud-rate ($16 f_b$), either 1.28 or 2.56 MHz in 80 or 160 kb/s modes, respectively. While it may be argued that this filtering could be performed in the receiver without any change in the overall transfer function there is a number of good reasons for doing it prior to transmission. First, the line driver must drive 800 Ω with a signal around 4 V peak to peak. The difference between the power in a square wave of this magnitude and the filtered signal is actually significant with respect to overall chip power consumption. Second, properties of the digital signal to be transmitted make it much easier to filter on the transmit side. The delay elements of the FIR structure can be flip-flops as opposed to analog sample-and-hold circuits. Finally, as a responsible user of the electromagnetic spectrum, it is undesirable to broadcast wide-band signals which could interfere with other systems.

The transmit filter is a linear-phase symmetrical-coefficient switched-capacitor FIR low-pass filter with zeros

located at the secondary spectral peaks of a random biphase signal. Fig. 2 shows simulated response in 160 kb/s mode, where the secondary lobes of the biphase spectrum are attenuated at least 35 dB with respect to the primary lobe without any phase distortion. In the passband a 4-dB pre-emphasis peak with respect to dc is included to offset line attenuation without distorting significantly the eye. Less emphasis would result in larger low frequency components giving a longer transhybrid response and poorer echo cancelling performance. Overemphasis results in gross amplitude differences between the two eye trajectories bringing the low frequency components closer to the noise floor. The 15-tap filter samples the digital input stream at $16 f_b$ so that the output is based on at most two bauds of information, such that unnecessary intersymbol interference in the transmitted eye is kept to a minimum.

The TX filter and indeed all switched-capacitor filters in the chip (Fig. 3) require the use of a high bandwidth, high slew rate operational amplifier to accomodate passbands on the order of the baud rate and clock frequencies as high as 2.56 MHz. The core amplifier is based on the folded cascode configuration [7] and is compensated by the capacitive load on the output. Internally compensated amplifiers must ensure that the frequency of the parasitic output pole is high enough not to cause instability by virtue of a low-output impedance. This is accomplished at the expense of supply current. In contrast, a transconductance amplifier takes advantage of the natural positions of the internal parasitic pole and dominant load pole with a high-output impedance, and achieves low-power consumption.

The on-chip voltage reference using a vertical n-p-n transistor-based bandgap circuit provides a -1.8-V reference with respect to an internally generated midrail for use in both the echo canceller and TX filter. The peak-to-peak echo which must be covered by the echo canceller D/A converter is a function of the TX filter output level. As both circuits share the same reference, the chip is self-compensating for variations in the voltage reference, reducing the requirement for a high precision trimmable circuit.

External line interface circuitry consists of a 2:1 transformer with a split winding on the line side to permit remote power feed and a line termination network, which

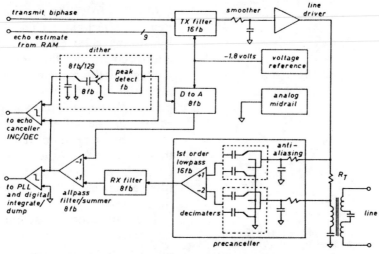

Fig. 3.   Block diagram of analog circuitry.

can be as simple as a single 400-Ω resistor, connected between the line driver and transformer. The composite line signal at the point between transformer and termination resistor, which contains both near- and far-end biphase signals, is routed back into the chip to the first stage of the receive path.

III.   RECEIVE PATH

The first stage of the receive path, shown in Fig. 3, is a precanceller realizing a fixed summation of the transmit and line signals to achieve limited cancellation of the TX component. Reducing the amplitude of the near-end signal improves the dynamic range of the receiver and the echo canceller. With a purely resistive termination, a worst-case precancelled signal −9 dB with respect to the line signal was observed over a variety of different lines. Precancellation allows an additional 9 dB of gain without risk of clipping the near-end signal. Clipping of the far-end signal superimposed on the near-end echo is acceptable but if the near-end signal itself were clipped the far-end data would be masked out. Included in this section is an anti-aliasing filter on each of the LIN and LOUT inputs consisting of a continuous-time single-order low-pass filter. These were realized as distributed RC's for fast rolloff having a −3-dB point at 3.2 f_b. Following these are switched capacitor decimators with an effective input sampling rate of 32 f_b for the suppression of the replicated lobe at 16 f_b. Overall anti-aliasing performance is such that frequencies above the Nyquist limit are attenuated at least 20 dB with respect to the passband.

A third-order switched-capacitor bandpass filter serves as a fixed equalizer optimized for best performance at long line lengths. It is comprised of a first order lowpass section, incorporated in the pre-canceller, clocked at 16 f_b and a bilinear bandpass biquad clocked at 8 f_b. The composite response of the receive path, shown in Fig. 4 for a 160-kb/s transmission rate, has a passband between 64 and 256 kHz with a dc zero and a bilinear notch at 640 kHz. When the transceiver is in the 80-kb/s mode the filter responses are compressed in frequency by a factor of two by scaling the

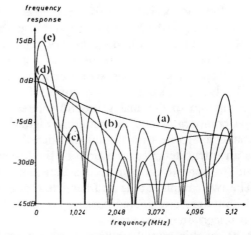

Fig. 4.   Receive path simulated frequency response. Progressive filtering is shown through: (a) Anti-aliasing filter, (b) Decimator, (c) Precanceller, (d) Bandpass Biquad, (e) All pass.

clocks. The time constant of the RC antialiasing filter is doubled by switching in additional resistance and capacitance. The receive filter emphasizes the higher frequency components of the biphase signal to offset the greater attenuation experienced by these components on twisted-pair cables. It also introduces delay which is greater for high frequencies to compensate for the dispersive effects of long lines. Fixed equalization in the receive filter is intended to optimize performance over long lines at the expense of short-line performance. A final filter section just before the point of echo cancellation adjusts short-line performance within acceptable limits.

An all-pass filter realized as a first-order switched-capacitor section clocked at 8fb has a pole–zero location of 0.8fb. Its primary purpose is to adjust the horizontal eye opening for short-line lengths. Fig. 5 shows the horizontal opening as a function of line length at the input and the output of the all pass. The effect for long lines is small while performance at 0 km is greatly improved. The all-pass filter is also the point of echo cancellation, where the analog echo estimate from the D/A converter is subtracted from the filtered composite line signal. Cancellation is

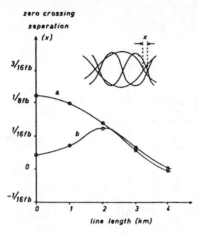

Fig. 5. Horizontal eye closure as a function of line length (24 gauge) for 160 kb/s transmission. (a) Without all pass section. (b) With all pass section.

performed after the receive filter so that the echo canceller can take advantage of the shortened impulse response, keeping the amount of TX data history required to a minimum. After removal of the near-end component, a decision is made as to whether a zero or a one has been received. Each analog sample at the output of the all-pass filter is compared to the analog ground reference with a high-speed comparator and the resulting digital information is made available to a digital decoder. This circuit performs an integrate and dump on 6 of the 8 samples in each baud to achieve an enhanced signal-to-noise ratio, (SNR). The two samples at the mid-baud zero crossing and the band boundary are ignored because the horizontal eye opening is uncertain in these regions.

Either the on-chip crystal oscillator or an external 10.24-MHz clock driving the crystal pin provides the transceiver with a reference timebase. In CO mode the PCM-highway and transmit path timebases are directly derived from this clock. The opposite SET device uses a digital phase locked loop to create transmit and receive timebases locked to the incoming biphase signal. In CO mode the phase locked loop output is used only for the digital part of the receive leg, while the local crystal oscillator supplies clocks to the sampled analog filters. Here, the DPLL accounts only for the varying time delay of the far-end signal over the range of line lengths, since the received signal originating from the SET is locked in frequency to the CO timebase. The digital phase locked loop applies 49 ns corrections to the local crystal oscillator frequency at a rate of 5 or 2.5 kHz for 160 and 80 kb/s, transmission, respectively. Required crystal tolerance is either 60 or 120 ppm.

IV. Echo Cancellation

The echo canceller is based on the memory-compensation principle using a sign algorithm adaptation process. Traditionally echo cancellers have used an adaptive filter, typically an FIR structure, which converges to the impulse response of the echo path. The filter performs a convolution of the transmitted data stream with the estimated impulse response to provide echo estimates which are

subtracted from the composite signal. The memory compensation technique circumvents the convolution process by storing echo estimates directly in a RAM which is addressed by the transmitted data stream. A major advantage of the memory compensation scheme is a simple algorithm and an associated reduction in hardware complexity. While a relatively large RAM is needed to store the echo responses of every possible transmit data sequence, the technique need not rely on the linearity of the channel to achieve perfect cancellation. The memory compensation principle provides enhanced performance over the FIR filter approach, by accounting for nonlinear distortion in the echo path.

The sign algorithm adaptation process is chosen for its simplicity and ease of implementation. This method uses the sign of the received signal after echo cancellation to increment or decrement the corresponding echo estimate, as opposed to the stochastic iteration algorithm [8] which adds or subtracts an amount proportional to the magnitude of the received signal. Instead of an A/D converter to digitize the echo cancelled signal, only a comparator is required. The sign algorithm represents the most extreme quantization of the error signal which may result in inaccurate cancellation in the presence of a far-end signal. Extra noise, in the form of a dither signal, is added to the signal before taking the sign, to make the average quantization smoother for accurate echo-cancellation. Without dithering, an echo estimate can experience wander about the correct value, equal in magnitude to the far-end signal, which may prevent convergence completely.

The major disadvantage of the sign algorithm is a relatively long convergence time, particularly when coupled with the memory compensation architecture which is slower than the adaptive filter approach. This can be overcome by an accelerated convergence option which modifies words in RAM by more than 1 LSB, or by providing continuous communication even during on-hook periods.

The echo canceller in the transceiver chip has been designed to accommodate 40 dB of cable attenuation. If we assume a required signal to noise ratio of 15 dB for a bit error rate of 10^{-7} and an analog hybrid attenuation of 10 dB, then we require 45 dB of echo cancellation. The echo canceller achieves this level of cancellation using 14-bit words in the RAM, 5-bit history with 8 samples per baud, a 9-bit D/A converter and an adaptive dither source that tracks the amplitude of the received signal. Holte and Stueflotten [4] have analyzed the noise performance of the sign algorithm memory compensation scheme. The major sources of internal noise are cancellation error, quantization noise, residual echoes and jitter noise. Steady-state cancellation error has been derived by considering the adaptation of a single memory register and is a function of the finite resolution in the register and the dither amplitude. The SNR due to cancellation error with dither amplitude 1.4 times the received signal amplitude and the 13-bit plus sign code equal in magnitude to the echo after 10 dB of precancellation is 25.6 dB [4, eq. (20)]. It should be noted that as an approximation, the received signal is con-

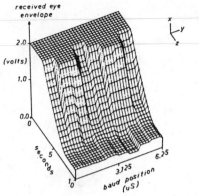

Fig. 6. Simulated echo canceller convergence in the presence of a far-end signal.

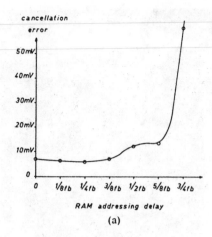

(a)

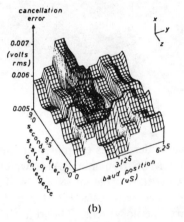

(b)

Fig. 7. (a) Worst-case rms cancellation error versus RAM addressing delay. (b) rms cancellation error with RAM addressing delay = 3/8 f_b.

sidered to be a binary signal free of intersymbol interference.

Although a 14-bit echo estimate is stored in RAM, only the nine most significant bits are sent to the D/A converter. This makes the effective D/A step size 32 times the RAM step size. The effect upon the adaptation may be neglected and the noise simply treated as quantization noise. For 40 dB of cable attenuation this results in a signal-to-quantization noise ratio of 28.9 dB [4, eq. (21)].

Another important source of noise in the echo canceller is that of residual echo. The echo path's impulse response has been assumed to be finite but it extends beyond the 5 bauds chosen to address the memory so that earlier pulses will contribute a noise in the form of an uncancelled residual echo. Since the impulse response is a strong function of the line type and the presence of bridge taps or gauge changes, it becomes difficult to predict the level of noise due to residual echo. Simulations and laboratory work with a variety of lines have shown that 5 bauds of echo history provide satisfactory performance.

The necessity of deriving a clock from the data creates further noise due to jitter. The transceiver uses a simple digital phase locked loop with a master clock of 10.24 MHz to provide discrete adjustments of approximatly 49 ns. It has been found that the noise due to jitter is at least 20 dB below the received signal.

Simulations provide a means of verifying an echo–canceller design based on Holte and Stueflotten's analysis. Fig. 6 shows a simulation of the dynamics of echo canceller operation in the presence of a far-end signal. Planes parallel to the $x - y$ axis hold the instantaneous peak envelope of the received eye as it would be seen on an oscilloscope. The three-dimensional surface represents this peak eye envelope as the echo canceller converges. At time zero the signal is completely saturated at the opamp clipping level. Then the eye envelope slopes towards zero with the profile of the as yet uncancelled near end signal in evidence. Finally, after about 7 s, the curve levels off in the form of the highly attenuated far-end signal. Convergence can be speeded up by a factor of 32 with the externally programmed accelerated convergence feature.

The echo canceller cannot account for echo unrelated to the RAM address, which consists of the most recently

transmitted 5 bits of data. In order to extend echo coverage as far as possible with the 5-bit prehistory scheme, the address to the RAM was delayed slightly with respect to the TX filter input to account for the delay inherent in the transmit and receive filters. It was determined that an addressing delay of 3/8 baud was optimum. Fig. 7(a) shows RMS cancellation error for 0 km line as a function of this delay. The curve starts to increase after 3/8 baud delay. The rms cancellation error as a function of baud position over time is shown in Fig. 7(b). One baud position shows a greater error which represents a small systematically uncancelled portion of the near-end signal which sneaks through before the RAM is even addressed. The relatively constant noise floor is due to short term wander of RAM coefficients, clock jitter and D/A converter truncation error. Considering the 15-dB receive path gain and 40 dB of line attenuation, the receiver SNR with 7 mV rms noise is better than 15 dB, which is close to what the theoretical analysis leads us to expect.

The D/A converter is an 8-bit plus sign binary weighted switched capacitor structure which uses an on chip reference and a 0.05-pF unit capacitor for a step size of 7 mV. Linearity is not so important in this application as monotonicity and low current consumption. The most significant 9 bits of the 14-bit wide RAM data bus are supplied to the D/A converter at a rate of 8 f_b to cancel the near end component on each sampled output of the receive filter.

Fig. 8.   Chip photomicrograph.

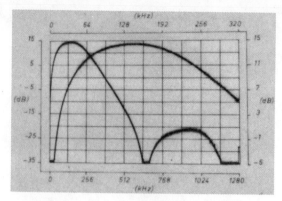

Fig. 9.   Receive filter response at 160 kb/s.

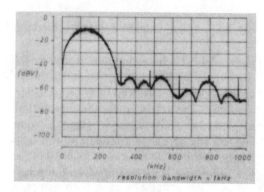

Fig. 10.   Transmit signal spectrum at 160 kb/s.

The echo canceller stores estimates of the line response in a 3584-bit CMOS dynamic RAM, which undergoes a READ–MODIFY–WRITE cycle at 8 f_b. A folded bit line architecture is used with each bit stored differentially in complementary memory cells. The refresh is interleaved with normal operation, so that all 32 rows are completely refreshed every 4 bauds. The 8-bit address consists of 5 bits of TX data history plus 3 bits indicating 8 positions within each baud. On each cycle an echo estimate is altered by one LSB depending on the result of a sign comparison between the RX signal and a dither signal having the desired statistical properties for uniform convergence.

The dither circuit generates a triangular waveform centered on analog ground with an amplitude related to the peak received far-end signal. It consists of a switched-capacitor peak detector clocked at the baud rate followed by a passive switched-capacitor ramp generator with an output frequency of 8 f_b/129 and a peak amplitude 1.4 times the peak detector output up to a maximum of 1 V. The frequency was chosen so as not to correlate with scrambled data from either end but high enough not to cause significant wander in the RAM coeffecients. On short lines the received signal will sometimes exceed the dither amplitude and will experience wander in the area outside the 1-V peak-to-peak range. This will not affect reliable data recovery.

V.   CHIP PERFORMANCE

Fig. 8 shows a photomicrograph of the integrated circuit with the major circuit blocks identified. The die which occupies 27.7 mm² (192 × 224 mils) of silicon is packaged in a 22-pin DIP with 0.4BSC spacing.

The receive filter frequency response and the transmit filter output spectrum are displayed in Figs 9 and 10,

respectively. Eye diagrams of the transmit, receive, and composite line signals for 160 kb/s transmission over 2- and 3.2-km 24 gauge lines are shown in Fig. 11. A small far-end component can be seen superimposed on the large near-end component for the 2-km composite line signal while at 3.2 km the far-end is attenuated to the point where it not easily visible on the oscilloscope. The received signal at both line lengths showing no transmit component demonstrates the functionality of the echo canceller. The effect of the fixed equalizer at 2 km is to cause multiple zero crossings which reduces the horizontal eye opening. At 3.2 km the line brings these zero crossings closer together, for improved opening.

Fig. 12 plots the signal-to-noise ratio for a fixed bit error rate of 10^{-6} over a range of 24 gauge plastic insulated cable lengths, at both 80 and 160 kb/s. Low-pass filtered white noise, having a bandwidth of the baud-rate, is injected onto the line at the receive transformer. The SNR is defined as the ratio of far-end signal to total noise, measured at the line input of the transceiver.

For zero line length the SNR is relatively high as a result of the fixed equalization optimized for long lines, and the dither signal which falls short of the far-end signal amplitude. The 160-kb/s curve approaches a minimum of 12 dB at medium length lines and becomes asymptotic after 4 km, indicating that transmission is impossible after this point. At 80 kb/s, SNR performance is marginally worse than that at 160 kb/s. This is primarily due to the line termination components which were optimized for 160-kb/s transmission. The 80-kb/s curve has not yet started to show the

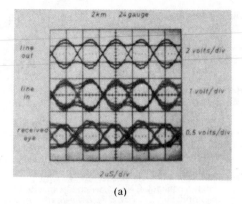

(a)

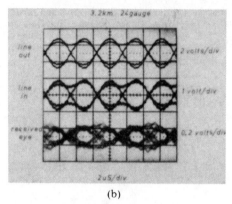

(b)

Fig. 11.   Transceiver signals. (a) 2-km line, (b) 3.2-km line.

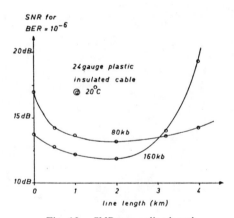

Fig. 12.   SNR versus line length.

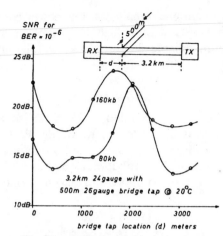

Fig. 13.   SNR versus bridge tap location.

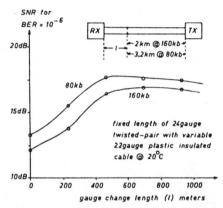

Fig. 14.   SNR versus gauge change location.

effects of line attenuation even at 4 km. Reliable transmission on up to 5 km is possible at this data rate before the SNR curve deteriorates.

The signal to noise performance was also measured as a function of bridge tap placement (Fig. 13). A 500-m open-circuit 26 gauge bridge tap was placed along a 3.2-km length of 24 gauge twisted pair cable between the SET and CO devices. When the bridge tap is positioned right at the RX position, where BER is measured, required SNR is high. This is due to gross line mismatch and the resulting clipping within the precanceller. As bridge tap distance from the receiving end increases, the SNR peaks and falls off once again. This phenomena can only be the result of imperfect echo cancellation after the 5-baud echo history, since ISI would be the same no matter where the bridge tap

was located. The peak occurs at the point where the combined effects of round trip echo delay and attenuation to the bridge tap create the largest echo unrelated to the 5-bit history. At no point does the bridge tap curve approach that of the straight line performance. This is due mainly to the ISI introduced into the far-end signal.

A short length of 22 gauge quad installation cable was connected in series with a 2- or 3.2-km 24 gauge twisted pair to examine the effects of gauge changes on performance (Fig. 14). Different lengths of 24 gauge cable were used for the 80- and 160-kb/s measurements so that line attenuation would be similar for the two baud rates. Both curves rise and then level off as a result of two different effects. Increasing the composite line length results in a continually rising curve. Reflections from the gauge change mismatch will cause a peaking curve, so that the resultant SNR levels off after 500 m.

Several key features of the IC provide enhanced performance without penalties in power or area. Biphase line-code tailored to the limited channel bandwidth permits the use of a simple fixed equalizer. RAM-based echo cancellation allows for simple line interfacing and is capable of handling a variety of line impedances, bridged taps, gauge changes, and nonlinearities. Symmetry between SET and CO devices, together with complete control of transceiver functions through external programming, makes the part very flexible.

This chip combines small size, low-power consumption, data transparency, and a high degree of integration in accomplishing reliable 80- or 160-kb/s full duplex data transmission for the PABX environment or the outside plant.

ACKNOWLEDGMENT

The authors wish to thank A/S Elektrisk Bureau of Norway, G. Aasen, P. Beirne, K. Buttle, R. Gervais, G. Reesor, D. Ribner, H. Schafer, and R. White for their substantial technical contribution as well as M. Foster and D. Brown for their support and encouragement in this project.

REFERENCES

[1] O. Agazzi, D. Hodges, and D. Messerschmitt, "Large scale integration of hybrid-method digital subscriber loops," *IEEE Trans. Commun.*, vol. COM-30, pp. 2074–2082, Sept. 1982.

[2] A. Komori, M. Furukawa, T. Sato, T. Komazaki, "A 200Kb/s burst mode transceiver with two-bridge tap equalizer," in *ISSCC Dig. Tech. Papers*, pp. 236–237, Feb. 1984.

[3] Y. Hino, T. Chujo, N. Ueno, K. Fujita, M. Yamamoto, K. Yamaguchi, and H. Kikuchi, "A burst-mode LSI equalizer with analog building blocks," in *ISSCC Dig. Tech. Papers*, pp. 238–239, Feb. 1984.

[4] N. Holte and S. Stueflotten, "A new digital echo canceller for two-wire subscriber lines," *IEEE Trans. Commun.*, vol. COM-29, pp. 1573–1581 Nov. 1981.

[5] U.S. Patent 4237463.

[6] K. Buttle, G. Aasen, R. Colbeck, R. Gervais, P. Gillingham, D. Ribner, H. Schaefer, and R. White, "A 160kb/s full duplex echo cancelling transceiver," in *ISSCC Dig. Tech. Papers*, pp. 152–153, Feb. 1985.

[7] T. Choi, R. Kaneshiro, R. Broderson, P. Gray, W. Jett, and M. Wilcox, "High-frequency CMOS switched-capacitor filters for communications applications," *IEEE J. Solid-State Circuits*, vol. SC-18, pp. 652–664; Dec. 1983.

[8] N. Verhoeckx, H. van den Elzen, F. Snijders, P. van Gerwen, "Digital echo cancellation for baseband data transmission" *IEEE Trans. Acoust. Speech, Signal Processing*, vol. ASSP-27, pp. 768–781, Dec. 1979.

A CMOS Switched-Capacitor Variable Line Equalizer

TOSHIRO SUZUKI, MEMBER, IEEE, HIROSHI TAKATORI, HIROTOSHI SHIRASU,
MAKOTO OGAWA, AND NOBUO KUNIMI

Abstract —This paper describes the design concept and experimental results for a CMOS switched-capacitor variable line equalizer to be used in time compression multiplexed (TCM) digital subscriber loop (DSL) transmission systems. The equalizer transfer function is optimized in the time domain to relax the filter complexity to half that required by the application of classical communication technique. In order to equalize wide bandwidth, high-speed digital data, a 50 MHz CMOS operational amplifier is proposed. The amplifier has a novel folded cascode and buffer structure to achieve good stability against load capacitance change. An experimental chip has been fabricated with 2.5 μm CMOS technology. The chip shows excellent characteristics for the equalization of 200 kbit/s data traveling through pair cables of 5 km and 0.4 mmφ dimensions.

Manuscript received May 9, 1983; revised July 7, 1983.
T. Suzuki, H. Takatori, and H. Shirasu are with the Central Research Laboratory, Hitachi, Ltd., Kokubunji, Tokyo, Japan.
M. Ogawa is with Totsuka Works, Hitachi, Ltd., Totsuka, Tokyo, Japan.
N. Kunimi is with the Device Development Center, Hitachi, Ltd., Kodaira, Tokyo, Japan.

INTRODUCTION

WITH the recent advances being made in digital telecommunication networks, or integrated services digital networks (ISDN's), the need has been increasing for a digital high-speed bidirectional transmission system that utilizes existing analog subscriber loops (DSL). In regard to this system, the most important technical requirement is monolithic implementation of line equalizers which will automatically compensate for high-frequency cable loss in accordance with cable length. Furthermore, equalizer power dissipation should be at a minimum in order to make remote power feeding from a local switching office through the subscriber cable itself feasible.

There are at least two conceivable approaches to realize bidirectional high-speed digital transmission: a hybrid method using an automatic wideband echo canceller (EC),

Reprinted from *IEEE J. Solid-State Circuits*, vol. SC-18, no. 6, pp. 700–706, Dec. 1983.

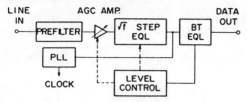

Fig. 1. Total equalizer configuration.

and a time compression multiplexing (TCM) transmission, the so-called ping-pong method [1]. The hybrid method has an advantage in its narrow transmission bandwidth. However, accuracy requirements for the EC are extremely stringent. Residual error must be at least 60 dB below the transmitting signal [2]. With the TCM approach, on the other hand, equalizer accuracy requirements are relaxed about a hundred times, but this is achieved at the price of higher frequency operation.

A skillful analog–digital combined echo canceller has been integrated on an NMOS test chip [3]. A monolithic line equalizer using a CCD delay line was fabricated as well [4]. Both chips, however, may have disadvantages in terms of power dissipation and chip complexity.

We thus concluded that adoption of CMOS switched-capacitor equalizers for TCM systems [5] seems to be the best way to go due to the possibility of achieving suitable accuracy without use of off-chip components and due to their low power properties [6]. The required gain–bandwidth product for the equalizer is close to a thousand times greater than that of previously developed codec LSI's [7]. Thus, a major technical problem affecting integration of the equalizer is the development of very high speed and low power CMOS switched-capacitor circuits.

This paper describes the design principle and results for a prototype CMOS switched-capacitor variable line equalizer. The equalizer was fabricated by means of 2.5 μm silicon gate process, and included 50 MHz operational amplifiers [8]. The results show that the equalizer can compensate for 200 kbaud data traveling through 5 km (0.4 mmϕ) subscriber pair cables.

TOTAL EQUALIZER CONFIGURATION

A total equalizer configuration is shown in Fig. 1 [6]. Here, the $\sqrt{f}$ equalizer compensates for high-frequency cable transmission loss. The subsequent bridge-tap (BT) equalizer can cancel echos generated by BT's, which are additional open-ended spare cables connected to the main cable. A digital PLL is used to extract optimized decision timing. As step intervals of the $\sqrt{f}$ step equalizer are relatively large, a fine gain adjusting AGC amplifier is employed. The gain of the $\sqrt{f}$ step equalizer and AGC amplifier is controled by level control logic.

Typical cable frequency response is shown in Fig. 2. As can be seen, the transmission loss increases drastically as frequency increases. The loss is approximately proportional to the $\sqrt{f}$. According to this system's specifications, the line baud rate is 200 kbaud. The data burst has 233 bit length for each direction and it is transmitted by every 2.5

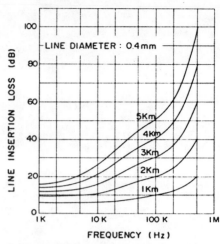

Fig. 2. Typical subscriber pair-cable frequency characteristics.

ms period. Therefore, 93.2 kbit data, which include 64 kbit voice, 16 kbit data, 8 kbit signaling, and 5.2 kbit frame synchronization informations can be transmitted for each direction. The maximum cable length is 5 km (0.4 mmϕ), and the transmission code is 50 percent AMI. Therefore, the equalizer must realize at least a 50 dB gain at 100 kHz, which is the Nyquist frequency for a 200 kbaud data rate. This means a 30 MHz gain-bandwidth product. Within the entire equalizer configuration, the $\sqrt{f}$ equalizer must realize most of this gain–bandwidth product. This is because the BT equalizer, which is realized by a decision feedback algorithm, has essentially no gain, while the AGC amplifier has only a small gain for fine gain adjustment. Therefore, the $\sqrt{f}$ equalizer prototype was fabricated prior to the other portions in order to check the feasibility of monolithic implementation of the entire equalizing system.

TRANSFER FUNCTION DESIGN METHOD

Generally, the $\sqrt{f}$ equalizer is composed of a roll-off filter and a $\sqrt{f}$ filter, as shown in Fig. 3. The purpose of the roll-off filter is to limit the transmission bandwidth in order to maximize the signal-to-noise ratio. In the meantime, the roll-off filter must minimize intersymbol interference as well. Therefore, most roll-off filters have a so called "full-cosine roll-off" transfer function $A(\omega)$, as is described by this [9]

$$A(\omega) = \frac{1}{2}\left(1 + \cos\frac{\pi\omega}{2\omega_1}\right), \qquad (1)$$

where ω_1 represents the Nyquist frequency.

Practically, this transfer function is realized by finite-order rational functions. Equation (1) can then be expanded as follows:

$$A(\omega) = \frac{1}{2}\left\{2 - \frac{1}{2!}\left(\frac{\pi\omega}{2\omega_1}\right)^2 + \frac{1}{4!}\left(\frac{\pi\omega}{2\omega_1}\right)^4\right.$$
$$\left. - \frac{1}{6!}\left(\frac{\pi\omega}{2\omega_1}\right)^6 + \cdots\right\}$$

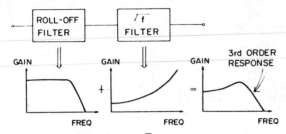

Fig. 3.   Conventional $\sqrt{f}$ equalizer organization.

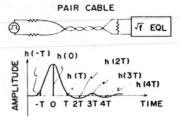

Fig. 4.   Isolated pulse response at $\sqrt{f}$ equalizer output.

$$= 1 - 0.61685 \left(\frac{\omega}{\omega_1} \right)^2 + 0.126835 \left(\frac{\omega}{\omega_1} \right)^4$$

$$- 0.01043 \left(\frac{\omega}{\omega_1} \right)^6 + \cdots \qquad (2)$$

If the error due to a limiting of the expansion length is to be less than ± 1 dB at Nyquist frequency ($\omega = \omega_1$), then at least the first three terms must be left. This means that the roll-off filter requires at least a fourth-order transfer function. Sometimes, additional filters are required to equalize phase error.

At the $\sqrt{f}$ filter, at least a second-order filter is required to compensate for the cable $\sqrt{f}$ characteristics shown in Fig. 2. As a result, the $\sqrt{f}$ equalizer itself might become a sixth-order filter. However, the high cut characteristics of the roll-off filter and high boosting characteristics of the $\sqrt{f}$ filter cancel each other out in the high-frequency portion of the passband, as Fig. 3 shows. If both the roll-off and $\sqrt{f}$ filters can be designed at one stroke, the order of the $\sqrt{f}$ equalizer might come down to half that with the conventional independent design method. This would be a very attractive result in regard to reducing power dissipation and chip size.

In order to realize this advantage, a time domain optimized total design method was employed. The purpose of the $\sqrt{f}$ equalizer is to minimize intersymbol interference D_n, which is defined as

$$D_n = \frac{\displaystyle\sum_{\substack{n = -\infty \\ n \neq 0}}^{\infty} |h(nT)|}{|h(0)|} \qquad (3)$$

where $h(nT)$ is an isolated pulse response at $t = nT$, as observed at the equalizer output node after a particular cable transmission. The nT represents nth decision timing as shown in Fig. 4. The $h(0)$ represents main pulse amplitude. Here the $h(nT)$ is sampled data decimated from

equalizer was 800 kHz; thus the T/T' was 4. The $h(mT')$ can be calculated as

$$h(mT') = \sum_{l = -\infty}^{\infty} y(lT') g(mT' - lT') \qquad (4)$$

where $y(lT')$ is an isolated pulse response at the cable output node. This response is determined by the transmitting waveform and cable transfer function. The $g(mT' - lT')$ is an equalizer impulse response derived from a series expansion of (5).

As shown in (3) and (4), the intersymbol interference D_n can be defined by cable conditions and an equalizer transfer function. This function, described as (5), was optimized so as to minimize D_n by means of a nonlinear programming (NLP) procedure [10]. In the optimization procedure, isolated pulse response at the cable output is simulated by a special-purpose cable simulator. The algorithm is equivalent to the so-called "zero forcing" widely used with automatic equalizers. As a result of this arrangement, the total transfer function of the $\sqrt{f}$ equalizer can be simplified to a third order Z-domain function for 0 to 5 km of line (0.4 mmϕ) length.

Another important advantage of this method is the automatic disappearance of any peculiar frequency characteristic distortion caused by an S-to-Z-domain transform. Thanks to this advantage, an equalizer sampling frequency can be reduced to only four times the data bit rate.

Circuit Configuration

The circuit configuration of the $\sqrt{f}$ equalizer is shown in Fig. 5. As shown, the equalizer is composed of a sample and hold circuit, a third-order roll-off, and a $\sqrt{f}$ filter. The filter is realized by a stray-free state-variable switched-capacitor circuit [11]. The transfer function can be described as follows:

$$F(Z) = \frac{C_1 Z^{-1/2} \left\{ C_4 C_6 + Z^{-1}(C_2 C_7 - C_5 C_6 - C_4 C_6) + Z^{-2}(C_5 C_6 - C_3 C_7) \right\} (C_{11} - Z^{-1} C_{12})}{C_0 \left\{ C_6 (C_{10} + C_9) + Z^{-1}(C_7 C_8 - C_6 C_9 - 2 C_6 C_{10}) + Z^{-2} C_6 C_{10} \right\} (C_{13} + C_{14} - Z^{-1} C_{14})} . \qquad (5)$$

$h(mT')$. T' is a sampling period for the $\sqrt{f}$ equalizer, and $m = nT/T'$. In our setup, the sampling frequency for the $\sqrt{f}$

An 800 kHz sampling clock was adopted after considering high-frequency crosstalk noise aliasing, and the settling

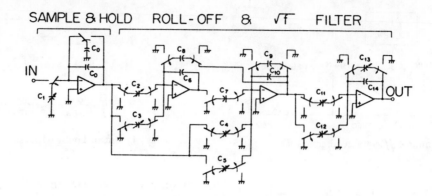

SAMPLE & HOLD   ROLL-OFF & √f   FILTER

—╫— : DIGITALLY PROGRAMMABLE
          CAPACITORS

Fig. 5.   Circuit configuration of $\sqrt{f}$ equalizer.

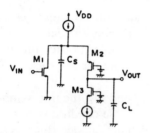

Fig. 6.   Equivalent circuit for folded cascode amplifier.

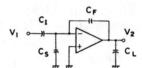

Fig. 7.   General circuit connection for switched-capacitor filter.

requirements for internal amplifiers.

Some capacitors in the equalizer are digitally program-mable to ensure step adaptability. The equalizer has 5 steps which correspond to a 1 to 5 km line (0.4 mmφ) length. The step interval is about 10 dB at the Nyquist frequency.

OPERATIONAL AMPLIFIER DESIGN

As has been described, the high-frequency operation is the most important property of this equalizer. The frequency limitation is mostly decided by the internal amplifier unity gain bandwidth. Rough estimation shows that close to 50 MHz is required. In order to realize this value, a 2.5 μm CMOS process was selected.

For this class of high-frequency amplifier, a folded cascode amplifier, as sketched roughly in Fig. 6, is the most feasible [12]. This amplifier has essentially a high second pole frequency, and two gain stages. Therefore, a high gain may be expected. The cascode configuration produces high gain, even when used with short channel devices which exhibit low output impedance due to short channel effects. The folded cascode amplifier, however, has some disad-vantages, as will be described below.

1) For conventional operational amplifiers, the transfer function $A(s)$ can be written as

$$A(s) = \frac{A_0}{1 + \dfrac{s}{\omega_p}}.$$ (6)

Here, A_0 is dc open-loop gain, and ω_p is dominant pole frequency. At the folded cascode connection, ω_p is the function of load capacitance C_L such that

$$\omega_p = \frac{G_1 G_2}{C_L Gm_2}.$$ (7)

Here, Gm_2 is the transconductance of transistor $M2$. G_1 and G_2 are the drain conductance of $M1$ and $M2$. There-fore, for a variable switched-capacitor filter as shown in Fig. 5, the amplifier transient response may no longer be stable against load capacitance change.

2) In the switched-capacitor circuits, the amplifier con-nection is generally described as shown in Fig. 7. Here, C_L and C_F represent load and feedback capacitance. C_I is input capacitance, which is in most cases connected through switches. C_S includes both stray capacitances generated at the amplifier input node, and other input capacitances. Using this connection, the transfer function between V_1 and V_2 can be defined as

$$\frac{V_2}{V_1} = \frac{C_I}{C_F} \cdot \frac{1}{1 + \dfrac{C_S + C_F + C_I}{C_F A_0 \omega_p} s}.$$ (8)

If ω_p is independently determined against external capaci-tances, the cutoff frequency ω_c can be calculated as

$$\omega_c = \frac{C_F A_0 \omega_p}{C_S + C_F + C_I}.$$ (9)

For the case of a folded cascode amplifier, C_F, C_I, and C_S, as well as C_L, must be regarded as being part of the load capacitance [13]. Therefore, an actual ω_p will be changed,

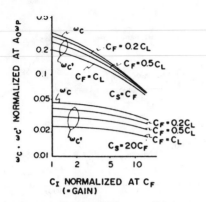

Fig. 8. Variation in operational amplifier cutoff frequency shown against gain change.

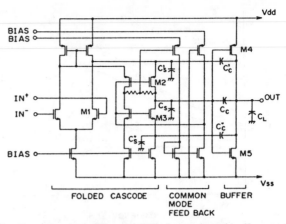

Fig. 9. Operational amplifier with folded cascode and buffer structure.

and the cutoff frequency ω_c' can be defined as

$$\omega_c' = \frac{C_F C_L A_0 \omega_p}{C_L(C_I + C_S + C_F) + C_F(C_F + C_S)}. \tag{10}$$

Comparing (9) and (10), it can be seen that the cutoff frequency for the folded cascode amplifier ω_c' is lower every time than ω_c, as shown in Fig. 8. This difference can be reduced using a very small C_F compared to C_L. However, there is a limit to the minimum value of C_F due to the filter coefficients, etching errors, and PSRR requirement.

3) Furthermore, gate-to-source bias voltages at output transistors $M2$ and $M3$ of the folded cascode amplifier may not be enough to cause a large drain current to flow due to power rail voltage splitting among stacked transistors. Therefore, the load drivability of a folded cascode amplifier is not high without use of large output transistors, which drastically degrade high-frequency characteristics because of their stray capacitance increment.

After considering these disadvantages, an output buffer stage was added. The total amplifier circuit is shown in Fig. 9. It can be seen that three internal compensation capacitors are used. In this circuit, the dominant pole, ω_p and second pole ω_2 can be written as

$$\omega_p = \frac{G_1 G_2}{A_B Gm_2 \left(C_c' + C_c''\right)} \tag{11}$$

$$\omega_2 = \frac{A_B Gm_2}{C_S + A_B C_c}. \tag{12}$$

Here, Gm_2 is transconductances for $M2$. G_1, and G_2 are, respectively, drain conductance of $M1$ and $M2$. A_B is gain for the buffer amplifier. If there is no C_c, ω_2 becomes higher than the second pole observed at a simple folded cascode amplifier, where the second pole is Gm_2/C_S'. In some cases a small C_c may be added to reduce peaking at ω_2.

In the folded cascode amplifier, a common mode feedback circuit was adopted to stabilize dc bias conditions against supply voltage and temperature change. As shown in Fig. 9, the common mode feedback circuit senses a difference between a common mode node voltage at cascode output and an external bias voltage which can be adjusted

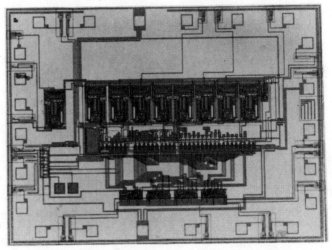

Fig. 10. Photograph of prototype $\sqrt{f}$ equalizer chip.

for stabilizing the dc operating points. The amplified output is fed back to the gates of cascode transistors. Furthermore, the common mode feedback circuit can cause the slew rate to increase. This is because the common mode feedback circuit causes all possible transistors to be in the active region, even if a large step signal is coming to amplifier input.

As a consequence of what has just been described, a 50 MHz unity gain bandwidth, 6.5 mW power dissipation for 5 V single power supply, 67 dB dc gain, and 110 ns 0.1 percent settling time can be achieved.

EXPERIMENTAL RESULTS

The equalizer circuit shown in Fig. 5 was realized in the form shown in Fig. 10 using 2.5 μm CMOS technology. The test chip had nine operational amplifiers, including five monitoring and testing amplifiers. The chip's active area was about 3.5 mm², which could be reduced to 2.2 mm² except for the extra amplifiers.

Measured frequency characteristics are shown in Fig. 11. As shown in the figure, the measured values show extremely good agreement with calculated values for all gain

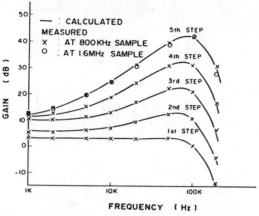

Fig. 11. Gain versus frequency characteristics for prototype chip.

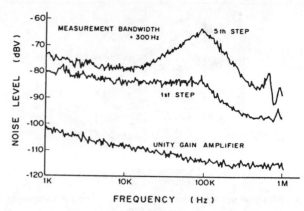

Fig. 13. Noise spectrum for operational amplifier and $\sqrt{f}$ equalizer test chip.

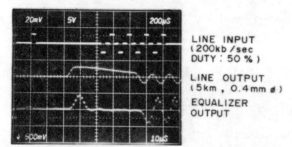

Fig. 12. Time domain response for chip with actual use of subscriber cable.

steps. Even when clocked at 1.6 MHz—double the design value—the measured errors are very small. This indicates that the internal amplifiers have enough speed.

Fig. 12 shows time domain response observed using an actual subscriber cable which was 5 km long and had an 0.4 mm diameter. As shown in the photograph, a distorted line output signal was perfectly equalized.

Fig. 13 shows the measured noise level for the amplifier and equalizer. Since the equalizer nominal output level was designed to be 1.3 V_{o-p}, and the equivalent noise bandwidth at the output to be 100 kHz, the signal-to-noise ratio at the fifth step becomes 42 dB. This is sufficiently small to keep the data error rate very low.

CONCLUSION

This paper has described a CMOS $\sqrt{f}$ variable line equalizer to be used in a DSL transmission system. The equalizer is realized by an 800 kHz sampled third-order switched-capacitor filter whose coefficients are determined by a newly designed time domain optimized algorithm. The operational amplifiers implemented in the equalizer have a novel folded cascode and buffer structure to stabilize cut-off frequency against load and input capacitance change, and to achieve high load drivability. A 50 MHz unity gain bandwidth can be achieved with only 6.5 mW power dissipation.

An experimental chip was fabricated using 2.5 μm CMOS technology. The chip shows excellent frequency character-

istics and indicates that a 200 kbaud signal can be equalized during 5 km subscriber cable transmission. The measured signal-to-noise ratio is low enough.

These results point out the fact that a total equalizing system for DSL can be integrated into a low power CMOS chip without any further technical breakthroughs.

ACKNOWLEDGMENT

The authors wish to thank K. Tamaki, T. Morishita, M. Tanaka, and Y. Kita of Hitachi, Ltd. for their valuable suggestions and continuous encouragement.

REFERENCES

[1] S. V. Ahamed, P. P. Bohn, and N. L. Gottfried, "A tutorial on two-wire digital transmission in the loop plant," *IEEE Trans. Commun.*, vol. COM-29, pp. 1554–1564, Nov. 1981.

[2] D. D. Falconer, "Adaptive reference echo cancellation," *IEEE Trans. Commun.*, vol. COM-30, pp. 2083–2094, Sept. 1982.

[3] O. Agazzi, D. A. Hodges, and D. G. Messerschmitt, "Large-scale integration on hybrid-method digital subscriber loops," *IEEE Trans. Commun.*, vol. COM-30, pp. 2095–2108, Sept. 1981.

[4] T. Enomoto, M. Yasumoto, T. Ishihara, and K. Watanabe, "Monolithic analog adaptive equalizer integrated circuit for wide-band digital communication networks," *IEEE J. Solid-State Circuits*, vol. SC-17, pp. 1045–1054, Dec. 1982.

[5] H. Ogiwara and Y. Terada, "Design philosophy and hardware implementation for digital subscriber loops," *IEEE Trans. Commun.*, vol. COM-30, pp. 2057–2065, Sept. 1982.

[6] T. Suzuki, H. Takatori, M. Ogawa, and K. Tomooka, "Line equalizer for a digital subscriber loop employing switched-capacitor technology," *IEEE Trans. Commun.*, vol. COM-30, pp. 2074–2082, Sept. 1982.

[7] K. Yamakido, T. Suzuki, H. Shirasu, M. Tanaka, K. Yasunari, J. Sakaguchi, and S. Hagiwara, "A single-chip CMOS filter/codec," *IEEE J. Solid-State Circuits*, vol. SC-16, pp. 302–307, Aug. 1981.

[8] T. Suzuki, H. Takatori, H. Shirasu, M. Ogawa, and N. Kunimi, "A CMOS switched capacitor variable line equalizer," in *ISSCC Dig. Tech. Papers*, Feb. 1983, pp. 74–75.

[9] W. R. Bennett and J. R. Davey, *Data Transmission*. New York: McGraw-Hill, 1965.

[10] R. Fletcher, "A new approach to variable metric algorithms," *Computer J.*, vol. 13, pp. 317–322, 1970.

[11] K. R. Laker and P. E. Fleisher, "A general active switched capacitor biquad topology for precision MOS filter," in *Proc. ISCAS'80*, pp. 304–308, Apr. 1980.

[12] T. C. Choi, R. T. Kaneshiro, R. Brodersen, P. R. Gray, W. Jett, and M. Wilcox, "High-frequency CMOS switched-capacitor filter for communication applications," in *ISSCC Dig. Tech. Papers*, Feb. 1983, pp. 246–247.

[13] K. Matsui, T. Matsuura, and K. Iwasaki, "2 micron CMOS switched capacitor circuits for analog video LSI," in *Proc. ISCAS'82*, pp. 241–244, May 1982.

A Single-Chip *U*-Interface Transceiver for ISDN

DANNY SALLAERTS, DIRK H. RABAEY, MEMBER, IEEE, RUDI F. DIERCKX, MEMBER, IEEE, JAN SEVENHANS, MEMBER, IEEE, DIDIER R. HASPESLAGH, AND BART J. DE CEULAER

Abstract—A single-chip *U* transceiver for ISDN based on echo-cancellation techniques is described. The circuit is realized in a 2-μm CMOS process with a die size of 45 mm^2.

I. INTRODUCTION

ISDN field trials and pilot services are currently set up worldwide either to probe for service potentialities or to investigate implementation alternatives. Whether the introduction of ISDN becomes a success or not largely depends on the cost effective availability of the two-wire transceiver for the so-called *U*-reference point.

Several considerations make it nonobvious to implement such a *U* transceiver:

a) the need for a net user bit rate of 144 kbit/s;
b) the evolutionary aspect of ISDN, which means transmission on almost 100 percent of the loops; and
c) the fact that it should be offered at a cost comparable to that of the plain old telephone set.

The paper discusses system and implementation aspects of a one-chip *U* transceiver based on echo-cancellation techniques using the MMS43 line code and a Barker synchronization word for bit and frame timing and including a 1-kbit/s transparant channel for maintenance purposes. The LSI is realized by a 2-μm CMOS, double-metal, double-poly technology shrinkable to a 1.5-μm technology and packaged in a 28-pin DIL package.

Most of the development effort was spent on three aspects of the implementation:

1) from a transmission point of view, the two- to four-wire conversion technique, the line code, and the synchronization procedures had to be defined and verified by simulations and hardware emulators;
2) the system architecture was studied in conjunction with the chosen transmission techniques; this included using an analog or digital summation point, echo-canceller and equalization techniques, number of taps, and coefficient word length; and
3) the optimal silicon implementation for digital signal

Manuscript received May 20, 1987; revised August 28, 1987.
The authors are with the Microelectronic Design Center, Alcatel-Bell Telephone Manufacturing Company, B-2018 Antwerp, Belgium.
IEEE Log Number 8717536.

TABLE I
ISDN *U*-INTERFACE TRANSCEIVER REQUIREMENTS

- full duplex transmission on 2 wire interface
- net bit rate : 144 kbit/s
- subscriber loops up to 8 km, 0.6 mm gauge
- BER $\leq 10^{-7}$
- 100 ms activation time.

processors and the high-performance analog front end was investigated.

II. SYSTEM DESCRIPTION

Systems aspects are treated in detail in [1]. Therefore we briefly highlight the most important items only. For the *U* interface the transmission method has to be defined by fulfilling the requirements given in Table I.

A. Transmission Performance

1. Range: For the transmission method the three key system aspects are: 1) the use of echo cancelling for the two- to four-wire conversion instead of time compression techniques; 2) the use of a well-known ternary code, MMS43, which offers advantages concerning crosstalk and spectral distribution; and 3) choosing a deterministic Barker synchronization word for secure and robust synchronization. To justify the system architecture, numerous simulations and tests using hardware emulators were performed [1], showing that the *U*-interface circuit (UIC) system can be used for ISDN on subscriber loops up to 4.2 km on 0.4-mm lines and 8.0 km on 0.6-mm lines with 10 μV/Hz as Gaussian noise.

2. Robustness: A digital transmission system must be at least as robust as speech transmission and signaling in a conventional analog telephone system. Short-term disturbances must not cause synchronization loss, equalizer divergence, or an excessive number of errors in the equalizer.

The exceptional robustness of the synchronization procedure is based on the fact that frame synchronization is independent of bit synchronization. Any correlation between the receive signal and the synchronization word is used by the receiver in independent control loops to derive the frame clock and to generate the symbol clock. Consequently, frame synchronization remains intact even if bit

Reprinted from *IEEE J. Solid-State Circuits*, vol. SC-22, no. 6, pp. 1011–1021, Dec. 1987.

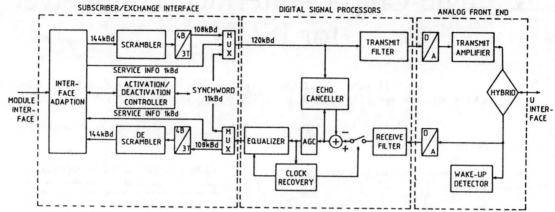

Fig. 1. Block diagram of the *U*-transceiver LSI.

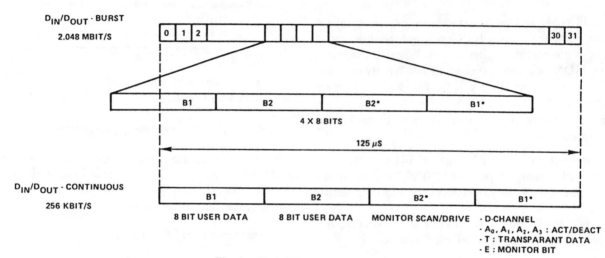

Fig. 2. Module interface configuration.

synchronization is lost, and the resynchronization time is minimized.

B. Block Diagram (Fig. 1)

The *U* transceiver includes all functions to perform the layer 1 function as defined by the OSI model. In addition to the transmission of the 144-kbit/s user information, activation, deactivation, and maintenance procedures are supported. Master and slave modes are implemented on the same device as well. A standard system interface, the so-called *V** interface, makes it possible to use the UIC LSI in different kinds of applications.

The *U* transceiver can be divided into three main functional parts: the system interface, the signal processors, and the analog front end.

1. The System Interface: The system interface converts the data presented on the module interface into the appropriate frame to be transmitted on the line. The module interface is a bidirectional four-wire interface for *RX* (-receive) and *TX* (-transmit) data, frame, and clock. Four bytes are used per basic access, containing (Fig. 2) B_1, B_1, D user data, control and indication channel C/I, transparent maintenance channel T, monitor-channel handshaking bit E, and monitor channel M.

After scrambling the MMS43 coder encodes groups of 4 bits into groups of three ternary symbols, hence reducing effectively the symbol rate from 144 kbit/s to 108 ksymbol/s. Then the transmit frame is formatted (Fig. 3) containing a total of 120 symbols in a 1-ms frame: 108 user data symbols, 11 symbols Barker synchronization, and one symbol for the transparent maintenance channel. For the receive direction, the mirror functions are implemented. Furthermore, the system interface includes an activation/deactivation controller, transmission-quality supervision, and looping functions.

2. The Signal Processors: The digital signal processors execute the computations for the echo canceller, the precursor and decision feedback equalizer, the digital correlation PLL, and the decimation filter for the 1-bit oversampling A/D convertor.

The echo canceller is a transversal adaptive filter consisting of 34 taps with 32-bit coefficients, which is clocked at the symbol rate of 120·ksymbol/s. The S/N ratio after the digital summation point reached by this configuration is better than 21 dB including residual echo signal, nonlinearities of the analog front end, and quantization noise. Therefore, for a worst-case rms insertion loss of 29 dB (corresponding to 35 dB at $f = 60$ kHz), assuming 6-dB echo attenuation at the analog hybrid and 12-dB margin, the echo canceller realizes 56-dB echo cancelling.

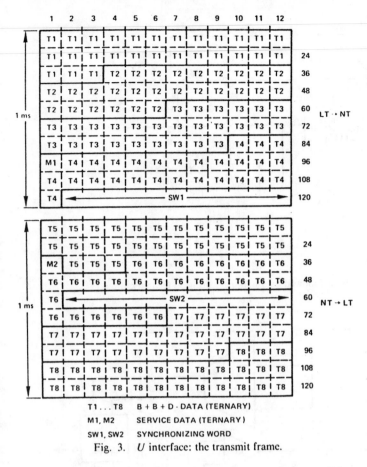

Fig. 3. *U* interface: the transmit frame.

The equalizer is implemented as a linear transversal filter followed by a nonlinear decision feedback filter (Fig. 4). The required bit error rate (BER) of 10^{-7} corresponds to an S/N ratio of 18.5 dB at the decision point for a ternary equiprobable code. The noise amplification introduced by the equalizers is less than 2.5 dB for a two-tap linear equalizer and a 24-tap decision feedback equalizer (DFE) using 16-bit coefficients.

An all-digital PLL has been developed [4] which performs both frame and bit synchronization. The signal processors perform the correlation by an 11-tap linear filter followed by a phase differentiating function with a low-pass phase filter to result in a steady-state jitter amplitude of maximum 65 ns.

3. The Analog Front End: The analog front end of the transceiver driving the twisted pair determines the general performance of the digital subscriber loop. A block diagram is given in Fig. 5. The A/D conversion of the receive signal and the D/A conversion of the transmit signal is done by oversampled pulse density modulation (PDM)-type A/D and D/A circuits clocked at 15.36 MHz. In the *RX* path, a second-order sigma–delta modulator ($\Sigma\Delta$M) is implemented. The second-order noise-shaping filter of the $\Sigma\Delta$M consists of two differential switched-capacitor low-pass stages with cascode-type operational amplifiers.

The PDM output signal of the $\Sigma\Delta$M is low passed and decimated by a third-order decimating filter yielding a 14-bit resolution at 120 ksymbols.

In the *TX* path, D/A conversion of the $4B/3T$ 120-kbit/s binary code to the ternary line signal is provided through a PDM signal stored in a ROM followed by a differential charge quantizing pulse generator and a first-order low-pass filter. The *TX* amplifier delivers 10 mW to the 150-Ω line with very low distortion over an output voltage swing of 4 V_{p-p}/n (n = transform ratio) on the transmission line or 8 V/n on the transmit amplifier output. To provide this output voltage swing in an amplifier powered by a single 5-V supply, the transmit amplifier uses two single-ended power operational amplifiers (Fig. 6).

The overall performance of the analog front end provides a signal to noise and distortion better than 70 dB required for 8-km loops with 0.6 mm wires to ensure a BER of better than 10^{-7}.

III. DIGITAL SIGNAL PROCESSORS

A. Architecture Alternatives

Because of the high amount of number crunching, two signal processor structures were considered: a time-distributed approach in which only a few arithmetic units process the data in a time serial way at a high speed, and a space-distributed approach in which many arithmetic units process the data at the same time at a much lower clock rate. The number-crunching requirements are imposed by the required transmission performance. For echo canceller, precursor equalizer, DFE, AGC, and DPLL, one needs to execute in one symbol cycle of 120 ksymbol/s 68 additions of 32 bit, 59 additions of 16 bit, three 16×16-bit multiplications, and a number of calculations for signal monitor applications. Since the maximum clock rate available is 15.36 MHz, one has 128 time slots to execute all arithmetic functions if one processor is used, or 256 if two are used.

The top level design of the digital signal processor based on two processors running with an instruction cycle of 60 ns on a Harvard architecture showed a required pipelining of six levels, a fast SRAM with READ/WRITE access within one 60-ns cycle, and a complex arithmetic unit with two's complement ADD, SUBSTRACT, MULTIPLY, and SHIFT as basic instructions.

In parallel, the space-distributed structure was applied starting from the basic systolic principles given by Kung [7] and it became rapidly clear that an optimized version for this application had many advantages. Since the number-crunching task is distributed over several processors, one processor performing only one task, the design complexity is reduced to a much smaller scale than it would be with one central processor. The clock rate is reduced, hence design can be done on a more straightforward basis. No extensive controller needs to be designed since control is generated within the cell itself. Finally, because of the serial structure, inherent serial test paths are available.

For all these reasons, the distributed processing structure was chosen to obtain a much faster and less complex signal processor.

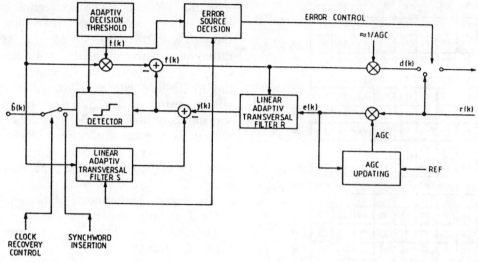

Fig. 4. The equalizer, implemented as a linear transversal filter followed by a nonlinear decision feedback filter.

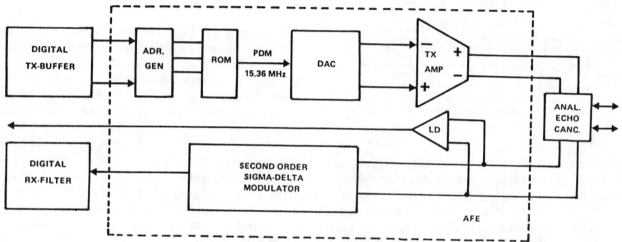

Fig. 5. Block diagram of the analog front end.

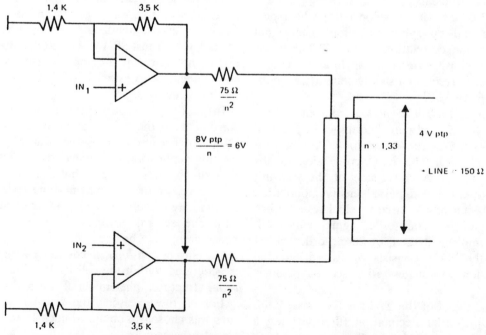

Fig. 6. The transmit amplifier driving the line transformer.

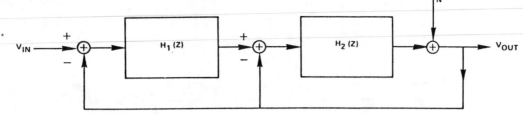

Fig. 7. Block diagram of a second-order $\Sigma\Delta$M.

IV. ANALOG FRONT END

The analog front end of the UIC consists of a receive path *RX* and a transmit path *TX* providing the full duplex analog interfacing to the 150-Ω twisted pair.

A. Analog-to-Digital Conversion (RX Path)

For A/D and D/A conversion, sigma–delta modulation appears to fit future MOS fabrication trends as it consists largely of technology-tolerant circuitry and can therefore scale efficiently as MOS feature sizes shrink. The analog components can be designed to accommodate large parameter tolerances associated with digital technologies. The design of the second-order $\Sigma\Delta$M was based on a physical model using idealized components to allow straightforward calculation of filter coefficients [5]. The design target is a $\Sigma\Delta$M with a specified SNR of at least 72 dB for an input dynamic range up to 2 V and a sampling frequency of 15.36 MHz. For a signal bandwidth smaller than 60 kHz this provides an oversampling factor of 260. By increasing the sampling frequency the number of quantization levels in the A/D and D/A conversion in the PDM loop is reduced to 1 bit (see Fig. 7).

A/D conversion is accomplished by a comparator at the differential output of the second-order low-pass filter and a clamp circuit adding up $+V_{\mathrm{ref}}$ or $-V_{\mathrm{ref}}$ to the summing nodes at the low-pass filter inputs.

From Fig. 7, the transfer functions can be derived for a second-order $\Sigma\Delta$M:

$$\mathrm{STF}_2 = \frac{H_1 H_2}{1 + H_2 + H_1 H_2}$$

$$\mathrm{NTF} = \frac{1}{1 + H_2 + H_1 H_2}.$$

If H_1 and H_2 are integrators (low-pass filters), the signal transfer function (STF_2) of a $\Sigma\Delta$M is of the low-pass type and the noise transfer function (NTF_2) is of the high-pass type.

Typical noise spectra of first- and second-order $\Sigma\Delta$M's are given in Fig. 8. Note that the effect of noise shaping is to move the quantization noise away from the baseband, at the expense of amplifying the total noise power by a factor *P*.

In the system the $\Sigma\Delta$M is followed by a digital low-pass decimation filter. The low-pass decimation filter cancels the high-frequency noise. A detailed description of the design methodology for the noise-shaping filter in the

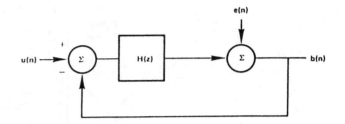

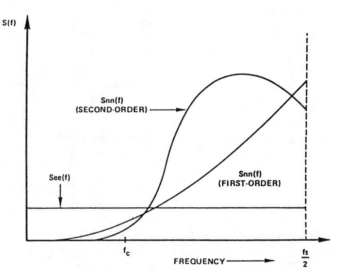

Fig. 8. Noise spectra of a first-order and a second-order $\Sigma\Delta$M.

$\Sigma\Delta$M is given in [5]. The design of the $\Sigma\Delta$M is determined by the system requirements on signal to noise plus distortion, the maximum signal amplitude in the linear region of the two operational amplifiers in the $\Sigma\Delta$M, and the stability of the second-order noise-shaping filter. In order to be stable, the poles of the transfer function should be inside the unity circle, so if the poles are $r \cdot e^{j\theta}$, then $r < 1$ is a necessary restriction for stability. When the poles are close to $z = 1$, it will also cause an early peaking of the NTF, requiring a complex subsequent decimation filter.

The noise peaking frequency is given by $f = (\theta/2\pi) \cdot f_s$. The criteria used for the circuit design can be summarized as follows:

1) the NTF pole angle θ has to be large enough to prevent early peaking ($> 5.6°$);
2) the pole amplitude $r < 0.75$;
3) $\mathrm{SNR} \geqslant 72$ dB; and
4) the main power gain $P \leqslant 3$.

In this design model, the output S/N, total noise amplification *P*, and the position of poles and zeros of the loop

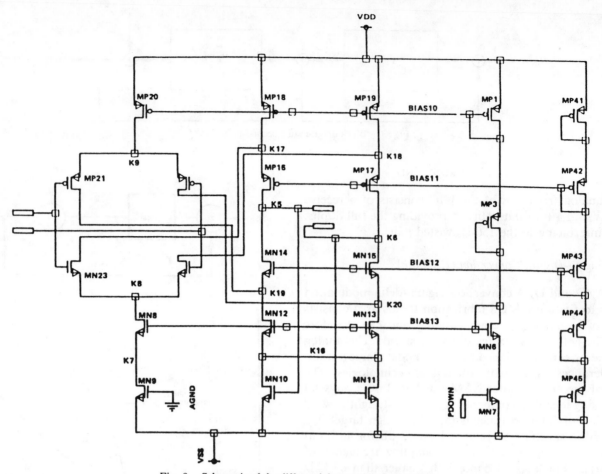

Fig. 9.   Schematic of the differential operational amplifier.

filter were calculated as function of the input and sampling frequencies and the capacitor ratios α_1, α_2, and β_1.

Following the design methodology of [5], the capacitor ratios of the $\Sigma\Delta M$ are defined within the following limits: $\beta_1 > 0.6$ and $0.2 < \alpha_2 < 0.6$. The linearity limits of the op amps are an additional criterion for the capacitor ratios. For loads up to 5 pF the unity-gain bandwidth is over 100 MHz for signals up to 2.5 V and the d.c. gain is 60 dB.

The loop amplifier used is shown in Fig. 9 [6]. Input FET's are chosen wide enough to minimize $1/f$ noise. PSRR problems are minimized by the differential approach, and the dynamic range is enhanced for the same noise characteristics. Common-mode stability is provided via feedback FET's MN10 and MN11. Additionally, common-mode stability is assured by providing a leaky integrator in the first stage of the loop filter. This operational amplifier was realized in a 2-μm CMOS process with threshold voltages of 0.7 and -0.7 V, respectively, for NMOS and PMOS transistors. Measurements show that it has a 100-MHz bandwidth and 35° phase margin.

The comparator shown in Fig. 10 basically consists of a differential input stage coupled to a strobed bistable flip-flop over the current mirrors MP9-MP9A-MN4A-MN4 and MP10-MP10A-MN5A-MN5. The complementary output signals K_{14} and K_{15} are buffered by two inverters and a flip-flop clocked in the same strobe sequence. The com-

parator operates with a 10-mV-wide death zone up to 20-MHz clock rate.

For the practical implementation, the second-order low-pass filter is realized in two switched-capacitor integrator stages. A switched-capacitor implementation was chosen because of the accuracy needed for the filter coefficients. The differential approach was found to be superior to the single-ended one because it reduces the signal-to-distortion degradation from second-order harmonics. It also provides an essential improvement in PSRR and dynamic range.

The theoretical second-order model used for calculation is depicted in Fig. 11:

$$H_1 = \frac{\alpha_2 \cdot z^{-1}}{1 + \alpha_1 - z^{-1}}$$

$$H_2 = \frac{\beta_1 \cdot z^{-1}}{1 - z^{-1}}.$$

The coefficients α_2 and β_1 are the amplification factors of the first and second stage, and α_1 is the loss coefficient of the first stage (resistive feedback). In this design approach, only quantization noise in the comparator is taken into account.

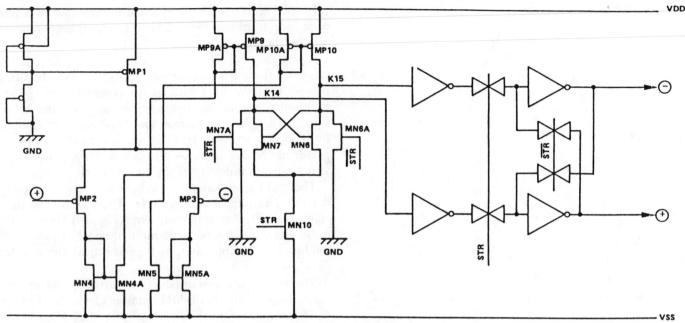

Fig. 10. Schematic of the high-frequency comparator.

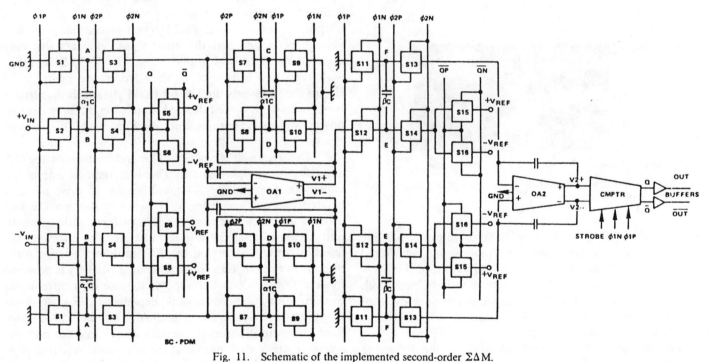

Fig. 11. Schematic of the implemented second-order $\Sigma\Delta$M.

Measurement results are shown in Fig. 12 and Fig. 13. From Fig. 12 it is clear that the $\Sigma\Delta$M reaches its optimum performance for approximately 2.5–3-V input voltage. This is what we expect for ± 1-V reference voltage. In the idle channel noise spectrum of Fig. 13, the typical second-order noise transfer function appears. The idle channel noise is below -80 dB in the low-frequency region. Above 300 kHz the noise peaks up to a maximum value of -45 dB at $f_s/4$ ($f_s = 15.36$ MHz). The idle channel noise was referred to the overload point of 4 V_{p-p}. The overload point corresponds to the maximum PDM signal being a monotonous stream of ones.

B. The Transmit Path: Digital-to-Analog Conversion and Transmit Amplifier

In the D/A converter the binary-coded ternary data (sign magnitude) are converted to a 120-kbit/s analog ternary signal. In the D/A converter the ternary impulses are shaped to minimize the signal bandwidth and reduce crosstalk on the line.

Each transition of the 120-kbit ternary signal is coded in ROM as a 128-bit PDM bit stream and shifted out at a rate of 15.36 MHz. Clamping the PDM stream and low passing the signal by a first-order 180-kHz *RC* filter

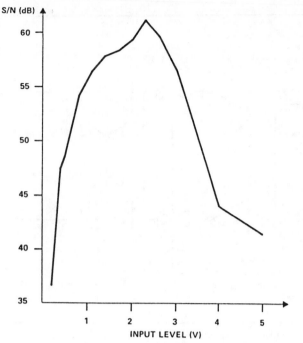

Fig. 12. Measured signal-to-noise plus distortion ratio of the $\Sigma\Delta M$ as a function of the input level.

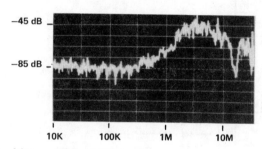

Fig. 13. Measured idle channel noise spectrum of the second-order $\Sigma\Delta M$.

delivers the analog ternary signal. To drive the 150-Ω line of the digital subscriber loop, the ternary signal is buffered in the transmit amplifier. To bring the output of the D/A converter and transmit amplifier to a well-defined level, independent of the supply voltage and the duty cycle of the PDM bit stream, the gain of the D/A converter is regulated by a reference loop.

The transmit path is designed to meet the following specifications: total distortion < −60 dB; PSRR > 40 dB ($f < 30$ kHz); output level control < ±7.5 percent; offset < 5 percent; and output impedance <10 Ω ($f < 120$ kHz).

1. The D/A Converter: The ternary line code is fed to the analog front end of the UIC in a 2-bit binary representation. The nine possible transitions of the ternary code ($-1/+1$, $-1/0$, $-1/-1$, $0/+1$, $0/0$, $0/-1$, $+1/+1$, $+1/0$, $+1/-1$) are programmed in ROM as a 128-bit PDM bit stream. The 2-bit binary representation of the current symbol and the previous one are used as the ROM address, to read the PDM bit stream of the corresponding transition. This PDM signal is shifted out with a clock rate of 15.36 MHz and fed into a clamp circuit. In the clamp circuit the amplitude of the PDM impulses is set to the reference level and the duty cycle is regulated in a reference loop to obtain the well-defined signal amplitude of 4 V_{p-p} on the line.

The clamp circuit, built up with transmission gates and powered from the 5-V power supply, drives a first-order passive low-pass filter with a cutoff frequency of 180 kHz, to eliminate the high-frequency content of the ternary signal. A single clamp circuit and low-pass filter combination on the 15.36-MHz PDM stream would give a bad performance due to duty-cycle dependency, power supply variations, nonlinearity of the clamp circuit itself, etc.

Therefore a reference loop is built to control the gain of the DAC through the duty cycle of the PDM bits in the clamp circuit. The reference loop (Fig. 14) consists of a pulse generator (monostable multivibrator), a pulse-width modulator, a comparator, and an identical clamp circuit plus low-pass filter.

The pulse generator produces a narrow pulse on every rising edge of the 15.36-MHz master clock. In the pulse-width modulator the duty cycle of this 15.36 MHz clock is controlled by the gain control voltage of the comparator.

The clamp circuit of the reference loop (AGC loop) receives a known PDM signal (overload = all ONE's). The output of the overload PDM signal is compared with a voltage reference and the error signal corrects the duty cycle of the pulse-width modulator. The 15.36-MHz clock with regulated duty cycle is then fed to the main clamp circuit that converts the 128-bit PDM stream to the ternary transitions with a well-defined amplitude, independent of power supply variations and duty cycle of the master clock.

2. The Transmit Amplifier: The architecture of the *TX* amplifier is straightforward. The first requirement is the high output swing: the *TX* amplifier has to drive a 150-Ω twisted pair with a differential swing of 4-V on the line using a line transformer with a transform ratio close to unity. Considering the line termination, the effective differential voltage swing on the *TX* amplifier is 6 V_{p-p} for a transform ratio of 1.33. Since all analog circuitry is powered from a single 5-V power supply, the use of a differential configuration is necessary as depicted in Fig. 6. Both amplifiers provide a swing of 3 V_{p-p} on the 150-Ω load. Based on output power considerations (60 mW for both op amps), a class-A/B output stage is a must. And the high output swing (3 V within the 5-V power supply) demands some gain, in the output stage to relax the swing, required in the gain stages.

An elegant manner by which to drive the push–pull output stage is to use two complementary gain stages, each driving a transistor of the push–pull output stage. These complementary gain stages are realized in a common-source configuration. Finally, for the input stage a folded-cascode amplifier is chosen for high gain and high bandwidth capabilities.

The transmit amplifier needs a high gain bandwidth product to satisfy the requirements on distortion and output impedance, for frequencies up to 60 kHz. The output impedance of the push–pull output stage is the

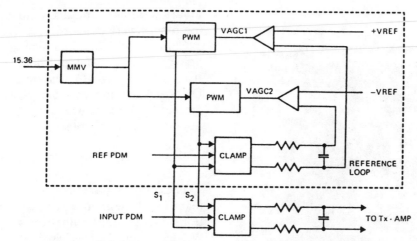

Fig. 14. Block diagram of the D/A convertor with the reference loop.

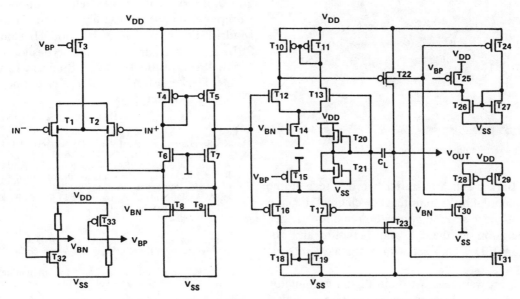

Fig. 15. Schematic of the operational amplifiers used in the transmit amplifier.

drain–source impedance of the transistors T_{22} and T_{23}. The loop gain reduces the output impedance below 10 Ω for the 120-kHz signal. The amplifier of Fig. 15 has a dc gain of 90 dB and a unity-gain bandwidth of 12 MHz.

To prevent excessive power consumption and crossover distortion, the quiescent current in the class-*A/B* output stage is controlled by the output bias circuits (T_{24}–T_{30}). In the crossover region the current of the push–pull output drivers T_{22} and T_{23} is measured in T_{24} and T_{31}, respectively, and mirrored to T_{26} and T_{28}. The current in T_{26} and T_{28} is balanced with the reference current of T_{25} and T_{30}. If a current unbalance occurs the gate voltages of the push–pull transistors T_{22} and T_{23} are driven to regain the balance. Consider an offset voltage in the differential stage driving T_{22} causing an excessive current in T_{22}. T_{24} will sense the excess of current in T_{22} and mirror it to T_{26}. This extra current in T_{26} causes a lower drive voltage on T_{23} and consequently the operational-amplifier output voltage will rise. In the application the amplifier is in a negative feedback loop. So from the error signal on the input IN the gain stages will drive the gates of T_{22} and T_{23} to restore

the output voltage. The same algorithm applies for an offset causing crossover. The quiescent current in T_{22} and T_{23} is determined in the design by choosing the drive current in T_{25} and T_{30} and by the current mirroring ratio in T_{24}, T_{27}, T_{26} and T_{31}, T_{29}, T_{28}. Mismatch in T_{25} and T_{30} will cause a slight offset in the amplifier. The output bias circuit keeps the quiescent current within a range of ±20 percent of the nominal value for 10-mV offset mismatch in the differential stages.

The stability of the amplifier for unity-gain feedback is obtained by the capacitive feedback from the output over C_c and the differential stages, introducing a dominant pole at 100 Hz. The transistors T_{20} and T_{21} provide the dc bias for the drivers of the differential stages in the compensation loop.

V. CONCLUSIONS

A single-chip *U* transceiver has been presented which will be used for pilot ISDN installations in Europe beginning in 1988. Transmission techniques are based on echo cancelling using the *4B3T* (-MMS43) line code, a DPLL at

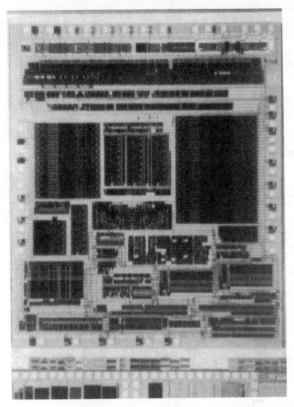

Fig. 16.   Photograph of the UIC chip.

the basis of a Barker synchronization word, and a frame format to include a 1-kbit/s maintenance channel. From a system point of view several application modes are included for operation at the LT and NT side and for interfacing to an ISDN chip set with a standard module interface.

Novel techniques were used for both digital and analog signal processing. To reduce analog processing a digital summation point was used, in conjunction with a DPLL. An optimized systolic processing scheme to reduce silicon area was developed. Oversampling A/D and D/A tech-niques were used and differential analog processing building blocks were designed.

Chip size is 6.20×7.30 mm^2 using a 2-μm CMOS technology, double metal and double poly for a 72 000-transistor device (Fig. 16). The transceiver is packaged in a 28 DIL package and uses 5-V power supply only. External devices required are a crystal, a line transformer, feed resistors, and line protection devices.

ACKNOWLEDGMENT

The authors want to thank all members of the project team for their important contributions on design, simulations, layout, and testing: for system design, Dr. F. Zapf, W. Beisel, and K. Széchényi, all of Alcatel-SEL, Stuttgart, Germany; for digital design, D. Van De Pol, E. Moerman, and J. Merino of Alcatel-Bell Telephone Manufacturing Company, Antwerp, Belgium, P. Pedron of Alcatel-SESA, Madrid, Spain, and P. Jedele and H. Conrad of Alcatel-SEL, Stuttgart; and for analog design, J. Schnabel of Alcatel-SEL, Stuttgart, and F. Bonjean, K. Callewaert, and E. De Decker of Alcatel-Bell Telephone Manufacturing Company, Antwerp.

REFERENCES

[1] K. Szechenyi, F. Zapf, and D. Sallaerts "Integrated full-digital U-interface circuit for ISDN subscriber loops," *IEEE J. Selected Areas Commun.*, vol. SAC-4, no. 8, pp. 1337–1349, Nov. 1986.
[2] R. A. McDonald, "Performance measurements of implementations of several line codes," ANSI contribution T1D1.3/86-154.
[3] R. A. McDonald, C. A. Larson, and W. Y. Barkley, "Field trial of MMS43 emulation in Garden City, NY," ANSI contribution T1D1.3/86-118.
[4] K. Szechenyi, "Anwending der Adaption Entzerrung und Echo-kompensation bei vollduplex 2-Draht-Übertragung unter Berück-zichtigung der Taktableitung," *Kleinheubacher Berichte*, Band 27, 1984.
[5] B. P. Agrawal and K. Shenoi, "Design methodology for $\Sigma\Delta M$," *IEEE Trans. Commun.*, vol. COM-31, no. 3, pp. 360–370, Mar. 1983.
[6] P. R. Gray and R. G. Meyer, "MOS operational amplifier design—A tutorial overview," *IEEE J. Solid-State Circuits*, vol. SC-17, no. 6, pp. 969–982, Dec. 1982.
[7] H. T. Kung, "Let's design algorithms for VLSI systems," presented at the CALTECH Conf. VLSI, Jan. 1979.

A CMOS Double-Heterodyne FM Receiver

BANG-SUP SONG, MEMBER, IEEE, AND JEFFREY R. BARNER

Abstract —Experimental results of a narrow-band, adjustment-free, double-heterodyne CMOS FM receiver with a high-Q switched-capacitor intermediate-frequency (IF) filter centered at 3 MHz are presented. The integration covers all the filtering and demodulation circuits from radio-frequency (RF) circuits (50–100 MHz) to the audio output. An experimental prototype FM receiver exhibiting a 5-mV input sensitivity and a -30-dB quieting level is implemented using a 1.75-μm double-poly CMOS technology. The chip occupies 7.7 mm^2 and dissipates 80 mW with a 5-V supply.

I. INTRODUCTION

THIS PAPER describes a way of integrating a narrow-band FM receiver applicable to receiver systems such as a pager, a toy radio, or a cordless telephone, as well as to TV audio and home and car FM receivers with modifications. In commercially available FM receiver integrated circuits (IC's), only bipolar transistors are integrated, and components such as an intermediate-frequency (IF) filter and a quadrature coil have to be supplied externally [1]–[7]. The experimental receiver described in this paper is a very simple and primitive example in the attempt to implement a whole receiver function from antenna to speaker on a single chip of Si. The main thrust toward an everything-on-Si approach is to minimize manufacturing costs by eliminating external components and adjustment procedures entirely. Additional components that are not integrated but are needed in the FM receiver are a tuned load, an automatic gain control (AGC) timing capacitor, a crystal or a reference clock, and a few power-supply decoupling capacitors if needed. No component adjustment is necessary because the whole receiver relies on a single reference clock.

The IF filtering in FM receivers has been performed directly at the 10.7-MHz IF frequency using highly selective *LC*, ceramic, or crystal filters. As a result, integration efforts have been focused on the implementation of high-frequency filters employing IC techniques [8], [9]. However, an actual realization has been delayed by problems associated with high-Q and high-frequency filtering [10]. An alternative approach is a modulation filter such as an *N*-path filter and a filter based on a single-sideband modulation, etc. [11]–[14]. The modulation filter shifts down the passband signal to dc or low frequencies for easy filtering, and the filtered signal is remodulated back to the

original passband or different frequencies. Because actual filtering is done at the baseband, the modulation filter can easily meet high-Q and high-frequency requirements and is also compatible with IC technology. However, a major drawback of the modulation filter is the inevitable generation of extraneous signals such as image and clock components inside the signal band as a result of the remodulation process.

This paper attempts to address the practical limits associated with a CMOS implementation of a conventional double-heterodyne FM receiver. In Sections II and III of this paper, IF filtering scheme as well as overall chip architecture of an experimental receiver are summarized, and in Section IV, a pulse-counting FM demodulation is discussed. Lastly, in Section V, three switched-capacitor filters including two IF filters are described, and experimental results of a prototype FM receiver are presented.

II. IF FILTERING

An IF filtering scheme for a double-heterodyne receiver is illustrated in Fig. 1. The receiver has a switched-capacitor all-pole bandpass filter (IF1) operating at 3 MHz with a Q of 55 as a first IF filter. From the viewpoint of feasibility, the 3-MHz IF is appropriate for receivers with switched-capacitor IF filters. The FM signal at 3 MHz is mixed down to the 72-kHz second IF as illustrated in Fig. 1. The second IF filtering at 72 kHz is also performed by a switched-capacitor elliptic bandpass filter (IF2). There are two advantages in double IF approaches. One advantage is that more channel filtering can be done at low frequencies, and the other is that the FM demodulation is easy at low frequencies.

In narrow-band FM receiver systems, assigned channel bandwidths are usually too narrow for monolithic IC filters. For example, to have a bandwidth of 20 kHz at 3 MHz, a selectivity Q of 150 is required, which is not feasible to meet using metal–oxide–semiconductor (MOS) technologies that have a typical 0.1–0.5-percent level of component matching characteristics. This high selectivity can be attained if modulation IF filtering is used [14]. However, the penalty is the performance degradation caused by spurious in-band signals generated during the remodulation process. Therefore, in conventional double-heterodyne receivers, additional channel filtering at the second IF will be of great help because the bandwidth of the first IF filter implemented using monolithic IC techniques will be much wider than the channel bandwidths of

Manuscript received April 22, 1986; revised July 14, 1986.
The authors are with AT&T Bell Laboratories, Murray Hill, NJ 07974.
IEEE Log Number 8610628.

Reprinted from *IEEE J. Solid-State Circuits*, vol. SC-21, no. 6, pp. 916–923, Dec. 1986.

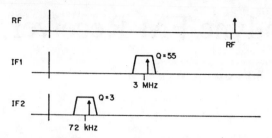

Fig. 1. IF filtering scheme for a double-heterodyne receiver.

narrow-band FM receiver systems. The Q of 55 in 3-MHz range seems to be the upper limit of active filters when a typical ± 0.5-percent center frequency accuracy of high-Q high-frequency filters is taken into consideration [10].

III. FM RECEIVER ARCHITECTURE

The schematic block diagram of an experimental receiver is illustrated in Fig. 2. The incoming radio-frequency (RF) FM signal (50–100 MHz) is amplified by a front-end RF amplifier, which is a simple differential pair amplifier with an open drain. An external tuned tank circuit must be added to give more image rejection as well as to improve the receiver sensitivity. The RF amplifier also has an AGC function to limit the amplifier gain under large-signal conditions. It is followed by a four-quadrant multiplier (MULTI1), which mixes the RF signal down to the 3-MHz first IF. The output of this balanced multiplier is band-limited to 8 MHz because only the frequency down-conversion is needed. The differential output of the multiplier drives a first-IF switched-capacitor bandpass filter (IF1). At this point, the signal is divided into two branches because the IF1 is a differential filter. Two sets of anti-aliasing and smoothing filters (ALIAS1 and ALIAS2) are used with this filter for the proper rejection of aliasing components at each of the clock multiples. Both filters are standard second-order maximally flat filters with a cutoff frequency of 8 MHz. Because of the front-end RF tuning in most RF receivers, the frequency band to be potentially aliased in the first IF filter is RF frequencies (50–100 MHz). These components are rejected by the balanced modulator (MULTI1) as well as by the anti-aliasing filter (ALIAS1). A carrier suppression of 30–40 dB can be achieved in typical four-quadrant multipliers [15]. Therefore anti-aliasing filter requirements are not stringent in typical RF receivers even though sampled-data IF filters are used.

The IF filter (IF1) is driven by a two-phase 12.4-MHz on-chip clock generator (CLOCK2). The filtered 3-MHz FM signal is mixed down again by the following multiplier (MULTI2) to the 72-kHz second IF. The signal level at this point is 130 mV for an RF input of 1 mV. Most signal gain is obtained in the IF filter, which has a gain of $Q/2$ corresponding to 28 dB at the center frequency. A switched-capacitor elliptic bandpass filter (IF2) for the second IF is a simple low-pass-to-bandpass transform of a third-order elliptic low-pass filter. This filter is driven by a

two-phase clock generator (CLOCK1). The input of CLOCK1 is a 775-kHz clock that is coming from an internal clock divider (DIVIDER). Two maximally flat anti-aliasing/smoothing filters (ALIAS3) with a 150-kHz cutoff frequency are used with this filter, and the FM signal is amplified by a 15.6-dB amplifier (AMP) to yield a 1.2-V signal for peak detection (PEAK DETECT). The peak of the signal at this point controls the gain of the front-end amplifier (RF) through AGC feedback. Following the second IF stage is a limiter for AM suppression (LIMIT). After the signal is limited, it is converted into a short pulse train (SHORT PULSE) to provide a reset pulse for a counter-type FM demodulator (FM DEMOD).

The output of the FM demodulator is a pulsewidth modulated signal whose duty cycle corresponds to the frequency deviation of the FM signal [16]. Therefore a simple low-pass filtering of this pulsewidth modulated signal will produce a demodulated voice output to drive an output speaker. A dc level shifter (LEVEL) shifts up the dc level of the pulsewidth modulated signal to the center of the supply voltage. This level shifter is an inverting differential amplifier with one input biased at $V_{DD}/4$. A switched-capacitor elliptic low-pass filter (LPF1) removes all high-frequency harmonic components from the pulse-width modulated signal. This low-pass filter has a 3.75-kHz cutoff frequency. In conjunction with this filter, two maximally flat anti-aliasing/smoothing filters (ALIAS4) with a 15-kHz cutoff frequency and a two-phase 194-kHz clock generator (CLOCK1) are used. As shown in Fig. 2, the overall system depends on the single 12.4-kHz reference clock. Therefore the receiver performance is insensitive to any misalignment, and any adjustment or trimming is absolutely unnecessary in this receiver.

IV. FM DEMODULATION

The FM demodulator is a simple counter chain composed of positive edge-triggering T flip/flop's (FT1N3D, CMOS polycell name in AT&T Bell Laboratories) as shown in Fig. 3. A very short pulse is generated using the FM pulse input signal and the 10-ns delayed pulse of its own. Whenever the counter chain receives this FM short pulse, it is reset to zero as illustrated in Fig. 4. After the short pulse goes down, the counter chain starts counting the incoming clock. Consider the output frequency of the counter to be f_N, which is simply $f_{clk}/2^N$ assuming the number of counters is N and the input clock frequency is f_{clk}. Then, the counter output will go up after the period of $1/f_{N-1}$, as shown in Fig. 4 where $f_{N-1} = f_{clk}/2^{N-1}$. The counter will be reset again at the next FM short pulse that has a period of $1/f_{FM}$, where f_{FM} is the FM signal frequency. Therefore the duty cycle of the demodulator output is

$$\text{duty cycle} = \frac{1/f_{FM} - 1/f_{N-1}}{1/f_{FM}} = \frac{f_{N-1} - f_{FM}}{f_{N-1}}. \quad (1)$$

Because the reference frequency f_{N-1} is fixed, the output

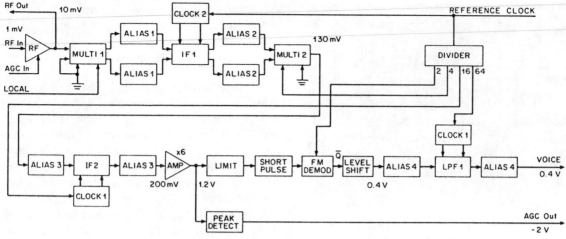

Fig. 2. Schematic diagram of a double-heterodyne FM receiver.

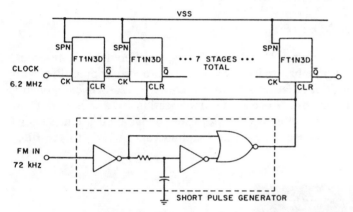

Fig. 3. Pulse-counting FM demodulator with a short-pulse generator.

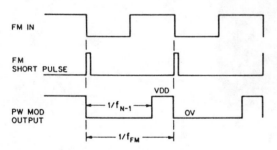

Fig. 4. Illustration of the FM demodulation process.

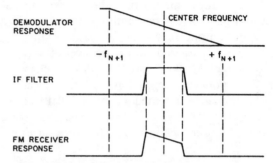

Fig. 5. Transfer characteristic of the FM receiver.

duty cycle is proportional to the frequency deviation of the FM signal.

A low-pass filter is needed to restore the dc component of this pulsewidth modulated pulse train. The resolution of this kind of demodulator is limited by the phase jitter of the pulsewidth modulated output because the FM short pulse and the reference clock are asynchronous. Therefore the uncertain rising/falling transitions of the output pulse appear as a wide-band phase jitter in the demodulated output. For example, the duty cycle of the pulsewidth modulated output has a maximum random variation corresponding to the period of the input clock. Therefore the maximum duty cycle error can be obtained by

$$\text{duty cycle error} = \pm 2 \times \frac{\text{FM input frequency}}{\text{input clock frequency}}. \quad (2)$$

Although a more elaborate synchronous one-shot oscillator and higher reference clock will provide cleaner demodulation, the values of $N = 7$ and $f_{clk} = 6.2$ MHz were chosen to obtain a moderate 30-dB level of resolution limited by the phase jitter in this experimental implementation. Using the relation of (2), the maximum jitter in this case can be estimated to be $\pm 2 \times (72 \text{ kHz}/6.2 \text{ MHz}) = \pm 2.3$ percent (-32.7 dB maximum).

The maximum voltage swing of the demodulator output is $\pm V_{DD}/4$, and the average dc voltage is $V_{DD}/4$, where V_{DD} is assumed to be the output pulse height. Because the demodulated output is sensitive to the pulse height, the actual pulse height is limited by the voltage drop of four MOS diodes ($4V_{GS} \approx 4.5$ V) rather than by the power supply voltage V_{DD}. To bring the dc output voltage back to the center of the power supply, the dc level shifter (LEVEL) is used before the low-pass filter (LPF1). The transfer characteristic of the demodulator is sketched in Fig. 5. It has a linear transfer characteristic within $\pm f_{N+1}$, where $f_{N+1} = f_{clk}/2^{N+1}$. Because the incoming signal is limited by the composite IF filter bandwidth (IF1 and IF2), which is narrower than the demodulator bandwidth, the resulting receiver frequency response will be as shown in the bottom sketch of Fig. 5.

Note that the demodulated output of the FM receiver becomes the lowest voltage for the signal outside the IF filter passband (signal below detection threshold) because of the polarity inversion in the level shifter (LEVEL). Actually, the pulsewidth modulated output rises to $V_{DD}/2$

Fig. 6.   Die photo of the FM receiver.

when there is no FM input signal to clear the counter chain. This in turn implies that the counter in the demodulator (Fig. 3) keeps on counting the reference clock. Therefore the output duty cycle of the demodulator is 50 percent, which corresponds to the dc level of $V_{DD}/2$. The passband slope of the receiver transfer characteristic changes its polarity depending on the local carrier frequency. The slope obtained when the local frequency is higher than the RF signal is the opposite to the slope obtained by a lower local carrier. Therefore, in Fig. 5, the

same slope is maintained before and after the polarity inversion for simplicity of explanation.

V. EXPERIMENTAL RESULTS

An experimental prototype FM receiver was fabricated using a 1.75-μm double-poly CMOS technology. All the blocks illustrated in Fig. 2 were integrated. The chip occupies 7.7 mm² and dissipates 80 mW with a 5-V supply. Half of the power is consumed in IF1 for high-frequency filtering. The die photo of the receiver is shown in Fig. 6. All measurements were performed with a 5-mV RF input applied after the front-end RF amplifier stage. Three switched-capacitor filters are used in the receiver. Two of them are bandpass filters, IF1 and IF2, and the other is a low-pass filter LPF1. Frequency responses of the three filters are shown in Fig. 7.

A. First IF Filter (IF1)

The first IF filter (IF1) is a differential, sixth-order, all-pole bandpass filter. It is a 0.3-dB equiripple filter with a Q of 55 at 3 MHz, and it rejects ± 120-kHz components by 40 dB. Since noise amplification by a factor of Q in high-Q filters is critical for low-signal applications such as in this receiver, input coupling capacitances to the filter

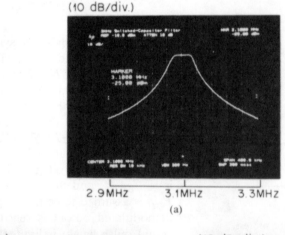

(10 dB/div.)

2.9 MHZ   3.1 MHZ   3.3 MHZ

(a)

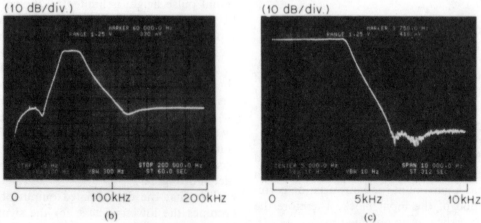

(10 dB/div.)

0   100 kHz   200 kHz

(b)

(10 dB/div.)

0   5 kHz   10 kHz

(c)

Fig. 7.   Frequency responses of the three switched-capacitor filters (10 dB/div., vertical): (a) IF1 (40 kHz/div., horizontal); (b) IF2 (20 kHz/div., horizontal); and (c) LPF1 (1 kHz/div., horizontal).

were intentionally made Q times larger than its nominal value to render this filter a gain of $Q/2$. Although this additional gain of 28 dB will limit the maximum input signal amplitude by the same amount, it is useful for IF filtering applications because the input signal from the RF front-end amplifier is very weak, and an AGC feedback will limit the maximum signal to the input stage. The frequency response pictured from a separate test chip with a higher clock rate is shown in Fig. 7(a), where the marker is shown at the center frequency of 3.1 MHz. Detailed experimental results of this filter are reported in [10]. Approximately, one pole in this filter requires 7.5 mW of power and 0.33 mm² of area.

B. Second IF Filter (IF2)

The second IF filter (IF2) is a sixth-order elliptic bandpass filter. It is designed to reject the sideband by 49 dB with a 0.3-dB passband ripple. The center frequency is 72 kHz with a Q of 3. IF2 is designed in a single-ended form, and the architecture of the filter is identical to that of the IF1 in the passband. It has a transmission zero for the sideband rejection. The frequency response of this filter is shown in Fig. 7(b), where the marker is at the center frequency. The center frequency of Fig. 7(b) is 60 kHz (not 72 kHz) because the picture was taken with a lower clock rate in a separate test chip. The same filter in a test chip worked up to 300 kHz by changing the clock frequency. This filter has an asymmetric frequency response because it is a direct low-pass-to-bandpass transform of a third-order low-pass filter. Therefore the sideband rejection at the higher frequency side is not as good as at the lower side. Although this asymmetry is not taken care of in this implementation, to achieve more rejection at the higher frequency side, a low-order low-pass filter can be added to this filter, or the bandpass filter itself can be implemented using a low-pass/high-pass combination. The measured performances of the two bandpass filters are summarized in Table I.

C. Voice-Band Low-Pass Filter (LPF1)

The low-pass filter LPF1 is a seventh-order elliptic ladder filter with a cutoff frequency of 3.75 kHz. It is designed to have a 0.1-dB passband ripple and an 80-dB rejection of frequency components higher than 6.25 kHz to suppress the harmonic components of the pulsewidth modulated signal from the FM demodulator. The input coupling capacitor is doubled to obtain the filter gain of unity. The filter is driven by a 194-kHz clock that is internally generated. The measured filter performance is summarized in Table II. It is designed in a single-ended form, and the frequency response of the filter is shown in Fig. 7(c).

D. Receiver Transfer Characteristics

The shape of transfer curve as predicted in Fig. 5 was observed in an actual experiment as shown in Fig. 8. The measured channel bandwidth of the filter, limited by the

TABLE I
MEASURED PERFORMANCES OF THE TWO BANDPASS FILTERS

Characteristics	IF1	IF2
Center Frequency	3 MHz	72 kHz
Selectivity Q	55	3
Passband Ripple	0.3 dB	0.3 dB
PSRR+	40 dB	35 dB*
PSRR−	35 dB	21 dB*
Dynamic Range**	46 dB	58 dB
Clock Frequency***	12.4 MHz	775 kHz
Passband Gain	27	1.5
Power	45 mW (5V Supply)	10 mW (5V Supply)
Area	2 mm²	0.5 mm²

* Measured at the center frequency.
** 1% intermodulation level.
*** Double sampling is used.

TABLE II
MEASURED PERFORMANCE OF THE LOW-PASS FILTER

Characteristics	LPF1
Cutoff Frequency	3.75 kHz
Passband Ripple	0.2 dB
Sideband Rejection	75 dB
PSRR+	36 dB*
PSRR−	25 dB*
Dynamic Range**	63 dB
Clock Frequency	194 kHz
Passband Gain	1
Power	1.4 mW (5V Supply)
Area	0.54 mm²

* Measured at the cutoff frequency.
** 0.1% harmonic distortion level.

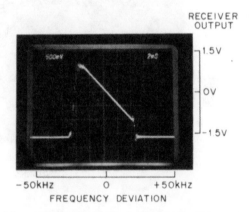

Fig. 8. Measured transfer characteristic of the FM receiver (receiver output versus input frequency deviation).

composite bandwidth of two IF filters, was approximately 28 kHz for a 10-mV RF input. For strong 100–500-mV RF signals, the channel bandwidth was increased to 35–45 kHz as shown in the example of Fig. 8. The maximum deviation of the FM signal in this experiment is 8 kHz with a 1-kHz modulating tone. The responses of the receiver to a sinusoidally modulated FM input are illustrated in Fig. 9(a) and (b). In Fig. 9(a), the modulating input frequency is swept down from approximately 10 kHz to almost dc in 10 ms as shown in the top trace. The output starts growing after the modulating frequency becomes lower than 3.75 kHz, which is the output low-pass filter bandwidth (LPF1). For the in-band FM signal, the sinusoidal modulating signal is reproduced at the receiver output as shown in Fig. 9(b). Other examples of triangle

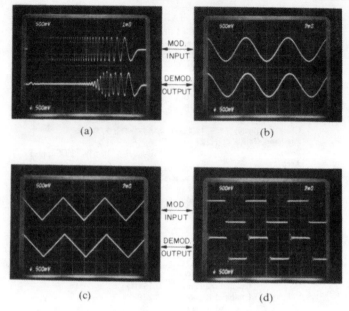

Fig. 9.   FM receiver responses: (a), (b) a sinusoidal modulation; (c) a triangle modulation; and (d) a pulse modulation.

and pulse modulations are shown in Fig. 9(c) and (d), respectively.

E. Distortion, AM Rejection, Intermodulation, and Capture Ratio

The output spectrums for measuring harmonic component, AM rejection, intermodulation, and capture ratio of the receiver are shown in Fig. 10. The harmonic components can be directly measured as shown in Fig. 10(a). The AM rejection is measured by applying a 50-percent AM-modulated 2.5-kHz FM signal to the receiver. Considering that the AM components are 12 dB down for 50-percent modulation, the measured AM rejection of the receiver is approximately 30 dB as shown in Fig. 10(b). An AM-modulated FM carrier is used as an AM-rejection test signal because the simultaneous AM/FM modulation provides the information about the AM suppression of the FM signal contaminated by the AM modulation. The intermodulation components of two in-band signals are rejected by more than 32 dB as shown in Fig. 10(c). The receiver can be captured by any signal approximately 4.4 dB larger than the others. The spectrum shown in Fig.

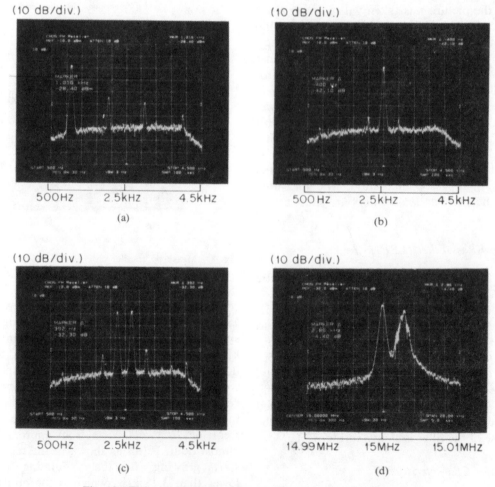

Fig. 10.   FM receiver measurements (10 dB/div., vertical): (a) distortion (400 Hz/div., horizontal); (b) AM rejection (400 Hz/div., horizontal); (c) intermodulation (400 Hz/div., horizontal); and (d) capture ratio (2 kHz/div., horizontal).

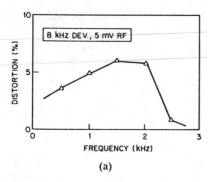

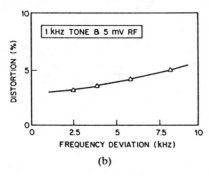

Fig. 11. Measured distortion characteristic of the FM receiver: (a) distortion versus frequency; and (b) distortion versus frequency deviation.

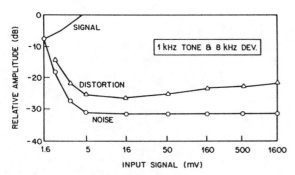

Fig. 12. Measured quieting and distortion characteristics of the FM receiver.

10(d) illustrates this capture condition. The clean signal can take over the FM receiver under the presence of interferences. This capture measurement was done at 15 MHz.

F. Receiver Quieting Characteristics

The distortion and noise characteristics of the receiver are illustrated in Figs. 11 and 12. Note that the input signal levels are the signal levels after the front-end RF amplifier of Fig. 2. The figures are self-explanatory, and measurement results are typical. It was observed that the harmonic distortion components of frequencies higher than 2 kHz are considerably attenuated by the output low-pass filter (LPF1) as shown in Fig. 11(a). The output noise is dominated by the phase jitter of the demodulator, and the receiver quieting level was approximately constant -30 dB for input signals ranging from 5 mV up to 1.6 V. However, in normal operations, the input signal should be limited to a 10-mV level by AGC because intermodulation is a

TABLE III
MEASURED PERFORMANCE OF THE FM RECEIVER

Characteristics	Measurement
Channel Bandwidth	28 kHz
Maximum Frequency Deviation	8 kHz
Input Sensitivity*	5 mV for 30-dB Quieting
Signal-to-Noise Ratio	31 dB
Harmonic Distortion at 1 kHz	5%
Intermodulation Distortion	2.5%
AM Rejection	30 dB
Capture Ratio	4.4 dB
Alternate Channel Selectivity	> 31 dB
Adjacent Channel Selectivity	> 31 dB
Audio Output	$1 V_{p-p}$
Power	80 mW (5V Supply)
Die Area	$7.7 mm^2$

* Without considering front-end RF amplifier gain.

potential problem under large input signal conditions. The measured performance of the receiver is summarized in Table III.

VI. CONCLUSIONS

In summary, an experimental CMOS double-heterodyne FM receiver exhibiting a 5-mV input sensitivity and a -30-dB quieting level has been demonstrated. The prototype receiver fully worked with a 3-MHz switched-capacitor first IF filter. In actual applications, an appropriate signal boost-up in an antenna matching network as well as in a front-end RF amplifier will help achieve a 50-μV level input sensitivity. The quieting level can easily be improved by an additional 6 dB if the number of counters in the FM demodulator is increased by one, and the distortion can be considerably reduced by using linear-phase filters for all IF filters. It should be admitted that there is enough room for further improvements in performance as well as for new system architectures suitable for MOS technology, and achieved performances are not yet comparable to those of commercial FM receivers at this time. However, the integration approach will surely be the most cost-effective way for miniature receiver applications as progress continues in the future.

ACKNOWLEDGMENT

The authors express their best thanks to M. F. Tompsett for supporting this work. The authors also appreciate valuable comments from N. Dwarakanath, D. Soo, J. Michejda, M. Banu, and two unknown reviewers about the manuscript.

REFERENCES

[1] "MC3362, low power dual conversion FM receiver," Motorola Semiconductors, Data Sheet.
[2] L. Blaser and T. Taira, "An AM/FM radio subsystem IC," *IEEE Trans. Consumer Electron.*, vol. CE-23, pp. 129–136, May 1977.
[3] W. Peil and R. J. McFadyen, "A single chip AM/FM integrated circuit radio," *IEEE Trans. Consumer Electron.*, vol. CE-23, pp. 424–429, Aug. 1977.
[4] O. L. Richard, "A complete AM/FM signal processing system," *IEEE Trans. Consumer Electron.*, vol. CE-24, pp. 34–38, Feb. 1978.
[5] W. Beckenbach, C. W. Malinowski, and H. Rinderle, "A new

design approach to digitally tuned radio receivers," *IEEE Trans. Consumer Electron.*, vol. CE-25, pp. 578–595, Aug. 1979.

[6] S. Miki, T. Nyuji, S. Ninomiya, M. Kaneko, Y. Daimaeu, and M. Fumoto, "Thin radio receiver by using PLL synthesizer digital tuning techniques," *IEEE Trans. Consumer Electron.*, vol. CE-25, pp. 597–603, Aug. 1979.

[7] T. Okanobu, T. Tsuchiya, K. Abe, and Y. Ueki, "A complete single chip AM/FM radio integrated circuit," *IEEE Trans. Consumer Electron.*, vol. CE-28, pp. 393–407, Aug. 1982.

[8] T. C. Choi, R. T. Kaneshiro, R. W. Brodersen, P. R. Gray, W. B. Jett, and M. Wilcox, "High-frequency CMOS switched-capacitor filters for communications application," *IEEE J. Solid-State Circuits*, vol. SC-18, pp. 652–664, Dec. 1983.

[9] C.-F. Chiou and R. Schaumann, "Design and performance of a fully integrated bipolar 10.7-MHz analog bandpass filter," *IEEE J. Solid-State Circuits*, vol. SC-21, pp. 6–14, Feb. 1986.

[10] B.-S. Song and P. R. Gray, "Switched-capacitor high-Q bandpass filters for IF applications," *IEEE J. Solid-State Circuits*, vol. SC-21, no. 6, pp. 924–933, Dec. 1986.

[11] L. E. Franks and I. W. Sandberg, "An alternative approach to the realization of network transfer functions, the N-path filter," *Bell Syst. Tech. J.*, pp. 1321–1350, Sept. 1960.

[12] A. Fettweis and H. Wupper, "A solution to the balancing problem in N-path filters," *IEEE Trans. Circuit Theory*, vol. CT-18, pp. 403–405, May 1971.

[13] M. B. Ghaderi, G. C. Temes, and J. A. Nossek, "Switched-capacitor pseudo N-path filter," in *Proc. IEEE Int. Symp. Circuits and Systems*, Apr. 1981, pp. 519–522.

[14] B.-S. Song, "A narrow-band CMOS FM receiver based on single-sideband modulation IF filtering," to be published in *IEEE J. Solid-State Circuits*.

[15] B.-S. Song, "CMOS RF circuits for data communications applications," *IEEE J. Solid-State Circuits*, vol. SC-21, pp. 310–317, Apr. 1986.

[16] S. Inoue and Y. Iso, "Super high quality FM detector and its development process," *IEEE Trans. Consumer Electron.*, vol. CE-24, pp. 226–234, Aug. 1978.

A Full Duplex 1200/300 Bit/s Single-Chip CMOS Modem

GARRY R. SHAPIRO, MEMBER, IEEE, AND ANDREW C. E. PINDAR

Abstract —The design of a Bell 212A (AT&T) compatible single-chip modem is described, and measured results presented. The IC (Fairchild μA212A) contains all signal-processing functions and supports all operating modes, including test modes and selection of either 1200-bit/s QPSK or 300-bit/s FSK operation. The modem offers such features as a coherent digital receiver, call-progress tone monitoring, 8–11-bit character lengths for asynchronous operation, and on-chip handshaking for the remote loopback test mode. Only connect and disconnect sequences are user-provided, typically by a (generic) microcontroller. Implemented in Linear-Compatible Dual-Poly CMOS (LCCMOS), the chip measures 57K mils^2 and is housed in a 28-pin package. The IC operates from $+/-5$ V_{dc}, and typically dissipates 35 mW.

I. INTRODUCTION

THE EXPLOSIVE GROWTH of data communications via the switched telephone network has created a demand for inexpensive integrated circuits which perform the modem function. In North America, this demand currently focuses upon 1200-bit/s full-duplex communications utilizing the Bell 212A protocol. Described here is a CMOS single-chip modem integrated circuit which contains all of the 212A signal-processing functions. The modem supports all standard 212A modes, as well as call-progress tone monitoring for auto-dialer support, 8–11-bit character asynchronous operation, and on-chip handshaking for the remote loopback test mode. Only connect and disconnect sequences are user-provided, typically by a (generic) microcontroller.

II. ARCHITECTURE

Fig. 1 is a block diagram of the modem IC. The major functional groups are the transmitter, including the high- and low-speed modulators, (bottom of figure), the receiver, including both demodulators (top of figure), and control and timing circuitry.

The modem employs frequency division multiplex (FDM) techniques to achieve full duplex operation. Two 900-Hz-wide channels are centered at 1200 and 2400 Hz, with the originating (calling) modem transmitting low. Most FDM-based modem designs utilize two complex bandpass filters, which are switched between transmitter and receiver according to whether the modem is originating or answering. In this IC, the use of separate programmable transmit

Manuscript received April 17, 1985; revised August 7, 1985.
The authors are with Fairchild Semiconductor Corporation, Advanced Signal Processing Division, Mountain View, CA 94039.

and receive filters allows optimization of each for the unique requirements of the transmitter and receiver. Except for the input anti-alias filter, all filters utilize switched-capacitor (SC) technique.

The modulation technique utilized for the 1200 bit/s mode is known as both Quadrature Phase Shift Keying (QPSK) and four-state Quadrature Amplitude Modulation (4-QAM). One may design the modulator and demodulator from either viewpoint, but certain advantages are associated with the 4-QAM approach. A coherent demodulator can provide up to 2.3-dB improvement in noise performance over a comparison or differential demodulator [1], [2]. In the classic QAM modulator of Fig. 2(a), baseband filters $F(\omega)$ provide both passband shaping and stopband attenuation, which relaxes the requirements on post-filter $G(\omega)$.

III. TRANSMITTER

The transmitter consists of a transmit buffer, scrambler and encoder, high-speed and low-speed modulators, and a transmit filter and line driver.

A. Transmit Buffer

In the high-speed asynchronous mode, start-stop characters enter the transmit buffer, which synchronizes the data to the internal 1200-bit/s clock. Bell-212A compatibility requires the buffer to handle intracharacter bit rates from 1170 to 1212 bit/s. Data which are underspeed relative to 1200 bit/s periodically have the last (stop) bit sampled twice resulting in an added stop bit. Similarly, overspeed input data periodically have unsampled—and therefore deleted—stop bits. A stop bit may be deleted no more than once in each nine consecutive characters.

The MOD1 and MOD2 pins choose 8-, 9-, 10-, or 11-bit character lengths, as in the CCITT V.22 specification; these consist of one start bit, one stop bit, and 6, 7, 8, or 9 data bits, including parity. The original Bell 212A offered only 9- and 10-bit character lengths, but many modem manufacturers currently offer the additional choices.

The buffer also services the break (all space) function. 212A compatibility requires a transmitted sequence at least $2M+3$ bits long, where M is the character length. If a space sequence longer than one character but less than $2M+3$ bits is presented at TXD (Transmit Data), the transmit buffer will extend its length to $2M+3$ bits. If the

Reprinted from *IEEE J. Solid-State Circuits*, vol. SC-20, no. 6, pp. 1169–1178, Dec. 1985.

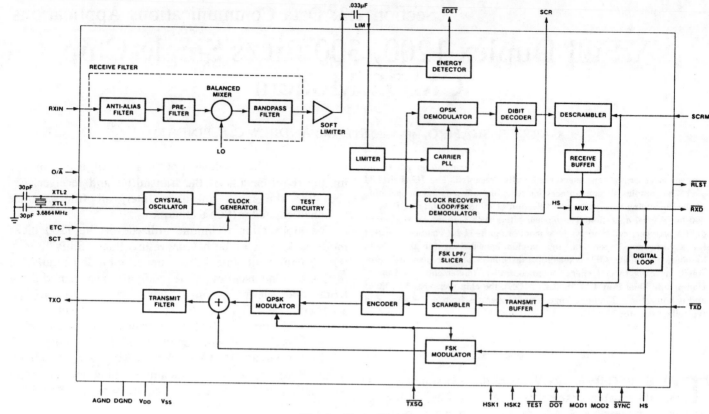

Fig. 1.   Block diagram.

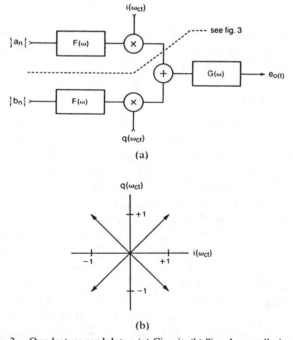

Fig. 2.   Quadrature modulator. (a) Circuit. (b) Signal constellation.

data source provides a character immediately following a break that is to short, that character will be lost.

B. Transmit Clocks

In synchronous mode the transmit buffer is bypassed. The transmitter clock may be internal, external or derived from the recovered received data (slave mode). When inter-

nal clock is chosen, the crystal-derived 1200-Hz clock is provided at the Serial Clock Transmit (SCT) pin to clock data from the synchronous source into the TXD pin. When external clock is chosen, the synchronous source provides a 1200-Hz clock to the External Transmit Clock (ETC) pin while providing synchronized data to the TXD pin. A digital phase-locked loop synchronizes the internal transmitter clock chain to the external clock. Similarly, in slave mode the transmitter clock chain is locked to the recovered receive clock, which is also presented at Serial Clock Receive (SCR).

C. Scrambler

A scrambler precedes encoding to ensure that the line spectrum is sufficiently distributed so as to not interfere with the in-band supervisory single-frequency signaling system employed in most Bell system toll trunks. The randomized spectrum also facilitates timing recovery in the receiver. An unscrambled, repeated 01 dibit encodes as a 0° phase shift, or unmodulated carrier. The absence of sidebands capable of generating a 600-Hz product may cause a loss of phase lock in the receiver clock recovery circuit.

The 17-bit scrambler is characterized by the following recursive equations:

$$Y_i = X_i + Y_{i-14} + Y_{i-17}$$

where X_i is the scrambler input bit at time i, Y_i is the scrambler output bit at time i, and + denotes the EXCLU-

SIVE-OR operation. Since the initial state of the register is not known, the first 17 bits serve to initialize the scrambler. A "lockout" counter is included to prevent scrambler lockup: if Y_{i-1} for 64 bits, the following bit is inverted.

D. Encoder

The 212-type modems achieve full-duplex 1200 bit/s operation by encoding transmitted data by bit pairs (dibits), thereby halving the apparent line data rate to 600 Bd (symbols/s). The resultant reduced spectral width allows both frequency channels to coexist in a limited bandwidth telephone channel with achievable levels of filtering. The four unique dibits thus obtained are gray coded and differentially phase modulated onto a carrier at either 1200 Hz (ORIGINATE mode) or 2400 Hz (ANSWER mode) as phase steps which are multiples of 90°. Gray coding means that adjacent dibits in the sequence change by only one bit [01, 00, 10, 11], compared to natural binary [00, 01, 10, 11]; a small but demonstrable improvement in bit-error-rate performance results [2]. Each dibit is encoded as a phase change relative to the phase of the preceding signal dibit element. This differential encoding is necessary because there exists no network-level timing reference with which to measure phase shift. Hence, by storing the current dibit for comparison to the next, a reference is provided.

Dibit encoding is as follows:

Dibit	Phase shift [deg]
00	+90
01	0
11	−90
10	180

The left-hand digit of the dibit is the one occurring first in the data stream as it enters the encoder after the scrambler.

E. 1200-bit/s modulator

The high-speed modulator design utilizes a novel 4-QAM approach, combining switched-capacitor and digital techniques [3]–[5].

In the classic QAM modulator of Fig. 2(a) the input sequences $\{a_n\}$ and $\{b_n\}$ are binary synchronous NRZ random sequences from the dibit encoder, while $i(\omega_c t)$ and $q(\omega_c t)$ are periodic quadrature functions at the desired carrier frequency. Constant (dc) values of the binary inputs provide the four steady-state phasors illustrated in Fig. 2(b), which are separated by 90° increments and are of nominally equal amplitude.

Filters $F(\omega)$ are identical baseband shaping filters, which provide passband shaping and attenuation of transmitter energy from the adjacent receive channel (750–1650 Hz). The post-modulator transmit filter provides only line equalization and suppression of the odd-harmonic carrier spectra resulting from square-wave carrier signals.

In the modulator of Fig. 2(a), practical performance limitations stem from the symmetry required in the input

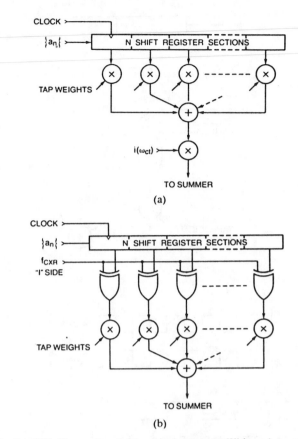

Fig. 3. FIR filter and modulator (single arm). (a) With analog post-multiplier. (b) Digital equivalent.

arms, and from dc offset in the modulating sequences. These result in unsuppressed carrier and sideband components. In this modem, a single finite impulse response (FIR) filter is realized with SC tap weights and multiplexed between the two input symbol sequences. The FIR filter offers the additional advantage of linear phase response and therefore adds no delay distortion.

Fig. 3(a) depicts one arm of the modulator of Fig. 2(a) realized with a shift-register-based FIR filter followed by an ideal multiplier; the outputs of the N-stage shift register are again binary NRZ with values $+/-1$. If $i(\omega_c t)$ is a square wave of equal constraint, the structure is equivalent to that of Fig. 3(b), where the analog post-multiplier is replaced by EXCLUSIVE-OR gates at each shift register output. The coefficient multipliers and summer are constructed using a summing SC integrator.

Quadrature modulation is achieved by multiplexing the shift registers of each input arm at twice the carrier rate, ahead of the XOR modulators, as shown in Fig. 4(a). Fig. 4(b) illustrates the resultant equivalent carrier component multiplying each set of shift register outputs. These are equivalent to sums of square waves 45° out of phase.

In the preceding discussion, every register tap will display a value of $+1$ or -1, resulting in the constant exercise of every tap multiplier. Substantial simplification results from exploitation of the synchronicity of $\{a_n\}$ and $\{b_n\}$: the sequences need only be sampled once per symbol interval and may be zero at all other times. Further, the

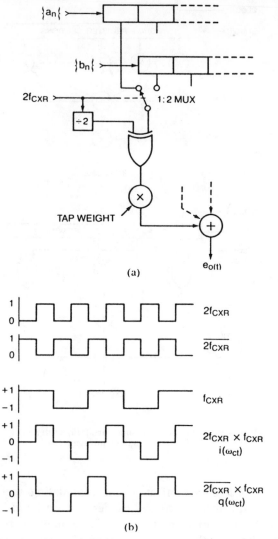

(a)

(b)

Fig. 4. Quadrature modulation. (a) Technique. (b) Equivalent carriers.

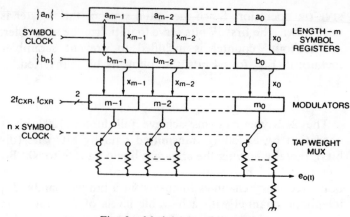

Fig. 5. Modulator architecture.

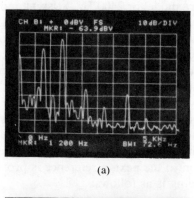

(a)

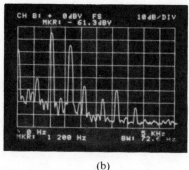

(b)

Fig. 6. Unscrambled 1200 bit/s output spectra. (a) +90° phase shifts at 1200 Hz. (b) 270° phase shifts at 1200 Hz.

resultant flattened input spectrum may be readily shaped by the filter to provide the desired passband and stopband. Only one register tap per symbol interval will then be active and nonzero. That is, for $N = mn + 1$, where

N = number of taps,
m = number of symbol intervals in the truncated (finite) impulse response,
n = number of taps/symbol

only m taps need be examined at any instant. For this modem IC, $N = 65$, $m = 4$, and $n = 16$.

Fig. 5 illustrates the simplified architecture, with resistor weighting shown for clarity. The 65-bit shift registers clocked at the sample rate have been replaced by 4-bit registers clocked at the symbol rate (600 Hz). The four outputs are modulated and passed to ganged commutating MUXes which select the 16 taps/symbol at the sample rate (9600 Hz).

Because of the symmetry of the discrete impulse response and finite capacitor ratios, only a fraction of the 65 taps are nonzero. The final design employs one operational amplifier and 15 switched and fixed capacitors.

The design provides a $\sim (\sin x)/x$-passband, a stopband loss ~ 50 dB in the adjacent channel, and excellent suppression of the carrier (> 55 dB) and unwanted sidebands (> 40 dB) in unscrambled patterns. Fig. 6 shows the transmitter output at 1200 Hz for two unscrambled repeated patterns: 90° phase shifts (00 dibits) in Fig. 6(a), and 270° phase shifts (11 dibits) in Fig. 6(b).

F. FSK Modulator

The low-speed modulator generates phase-coherent FSK using a programmable counter clocked at 1.2288 MHz. The counter modulus is selected by the ORIGINATE/ANSWER (O/Ā) and transmit data (TXD) pins. The counter completes a count cycle before responding to a change in TXD; this distortion of the bit length is reduced by a "divide-by-8" counter. Modulator output frequencies (in Hz) are as shown in Table I. Note that ANSWER mode mark

TABLE I
FSK MODULATOR FREQUENCIES

Mode	OAB	TXD	F_{nom}	F_{actual}
ORIG	1	0	1070	1066.67
ORIG	1	1	1270	1269.42
ANS	0	0	2025	2021.05
ANS	0	1	2225	2226.09

(TXD = 1) is also utilized as ANSWER TONE (2225 Hz) in both low- and high-speed operation, and can be forced with either the 1200 bit/s or 300 bit/s receiver engaged.

G. Transmit Filter

The QPSK and FSK modulator outputs are summed at the transmit filter input.

Since the QPSK modulator provides passband shaping and stopband rejection, the transmit filter provides only harmonic suppression and channel equalization.

Due to the equivalent square-wave nature of the carrier in both the FSK and QPSK modulators, the transmitter output spectra are replicated at the odd harmonics of the carrier, the harmonic spectra decreasing in amplitude as with a square wave. The transmit filter reduces the out-of-band components of this harmonic energy to levels below those allowed by AT&T on its lines. The filter also attenuates in-band harmonics which would otherwise interfere with modem operation, e.g., the third harmonic of the 1200-Hz QPSK signal, which would place considerable energy within the high band.

The transmit filter passband response provides the transmitter's half of the modem's fixed compromise line equalization and adjusts gain to provide the nominal output level for each band and mode. The filter consists of two fixed second-order SC delay equalizer sections, and a 3-pole, 2-zero SC lowpass filter, configured by the O/$\overline{A}$ pin. Since both modulators are synchronized to the 153.6-kHz filter clock, no anti-alias filter is required. Fig. 7 shows the transmit filter gain responses for the ORIGINATE and ANSWER modes.

Following the filter is a unity-gain buffer amplifier possessing low Z_{out}. The buffer provides ~ 0.7-V rms at the TXO pin; this will normally be more than enough to provide −9 dBm to the line, considering insertion loss in the line-connect circuit, line (load) impedance, and the source impedance provided between TXO and the load.

IV. RECEIVER

The receiver is comprised of the receive filter, two-stage limiter and energy detector, high-speed demodulator, descrambler and decoder, receive buffer (sync-to-async converter), and the low-speed receiver.

A. Receive Filter

There are three major functions for the receive filter:

1) to equalize gain and delay distortion introduced by the telephone channel;

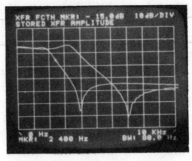

Fig. 7. Receive filter group delay.

2) to provide signal shaping to convert a ~ $(\sin x)/x$ input spectrum to a raised-cosine spectrum to reduce intersymbol interference;

3) to reject both the echo presented to the adjacent frequency band and the noise outside the passband, thus improving the signal-to-noise ratio of the received signal.

To achieve full-duplex communication, two channels are defined within the 4-kHz baseband telephone channel. When the modem is in ANSWER mode, it receives on the low band (750–1650 Hz) and transmits on the high band (1950–2850 Hz). Conversely, when the modem is in ORIGINATE mode, it receives high and transmits low. In most modem designs, two filters are used, with one employed as a transmit filter, selected by the chosen mode. In this IC, a heterodyne approach is utilized: a prefilter and mixer are placed in front of a 1200-Hz bandpass (BP) filter. In ANSWER mode, the mixer operates "straight through," i.e., with a local oscillator (LO) frequency of 0 Hz, thus passing the entire received signal unaltered to the following 1200-Hz bandpass filter. When the modem is in ORIGINATE mode, the mixer LO frequency is 3600 Hz. The high-band received signal is downconverted to the low band; the desired spectrum—now centered on 1200 Hz (in high-speed mode)—is again passed to the following 1200-Hz BP filter. In LOW-SPEED ORIGINATE mode, the center of the received spectrum is at 2125 Hz rather than 2400 Hz; a 3200-Hz LO is employed to more closely center the received signal in the 1200-Hz passband.

The heterodyne architecture, together with the 1200-bit/s modulator design, allows simplification of the transmit filter (Section III-G), elimination of the higher-Q 2400-Hz bandpass filter, and optimization of the receive filter transfer function for the demodulator. Thirty poles of SC filtering (7 transmit, 23 receive) are employed, compared with at least 40 poles for the two-filter approach.

The receive filter consists of five sections:

1) anti-alias filter;
2) prefilter;
3) mixer;
4) delay equalizer;
5) bandpass filter.

1) Anti-alias filter: The received signal from the RXIN pin passes first through a Sallen and Key second-order low-pass section. This active-*RC* section is maximally flat in the baseband with a −3-dB frequency of 28 kHz and unity gain. It provides ~ 45-dB attenuation at 460.8 kHz, which is the effective sampling frequency of the following switched-capacitor section.

2) Prefilter: The prefilter is a 3-pole, 2-zero low-pass filter which drives the mixer directly in the low-band (ANSWER mode) and through a second-order equalizer section in the high-band (ORIGINATE mode). The prefilter provides:

a) Bandlimiting of the input signal: The gain function rolls off above ~ 4 kHz, with the transmission zero at 7 kHz. This reduces the image response of the mixer to noise and spurious energy above the LO frequency.

b) Gain: The first section (real pole) of the prefilter provides a net gain of 3.5 (~11 dB). The remainder of the receive filter operates at unity gain. For a (conservative) maximum filter output level of 3.5-V peak (for which the input signal is 1.5-V peak) and a typical input threshold of 6.5-mV rms, one obtains a minimum dynamic range of 35 dB.

c) Compromise equalization: The prefilter gain function provides compromise gain equalization below 3 kHz. For a representative middle-distance circuit, relatively flat response in the low channel and a rising characteristic in the high passband are required. The primary group delay equalization is performed in the 1200-Hz delay equalizer and is explained in that section. The prefilter essentially optimizes that characteristic for each passband by selecting or deselecting the equalizer section.

3) Mixer: There are three operational modes for the mixer:

a) ORIGINATE mode: The mixer operates with a local oscillator (LO) frequency of 3600-Hz (high-speed mode) or 3200 Hz (low-speed mode), down-converting the received high band signal to the 1200-Hz band for processing by the 1200-Hz bandpass filter. The primary SC clock for the receive filter is 57.6 kHz, which is an even multiple of both frequencies.

b) ANSWER mode: The LO frequency is 0 Hz. The mixer operates straight through as a simple low-pass filter section with a very high cutoff frequency.

c) DIALER mode: The receive filter clocks are scaled by half to receive the call-progress tone band (350–620 Hz). The energy detector is reconfigured to provide call-progress tone detection to the microcontroller. On and off response times are made relatively short so that the EDET pin follows the ON/OFF profile of the respective call-progress tone pairs; no frequency discrimination is performed.

In ORIGINATE mode, the most critical parameter for the mixer is input feedthrough suppression. Talker echo (reflected transmitter energy) at 1200 Hz can exceed the desired 2400-Hz signal by > 30 dB, and, if fed through, can interfere with the downconverted 2400-Hz received signal. The SC mixer must suppress this component by at least 55 dB.

In ANALOG LOOPBACK mode, the receiver ORIGINATE and ANSWER mode assignments are inverted, which forces the

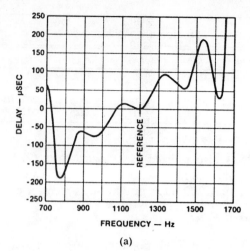

(a)

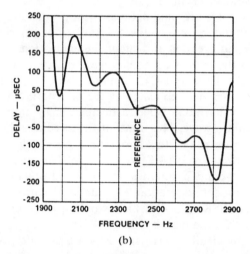

(b)

Fig. 8. Transmit filter gain functions. (a) 1200-Hz channel. (b) 2400-Hz channel.

receiver to operate in the transmitter frequency band.

4) Delay Equalizer: The delay equalizer precedes and functionally is part of the 1200-Hz bandpass filter. Its purpose is to convert the group delay response of the bandpass filter to that which will equalize (i.e., flatten) the delay profile of a chosen typical telephone channel. Since the channel delay varies considerably within often broad limits, such fixed equalization is a compromise, and is so called.

The chosen group delay characteristic to be equalized represents a medium-distance path as reported by AT&T [6]; it is essentially parabolic with its vertex (minimum) between the high and low channels. Thus the compensating characteristic increases with frequency in the low band and decreases in the high band. Since the down-conversion of the high band inverts the passband, in a left-right sense, the correct sense for both channels is provided by one filter/equalizer combination. The difference between the required responses for each band is adjusted by the prefilter. Four cascaded second-order equalizers are employed.

Fig. 8(a) is the receive filter group delay response for the low band, while Fig. 8(b) shows the high-band response. Note that the transmitter and receiver each provide half of the total required channel equalization.

5) 1200-Hz Bandpass Filter: The 1200-Hz bandpass filter passes the desired received signal while attenuating the adjacent transmitted signal component reflected from the line (talker echo). The chosen passband shape and accompanying delay equalization convert the $\sim \sin x/x$ spectral shape of the received high-speed signal to a 100-percent raised cosine (cosine-squared) spectrum (exclusive of line equalization). Such Nyquist filtering minimizes intersymbol interference in the recovered data. The passband over which this approximation holds is, considering both gain and delay, from 750 to 1650 Hz.

The filter stopband rejects transmitter energy in the adjacent channel which is reflected from the imperfect line interface (talker echo). The required stopband attenuation is that which reduces adjacent channel energy sufficiently to have an insignificant effect upon performance at minimum receiver input levels. The receive filter provides > 55-dB rejection from 1950 to 2850 Hz.

The bandpass filter transfer function contains five pole pairs and four finite transmission zeros, three of which form the asymmetric stopband. The fifth section is bandpass. The transfer function was realized with a leapfrog structure to minimize passband sensitivity.

Fig. 9 shows the receive filter gain function for the low band.

B. Limiter Blocks

The limiter functional blocks are the soft limiter, limiter, and energy detector. The major functions of the limiter blocks are:

1) to perform accurate zero-crossing detection on the receive filter output;
2) to provide a logic-level signal to the digital demodulator;
3) to provide accurate and repeatable threshold measurements of energy levels in the receive passband.

Although the analog output of the receive filter contains both amplitude and phase information, the carrier phase information is sufficient for recovery of both carrier and dibit clock, and decoding of the modulation. Hence, if the output of the receive filter is $e(t)$, it is sufficient to extract the zero crossings of $e(t)$, that is, to obtain $\text{sgn}[e(t)]$. A limiter performs this zero-crossing detection.

The modem utilizes a two-stage limiter to achieve at least 40 dB of usable limiter dynamic range. The first or soft limiter is a linear dc amplifier with a gain of 35 dB which drives both the energy detector and a hard limiter. For a threshold signal at RXIN (6.5-mV rms), the soft-limiter output level is 1.3-V rms, well within its linear region, and a convenient level for the energy detector. Signals above threshold drive this amplifier into clipping. However, its high slew rate and short delay enable quick recovery when returning to the linear region, thus preserving the accuracy

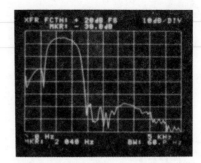

Fig. 9.   1200-Hz receive filter gain function.

of the zero crossings. A 0.033-uF capacitor from SLIM to LIM (the hard limiter input) removes the offset component.

The hard limiter is a single comparator cell. Since the limiter input-referred offset is <1 percent of the minimum signal of interest, this offset has little effect on the zero-crossings.

The energy detector provides a digital indication at EDET that energy is present within the filter passband at a level above a preset threshold. Nominal input-referred threshold levels are 6.5-mV rms OFF/ON and 4.6-mV rms ON/OFF. Note that ~ 3-dB of hysteresis is provided between ON and OFF levels to prevent ratcheting of the detector output. Assuming ~ 2-dB insertion loss for the data access arrangement (DAA) and ~ 6-dB gain in a nominal line hybrid, the nominal thresholds are -45.5 dBm OFF/ON and -48.5-dBm ON/OFF, referred to the line input. Variations in line coupling circuitry and in line impedance will affect these figures.

The energy detector consists of a peak detector with hysteresis driving timing counters. In data mode, EDET goes low 105–205 ms after receipt of a passband signal, and high after 10–24 ms of signal loss. Note that energy detection is only part of the carrier detect function: carrier detect represents, in addition to energy detection, the timing provided for clock and carrier acquisition, and line transient damping. The OFF/ON timing provided on-chip represents carrier detect for the low-speed mode, which is shorter than the carrier detect interval required for high-speed carrier detect. In high-speed mode, the remainder of the nominal required 270 ms is provided by the microcontroller. Received data (RXD) are held high (clamped to MARK) by EDET when EDET is high.

In DIALER mode, EDET provides a logic-level indication of the presence of call-progress tone pairs (i.e., dial tone, busy, ringback). The timing intervals (33-ms OFF/ON, 40-ms ON/OFF) are long enough to be nonambiguous but short enough to allow accurate detection of "trunk busy" (120 intervals/min).

C. 1200 bit/s Demodulator

The 1200-bit/s coherent demodulator implementation is digital, and includes dibit clock recovery circuitry, a carrier recovery PLL, the QAM demodulator, and a decoder.

Clock Recovery. The clock recovery circuitry consists of

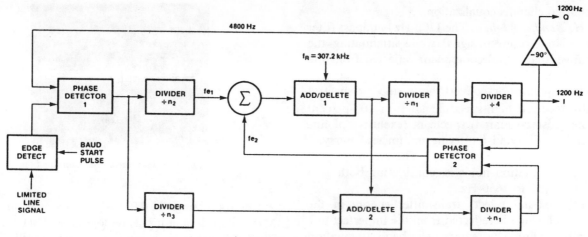

Fig. 10.   Carrier recovery loop.

a delay line detector [7], followed by a first-order digital phase locked loop (DPLL). In the delay line detector, dibit transitions, i.e., carrier phase discontinuities, are detected by correlating the tap outputs of a shift register whose taps are spaced 180° relative to the 600-Hz symbol rate. Two pairs of taps separated by 33.75° provide immunity to erroneous transitions due to noise. A train of baud transition pulses generated by the detector is the coarse recovered clock and is the input to the following PLL. No pulses appear for 0° phase transitions. Since this occurs, on average, one-fourth of the time for properly scrambled data, a factor of 3/4 must be included when computing the loop gain.

The "cleanup" PLL which follows is a first-order digital loop, which locks on the average value of the input pulse train. The loop outputs are both the 600-Hz dibit clock required by the decoder, and the 1200-Hz recovered receive clock (SCR) which accompanies received data (RXD) in synchronous mode. The loop bandwidth (0.56 Hz) easily tracks the $+/-0.06$-Hz allowed clock frequency offset, and provides adequate noise immunity at the expense of a long pull-in time. Pull-in is partially compensated by additional circuitry to ensure that the loop starts up almost perfectly locked.

Carrier Recovery. The carrier PLL is a second-order design operating at 307.2 kHz [8] and is illustrated in Fig. 10. The phase detector inputs are the filtered, limited line signal and a corrected 4800-Hz clock. The four possible transitions of the 1200-Hz carrier are all multiples of 360° at 4800 Hz. The 4800-Hz clock is therefore essentially free of phase transitions, as is loop operation. Comparison is performed once in each dibit period. The secondary phase detector inputs are the phase-locked 1200-Hz recovered carrier, and a corrected 1200-Hz clock driven by the secondary ADD/DELETE circuit from the same error reference. When the loop is locked, the two clocks will be in quadrature. Corrections from the two phase detectors feed the primary ADD/DELETE circuit, from which are generated the in-phase and quadrature carrier clocks required for demodulation. The loop response is critically damped with a bandwidth of 4.7 Hz and a capture range of 9.4 Hz,

sufficient for the carrier offsets encountered in the domestic switched telephone network. Maximum pull-in time is approximately 75 ms.

Demodulator/decoder. Demodulation of the limited line signal is achieved with EXCLUSIVE-OR gates supplied with the in-phase and quadrature carrier clocks from the carrier recovery PLL. The demodulated 600 Bd dibit streams are then decoded to produce a serial bit stream at 1200 bit/s, and descrambled. In synchronous mode, the descrambled 1200-bit/s data are presented at the RXD (Received Data) pin.

D. Receive Buffer

The receive buffer complements the previously described transmit buffer. It operates when the modem is in high-speed asynchronous mode and is bypassed in synchronous mode. Its primary function is to detect and replace stop bits which were deleted at the far-end transmitter. The buffer clocks synchronous data from the demodulator to the RXD pin in character asynchronous format at an intracharacter rate of 1219.05-bit/s (64/63*1200 bit/s). As in the transmit buffer, the receive buffer handles 8–11-bit characters. Reception of a "break" (at least two characters of "all zeros") is detected in order to disable stop bit replacement.

E. FSK Receiver

Assertion of the 0–300-bit/s FSK mode reconfigures the delay line detector, used for coarse clock recovery in the 1200-bit/s QPSK mode, to a simple FM comparison demodulator [1] (see Fig. 11). For each mode, the shift register delay is selected to be 270° at the nominal carrier frequencies. Resolution error due to the finite clock frequency employed in the delay register is traded off against register length. The XOR output contains baseband and double-frequency carrier components and yields a linear discriminator characteristic with a zero at the carrier frequency. The carrier components are attenuated by a

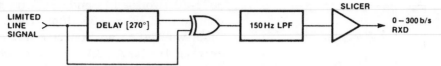

Fig. 11. 300-bit/s FSK receiver.

Fig. 12. Chip photomicrograph.

3-pole, 2-zero SC lowpass filter which provides a linear phase passband and a 150-Hz nominal cutoff. The resultant recovered baseband signal drives a comparator which serves as a zero-crossing detector and provides reconstituted data to the RXD pin.

V. Control and Timing

On-board clocks and timing signals are derived from either an on-board 3.6864-MHz crystal oscillator or from an external logic-level source. As previously discussed, the 1200-bit/s transmitter can be locked to an external clock or slaved to the receiver when in synchronous mode. Connect and disconnect sequences are typically provided by an external microcontroller. A notable exception is the elaborate handshake for the remote digital loopback (RDL) test mode, which has been included on-chip to simplify user programming requirements. This mode is activated by assertion of one control state and monitoring of one status pin.

VI. Fabrication and Performance

The modem is implemented in 3.75-μm Linear-Compatible Dual-Poly CMOS (LCCMOS), measures 57K mil^2 and is housed in a 28-pin package. It typically requires 5 V at 4.5 mA and -5 V at 2.5 mA, for a dissipation of 35 mW.

Fig. 12 is a photomicrograph of the chip. Due to extensive use of digital techniques, less than 35 percent of the chip area is analog. The design was based upon a standard cell system: the analog circuitry was manually placed and interconnected, while the digital sections were automatically placed and routed.

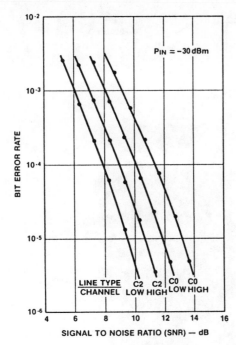

Fig. 13.   Bit-error rate versus signal-to-noise ratio.

3002-C∅ (worst-case) and −C2 (average) lines for both channels, at $P_{in} = -30$ dBm. These measurements were "four-wire" (i.e., no "talker echo" or reflected transmit energy). The AEA S3A channel simulator provided 5-kHz flat noise, measured in the 3-kHz bandwidth normally used for such tests.

ACKNOWLEDGMENT

The authors wish to acknowledge the contributions of H. El-Sissi, J. R. Cressey, P. T. Kwok, C. S. Meyer, C. W. Papalias, L. van Allen, and J. R. Woodbury.

REFERENCES

[1] R. W. Lucky, J. Salz, and E. J. Weldon, Jr., *Principles of Data Communication.* New York: McGraw-Hill, 1968, ch. 9.
[2] W. R. Bennett and J. R. Davey, *Data Transmission.* New York: McGraw-Hill, 1965, ch. 10.
[3] G. R. Shapiro, A. C. E. Pindar, and P. T. Kwok, "A full duplex 1200/300 b/s single-chip CMOS modem," in *Proc. ISSCC*, 1985, pp. 288–289.
[4] G. R. Shapiro "FIR switched-capacitor modulators for discrete QAM," in *Proc. 1985 ISCAS* (Kyoto, Japan), pp. 1629–1632.
[5] G. R. Shapiro and C. S. Meyer, patent pending.
[6] F. P. Duffy and T. W. Thatcher, Jr., "Analog transmission performance on the switched telecommunications network," BSTJ 50/4, Apr. 1971, pp. 1311–1347.
[7] V. K. Bhargava, D. Haccoun, R. Matyas, and P. Nuspi, *Digital Communications by Satellite.* New York: Wiley, 1981, pp. 151–153.
[8] Case-Rixon, Inc., patent pending.

Transmitter and filter performance are illustrated by Figs. 6, 7, and 9. Measured bit-error-rate (BER) performance of the 1200 bit/s QPSK receiver versus signal-to-noise ratio (SNR) is shown in Fig. 13 for unconditioned

Line and Receiver Interface Circuit for High-Speed Voice-Band Modems

JONATHAN H. FISCHER, JEFF L. SONNTAG, JAMES S. LAVRANCHUK, MEMBER, IEEE,
DONALD P. CIOLINI, A. GANESAN, MEMBER, IEEE, DOUGLAS G. MARSH, WILLIAM E.
KEASLER, MEMBER, IEEE, JOE PLANY, AND LAWRENCE H. YOUNG

Abstract —This paper describes a single-chip implementation of the analog signal processing functions required for full-duplex voice-band modem operation over twisted pair wires. Echo path signal-to-total harmonic distortion plus noise exceeds 65 dB. The device is implemented in a 3.5-μm twin-tub CMOS process, and typically dissipates 180 mW.

I. INTRODUCTION

TODAY, the switched voice telephone network carries a significant amount of data traffic. Full-duplex data rates above 2400 bit/s over unconditioned two-wire twisted pair subscriber loops require some form of echo cancellation. An earlier paper described an analog echo cancellation technique [1]. Fig. 1 is a simplified block diagram of a full-duplex high-speed modem (CCITT standards V.26 and V.32) that uses digital echo-cancellation techniques. This paper discusses a line and receiver interface circuit (LARIC) which performs the front-end analog signal processing functions enclosed in the dashed box in Fig. 1. For V.32 9600-bit/s full-duplex transmission, 65 dB of echo suppression is required for proper operation under worst-case hybrid feed and weak receive signal conditions. This requires a local transmit-to-receive path signal-to-total harmonic distortion plus noise ($S/(D+N)$) ratio of at least 65 dB for large input signals.

This paper describes the LARIC beginning with the transmit path, followed by the receive path, baud clock recovery path (top of Fig. 1), the A/D and D/A converters, amplifiers, and layout notes.

II. TRANSMIT PATH

The transmit path is shown in Fig. 2. The digital data to be transmitted are decoded by the D/A converter and shaped by the gain/slope equalizers (EQUZ1 and EQUZ2) to compensate for local loop passband gain slope impairments. With single-ended filters, noise energy at high frequencies (from switching power supplies and digital circuit noise coupling into analog supplies) is sampled by the switched-capacitor filters and aliased into the passband. Balanced differential filters are used to take advantage of their power supply rejection characteristics, and the balanced structure also helps to minimize nonlinear signal distortion [2]. The low-pass notch (LPN) and RC low-pass filter (RC LPF) band-limit the output to conform with FCC Part 68 specifications on energy above 4 kHz on the public switched network.

The gain/slope equalizer has four compromise shapes (set by the user) to compensate for roll-off in the 1–3 kHz band of the data signal between the modem and the central office. The equalizer shapes are designed so that the total transmitted power is independent of equalizer setting.

The equalizer is implemented in two biquad blocks [3]. EQUZ1 (Fig. 3) is used for 0- and 14-dB shapes; EQUZ2 is similar and is used for 4- and 7-dB shapes. The equalizer is separated into two circuits to reduce the number of programming switches required, which saves more area than would be used by the additional amplifiers. As shown in Fig. 3, 12 programming switches are needed for two equalizer settings.

The LPN filter is also a biquad section, and attenuates the transmitted energy above 4 kHz.

The final stage of the transmit filter is a second-order RC LPF design (Fig. 4) which attenuates signals above 20 kHz. The nominal filter corner frequency of 20 kHz allows a 4:1 RC variation over process and temperature without causing excessive passband droop or insufficient rejection at the LPF sample rate (153.6 kHz). To balance supply feeds, C_1 has been split into two parallel capacitors (C_{1A} and C_{1B}).

The balanced RC LPF signal is converted to a single-ended output by a differential amplifier (Fig. 2). The signal then passes through a muting switch and a voltage-follower output buffer.

The transmit-path amplitude response is shown in Fig. 5 for the four equalizer settings. The blip at 4800 Hz occurs

Manuscript received April 22, 1987; revised July 21, 1987.
J. H. Fischer and D. G. Marsh are with AT&T Bell Laboratories, Holmdel, NJ 07733.
J. L. Sonntag is with AT&T Bell Laboratories, Reading, PA 19603.
J. S. Lavranchuk and L. H. Young are with AT&T Bell Laboratories, Morristown, NJ 07960.
D. P. Ciolini and W. E. Keasler are with AT&T Communications/Information Systems, Middletown, NJ 07748.
J. Plany is with AT&T Bell Laboratories, Murray Hill, NJ 07974.
A. Ganesan was with AT&T Information Systems, Middletown, NJ 07748. He is now with Analog Devices Inc., Wilmington, MA 01887.
IEEE Log Number 8716802.

Reprinted from *IEEE J. Solid-State Circuits*, vol. SC-22, no. 6, pp. 982–989, Dec. 1987.

443

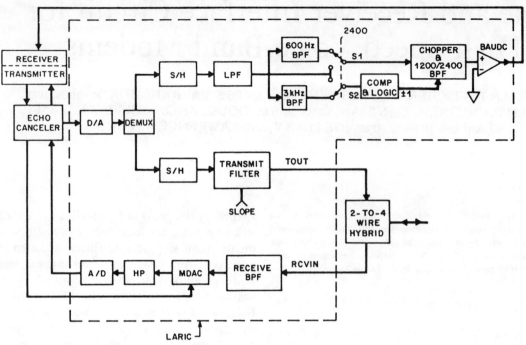

Fig. 1. Block diagram of a digital echo canceler with the analog front end (including timing recovery) for a 9600-bit/s modem enclosed in the dashed box.

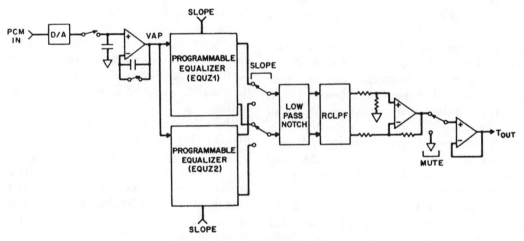

Fig. 2. Transmit-path block diagram.

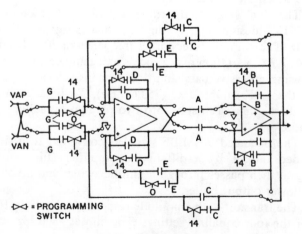

Fig. 3. EQUZ1 circuit.

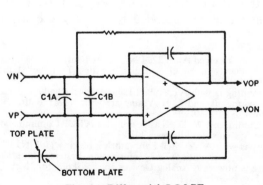

Fig. 4. Differential *RC* LPF.

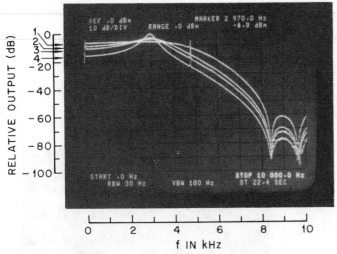

Fig. 5. Transmit-path frequency response: (1) 0-dB, (2) 4-dB, (3) 7-dB, and (4) 14-dB equalizer settings. The glitch at 4800 Hz is from the D/A, sampling at a 9600-Hz rate, interacting with the spectrum analyzer.

at the half sample rate of the D/A converter. It is an artifact of the spectrum analyzer interacting with the D/A aliases.

III. RECEIVE PATH

The receive signal path band-limits the signal from the hybrid, performs a coarse AGC function, and digitizes the analog signal for the digital echo canceler. A receive path block diagram is shown in Fig. 6. As with the transmit path, it is critical that the receive path $S/(N + D)$ ratio exceeds 65 dB for large signals. Again, a fully balanced differential topology is used for the receive filter to maximize PSRR and to improve filter linearity.

The high-frequency PSRR of the single-to-differential converter circuit is not critical because any output noise is band-limited by the 20-kHz antialiasing RC LPF. The fifth-order LPF removes signal energy above the voice band prior to the 9.6-kHz sampling of the 60-Hz second-order high-pass notch (HPN) and A/D converter.

The multiplying D/A converter (MDAC) (Fig. 6) implements an AGC function with 18-dB range and 1.5-dB gain steps. This ensures that the signal level is always near the A/D full-scale range to maximize $S/(D + N)$ performance.

The first-order high-pass section (see Fig. 6) follows the MDAC to block the MDAC output dc offset from the A/D converter. It also serves as sample and hold for the A/D. A differential amplifier converts the filter output into single-ended form and drives the A/D.

The frequency response of the receive path is shown in Fig. 7.

IV. CLOCK RECOVERY PATH

The clock recovery path (Fig. 1) is used to extract the 1200- or 2400-Hz symbol (baud) clock from the data signal

after the echo has been removed from the received signal.

The PCM signal from the echo canceler is reconstructed by the D/A and low-pass filtered by the second-order LPF with a 3-kHz cutoff frequency. To conserve area, the transmit D/A is time multiplexed and is used here. Transmit-clock recovery path crosstalk at the D/A output is avoided by using separate S/H blocks for each path. Since PSRR and linearity requirements are not stringent in the clock recovery filters, single-ended structures are used. The rise and fall times of the clock bus drivers are controlled (200 ns typical) to reduce charge injection offsets.

The data signal (9600 bit/s) is confined to the 600-Hz to 3-kHz frequency band (CCITT V.32). The 2400-Hz symbol (baud) clock is recovered by modulating the energy in the 600-Hz lower band edge by a logic signal derived from the 3-kHz upper band-edge energy. This is commonly referred to as band-edge timing recovery. Both the 600-Hz BPF and 3000-Hz BPF are conventional high Q biquad structures [3]. Alternately, the 1200-Hz clock is recovered by feeding the data signal directly into the modulator (chopper), effectively full-wave rectifying the input. A programmable 1200/2400-Hz BPF passes only the energy at the symbol clock, which is then squared up by the limiter.

Fig. 8 shows the chopper and baud clock filter details. The signal enters at the left (V_{in}) and goes through a gain switch to compensate for the difference in the signal levels between the 1200- and 2400-Hz modes. The chopper multiplies the input by ± 1 depending on the relative switch phasing of the chopper input switches. The switch phasing is selected by the S_{CMP} switch under comparator control. The baud clock is recovered by the second-order biquad bandpass filter. In the 2400-Hz mode, the switches are as shown, and the filter samples at a 153.6-kHz rate. In the 1200-Hz mode, the filter sample rate is halved, and the E and J capacitors are changed to keep the filter bandwidth the same in both modes.

Fig. 9 shows the spectrum at the baud clock BPF output. Fig. 9(a) shows the output (in the 2400-Hz mode) for a 600-Hz and 3-kHz two-tone input. Fig. 9(b) shows the 1200-Hz mode response to a 600-Hz input tone. In both cases, the output at the filter center frequency is more than 30 dB greater than that at any other frequency, ensuring a strong baud clock for data recovery.

V. A/D AND D/A

Referring back to the chip diagram (Fig. 1), the D/A and A/D are in the near-end echo path. To achieve 65-dB $S/(N + D)$ for large signal levels in this path, the converters need at least 11 bits of linearity. Since they are in a feedback loop, a monotonic response is needed. The system signal level ranges dictate 14 bits of A/D and D/A resolution. An additional constraint on the D/A is that the transmit-to-clock recovery path isolation exceed 50 dB for proper system operation.

As implemented, the A/D and D/A are 15-bit uniform quantization converters with 11-bit linearity. A conven-

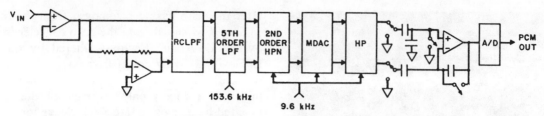

Fig. 6. Receive-path block diagram.

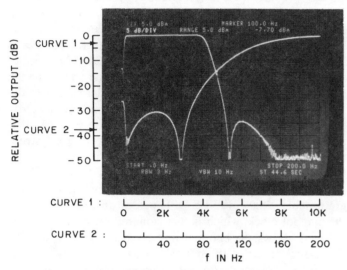

CURVE 1 :

0	2K	4K	6K	8K	10K

CURVE 2 :

0	40	80	120	160	200

f IN Hz

Fig. 7. Receive-path magnitude response.

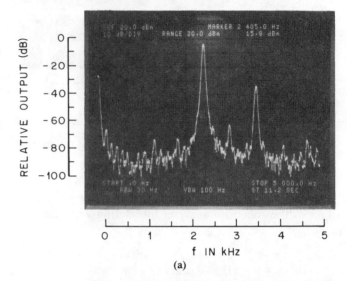

(a)

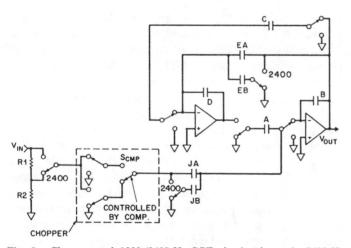

Fig. 8. Chopper and 1200/2400-Hz BPF circuit (shown in 2400-Hz mode).

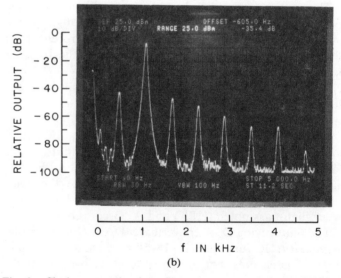

(b)

Fig. 9. Clock recovery bandpass filter output spectrum for (a) 2400-Hz mode with 600- and 3000-Hz input tones; and (b) 1200-Hz mode with a 600-Hz input tone.

tional binary-weighted array requires 1/2-LSB accuracy of the MSB to ensure monotonicity. One way to avoid this requirement is to form the N high-order bits with an array of 2^N equal elements and to feed the remaining lower LSB's in from a second D/A, as shown in Fig. 10 [4]. Each higher output level is the sum of all the levels below it, so the monotonicity of this array is independent of the weighting values. The levels between the high-order bits are generated by weighting V_R by a binary-weighted LSB D/A and feeding the output (V_R^*) into the MSB array [5].

A simplified schematic of the A/D circuit is shown in Fig. 11. The D/A is very similar; the comparator is replaced by a voltage follower. The MSB array implements the top 5 bits using 32 unit capacitors. The monotonicity is obtained by feeding V_R^* through C_{k+1} for levels in the range

$$W_k \leqslant \text{input} < W_{k+1}.$$

The coder output level when C_{k+1} is connected to V_R will always be greater than when C_{k+1} is connected to V_R^*, since V_R is greater that V_R^*. The only requirement for MSB array monotonicity is that the LSB array buffer offset be

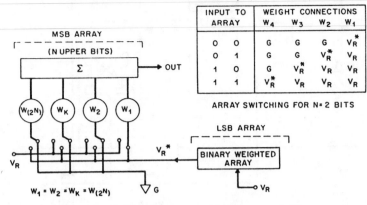

EXAMPLE, LET N = 2 BITS;

INPUT TO ARRAY		WEIGHT CONNECTIONS			
		W_4	W_3	W_2	W_1
0	0	G	G	G	V_R^*
0	1	G	G	V_R^*	V_R
1	0	G	V_R^*	V_R	V_R
1	1	V_R^*	V_R	V_R	V_R

ARRAY SWITCHING FOR N = 2 BITS

Fig. 10. DAC architecture and MSB array switching.

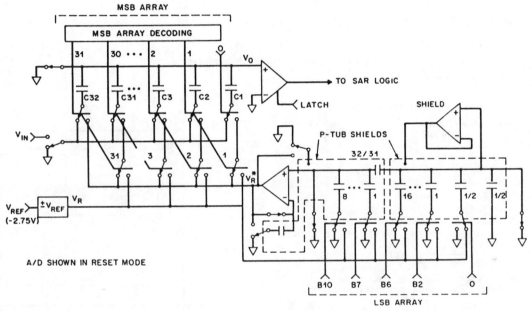

Fig. 11. Simplified A/D circuit.

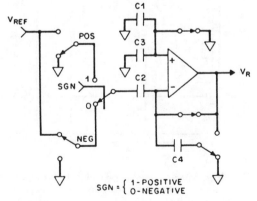

Fig. 12. $\pm V_{REF}$ circuit (shown in reset mode).

smaller than 1/2 LSB of the LSB array. This is insured by nulling the buffer offset [6].

A coupled array is used for the LSB array since this uses a less chip area than a straight binary-weighted array. The LSB array top plate-to-substrate parasitic capacitance is neutralized by placing bootstrap-driven P-tub shields un-

der the array sections, and driving them from buffers that track the top-plate voltages. The dc offset of the shield amplifier is not important as long as the amplifier stays in its linear operating range. Small input devices minimizing the array top-plate loading by the amplifier.

The bipolar V_R signal is generated by the $\pm V_{ref}$ circuit of Fig. 12. C_1–C_4 and the op amp implement a unity-gain amplifier with the SGN switch determining the sign of the gain. If the $\pm V_R$ voltage magnitudes are not matched, second harmonic distortion will result. The dominant source of this imbalance is dc offset caused by parasitic mismatch in the C_1, C_3 and C_2, C_4 branches.

Fig. 13 shows the comparator circuitry. The input is applied to the input differential amplifier ($M_1 - M_5$). The single-ended output is level shifted by the M_5 voltage follower and fed into the cascode amplifier latch ($M_6 - M_{15}$). A cascode amplifier is used to isolate the M_5 level shifter from latch transients. The comparator frequency response is dominated by the input stage, with an overall -3-dB gain frequency of about 1.5 MHz and unity gain at about 35 MHz.

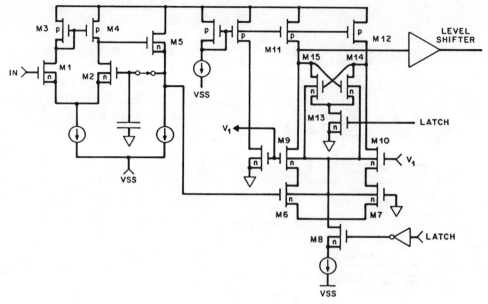

Fig. 13. Simplified A/D comparator circuit.

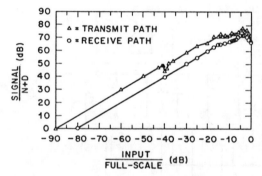

Fig. 14. Transmit and receive path $S/(N+D)$ results. The receive-path MDAC gain is set to unity. The transmit equalizer is set to 0 dB.

Fig. 14 shows the $S/(N+D)$ versus signal level performance of the complete transmit and receive paths. Note that 65-dB $S/(N+D)$ is maintained over the top 10–15 dB of signal range. The dip in the transmit path curve (40 dB below full scale) is caused by a capacitor etching error in the D/A LSB array.

VI. Amplifiers

Several versions of a basic amplifier [6], [7] are used in this chip, each optimized for speed, noise, or output drive capability.

To meet the 65-dB S/D target for this chip, the A/D array amplifiers V_R, V_R^*, and V_{shield} of Fig. 11 should settle to 0.05 percent for each conversion cycle. Large output devices and high idle current are used to provide the low output impedance to drive the worst-case capacitive loads with acceptable phase margin.

The voltage reference V_{ref} is generated by a threshold differencing amplifier [8]. The amplifier is a standard CMOS design except that the threshold of one of the n-channel input devices has been shifted negative. The

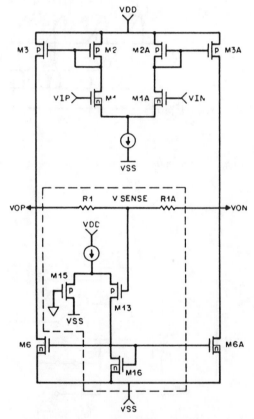

Fig. 15. Simplified balanced amplifier circuit. The boxed-off circuitry is the common-mode control loop.

output stage is able to drive the $\pm V_{ref}$ input capacitor array and recover quickly. The bias current source and input differential pair device sizes have been designed for a low overall temperature coefficient. The typical measured voltage reference temperature coefficient is 1 mdB/C over the 0–85°C temperature range. To minimize the number of trim operations, only one reference is used. The trimmed

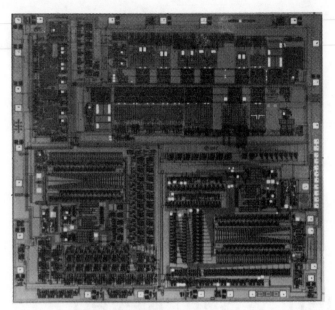

Fig. 16. LARIC chip microphotograph.

TABLE I
LARIC CHARACTERISTICS

Technology	3.5μm twin tub CMOS
Power Supply: Voltage Power	±5V ±5% 180 mW typical
Maximum input (Receive path)	1.75Vpeak
Maximum output (Transmit at 1kHz, equalizer set to 0)	1.095Vpeak
Signal to Distortion for maximum signal swing (transmit and receive)	≥68 dB
Idle channel noise Transmit (0 dB setting) Receive (input referred) MDAC−1 MDAC−7	 0 dBrnC 14 dBrnC 4 dBrnC
Crosstalk: Receive-Transmit Receive-Clock Recovery Transmit-Clock Recovery	 -70 dB -60 dB -60 dB
Die Size	7.50 mm × 7.03 mm

V_{ref} voltage is then separately buffered to both the A/D and D/A.

Fig. 15 is a simplified schematic of the balanced filter amplifier [9]. The input stage current from M_1 and M_{1A} is mirrored about V_{DD} by M_2, M_{2A}, M_3, and M_{3A}, and drives the load. The balanced filter requires tight control of the common-mode signal to meet the LARIC S/D target of 65 dB. The common-mode signal is sensed at the junction of R_1 and R_{1A}. The differential amp ($M_{13}-M_{15}$) adjusts the bias point of M_6 and M_{6A} until $V_{\text{sense}} \approx 0$. To drive resistive loads (such as in the RC LPF block), source followers are added to the outputs. Compensation capacitors are placed between V_{OP} and V_{ON} to stabilize the amplifier when the source followers are added [9].

VII. LAYOUT NOTES

Fig. 16 is the die photomicrograph. To meet the 65-dB S/D requirement, special care was exercised in laying out the MSB and LSB arrays of the A/D and D/A. The unit capacitors of the MSB array consist of two capacitors that are placed diametrically about the center line of the array (common centroid) in order to minimize the first-order process variation effects. Similarly, the LSB array is a binary-weighted coupled common centroid array. The balanced structure of the receive and transmit paths is also evident.

VIII. SUMMARY

A single chip implementing the analog functions for a 9600-bit/s full-duplex voice-band modem has been described. Key device parameters for the LARIC chip are shown in Table I. The 65-dB $S/(N + D)$ performance for large signal levels and the low idle channel noise of the near-end echo path makes it possible to run 9600-bit/s full-duplex data transmission over unconditioned two-wire twisted pair. The transmit-clock recovery path isolation exceeds the 50 dB required for proper data recovery.

ACKNOWLEDGMENT

The authors wish to express their thanks to R. Bell for the test program; R. S. Shariatdoust for engineering support; G. Malek for system support; C. A. Adams, T. L. Jebb, L. E. Pupa, and M. J. Tarsia for the chip layout; and C. A. Bollinger, E. P. Eberhardt, R. W. Gregor, and R. Johnson for wafer fabrication.

REFERENCES

[1] J. C. Bertails, C. Perrin, L. Tallaron, L. Mary, and C. DeLange, "A full duplex analog front-end chip set for split-band and echo-canceling modems," in *ISSCC Dig. Tech. Papers*, Feb. 1986, pp. 174–175.
[2] M. Banu and Y. Tsividis, "Fully integrated active RC filters in MOS technology," in *ISSCC Dig. Tech. Papers*, Feb. 1983, pp. 244–245.
[3] P. E. Fleischer and K. R. Laker, "A family of active switched capacitor biquad building blocks," *Bell Syst. Tech. J.*, vol. 58, pp. 2235–2269, Dec. 1979.
[4] J. A. Schoeff, "An inherently monotonic 12-bit DAC," *IEEE J. Solid-State Circuits*, vol. SC-14, no. 6, pp. 904–911, Dec. 1979.
[5] Y. P. Tsividis, P. R. Gray, D. A. Hodges, and J. Chacko, Jr., "A segmented μ-255 law PCM voice encoder utilizing NMOS technology," *IEEE J. Solid-State Circuits*, vol. SC-11, pp. 740–744, Dec. 1979.
[6] D. G. Marsh, B. K. Ahuja, T. Misawa, M. R. Dwarakanath, P. E. Fleischer, and V. R. Saari, "A single-chip CMOS PCM codec with filters," *IEEE J. Solid-State Circuits*, vol. SC-16, no. 4, pp. 308–315, Aug. 1981.
[7] V. Saari, "Low-power high-drive CMOS operational amplifiers," *IEEE J. Solid-State Circuits*, vol. SC-18, no. 1, pp. 121–127, Feb. 1983.
[8] B. S. Song and P. R. Gray, "Threshold-voltage temperature drift in ion-implanted MOS transistors," *IEEE J. Solid-State Circuits*, vol. SC-17, pp. 291–298, Apr. 1982.
[9] J. A. Klecks and C. F. Rahim, private communication, 1984.

A Single-Chip Frequency-Shift Keyed Modem Implemented Using Digital Signal Processing

RUSSELL J. APFEL, STEVAN EIDSON, AND DAVID M. TAYLOR

Abstract —Modern day communications systems use modulator/demodulators (modems) to convert digital data into analog signals capable of being transmitted on the public telephone system. This paper will describe the philosophy and design of a single-chip modem that employs digital signal processing to perform the modem functions. The DSP modem[1] is a complete single-chip modem which handles all of the popular international standards for frequency-shift keyed (FSK) modems. These specifications include the Bell 103 and 202 specifications for North America and the CCITT V.21 and V.23 specifications in Europe. The paper will describe modem system aspects, the architecture and philosophy of design of this modem, and some of the special features that are incorporated into the modem through the use of digital signal processing.

I. INTRODUCTION

A modem is a device which acts as the interface between two dissimilar media. In this paper, it refers to the modulation and demodulation which must take place for transmission and reception of digital data over the public switched telephone network (PSTN). The telephone network is a 4 kHz band-limited medium. Because of amplitude and group delay distortions present at both the high- and low-frequency portions of this spectrum, the usable bandwidth is approximately 300–3000 Hz. Tele-

phone companies band-limit the signal spectrum to less than 4 kHz because each channel may be multiplexed using frequency-division multiplexing (FDM) techniques or sampled and spaced in time using time-division multiplexing (TDM) techniques.

Square-wave digital data contain an infinite spectrum of odd harmonics at a relatively high amplitude level. Hence, square waves would not be suitable for transmission over the telephone network because of low- and high-frequency harmonic content. This digital waveform must be modulated to a suitable carrier frequency and then filtered to meet bandwidth requirements prior to transmission over the network.

There are three primary modulation techniques used for digital data transmission over the telephone network: frequency-shift keying (FSK), phase-shift keying (PSK), and quadrature-amplitude modulation (QAM) [1]. This DSP modem implements several FSK modem specifications. FSK modulation is a digital form of frequency modulation (FM) where the output frequency is a function of the baseband input signal. An example of a time-domain FSK waveform is shown in Fig. 1. In the modem transmitter, the input digital data control the frequency which is generated and transmitted over the network. In the modem receiver, digital data are recovered as a function of the frequency of the instantaneous received carrier.

Manuscript received May 30, 1984.
The authors are with Advanced Micro Devices, Sunnyvale, CA.
[1]Advanced Micro Devices Am7910.

Reprinted from *IEEE J. Solid-State Circuits*, vol. SC-19, no. 6, pp. 869–877, Dec. 1984.

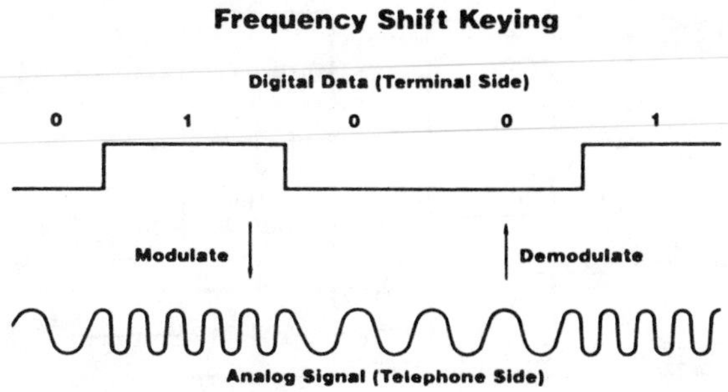

Frequency Shift Keying

Fig. 1. Frequency-shift keying is a modulation technique that utilizes two different frequencies to represent the two binary states. This allows digital data to be transmitted over an analog telephone network.

Modem implementation complexity and costs rise with the increase in transmission data rates provided by PSK and QAM modems. FSK modems are generically known as low-speed modems, PSK modems are medium speed, and QAM modems are high-speed modems.

PSTN modems may operate in one of three modes on the network: simplex, full duplex, or half duplex. Simplex operation means that data are transmitted in only one direction from one modem to another; data are never transmitted in the opposite direction.

Full duplex means that data may be transmitted in both directions simultaneously on the network. For low-speed modems, this is accomplished by using a frequency-division multiplexing (FDM) technique. Each modem makes use of nearly half of the available bandwidth on the network for its transmission. The modem originating the data call will transmit data to the answering modem on the low-frequency channel, while the answering modem will transmit to the originating modem on the high-frequency channel.

Half-duplex operation means that data may be transmitted in both directions on the network, but in only one direction at a time. One modem will transmit in one direction until finished. Then the transmission channel will be "turned around" for data transmission in the opposite direction. Thus, almost the entire available bandwidth which the network provides is used for this unidirectional transmission.

There is, however, an exception to the definition just provided for half-duplex operation. In many instances, a lower data rate "back channel" is provided in half-duplex modems. The back channel allows data to be transmitted in the opposite direction of the higher speed transmission which occurs on the "main channel." Back channel data transmission occurs simultaneously with main channel transmission, so it achieves "split rate" operation. Back channel data rates for PSTN modems vary; usually they are between 5 and 150 bits/s for FSK modems.

The U.S. standardized FSK modems are the Bell 103 and the Bell 202. The Bell 103 [2] is capable of operation at 300 bits/s full duplex. The Bell 202 [3] is a 1200 bit/s half-duplex modem which contains an ON/OFF-keyed back channel. The back channel operates at 5 bits/s, too slow for much data transmission to be accomplished, but adequate for signaling or indicating the status of received data

to the modem transmitting on the main channel. The CCITT (Consultive Committee for International Telephony and Telegraphy) has adopted two equivalent FSK modem specifications. These are the CCITT V.21 and CCITT V.23. These CCITT modem specifications are very similar in nature to the Bell modems discussed previously. The V.21 [4] is a 300 bit/s full-duplex FSK modem, although the V.21 modulating frequencies differ from the Bell 103 frequencies. The V.23 [5] is a 1200 bit/s half-duplex modem with a 75 bit/s FSK back channel. It also uses different modulating frequencies from the Bell 202 and has a full FSK back channel compared to the Bell 202 signaling back channel.

The interface to all four of the modem specifications (Bell and CCITT) is also quite similar. In the U.S., the RS-232-C is the most common interface. It contains all of the status, command, and data pins to connect the modem directly to a terminal or computer. The CCITT has adopted the V.24 specification which is very similar in nature to the RS-232-C. Nearly all pins are identical with the RS-232-C; however, in certain cases, the V.24 pin names were changed by the CCITT. Differences are limited primarily to timing and minor protocols.

There are additional factors to consider when designing modems besides simply choosing the modulation technique and interface. The telephone network is a harsh environment for data transmission. Impairments such as amplitude and group delay distortion tend to distort the data spectrum and may cause demodulation problems for the receiver. In addition, impairments such as phase jitter, frequency offset, amplitude and gain hits, and switching noise created by central office switching equipment may cause havoc with a modem signal. The human ear is quite tolerant of many of these distortions, but data modems are not as impervious to poor line conditions. In order to increase transmission data rates, modulation complexity must increase, and telephone environmental conditions increasingly affect modem performance.

II. Goals and Objectives

In developing the DSP modem, some key goals and objectives were established to guide the project. The overriding goal was to produce a cost-effective single-chip modem which provided a high level of performance. This meant integrating many interface and control features into the chip to simplify the user's system design. In addition, the features had to have truly international applications for the modem to be effective.

Among these items, are the inclusion of the following.

1) Pin programmable selection of all of the widely used FSK modem specifications (CCITT V.21, CCITT V.23, Bell 103, and Bell 202). These four modem specifications may be operated in nine different data mode configurations.

2) All traditional modem functions such as modulation, demodulation, and filtering. It was important that the modem be a true single-chip solution requiring that no

critical external components be supplied by the user—particularly if they were required for implementation of the traditional modem functions.

3) All analog-to-digital and digital-to-analog conversion on chip.

4) The traditional RS-232-C/V.24 functional interface for modem control; this allows the user to design a simple interface for the DSP modem using a standard with which he may already be familiar.

5) The ability to perform loopback testing of the modem for all of the modes in 1) above; loopback modes provide ten additional operational configurations for testing the nine data modes, for a total of 19 different operational modes.

6) The ability to assist the user in an auto-answer mode for automatic answering applications; this relieves the user of the need to take an active part in the automatic answering process.

The end modem user also should not be required to possess intimate knowledge about modems or the PSTN. The digital logic designer should be able to design the traditionally analog modem function into his system without additional knowledge. Continuing along this line, the modem should require as few external parts as possible and no active external parts. A crystal and a single resistor and capacitor required by the DSP modem meet this criterion.

III. DIGITAL SIGNAL PROCESSING

This DSP modem makes extensive use of digital signal processing as opposed to to other designs which use analog signal processing. Digital signal processing offers many advantages in performance and design capability over traditional analog processors. Digital signal processing provides higher performance because there is no drift with time or temperature. Once the modem is designed, the filter characteristics are fixed because the coefficients are precisely known; there is no unit-to-unit variation. Digital signal processing also allows the use of a wide range of filter types, both infinite impulse response (IIR) and finite impulse response types (FIR) [6].

The DSP modem makes extensive use of linear phase finite impulse response filters [6]. FIR filters are the only type of filters which can achieve linear phase characteristics. This means a constant delay characteristic which is very important to achieve high performance in modems. The digital signal processing system also allows the designer a much greater range of flexibility and programmability since the coefficients are stored in ROM. Selecting different sections of ROM provides the opportunity to easily change filter characteristics.

Another advantage of DSP is that a core signal processor can be multiplexed and used to perform many different tasks. There is the attendant overhead of putting together the basic signal processor that includes the arithmetic and logical unit (ALU), the RAM and ROM, addressing capabilities, and the analog-to-digital and digital-to-analog con-

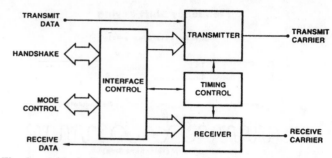

Fig. 2. A block diagram of the DSP modem shows three major blocks and the timing control among them. The transmitter and the receiver use DSP to modulate and demodulate data, while the interface control determines the configuration and state of the modem.

version blocks. However, once this overhead is paid for, the incremental cost of adding extra stages of filtering or extra complexity is relatively low.

The other major advantage of the digital signal processor is the fact that it can be built with standard microprocessor technology without special processing steps; manufacturing yields are comparable to those of microprocessors. The microprocessor and static memory technology is the most advanced, high-volume, production-oriented technology where advances in line widths and reduction in silicon area can be piggy-backed onto those already being generated. Modern microprocessors have moved rather rapidly from 5 μm technologies down to 2 μm technologies with dramatic reduction in silicon area. These types of advances will happen within digital signal processing products as well.

IV. SYSTEM ARCHITECTURE

Fig. 2 shows a functional block diagram of the entire DSP modem. Notice that the transmitter and receiver are shown as separate blocks in the diagram. This is also how they are implemented in silicon. The transmitter and receiver each contain their own instruction ROM, RAM, and ALU for performing the signal processing. There is no interaction between the two processors. The interface block controls the operation of both the transmitter and receiver as a function of the programming of the mode control pins provided by the user and the states of the handshake (RS-232-C or V.24) pins. The transmit data provided by the user are a direct input to the transmitter, and the demodulated data are provided as an output directly from the receiver.

Fig. 3 shows a block diagram of the transmitter signal processing blocks implemented within the DSP modem. The frequency of the carrier to be generated is a function of the modem type selected on the mode control pins and the state of the transmit data (*TD*) pin. The phase constant selection, serial adder, and phase-to-amplitude conversion blocks comprise the frequency-shift sine wave generation block.

The phase of a sine wave is linear with time, while the amplitude is nonlinear. To generate the sine amplitude, the sine phase is accumulated in a serial adder, with a frequency-selectable phase constant being the incremental

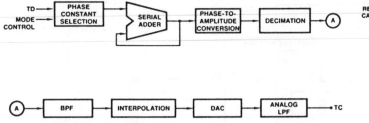

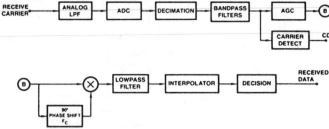

Fig. 3. The block diagram of the modem transmitter can be divided into three types of hardware: sine synthesis, filtering, and digital-to-analog conversion. Sine synthesis is done in dedicated hardware outside the DSP ALU, while filtering is accomplished wholly in the ALU.

Fig. 4. The modem receiver has several pieces of dedicated hardware to perform demodulation. Carrier detect interfaces to an external pin to control the receiver state. The multiplier takes control of the ALU to perform the Booth algorithm. The interpolator hardware linearly divides the low-pass filter data to decrease jitter distortion.

value. The accumulated phase value then addresses a sine ROM that provides the amplitude of the sine wave for that phase angle [7]. Because of the symmetry of a sine wave, only a single quadrant of amplitudes is stored in the ROM. The remaining three quadrants may be derived from the stored quadrant by routing the two MSB's of the phase address as control for quadrant steering. When a frequency shift is desired, a different phase constant representing the new frequency is provided as input to the serial adder. This creates a continuous-phase FSK signal.

The modem performs the sine wave generation at a high sampling rate (122 kHz). This high sampling rate is used to accurately locate the instants of change in frequency (provided by the user on the *TD* input pin). The high sample rate signal must have its sample rate decimated (reduced) down to a value near 8 kHz prior to bandpass filtering. The bandpass filters consist of a cascade of group delay-equalized elliptic second-order sections. Decimation is accomplished with a series of low-pass FIR filters [8]. Bandpass filtering of the FSK waveform is required to remove high-frequency components which exceed telephone company requirements. If filtering was performed at 122 kHz, the speed requirements of the ALU would be too great to allow any significant signal processing to be performed. Generation of the FSK waveform inherently produces an infinite spectrum. However, in comparison to the spectrum of the square wave discussed earlier, the FSK spectrum is much more compatible with telephony requirements.

After bandpass filtering, the FSK waveform is interpolated [8] (sample rate is increased) up to a high sampling rate again (122 kHz) for output to the DAC. The interpolation is performed to allow a simple analog low-pass filter to be used as the reconstruction filter on the analog output. If the inputs to the DAC were provided at the low sampling rate of the ALU, the analog filtering requirements of the reconstruction filter would dictate a high-order filter to eliminate harmonics that are generated at multiples of the sampling rate. It is not an easy task to design high-performance, high-order analog filters in NMOS.

Fig. 4 shows a block diagram of the DSP modem receiver. The analog low-pass filter eliminates energy which may alias back into the passband after sampling by the analog-to-digital converter. Similar to the transmitter DAC operation, the receiver ADC operates at a very high sampling rate (496 kHz). This reduces the burden on the analog

prefilter. In fact, the prefilter is a single-pole (−6 dB/octave) *RC* low-pass filter.

The receiver decimators reduce the sampling rate down to the basic ALU signal processing rate. Decimation needs to be performed within the receiver for precisely the same reason that it was performed within the transmitter—to reduce the sampling rate in order to allow more signal processing to be performed.

Following the decimation, the received signal is filtered to remove out-of-band energy which may corrupt the demodulation process. The bandpass filtering is performed entirely within the receiver ALU. The filters are cascaded elliptic second-order stages which are group-delay equalized. Fig. 5 shows two typical amplitude responses of bandpass filters within the DSP modem. After filtering, the automatic gain control (AGC) block ensures that the received signal is of the proper signal strength to perform the demodulation. A low-level received signal is amplified to a higher level to maintain the largest dynamic range possible within the ALU.

The carrier detection block within the DSP modem determines if the received signal contains sufficient energy to indicate a valid carrier. If a valid carrier is detected, the $\overline{CD}$ pin will go LOW; otherwise, it will be HIGH. If a valid carrier is not being received, the remainder of the interface pins are either ignored if they are input pins or set to an appropriate initialized logic state if they are output pins.

The DSP modem demodulator is known as a product demodulator. The major advantage of this type of demodulator is that it is less susceptible to phase distortions present on the telephone network relative to other types of demodulation schemes. It consists of phase shifting the incoming signal by 90° at the center frequency between the two frequency shifts and multiplying this phase-shifted signal with the incoming signal. The result is a baseband term containing the desired digital data information, and a double-frequency term containing no useful information. This double-frequency term is filtered out by the low-pass filter following the demodulator.

The receive interpolator has the job of accurately determining where the zero-crossing transitions of the demodulated and low-pass filtered signal occur. These zero crossings determine changes in the state of the data. The zero crossings should be accurately defined to reduce the

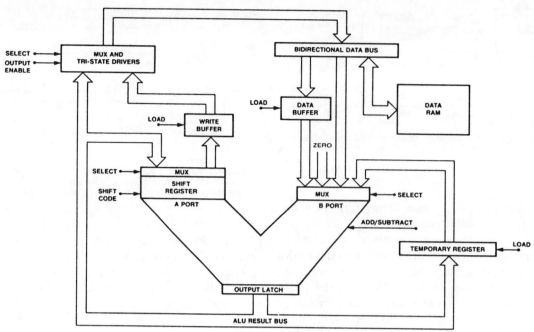

Fig. 5. A block diagram of the DSP modem ALU reveals a structure that can be used for both the receiver and the transmitter. Since coefficients are represented as shifts and adds, the shift register and adder must be able to return a result within one microinstruction.

amount of jitter observed in the received data waveform. The signal processor provides approximately seven samples/bit interval (1200 bits/s) or 26 samples/bit interval (300 bits/s) out of the low-pass filter. If a zero crossing occurs between two of these samples, the jitter out of the modem contributed by the decision process alone would be unacceptable. (Jitter is defined as the deviation from the ideal change of state instant by the received data out of the modem.)

The interpolator divides the interval between signal processing samples into eight equal smaller samples. If a zero crossing occurs between signal processing samples, the interpolator chooses which of the eight smaller intervals lies closest to the zero crossing. Hence, interpolation reduces the percentage of jitter contributed by the decision process by a factor of eight. The decision process then is simply a comparison process to change the state of the received data when a sign change is detected from the interpolation block.

The function of the decision block is simply to restore the sharp edges to the demodulated waveform. The decision block samples the output of the interpolator. If a zero crossing is detected, the state of the received data pin is changed to reflect the new data.

The interface control (refer to Fig. 2) is a state machine which provides the control to the transmitter and receiver as a function of the mode programming by the user. The five mode control pins select the operational mode of the modem, while the handshake pins control the dynamic modem operation. The handshake pins provided are: $\overline{DTR}$, main and back channel $\overline{RTS}$, main and back channel $\overline{CTS}$, main and back channel $\overline{CD}$, main and back channel $\overline{TD}$, and main and back channel iR D. The back channel signals

are used only by the half-duplex modems; they are ignored for full-duplex operation.

V. DIGITAL CIRCUIT DESIGN

The DSP modem is a dual processor architecture; both the receiver CPU and the transmitter CPU process data independently at 2.4576 MHz The two central processors have an identical architecture which is shown in Fig. 6. The architecture consists of a two-port ALU, three data buffers, an ALU result bus, and a bidirectional bus for access to data RAM. After data are read from RAM onto the bidirectional bus, they can be read into the B port of the ALU or into the data buffer. The data buffer has three output paths: one to the write buffer, and one each to the A and B ports of the ALU. The A port has a shift register which is controlled by the microcode word. Data at the input ports of the ALU are added or subtracted on every instruction cycle. The result of this arithmetic operation is latched onto the ALU result bus. The ALU result bus has return paths to the A port of the ALU and to the bidirectional bus. In addition, the ALU result can be transferred into the temporary buffer. The timing of the ALU bus and the temporary buffer is such that results can be passed through the buffer to the ALU B port on a single microcode cycle. Data can also be read onto the bidirectional bus from the write buffer.

By using the same filter and demodulator structures for every modem type on the DSP modem, it is possible to use the same microcode instruction set in every case. Only the filter coefficients are changed from modem type to modem type. Two microcode ROM's are employed by each CPU. The first ROM contains the microcode for the filter and

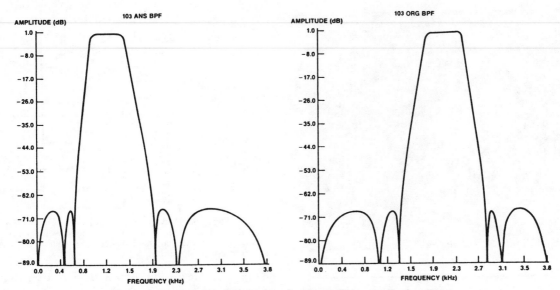

Fig. 6. A graph of the receiver bandpass filter amplitudes versus frequency for the Bell 103-type modems. Note that the passbands are flat since FSK demodulators are quite sensitive to amplitude variations.

demodulator structures, while the second ROM contains the filter coefficients for each modem type. By separating the microcode into two ROM's, the DSP modem is able to operate in nine different modem configurations, with only 10 percent of the die dedicated to the modem-type-dependent coefficient ROM. Some microcode blocks such as the decimators, interpolators, and multiplier are common to all modem types. When the address of these common blocks is decoded, coefficients for the filters are not chosen from the coefficient ROM; they come from an extra set of coefficients stored in the instruction ROM. This technique of coefficient selection reduced the size of the coefficient ROM by approximately 100 words per modem type.

Because of the limited number of microcode instructions, it was necessary to place the first receiver decimator in a separate ALU from the main CPU. Samples from the hardware decimator ALU are strobed onto the bidirectional bus by the main CPU.

A single multiply is necessary to demodulate the input receive carrier. The basic DSP CPU does not have the capability to do a multiplication on two RAM values since its filter coefficients are stored as shift-and-add codes in ROM. For this reason, some hardware modifications were made to the receiver CPU to allow multiplication. By using a modified Booth multiplication algorithm and controlling the multiplier with strobes derived from unused RAM locations, an 18 bit product is accumulated from two nine-bit operands. Only 52 words of the possible 64 data RAM locations have RAM cells in the receiver CPU. Four of the nonexistent locations are decoded by the CPU as multiplier strobes. The first strobe takes the absolute value of both operands and saves both signs for the final product. The second strobe clears the lower half of ALU result bus so the two operands can both be stored as a single 18 bit number on the ALU bus. The third strobe takes control

of the ALU and performs a repetitive shift and add on the 18 bit ALU number. After nine shift and add operations, the fourth multiply strobe corrects the sign of the final product.

The method of sine wave generation on the DSP modem employs two ROM's, a serial adder, and an accumulator external to the transmitter CPU. The mode control pins and the transmit data pin select a value in a 28-word phase ROM. The output of the phase ROM is serialized and fed into the serial adder. The other input to the serial adder comes from the phase accumulator. The sum of the phase constant and the phase accumulator is stored in the phase accumulator. The accumulated phase is used to address a sine ROM which converts the phase to an amplitude. Due to the symmetry of the sine function, it is necessary to store amplitudes for only one quadrant of a sine wave. The two most significant bits of the phase accumulator are used to control the quadrant function of the sine wave. The output of the sine wave generator is strobed into the transmitter CPU by setting a microcode bit.

Approximately 10 percent of the die area was dedicated to test logic. The RAM's, ROM's, ALU's, and analog sections are separated into components for test procedures. The test logic allowed each component of the DSP modem to be tested individually without resorting to time-consuming system tests.

Circuit functions developed external to the main CPU structure (such as the multiplier and data interpolation blocks) could have added a large amount of unnecessary circuit complexity. By tailoring each function to work with the CPU (by providing strobes generated from nonexistent RAM locations, for instance), it was possible to implement the system without an unwieldy circuit. Fig. 7 shows a die photo and approximate layout of the circuit functions of the DSP modem.

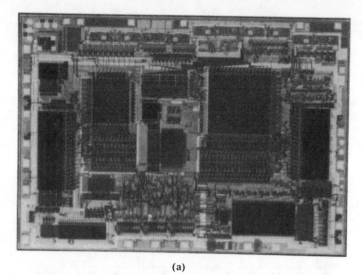

(a)

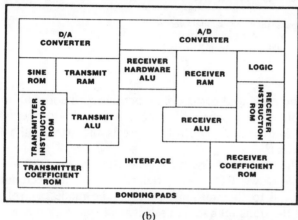

(b)

Fig. 7. (a) A photograph of the DSP modem showing the placement of the major architectural circuit blocks. (b) Layout of FSK modems, Am7910.

VI. ANALOG-TO-DIGITAL CONVERSION FOR DSP MODEMS

The receiver of the digital signal processing modem requires a specialized type of analog-to-digital converter. Since the modem is a sampled data system, the ADC must band-limit the signal prior to A/D conversion to meet the Nyquist criterion. To simplify the signal processing and the anti-aliasing filter, the A/D converter must operate at a high sampling rate. The higher the sampling rate, the simpler the requirements of the anti-aliasing filter. The highly sampled digital signal is later decimated down to a rate that is more conducive to the bulk of the signal processing.

The A/D converter chosen for this DSP modem is an interpolative coder [9]–[11]. An interpolative coder, as shown in Fig. 8, is a modified form of sigma–delta modulator. The main components of the ADC are the integrating amplifier, the decision comparator, the feedback control logic, and the feedback digital-to-analog converter. The DAC generates a quantized approximation of the analog input signal. This quantized approximation is subtracted from the input signal and the difference is integrated

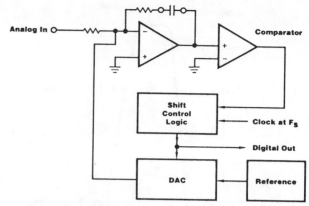

Fig. 8. The DSP modem's interpolative A/D converter. The comparator decides whether to increase or decrease the DAC output voltage based on the integrated difference between the last DAC output and the present analog input.

using the integrating amplifier. The decision comparator examines the output of the integrating comparator and decides whether it is greater than or less than zero. A decision is made to increase or decrease the magnitude of the quantized signal based on this information. The comparator output is fed into the control logic which then adjusts the magnitude, and on occasion the sign, of the quantized signal and generates a new sample.

The interpolative coder operates at a highly oversampled rate which performs quantization with very coarse levels in the A/D converter. The digital-to-analog converter used has 11 positive and 11 negative exponentially scaled states, each increasing positive state being exactly twice the previous state. On every A/D sample, the value of the DAC output voltage is doubled or halved, depending on the decision comparator output.

The accuracy and dynamic range of the ADC is a function of the sampling rate, the number of levels allowed in the digital-to-analog converter, and the algorithm chosen. The accuracy increases as the sampling rate increases because there are more sample values that will be averaged during decimation. An increased number of DAC levels relative to full scale allows finer resolution at low signal levels. But as the number of DAC levels increases, it becomes increasingly difficult to traverse the entire range of values while tracking a rapidly changing signal. There is an apparent tradeoff between dynamic range and the amount of frequency response to the system.

This "slope overload" problem is obviated by the DSP modem control algorithm. The algorithm automatically removes levels around 0 for high-level signals and reinserts them for low-level signals. Because the lowest levels of a full-scale signal are a small percentage of the whole, it is not necessary for a high-level signal to have fine accuracy near zero. Waveforms also have the highest rate of change near zero, and therefore need to approach zero much more rapidly. The control algorithm keeps track of the analog-to-digital converter peak level during any half cycle. When the magnitude is decreased to 1/32 of that peak level (i.e., five level changes), the next reduction of magnitude changes the sign bit rather than reducing the magnitude further.

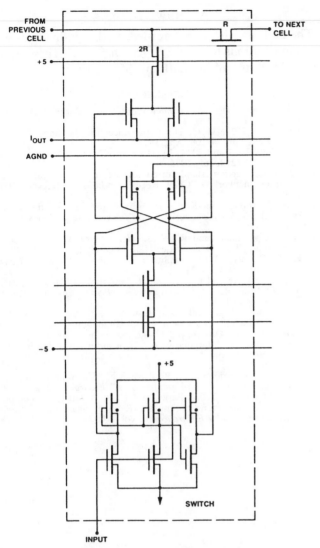

Fig. 9. A DAC cell from the DSP modem. Note that the $R/2R$ resistor ladder is built using MOS transistors.

the output is always doubling; the first or 0 level is 1 LSB, a code of 1 gives you 2 LSB's, 11 gives you four times the LSB, 111 gives you eight times the LSB, etc. Note that the weighting of the bits into the DAC is not interpreted directly as a binary weighting. Instead, a shift left of 1 bit doubles the DAC current. A "1" in the next left position turns on another branch of current which is exactly double the last current.

Notice that accuracy is also improved in this scheme because as we double the level, we add a new current which is double the previous value. Half of the new current is the current which was already summed into a summing node for the lower level. This reduces the accuracy requirements of the analog components by a factor of two. The DAC implemented on the DSP modem does not use resistors or capacitors. The resistor structure is simply the ON resistance of MOS transistors as shown in Fig. 9. The unit resistance is built out of an MOS transistor which is approximately 20 μm on the side and has value of 1 kΩ.

VII. MODEM ALGORITHMS

The DSP modem uses both IIR and FIR filters The IIR filters are implemented as standard direct form II second-order sections [6], while the FIR filters are simply linear phase filters developed using the McClellan–Parks algorithm [12]. Each IIR second-order section requires two memory locations and has five coefficients. One memory location and one coefficient are required per filter order for the standard ladder implementation of the FIR filters. The memory locations are allocated from data RAM on the CPU, but the filter coefficients are stored in the instruction and coefficient ROM's of the CPU. The multiplication of data is simulated by adds and shifts in the ALU, but the DSP modem also has the ability to represent coefficients by subtraction. A canonic sign-digit representation was chosen for the coefficients in ROM. Canonic sign digit allows the grouping of bits in a binary representation to be replaced by one addition and one subtraction. An example of an 8 bit coefficient is given below in both binary and canonic sign-digit.

$$-0.11100111_2 = -0.90234375_{10}.$$

In canonic sign-digit:

$$-1.00\bar{1}0100\bar{1} = -0.90234375_{10}$$

where $\bar{1}$ in canonic sign-digit means subtract.

On the DSP modem:

$$-2^0\left(1-2^{-3}\left(1-2^{-2}\left(1-2^{-3}\right)\right)\right) = -0.90234375_{10}.$$

From the above example, the exponents are the shift codes for the ALU. Note that the binary representation would have required six add-and-shift operations, while the canonic sign digit representation takes only four operations.

Therefore, for low peak amplitudes, lower levels (still 1/32 of the peak value) are used before the sign bit is changed.

A decay algorithm in the control allows the system to retreat back to lower levels after it has been at a peak level. This decay algorithm is a dither signal which reduces by one the peak level stored during each sample period. This dither signal additionally provides an out-of-band noise source which prevents the buildup of any periodic type of folded signal from being created within the A/D converter.

The heart of the ADC is the feedback digital-to-analog converter. This is a unique DAC in that it only puts out levels which are binarily weighted. This is accomplished by using a standard $R/2R$ array (or ladder), with the final termination being two $2R$ resistors in parallel. Instead of terminating the unused current into ground, it is taken into the summing node of the D/A converter. Thus, with a 0 code input to the DAC, there is still 1 LSB of current flowing in the summing node. The D/A converter is now driven with a code of 1. Next, the 1 is shifted left so that we have a code of 1, 11, 111, etc. It can be observed that

IX. Conclusion

The DSP modem discussed in this paper is truly a "world-chip" modem by being able to operate in almost all FSK modem applications with a standard RS-232-C type of interface. The few external components required ensure a simple and cost-effective modem implementation.

The inclusion of all of the modem operating modes is made possible by the use of digital signal processing. In addition, the DSP approach ensures stable, repeatable, and predictable performance across all units. The filter structures for all of the modems are identical, so implementing one of the four different modem specifications is simply a matter of switching in a different set of filtering coefficients. In fact, the inclusion of all of the operating modes consumes only an approximate 10 percent increase in die size relative to implementing just a single modem.

This DSP modem is an example of the first generation of single-chip modems to appear on the market. In the next few years, more sophisticated single-chip modems using PSK and QAM modulation techniques will begin to appear which contain higher data rates. As the complexity of these modems increases, other semiconductor manufacturers will be forced to use DSP approaches because these modem types require advanced DSP techniques such as adaptive equalizers.

Acknowledgment

The authors wish to acknowledge other individuals with key contributions to the DSP modem project. These people are D. Mullins and his staff and M. Stauffer.

References

[1] J. G. Proakis, *Digital Communications.* New York: McGraw-Hill, 1983, pp. 116–117, 162–171, 183–190.

[2] Data Sets 103J, 113C, 113D, Interface Specification, *Bell Syst. Tech. Ref.*, Publ. 41106, Apr. 1977.

[3] Data Sets 202S and 202T, Interface Specification, *Bell Syst. Tech. Ref.*, Publ. 41212, July 1976.

[4] "200-baud modem standardized for use in the general switched telephone network," CCITT Recommendation V.21, vol. VIII.1, 1976.

[5] "600/1200-baud modem standardized for use in the general switched telephone network," CCITT Recommendation V.23, vol. VIII.1, 1976.

[6] A. V. Oppenheim and R. W. Schafer, *Digital Signal Processing.* Englewood Cliffs, NJ: Prentice-Hall, 1975, ch. 5.

[7] H. W. Cooper, "Why complicate frequency synthesis?," *Electron. Design*, July 19, 1974.

[8] R. E. Crochiere and L. R. Rabiner, "Interpolation and decimation of digital signals—A tutorial review," *Proc. IEEE*, vol. 69, pp. 302–331, Mar. 1981.

[9] J. C. Candy, W. H. Ninke, and B. A. Wooley, "A per channel A/D converter having 15-segment u-255 companding," *IEEE Trans. Commun.*, vol. COM-24, pp. 33–42, Jan. 1976.

[10] G. Ericson, "An interpolative A/D converter with adaptive quantizing levels," in *Nat. Telecommun. Conf. Rec.*, 1980, pp. 56.6.1–56.6.6.

[11] R. Apfel, H. Ibrahim, and R. Ruebusch, "Signal-processing chips enrich telephone line-card architectures," *Electronics*, May 5, 1982.

[12] J. H. McClellan, T. W. Parks, and L. R. Rabiner, "A computer program for designing optimum FIR linear phase digital filters," *IEEE Trans. Audio Electroacoust.*, vol. AU-21, pp. 506–526, Dec. 1973.

A CMOS Ethernet Serial Interface Chip

Haw-Ming Haung, Dado Banatao, Gust Perlegos, Tsing-Ching Wu, Te-Long Chiu

SEEQ Technology, Inc.

San Jose, CA

A CMOS SINGLE CHIP SOLUTION, developed for the Ethernet Serial interface function, will be described. The chip* provides the interface between the data link controller and the transceiver, and represents a second step toward total integration of the Ethernet connection hardware. The chip performs the 10MHz Manchester encoding/decoding and detects the differential collision signal sent by the transceiver indicating collision of data on the cable. Functionally, the chip is divided into two sections: namely, the transmitter and receiver.

Figure 1 shows the receiver input circuit. It has a two-stage differential amplifier and a CMOS level converter with a total delay 12ns. The amplifier has P channel load devices which are biased in the linear region to obtain a gain of 5 per stage. To provide the static noise margin, a squelch circuit is used which has a -200mV dc threshold during idle and is disabled after the first valid transition. A digital noise rejection filter is also provided to prevent detection of spurious signals. The collision detector uses a similar input circuit.

In Manchester coding, the first half of the bit cell contains the complementary data and second half contains the true data. Thus, there is always a transition in the center of the bit cell. At the beginning of a frame, a series of alternating '1's and '0's (preamble) signal is sent which has no boundary transitions.

The PLL uses this signal to lock-in and recovers the data and clock.

The PLL, which does not require external components, is shown in Figure 2. During idle, the PLL locks to TXC and the VCO runs at 20MHz. Once incoming data is active, which asserts CRSN, the PLL adjusts its phase to be synchronous with the data. A phase/frequency comparator generates the UP and DOWN signals to adjust the phase or frequency. Two 10MHz clocks are generated with a 90° phase relationship. The data transitions are compared with ϕ, while $\phi + 90°$ is used to window out the boundary transitions and sample the data. Different charge-pumps are used in idle or active (preamble and data) periods. The PLL uses a second order low-pass loop filter. By changing the resistor value, the filter changes the damping during preamble or data periods. A delay circuit in the ϕ signal path is used to match the delay in the multiplexer and transition detector. The RXC is switched from the VCO clock to TXC clock during idle and preamble. This method provides an optimum RXC clock duty cycle. Using this approach, the PLL can lock-on to the preamble with a jitter of ≤12ns within 12 bit times and can sample the incoming data having ≤18ns of jitter. Figure 3 shows the received waveform.

The transmitter has an on-board 20MHz crystal oscillator which is used to generate the 10MHz TXC signal. The encoding is accomplished by exclusive ORing the clock and data. The output driver shown in Figure 4 converts the CMOS level to a differential level of 850mV which drives a 78Ω transceiver cable. The output waveform has a 4ns rise and falltime and a skew of less than 0.5ns. A pair of 240 ohm external resistors are used as loading. During idle, the differential output is returned to near zero with a backswing voltage of less than 100mV to eliminate dc current in the coupling transformer. To achieve this, signals IDLE1 and IDLE2 force TX- to 90% of TX+ voltage 200ns (minimum) after the last transition, and then TX- slowly approaches the TX+ level. Figure 5 shows the transmitter waveform.

The chip (99 x 115 mil^2) which is shown in Figure 6 was fabricated using a 2μm, 400Å N-well CMOS technology, and consumes 150mW of power.

Acknowledgments

The authors wish to thank B. Gunning and D. Hodges for their technical advice, G. Perlegos for overall direction and L. Whitcomb for test support.

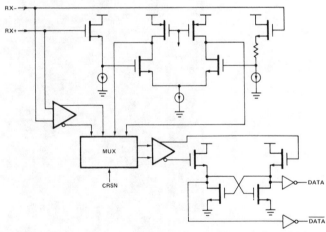

FIGURE 1—Receiver differential amplifier and squelch circuit.

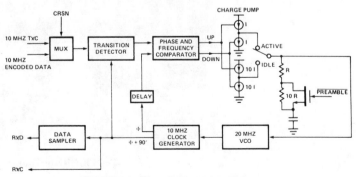

FIGURE 2—Phase-locked-loop diagram.

---

*Device meets specifications set in the IEEE 802.3 standards.

Reprinted from *1984 IEEE Int. Solid-State Circuits Conf.*, pp. 184–185, Feb. 1984.

RX+

CRSN

RXD

RXC

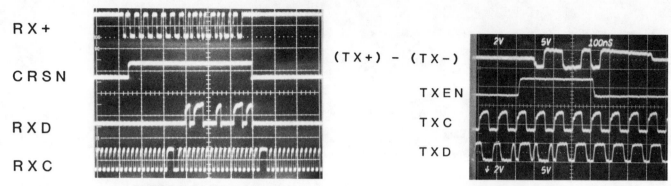

FIGURE 3—Receive waveform.

(TX+) - (TX-)

TXEN

TXC

TXD

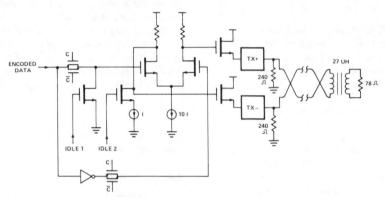

FIGURE 5—Transmit waveform.

FIGURE 4—Differential output driver.

FIGURE 6—Photograph of chip.

A Single Chip NMOS Ethernet Controller

Allen G. Bell, Gaetano Borriello

Xerox Research Center

Palo Alto, CA

THIS PAPER WILL COVER the design of a single-chip NMOS integrated controller* for a 10MHz Ethernet. The chip implements the physical layer and low-level data-link layer functions of the Ethernet protocols[1]. It interfaces directly to the transceiver drop cable and provides a byte-wide data-path to a host dependent interface. A novel self-calibrated tapped delay line concept is central to this chip. The chip is 6.6mm by 4mm, is composed of 4,500 transistors using a 4μ Si-gate NMOS process ($\lambda = 2\mu$), and dissipates approximately 0.9W.

This chip comprises four subsystems; Figure 1. The signal-handling transmitter (SHTx) encodes the data into Manchester format, then drives the transceiver with differential ECL logic levels. The signal-handling receiver (SHRx) translates differential ECL input signals into clocked NMOS signals by extracting the serial data from the Manchester encoded waveforms. The data-handling transmitter (DHTx) takes as input a byte stream from the host-dependent interface and drives SHTx with a preamble, the serialized data and a generated cyclic redundancy checksum, performing interpacket spacing and exponential backoff on collision. The data-handling receiver (DHRx) reformats the data decoded by SHRx, stripping off the leading preamble, performing the cyclic redundancy check, and presents a byte stream to the host-dependent interface.

The self-calibrating tapped delay line is crucial to the signal-handling receiver. This subsystem utilizes an automatic frequency control loop which adjusts the control voltage of a VCO so that its frequency equals a reference frequency. This control voltage is then used to regulate the delay through a series of delay stages identical to those in the VCO. By tapping this delay line at appropriate points, one can obtain delayed versions of the input; Figure 2. Arbitrary waveforms, with edge resolution equal to the delay of one stage, can be generated by applying logic functions to the outputs of these taps. The error of the delay elements has been measured to be 0.1% per stage.

The automatic frequency control loop[2] is a charge-pump, phase-lock loop with a second-order loop filter; Figure 3. Since its only function is to adjust the VCO frequency to the reference frequency, the loop can be made highly process-independent, utilizing standard NMOS constructs. The loop is composed of four elements: the VCO the frequency division logic, a phase-frequency detector used in

generating *up* and *down* pulses for the charge-pump, and a second-order loop filter with an inverting level shifter. The output of the loop is a control voltage which is applied to the delay line elements, so acquisition time and loop bandwidth are not critical parameters. The loop must be stable and have a relatively large time constant.

To implement the signal-handling receiver (Figure 4), a delay line calibrated at 2.5ns per stage is used to generate a 25ns-wide pulse at every transition of the input. Delayed versions of this pulse (after masking the extraneous, between-bit transitions — Figure 5) become the sampling signal for the data, the sampling pulses and clock for data valid logic, and the inputs to the receive clock flip-flop. The receiver has a noise immunity of 450mV peak-to-peak and is capable of handling bit jitter of up to 10ns on every edge of the Manchester encoded data.

More broadly, in addition to the Ethernet application, the self-calibrating delay line concept can find a more general application as a versatile, on-chip clock generator. This subsystem enables the partitioning of an input signal into arbitrary sections by appropriate taps along the delay line. It can provide asynchronous clocking for many subsystems on a single chip, in some applications avoiding the use of synchronizers and the problems of meta-stable states.[3] Other uses of this subsystem include generation of arbitrary waveforms with an edge resolution as high as 2.5ns for use in, for example, specialized device controllers and memory interfaces.

Acknowledgments

The authors wish to thank R. Lyon for his many contributions to the concepts presented in this paper and L. Conway for supporting our work and providing the tractable design methodology that made this design possible. We are also indebted to R. Garner for creating the demonstration vehicle for the chip, to A. Paeth for providing the implementation service, and to the Xerox Integrated Circuit Laboratory led by W. Spencer for the fast-turna-round fabrication.

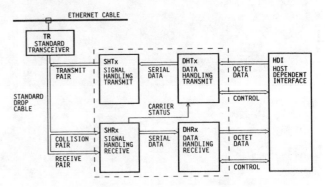

FIGURE 1—Functional units of the integrated Ethernet controller.

---

*Patents applied for.

[1] Xerox Corp., Digital Equipment Corp. and Intel Corp., "The Ethernet", *A Local Area Network, Data Link Layer and Physical Layer Specifications*, Version 1.0; Sept., 1980.

[2] Gardner, F.M., "Charge-Pump Phase-Lock Loops", *IEEE Transactions on Communications*, p. 1849-1858; Nov., 1980.

[3] Mead, C. and Conway, L., "Introduction to VLSI Systems", *Addison-Wesley Publishing Company*; 1980.

Reprinted from *1983 IEEE Int. Solid-State Circuits Conf.*, pp. 70–71, Feb. 1983.

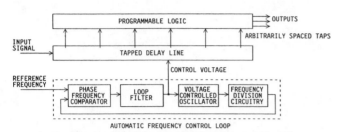

FIGURE 2—Self-calibrating tapped delay line with automatic frequency control loop.

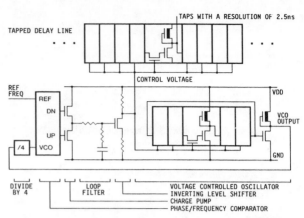

FIGURE 3—Automatic frequency control loop implementation.

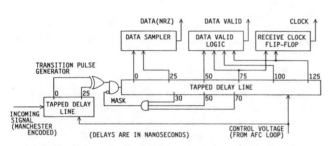

FIGURE 4—Use of the tapped delay line into the receiver of the integrated controller.

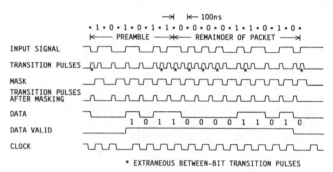

FIGURE 5—Sampling pulse generation by masking transition pulses.

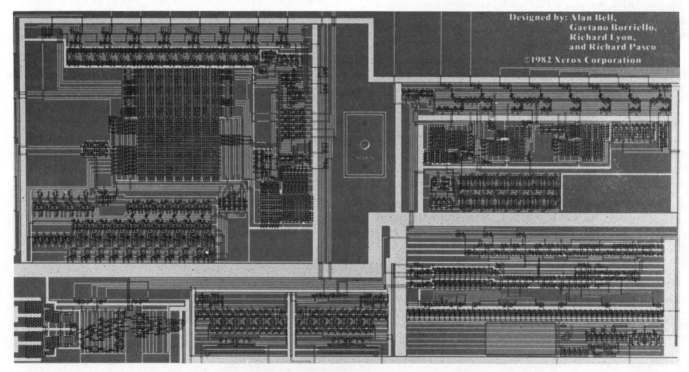

FIGURE 6—A single chip NMOS Ethernet controller.

A Monolithic Line Interface Circuit for T1 Terminals*

Kenneth J. Stern, Nav S. Sooch, David J. Knapp, Michael A. Nix

Crystal Semiconductor Corp.

Austin, TX

ONCE THE DOMAIN OF BIPOLAR DEVICES, the low-impedance line interface requirements of the 1.544 MBPS T1 terminals can now be addressed by the $3\mu m$ CMOS device to be described. A block diagram appears in Figure 1. The linear, fully differential line driver operates with a 25Ω equivalent load with a 10MHz f_T using a single 5V supply. The device provides the line-length-selectable pulse shaping necessary to meet T1 requirements. The receive side has adaptive slicing levels and a frequency-phase-lock-loop (FPLL) for clock and data recovery. The circuit interfaces between TTL or CMOS logic levels and 100Ω twisted pair, allowing transmission of 1.544MBPS data at distances of up to 1500'. It is implemented in a double polysilicon, single metal, bulk CMOS process.

T1 terminals are required to transmit alternate bipolar pulses that conform to a specified template at the DSX-1 cross-connect, a variable distance down the line[1]. The fast rolloff of twisted pair cable requires that the pulses be equalized prior to transmission. Differences in cable length require variable equalization to meet the template in all cases. Measured results at the input of the cable and at the cross-connect are shown in Figure 2 for medium and maximum length settings. Note the similarity of the two pulses after the respective cable lengths.

Pulses are created with four voltage levels by a controlled-slew switched-capacitor DAC. The coefficients for the voltage levels and the slew are stored in ROM, grouped by line length, and accessed sequentially. Slew control is continuously calibrated via a bandgap reference and a crystal timebase; both circuits are on-chip.

The line driver (Figure 3) interfaces with the 100Ω cable via a 2:1 step-up transformer. The resulting 25Ω load is driven with peak currents of 120mA by four $10,000\mu m/3\mu m$ devices (M3a-b, M4a-b). These transistors are adaptively biased at the threshold of class A-B operation by transistors M5a-b, M6a-b and switches S1a-b, S2a-b. Current sources (I) reduce quiescent power while maintaining high slew current, and the entire driver is powered down except when a pulse is to be transmitted. The gain of the driver stage is held fixed by differential resistive feedback. Closed-loop bandwidth greater than 10MHz allows the driver to follow the 40ns risetime required of the pulse. Fully differential operation allows the AMI output to reach $6V_{p-p}$.

The receiver uses a peak detector to set slicing levels for the incoming pulses. Comparator outputs for both positive and negative pulses go to the FPLL (Figure 4) for clock extraction. The center frequency of the oscillator is calibrated at power-up or at user command to 1.544MHz using the crystal timebase. The oscillator is then constrained within 6% of this value, preventing a false lock at 7/8 and 9/8 frequencies. Loop parameters are chosen so that worst-case static phase error is less than 23°, similar to classical LC-tank-based clock recovery approaches[2]. Jitter tolerance results appear in Figure 5.

A die photo appears in Figure 6. Automatic placement and routing techniques have been used extensively. The standard cell digital blocks and the analog cells were autorouted, and the interface between them was then generated for the most part automatically.

Acknowledgments

The authors wish to thank E. Swanson and R. Bridge for their continued support, and also D. Caldwell for his invaluable assistance during characterization.

---

*The 1.544 MBPS primary transmission rate in the telephony hierarchy.

[1] AT&T, "Interconnection Specification for Digital Cross-Connects — Issue 3", *AT&T Technical Advisory No. 34*, Oct., 1979.

[2] Mayo, J.S., "A Bipolar Repeater for Pulse Code Modulation Signals", *The Bell System Technical Journal*, p. 25-97; Jan., 1962.

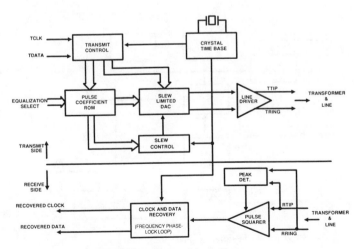

FIGURE 1—System block diagram.

Reprinted from *1987 IEEE Int. Solid-State Circuits Conf.*, pp. 292–293, 432, Feb. 1987.

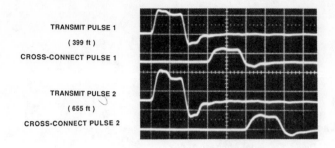

TRANSMIT PULSE 1
(399 ft)
CROSS-CONNECT PULSE 1

TRANSMIT PULSE 2
(655 ft)
CROSS-CONNECT PULSE 2

FIGURE 2—Pulse shapes at cable interface and at cross-connect.

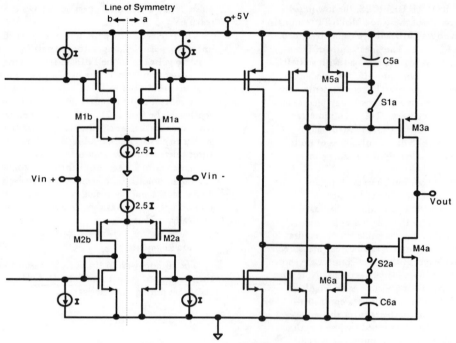

FIGURE 3—Driver partial schematic.

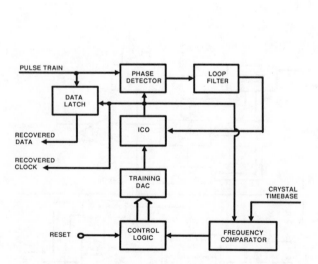

FIGURE 4—Frequency-phase-lock-loop.

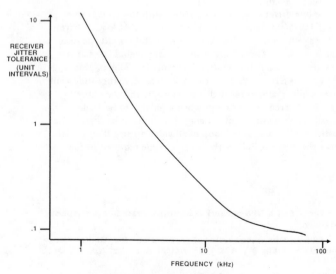

FIGURE 5—FPLL jitter tolerance.

464

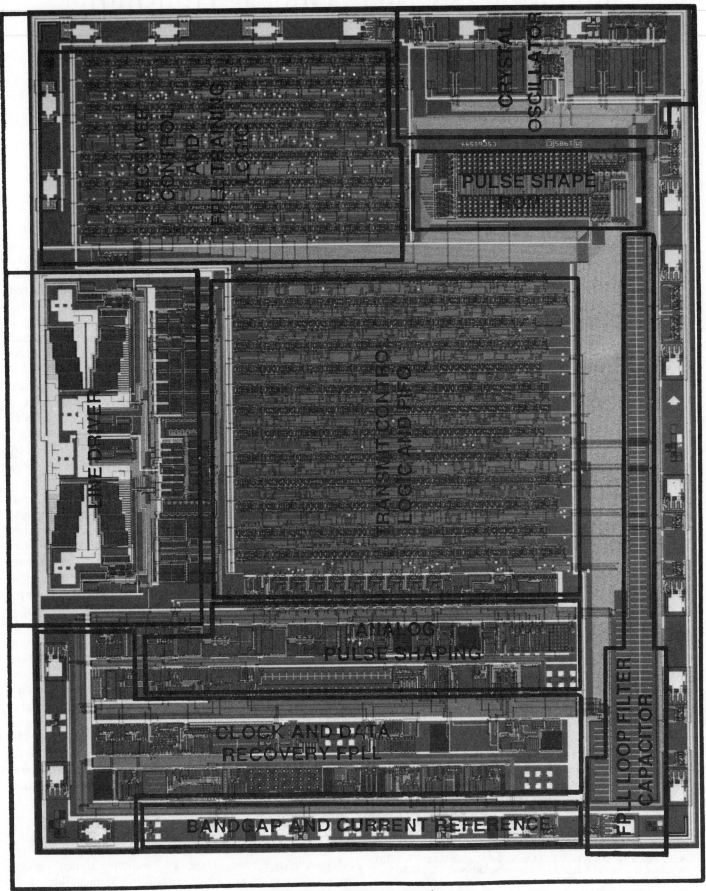

FIGURE 6—Chip die photograph.

465

A 50-Mbit/s CMOS Optical Transmitter Integrated Circuit

AARON L. FISHER, MEMBER, IEEE, AND N. LINDE

Abstract—An integrated circuit for an optical transmitter has been designed in a standard digital 1.5-μm CMOS technology. This production chip features a 50-MHz digital data scrambler, a trimmable temperature-compensated output current of between 50 and 80 mA, and switching times of < 2 ns. Asynchronous speeds of over 200 Mbit/s have been achieved. This paper describes the design objectives of this integrated circuit as well as a detailed design description followed by electrical and optical measured results.

I. INTRODUCTION

AS THE frequency of operation of VLSI devices continues to increase, the problems associated with high-frequency digital communication between chips, boards, and systems grow more severe. Transmission line effects and electromagnetic interference are two of the many challenges encountered in high-frequency signal distribution. One method of avoiding these problems is through the use of an optical data link. By transmitting data over an optical fiber, issues such as line termination and shielding are eliminated. In addition, optical fiber is less bulky than most coaxial cable and is therefore more desirable in systems where size and weight are important considerations.

In the past, however, several difficulties have prevented the widespread use of these data links. The need for multiple power supplies, extensive filtering, and external components increased the cost to the user. Also, the lack of standard packaging and pin placement has made incorporation of the data-link modules onto printed circuit boards more difficult. Finally, in order to simplify timing recovery as well as eliminate low-frequency signal components which are incompatible with many high-speed data links, it was necessary for the user to implement digital preprocessing circuitry. Consequently, the overall cost of using these links was often prohibitive.

This paper describes an integrated circuit [1] designed for the transmitter module of a new high-performance optical data link [2]. In Section II of this paper, the design objectives used to define the transmitter module and its circuitry are described. In Section III a detailed description of the circuit and system design can be found. Finally,

in Section IV the results of electrical and optical measurements are presented.

II. PRODUCT DESIGN CONSIDERATIONS

A. Design Objectives and Features

The desire to solve the problems associated with using optical data links resulted in four major design objectives for the transmitter IC described by this paper.

The first objective was to obtain a higher level of integration in this design than in previous ones. Traditionally, lightwave transmitters have consisted of a driver circuit and an optical source (an LED or a LASER). Any preprocessing of the serial digital data was done outside of the transmitter itself. In this design a higher level of integration was achieved by the incorporation of a 50-MHz synchronous data scrambler directly onto the integrated circuit with the driver. This design feature allows many overall performance enhancements to be realized. These enhancements will be described in depth in Section III.

The second objective was associated with ease of product use. The goal was to design a part that would be digital in appearance. In this design the transmitter module itself is completely TTL compatible requiring no special filtering, power supplies, input levels, or external components. In addition, the package has been designed in such a way that the pins conform to the footprint of a standard 16-pin DIP. Therefore this product is easier for the customer to design into future as well as present digital systems. In addition, by making the chip and package digital in appearance, they require less unique handling and testing than a specialized analog product. Consequently, the transmitter IC is less costly to manufacture. In fact, the chip and the package are tested on standard high-speed digital test sets.

The third objective was to design a flexible product that would have broad application. It was necessary that the transmitter operate either synchronously or asynchronously and that the user be able to control the mode of operation. It was also desirable to have the actual output current delivered to the LED be programmable, although not necessarily by the customer. This would extend the IC's lifetime allowing it to adapt to changing LED electrical characteristics brought about by evolving optical component technology as well as extend the applicability of the

Manuscript received May 23, 1986; revised August 5, 1986.
A. L. Fisher is with AT&T Bell Laboratories, Allentown, PA 18103.
N. Linde is with AT&T Bell Laboratories, Orlando, FL 32809.
IEEE Log Number 8610947.

Reprinted from *IEEE J. Solid-State Circuits*, vol. SC-21, no. 6, pp. 901–908, Dec. 1986.

chip design to various data-link products with different optical sources.

Various features were incorporated into the design in order to achieve this flexibility. Two modes of operation are available and the mode is pin selectable by the user. A synchronous mode is obtained via use of the data scrambler and an asynchronous mode is available which bypasses the data scrambler. The output current value is also programmable, although not by the user. This programmability involves a fusible-link meltback trim performed at wafer probe. By use of this trim technique, the output current can be programmed in the range of 50–80 mA.

Finally, the fourth objective was to achieve a low-cost product without sacrificing performance. The goal was to reduce the cost of the silicon to a negligible fraction of the total cost of the data link itself. This translated into using a standard digital technology while keeping die size and test time to a minimum.

B. Technology Choice

In order to achieve the product objectives described above, it was determined that the technology of choice would be a standard 1.5-μm digital CMOS technology [3]. The use of a digital technology eliminated most of the problems associated with designing a part "digital in appearance." The design of TTL input buffers was well understood, as was the design of high-speed static CMOS logic. The only unknown was the design of low-power-supply, high-current driver circuits in CMOS. Finally, since the technology was a mainstream digital process, the benefits associated with yield improvement, modeling, process support, and other advantages associated with economy of scale could be capitalized upon.

III. DESIGN DESCRIPTION

A. Overview

A block diagram of the optical transmitter chip is shown in Fig. 1. All of the blocks shown in the figure are contained on the CMOS die except for the GaAlAs LED which is contained within the transmitter package itself. The three blocks on the right side of the figure, the driver, current source, and proportional-to-absolute-temperature (PTAT) current reference, are high-performance analog circuit designs. The remaining blocks are high-performance digital circuit designs.

Operation of the chip is available in two modes, synchronous and asynchronous. The mode of operation is selected via the *ENABLE SCRAMBLER* pin. This pin controls the state of a selector circuit which is implemented as a transmission gate 2:1 multiplexor. When in one state, the multiplexor selects asynchronous data directly from the *DATA* input buffer and directs it to the driver circuit. When in the other state the digital data to be presented to the driver comes from the output of the synchronous data scrambler, whose output is under the

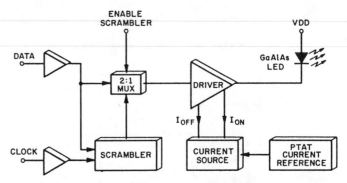

Fig. 1. Transmitter IC block diagram. All blocks are contained on the chip except the GaAlAs LED.

control of both the *DATA* input as well as the *CLOCK* input.

The driver circuit presents the LED with one of two currents, depending on whether the LED should be emitting light (digital ONE) or not (digital ZERO). When producing a ONE the LED is driven with a high-valued current (between 50 and 80 mA), labeled I_{on} in Fig. 1. When producing a ZERO the LED is driven with a small trickle, or prebias current I_{off}. This prebias current is low enough that no appreciable amount of light is emitted from the LED but that the diode's forward voltage drop is maintained. This limits the overall voltage excursion on the output node of the driver and thereby increases the speed at which that node can be switched.

The two currents I_{on} and I_{off} are provided from a current source that receives its input from a PTAT current reference. This current reference provides the positive temperature coefficient of current necessary in order to cancel the negative temperature coefficient of emitted optical power which is characteristic of high-performance LED's [4]. In addition to providing a PTAT current, the reference demonstrates a high degree of power-supply rejection and therefore provides isolation from supply noise caused by high-speed high-magnitude demand current transients typical of optical transmitters. The design of the PTAT current reference will now be described.

B. PTAT Current Reference

Fig. 2 shows a simplified schematic of the PTAT current reference. It is based on a bipolar thermal voltage reference and one form of it has been previously described [5].

The bottom current mirror consisting of devices M_1 and M_2 utilizes substrate p-n-p devices Q_1 and Q_2 connected as diodes. Devices M_1 and M_2 force the voltages at nodes A and B to be equal causing a logarithmic relationship between I_1 and I_2 as given below:

$$I_2 R_{\text{REF}} = V_T \ln\left(\frac{I_1 I_{S2}}{I_2 I_{S1}}\right) \tag{1}$$

where I_{S1} and I_{S2} are the saturation currents of Q_1 and Q_2, respectively, and $V_T = kT/q$.

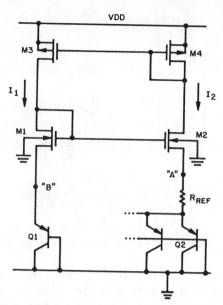

Fig. 2. PTAT current reference. Simplified schematic.

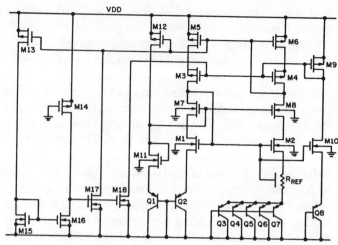

Fig. 3. Full PTAT current reference schematic.

The top current mirror M_3, M_4 sets

$$I_1 = I_2. \qquad (2)$$

Solving (1) and (2) yields

$$I_1 = I_2 = \frac{kT}{qR_{REF}} \ln(a) \qquad (3)$$

where $a = I_{S2}/I_{S1}$. Therefore I_1 and I_2 are PTAT. Perfect compensation of the LED's negative optical temperature coefficient would require matching of individual chips to individual LED's. This operation is extremely costly and not feasible in a high-volume production environment. It has been found that first-order cancellation is obtained by the use of this PTAT reference without regard to specific LED characteristics.

Although the circuit of Fig. 2 will function, several nonidealities exist which mandate a more complex design. The first problem with the circuit is due to the finite output conductance of the MOS device. Since the pairs of devices M_3, M_4 and M_1, M_2 have different drain-to-source voltages, significant error currents are present. The second problem is due to MOS threshold voltage dependence on drain-to-source voltage. The voltages at nodes A and B are equal only if the threshold voltages of M_1 and M_2 are equal. Since their V_{DS} terms are different, this results in an error term ΔV_{TN} being applied across resistor R_{REF}. This error can be as large as ± 10 percent.

The final problem with the circuit of Fig. 2 is inherent to most power-supply-independent biasing schemes. That is, the circuit has two valid operating points. One is at the desired current value and the other is a zero current state. This mandates that a start-up circuit be included in order to insure proper operation.

Fig. 3 shows the final full reference circuit developed. The first items of interest are the current mirrors. The

simple PMOS and NMOS current mirrors of Fig. 2 (M_1, M_2 and M_3, M_4) have been replaced with cascode current mirrors. Devices M_3–M_6 form the PMOS cascode current mirror and devices M_1, M_2, M_7, and M_8 form the NMOS cascode current mirror. Due to the cascode configurations, the output conductance mismatch problem described above is significantly reduced. The cascode configurations also serve to improve the power-supply rejection of the overall circuit. In addition the drain-to-source voltages of M_1 and M_2 are now forced to be equal, eliminating the errors associated with threshold dependence upon drain-to-source voltage. In order to improve device matching, M_5, M_6 and M_1, M_2, M_7, and M_8 have been laid out using common centroid geometries. Finally, all cascode devices have longer than minimum channel lengths which results in lower output conductances and therefore better supply rejection.

Unlike the circuit proposed in [5], this reference does not utilize diode-connected MOS devices. In the simplified circuit of Fig. 2, devices M_1 and M_4 are diode connected which is common for current mirrors. Cascode current mirrors often contain diode-connected devices as well. It was found, however, that a diode-connected circuit required about 5 V of power-supply headroom to maintain all devices in the saturated region of operation. This situation resulted in marginal circuit performance under worst-case processing and environmental conditions. It was found that variations in power-supply voltage produced variations in output current that were unsatisfactory. Therefore a ratio cascode biasing technique was utilized. This biasing scheme [6] significantly extends the power-supply range of the circuit and has been used extensively throughout this chip in order to obtain high-accuracy high-magnitude current sources with a minimum of voltage headroom required.

The PMOS cascode mirror is biased by the bias leg consisting of M_9, M_{10}, and Q_8. The NMOS mirror is similarly biased by devices M_{12}, M_{11}, and Q_1. This technique insures that devices M_1, M_2, M_5, and M_6 all remain in the saturation region of operation and yet have the minimum drain-to-source voltages possible. The use of this biasing technique has extended the low-voltage region of operation of the circuit by about 2.5 V.

As mentioned above, the reference requires a start-up circuit. Devices $M_{13}-M_{18}$ form the start-up circuit for the reference and the operation is as follows.

At power up no current is flowing in any of the legs of the circuit. Therefore the gates of M_{17} and M_{18} get charged to the value of the power supply by device M_{14} which is in the triode region of operation. M_{17} and M_{18} then conduct current pulling the gates of the PMOS cascode current mirror low. This snaps the circuit to the desired operating point causing current to flow in device M_{13}. This current is mirrored to M_{16} which turns off M_{17} and M_{18} disconnecting the start-up circuit from the reference.

The final item of interest in the reference circuit is the resistor R_{REF}. It is constructed of p$^+$ material in an n-type tub. As mentioned in Section II-A, one of the design objectives was to be able to adjust the value of the output current in order to adapt to changes in optical component requirements. The values of I_{on} and I_{off} are derived from this current reference. By performing a 3-bit fusible-link meltback trim on the resistor R_{REF} in the reference circuit, the final output current value can be varied between 40 and 120 mA. However, sheet resistance variations and metal migration related reliability considerations limit the realistic trim range to between 50 and 80 mA.

C. High-Current Driver

The design of the high-current driver circuitry was influenced by several considerations. The first consideration was frequency of operation. In order to produce a high-quality signal at 50 Mbit/s, the electrical rise/fall times for the high-current waveform were required to be 3 ns or less. Since functions such as switching 80 mA in 3 ns were previously thought to be unsuitable for digital CMOS, this requirement influenced the driver design to the largest extent.

The second consideration was pulsewidth distortion (PWD). PWD is defined as the difference in delay between a ONE going through the circuit and a ZERO going through the circuit. Since both the driver circuit and the LED contribute to the optical transmitter's overall PWD performance, it is typical to specify a tighter PWD on the driver than on the overall transmitter. Therefore one of the most critical specifications for LED driver circuits is the PWD inherent in the circuitry. In addition, since the data can come from two different sources (either the asynchronous data input or the scrambler), it was necessary that each of the two individual paths be separately balanced. For this particular chip design, the PWD specification is ±1 ns or less.

The final design consideration involved the reduction of high-frequency noise generated by the transmitter module itself. In a simple driver circuit high-valued demand current transients are produced. These transients induce noise on any stray series inductance and the resultant power-supply noise generated can disrupt other circuit board

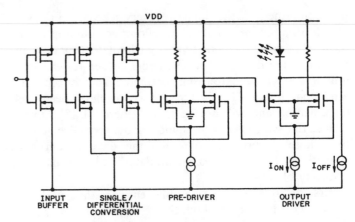

Fig. 4. Driver circuit schematic. Shown is the asynchronous data path without the 2:1 multiplexor.

components. Typical designs commonly require either extensive filtering of the transmitter module to prevent this noise, or provide this filtering inside the transmitter package. In order to facilitate ease of use as well as to maintain the "digital appearance" of the product, it was desirable to require only customary TTL filtering for the module. Therefore the reduction of power-supply-induced noise was an important design consideration.

The final driver design addresses these three considerations and a schematic can be found in Fig. 4. The circuit consists of four stages of gain, including the TTL input buffer. Fig. 4 is representative of the asynchronous data path and does not include the 2:1 multiplexor.

The first stage, the TTL input buffer, consists of a standard CMOS inverter with its device sizes set such that its threshold is at about 1.4 V. This provides TTL input level compatibility. The second stage performs single-ended-to-differential conversion. This is realized by providing the predriver with the digital signal at one input and a constant voltage at the other. By adjusting the device sizes of this stage, the relative delays for a ONE and a ZERO can be varied. These device sizes have been chosen in order to balance the logic delays for a ONE and for a ZERO in order to provide the optimum PWD performance.

The final two stages, the predriver and output driver, are fully differential in configuration. This results in faster switching than standard CMOS logic because of the reduced voltage swing on the differential nodes. Also, the use of differential switching results in more balanced operation which results in better PWD performance. Finally, since differential switching is a current steering technique, the demand current transients imposed on the power supply are greatly reduced; this results in less noise induced on the power supply. Therefore the filtering requirements for the chip can be significantly reduced.

D. Digital Data Scrambler

In most applications of optical data links, coding is performed on the data before transmission. By the use of digital coding, the user can insure that the serial bit stream contains an adequate number of data transitions to satisfy the requirement of the timing recovery scheme being used. In addition, coding of the data allows ac coupling of the receiver because the code typically eliminates any dc component in the data stream. The coder chosen for this data link was a self-synchronizing scrambler. The concepts behind the design of this type of coder have been well documented and will not be covered here [7], [8]. Simply stated, the scrambler is a pseudorandom sequence generator whose purpose is to enrich the data stream with transitions. This improves overall data-link performance in a variety of ways.

First of all, since the scrambler enriches the serial data stream with transitions, both the receiver design as well as the timing recovery design become less complex. By in effect limiting the dc component in the data stream, the receiver can be ac coupled. The timing recovery scheme can have a tighter acquisition range because a minimum low-frequency component is insured. Therefore the overall link performance is isolated from the particular statistics of the data. This allows the link to be used with data containing varying frequencies, such as burst-mode data. Burst-mode data are characterized by long periods with no data (all ZEROS or all ONES) and then short periods with high-frequency bursts of data. By using the scrambler and clocking it at the data rate, the minimum frequency that the data link sees will be limited.

Use of the scrambler improves performance in another way. Since the scrambler is a synchronous circuit, the data are retimed on the chip. Therefore if the incoming data have a great deal of PWD, the retiming function provided by the scrambler will improve that distortion. Therefore the PWD of the output data stream will not be dependent on the input signal. In the asynchronous mode of operation the driver will reproduce the input signal PWD with the ±1-ns figure added to it. Consequently the user can "clean up" degraded data signals by use of the synchronous mode of operation.

A logic schematic of the scrambler design can be found in Fig. 5(a). Fig. 5(b) shows the descrambler circuit which is contained on a companion retiming IC as described in [2]. The scrambler is a seven-stage tapped shift register with feedback. One of the key design considerations was the length of the shift register. As is discussed in [7] and [8] the length of the shift register is also the maximum number of consecutive bit positions without a transition that the scrambler will produce given constant input data. Therefore a short shift register is desired. However, one disadvantage of this type of scrambler is that there is a finite probability that a particular input data sequence could null the output. This probability is decreased by increasing the length of the shift register. Therefore the design goal was to reduce this probability to a value less than the bit error rate of the data link. This was achieved by including a lock-up detection circuit. This circuit monitors an internal node of the scrambler and counts the number of consecutive bit periods without a transition. When that number exceeds a certain value, the detection circuit reinitiates the scrambling sequence. The choice of that value

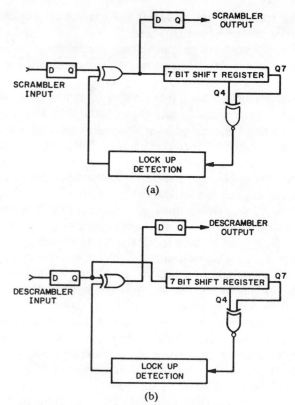

Fig. 5. (a) Scrambler logic schematic. (b) Descrambler logic schematic.

determines the probability that the output will be nulled (the equations that relate the two can be found in [7]).

The implementation of the scrambler was realized with static CMOS logic circuits and was relatively straightforward. The circuits used were from a standard cell library but the cells were hand sized, placed and routed in order to insure worst-case operation at 50 MHz. Included was a testability feature which allowed the contents of the shift register to be cleared. This allowed testing to be done with a shorter number of vectors.

E. Layout Considerations

Since the devices in the driver section of this chip switch currents with magnitudes on the order of tens of milliamperes, specifics concerning device design, layout, and arrangement are critical to obtaining a successful design. In addition, since the driver devices are extremely large (the source-coupled pair in the output stage has channel widths of 3000 μm), the transient charge/discharge currents on their gate nodes are also in the range of tens of milliamperes. Therefore issues such as series contact resistance, electromigration, and power/ground voltage drops had to be carefully considered and factored into the overall physical layout.

Fig. 6 shows a photomicrograph of the optical transmitter chip. The driver section can be clearly seen in the upper right corner. Attached to the output pads, VDD and LED are the output driver source-coupled pair. For this differential stage as well as the one used in the predriver an interdigitated multiple-stripe geometry was used. Fig. 7

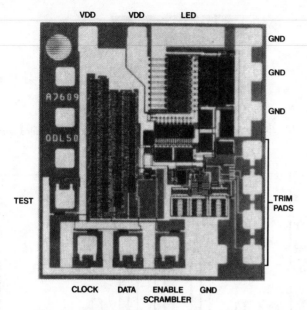

Fig. 6.   Optical transmitter chip photomicrograph.

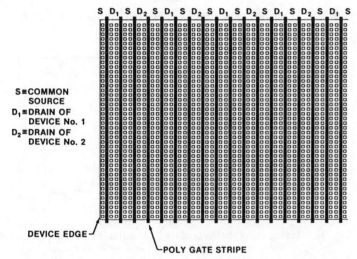

S ≡ COMMON
SOURCE
D_1 ≡ DRAIN OF
DEVICE No. 1
D_2 ≡ DRAIN OF
DEVICE No. 2

DEVICE EDGE

POLY GATE STRIPE

Fig. 7.   Details of the interdigitated multistripe source-coupled pair layout.

shows a section of the layout of the output driver. Every other diffusion stripe is the common-source node and the drain nodes alternate from one of the differential devices to the other. This distributes both devices over the entire structure and consequently improves overall device matching.

An examination of the photomicrograph shows that many of the devices on the chip utilize multistripe layout configurations. In order to facilitate quick layout implementations of these devices, a programmable layout generator was developed. This generator was implemented using a feature available in the graphics layout editor [9]. By specifying the overall device width in micrometers and the number of poly gate stripes desired, the generator produced the required layout. Since this generator allowed the layout to easily adopt to changes in design rules, the design of this chip was able to take place early in the process development schedule.

Also shown in Fig. 7 is the contact scheme used for the output driver. In order to avoid violating electromigration guidelines and thereby compromising the reliability of the chip, a double row of contact heads is provided on each of the diffusion regions in the output driver. In fact, the dominant consideration in determining the layout of the interconnect in the driver section was electromigration. This is unlike logic layouts where parasitic capacitance is usually the most important layout consideration. In this design the capacitances associated with the large driver devices dominate the layout parasitics. Therefore the widths of the metal interconnect are determined primarily by electromigration considerations.

The final item of interest concerning layout is the implementation of the meltback trim. The lower right four pads of the chip are the meltback trim nodes. Three minimum width metal fuses are located between the pads and these fuses short out three binary weighted resistors in series with the main resistor in the PTAT current reference. These fuses are blown at wafer probe with high-current pulses from the digital test set. By using this trim technique, a range of currents is available for use and the absolute accuracy of the current can be better controlled.

IV. EXPERIMENTAL RESULTS

This section presents experimental measurements on the performance of the transmitter IC. First described are the results from bench measurements on the PTAT current reference. Next shown are results derived from measurements on the high-current driver. These results are both from laboratory bench testing as well as compilations of data from automatic digital test set measurements.

Fig. 8 shows a plot of simulated and measured results for the PTAT current reference. Comparing the two we find a difference of about 7.5 percent between the simulated and measured results. It is believed that the inaccuracy is partially due to device mismatch in the current mirrors in the reference. Also suspected is the variation of the temperature coefficient of the current-controlling resistor in the reference. By differentiating (3) with respect to temperature, one can show that the temperature coefficient of the current TCI at a temperature T_0 is given by

$$TCI = \frac{1}{T_0} - TCR \qquad (4)$$

where TCR is the temperature coefficient of the resistor. Therefore variation of the TCR of the resistor could account for the difference.

Another important figure of merit for the reference is its power-supply regulation. Simulated results indicate proper operation at values of V_{DD} as low as 2.5 V. Fig. 9 shows the measured power-supply performance for the reference. The figure shows three curves representative of a nominal site, a high-current site, and a low-current site. The variation in absolute value of the current is due to variations in sheet resistance of the current-setting resistor in the refer-

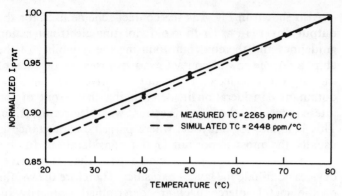

Fig. 8. Measured and simulated performance of the PTAT current reference.

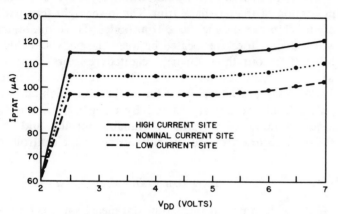

Fig. 9. Measured power-supply regulation of the PTAT current reference.

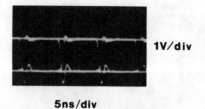

Fig. 10. Electrical switching waveform of the LED driver loaded with a 25-Ω impedance.

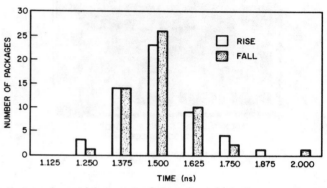

Fig. 11. Histogram of the electrical rise and fall times of the LED driver. Load impedance is 25 Ω.

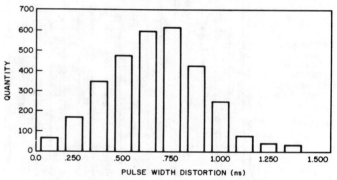

Fig. 12. Histogram of the PWD of the LED driver. Measurements were made automatically on roughly 3000 wafer sites.

ence. Measured power-supply performance of the reference indicates a ± 0.175-percent change in output current for a change in power supply of ± 10 percent (4.5–5.5 V). As the measurements demonstrate, the reference maintains regulation at supply values as low as 2.5 V.

Fig. 10 shows an oscilloscope photograph of the output node of the driver with a 25-Ω resistive load. The input signal was generated from a random sequence generator and the mode of operation in this case is asynchronous. The oscilloscope was triggered synchronously to the input data thus resulting in the eye diagram as shown. As can be seen from the figure, rise/fall times are on the order of 1.5 ns and PWD is on the order of 1 ns. It should be noted that this measurement was taken on a device before the trim procedure was performed and that this particular driver was switching 100 mA. The effect of a resistive load (as opposed to an LED) is to increase the voltage excursion on the output node of the driver. In this case it is on the order of 2.5 V. With an LED as the load, the voltage swing is much smaller, on the order of 0.5 V. Therefore the crossing of the rising and falling edges in the eye diagram, which appears near the bottom of the swing, is at about the 50-percent switching point when the driver is loaded by an LED.

By use of the digital test set, large numbers of devices can be accurately characterized. The chip is tested at full speed for both the wafer and package test. Since ac parametric measurements such as rise/fall times and PWD are

relative measurements (i.e., the quantities being measured are actually the differences of two measurements), an overall timing accuracy of 0.125 ns is obtained. Fig. 11 shows histograms of the electrical rise and fall times for 50 transmitter packages selected at random from a recent production lot. As the figure demonstrates, both rise and fall times are centered around 1.5 ns which is well within the 3-ns absolute maximum specification. A similar curve of PWD, measured on several thousand wafer sites, is shown in Fig. 12. As these data show, the PWD is centered at about 0.75 ns which is under the 1-ns absolute electrical maximum.

Although the chip itself is evaluated electrically, the bottom line for an optical product is the optical performance. Fig. 13(a) and (b) shows the measured optical performance of the transmitter in two experiments. In the first, a random sequence of 90 Mbit/s was used as an asynchronous input and the optical output was measured by using a short length of fiber and a wide-bandwidth avalanche photodiode as a detector. As the figure shows,

IEEE JOURNAL OF SOLID-STATE CIRCUITS, VOL. SC-21, NO. 6, DECEMBER 1986

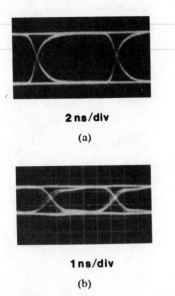

2 ns/div

(a)

1 ns/div

(b)

Fig. 13. Optical waveform measurements. (a) GaAlAs short-wavelength LED with a 90-Mbit/s pseudorandom word ($I_{on} = 80$ mA). (b) In-GaAsP long-wavelength LED with a 212-Mbit/s pseudorandom word ($I_{on} = 120$ mA).

the high-speed high-current capability of the technology and provides an existence proof for future higher performance driver designs.

V. CONCLUSIONS

This paper has described the design of a 50-Mbit/s optical transmitter IC. The chip contains both the high-current driver and a 50-MHz data scrambler and has been designed in a standard 1.5-μm digital CMOS technology. The work presented in this paper demonstrates the ability of digital CMOS to perform both high-speed logic functions as well as high-current driver functions which will be necessary for data-link products of the future.

ACKNOWLEDGMENT

The authors would like to acknowledge the valuable contributions of D. Morrison, D. Koehler, and L. Ackner as well as the support of J. Henry and D. Pedersen.

rise/fall times are on the order of 3 ns which is completely consistent with the performance expected of this LED. PWD as measured at the 50-percent point is almost indiscernable. In the second experiment, Fig. 13(b), a random word of 212 Mbit/s was provided as asynchronous input to the chip. By selecting a chip whose untrimmed value of output current was about 120 mA, a wider bandwidth LED could be used. The purpose of this experiment was to push the chip to its maximum asynchronous frequency of operation. Although this objective was not achieved (the maximum speed of the test equipment was first reached), the experiment demonstrated that this driver could perform at speeds exceeding 200 Mbit/s. Examination of the figure shows that the eye diagram still has room for closure, and therefore the driver itself might operate as much as 50 percent faster. This experiment demonstrates

REFERENCES

[1] A. L. Fisher and N. Linde, "A 50 Mbit/sec CMOS LED driver circuit," in *ISSCC Dig. Tech. Papers*, 1986 p. 58.
[2] D. P. Morrison, "A 50 Mbit/sec CMOS fiberoptic datalink," in *Proc. 1984 Fiber Optic and Local Area Network Conf.*
[3] J. Agraz-Guerena *et al.*, "Twin tub III—A third generation CMOS technology," in *Tech. Dig. 1984 IEDM*, pp. 63–66.
[4] H. Kressel *et al.*, *Semiconductor Devices for Optical Communication*. New York: Springer-Verlag, 1982, p. 40.
[5] P. R. Gray and R. G. Meyer, *Analysis and Design of Analog Integrated Circuits*. New York: Wiley, 1984, p. 735.
[6] E. J. Swanson, *Design of MOS VLSI Circuits for Telecommunications*, Y. Tsividis and P. Antognetti, Eds. Englewood Cliffs, NJ: Prentice-Hall, 1985, p. 560.
[7] J. E. Savage, "Some simple self-synchronizing digital data scramblers," *Bell Syst. Tech. J.*, Feb. 1967.
[8] D. G. Leeper, "A universal digital data scrambler," *Bell Syst. Tech. J.*, Dec. 1973.
[9] D. H. Potter, J. D. Tauke, and A. A. Yiannoulos, "Parameterized procedural module generation for physical integrated circuit design," in *Proc. 1983 IEEE Int. Conf. Computer Design: VLSI in Computers*, p. 290.

A 2Gb/s Silicon NMOS Laser Driver

Robert G. Swartz, Alexander M. Voshchenkov, Gen M. Chin, Sean N. Finegan Maureen Y. Lau

Mark D. Morris, Vance D. Archer

AT&T Bell Laboratories

Holmdel, NJ

Ping K. Ko

University of California

Berkeley, CA

THIS PAPER will describe a laser modulation circuit fabricated in a 1.5μm design rule, 1μm effective channel length (L_{eff}) NMOS process. The circuit is intended for application in optical communications and has been tested at a maximum bit rate of 2Gb/s. Ring oscillators with fan-in/fan-out = 1 fabricated in this process have previously demonstrated a 65ps gate delay with 1μm L_{eff} and 32ps delay with L_{eff} = 0.5μm[1]. Other reported circuits include 1.3GHz bandwidth amplifiers, 3.5GHz bandwidth buffer/line drivers and latched comparators with a 4b resolution at 750Msamples/s[2].

Figure 1 shows the circuit diagram for the laser driver which is configured as a direct-coupled amplifier with three differential amplifier stages and two pairs of source follower buffers. Diode connected (gate wired to drain) MOSFETs are used to set the voltage gain of the input stage to approximately 4 and to bias the succeeding stages properly. Depletion load MOSFETs are wired in parallel with the diode-connected MOSFETs to increase the transconductance of the input drivers by augmenting their drain current.

Following a buffer, the second gain stage is a cascode-differential pair and another set of diode-connected/depletion load MOSFETs. The final differential pair is buffered by a second pair of source followers. Output drive current is adjustable from 0mA to over 100mA by varying the gate voltage of the output current source. All internal circuit nodes have RC time constants of 100ps or less. Large transistors are assembled from an array of smaller modules each containing a gate and source/drain stripe. These modules are designed with a W/L ratio of 10μm/1μm, chosen as a compromise between design convenience and the need for minimizing RC gate delay. Modules are wired in parallel so that sources and drains are each shared by two gates, thus reducing sidewall junction capacitance. The chip dimensions of the completed laser driver circuit (Figure 2), including bonding pads are 1.0mm x 1.0mm.

Device gate/drain overlap capacitance is reduced to 0.15fF/μm of gate width by means of a thermal oxide spacer on the sidewall of the poly-silicon gate. A double profile source-drain composed of a shallow arsenic-implanted region and a deeper phosphorus-diffused contact region minimize short channel effects, while substantially decreasing drain/source contact resistance and junction capacitance (the latter to 0.2fF/μm^2).

Low-frequency wafer probing established a peak voltage gain of approximately 22dB (50Ω in/out) and a typical maximum output drive current of 210mA into 25Ω with a 0.8V single-ended ECL-level input. Nominal current draw for both the positive and negative supplies was about 125mA with $\pm$5V supplies, exclusive of the final stage (laser drive) current. For microwave testing, individual die were bonded to alumina substrates with connections to external SMA connectors via 50Ω microstrip. Nominal 10%-90% rise and fall times were approximately 320ps at maximum output into 25Ω with an ECL-level 0.65V input. These times were improved to 200ps by reducing the output current or, increasing the input drive; Figure 3. An error rate test set* was able to distinguish a 1 from a 0 with zero errors during 300ps of the 500ps wide bit transmission period at 2Gb/s.

A plot of peak output current vs. frequency for various input amplitudes is shown in Figure 4. The 22dB gain figure obtained in dc on-wafer measurements could not be repeated in testing with packaged die. We attribute this to inadequate heat sinking in the microwave packages. It is probable, therefore, that further attention to packaging will yield substantially improved high frequency performance. Figure 5 shows the detected output of a semiconductor laser in response to a pseudorandom NRZ input at 880Mb/s and 1.7Gb/s. External electronics were used to provide a feedback stabilized dc bias current.

Acknowledgments

The authors wish to acknowledge the assistance of K. Yanushefski and S. Lumish in performing the measurements described. They also thank P.K. Tien for his support throughout this project.

---

*Anritsu MS65A.

[1] Ko, P.K., Voshchenkov, A.M., Hanson, R.C., Grabbe, P., Tennant, D.M., Archer, V.D., Chin, G.M., Lau, M.Y., Soo, D.C. and Wooley, B.A., "SiGMOS — A Silicon Gigabit/s NMOS Technology", *IEDM Technical Digest*, p. 751-753; Dec., 1983.

[2] Soo, D.C., Voshchenkov, A.M., Chin, G.M., Archer, V.D., Lau, M.Y., Morris, M.D., Wooley, B.A., Ko, P.K. and Meyer, R.G., "A 750MS/s NMOS Latched Comparator", *ISSCC DIGEST OF TECHNICAL PAPERS*, p. 146-147; Feb., 1985.

Reprinted from *1986 IEEE Int. Solid-State Circuits Conf.*, pp. 64-65, 308, Feb. 1986.

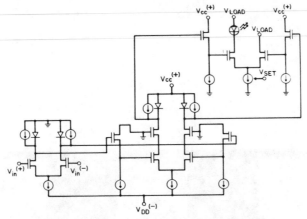

FIGURE 1—Laser driver circuit diagram.

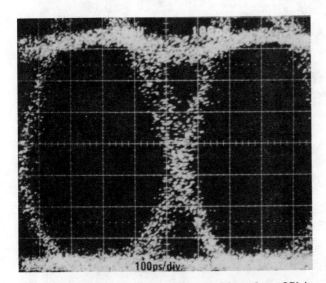

FIGURE 3—Current output with pseudorandom 2Gb/s, 1.4V NRZ input. Vertical scale: 8mA/div.

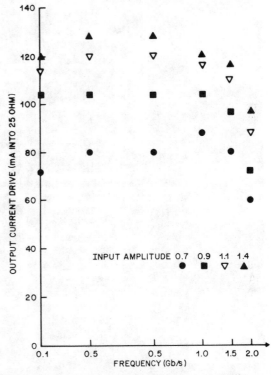

FIGURE 4—Maximum output drive current vs. frequency for input amplitudes of 0.7V, 0.9V, 1.1V, and 1.4V.

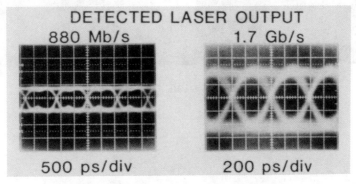

FIGURE 5—Laser output at 880Mb/s and 1.7Gb/s.

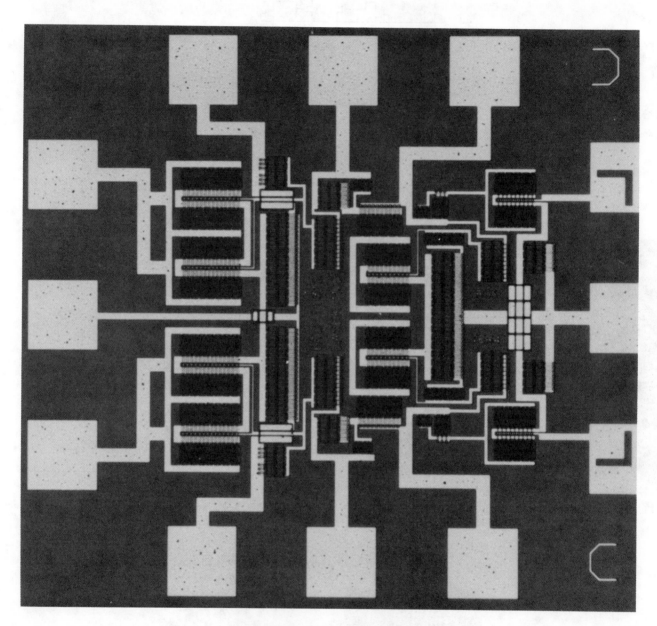

FIGURE 2—Chip photo.

A 50Mb/s CMOS Optical Data Link Receiver Integrated Circuit

John M. Steininger, Eric J. Swanson

AT&T Bell Laboratories

Reading, PA

FINE-LINE HIGH-SPEED MOS devices are allowing the fabrication of wide bandwidth circuits that have traditionally been a bipolar stronghold. The chip to be described exploits the fast devices in a 1.5μm CMOS technology with several circuit design techniques to produce a complete, high sensitivity optical receiver sub-system on a single chip. Mated with a custom low voltage PIN diode and housed in a compact package, the result is a low cost 50Mb/s lightwave receiver.

Except for power supply bypassing, the 5V-only part needs no external components for operation. All ac coupling and AGC time constants are executed on the IC. A quiet PIN diode bias supply, featuring a 1000pF capacitor and average diode current monitor is supplied from the chip. A bandgap derived current source using parasitic bipolar devices biases the analog sections for supply and temperature insensitive operation.

A low-noise front end enables the receiver to operate at input currents from the PIN diode as low as 100nA rms. A two-loop AGC system produces a 60dB dynamic range and drives a mute circuit. Without a valid input signal, the mute circuit pulls a Link Status Flag high and, when tied to an Output Disable, mutes the data outputs.

A standard cell layout approach has been used and all non-critical interconnect was machine routed. Over 100dB of input to output isolation at 25MHz has been achieved. With no input signal, no oscillations are observed even with the output disable defeated.

Block diagram for the circuit is in Figure 1. The external low voltage PIN diode is attached between the PIN bias supply and preamp input to complete the optical receiver.

A transimpedance preamp is a three-stage amplifier with a resistively-biased MOSFET connected from input to output; Figure 2*. The amplifier stages are common source with a diode connected device used as a load. These g_m/g_m (g_m of M1/g_m of M2) gain amplifiers have wide bandwidths and are process and temperature insensitive. Figure 3 shows the transimpedance and output noise response with an optical input.

Surrounding the preamp is its own AGC system. A preamp AGC processor peak-to-peak detector monitors the output of the preamp and compares it to a derived dc voltage. Loss can be applied to the preamp by controlling the gate of a shunt MOSFET (M16 in Figure 2) that steals diode current from the preamp input.

Signal from the preamp is ac coupled into the variable gain amplifier (VGA). This differential in/differential out stage has an adjustable gain from +2 to –40dB and uses a current steering/cancellation to achieve variable gain. Following the VGA are two stages of postamplification. These differential input/output amplifiers have a fixed gain of 20dB and 3dB bandwidths of 50MHz with a linear output swing of 1V. These postamplifiers (Figure 4) also have a gain which is set by a ratio

of transconductances. In this circuit the load device is a P-channel MOSFET, M13 (M14), and the gain device is M1 (M2).

A second AGC loop surrounds the VGA and postamp chain. The AGC processor is identical to the preamp AGC processor and controls the gain of the VGA to hold the output of the second postamp to 0.5Vp-p.

The output of the second postamp is ac coupled into a fast comparator where the analog signal is quantized into a digital output. The comparator uses a balanced architecture similar to the postamps, but with an added push-pull output stage and diode clamp to achieve a high open loop gain (100dB) and rail-to-rail output swing.

An eye diagram of the output data at 50Mb/s with an input signal well above the noise floor, is shown in Figure 5. The residual jitter or pulsewidth distortion at 50Mb/s is less than 2ns. The input power necessary to achieve a bit error rate of 10^{-9} at 50Mb/s is $\eta P = -40$dBm, corresponding to a minimum input signal of 100nA rms. A photomicrograph of the die on the package header is shown in Figure 6.

Acknowledgments

The authors wish to thank D. Sherry, D. Morrison, B. Del Signore, J. Hein, N. Sooch and G. Williams for their valuable contributions and support.

---

*U.S. Patent 4,540,952 issued to G.F. Williams; Sept. 10, 1985.

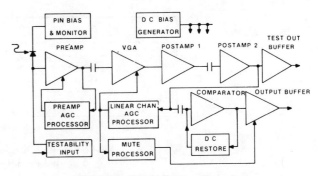

FIGURE 1—Block diagram of receiver.

Reprinted from *1986 IEEE Int. Solid-State Circuits Conf.*, pp. 60–61, 304, Feb. 1986.

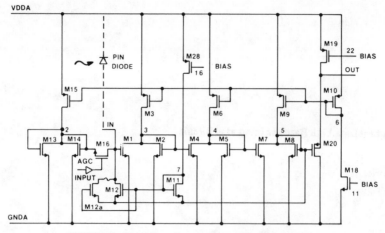

FIGURE 2—Transimpedance preamplifier schematic.

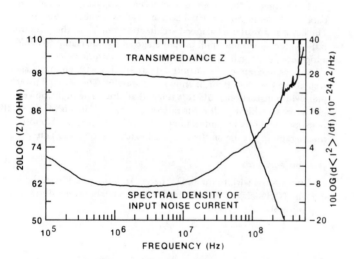

FIGURE 3—Frequency response and noise spectral density of preamplifier.

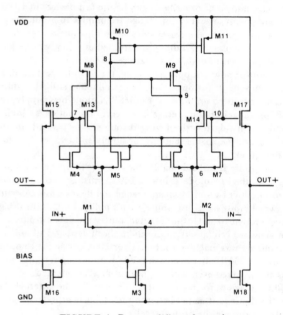

FIGURE 4—Postamplifier schematic.

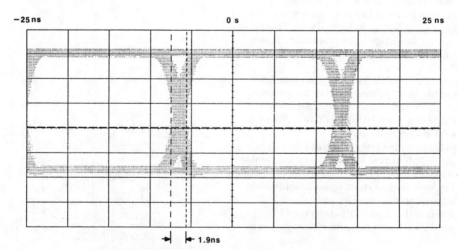

FIGURE 5—Eye diagram of 50Mb/s data output. Residual pulsewidth distortion is less than 2ns.

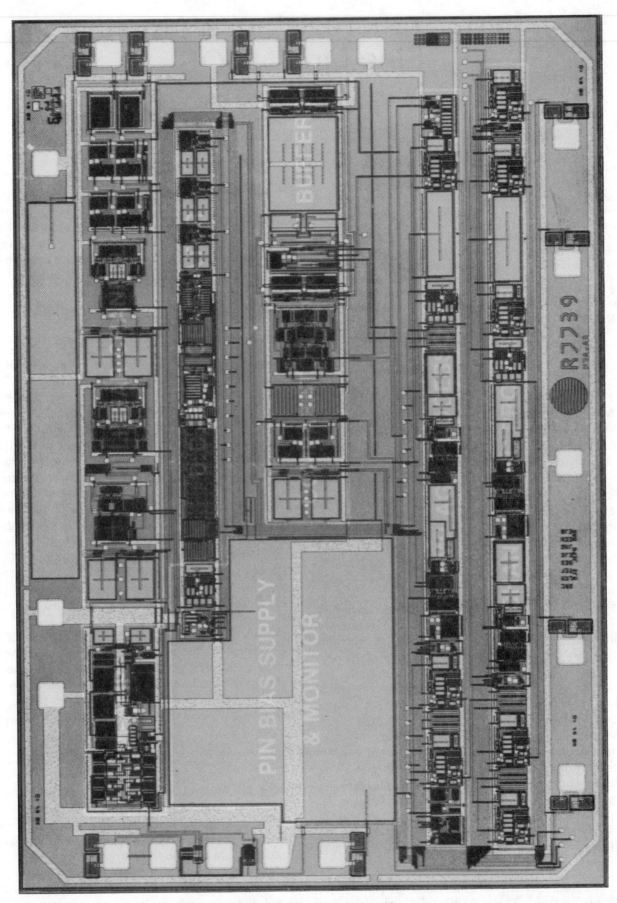

FIGURE 6—Photomicrograph of IC.

Gigahertz Transresistance Amplifiers in Fine Line NMOS

ASAD A. ABIDI

Abstract —Two front-end amplifiers for fiber optics applications at gigabit data rates were fabricated in fine line NMOS. One was designed for operation at 1 Gbit/s, and a simpler circuit was designed for 1.7 Gbits/s, with the function of amplifying a photocurrent into an output voltage. An AGC optionally follows the first amplifier. The bandwidth of the first amplifier was measured to be 920 MHz, with optical operation at 800 Mbits/s at an optical sensitivity of − 28 dBm using a pin detector for light at 1.3 μm wavelength. The bandwidth of the second amplifier was measured to be 1260 MHz. The AGC had a maximum gain of 32 dB at a bandwidth of 870 MHz. Both circuits operated at less than their designed frequency because of discrepancies between expected and measured device transconductances.

I. Introduction

WITH recent advances in fine line lithography for MOS fabrication [1], it has become possible to build circuits which approach the speeds of the traditional high-speed processes like silicon bipolar or GaAs MESFET's while retaining the low cost and high yield manufacturability of silicon MOS processing. An emerging application for high-speed circuits combining analog and digital functions is that of fiber optics signal regeneration at gigabit-per-second (Gbit/s) data rates. Regenerators are placed periodically along an optical fiber link, typically every few tens of kilometers with present day low-loss fibers. In each regenerator, the optical signal emerging from the fiber is converted to an electrical current by a photodiode, which is then amplified by a low-noise transresistance preamplifier. This is followed by a linear channel consisting of a variable gain amplifier, and possibly some filtering. A decision making circuit then converts this signal to a digital bit stream, digital circuits re-time the signal and check for parity errors, and finally, this regenerated data stream is applied to a high current driver device to modulate a semiconductor laser or LED.

High-speed trunk fiber optics links will become most practical when all the functions of regeneration can be fabricated on one or two reliable integrated circuits. The preamplifier and the linear channel are difficult elements to design because both are faced with the stringent requirements of low input noise, a very low ripple passband, and wide dynamic range. The noise requirement for the preamplifier is specified as an optical sensitivity, which is the minimum received optical power incident on a perfectly efficient photodiode connected to the amplifier input so that the presence of the amplifier noise corrupts only one bit per 10^9 bits of incoming data on average. The optical sensitivity (η) is related to the rms equivalent input noise current (i_n) of the amplifier as follows [2]:

$$\eta = 10\log_{10}\left(6i_n \times \frac{hc}{q\lambda \times 10^{-3}}\right) \text{dBm} \qquad (1)$$

where h is Planck's constant, q is electronic charge, λ is the wavelength of light (1.3 μm in high-speed optical fibers), and c is the speed of light. The best reported sensitivity to date for a preamplifier operating at gigabit data rates is − 34 dBm [3] at 1.8 Gbits/s. The circuit was a silicon bipolar amplifier with a bandwidth of 1.3 GHz using an avalanche photodetector connected at the input.

The bandwidth of the preamplifier is determined by how the data is encoded. If it is encoded as a nonreturn to zero (NRZ) waveform, the maximum fundamental frequency is at half the bit rate, so a preamplifier of this bandwidth suffices. Ripple in the passband of the amplifier should be less than 1 dB to avoid degrading the sensitivity. Following the preamplifier, an automatic gain control (AGC) amplifier with a variable gain of 40 dB is desired because the distance between regenerators may vary widely in practice. This adjusts the signal to a nominal level before it is fed to a decision-making circuit.

Two such preamplifiers are described in this paper. One was designed to operate at 1 Gbit/s, and the other at 1.5 Gbits/s data rates. An NMOS process with a one micron minimum feature size was used to fabricate them. A preamplifier operating at 45 Mbits/s and 90 Mbits/s fabricated in this process has been reported previously [4].

II. Amplifier Circuits

There are four main issues in designing an optical fiber preamplifier. It should be designed to minimize the equivalent input noise current: the bandwidth should be wide enough for operation at the desired data rate, the amplifier should produce a sufficiently large output voltage, and the operation of the amplifier should be stable over temperature and process fluctuations. The first three requirements are not mutually exclusive because bandwidth can often be

Manuscript received April 20, 1984; revised October 30, 1984.
The author is with AT&T Bell Laboratories, Murray Hill, NJ 07974.

Reprinted from *IEEE J. Solid-State Circuits*, vol. SC-19, no. 6, pp. 986–994, Dec. 1984.

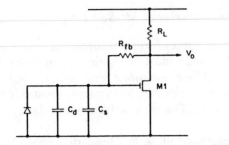

Fig. 1.   A prototype MOS transresistance amplifier.

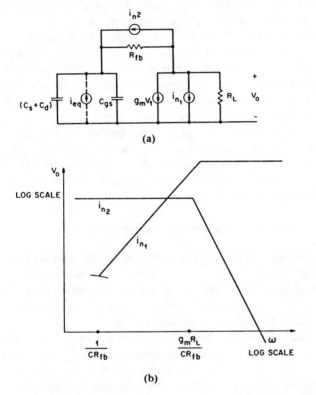

(a)

(b)

Fig. 2.   (a) The small-signal model for the transresistance amplifier. (b) Spectra of noise contributions at output of amplifier.

increased at the expense of noise, and gain can be increased at the expense of bandwidth. This shall become evident as each issue is discussed.

A. The Noise Optimum for Transresistance Amplifiers

A simplified MOSFET transresistance amplifier (Fig. 1) consists of a single transistor ($M1$) with a shunt feedback resistor (R_{fb}). At its input is a photodiode current source, whose capacitance C_d is in parallel with some stray capacitance C_s. Ignoring noise in the diode, an assumption justified by the very low leakage currents of modern devices, the two main contributors to the input noise of the amplifier are the thermal noise i_{n1} in the channel of MOSFET $M1$ and the thermal noise i_{n2} in resistor R_{fb}. The combined effect of these two can be expressed as the equivalent current noise (i_{eq}) at the amplifier input [Fig. 2(a)]. To calculate i_{eq}, note that the transresistance bandwidth of this amplifier is determined by the pole at the amplifier

input, that is,

$$\omega_{-3\,\mathrm{dB}} = \frac{(1 + g_m R_L)}{R_{fb}\left(C_{gs} + (1 + g_m R_L)C_{gd} + C_s + C_d\right)}$$

$$\simeq \frac{g_m R_L}{R_{fb}\left(C_{gs} + g_m R_L C_{gd} + C_T\right)} \tag{2}$$

where $g_m R_L \gg 1$, $C_T = (C_s + C_d)$. For operation at a data rate of B bits/s, and with NRZ waveforms, $\omega_{-3\,\mathrm{dB}} = \pi B$ is sufficient.

The combined effect of i_{n1} and i_{n2} at the output is as follows:

$$V_0 = \frac{1}{g_m} \times \frac{1 + j\omega\left(C_{gs} + C_{gd} + C_T\right)R_{fb}}{1 + j\omega\dfrac{\left(C_{gs} + g_m R_L C_{gd} + C_T\right)R_{fb}}{g_m R_L}} i_{n1}$$

$$+ R_{fb} \times \frac{1}{1 + j\omega\dfrac{\left(C_{gs} + g_m R_L C_{gd} + C_T\right)R_{fb}}{g_m R_L}} i_{n2}. \tag{3}$$

These two contributions are plotted in Fig. 2(b). The contribution of i_{eq} at the output is

$$V_0 = i_{eq} R_{fb} \times \frac{1}{1 + j\omega\dfrac{\left(C_{gs} + g_m R_L C_{gd} + C_T\right)R_{fb}}{g_m R_L}}. \tag{4}$$

Comparing (3) and (4),

$$i_{eq} = \frac{1}{g_m R_{fb}}\left(1 + j\omega\left(C_{gs} + C_{gd} + C_T\right)R_{fb}\right)i_{n1} + i_{n2}. \tag{5}$$

i_{n1} and i_{n2} are statistically independent, so (5) simplifies to the following at high frequencies near the amplifier bandedge

$$\overline{i_{eq}^2} = \omega^2 \frac{\left(C_{gs} + C_{gd} + C_T\right)^2}{g_m^2}\,\overline{i_{n1}^2} + \overline{i_{n2}^2}. \tag{6}$$

Due to the presence of the ω^2 coefficient, the transistor noise $\overline{i_{n1}^2}$ will dominate the total noise $\overline{i_{eq}^2}$ at high bit rates. In further simplifying (6), observe that $g_m = 2\pi f_T(C_{gs} + C_{gd})$ for the transistor, where f_T is a constant depending on the process. Therefore, (6) simplifies to

$$\overline{i_{eq}^2} \simeq \omega^2 \frac{\left(C_{gs} + C_{gd} + C_T\right)^2}{g_m^2} \times \overline{i_{n1}^2}$$

$$= \omega^2 \frac{\left(C_{gs} + C_{gd} + C_T\right)^2}{g_m^2} \times \alpha k T g_m$$

$$= \frac{\omega^2}{2\pi f_T} \frac{\left(C_{gs} + C_{gd} + C_T\right)^2}{C_{gs} + C_{gd}} \times \alpha k T \tag{7}$$

where the expression $\overline{i_{n1}^2} = \alpha k T g_m$ for the MOSFET channel noise is substituted and α is a constant. The only

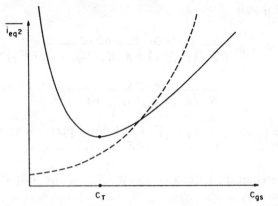

Fig. 3. Equivalent input noise as a function of input device capacitance with contribution of input stage (solid line) and of second and further stages (dotted line).

independent variable on the right-hand side of (7) is $(C_{gs} + C_{gd})$, which is determined by the transistor size; $\overline{i_{eq}^2}$ attains a minimum value when $(C_{gs} + C_{gd}) = C_T$ [2] (Fig. 3). Thus, for high bit rate transresistance amplifiers, the input device size is dictated by the photodiode and stray capacitances, and the optimum sensitivity of the amplifier improves inversely as the square root of this total capacitance.

The analysis above has assumed that noise due to the circuits following the first stage makes a negligible contribution to i_{eq}. High frequency amplifiers, however, tend to have a low gain per stage, so the noise of the following stages may become important. If the stage following $M1$ is a voltage amplifier, with equivalent input noise V_{n2}, then (7) is modified to

$$\overline{i_{eq}^2} = \frac{\omega^2}{2\pi f_T} \frac{(C_{gs} + C_{gd} + C_T)^2}{C_{gs} + C_{gd}} \alpha kT + \frac{\overline{V_{n2}^2}}{R_{fb}^2}. \quad (8)$$

From (2),

$$R_{fb} = \frac{g_m R_L}{\pi B (C_{gs} + C_{gd} + C_T)},$$

so if $g_m R_L$, the voltage gain of the input inverter, is assumed fixed, then

$$\overline{i_{eq}^2} = \frac{\omega^2}{2\pi f_T \nu} \frac{(C_{gs} + C_{gd} + C_T)^2}{C_{gs} + C_{gd}} \alpha kT$$
$$+ \frac{\overline{V_{n2}^2}}{(g_m R_L)^2} (\pi B)^2 (C_{gs} + C_{gd} + C_T)^2. \quad (9)$$

The second term in (9) increases monotonically with $(C_{gs} + C_{gd})$, so the noise minimum will occur at some $(C_{gs} + C_{gd}) < C_T$ (Fig. 3), depending on the magnitude of V_{n2}.

B. Additional Sources of MOSFET Noise

There are some further sources of noise special to MOSFET's which must be minimized in low noise amplifier design. These are noise due to gate resistance and possible excess noise due to substrate resistance and short-channel effects. The latter effects are under current investigation.

Gate resistance noise has been discussed elsewhere [5]. The thermal noise in the gate material induces noise in the MOSFET channel, thus increasing the total channel noise. This can be minimized by interdigitating the device gates, and most significantly by applying a silicide over the polysilicon layer; if tantalum silicide is used, the poly layer resistance is reduced by a factor of 10 to about 3 Ω/square.

The circuits were fabricated in a 5 μm deep n-type epitaxial layer on a highly conductive bulk material to reduce adverse effects of substrate resistance. Furthermore, the circuits were designed for operation from low power supply voltages: full gain could be obtained from 4.5 V supplies, but operation at 3 V was possible with reduced gain. A low power supply allows reduction in the peak electric field in the MOSFET channel, important in minimizing any excess noise.

C. A High-Speed Transresistance Circuit

For operation with a 4–4.5 V power supply, cascode stages were avoided in the amplifier. Instead, the circuit consists of three inverters with a local shunt feedback loop, and overall resistive feedback (Fig. 4). This circuit is henceforth referred to as amplifier A.

The size of the input device was chosen as 1200 μm to obtain the noise minimum given by (9); $C_{gs} + C_{gd} \simeq 0.75$ pF for this device. The shunt feedback on the second inverter acts to broadband the frequency response of the forward path. An approximate analysis is now done for the open-loop preamplifier, that is, with the outer feedback resistor device $M8$ removed. The pole at the output of the first inverter is

$$\omega_{p1} = \frac{g_{d1} + g_{mb2} + g_{d7}\left(1 + \frac{g_{m3}}{g_{d3} + g_{mb4}}\right)}{C_{gd1} + C_{db1} + C_{sb2} + C_{gd2} + C_{gs3} + C_{gd3}\left(1 + \frac{g_{m3}}{g_{d3} + g_{mb4}}\right) + C_{gs6} + C_{gd6}} \quad (10)$$

where C_{db}, C_{sb} are drain and source to substrate capacitances, and g_{mb} is the back-gate modulation transconductance. g_{d7} and C_{gd3} in (10) above are shown as Miller multiplied. The relative sizes of each driver and load device are chosen such that

$$1 + \frac{g_{m3}}{g_{d3} + g_{mb4}} = 10 \quad (11)$$

for every inverter. Substituting typical values of the small-signal parameters for the devices used,

$$\omega_{p1} = 6.8 \text{ grad/s } (=1.1 \text{ GHz}). \quad (12)$$

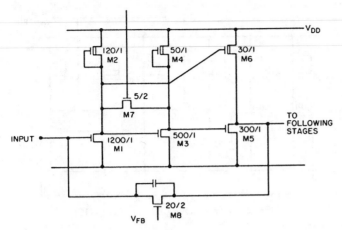

Fig. 4.   Transresistance amplifier A.

D. Choice of Bias Point and Bias Stabilization

The operating point of the amplifier stages was determined by two factors: the ability of the amplifier to operate with a 4.5 V power supply, and for each inverter to have a voltage gain of about 10. These two factors are not in conflict because a small depletion load for each enhancement device will supply a small current, thus requiring a small $V_{GS} - V_T$, while the gain per inverter shall be determined by g_m/g_d of the enhancement driver, which for the 1 μm NMOS process is about 10.

The transresistance amplifier (Fig. 4) is self-biasing. $M1$ is biased through the feedback resistor $M8$, which nominally has $V_{DS} = 0$. The ratio of drivers to loads in all the stages is identical, so all drain voltages are equal. This

Similarly, the pole at the output of the second inverter is

$$\omega_{p2} = \frac{g_{m3} + g_{d3} + g_{mb4}}{C_{gd1} + C_{db1} + C_{sb2} + C_{gd2} + C_{gs3} + C_{db3} + C_{sb4} + C_{gd4} + C_{gs5} + \left(1 + \dfrac{g_{m5}}{g_{d5} + g_{m6} + g_{mb6}}\right)C_{gd5}}. \tag{13}$$

Substituting approximate values,

$$\omega_{p2} = 31.6 \text{ grad/s } (= 5.0 \text{ GHz}). \tag{14}$$

The load of the third inverter is connected so that $V_{GS}(M6)$ $= 0$ without having to short the gate to the source. This ensures that all inverters are biased at exactly the same voltage, but the gain of the last inverter is reduced by about 3, and its output resistance is proportionately decreased. The lowered gain is necessary to ensure stability of the larger feedback loop where $M8$ feeds back around three inverters.

The pole of the output of the third inverter is

$$\omega_{p3} = \frac{g_{d5} + g_{m6} + g_{mb6} + g_8}{C_{gd5} + C_{db5} + C_{sb6} + C_{gs6} + C_{\text{(next stage)}}}. \tag{15}$$

Substituting approximate values,

$$\omega_{p3} = 42.5 \text{ grad/s } (= 6.7 \text{ GHz}).$$

The value of g_m used above is 90 mS for a 1 mm wide device with 1 μm coded gate length, and with an effective channel length of about 0.55 μm.

When the outer feedback loop is closed by resistor $M8$, a fourth pole due to the input capacitance is introduced. Computer simulations were used to choose the widths of $M3$ and $M5$ such that the closed loop response was maximally flat. If the voltage gain of the forward path is A_F, and the transresistance R_F, then the input pole is

$$\omega_{\text{in}} = \frac{(1 + A_F)}{R_F(C_T + C_{G1})} \tag{16}$$

where C_{G1} is the total capacitance to ground at the input due to the amplifier. As $C_T \simeq 1$ pF, and $C_{G1} \simeq 1.2$ pF after Miller multiplication, then for $A_F \simeq 30$, $R_F = 2.6$ kΩ.

implies $V_{DS}(M7) = 0$, forcing $M7$ into the triode region, which is the desired operating point for resistive shunt feedback. Finally, because all drain voltages have the same quiescent point, the gate connection of $M6$ does not upset the overall biasing.

The resistance of $M7$ and $M8$ is determined by their gate voltages, V_G and V_{FB}, respectively. Once the resistance of these devices has been chosen for the optimum performance, it is desirable to stabilize these characteristics against variations in chip temperature and against threshold voltage variations from one process lot to another. More precisely, the *resistance* of $M8$ and the *product* of the $g_m(M1)$ and the resistance of $M7$ should be so stabilized, the latter because $M7$ determines the gain of the forward path.

A resistor bridge circuit is used to generate V_{FB} [Fig. 5(a)]. R_1, R_2, and R_3 are off-chip low-temperature coefficient precision resistors, and the op amp is also external to the MOS chip. R_3 is chosen large so that a few tens of microamps flow through it, thus making $M22$ operate in the triode region. The op amp balances the bridge by forcing its differential input voltage to zero, so

$$g_d(M22) = \frac{R_1}{R_2} \times \frac{1}{R_3}. \tag{17}$$

By choice of device sizes, $V_{DS}(M20) = V_{GS}(M1)$, that is, $V_S(M22) = V_S(M8)$. When the op amp output is connected to the gate of $M8$,

$$g_d(M8) = \frac{W_8}{W_{22}} g_d(M22) \tag{18}$$

which, from (17), is completely determined by ratios of precision resistors, and is therefore stable.

The circuit to determine V_G is more complex [Fig. 5(b)], and relies on the near linear transfer characteristics of

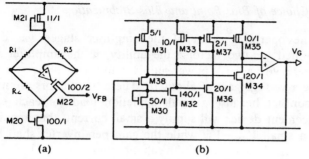

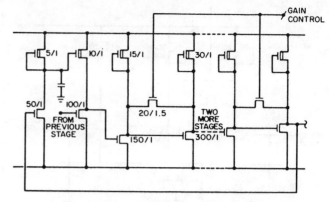

Fig. 5. (a) Transresistance stabilization circuit. (b) Gain stabilization circuit.

Fig. 6. Controllable gain amplifier (AGC).

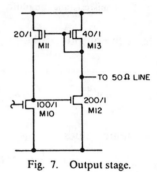

Fig. 7. Output stage.

short channel MOS devices [6],

$$I_D = C_{ox} W v_{sat}(V_{GS} - V_T) \qquad (19)$$

where C_{ox} is the capacitance per unit area of the gate oxide, W is the gate width, and v_{sat} is the saturation drift velocity of electrons in the channel. When the op amp has forced the voltage at its two inputs equal, $I(M32) = I(M34)$ because $M33$ and $M35$ are equal sized devices. Under these conditions

$$V_{GS}(M34) - V_{GS}(M32) = \frac{I(M32)}{C_{ox}v_{sat}}\left(\frac{1}{W_{34}} - \frac{1}{W_{32}}\right). \qquad (20)$$

Now,

$$I(M32) = I(M30) \times W_{32}/W_{30}$$

and

$$V_{DS}(M38) = V_{GS}(M34) - V_{GS}(M32)$$

$$= I(M30)/g_d(M38),$$

so

$$\frac{I(M30)}{g_d(M38)} = \frac{W_{32}}{W_{30}} \frac{I(M30)}{C_{ox}v_{sat}}\left(\frac{1}{W_{34}} - \frac{1}{W_{32}}\right). \qquad (21)$$

If the output of the op amp is applied to the gate $M7$ in the amplifier (Fig. 4), and since $V_s(M38) = V_s(M1)$,

$$\frac{g_m(M1)}{g_d(M7)} = \frac{g_m(M1)}{g_d(M38)} \times \frac{W_{38}}{W_7}$$

$$= \frac{W_{38}}{W_7} \times \frac{W_{32}}{W_{30}} \times \left(\frac{1}{W_{34}} - \frac{1}{W_{32}}\right) \qquad (22)$$

from (22) and (19). The forward gain of the first two inverters of the amplifier thus depends only on device widths, and so is stable. The gain of the third inverter is roughly $g_m(M5)/(g_m(M6) + g_{mb}(M6))$, and is also relatively stable.

E. Voltage Controllable Gain Amplifier and Output Stage

An AGC amplifier with a range of 40 dB is required to follow the transresistance amplifier described above. Such an amplifier was designed using a cascade of four shunt feedback pairs (Fig. 6), where the gain is varied by changing the gate voltages of the shunt feedback resistors. The device sizes were chosen to produce a broadband response, as well as to avoid loading the output of the previous stage. Computer simulations indicated that the maximum gain of each shunt feedback pair was 10 dB with a bandwidth of 1 GHz, thus producing a total gain of 40 dB.

The dc feedback loop corrects for threshold voltage mismatches in the devices which may otherwise saturate the AGC output level at high gains. The action of the loop is straightforward (Fig. 6), and it biases all the drain voltages at the same value. Feedback of the ac signal is disabled by the external high-frequency decoupling capacitor.

The output stage consists of a series feedback pair (Fig. 7), with the output resistance nominally adjusted to 50 Ω by the g_m of $M12$. This reduces ripple in the amplifier passband due to mismatches when driving a 50 Ω transmission line. This stage has a gain of about 2 when driving a 50 Ω load.

F. A 1.5 Gbit/s Circuit

A simplified version of the previous circuits was simulated for possible operation at 1.5 Gbit/s data rates, requiring a 3 dB bandwidth of about 1.0 GHz. An important component in the modeling of this circuit was the bonding wire inductances associated with the ground lines. A typical bonding wire has a self inductance of 3 nH which can cause serious peaking or even instability if it is not included in the simulation of a high-frequency amplifier circuit. In the design of the following circuit, peaking due

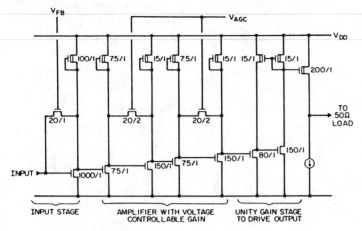

Fig. 8. Transresistance amplifier *B*.

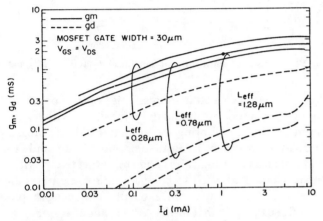

Fig. 9. Measured g_m and g_d of test devices.

to bond wire inductances was used to broadband the circuit up to 1.5 GHz.

This amplifier, henceforward referred to as amplifier *B*, is composed of a single inverter transresistance stage followed by two shunt feedback variable gain stages and an output stage (Fig. 8). By using multiple bond wires in parallel, the ground line bonding inductance is reduced to about 0.5 nH in the first stage. On the other hand, a single ground line inductance of 3 nH is used beneficially as an additional coupling path between the two inverters of every shunt feedback pair to produce some peaking at the bandedge. The transresistance of this circuit is about 530 Ω at the desired bandwidth, and the maximum gain of the voltage controlled gain section is simulated as 12 dB at 1.5 GHz. The output stage provides a gain of 3 into 50 Ω, and consists of a series feedback stage driving a source follower.

III. EXPERIMENTAL RESULTS

A. Measurements on Single Transistors

Before discussing the detailed performance of the amplifiers, the measured small-signal characteristics of single transistors fabricated on the same wafer are presented. It was found that the g_m of the devices was about a factor of 2 smaller than the simulated value (Fig. 9). The simulations of the amplifier had been based on an overly optimistic value of g_m derived from large-signal characteristics of the transistors.

The simulated g_m of about 90 mS/mm may be obtained if the effective channel lengths of the devices are shorter, and they are operated at 2–3 times the current density they were designed for (Fig. 9). However, for channel lengths less than 0.3 μm, short-channel effects produce a sudden decrease of g_m/g_d, resulting in a lowered gain per inverter.

B. Frequency Response and Noise Measurements on Amplifier A

The transresistance amplifier *A* (Fig. 10) was tested both in a 50 Ω network analyzer and on an optical bench with a photodiode connected at the input. The network analyzer measurement indicated a low frequency $|S_{21}| = 22$ dB with a bandwidth of 920 MHz (Fig. 11). This compares favorably with the simulated response, where the transconductances and channel lengths in the device model have been altered to match the measured values. The typical bandwidths for other units ranged from 750–850 MHz at comparable values of $|S_{21}|$. The maximum power delivered to the output load before measurable distortion sets in was about −5 dBm.

The output stage response when driving 50 Ω was determined separately from the rest of the circuit. The gain was measured to be 2 dB, and the bandwidth was in excess of 970 MHz.

Optical measurements were performed on the best unit using an Anritsu MG642A pseudorandom word generator driving a laser, whose light output was focused on a photodiode at the amplifier input. The transresistance stage output was further amplified before being applied to an Anritsu MS65A bit-error rate (B.E.R.) counter. Error rate measurements were done at data rates of 850 Mbits/s and 1 Gbit/s using an InGaAs pin photodetector, and at 850 Mbits/s, 1 Gbit/s, and 1.3 Gbits/s using a Ge avalanche photodiode (APD). The results of these measurements are summarized in Fig. 12. The simulated sensitivity at 1 Gbit/s with a pin photodetector at 10^{-9} bit-error rate was −32 dBm for this amplifier. The best sensitivities (Fig. 12) for amplifier *A* were measured at a power supply of only 3 V and a substrate bias of −2 V, with a considerable degradation in sensitivity observed at higher supply voltages. This suggests that there may be excess noise in these MOSFET's dependent on V_{DS}.

Amplifier *A* was fabricated with devices of $L_{eff} = 0.85$ μm and $L_{eff} = 0.45$ μm. The sensitivities in Fig. 12 are for the latter. The former version could only be operated up to 800 Mbits/s, where its sensitivity at 10^{-9} B.E.R was measured to be −28 dBm using a pin photodiode. This was 2 dBm better than the corresponding sensitivity in Fig. 12. As the measurements were done on two different test sets, it is difficult to establish if the shorter channel length is responsible for the worsened sensitivity.

The photodetector driving amplifier *A* is a capacitive signal source, which decreases the amplifier bandwidth

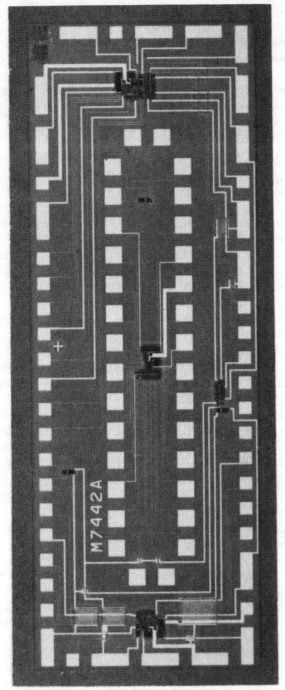

Fig. 10. Chip photograph of amplifier A.

SILICON MOS FIBER OPTICS PREAMPLIFIER

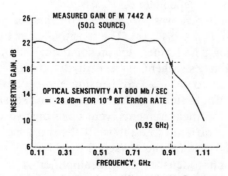

Fig. 11. Measured gain of amplifier A in 50 Ω system.

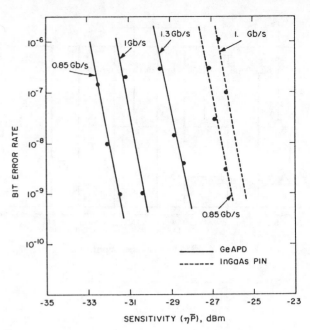

Fig. 12. Measured sensitivities of amplifier A at 0.85, 1.0, and 1.3 Gbit/s data rates.

when compared to a 50 Ω source. This "optical bandwidth" can be measured by modulating the laser with a sweep generator, and monitoring the amplifier output. Measured bandwidths with an InGaAs pin photodiode and a Ge APD are shown in Fig. 13(a) and (b), respectively, with the transresistance adjusted to compensate for the photodiode capacitance. The capacitance of the unpackaged pin is about 0.3 pF, while that of the APD is about 1.3 pF.

The effectiveness of the bias stabilization circuits was tested in the temperature range of 20 to 90°C. The amplifier characteristics with the control voltages V_G and V_{FB} connected to the stabilization circuits were compared with the characteristics if these voltages were held constant. With the stabilization circuits in place, the temperature coefficient of bandwidth was reduced by about 30 percent [Fig. 14(a)], while the temperature coefficient of gain remained approximately the same [Fig. 14(b)]. The net temperature coefficient of the amplifier gain was attributed to the output stage, whose transfer function is not stabilized in any way, and which, when measured separately, has the same temperature coefficient as the complete amplifier. Interestingly, the temperature coefficient of the gain is about the same in the absence of the stabilization circuits. The effect of these circuits is more apparent in the temperature coefficient of bandwidth, where V_G stabilization reduces the resultant drifts in the impedances of the internal nodes which determine the amplifier bandwidth. However, the stabilization is slight, and is restricted to only part of the temperature range.

The amplifiers operated at maximum gain with $V_{DD} = 4.5$ V and a power supply current of 60 mA.

C. Frequency Response Measurements on Controllable Gain Amplifier

Network analyzer measurements were done on the controllable gain amplifier. The maximum gain at which the

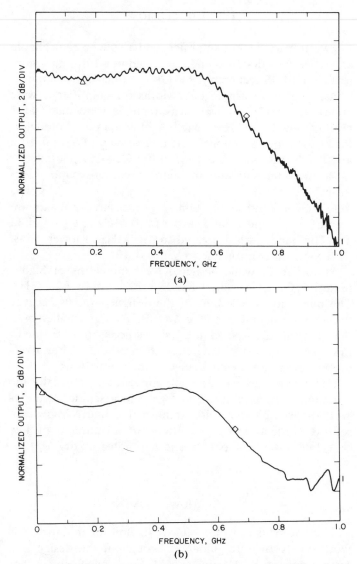

(a)

(b)

Fig. 13. (a) Measured optical frequency response of amplifier A with pin photodiode at input. (b) Measured optical frequency response of amplifier A with Ge avalance photodiode at input.

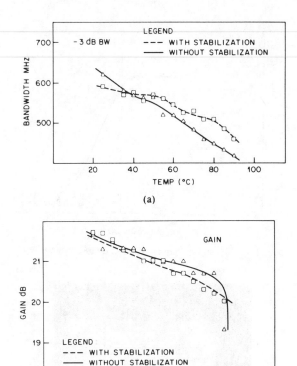

(a)

(b)

Fig. 14. (a) Measured temperature dependence of bandwidth, with and without stabilization circuit. (b) Measured temperature dependence of gain, with and without stabilization circuit.

circuit operated stably was 40 dB, with a bandwidth of 450 MHz. At 26 dB gain, the maximum bandwidth of 810 MHz was obtained (Fig. 15). The gain could be reduced to 0 dB; however, the bandwidth decreased to 650 MHz for low gains. The bandwidth attains a maximum value at medium gains due to peaking produced by the intrastage ground wire inductance, as predicted by simulations.

Again, a uniformly high bandwidth would have resulted had the device transconductances been closer to the simulated values.

D. Frequency Response Measurements on Amplifier B

The best measured frequency response of amplifier B (Fig. 16) was 1.26 GHz with a gain of 20 dB (Fig. 17). To decide whether the input transresistance stage or the gain stages following it was the main cause of the early rolloff, the bandwidth for varying transresistances and varying second stage gains was measured. The results suggest that the input stage imposes a dominant pole in the signal path. The amplifier operated at $V_{DD} = 4.5$ V with supply currents of 35–40 mA.

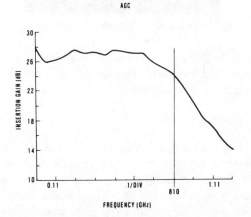

Fig. 15. Measured frequency response of AGC at 26 dB gain.

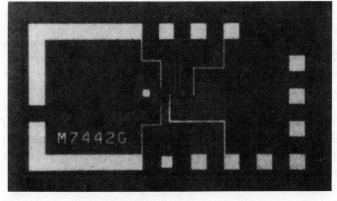

Fig. 16. Chip photograph of amplifier B.

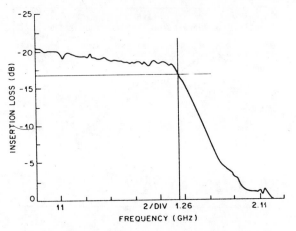

Fig. 17.   Measured frequency response of amplifier *B*.

Fig. 18.   Ceramic package used for testing circuits.

IV.   Conclusions

Two transresistance amplifiers and a voltage controllable amplifier were designed and fabricated in a 1 μm minimum feature NMOS technology for Gbit/s fiber optics applications. Typical measured bandwidths for amplifier *A* were about 920 MHz. Optical measurements were made at 1 Gbit/s with 1.3 μm wavelength light and a pin photodetector at the input of amplifier *A*; a sensitivity of -25.4 dBm at 10^{-9} B.E.R. was obtained. With a Ge APD photodetector at the same data rate, the sensitivity improved to -30.3 dBm.

A voltage controllable gain stage measured a maximum gain of 40 dB and a bandwidth of 810 MHz at a gain of 32 dB. The bandwidth decreased to 650 MHz at lower gains, with the gain controllable to below 0 dB.

Amplifier *B*, while designed for bandwidths of higher than 1.5 GHz, measured bandwidths of only 1.26 GHz. The bandwidths fell short of the designed values because the transconductance of the MOSFET's was smaller by 2 than the values assumed in the simulations; this was due to the operating points being close to threshold and an overestimation of transconductance in the simulations.

For many high-speed analog applications, normally the domain of bipolar silicon and gallium arsenide technologies, these circuits should establish fine line NMOS as a serious contender with its attendant advantages of low cost, low power dissipation, and high integrability.

Again, when the correct transconductances are used in the model, the measured response of the amplifier compares well with simulations.

E.   Packaging of Circuits

Careful packaging is important in making meaningful measurements on high-frequency circuits such as these. A ceramic substrate designed to accommodate all these amplifiers was used. One of its important features was a ring which encircled the chip and connected by via holes to the back side ground plane. Short bonding wires could thus be attached from the ground pads on the chip to this ring to minimize inductance to ground. High-frequency chip capacitors were mounted on the supply lines close to the circuit to bypass signal currents to ground, and it was found that effective bypassing of the substrate terminal was essential to avoid a characteristic dip in the response between 100 and 200 MHz.

The ceramic substrate was attached to a brass block (Fig. 18) and coaxial launchers screwed on to the block to connect signal cables on to the lines on the ceramic. All these lines were designed for a characteristic impedance of 50 Ω.

V.   Acknowledgment

The author is most grateful to H. Boll for initiating this work and providing guidance throughout the design and testing phases. B. Bayruns and M. Fang helped the author in the process of learning about high-frequency measurements. Chips were fabricated by E. Hofstatter, D. Kushner, and K. Watts. J. Kearney did most of the packaging in the latter part of the project. L. Nagel contributed by expediting the process flow and editing this manuscript. S. Lumish and R. Panock graciously agreed to do the optical measurements. D. Fraser made helpful comments on the manuscript.

References

[1] M. P. Lepselter *et al.*, "A systems approach to 1 micron NMOS," *Proc. IEEE*, vol. 71, May 1983.

[2] R. G. Smith and S. D. Personick, "Receiver design for optical fiber communication systems," in *Semiconductor Devices for Optical Communication.* Berlin, Germany: Springer Verlag, 1980.

[3] K. Iwashita *et al.*, "Monolithic integrated amplifiers for a gigabit optical repeater," *Electron. Lett.*, vol. 20, no. 11, May 24, 1984.

[4] D. L. Fraser *et al.*, "A single chip NMOS preamplifier for optical fiber receivers," in *Dig. Int. Solid-State Circuits Conf.*, Feb. 1983.

[5] K. K. Thornber, "Resistive gate induced thermal noise in IGFET's," *IEEE J. Solid-State Circuits*, vol. SC-16, Aug. 1981.

[6] S. M. Sze, *Physics of Semiconductor Devices*, 2nd ed.   New York: Wiley, 1981.

Author Index

Subject Index

493

494

Paul R. Gray (S'65–M'69–SM'76–F'81) was born in Jonesboro, Arkansas, on December 8, 1942. He received the B.S., M.S., and Ph.D. degrees from the University of Arizona, Tucson, in 1963, 1965, and 1969, respectively.

In 1969 he joined the Research and Development Laboratory, Fairchild Semiconductor, Palo Alto, California, where he was involved in the application of new technologies for analog integrated circuits, including power integrated circuits and data conversion circuits. In 1971 he joined the Department of Electrical Engineering and Computer Sciences, University of California, Berkeley, where he is now a Professor. His research interests during this period have included bipolar and MOS circuit design, electrothermal interactions in integrated circuits, device modeling, telecommunications circuits, and analog–digital interfaces in VLSI systems. During year-long industrial leaves of absence from Berkeley, he served as Project Manager for Telecommunications Filters at Intel Corporation, Santa Clara, California, in 1977–78, and as Director of CMOS Design Engineering at Microlinear Corporation, San Jose, California, in 1984–85.

Dr. Gray is the coauthor of a college textbook on analog integrated circuits. He has been corecipient of best-paper awards at the International Solid-State Circuits Conference and the European Solid-State Circuits Conference, and was corecipient of the IEEE R. W. G. Baker Prize in 1980, the IEEE Morris N. Liebmann Award in 1983, and the IEEE Circuits and Systems Society Achievement Award in 1987. He served as Editor of the *IEEE Journal of Solid-State Circuits* from 1977 through 1979, and as Program Chairman of the 1982 International Solid-State Circuits Conference. He currently serves as President of the IEEE Solid-State Circuits Council.

Dr. Gray is married and has two sons.

Bruce A. Wooley (S'62–M'70–SM'76–F'82) was born in Milwaukee, Wisconsin, on October 14, 1943. He received the B.S., M.S., and Ph.D. degrees in electrical engineering from the University of California, Berkeley, in 1966, 1968, and 1970, respectively.

From 1970 to 1984 he was a member of the research staff at Bell Laboratories in Holmdel, New Jersey. In 1980 he was a Visiting Lecturer at the University of California, Berkeley. In 1984 he assumed his present position as Professor of Electrical Engineering at Stanford University, Stanford, California. His research is in the field of integrated circuit design and technology where his interests have included monolithic broadband amplifier design, circuit architectures for high-speed arithmetic, analog-to-digital conversion and digital filtering for telecommunications systems, tactile sensing for robotics, high-speed memory design, and circuit techniques for video A/D conversion and broadband fiber-optic communications.

Dr. Wooley is a member of the IEEE Solid-State Circuits Council, and he is the current Editor of the *IEEE Journal of Solid-State Circuits*. He was the Chairman of the 1981 International Solid-State Circuits Conference. He is also a past Chairman of the IEEE Solid-State Circuits and Technology Committee and has served as a member of the IEEE Circuits and Systems Society Ad Com. In 1986 he was a member of the National Science Foundation-sponsored JTECH Panel on Telecommunications Technology in Japan. He is a member of Sigma Xi, Tau Beta Pi, and Eta Kappa Nu. He received an Outstanding Panelist Award for the 1985 International Solid-State Circuits Conference. In 1966 he was awarded the University Medal by the University of California, Berkeley, and he was the IEEE Fortescue Fellow for 1966–67.

Robert W. Brodersen (M'76–SM'81–F'82) received B.S. degrees in electrical engineering and in mathematics from California State Polytechnic University, Pomona, in 1966. From the Massachusetts Institute of Technology, Cambridge, he received the Engineers and M.S. degrees in 1968, and a Ph.D. in 1972.

From 1972 to 1976, he was with the Central Research Laboratory at Texas Instruments, Inc., Dallas, where he was engaged in the study of operation and applications of charge coupled devices. In 1976, he joined the faculty of the University of California, Berkeley, where he is a Professor. In addition to teaching, he is investigating the use of MOS technology for signal processing applications in the areas of speech recognition and synthesis, telecommunications, image processing, control systems, and robotics. Other projects involve the design of analog interface circuits and the development of a rapid prototyping environment based on integrated circuit technology. He is also an industrial consultant, working primarily on integrated circuits and the development of a rapid prototyping environment based on integrated circuit technology.

Among the awards he has received are the 1982 Morris Liebmann Award of the IEEE, the 1979 R. W. G. Baker Award for the outstanding paper in the IEEE journals and transactions, and in 1987 a Technical Achievement award from the Circuits and Systems Society. He has won conference best-paper awards at EASCON (1973), the International Solid-State Circuits Conference (1975), and the European Solid-State Circuit Conference (1978). He received a best-panel award at the International Solid-State Circuits Conference (1985). In 1986, he received the IEEE Circuits and Systems Society's best-paper award for the CAD transactions. He was named the 1978 outstanding engineering alumnus of California State Polytechnic University. In 1982 he became a Fellow of the IEEE. On September 25, 1988, he was officially inducted as a member of the National Academy of Engineering in Washington, D.C.